Dynamic Spectrum Access Decisions

Dynamic Spectrum Access Decisions

Local, Distributed, Centralized, and Hybrid Designs

George F. Elmasry
Rockwell Collins Advanced Technology Center
USA

Registered Offices
John Wiley & Sons, Inc., 111 River Street, Hoboken, NJ 07030, USA
John Wiley & Sons Ltd, The Atrium, Southern Gate, Chichester, West Sussex, PO19 8SQ, UK

Editorial Office
The Atrium, Southern Gate, Chichester, West Sussex, PO19 8SQ, UK

For details of our global editorial offices, customer services, and more information about Wiley products visit us at www.wiley.com.

Wiley also publishes its books in a variety of electronic formats and by print-on-demand. Some content that appears in standard print versions of this book may not be available in other formats.

Library of Congress Cataloging-in-Publication Data

Names: Elmasry, George F., author.
Title: Dynamic spectrum access decisions : local, distributed, centralized, and hybrid designs / George F. Elmasry.
Description: Hoboken, NJ, USA : Wiley, 2020. | Includes bibliographical references and index.
Identifiers: LCCN 2020017306 (print) | LCCN 2020017307 (ebook) | ISBN 9781119573760 (hardback) | ISBN 9781119573777 (adobe pdf) | ISBN 9781119573791 (epub)
Subjects: LCSH: Radio resource management (Wireless communications) | Cognitive radio networks.
Classification: LCC TK5103.4873 .E46 2020 (print) | LCC TK5103.4873 (ebook) | DDC 621.384–dc23
LC record available at https://lccn.loc.gov/2020017306
LC ebook record available at https://lccn.loc.gov/2020017307

Cover Design: Wiley
Cover Images: Spectrum access tools hybrid design courtesy of James Stevens, Abstract background© Studio-Pro/Getty Images

Set in 9.5/12.5pt STIXTwoText by SPi Global, Chennai, India

Printed and bound by CPI Group (UK) Ltd, Croydon, CR0 4YY

10 9 8 7 6 5 4 3 2 1

I would like to thank the IEEE for allowing the DySPAN-SC material to be included in this book. I also would like to thank Dr. James Stevens, the networking fellow at Collins Aerospace, and Dr. Thomas Tapp, the product owner of DSA capabilities at Collins Aerospace for their valuable input to this book. The DySPAN-SC material made this book a comprehensive reference for engineers developing DSA systems and Dr. Stevens's and Dr. Tapp's reviews and suggestions added so much value to this text.

Contents

About the Author

Dr. George F. Elmasry was born in Egypt and received a Bachelor of Science in Electrical Engineering and Electro-physics from Alexandria University, Egypt in 1985. He then went on to receive a Master of Science and a PhD in Electrical Engineering from the New Jersey Institute of Technology (NJIT) in 1993 and 1999, respectively. Dr. Elmasry has over 25 years of industrial experience in commercial and defense telecommunications and is currently part of the Advanced Technology Center of Collins Aerospace, which is a forward-looking research and development organization for defense and commercial aerospace telecommunications systems.

Dr. Elmasry has an interdisciplinary background in electrical and computer engineering and computer science. He is active in research, patenting, publications, grant proposals as well as system engineering of defense and commercial aerospace wireless communications systems. He has experience with technical task leads, team building and management. Dr. Elmasry has over 50 peer-reviewed publications and countless patents that pertain to network resource management, network management, network operation, software defined radios, cognitive networking, resilient communications and network and transport layers algorithms.

Dr. Elmasry has authored a book on tactical wireless communications and networking that has been used as a textbook for graduate and senior level courses in different universities in the United States and internationally. He also authored part of the McGraw-Hill encyclopedia on the history of military radios.

In addition to his publications and patents activities, Dr. Elmasry has been a technical member of the annual Military Communications Conference (MILCOM) since 2003, where he has participated in session organization, paper reviews and session chairing. Dr. Elmasry is also a senior member of the Institute of Electrical and Computer Engineering (IEEE), a member of the Armed Forces Communications and Electronics Association (AFCEA) International, a Member of Sigma Xi and a member of Alpha Epsilon Lambda, the NJIT graduate-student honor society. Dr. Elmasry holds countless awards, including the prestigious Hashimoto Award for achievement and academic excellence in electrical and computer engineering.

Preface

The coupling of spectrum sensing and dynamic spectrum access (DSA) solely with cognitive radios can do a disservice to the wireless communications community. Long before the concept of cognitive radios was coined, the defense community had elaborate spectrum sensing techniques. Since World War II, the defense community have had signal intelligence (SigInt) capabilities that are decades ahead of the commercial world's sensing capabilities. In today's commercial wireless system, we have the concept of finding the geolocation of a spectrum emitter through using multiple receivers (sensors) that can estimate the geolocation of the emitter and the direction of the emitter's beam. The defense community had this concept implemented in SigInt decades before the commercial world understood it and without having preknowledge of the emitted signal characteristics. SigInt capabilities used spectrum sensors extensively as part of the cat-and-mouse game of detecting the enemy's spectrum emission and overcoming the enemy's jammers.

As dynamic spectrum access capabilities are quickly evolving in the commercial wireless arena, one can see now how spectrum sensing leading to dynamic spectrum access ought to be decoupled from building the cognitive capabilities. As this book shows, we can have wireless communications systems that are not necessarily defined as cognitive systems, yet these systems can perform spectrum sensing, can analyze the sensed information and can use a form of dynamic spectrum access. Even with cognitive radios, breaking the problem domain to separate the spectrum sensing aspects from the cognitive capabilities can be helpful. When building communications systems, considering what type of spectrum sensing capabilities to use, how to process spectrum sensing information and how to make dynamic spectrum access decisions separate from designing and building a cognitive engine can be critical to optimizing these systems.

There are multiple goals behind this book from my own experience with commercial and military wireless communications systems. The first goal is to have the reader first focus on spectrum sensing and DSA approaches apart from building a cognitive engine. The second goal is to cover aspects of spectrum sensing without being tied to a specific system. The third goal is to create a reference and senior level or graduate school level textbook for a course in dynamic spectrum access. A course in cognitive radios and cognitive networks for electrical and computer engineering students should focus on the aspects covered in this book. The book considers cognitive engine design secondary to the digital communications aspects of DSA.

The book is divided into four parts. The first two parts can be used as a textbook, with chapters including exercises. The first part represents digital communications theoretical bases and concept descriptions of DSA that can apply to any system. The second part includes some case studies for designing DSA capabilities as a set of cloud services. The second part does not cover every possible application that can use DSA, but these examples should be eye-opening for any engineer who is looking to design DSA capable systems. The third part of the book includes the public domain publication of the US Army *Techniques for Spectrum*

Management Operation, which provides excellent information on the US Army doctrine for spectrum management. The reader can see the involvement of the military domain use of DSA and see the challenges facing military applications of DSA. The fourth part is the established DySPAN standardization, also known as the IEEE P1900.

George F. Elmasry

List of Acronyms

3GPP	third-generation partnership project
4G	fourth generation
5G	fifth generation
ACES	automated communications engineering software
ADC	analog to digital converter
AESOP	afloat electromagnetic spectrum operations program
AJ	antijamming
AM	amplitude modulation
AOC	area of coverage
API	application program interface
AWGN	additive white Gaussian noise
BGP	border gateway protocol
bps	bits per second
CBRS	Citizens' Broadband Radio Service
CEMA	cyber electromagnetic activities
CDMA	code division multiple access
CFAR	constant false alarm rate
CJCSM	Chairman of the Joint Chiefs of Staff manual
CJSMPT	Coalition Joint Spectrum Management Planning Tool
COA	course of action
COMSEC	communications security
COP	common operational picture
CROC	complementary receiving operating characteristics
CSDF	cyclic spectral density function
CSMA	carrier sense multiple access
CSV	comma separated values
DA	Department of the Army
DAC	digital to analog converter
DARPA	defense advanced research projects agent
dB	decibells
dBm	decibells relative to 1 mW
DD	Department of Defense
DF	decision fusion
DFC	decision fusion center
DFT	discrete Fourier transform
DL	down link
DLEP	dynamic link exchange protocol
DOD	Department of Defense
DSA	dynamic spectrum access/dynamic spectrum awareness
DSM	dynamic spectrum management
DTD	device-to-device

DTV	digital television
DySPAN	dynamic spectrum access network
EA	electronic attack
EE	energy efficiency
EMI	electromagnetic interference
EMOE	electromagnetic operational environment
EP	electronic protection
EW	electronic warfare
EWO	electronic warfare officer
FCC	federal communications commission
FD	full duplex
FDMA	frequency division multiple access
FEC	forward error correction
FFT	fast Fourier transformation
FSPL	free space path loss
FT	Fourier transform
G-2	assistant chief of staff for intelligence
G-3	assistant chief of staff, operations
G-6	assistant chief of staff, signal
G-7	assistant chief of staff, information engagement
GEMSIS	Global Electromagnetic Spectrum Information System
GHz	gigahertz
GPP	general purpose process
GPS	global positioning system
HAP	high altitude platform
HERF	hazards of electromagnetic radiation to fuels
HERO	hazards of electromagnetic radiation to ordnance
HERP	hazards of electromagnetic radion to personnel
HF	high frequency
HN	host nation
HNSWDO	Host Nation Spectrum Worldwide Database Online
Hz	Hertz
IaaS	infrastructure as a service
ICD	interface control definition
IP	internet protocol
IPSec	internet protocol security
J-3	operations directorate of a joint staff
J-6	communications system directorate of a joint staff
JACS	joint automated communications electronics operation instructions system
JCEOI	joint communications-electronics operating instructions
JFMO	joint frequency management office
JRFL	joint restricted frequency list
JSC	joint spectrum center
JSIR	joint spectrum interference resolution
JSIRO	joint spectrum interference resolution online
JSME	joint spectrum management element
kHz	kilohertz
LBT	listen before talk
LTE	long-term evolution
LPD	low probability of detection
LPI	low probability of interception
MA	multiple access
MAC	medium access control

MAC	multiple access channel
M-AM	M-level amplitude modulation
MANCAT	multispectral ambient noise collection and analysis tool
MANET	mobile ad hoc networks
MCEB	Military Communications Electronics Board
MDMP	military decision-making process
MHz	megahertz
MIMO	multiple-input multiple-output
MU MIMO	multi-user MIMO
mW	milliwatt
NCE	network-centric environment
NG	National Guard
NIPRNET	Nonsecure Internet Protocol Router Network
NIST	National Institute of Standards and Technology
NORM	nack oriented reliable multicast
NSA	national security agent
NTIA	National Telecommunicatoins and Information Administration
OFDMA	orthogonal frequency division multiple access
OPLAN	operation plan
OPORD	operation order
OSI	open systems interconnection
OTA	over-the-air
OWA	open wireless architecture
PAC	parallel access channel
PDF	probability distribution function
PDF	probability density function
PPPOE	point-to-point protocol over Ethernet
PSK	phase shift keying
PU	primary user
QAM	quadrature amplitude modulation
QoS	quality of service
R2R	router-to-radio
RAT	radio access technology
RF	radio frequency
ROC	receiver's operator characteristics
RS	Reed-Solomon
RSSI	received signal strength indication
S-2	battalion or brigade intelligence staff officer
S-3	battalion or brigade operations staff officer
S-6	(Army) battalion or brigade signal staff officer
S2AS	Spectrum Situational Awareness System
SA	spectrum awareness
SA	spectrum agent
SAS	spectrum access system
SDD	spectrum dependent devices
SDN	software defined network
SDR	software defined radio
SFAF	standard frequency action format
SI	self-interference
SigInt	signal intelligence
SINCGARS	single-channel ground and airborne radio system
SINR	signal to interference plus noise ratio
SIPRNET	SECRET Internet Protocol Rourter Network

SIR	signal to interference ratio
SiS	signal in space
SISO	single-input single-output
SMO	spectrum management operations
SNR	signal-to-noise ratio
SNIR	signal-to-noise interference ratio
SOI	signal operating instructions
SON	self-organized network
SPEED	systems planning, enginnering, and evaluation device
SSRF	standard spectrum resource format
SRW	soldier radio waveform
SU	secondary user
SWaP	size weight and power
TDMA	time division multiple access
TGP	terrestrial geolocation protocol
TOS	type of service
transec	transmission security
TTI	transmission time interval
UE	user equipment
UHF	ultra-high frequency
UL	up link
US	United States
VHF	very high frequency
WiFi	wireless fidelity (the term is trademarked for the IEEE 11.x)
WINNF	wireless innovation forum
WNW	wideband networking waveform
XG	next generation
XML	extensible markup language

About the Companion Website

Don't forget to visit the companion website for this book:

www.wiley.com/go/elmasry/dsad

There you will find valuable material designed to enhance your learning, including:

- Solution Manual

Scan this QR code to visit the companion website

Part I

DSA Basic Design Concept

1

Introduction

This book targets the field of dynamic spectrum access (DSA), which can also be referred to as dynamic spectrum awareness, dynamic spectrum management (DSM), or cooperative spectrum management. The book does not attempt to explain or summarize what is already established in standardization efforts, such as the dynamic spectrum access network (DySPAN), also known as the Institute of Electrical and Electronics Engineers (IEEE) P1900,[1] or the Wireless Innovation Forum (WINNF) Spectrum Access System (SAS). Rather, it's goal is to help engineers design the most suitable DSA approach for whatever wireless communications system is being built. DSA is needed for a wide range of civilian and military communications systems to dynamically optimize spectrum use. A form of DSA can be used for licensed and unlicensed spectrum bands in a wide variety of systems. DSA is presented in this book with a wider context than cognitive radios. There are many commercial and military communications systems that are not necessarily categorized as cognitive systems, but use these techniques to dynamically manage scarce spectrum resources. In today's ever-increasing appetite for bandwidth, every extra Hertz a wireless system can use means an increased rate of transmission in bits per second (bps) over the air (OTA). Different types of communications systems are evolving to add incremental DSA capabilities. For example, military communications systems are moving towards DSA with a mixed use of new cognitive waveforms and legacy noncognitive waveforms.

DSA is being deployed and enhanced in many commercial and defense systems. Cellular long-term evolution (LTE) and fifth-generation (5G) literature shows myriad DSA approaches that are combined with beam forming, multiple input and multiple output (MIMO) antennas, and arbitration by the base station for efficient spectrum use. The United States (US) Defense Advanced Research Projects Agent (DARPA) next-generation (XG) program is one example of defense initiatives for developing cognitive DSA radios. In both commercial and defense systems, DSA techniques are still evolving. There will always be room for further enhancement of DSA techniques in existing systems and for the development of better DSA techniques in new systems.

One goal of this book is to decouple DSA from cognitive radios and cognitive networks as covered in Part 1. The literature often presents DSA as the only drive behind cognitive radios. Even the US Federal Communications Commission (FCC) early definition of cognitive radio is based on the radio being able to opportunistically use unlicensed spectrum. There could be systems that lack the definition of being cognitive systems, which can use a form of DSA. Also, as cognitive wireless communications systems evolve, they should not be viewed as solely DSA systems. Chapter 4 explains the use of a form of "cognitive" reactive routing, which is built on top of DSA. The use of this reactive routing technique with military wireless communications systems enhances these systems' low probability of detection (LPD) and low probability of interception (LPI) capabilities.

This book is meant for a reader who has basic knowledge of wireless communications and wireless networks, and has an interest in the design and implementation of the physical

---

1 Part 4 of this book contains the DySPAN standards. DySPAN is selected over SAS in this book because it contains API design approaches that complement the goals of this book.

Dynamic Spectrum Access Decisions: Local, Distributed, Centralized, and Hybrid Designs, First Edition. George F. Elmasry.
© 2021 John Wiley & Sons Ltd. Published 2021 by John Wiley & Sons Ltd.
Companion website: www.wiley.com/go/elmasry/dsad

layer and medium access control (MAC) layer of wireless communication systems to include cognitive radios and cognitive networks. Regardless of whether the system under design is being targeted to use a licensed spectrum band or an opportunistic spectrum band, there is a need to consider DSA. It is more obvious that a system built to use opportunistic spectrum would need DSA capabilities. However, the design of systems such as cognitive mobile ad-hoc networks (MANETs) with allocated frequency bands has to consider the cooperative use of the allocated frequency bands to maximize the effective bandwidth of the formed network.

This book presents the most *generic* model to consider for DSA design. This model represents a large-scale collection of heterogeneous hierarchical MANETs that use a mix of licensed and unlicensed spectrum bands. With this model, DSA becomes a set of cloud services that can span from the network edge to the network core and to a single centralized point (root) in the network core. With this generic model, DSA decisions can be made anywhere in the network. An entity making a DSA decision can use local information, information shared with and obtained from peer nodes in a cooperative distributed manner, or information obtained from lower or higher hierarchical entities. With this model, the most studied comprehensive case of DSA, which is the cellular 5G DSM, can be seen as a special case of this generic model. The first chapter in the second part of this book introduces this *generic* model followed by a chapter dedicated to 5G DSM.

1.1 Summary of DSA Decision-making Processes

One aspect of optimizing DSA performance is to turn every node in every network into a spectrum sensor. The technology of spectrum sensing has evolved very well lately where a small size chip can perform comprehensive spectrum sensing techniques with minimal requirements on the node size, weight, and power (SWaP). The IEEE DySPAN standardization has a working group that defined the interface between a sensing hardware and the node module that is responsible for interfacing to the sensing hardware. This node module can be a distributed cognitive agent or a mere information collection agent.

Spectrum sensing can be tabulated under two main categories. The first category is *augmented sensing* where specialized spectrum sensing hardware/software is used as mentioned above. The second category is *same-channel in-band sensing*. With same-channel in-band sensing, the physical layer metrics of a received communication signal are leveraged to generate spectrum sensing information. This is a form of piggybacking of spectrum sensing over the ongoing communications signal, which should only require some processing of the physical layer metrics to obtain valuable spectrum sensing information. A comprehensive DSA solution may rely on both augmented sensing and same-channel in-band sensing. The advantages of same-channel in-band sensing, even in the presence of a specialized spectrum sensing hardware for augmented sensing, are detailed in this book.

There are two main components of the DSA design to consider. The first component pertains to obtaining spectrum sensing information from a sensor and being able to configure the sensor on what frequency bands to sense and what parameters to send to the distributed agent interfacing to the sensor. The second, and more challenging, DSA component is what to do with the obtained spectrum sensing information. This is sometimes referred to as decision fusion (DF). DF is a critical part of designing DSA capabilities where the design has to consider the following decision-making types:

1. *Local decisions*. With this decision type, an agent can make a local decision to overcome interference detected on a utilized frequency band. This agent can make decisions such as increasing transmission power or switching to a different frequency band. In a MANET, a node can suggest to peer nodes to switch to the new frequency band based on its local DF using some OTA protocol. The switch to the new frequency occurs without relying on

any external information that can be obtained from a higher or a lower hierarchal entity. Peer nodes can also be listening on a group of frequencies and synch to the new frequency without the need for any OTA control traffic transmission.

2. *Distributed cooperative decisions.* With this approach, a distributed cognitive agent will share spectrum sensing information, or a subset or processed version of it (fused spectrum sensing information), with peer distributed agents in the same network and cooperatively make a distributed decision to avoid a frequency band or to use a new frequency band. These cooperative decisions take into consideration that all nodes in the network have to synchronously switch to the new frequency.

3. *Centralized decisions.* With this decision type, spectrum sensing information is forwarded to a centralized entity (e.g., a network manager or a spectrum allocation arbitrator). The decision to use a frequency band or to stop using a frequency band is made with a bird's eye view (global view) of the status of spectrum use. This aspect of DSA is specifically needed when we have heterogeneous networks and there is no solution to create an equilibrium, using gaming theory implementation, between distributed agents' request for increased spectrum use. Without a centralized arbitrator, some networks can be making noncooperative decisions with respect to external networks that can result in "spectrum hugging". With this case, the spectrum arbitrator is more suited to ensure spectrum usage is fairly optimized with respect to a large-scale deployment of heterogeneous networks.

4. *Hybrid decisions.* With this approach, a mix of the above three decision types is considered in the DSA design. The balance between how much of each of the above three decisions types to use in a hybrid DSA design depends on the systems under design. As this book shows, DSA decisions can become a set of cloud services. The designer has to consider that the best approach to create DSA services[2] in a large system is to use hybrid DSA decisions that adapt to the current state of the network of networks. This hybrid approach should make DSA services always available at any network entity regardless of the conditions of the control plane used to communicate DSA control traffic.[3]

Figure 1.1 shows a notional view of this decision-making hierarchy. A distributed DSA decision has to consider that multiple local DSA processes can cooperate to dynamically solve a spectrum access challenge in near real time. Consider the case of a distributed MANET network where the nodes' distributed agents can make local decisions such as listening on a different frequency band while the distributed agents make cooperative decisions regarding DSA aspects such as beam forming, power increase/decrease, and changing the error control coding mode. In addition, multiple networks can have local and distributed decisions but these networks are also part of a centralized decision-making process that is running on a centralized network manager.

As the reader proceeds through this book, the notional view in Figure 1.1 will be seen as an oversimplification. It is meant to introduce a concept. The reader will see how in a set of heterogeneous MANETs, network gateways can perform a global cooperative distributed decision-making protocol that is different from the distributed cooperative decision making within a network (local to a network).

Using machine learning based engines to make these decisions can also be local, distributed, centralized or hybrid implementation. The designer of a DSA system should not limit machine learning approaches to a specific area although the design has to keep in mind that machine

---

2 There are different types of DSA services that can be offered in a set of DSA cloud services. Part 2 of this book shows that co-site interference avoidance can be a collection of DSA services that incrementally increases the efficiency of frequencies assignments. The consideration of blanking signal, which ensures the accuracy of sensing information, can be another type of DSA service.

3 DSA decisions are often needed when the radio frequency signal is compromised. The DSA control plane conditions may or may not be compromised at that time. When the DSA control plane is compromised, DSA services should be available. The response time between requesting a DSA service and granting the service should not be dependent on the DSA control plane conditions.

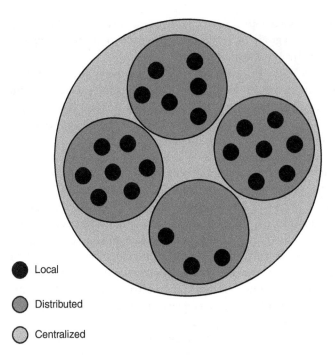

Figure 1.1 Hybrid DSA decision making.

learning techniques should be used when they are likely to produce better decisions than stochastic model decisions. The design of a point-to-point link operating at a cutoff threshold of a signal-to-noise interference ratio (SNIR) relying on a local spectrum sensing to avoid using a jammed frequency may not need a machine learning technique. This is because the cutoff SNIR is based on the physical layer stochastic models and the best machine learning approach will perform as good as a stochastic decision-making process in this simple case. There is a golden rule regarding the use of machine learning approaches: If a stochastic model gives the same performance as a machine learning technique, why complicate a design? The design should consider the stochastic model. As a rule of thumb, machine learning based techniques perform better as the number of factors contributing to a decision increases and as the uncertainty and change of behavior of the formed network change based on many surrounding factors. In most cases, DSA design should use a cognitive engine approach while relying on stochastic models for processing the raw physical layer metrics.

Sharing information in a distributed or centralized manner combines spectrum sensing information results from a much larger sample of measurements than a local node has. This information sharing can reduce noise uncertainty and overcome other signal distortion challenges. The downside of sharing sensing information is the need to develop mechanisms for information sharing that minimize bandwidth consumption from spectrum sensing information control traffic. The design has to consider a tradeoff between the gain obtained from DSA capabilities and the loss of bandwidth used by the DSA control traffic.

Figure 1.2 shows a conceptual view of the DSA key processes regardless of how and where it is implemented. Local, distributed, and centralized aspects of the DSA processes have to follow the theme of observe, orient, decide, and act. The observation process can include obtaining spectrum sensing information coupled with position location information [e.g., global positioning system (GPS) locations]. The orientation process can include processing of local spectrum sensing information and adding time stamps to the spectrum sensing messages distributed to peer agents and to the centralized arbitrator. The orientation process can also include fusing spectrum sensing information received from peer agents to gain a more comprehensive view of the spectrum use status. The decision process can be local distributed or centralized decisions and it can include frequency band change or changing the communications mode of a waveform.[4] The action process has to ensure synchronization of frequency change such that all the nodes in the network switch to the new frequency seamlessly and without losing any OTA transmitted frames.

It is important to note that spectrum sensing information has to be tied to both time and location. It has to be time stamped before distribution to peer agents or to the central arbitrator. Spectrum sensing information has to have geolocation information of the sensing node. Some rudimentary centralized DSA techniques can use location and time stamp information in the absence of spectrum sensing information to assign nonconflicting bands to a large-scale set of heterogeneous networks. This approach can be used when dedicated frequencies are assigned to the set of heterogeneous networks and assuming the absence of external users of these frequency bands. Frequency reusability with this case is purely spatial.

Software-defined radios (SDR) and software-defined networks (SDN) helped us do away with the old open systems interconnection (OSI) model. The line between the MAC and network layers is now blurred. When we think of machine learning based decision making, we can put it in the context of a MAC layer or a network layer. The 5G standardization uses the context of an open wireless architecture (OWA) where the system designer has the flexibility to add these machine learning techniques to the OWA layer as software modules. Thus, this book will not attempt to label DSA techniques as internet protocol (IP) layer techniques or MAC layer techniques. Rather, the book attempts to guide the reader to consider how to make the best out of the spectrum sensing assets and to decide what decisions can be made locally, what decisions can be made in a distributed manner, and what decisions can be made in a centralized location.

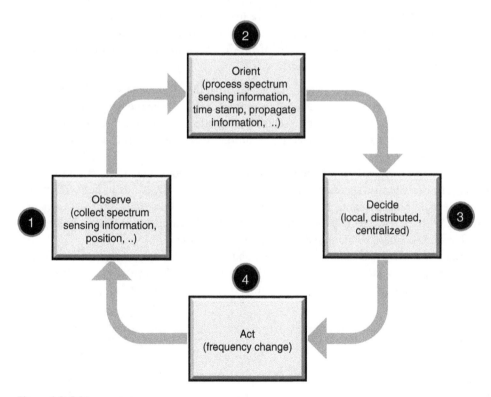

Figure 1.2 DSA processes.

---

4 Some military networks can have an antijamming mode. A DSA decision can be switching from a high bandwidth normal mode to a lower bandwidth antijamming mode instead of changing the operating frequencies.

1.2   The Hierarchy of DSA Decision Making

When a dynamic spectrum access decision is made locally, it can be made quickly. If we are considering a point-to-point link that has the flexibility to choose between different frequencies and its spectrum sensing has a simple SNIR threshold to compare to when using one frequency, it can simply decide that SNIR on a used frequency is low and switch to a different available frequency. In a distributed MANET, where we have to make cooperative distributed DSA decisions, propagation time of spectrum sensing information, processing time, and the need for making synchronized decisions will produce a relatively slower decision-making process. In a centralized decision-making construct, the decision time will even be longer as the spectrum allocation arbitrator has to obtain information from a large-scale set of heterogeneous networks and make slower, more lasting decisions. Centralized decision making is global, has to coordinate between all networks, has to consider geographical separation between networks, and cannot be performed every short period of time.

Figure 1.3 shows a conceptual view of this decision-making process timing where T_1 represents local decision-making time, T_2 represents distributed decision-making time, and T_3 represents centralized decision-making time. In Figure 1.3, T_1 T_2 and T_3 are expressive of both the speed of making DSA decisions and the time periods between making DSA decisions. Figure 1.3, by no means uses an accurate time scale but is meant to illustrate that in a hybrid DSA decisions system many local decisions can occur before a distributed cooperative decision can occur, and many distributed cooperative decisions can occur in different networks before a centralized arbitrator decision can occur.

It is important to consider this hierarchy of decision making when designing a hybrid DSA system. Local DSA decisions have to be quick, distributed DSA decisions have to be more insightful, and centralized DSA decisions have to be long lasting in order for the design to achieve the aspects described below and avoid the pitfalls described in Section 1.5.

Taking into consideration the hierarchy of DSA decision making, a comprehensive DSA design will consider these important aspects:

1. *Avoid rippling effects.* Rippling effects here means that the machine-based decision-making processes switches a user or a network from frequency f_1 to frequency f_2 only to decide quickly to switch back from f_2 to f_1. This rippling effect can reduce the throughput efficiency of the network and diminish the optimized use of the spectrum resources pool of frequency.
2. *Consider traffic demand.* In a large-scale set of heterogeneous networks, centralized DSA decisions can be more successful in allocating more frequency bands to networks where there is high traffic demand. This will increase the overall throughput of the managed networks. With DSA, spectrum is a commodity that can be distributed to overcome interference and resources should be increased where demand increases.
3. *Consider secondary user rules.* If the system being designed is for an opportunistic use of spectrum as a secondary user, the design has to adhere to the secondary user rules. Secondary user rules give the primary user the first right to spectrum use and spectrum sensing

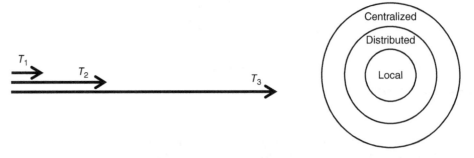

Figure 1.3 Conceptual view of DSA decision-making process timing.

has to always be on the lookout for primary user activities. Secondary user rules are discussed in more details in Chapter 2.

4. *Consider hidden node effects.* The challenges associated with hidden nodes are discussed in the next chapter. Being able to sense spectrum use from different locations helps avoid this challenge. In a hybrid system with a centralized spectrum arbitrator and using a large set of spectrum sensors, the hidden node challenge can be mitigated effectively.

5. *Consider the hierarchy of response times mentioned above.* Local decisions are fast, distributed decisions are slower and more insightful, centralized decisions are intended to be long lasting. A hybrid DSA decision system design has to consider making the different DSA hierarchal levels work in harmony.

6. *Avoid conflicts with other decision-making processes.* This challenge can surface as SDR and SDN allow the incremental addition of software modules that can be prone to making contradicting decisions. Consider this example: a software module makes adaptive power control decisions. This module can obtain SNIR information from the physical layer and decides to increase power to overcome low SNIR assuming that the peer node, from which it received a signal with a low SNIR, is suffering from the same degradation reciprocally. In the meantime, a DSA agent obtaining the same SNIR metric attempts to switch frequency to overcome the low SNIR. There are pros and cons for each of these decisions. Increasing spectrum emission power can quickly fix the low SNIR challenge but may reduce frequency reusability in a large-scale set of networks. Switching to a new frequency can also address the low SNIR challenge radically but can be a greedy decision. The new frequency may have been more needed by a neighboring network. The main point here is for the designer to consider how SDN and SDR can have other software modules that address spectrum conflicts from different angles and make sure DSA decisions are made in harmony between all spectrum access techniques. Chapter 5 explains the value of using a generic cognitive engine skeleton at all entities of DSA decision making with a single information repository and a single decision maker. This skeleton design ensures that a single process makes all the DSA-related decisions (i.e., offers all DSA services) at any given DSA service entity.

Clearly, DSA decisions are not "one size fits all". Depending on so many factors and other design decisions, one has to articulately consider all the design aspects of DSA. A comprehensive DSA system may have a mix of local, distributed, and centralized decision making and a good design must consider the order of time[5] these decisions ought to take, which decision to make at which area, and how to coordinate between all DSA-related decision-making capabilities.

1.3 The Impact of DSA Control Traffic

As the old saying goes "there is no free lunch". With DSA, the amount of control traffic could increase. In order to create distributed and centralized aspects of DSA, spectrum sensing information has to flow from sensor nodes to other sensor nodes or to the centralized arbitrator. Negotiation between distributed agents could use control plane bandwidth as well. Configuring sensing nodes on what frequency bands to sense will also require messaging over the air. The tradeoff between the increase in control traffic and the gain obtained from DSA optimizing the available spectrum resources must be quantified. In a network that operates on a narrowband of frequency, which results in a low bit per second transmission rate, the amount of control traffic increase may render the gain from DSA techniques worthless to this network. In a heterogeneous set of networks, such a low bandwidth network may be allocated a fixed frequency and the global DSA design may have to work around this fixed

---

5 Order of time here means response time and the gap of time between two consecutive DSA service requests.

frequency assignment when optimizing spectrum resources for the rest of the heterogeneous networks.

Even when we are certain that DSA gain exceeds the impact of DSA control traffic, DSA design has to find ways to reduce the impact of DSA control traffic through techniques that do some local processing (fusion) of spectrum sensing information and abstract them before sending them OTA such that we do not compromise spectrum sensing information relevance while we minimize the use of OTA resources for DSA control traffic.

1.4   The Involvedness of DSA Decision Making

Figure 1.4 shows how the DSA decision-making process has to consider many factors. Notice that Figure 1.4 does not attempt to put the DSA decision-making process in the context of a cognitive engine, which can be in a cognitive radio or a cognitive network node. Rather, the figure shows the place of the DSA decision-making process independent of how it is implemented and without reference to implementation.

As Figure 1.4 shows, in addition to spectrum sensing information, the DSA decision-making process involves the following:

1- *Spatial and time information.* Geolocation information is critical to most DSA decision-making processes. Cooperative distributed agents have to pass geolocation information to each other for spatial reference. As for time, sensing information has to have time reference so that the DSA decision-making agent can utilize the most relevant measurements and estimate tendencies.

2- *Traffic demand.* As mentioned above, the DSA decision-making process should be *fair* in terms of considering traffic demand. More spectrum resources can be dynamically allocated where there is more traffic demand.

3- *Constraints.* These include security constraints, rules, and policies. It is critical for the DSA decision-making process to adhere to security constraints. For example, encryption of geolocation information can be a security constraint. Secondary user rules are one example of DSA rules. Chapter 5 is dedicated to DSA as a set of cloud services that relies on policies, rule sets, and configuration parameters to constrain the DSA cognitive engine evolver. DSA has to evolve within the intended use of the managed networks. In military communications, it is critical to reflect the commander's intent in the DSA policies, rule sets, and configuration parameters in order to meet the mission needs.

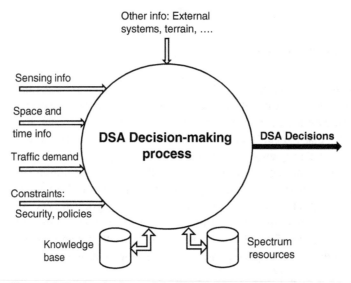

Figure 1.4 The involvedness of the DSA decision-making process.

4- *Other information*. This can include external systems and their use of spectrum, and terrain information. Terrain information can help the DSA decision-making process differentiate between interference from other systems versus degraded signal due to terrain effects as well as evaluating a spectrum assignment before it is executed.

The DSA decision-making process can use two important categories of information in its information repository among other information categories. The first category of information is the spectrum resources pool, which can be changed dynamically with time. The DSA decision-making process uses and updates the spectrum resources pool according to spectrum assignments. The second category of information is the knowledge base obtained from fusing spectrum sensing information. A good design of DSA as a set of cloud services should be able to use objective DSA metrics to measure the effectiveness of past decisions and should be able to adapt future decisions based on the behavior measured through the DSA metrics. Adapting future decisions will rely on ubiquitous changes in policies, rule sets, and configuration parameters, as further explained in Chapter 5.

The cognitive engine implementation of the DSA decision-making process can perform many simultaneous tasks. Consider, for example, the presence of different propagation models within the information repository. A process within the cognitive engine can use geolocation information and terrain data as shown in Figure 1.4 while analyzing spectrum sensing information to decide the best propagation model that can be used at a given time. As geolocation information is updated, another part of the terrain data may be used, which may require changing the propagation model. The cognitive engine continues to recommend the best propagation model to be used to analyze the effectiveness of a potential spectrum assignment before it is assigned. This makes the DSA decision-making process able to have a level of confidence for any spectrum change and be able to recommend the best spectrum assignment in cases where more than one option of spectrum assignment can be considered.

1.5 The Pitfalls of DSA Decision Making

DSA design must avoid some pitfalls that can render the design suboptimal. These pitfalls include the following:

1. *Failure to see a complete picture of the involvedness of the DSA decisions*. DSA designs that ignores the role of important information, constraints, and the intended use of the managed networks can be suboptimal.
2. *Failure to consider the hierarchy of DSA decisions*. This can result in a rippling decision-making process and diminishing throughput efficiency.
3. *Failure to take advantage of all available information*. The presence of augmented sensing hardware that gives rich information on wideband spectrum sensing can lure the designer to ignore the value of other information such as same-channel in-band sensing. Also, one must be cognizant of the relevance of the collected spectrum sensing information. Old sensing information must be discarded as it can be harmful to use since it does not reflect the current state of spectrum usage. As Chapter 8 shows, making co-site interference information available to the DSA decision-making process and taking into consideration the different types of co-site interference information can decrease rippling and ensure better optimization of the spectrum resources utilization.
4. *Overcomplicating the decision-making process*. The use of cognitive techniques is powerful and is needed in the DSA design. However, developing a complex machine learning technique where stochastic models can give the best decision is a pitfall that must be avoided. This type of overcomplication can occur in the local decision-making process where physical layer metrics are the reference for some decisions. Using stochastic models can render the best results a machine learning approach can give. Reducing the complication of DSA decision making can also include overlaying or superimposing decisions in the order of their

precedence. As Chapter 8 explains, co-site interference information can be considered as a second order of spatial separation that is only considered after the first order of spatial separation of frequency assignments is completed. This overlay of decision making reduces complexity while using later, less important decisions to fine-tune earlier more important decisions.

5. *Design rigidity.* DSA is evolving and a good design should consider incremental increases in DSA capabilities. One important approach is to leverage standardized interfaces such as the DySPAN standards covered in Part 4 of this book instead of utilizing proprietary interfaces. Standardized interfaces allow leverage of new capabilities with minimal integration efforts. Also, the overlay of decision-making techniques explained above can lead to reducing design rigidity. DSA as a set of cloud services can add new services that fine-tune the outcome of existing services for better optimization of spectrum resources without the need to modify earlier capabilities.

6. *Ignoring the role of DSA policies.* Overlooking how to tie cognitive decisions to policy automation can render a system unusable. This pitfall can result in violations of established rules and the cognitive engine making some decisions that should be avoided. This is specifically critical with military communications systems where DSA policies have to reflect the commander's intent while being ubiquitous.

7. *Not considering the need for decision hierarchy.* Very often designers of a distributed MANET focus on cooperative decisions and pay little attention to decisions that can be or should be made locally. Similarly, a designer of heterogeneous networks DSA may ignore the need for a centralized arbitrator.

8. *Ignoring decision fusion and abstraction.* When spectrum sensing information is shared in a distributed or centralized manner, fusing and abstracting the shared information is needed to reduce the control traffic volume. With distributed cooperative and centralized decision-making solutions, locally made decisions can be shared with an abstracted version of spectrum sensing information if needed.[6]

The chapters of this book are organized to help the reader make design decisions that can optimize the performance of DSA services and avoid the pitfalls mentioned above. For example, if you are designing a distributed cooperative DSA system, there could be room for local decisions and optimum local fusion and optimization of DSA control traffic volume. If you are designing a centralized DSA system, optimum hierarchical decision making can lead to more efficient spectrum management of heterogeneous networks than relying mainly on centralized decisions.

1.6 Concluding Remarks

DSA solutions can differ drastically from one system to another. However, there are common foundations that can be used in any DSA design approach. The goal is always the same: optimize the use of a given pool of spectrum resources dynamically and react to changes in environments. With this book laying some theoretical foundations of DSA design, addressing the most generic model of DSA, and then showing with case studies how this generic model can be applied to different cases, the reader should be able to obtain knowledge on how to approach DSA and how to create design concepts for any system under consideration. The separation of the physical layer aspects of DSA from the cognitive engine aspects in this book should help the reader address the design aspects of the physical layer separate from the design aspects of cognitive engines such that DSA can be applied to noncognitive systems and cognitive systems and the design can address any system requirements and boundaries.

As the reader goes through the rest of this book, the many facets of DSA will become clearer and the reader will see how a large system of heterogeneous networks may require using

---

6 Abstracting spectrum sensing information is covered in detail in Chapter 4.

heuristic approaches, overlaying of decisions-making processes, and considering many aspects and tradeoffs to design an effective system.

Exercises

1. Consider a hybrid DSA system you are designing with 10 mins minimum elapsed time before changing frequency assignment for a large-scale set of heterogeneous networks. One of these heterogeneous networks has distributed agents in its nodes that make DSA decisions for the network and these decisions can take up to 200 ms due to spectrum sensing information propagation time and processing time. If you are asked to come up with a minimum time interval for a local agent to make a local DSA decision (time elapse after making a local DSA decision before you can consider making another one), which of the following time intervals would you choose?

    A. 200 ms

    B. 100 ms

    C. 20 ms

    Explain the reason for your choice.

2. In a hybrid DSA system, would you want a local DSA decision (e.g., power increase) to be propagated to peer agents that make distributed decisions? Explain your reason.

3. You are given a set of heterogeneous wireless networks with one network that operates at a narrowband of 100 kbps. Your analysis of DSA shows that DSA control traffic (e.g., spectrum sensing information and other DSA configuration parameters) would use about 30 kbps over this narrowband network. Would you proceed with your DSA design or consider assigning the narrowband network a fixed frequency band to operate at? Explain the reason for your choice.

Bibliography

Gandetto, M. and Regazzoni, C., Spectrum sensing: A distributed approach for cognitive terminals. *IEEE Journal on Selected Areas of Communications*, vol. 25, no. 3, pp. 546–557, April 2007.

Haykins, S., Cognitive radio: brain empowered wireless communications. *IEEE Journal on Selected Areas in Communications*, pp. 201–220, February 2005.

Hoffman, H., Ramachandra, H.P., Kovács, I.Z. et al., Potential of dynamic spectrum allocation in LTE macro networks. *Advances in Radio Science Open Access Proceedings*, vol. 13, pp. 95–102, 2015.

Hossain, E., Niyato, D. and Han, Z., *Dynamic Spectrum Management in Cognitive Radio Networks*. Cambridge University Press, 2009. ISBN: 978-0-521-89847-8.

Hu, R. and Qian, Y., An energy efficient and spectrum efficient wireless heterogeneous network framework for 5G systems. *Proceedings of the IEEE Communications Magazine*, vol. 52, issue 5, May 2014.

McHenry, M, Livsics, E., Nguyen, T., and Majumdar, N., XG dynamic spectrum sharing field test results. In *Proceedings of the IEEE International Symposium on New Frontiers in Dynamic Spectrum Access Networks*, Dublin, Ireland, April 2007, pp. 676–684.

Mitola, J., *Cognitive Radio Architecture, in Cognitive Networks: Towards Self-Aware Networks* (ed. Q.H. Mahmoud). John Wiley & Sons, Chichester, 2007. doi: 10.1002/9780470515143.

Srinivasa, S. and Jafar, S., Cognitive radios for dynamic spectrum access – The throughput potential of cognitive radio: A theoretical perspective. *IEEE Communications Magazine*, vol. 45, issue 5, May 2007.

http://grouper.ieee.org/groups/dyspan/ .

http://www.sharedspectrum.com/resources/darpa-next-generation-communications-program/ .

https://www.wirelessinnovation.org/.

2

Spectrum Sensing Techniques

In order for a DSA wireless system to observe, orient, decide, and act as explained in the previous chapter, it must be aware of the spectrum sensing parameters and how they relate to the sensed frequency band characteristics. This chapter addresses the different spectrum sensing techniques that can be utilized. The spectrum sensing techniques covered in this chapter are presented in a generic way while pointing to which techniques can be implemented on specialized hardware and which techniques can be implemented as same-channel for in-band sensing. Notice that the focus of this chapter is not limited to spectrum sensing performed for a secondary user to use some spectrum bands opportunistically. As the previous chapter explained, there are many DSA cases in defense and commercial applications that go beyond the secondary user scope and are not necessarily in the cognitive radio domain. The focus of this chapter is also not limited to spectrum sensing techniques that can be developed for distributed cooperative MANETs. The spectrum sensing techniques explained here are generic and can be used for local decisions, distributed cooperative decisions, centralized decisions, or hybrid decisions, and for different types of communication systems.

2.1 Multidimensional Spectrum Sensing and Sharing

Spectrum sensing provides multidimensional spectrum usage characteristics over time, space, and frequency bands. Spectrum sensing can range from simple techniques such as measuring a signal energy or the SNIR, to learning about the types of signals occupying the frequency band while analyzing the signal specifics such as modulation techniques, utilized bandwidth, and carrier frequency. Naturally, the more complex the sensing techniques are, the more computationally intensive they are. One of the most important advantages of using specialized hardware for spectrum sensing is the ability to sense a wide band of frequencies through parallel processing of different segments of the sensed wideband spectrum. Specialized hardware can be built to process algorithms such as fast Fourier transformation (FFT) in real time.

There are different spectrum sensing techniques that are covered in this chapter. Notice that designing a system that utilizes spectrum sensing can mix and match any of the below techniques and shouldn't necessarily use a single sensing technique. Spectrum sensing techniques can include the following:

1. *Time, frequency, and power sensing.* This spectrum technique focuses on detecting the presence of signal energy on certain frequency bands based on factors such as time intervals and signal power measured within a frequency band.
2. *Energy detection.*[1] This is the simplest and most commonly used spectrum sensing technique as it does not require prior knowledge of the signal being sensed. This technique simply measures the received signal power over a given sensing time. Energy detection can be used in conjunction with many of the other spectrum sensing techniques explained

---

1 Please reference Appendix 2A for an explanation of the use of the term "energy detection" in spectrum sensing.

Dynamic Spectrum Access Decisions: Local, Distributed, Centralized, and Hybrid Designs, First Edition. George F. Elmasry.
© 2021 John Wiley & Sons Ltd. Published 2021 by John Wiley & Sons Ltd.
Companion website: www.wiley.com/go/elmasry/dsad

below. Some sensing techniques perform energy detection before exploring some of the signal characteristics explained below.

3. *Signal characteristics*. With this spectrum sensing technique, the sensor attempts to synthesize the sensed signal to find specific details about it. This spectrum sensing approach can start with technique such as:
   a. matched filter based spectrum sensing
   b. autocorrelation based spectrum sensing.
   After sensing a specific signal presence, the sensor can improve signal detection accuracy using signal characteristics based spectrum sensing such as:
   c. spreading code spectrum sensing
   d. frequency hopping spectrum sensing
   e. orthogonality based spectrum sensing.
   With some known signals, the sensor can perform more signal-specific spectrum sensing techniques such as:
   f. waveform based spectrum sensing
   g. cyclostationarity based spectrum sensing.
4. *Euclidean space based detection*. As explained in Chapter 1, geolocation and timing are an essential part of comprehensive spectrum sensing. DSA can include:
   a. geographical space spectrum sensing
   b. angle of radio frequency (RF) beam spectrum sensing.

Depending on the system under design, certain sensing techniques would have more importance than others. For example, in typical commercial opportunistic spectrum use applications (e.g., a secondary user using a primary user spectrum opportunistically), time, frequency, and power techniques combined with Euclidean distance based techniques may be the most important sensing techniques to consider. In some defense applications, hopping pattern spectrum sensing can be critical for some SigInt needs. In commercial and defense applications, finding the modulation technique and spread spectrum codes can be important to identify the characteristics of a known signal occupying a certain bandwidth (e.g., a commercial cellular signal). While in defense applications this identification of a known commercial signal presence can rule out the presence of malicious users (e.g., jammer), in some commercial and defense systems this signal identification can mean the ability of the secondary system to use the sensed frequency band with a signal that appears as background noise to the primary user (this shared spectrum use is known as underlay transmission) or it can mean ruling out using this frequency band altogether.

Before proceeding with spectrum sensing techniques, let us reference some definitions that are common in the literature:

- *A primary user* is the system that has rights to the bandwidth. This system can be oblivious to the presence of a secondary user.
- *A secondary user* is the system that needs to detect the activity of the primary user and can only take advantage of the unused parts (holes) of the primary user spectrum. In some cases, the secondary user may use the sensed spectrum with a very low power signal that ensures the primary user rights are not compromised.
- *Spectrum opportunity* is a band of frequencies that are not being used by the primary user at a particular time in a particular geographic area.

This book's approach to DSA is not limited to the case of a primary user and a secondary user. Even with communication systems that have allocated spectrum to share with no primary and secondary users rules, DSA can allow a set of heterogeneous networks to dynamically share the allocated spectrum based on up-to-date spectrum sensing information.

DSA can result in spectrum sharing, which can take different forms as follows:

- *Underlay transmission*. With this transmission, the secondary user may transmit in parallel to the primary user if the secondary user signal does not interfere with the primary user signal. The definition of interference here means a certain threshold that the primary user can tolerate. Typically, a secondary user can use spread spectrum with large spreading codes (and lower bit rate) to ensure that the secondary user signal appears as background noise to the primary user signal. Notice that the underlay transmission concept can be used with defense applications to generate signals with low probability of detection and low probability of interception (LPD/LPI). There is a proposed rule before the FCC that makes the underlay transmission party "hold harmless" the primary user. In the commercial world, a primary user can file a claim against an underlay transmission party for harming its OTA use. In the defense world, and during war time, this is a hard claim to make.

- *Overlay transmission.* With this transmission, the primary user and the secondary user may cooperate in a way that allows the secondary user to use the frequency band under certain conditions. Notice that in defense applications overlay transmission can have a different meaning. Defense applications can become a secondary user of commercial frequency bands in peace time using underlay transmission rules, but in war time, the same systems can overlay the defense signal over the commercial signals with or without cooperation with the commercial systems.

- *Opportunistic transmission.* With this transmission, the secondary user can only use the primary user frequency band if there is no primary user transmission.

- *Cooperative transmission.* Within a MANET, nodes can share a frequency band divided into sub-bands (slots) in a cooperative manner relying on geographical separation rules and spectrum sensing. In a large-scale set of heterogeneous networks, networks can also share different frequency bands relying on distributed cooperative techniques and/or a centralized DSA arbitrator decision-making process. Other known concepts of cooperative transmission include:
  - ➤ time division multiple access (TDMA), where multiple users share the same frequency band cooperatively in different time slots
  - ➤ frequency division multiple access (FDMA), where multiple users share the same frequency band cooperatively in different frequency sub-bands
  - ➤ carrier sense multiple access (CSMA), where nodes sense if the carrier is busy with transmission and only transmit if the carrier is available
  - ➤ code division multiple access (CDMA), where different simultaneous transmissions use different spreading codes that makes a communicating pair of nodes see the other communications as background signals
  - ➤ orthogonal frequency division multiple access (OFDMA), where interference between orthogonal signals is minimized.

This chapter focuses on spectrum sensing techniques independent of how spectrum sharing occurs. Certainly, one must consider the importance of the information obtained from a spectrum sensing technique versus the complexity of the technique. With the existence of different options for dedicated spectrum sensing hardware, the system designer must understand all the aspects of spectrum sensing techniques before making an educated decision on which hardware to include in a system design, how to configure it, and how to interpret the spectrum sensing information obtained from that hardware. The system designer also has to be able to make the most out of same-channel in-band sensing. This chapter addresses the first steps towards grasping the many facets of spectrum sensing through covering the foundations of many of the known techniques to help the system designer select the most appropriate sensing technique for a given system.

Figure 2.1 illustrates how sensing a wideband of frequency can be expressed as spectrum usage in terms of frequency and amplitude. A spectrum sensor can divide this wideband into small sub-bands, with each sub-band defined by a frequency range where the spectrum sensor performs sensing over all of the sub-bands in parallel. An important objective of spectrum sensing is the ability to sense a wideband of frequency divided into sub-bands so that the information output of the spectrum sensor can help the DSA techniques find different spectrum opportunities.

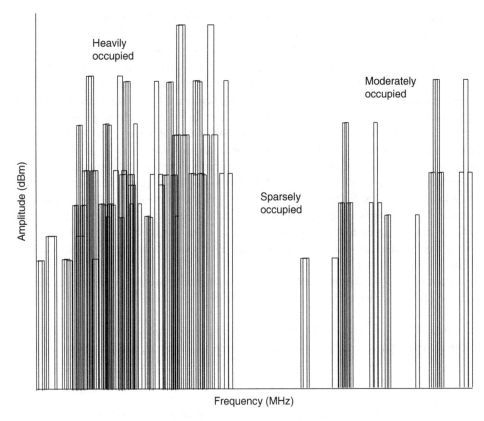

Figure 2.1 Sensing a wide band of spectrum.

Notice how the vertical access unit in Figure 2.1 is dBm, which stands for decibels (dB) below one milliwatt (mw). This unit is commonly used for energy detection which uses the one milliwatt as a reference and bridges the gap between SNIR, which is in dB,[2] and energy detection in watts. With this normalization, a signal received at 1 mw maps to 0 dBm while a signal received at 0.1 mw maps to –10 dBm. Energy detection can be expressed as received signal strength indication (RSSI), which uses the unit dBm. This conversion uses the formula $dBm = 10 \times \log(P/1 \text{ mw})$, where P is signal power.

2.2 Time, Frequency, and Power Spectrum Sensing

Although energy detection is the most common spectrum sensing technique, time, frequency, and power spectrum sensing are covered first in this chapter to emphasize the multidimensional aspects of spectrum sensing. The idea behind this spectrum sensing technique is to create multidimensional spectrum awareness. The simplest form of this spectrum sensing technique is a two-dimensional spectrum sensing that uses the frequency and time dimensions. Figure 2.2 shows this two-dimensional spectrum sensing where the spectrum sensor looks for occupancy of certain frequency bands at certain times.

A secondary user can use this spectrum sensing technique where it can hop to a different frequency band once it detects another user on a frequency it is using. This technique does not consider the signal power and relies on DSA defining a cutoff RSSI level to consider a frequency band as occupied or can be opportunistically used. The cutoff RSSI level may also be an estimation of additive white Gaussian noise (AWGN) without the presence of any signal.

If a communications system is to consider underlay or overlay transmission, the two-dimensional spectrum sensing in Figure 2.2 can be turned into three-dimensional spectrum

---

2 Notice that dB is a logarithmic scale that is used to describe the ratio of signal to noise.

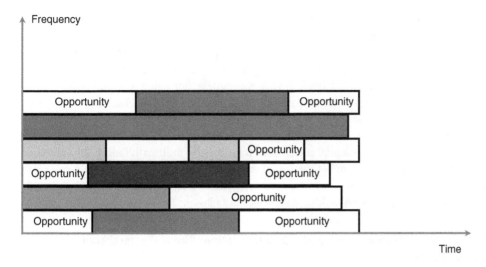

Figure 2.2 Two-dimensional spectrum sensing.

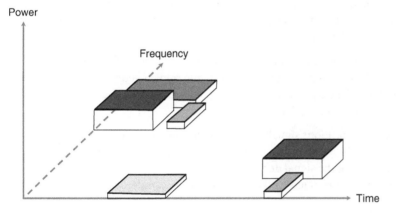

Figure 2.3 Three-Dimensional spectrum sensing.

sensing as shown in Figure 2.3. In this case, the power dimension is added. In Figure 2.3, the unoccupied areas are often referred to as *spectrum holes* or *white spaces*.

Notice that more dimensions can be added to this multidimensional spectrum sensing technique. For example, the spreading code can be sensed and made a fourth dimension. Spreading code sensing can show opportunistic transmission based on using specific spreading codes. Spectrum sensing of spreading code is covered in Section 2.3.3 as a signal characteristic.

2.3 Energy Detection Sensing

This is the most common spectrum sensing approach used today. As explained in Chapter 3, the receiver's operator characteristic (ROC) function makes good use of this simple energy detection approach. Chapter 3 covers how same-channel in-band sensing can use energy detection sensing with minimal requirements on the receiver to hypothesize the presence of interference.

RSSI is a common expression of energy detection. RSSI is so common that you can look at your phone while having wireless fidelity (WiFi)[3] connectivity and count the WiFi connectivity indicator lines on the top of your screen to see how RSSI is commonly mapped to about four levels (no connectivity, low connectivity, medium connectivity, and good connectivity). Laptop

---

3 WiFi is a trademarked phrase that means IEEE 802.11.

WiFi connectivity indicators typically map the WiFi RSSI to five or six levels. Other devices illustrate RSSI in more or less this number of levels.

A simple receiver can collect the energy received on the antenna in a certain frequency band and quantize it. Low computational complexity and simple implementation are what makes energy detection commonly used. In Figure 2.3, energy detection below a certain power threshold constitutes an opportunity for a spectrum band use (e.g., by a secondary user[4]). On the other hand, energy detection above that threshold constitutes an occupied band. In Figure 2.3, energy detection is the power axis. Deciding the value of that cutoff threshold can be challenging as the primary user signal may suffer from interference, multipath fading, and jamming among other factors that can affect the signal strength. Energy detection becomes especially more challenging when sensing spread spectrum signals that tend to have low energy. Chapter 3 is dedicated to DSA decision making and will discuss how cutoff thresholds can be used.

2.3.1 Energy Detection Sensing of a Communications Signal (Same-channel in-band Sensing)

Let us start from the unit of energy of a communications signal $\Phi_j(t)$, which can be defined as follows: If the receiver detects a 1 V signal across a 1 Ω resistor, the integration of the square value of signal voltage over a specific time period (T_g, T_f) is 1, that is, the receiver has detected one unit of energy.[5] Notice the following:

1. The signal can be constructed in a multidimensional signal-in-space (SiS) as a vector.[6]
2. The time $T = T_f - T_g$ is a critical factor in detecting the signal energy.[7] If the signal is too weak, the integration of the square value of the signal voltage may need a long period of time to yield reliable energy detection.

The receiver of a communications signal detects a multicoefficient signal in N-dimensional SiS and attempts to match the received signal with one of M signals.[8] The energy detector cares only for the signal energy not the signal decoding.

The signal's ith dimension projected on the kth base can be expressed as follows:

$$S_{ik} = \int_{T_g}^{T_f} S_i(t) \, \Phi_k(t) \, dt, \tag{2.1}$$

where $\Phi_k(t)$ is the signal basis per coefficient.

Notice that both the signal receiver and the spectrum sensor performing energy detection need to carry similar steps to calculate the received signal energy. Chapter 3 covers how to use same-channel in-band sensing to hypothesize the presence of an interfering signal. This section is intended to show how to piggyback on the communications receiver's energy calculation to create same-channel in-band energy sensing.

Figure 2.4 shows how a communications receiver recovers a signal $S_i(t)$ based on knowing its base per each dimension. The projection of the signal per each dimension is expressed in Equation (2.1). The baseband signal $S_i(t)$ is then mapped to a point in the N-dimensional SiS. The square distance from the origin to the projected point from Figure 2.4, $\sum_{m=0}^{N-1} S_{im}^2$, is simply the received signal energy. While the communications receiver takes further steps to decode the projected signal based on the squared intra-signal distances, an outcome of

---

4 Opportunistic use of a spectrum by a secondary user is one example of opportunistic spectrum use. The more common opportunistic spectrum use is with unlicensed frequency bands where any user looks for spectrum holes to transmit on.

5 This one unit of energy can be dependent on the sensing time period.

6 The reader can refer to digital communications specialized references on how a signal is constructed in multidimensional SiS.

7 With same-channel in-band sensing, T is well correlated to the *dwell time*, as explained in Chapter 3.

8 M is typically a power of 2. With a binary antipodal signal, $M = 2$. With a 4-ary phase shift keying signal, $M = 4$. With 8-ary phase shift keying, $M = 8$. QAM signals can be 16, 32, 64, etc. Except for the binary antipodal signal, which has one direction, all of these examples have a two-dimensional signal (i.e., $N = 2$). Orthogonal signals can have higher dimensions.

Figure 2.4 Leveraging signal receiver reconstruction of the received signal for same-channel in-band spectrum sensing.

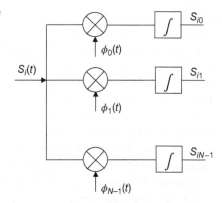

the communications signal receiver, $\sum_{m=0}^{N-1} S_{im}^2$, is energy detection that can be utilized for same-channel in-band sensing. Calculating $\sum_{m=0}^{N-1} S_{im}^2$ is the common process between signal decoding and signal energy detection. As will become clear in Chapter 3, the purpose of same-channel in-band sensing is to decide if the communications signal utilized frequency band suffers from an interference level that requires using a different frequency band or if the interference level is tolerable.

A communication receiver also calculates SNIR. The SNIR calculation is based on the energy detection of the signal and the energy detection of the noise (known as the noise floor). Signal energy detection and noise floor energy detection are metrics that can be leveraged for same-channel in-band spectrum sensing because they are common between signal decoding and same-channel in-band sensing. The value of using SNIR in same-channel in-band sensing will become clear in Chapter 3.

While energy detection is a natural outcome of signal decoding, building specialized hardware for spectrum sensing can perform energy detection in different ways. This specialized hardware, which is sometimes referred to as an augmented sensor, may or may not have prior knowledge of the signal bases. The following subsections show different methods that can be utilized by the augmented sensors to perform energy detection.

2.3.2 Time Domain Energy Detection

With this technique, the spectrum sensor has to rely on using bandpass filters. The spectrum sensor is given a center frequency f_0 and a bandwidth W to define the frequency range to sense. The spectrum sensor inputs the signal $s(t)$ through the bandpass filter followed by a squaring device and an integrator, as shown in Figure 2.5. The details of how to build such a sensor are beyond the scope of this book. However, these sensors have the ability to define bandpass filters for any given center frequency f_0 and any given bandwidth W in the broad-spectrum range they are designed for. These sensors also have the ability to perform energy detection on a wide band of frequency divided into smaller sub-bands in parallel.

Notice the importance of T in the integrator in Figure 2.5. A signal with weak power spectral density such as a spread spectrum signal would needs a longer time period T. With augmented sensors, the bandpass filter has a critical transfer function that can be expressed as follows:

$$H(f) = \begin{cases} \dfrac{2}{\sqrt{N_0}}, & |f - f_0| \le \dfrac{W}{2} \\ 0, & |f - f_0| > \dfrac{W}{2} \end{cases} \tag{2.2}$$

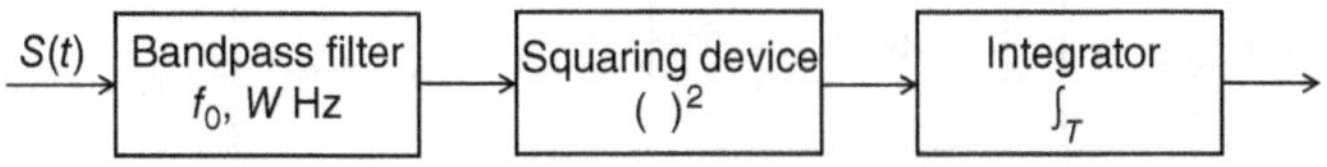

```
S(t) → | Bandpass filter |  → | Squaring device | → | Integrator |  →
         |   f₀, W Hz     |    |      ( )²        |   |     ∫_T      |
```

Figure 2.5 Time domain energy detection.

The reason for emphasizing Equation (2.2) is that the augmented sensor needs to estimate the noise one-sided power spectral density N_0, which is a challenge when the augmented spectrum sensor may have no prior knowledge of the signal it is sensing and hence has no means to directly estimate the noise floor, as in the case of same-channel in-band sensing. Instead, the augmented sensor relies on normalizing for the noise power and uses this normalization to compute the probability of a false alarm and the probability of detection as detailed in Chapter 3. If we conceptualize how the sensor creates an energy detection sample every T seconds, then in a large number of samples we have a high probability that the signal being sensed was not transmitted during the entire time period T. Thus, the noise floor becomes the energy sample collected with the minimum energy detection. The importance of Equation (2.2) is that it expresses the noise one-sided power spectral density that correlates to the noise floor.

In time domain spectrum sensing, the time duration that the sensed signal remaining in a particular state can affect the outcome of the spectrum sensor. This time duration is referred to as the *dwell time*. The spectrum sensor observation time length should correlate to dwell time. Chapter 3 shows that one advantage of same-channel in-band sensing is that the in-band signal is sensed during a known state and the sensing technique does not observe the signal during multiple states within a single sensing window. That is, the sensing technique can have knowledge of whether the transmitted signal is present or not during the entire sensing window. On the other hand, augmented sensors using time domain energy detection, where the sensed signal may change state during the observation time, can lend a higher probability of false alarm.

2.3.3 Frequency Domain Energy Detection

With this energy detection technique, the sensor also has to be configured for a center frequency f_0 and a bandwidth W to define the frequency range to sense. The sensor uses a bandpass filter, as with time domain energy detection, followed by an analog-to-digital convertor (ADC) to digitize the signal and FFT to convert the signal to the frequency domain (Figure 2.6). The squaring device calculates the energy per each frequency coefficient and the mean value stage is used to calculate the average energy over the observed frequency band.

As with time domain energy detection, frequency domain energy detection has to consider the presence of noise. The method used to estimate the noise power spectral density can rely on discrete Fourier transformation (DFT) where the digitized data is divided into segments and a sliding window is used to estimate the average noise spectral density. One reason to choose frequency domain energy detection over time domain energy detection in augmented sensors is the higher accuracy of noise estimation but the price for that is the need for more computational power.

Notice how with the three energy detection techniques covered so far, the outcome is simple:

1. Signal energy level at the defined carrier frequency f_0 and bandwidth W, and
2. Noise floor energy at the same carrier and bandwidth.

It is important to note that the hypothesizing and decision-making processes covered in Chapter 3 can be tricky under certain circumstances, such as fading channels. While frequency domain energy detection can implement good techniques such as the sliding window explained above, distributed and centralized DSA techniques can have a view of spectrum sensing that is more comprehensive than a local node. Distributed and centralized DSA techniques can analyze spectrum sensing information per RF neighbor and further overcome the uncertainty that can result from fading channels.

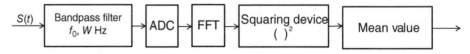

Figure 2.6 Frequency domain energy detection.

While Section 2.3.1 introduced same-channel in-band sensing, Sections 2.3.2 and 3.3.3 introduced the most common forms of augmented sensing where the augmented sensor can be configured to sense a frequency band defined by the carrier frequency f_0 and bandwidth W as illustrated in Figures 2.4 and 2.5. Augmented sensors can be built to sense a wide band of frequencies and sense multiple sub-bands each defined by a carrier frequency f_0 and a bandwidth W simultaneously. This simultaneous sensing utilizes parallel paths where each path starts with a bandpass filter configured for the carrier frequency f_0 and bandwidth W for one of sub-bands being sensed.

2.4 Signal Characteristics Spectrum Sensing

There are different signal characteristics that a spectrum sensor can detect. Here, we go beyond simple energy detection with no prior knowledge of the signal being sensed to having some prior knowledge of the signal and the ability to synthesize the detected signal to extract more information.

2.4.1 Matched Filter Based Spectrum Sensing

This technique requires pre-knowledge of many aspects of the sensed signal such as bandwidth, operating frequency, modulation type and order, pulse shaping, and frame format. The spectrum sensor can quickly detect the presence of the sensed signal with high accuracy. This technique can be used before discovering more detailed signal characteristics such as spreading code and hopping pattern.

The matched filter will accentuate the targeted signal $S(t)$ and will suppress other signals and noise. Notice that signals other than the targeted signal $S(t)$ are essentially noise with respect to $S(t)$. The impact of the suppressed signals and noise are referred to as $W(t)$. The design of this matched filter includes:

1. Creating a contrast between $S(t)$ and $W(t)$ such that when $S(t)$ is present at a time t, the output of the filter will have a large peak
2. Minimizing the probability of error. This can be achieved by considering the energy of the signal and the energy of the noise over a time T instead of considering the signal and noise amplitude. Energy calculation uses the square of the amplitude.

Notice that with wireless communications systems where we decode symbols, minimizing the probability of symbol error also uses signal and noise energy. However, the probability of error in spectrum sensing has two folds. With spectrum sensing, we have a probability of false alarm where the matched filter decides that $S(t)$ is detected but $S(t)$ was absent and the probability of misdetection where the matched filter decides that $S(t)$ is absent but $S(t)$ was present.[9]

Figure 2.7 shows the use of a matched filter to detect the presence of a signal $S(t)$. Notice that a spectrum sensor can sense more than one signal type using a bank of matched filters. Once the spectrum sensor decides the signal type, other signal synthesizing techniques can be used to discover characteristics such as spreading code or orthogonality.

Some of the disadvantages of the matched filter sensing technique include the following:

- The implementation complexity may not be practical to implement for a large set of signals. Consider the detection of all types of commercial cellular signals and other known commercial but not cellular signals.
- Large power consumption is needed to execute the various receiver algorithms.

---

9 Statistical decisions in symbol detection also have a probability of false alarm and a probability of misdetection. However, both probabilities result in detecting the wrong symbol and hence most signal detection techniques are interested in the probability of symbol error without the distinction between the probability of false alarm and the probability of misdetection.

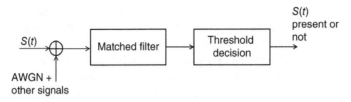

Figure 2.7 Signal detection using matched filters.

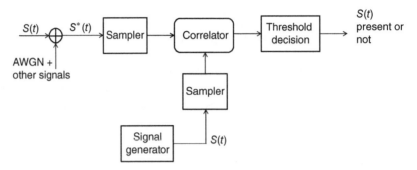

Figure 2.8 Signal detection using autocorrelation.

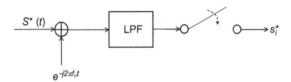

Figure 2.9 Signal sampling before autocorrelation.

2.4.2 Autocorrelation Based Spectrum Sensing

This spectrum sensing approach has more advantages than the matched filter approach described in the previous section when attempting to detect the presence of a narrow band signal. Autocorrelation estimation can result in a constant false alarm rate (CFAR) by employing techniques that reduce the dependency on noise power. With this sensing technique, the sensor can have a stored representation of the signal as samples $S(t) = (s_0, s_1, \ldots, s_{N-1})$ or has a signal generator that can generate a copy of the sensed signal and sample it. The sensed signal $S^*(t)$ is sampled to produce $S^*(t) = (s_0^*, s_1^*, \ldots, s_{N-1}^*)$. The result of the correlation process[10] between $S(t)$ and $S^*(t)$ is compared to threshold to decide if the sensed signal matches $S(t)$ or not, as shown in Figure 2.8.

If the sensor uses a signal generator and a sampler instead of stored samples, as in Figure 2.8, the sampler can be broken down as shown in Figure 2.9 where down conversion and a low pass filter (LPF) is used before time sampling.[11] In Figure 2.9, the signal is a complex baseband signal with center frequency f_c. The LPF has a bandwidth of $(-f_{bw}, f_{bw})$ Hz and the sampling rate is $T_s \overset{\Delta}{=} (2f_{bw})^{-1}$.

The correlation technique attempts to produce unbiased estimation of the signal that can be expressed in terms of a variable l that is adaptively chosen to reduce noise power dependency. Different techniques can be employed to incorporate l in the correlation function, such as estimating a covariance matrix.

Autocorrelation based spectrum sensing can use a bank of signal generators, samplers, and correlators to detect the presence of multiple signals.

---

10 The differences between $S(t)$ and $S^*(t)$ are two-fold. The first difference is the added noise in $S^*(t)$. The second difference, which can have a larger impact, is the time lag between the samples of $S^*(t)$ and $S(t)$. If the time lag is zero, $S^*(t)$ is aligned to $S(t)$ and this process becomes a convolution process.
11 Signal detection using autocorrelation can also happen in frequency domain where noise covariance can be better estimated.

Figure 2.10 Direct sequence spread spectrum modulation.

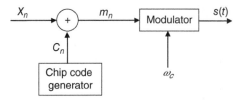

Figure 2.11 Frequency-hopping spread spectrum signal modulation.

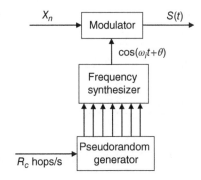

Notice the difference between time-domain cyclic marks and frequency domain cyclic marks. OFDM signals use a frequency-domain cyclic prefix, which protects the OFDM signals[12] from inter-symbol interference. This cyclic prefix can be utilized in a frequency-domain based correlation technique to affirm the presence of the targeted signal. If the targeted signal uses a preamble of symbols, this preamble can be utilized to affirm the presence of the signal in time-domain correlation. Both cyclic prefix and time-domain preamble can help the autocorrelation capable sensor decrease the probability of misdetection and the probability of false alarm when making a decision regarding the presence of the targeted signal.

2.4.3 Spreading Code Spectrum Sensing

Spread spectrum is implemented differently in commercial signals than in military signals. Commercial signals tend to use direct sequence spread spectrum, which is easy to sense, while military signals tend to use frequency hopping spread spectrum, which is intentionally made hard to detect.

Figure 2.10 shows the commercial case where a chip code C_n is module-2 added to the signal binary stream X_n before modulation. The resulting binary stream m_n is a pseudo random sequence. The modulated signal $s(t)$ is a spread spectrum signal.

Figure 2.11 shows a frequency hopping spread spectrum, which divides the available spectrum into slots. The carrier frequency then hops among these slots at a rate that is known as the frequency hopping rate. The hopping range of W Hertz is divided it into N slots with each slot size being Δf. The carrier frequency is always decided by the frequency synthesizer at each moment. The synthesizer is driven by a pseudorandom generator which generates unique N input vectors to the frequency synthesizer.

With Figure 2.11, a pseudorandom generator, clocked at the hopping rate R_c, feeds the frequency synthesizer. The frequency synthesizer generates the hopping carrier frequency for the modulator expressed as $\cos(\omega_i t + \theta)$. Notice that the binary stream X_n is modulated over the carrier with frequency ω_i, which corresponds to the ith slot of the N slots available for hopping. The hopping rate is determined by R_c.

When autocorrelation detection is used and the presence of the targeted signal is hypothesized to exist, the spectrum sensor can use a chip code generator to generate all the chip codes the targeted signal is known to use. One of the chip codes that is used by the targeted signal

---

12 OFDM is used in both commercial and defense signals. Commercial signals include WiMAX, LTE 5 MHz, LTE 20 MHz, and 5G. Defense signals include the wideband networking waveform (WNW) and the soldier radio waveform (SRW).

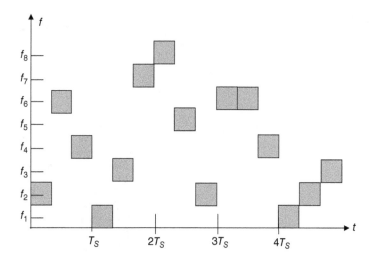

Figure 2.12 Fast hopping where three hops occur during the modulation of one symbol.

will result in the highest correlation with the sensed signal. The spectrum sensor may be able to help the local DSA decision fusion agent pinpoint which chip code is not used in a certain vicinity at a certain time, allowing for the opportunistic use[13] of the sensed frequency band only with an unused chip code.

2.4.4 Frequency Hopping Spectrum Sensing

An example of frequency hopping spread spectrum is illustrated in Figure 2.12, where three hops can occur during the modulation duration time, T_s, of a single symbol. In reality, defense signals tend to create fast hopping such that T_s is as small as possible.

With frequency hopping spread spectrum, both the hopping pattern and the chip code vectors are not known to an external spectrum sensor. External commercial spectrum sensors that sense a military signal should rely on spectrum sensing techniques that do not utilize frequency hopping and spread spectrum detection (i.e., not attempt to find the spreading code and the hopping pattern of the military signal). A benign method could be simple energy detection. In defense applications, a spectrum sensor can detect the presence of a malicious signal that attempts to jam the used defense signal through different means, including jamming a subset of the frequency slots ($f_1 - f_8$ in Figure 2.12) continually.[14] This type of spectrum sensing is performed by the military system to overcome malicious jammers when the defense application signal is not a secondary user.

2.4.5 Orthogonality Based Spectrum Sensing

This type of spectrum sensing is common among cooperative spectrum users of the same signal. With orthogonal spectrum sensing, the decision-making entity can be distributed, centralized or hybrid, as introduced in Chapter 1. The centralized decision-making entity is sometimes referred to as the decision fusion center (DFC). The decision-making process attempts to exploit signal orthogonality for cooperative spectrum use while mitigating the effect of fading, shadowing, out-of-range, and other factors that can increase the probability of false alarm and the probability of misdetection.

Orthogonal cooperative spectrum sensing communications systems have to take into consideration the use of multiple-input multiple-output (MIMO) antennas where multipath fading

---

13 This opportunistic use can be by another commercial signal or a military signal.
14 All the jammer needs to do is to jam enough frequency slots to cause error patterns that make the military signal error correction coding fail to correct the error patterns. This type of spectrum sensing is a cat and mouse game. The defense signal platform spectrum sensing can monitor if the jammer succeeds in overcoming the defense signal before switching the signal to a different mode or using another signal type.

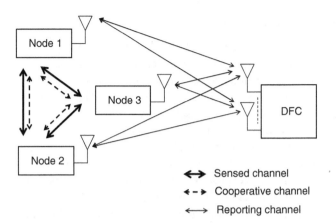

Node 1

Node 3

Node 2

DFC

⟷ Sensed channel
◄ - ► Cooperative channel
⟷ Reporting channel

Figure 2.13 Cooperative spectrum sensing with MIMO DFC.

is a critical factor. Some implementations of this spectrum sensing technique use a hybrid approach between a local, distributed, and centralized decision-making processes. In a typical system, referred to as the standard centralized fusion model, each node transmits its local decision outcome to a centralized DFC or a peer node. Spectrum sensing traffic, which include this reporting of node decision, can use a standalone channel known as parallel access channel (PAC) or can use one of the orthogonal signal multiple access channels (MACs). Reporting to peer nodes can use a separate channel from reporting to a centralized DFC, which is often referred to as the cooperative channel. Figure 2.13 shows three types of channels: (i) the sensed communications channel indicated by the thick lines; (ii) the reporting channel to the centralized DFC indicated by the thin lines; and (iii) the cooperative channel for peer-to-peer reporting of spectrum sensing decisions indicated by the dashed lines.

Notice in Figure 2.13 that all the nodes can have MIMO antennas (not just the DFC centralized entity). This model can work with a single-input single-output antenna for each node or a MIMO for each node. In either case, the centralized location must have a massive MIMO antenna in order to account for multipath fading.

With this cooperative mode, the ROC decision-making process explained in the next chapter is altered to a complementary receiving operating characteristics (CROC) decision-making process, which takes into consideration the difference in spectral efficiency.

2.4.6 Waveform Based Spectrum Sensing

This spectrum sensing approach relies on pre-knowledge of the signal to be sensed. Some commercial wireless signals use known synchronization patterns to align the receiving node processing to the received signal. These patterns can be exploited by the spectrum sensor to hypothesize the presence of the sensed signal. Signal synchronization patterns can include preambles, mid-ambles, regularly transmitted pilot patterns, spreading sequences,[15] etc.[16] These patterns allow the spectrum sensor to correlate the received signal with a known copy of itself (it is essentially a form of coherent detection). This correlation process leads to a spectrum sensing result that outperforms energy detector based sensing. The reliability of the correlation process increases when the known signal length increases. Waveform-based detection is used with known signals such as IEEE 802.11 signals.

While the autocorrelation based signal detection explained in Section 2.4.2 can be influenced by noise and the time lag between the samples, waveform based spectrum sensing is only affected by the presence of noise as the signal patterns align with the correlation process. Chapter 3 shows how the decision-making process of waveform based spectrum sensing may differ from that of simple energy detection spectrum sensing.

---

15 Spread spectrum based sensing was covered earlier in this chapter.
16 The preambles mentioned here are different from the time domain preamble, which is a sequence of symbols in the transmitted frame. The preambles here are frequency domain preambles.

2.4.7 Cyclostationarity Based Spectrum Sensing

With some commercial OFDM signals, waveforms are altered by the transmitter to add signatures in the form of cycle frequencies at certain frequencies. These signatures can increase the robustness against multipath fading. Spectrum sensors can leverage these features for signal sensing. These signatures introduce periodicity features. The introduced cyclic frequencies and the periodicity make the signal cyclostationary. Cyclostationary signals follow a spectral density (cyclic spectral density function, CSDF) that is leveraged by the detection process and is used to differentiate noise from the sensed signal. This differentiation happens because the modulated signal has cyclostationary characteristics while the noise has wide-sense stationary characteristics with no correlation. Cyclostationarity characteristics can also be used for distinguishing among different types of sensed signal.

Chapter 3 shows a type of cyclic autocorrelation function used with same-channel in-band signal sensing which estimates the noise spectral density separate from estimating the in-band signal spectral density.

2.5 Euclidean Space Based Detection

As explained in Chapter 1, DSA involves many factors other than spectrum sensing. When propagating spectrum sensing information to peer nodes or to a centralized location, the geolocations of the sensors must be attached to the spectrum sensing information. This allows a distributed or a centralized DSA process that fuses spectrum sensing information from different sensors to create a comprehensive view of spectrum use (spectrum map) in a given area of operation.[17] This comprehensive view can find each sensed signal's area of coverage (AOC) to create spectrum opportunities based on locations as well as find directional beams that can show more spatial opportunistic use.

2.5.1 Geographical Space Detection

When spectrum sensing information is propagated with location information defined by the sensor's latitude, longitude, and elevation, the fused information can create more spectrum use opportunities. The fusion process can show how at any given time, spectrum opportunity can be available in some parts of the area of operation while being fully occupied in other parts. The geographical space dimension helps the fusion process estimate propagation loss (path loss) in space to further ensure that spectrum reuse will not interfere with the sensed signals. Figure 2.14 shows an example of geographical separation creating opportunistic spectrum use in the case of a secondary user opportunistically using a primary user's spectrum. Figure 2.15 illustrates the case of a set of heterogeneous MANETs where DSA allows them to cooperatively and dynamically share a set of frequency bands (f_1, f_2, and f_3) and an area of operation indicated by the dashed rectangle while avoiding interference. Notice that the illustration in Figure 2.15 differs from dynamic use of frequency slots within a single network. With Figure 2.15, a centralized entity may be utilized to fuse spectrum sensing information collected from sensors dispersed geographically through the area of operation to dynamically use a set of predefined frequency bands.

Figure 2.16 illustrates the advantage of having a centralized entity to arbitrate DSA. If the spectrum reuse decision is made at the local node, there remains the potential of interfering with a hidden primary user. This issue can occur if there is severe multipath fading or shadowing of the RF signal as it is emitted from the primary user while the sensor targeting the primary user's transmissions fails to sense the primary user signal. With Figure 2.16, the circles on the right-hand side represent the AOC of the primary user. One can see how the secondary user

---

17 In defense applications, this area of operation is referred to as the theater of operation.

Figure 2.14 Geographical separation creating opportunistic spectrum use for a secondary user.

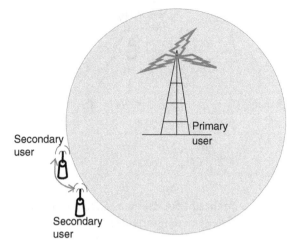

Figure 2.15 Geographical separation creating the DSA of a limited set of frequency bands for a set of heterogeneous MANETs.

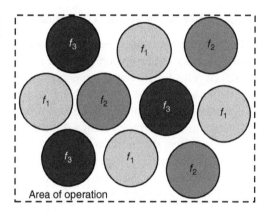

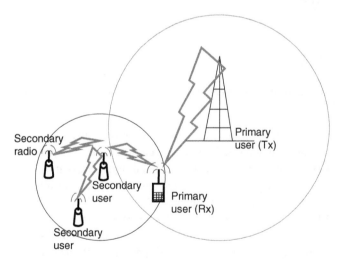

Figure 2.16 The hidden node problem.

AOC depicted by the circle on the left-hand side causes unintended interference with the primary user's receiver since the transmitting signal from the primary user could not be detected. One can see how distributed cooperative sensing or the use of a centralized arbitrator can better address the hidden primary user problems. In order to reduce the chance of encountering a hidden node problem, one can rely on:

1. geographically dispersed spectrum sensors
2. a centralized DSA decision process, which can have a bird's eye view of the area of operation

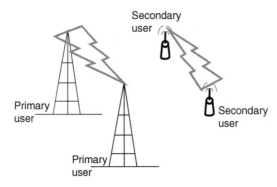

Figure 2.17 Directional secondary user's leveraging of the primary user's beam angle to create spectral opportunity.

3. the centralized DSA decision-making process having an algorithm that can estimate the AOC of the primary user based on multiple sensor input.

2.5.2 Angle of the RF Beam Detection

When primary users use directional antennas such as microwave links, the locations of the primary users and the directions of their RF beam – which include the azimuth and elevation angle – can be detected by the geographically dispersed spectrum sensors and leveraged by the secondary user's DSA decision-making process to create opportunistic spectrum use. The spectrum sensing techniques will estimate the geolocation or position of the primary users and the direction of their beam. With a directional antenna, if a primary user is transmitting in a specific direction, the secondary user can transmit in the opposite direction without creating interference, as shown in Figure 2.17.

Obviously, the case illustrated in Figure 2.17 points out to the importance of having a centralized arbitrator or the secondary user having knowledge of peer nodes location and the primary user nodes locations and signal characteristics.

2.6 Other Sensing Techniques

There are many other sensing techniques that have been proposed in the literature and are not detailed here, including the following:

- *Multitaper spectral estimation.* This technique is widely used in neuroscience and other biomedical engineering applications. It estimates the power spectrum of the signal that is a stationary ergodic random process with finite variance (wideband signals can carry these characteristics). It detects contiguous realization of the signal and uses a maximum likelihood estimator to calculate the signal's power spectral density.
- Random Hough transform. While most non time domain spectrum sensors use Fourier transform (FT), this approach proposes the use of a different transform domain for signal detection. The Hough transform domain suits signals with periodic patterns where it exploits the statistical covariance of noise and signal. This spectrum sensing technique can be effective at detecting digital television (DTV) signals.
- Wavelet transform based estimation. This approach uses another transform domain. Wavelet[18] domain is used for detecting edges in the power spectral density. These edges can result from the transition from the occupied band to the empty band and vice versa. Analog implementation of wavelet transform based sensing have the advantage of needing low power consumption and it can be implemented in real time.

---

18 Wavelet transform is widely used in digital image processing. With spectrum sensing, it can be used with sensing signals that use a wideband.

2.7 Concluding Remarks

Regardless if one is building spectrum sensing capabilities from the bottom-up or utilizing an existing spectrum sensing technology, one must understand the different spectrum techniques that can be used as reviewed in this chapter. The decision-making process using spectrum sensing information is covered in the next chapter.

When designing a system that uses DSA, one may build the best spectrum sensing capabilities and choose the best spectrum sensing hardware and configure it for the appropriate bands to sense and use the sensing parameters appropriately; however, if some critical factors are ignored, the design can be way suboptimal. One of the critical factors worth mentioning at the conclusion of this chapter is making sure a concept called *blanking signal* is applied. On a platform that has both a communications waveform and a spectrum sensor, it will be important for the communications waveform to inform the spectrum sensor of transmission time intervals. During these intervals, the spectrum sensor should refrain from collecting spectrum sensing information since spectrum sensing will be dominated by the emitted communications signal. Even if sensing is at a different frequency band from the transmission band, frequency domain harmonics can impact the sensing accuracy, as explained in Chapter 8.

Another critical factor worth mentioning is the relationship between *dwell time* and the signal characteristics. When performing same-channel in-band sensing, many aspects of the signal characteristics are known. For example, a frame transmitted over the air can have a preamble. The presence of the preamble can inform the energy detection process that the sensed energy level includes the presence of the communications signal. The absence of the preamble can inform the energy detection process that the sensed energy level is for noise or noise plus interfering signal. One can map dwell time to the frame time, taking multiple samples from the frame, or one can sample multiple frames in tandem with a larger dwell time. This can yield a good estimation of the noise floor when sensing the in-band frequency. With augmented sensors, performing energy detection is separate from the demodulation process. Cyclostationary characteristics can also help define a dwell time that informs the energy detection process whether or not the sensed energy level includes the presence of the sensed signal. Having a dwell time that can allow the energy detection process to overlay noise, in-band signal, and interfering signal can lead to a lower probability of misdetection and a lower probability of false alarm.

There are other critical factors to consider as discussed in the following chapters and introduced in Chapter 1, such as abstraction, the value of same-channel in-band sensing, making the best out of local, distributed, centralized, and hybrid decisions, the reduction of spectrum sensing control traffic, the powerful features of augmented sensing, and the role of policies and security with DSA systems.

Exercises

1. Create a table of 11 entries that map dBm values to milliwatt values in the range of 1–50 mW. Use this table to create six approximate RSSI thresholds to express:
   1. Undetectable signal
   2. Very weak signal
   3. Weak signal
   4. Good signal
   5. Very good signal
   6. Excellent signal

2. State some of the dimensions you would want a spectrum sensor to consider in sensing a 5G commercial cellular signal.

3. Consider the case of M-AM (M-level amplitude modulation, where M is an even number) where the number of bits transmitted per symbol is ($\log_2 M$). The transmitted signal is related to a data symbol $x_i \in \{0, 1, ..., M-1\}$ by $s_i(t) = (2x_i - M + 1)\,\phi(t)$, where $\phi(t)$ is the common signal shape to all signals. This is essentially a one dimensional signal in space. Assume the net amplitude modulation is symmetric at each AM constellation point. Arrange the M constellation points on a straight line separated by a distance of $d = 2a$ and answer the following:

   (a) Draw the one-dimensional signal space constellation for this case from s_0 to s_{M-1} symmetrically around the zero-power point.
   (b) Calculate the ideal signal energy for the inner six transmitted symbols in terms of a.
   (c) Is simple energy detection ideal for this type of signal? Why?

4. Consider the 32-QAM constellation case shown below. QAM signals are commonly used with microwave links. This is a two-dimensional signal in space. In QAM, inner constellation points have four nearest neighbors, the edge points have three nearest neighbors, and the corner points have two nearest neighbors.

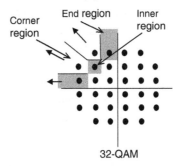

32-QAM

   (a) Assuming signal spacing is $d = 2a$ in each dimension, what are the various energies for the different points of the constellation?
   (b) How many instantiations are there for each energy level?
   (c) Is this type of signal a better candidate for energy detection than the one-dimensional case in Problem 3? Why?
   (d) What would you consider as another important metric in addition to energy detection when detecting microwave signals?

5. Comparing to AM and QAM, what do you think of the suitability of 4-ary PSK and 8-ary PSK signals for energy detection?

Appendix 2A: The Difference Between Signal Power and Signal Energy

There is a significant difference between energy and power when analyzing a signal. A signal can be categorized in different ways, as shown in Table 2A.1, including an energy/power signal category. Notice that in Table 2A.1 the signal can be either category A *or* category B.

A signal is time varying and can convey information or not convey information. Noise is a type of signal that does not convey information. Some jammers can also be a type of signal that does not convey information. A signal can be a function of time and a function of other independent variables.

A signal is categorized as an energy signal if it has finite energy, that is, $0 < E < \infty$. When the signal is analyzed over a given time duration and the signal energy can be measured, it is an energy signal. An example of that signal is a pulse signal where the pulse is either positive or negative. A decaying pulse signal is also an energy signal because its energy can be measured

Table 2A.1 Different categories of signals

Signal category	Category A	Category B
1	Continuous time	Discrete time
2	Deterministic	Random
3	Periodic	Aperiodic
4	Even	Odd[19]
5	Energy	Power

over a given time duration. On the other hand, a sinusoidal signal energy cannot be measured in time domain, and hence it cannot be categorized as an energy signal. The sinusoidal signal is a power signal. Moving from time domain to frequency domain, the sinusoidal signal power spectral density can be measured. The sinusoidal signal is a power signal over infinite time. A signal can be categorized as a power signal when it has finite power without time limitation.

Notice that with spectrum sensing, we may sense a modulated signal over a sinusoidal wave (carrier). We look at frequency bands of carrier frequencies and hence we measure the signal power. The term "energy detection" is used loosely with spectrum sensing and it means integrating the measured signal power over a limited time period (time of sensing or dwell time). With spectrum sensing, the correct term for energy detection should be *power integration over a finite time*. The term "energy detection" is widely used because spectrum sensors integrate the sensed power spectral density over the sensing time period and the process of integration over time leads to using the term energy detection. In spectrum sensing references, spectrum sensing is a multidimensional process that considers time, frequency, and power. Power here can be power spectral density measured over limited time. The simplest way of spectrum sensing is known as energy detection, which integrates the power spectral density measured by the spectrum sensor, in frequency domain, over a given sensing period.

Bibliography

Abdulstar, M. and Hussein, Z., Energy detection techniques for spectrum sensing in cognitive radio: A survey. *International Journal of Compute Networks and Communications*, vol. 4, no. 5, September 2012.

Cabric, D.B., *Cognitive radios: System design perspective*. University of California, Berkeley, UMI Microform 3306077.

Cabric, D., Tkachenko, A., and Brodersen, R., Spectrum sensing measurements of pilot, energy, and collaborative detection. In *Proceedings of the IEEE Military Communications Conference*, Washington, D.C., USA, October 2006.

Cordeiro, C., Challapali, K., and Birru, D., IEEE 802.22: An introduction to the first wireless standard based on cognitive radios. *Journal of Communications*, vol. 1, no. 1, April 2006.

Digham, F., Alouini, M., and Simon, M., On the energy detection of unknown signals over fading channels. In *Proceedings of the IEEE International Conference on Communication*, vol. 5, Seattle, Washington, USA, May 2003, pp. 3575–3579.

Elmasry, G.F., *Tactical Wireless Communications and Networks, Design Concepts and Challenges*. Wiley, 2012. ISBN: 978-1-1199-5176-6.

Federal Communications Commission, Notice of proposed rulemaking and order: Facilitating opportunities for flexible, efficient, and reliable spectrum use employing cognitive radio technologies. ET Docket No. 03-108, February 2005.

Gandetto, M. and Regazzoni, C., Spectrum sensing: A distributed approach for cognitive terminals. *IEEE Journal on Selected Areas of Communications*, vol. 25, no. 3, pp. 546–557, April 2007.

---

19 An even signal is symmetric around the vertical axes. An odd signal is symmetric about the origin.

Geirhofer, S., Tong, L., and Sadler, B., A measurement-based model for dynamic spectrum access in WLAN channels. In *Proceedings of the IEEE Military Communications Conference*, Washington, D.C., USA, October 2006.

Geirhofer, S., Tong, L., and Sadler, B., Dynamic spectrum access in the time domain: Modeling and exploiting white space. *IEEE Communications Magazine*, vol. 45, no. 5, pp. 66–72, May 2007.

Ghasemi, A. and Sousa, E.S., Asymptotic performance of collaborative spectrum sensing under correlated log-normal shadowing. *IEEE Communications Letters*, vol. 11, no. 1, pp. 34–36, January 2007.

Guzelgoz, S., Celebi, H., and Arslan, H., Statistical characterization of the paths in multipath PLC channels. *Proceedings of the IEEE Transactions on Power Delivery*, vol. 26, no. 1, pp. 181–187, January 2011.

Haykins, S, Cognitive radio: brain empowered wireless communications. *IEEE Journal on Selected Areas in Communications*, pp. 201–220, February 2005.

Hu, W., Willkomm, D., Abusubaih, M., *et al.*, Dynamic frequency hopping communities for efficient IEEE 802.22 operation. *IEEE Communications Magazine*, vol. 45, no. 5, pp. 80–87, May 2007.

Lehtomäki, J., Vartiainen, J., Juntti, M., and Saarnisaari, H., Spectrum sensing with forward methods. In *Proceedings of the IEEE Military Communications Conference*, Washington, D.C., USA, October 2006.

Leu, A., McHenry, M., and Mark, B., Modeling and analysis of interference in listen-before-talk spectrum access schemes. *International Journal of Network Management*, vol. 16, pp. 131–147, 2006.

Marcus, M., Unlicensed cognitive sharing of TV spectrum: the controversy at the federal communications commission. *IEEE Communications Magazine*, vol. 43, no. 5, pp. 24–25, 2005.

Mathur, C.N. and Subbalakshmi, K.P., Digital signatures for centralized DSA networks. In *First IEEE Workshop on Cognitive Radio Networks*, Las Vegas, Nevada, USA, January 2007, pp. 1037–1041.

Mitola, J., *Cognitive Radio An Integrated Agent Architecture for Software Defined Radio*. PhD thesis, KTH Royal Institute of Technology, Stockholm, Sweden, 2000.

Mitola, J., *Cognitive Radio Architecture, in Cognitive Networks: Towards Self-Aware Networks* (ed. Q.H. Mahmoud). John Wiley & Sons, Chichester, 2007. doi: 10.1002/9780470515143.

Naraghi-Pour, M. and Ikuma, T., Autocorrelation-based spectrum sensing for cognitive radios. *Proceedings of the IEEE Transactions on Vehicular Technology*, vol. 59, no. 2, February 2010.

Papadimitratos, P., Sankaranarayanan, S., and Mishra, A., A bandwidth sharing approach to improve licensed spectrum utilization. *IEEE Communications Magazine*, vol. 43, no. 12, pp. 10–14, December 2005.

Pawełczak, P. Janssen, G.J., and Prasad, R.V., Performance measures of dynamic spectrum access networks. In *Proceedings of the IEEE Global Telecommunications Conference (Globecom)*, San Francisco, California, USA, November/December 2006.

Rossi, P., Ciuonzo, D., and Romano, G., Orthogonality and cooperation in collaborative spectrum sensing through MIMO decision fusion. *Proceedings of the IEEE Transactions on Wireless Communications*, vol. 12, no. 11, November 2013.

Shobana, S., Saravanan, R., and Muthaiah, R., Matched filter-based spectrum sensing on cognitive radio for OFDM WLANs. *International Journal of Engineering and technology*, vol. 5 no 1, February 2013.

Sun, Z., Bradford, G., and Laneman, J., Sequence detection algorithm for PHY-Layer sensing in dynamic spectrum access networks. *Proceedings of the. IEEE Journal of Selected Topics in Signal Processing*, vol. 5, no. 1, pp. 97–109.

Yucek, T. and Arslan, H., A survey of spectrum sensing algorithms for cognitive radio applications. *IEEE Communications Surveys and Tutorials*, vol. 11, no. 1, first quarter 2009.

3

Receiver Operating Characteristics and Decision Fusion

Receiver operating characteristic (ROC) is not a unique methodology to DSA decision making. It is widely used in many areas where statistical decisions are adaptively made based on myriad metrics. ROC is a generic approach developed for low computational and implementation complexities. This chapter covers different DSA scenarios that rely on using ROC models. A ROC model can be used when probing a frequency band to discover if it is occupied or not at a certain geographical location (e.g., a secondary user is sensing if a primary user is using this frequency band or not). Another scenario that uses a different ROC model is the case of same-channel in-band sensing where the ROC model can hypothesize the presence or absence of an interfering signal. The detected energy of the sensed frequency band is compared to an adaptive threshold to hypothesize the presence of the interfering signal. This threshold adaptation is highly dependent on noise estimation and is decided based on tradeoffs driven by the design needs. Estimating ROC models' thresholds can be challenging when noise variance increases and when signal power is too low.[1] The spectrum sensor can be looking into interference from other signals overlaid with the noise without being able to distinguish between the interfering signal power and the noise power. In other cases, the spectrum sensor can be looking at an overlay of different signals occupying the same frequency band. This can mislead the decision-making process and result in increasing the probability of false alarm and of misdetection.[2]

With DSA, the ROC methodology can be implemented in different approaches depending on the sensing metrics and where the decision is made. This chapter will start with the generic aspects of the ROC hypothesizing process in DSA applications and present simple ROC-based decision fusion cases while gradually moving to the harder cases. Statistical decision models in modulation and coding are well studied and well presented in textbooks. This chapter covers decision models for spectrum sensing pointing to the similarities and difference with modulation and coding models. While a demodulator may use fixed thresholds and rely on well-known statistical models such as AWGN and the communication signal known power spectral density characteristics to decode a symbol, spectrum sensing models use adaptive thresholds and machine learning techniques to account for the many factors that can compound the spectrum sensing hypotheses. If the reader is not familiar with the ROC models, the reader is encouraged to refer to Appendix A of this chapter to get some basic understanding of the ROC methodology.

3.1 Basic ROC Model Adaptation for DSA

This ROC model is the most basic model where a sensor is probing a frequency band to check for the presence or absence of a communications signal. This basic model relies on energy

---

1 Spread spectrum signals are especially hard to detect with simple energy detection.
2 The definitions of false alarm and misdetection depend on the spectrum sensing goal, as clarified later in this chapter.

Dynamic Spectrum Access Decisions: Local, Distributed, Centralized, and Hybrid Designs, First Edition. George F. Elmasry.
© 2021 John Wiley & Sons Ltd. Published 2021 by John Wiley & Sons Ltd.
Companion website: www.wiley.com/go/elmasry/dsad

detection and can assume that the noise is AWGN such that the signal received by the sensor can be expressed as follows:

$$y(n) = s(n) + w(n). \tag{3.1}$$

In Equation (3.1), $s(n)$ is the sensed signal, $w(n)$ is the AWGN, and n is the sampling index. If the sensed frequency band has no signal occupying it, then $s(n) = 0$ and the sensing process will detect the energy level of the AWGN.

The sensed signal energy can be expressed as a vector of multiple sampling points as follows:

$$M = \sum_{n=0}^{N-1} |y(n)|^2, \tag{3.2}$$

where N is the size of the observation vector.

The value of N and the definition of sampling points can differ from one sensor to another and any pre-knowledge of the sensed signal waveform characteristics can guide the sensor into creating more optimal sampling points.

The energy detection process can compare the decision metric M from Equation (3.2) against a fixed threshold λ_E. This processes needs to distinguish between two hypotheses, one hypothesis is for the presence of only noise and the other hypothesis is for the presence of signal and noise. These two hypotheses are:

$$\mathcal{H}_0: \qquad y(n) = w(n), \tag{3.3}$$

$$\mathcal{H}_1: \quad y(n) = s(n) + w(n). \tag{3.4}$$

The spectrum sensor detection algorithm can successfully detect the sensed frequency with probability P_D and the noise variance can cause a false alarm[3] with a probability of P_F. The detection problem can be expressed as:

$$P_D = \Pr(M > \lambda_E | \mathcal{H}_1), \tag{3.5}$$

$$P_F = \Pr(M > \lambda_E | \mathcal{H}_0). \tag{3.6}$$

Equations (3.5) and (3.6) can be illustrated as shown in Figure 3.1 where selecting an energy threshold λ_E can deviate from the optimum threshold. The optimum threshold is not known at any given instant and would have resulted in $P_D = 1$ and $P_F = 0$. The estimated λ_E can either be intentionally shifted to the right or shifted to the left as shown by the arrows at the bottom of Figure 3.1. If it is shifted to the left, the ROC model would increase the probability of hypothesizing $\mathcal{H}_1$, leading to a higher false alarm probability. If it is shifted to the right, the ROC model would increase the probability of hypothesizing $\mathcal{H}_0$, leading to a higher misdetection probability. This intentional shifting of the threshold depends on the communications system being designed. The system requirements can lead to shifting the threshold either to the left or to the right within a range, as indicated by the dashed arrow at the top of Figure 3.1.

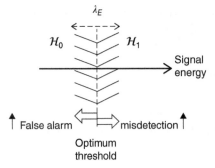

Figure 3.1 Single-threshold ROC model leading to false alarm and misdetection.

---

3 As mentioned earlier, the definition of the probability of detection P_D and the probability of false alarm P_F are case dependent. In the case presented here, P_D is the probability of hypothesizing the signal presence where the signal is actually present while P_F is the probability of hypothesizing the signal presence but there was only noise power present.

Maximum likelihood decisions can be applied to the decision threshold λ_E. A key factor in selecting λ_E is the estimation of noise power. Also, estimating the signal power by the sensor can be difficult since it can change due to propagation environments, and the distance between the transmitting node and the sensor. A good approach to select λ_E is to balance P_D and P_F based on given requirements. The threshold λ_E can be chosen to meet a given false alarm rate that can be deemed acceptable for the system under design. This makes it sufficient to model the noise variance, assuming a zero-mean Gaussian random variable with variance σ_w^2. With this noise model, we can express the noise as $w(n) = (0,\ \sigma_w^2)$. We can also simplify how we model the signal $s(n)$ in order to make this analysis possible.[4] We can assume that there is no fading and express the signal as a zero-mean Gaussian variable making $s(n) = (0,\ \sigma_s^2)$.[5] These assumptions allows us to express the decision metric M as a Chi-square distribution with $2N$ degree of freedom (χ_{2N}^2) giving:

$$
M = \begin{cases} \dfrac{\sigma_w^2}{2}\chi_{2N}^2 & \mathcal{H}_0, \\[3mm] \dfrac{\sigma_w^2 + \sigma_s^2}{2}\chi_{2N}^2 & \mathcal{H}_1. \end{cases} \tag{3.7}
$$

Thus, the ROC model can calculate P_D and P_F as follows:

$$
P_F = 1 - \Gamma\left(L_f L_t, \frac{\lambda_E}{\sigma_w^2}\right), \tag{3.8}
$$

$$
P_D = 1 - \Gamma\left(L_f L_t, \frac{\lambda_E}{\sigma_w^2 + \sigma_s^2}\right), \tag{3.9}
$$

where $\Gamma(a, x)$ is the incomplete gamma function and L_f and L_t are the associated Laguerre polynomials.

The DSA decision fusion process can use the ROC model to compare the performances for different threshold values. ROC models are a set of convergence curves that explore the relationship between the probability of detection and the probability of a false alarm for a variety of different thresholds. Based on given requirements and machine learning techniques that count for the dynamics of the sensed environments, a close-to-optimal threshold can be reached. Figure 3.2 exemplifies different ROC curves for different SNIR values using the equations above. SNIR is defined as the ratio of the sensed signal power to noise power $\frac{\sigma_s^2}{\sigma_w^2}$.

One can see from Figure 3.2 that as SNIR increases,[6] we can achieve higher P_D at lower P_F. While one can see the same tendency in demodulation and decoding of a signal where higher SNIR results in less symbol error probability, it is critical to understand that this energy detection process relies on estimating the noise variance. Noise power estimation error can cause significant performance loss (significant shift in λ_E in Figure 3.1). Noise level can be estimated dynamically and more accurately if the spectrum sensor is able to separate the noise subspace from the signal subspace. Some sensors estimate noise variance as the smallest eigenvalue of the sensed signal's autocorrelation. This estimated noise variance can then be used to find the decision threshold λ_E that satisfies the requirements for a given false alarm rate. This noise estimation algorithm is applied iteratively, where N can be a moving average window that continuously normalizes the noise power.

---

4 Notice that using the ROC model for DSA can require considering two main aspects. The first aspect is to estimate the decision threshold λ_E through analysis. The second aspect is the use of a DSA cognitive algorithm that can fine-tune the value of λ_E based on deployment dynamics and design requirements relying on measured metrics.

5 It is important to note how the ROC model uses energy instead of signal and noise amplitude sampling. Energy detection uses square values which extends the scale of decision making, leading to more accuracy, but creates a one-dimensional positive axis only scale, as shown in Figure 3.1.

6 For signals with low energy such as spread spectrum signals, SNIR is inheritably low, making energy detection based spectrum sensing more challenging.

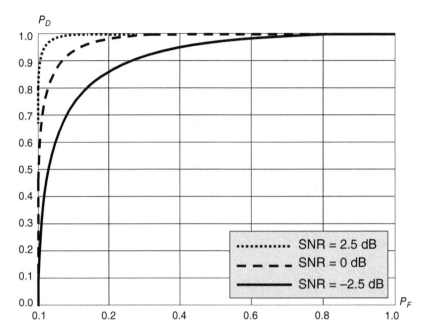

Figure 3.2 Different ROC curves for different SNIR (not to scale).

In addition to noise power estimation, a machine learning technique[7] that uses the ROC model can leverage the following techniques to tune the decision threshold:

1. Measure the success of its own decisions.[8]
2. Take into consideration external variables such as emitter power, emitter distance to the sensor, terrain, rain, and fog that can affect SNIR.
3. Increase accuracy by increasing the number of decision samples. Cooperative distributed DSA and centralized DSA techniques can be looking at more comprehensive information than a single node to make the ROC estimation more accurate.

The purpose of using the above three techniques is to make the DSA system able to adapt the decision threshold to adhere to the same P_D at the same given requirement of P_F even with the increase of uncertainty.

Example: Evaluation Metrics and ROC Design for Different Applications

Equations (3.5) and (3.6) express the probability of detection and the probability of false alarm, respectively, for a single threshold ROC model. A third probability calculation could be the probability of misdetection. If we are to evaluate the accuracy of this hypotheses-based decision making, we could create the following three metrics:

$$P_D = P(H_1/H_1), \tag{3.10}$$

$$P_F = P(H_1/H_0), \tag{3.11}$$

$$P_m = P(H_0/H_1), \tag{3.12}$$

where P_D is the probability of hypothesizing the presence of the sensed signal given that the sensed signal is present, P_F is the probability of hypothesizing the presence of the sensed signal given that the sensed signal was not present, and P_m is the probability of hypothesizing the absence of the sensed signal given that the sensed signal was present.

---

7 This machine learning technique can be local, distributed using cooperative techniques or centralized.
8 Chapter 5 explains how to create metrics that measure the performance of DSA decision fusion results.

Table 3.1 Signal presence versus hypotheses.

Signal presence	Hypotheses	Evaluation metric
Y	Y	P_D
Y	N	P_F
N	Y	P_m
N	N	N/A

Notice that there is a fourth possibility that is irrelevant to performance evaluation. Table 3.1 shows the four cases of signal presence and absence versus hypotheses with the "N, N" case (the signal is not present and the sensor did not detect it) being irrelevant.[9]

The P_D, P_F, and P_m metrics can be used to measure the efficiency of the decision-making process given some design requirements. Notice that:

$$P_D + P_F + P_m = 1. \tag{3.13}^{10}$$

Although Equations (3.10)–(3.13) can apply to different systems, the system under design should influence how a machine-learning algorithm would estimate λ_E. Let us consider the following two cases:

Case 1: A commercial communication system of a secondary user attempting to opportunistically use the primary user spectrum. In this case, a higher probability of false alarm can be acceptable as the higher probability of misdetection can cause the secondary user to interfere with the primary user. With this case, the design of the machine learning algorithm would accept a higher probability of false alarm to minimize the probability of misdetection.

Case 2: A military MANET system that can operate in an antijamming mode and the formed MANET can switch to a different waveform type only if the interference level is too high. With this case, a higher probability of misdetection may be acceptable since the antijamming waveform can operate in the presence of some level of interference. With this case, the design of the machine learning algorithm may target a higher probability of misdetection to minimize the probability of false alarm.

3.2 Adapting the ROC Model for Same-channel in-band Sensing

Same-channel in-band sensing use of the ROC model has some factors that can make the hypothesizing process more accurate but also has its own challenges. As mentioned in Chapter 2, with same-channel in-band sensing, the receiver can have a clear dwell time in the presence of the communications signal and a clear dwell time in the absence of the communication signal. For example, if we have a time-domain preamble and a fixed over-the-air frame

---

9 Sometimes it is difficult for those who know digital communications and understand decision theory as it applies to symbol decoding to see how the fourth case in Table 3.1 is irrelevant. A simple way to conceptualize the ROC model, while being cognizant of the difference with symbol decoding, is to map the second and third rows in Table 3.1 to the case of symbol error decoding. The ROC model distinguishes between these two error types while the digital communications decoding model combines them as symbol error probability. The first row is simply mapped to the probability of decoding the correct symbol. The fourth column is irrelevant in both cases. With symbol decoding, not decoding a symbol that was never sent or the ROC model not hypothesizing the presence of a signal that is not present are both irrelevant cases.
10 Following the same theme of correlating to symbols decoding: in symbol decoding, the probability of correctly decoding a symbol plus the probability of error decoding add to one. The reader has to be careful in using this resemblance between energy detection and symbol decoding as energy detection has no consideration of the signal dimensions and symbol decoding error can either be an erasure or can produce another known symbol.
11 The FCC regulation states that "no modification to the incumbent system should be required to accommodate opportunistic use of the spectrum by secondary users".

size, a sampling point can be an instant detection of energy within the length of the frame and N in Equation (3.2) can be selected for the number of samples (instants) collected during the frame time. N should be large enough to smooth the effect of noise spikes. This sampling can occur when the preamble is acquired and when the preamble is not acquired separately.

The goal of same-channel in-band sensing has two folds. The first is to hypothesize for the presence of an interfering signal plus noise or hypothesize the presence of only noise when the communications signal is known to be absent. The second is when the communications signal is known to be present, the ROC model would then hypothesize if the interfering signal and noise power are too high to warrant a change of the operating frequency. The ROC model explained in the previous section has to be adapted to consider two different thresholds instead of one threshold. With same-channel in-band sensing, Equation (3.1) becomes:

$$y(n) = s(n) + [r(n) + w(n)], \tag{3.14}$$

where $r(n)$ is the interfering signal.

When the same-channel in-band demodulation finds a preamble and it is demodulating and decoding an over-the-air frame, the same-channel in-band energy detection process is faced with the following two hypotheses:

$$\mathcal{H}_0: \quad y(n) = s(n) + r(n) + w(n), \tag{3.15}$$

$$\mathcal{H}_1: \quad y(n) = s(n) + w(n). \tag{3.16}$$

When the same-channel in-band demodulation does not find a preamble and it is not demodulating and decoding an over-the-air frame, the same-channel in-band energy detection process is faced with the following two hypotheses:

$$\mathcal{H}_2: \quad y(n) = r(n) + w(n), \tag{3.17}$$

$$\mathcal{H}_3: \quad y(n) = w(n). \tag{3.18}$$

Same-channel in-band sensing and the presence of the communications signal's marks such as preambles should lead the ROC model to minimizing P_F, with very low probability that the receiver falsely decides that the preamble exist when it does not exist.[12] With same-channel in-band detection, P_F is the probability of deciding an interfering signal r exists when it does not exist. The key here is to have a good estimation of the noise floor energy and a good estimation of the communication signal energy in order to effectively hypothesize the presence or the absence of an interfering signal.

Let us consider one of the most suitable signal types for same-channel in-band sensing with minimal computational power requirements. This is the n-ary phase shift keying (PSK) signal type[13]. This signal can be used with OFDM as 4-ary PSK when noise level is high to increase range and reduce data rate, or it can be used as 8-ary PSK when noise level is low and the signal needs to achieve a higher transmission rate.[14] This signal is depicted in Figure 3.3 with the 4-ary case encoding two bits per symbol and the 8-ary case encoding three bits per symbol.

Notice that using two-dimensional signal space as in Figure 3.3 is proven to reduce computational complexity with signal encoding and decoding. Also, PSK makes the signal amplitude change due to noise or interfering signals, not affect the symbol decoding process as shown in Figure 3.4 where the decision zones are depicted with dashed lines. Figure 3.5 shows the probability distribution function (PDF) contours of one of the signal symbols, S_1, in relation to the decision line (perpendicular bisector) between S_1 and S_0. Notice in Figures 3.4 and 3.5 how the received signal vector, v, minor phase shifting will not cause a probability of a symbol error and how amplitude increase or decrease will never affect symbol errors. With this signal type, symbol error occurs only with considerable phase shifting. This modulation technique is

---

12 This can only happen if the correlation process output in the absence of a preamble is high for some odd reason.
13 Other constant envelope signals such as minimum shift keying signals have the same advantage.
14 Other higher order PSK can also be used to increase the number of bits per symbol.

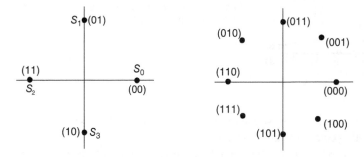

Figure 3.3 4-ary PSK and 8-ary PSK.

Figure 3.4 Decisions zones for 8-ary PSK.

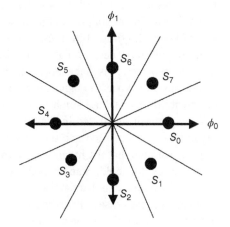

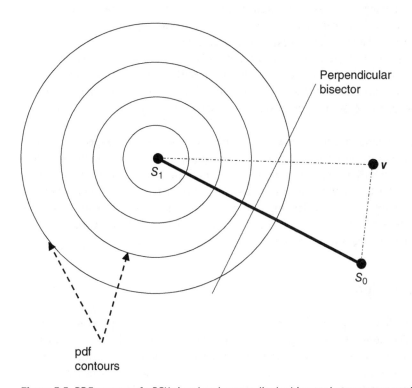

Figure 3.5 PDF contour of a PSK signal and perpendicular bisector between two symbols in signal space.

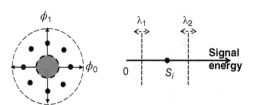

Figure 3.6 Hypothesizing the presence of noise and interfering signal with PSK signals.

widely used with military communications signals and can be useful in adding same-channel in-band sensing to the existing communications signal.

Same-channel in-band sensing of n-ary PSK signals leverages the signal characteristics where the hypotheses in Equations (3.15)–(3.18) can be simplified. The left-hand side of Figure 3.6 shows how the inner grey circle can define a noise floor (similar to a decision zone for the presence of only noise) and how a measured signal power outside of the outer circle can define an interfering signal that drastically affected the signal amplitude. The right-hand side of Figure 3.6 shows how these circles can be projected into decision lines (thresholds) in terms of the energy detection. Keep in mind that a major difference between symbol decoding and energy detection is that energy detection projects the signal vector into a one-dimensional positive axis shown as the signal energy axis on the right of Figure 3.6, while symbol decoding deals with the signal based on it SiS dimensions and characteristics.

The approach illustrated in Figure 3.6 allows for the estimation of noise floor when the preamble is not acquired and the energy level is low. Noise floor estimation can be a moving average such that the inner dashed line on the left of Figure 3.6, which maps to λ_1 for the noise energy threshold, is an adaptable threshold. Similarly, the λ_2 threshold defines the separation between a signal and noise versus a signal plus noise plus an interfering signal. λ_2 can be adapted as the noise floor estimation changes and as the estimated received signal power is changed.[15] Notice that if the communications waveform has an adaptable power control feature, knowledge of the signal transmission power, the distance between the transmitter and the sensor, and the terrain type can help decide where λ_2 changes adaptively.

Equations (3.15)–(3.18) and Figure 3.6 explain an overlay concept of the noise, in-band signal, and interfering signal. If the noise floor estimation is accurate, the sensor can subtract $w(n)$ from Equations (3.11)–(3.14). If the sensor has information regarding the transmission power of the in-band signal, the distance to the emitter and terrain information, then $s(n)$ can be estimated, allowing the sensor to hypothesize the presence of an interfering signal in a close to optimal way.[16]

It is critical to understand the importance of collecting large samples by the spectrum sensor to make the ROC model viable in implementation. More importantly, noise and the interfering signal are manifested not by a simple increase in energy detection level, but by an increase in the variance of the collected samples. Relying on a small sample can lead to suboptimal results as the set of small samples can be misleading. Estimating the *deviation* in the energy samples is what accurately reflects the impact of noise and interfering signals and what should be used for dynamically adapting the thresholds.

3.3 Decision Fusion

The ROC model implementation at the sensing node could be the first step towards making spectrum sensing decisions. The next step is referred to as decision fusion (DF), which uses the ROC model hypotheses outcome to make more comprehensive spectrum sensing decisions. This section presents local, distributed, and centralized decision fusion approaches to help the reader decide the most suitable place to make a DSA decision in a hybrid DSA system.

---

15 The estimated received signal power can change due to factors such as mobility and adaptive power control.
16 Problem 2 in the exercise section should lead the reader to approach this ROC model as an overlay case instead of attempting to create a complex ROC model analysis with two thresholds.

3.3.1 Local Decision Fusion

With a spectrum sensor performing a simple energy detection decision, this may be the end of the decision-making process that can be made locally. The local decision fusion process would rely on the local hypotheses that differentiate if the frequency band being sensed is occupied or not. If a hypothesis is persistent for the presence or absence of a signal, the decision fusion will turn the hypotheses into a decision. However, if an augmented sensor is able to utilize a multi-sector antenna or antenna arrays, there could be further fusion steps before making a decision. An example of a further fusion step is to identify the direction of the interfering signal when the local process hypothesizes the presence of interference relying on the difference in the energy received per sector. This case is covered in Section 3.3.1.2. The more common case to perform further local decision fusion is for the same-channel in-band sensing in a MANET where the reception of the sensed communications signal can be mapped to an RF neighbor. This can make the spectrum sensor in the MANET node able to create a more detailed spectrum map (i.e., identify interference directionality) without using sectored antennas,[17] as explained in Section 3.3.1.1.

Notice that if the local fusion process stops without further fusion of spectrum sensing information, the higher hierarchical levels of decision making (e.g., distributed cooperative or centralized decision fusion) can make DSA decisions that are more optimum than that of the local decision-making process. With hybrid DSA designs, fusion at the lower hierarchical level can always help reduce control traffic volume and make the overall decision-making process more accurate even if the final decisions are left for the higher hierarchical level.[18]

3.3.1.1 Local Decision Fusion for Same-channel in-band Sensing

With same-channel in-band sensing, the sensed signal can be associated with an RF neighbor in a MANET without the need to use sectored antennas. The sensing node may have information about the geolocation of its RF neighbors. This will allow a local decision-making process to map interference to its RF space. Consider Figure 3.7 where node 0 is the sensing node and there is an external RF emitter of the same sensed frequency, as illustrated by the small

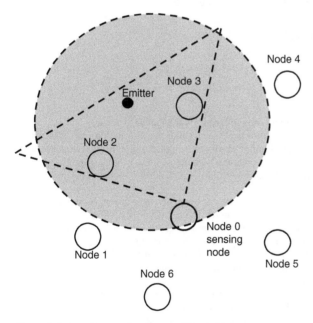

Figure 3.7 Interference from some RF neighbors.

---

17 The MANET nodes can be using omnidirectional antennas.
18 Chapter 5 explains how DSA can become a set of cloud services that can be offered at any hierarchical entity in a large-scale set of heterogeneous networks.

black circle. This emitter can be covering the gray circular area but the sensor may perceive the direction of the emitter as shown by the dashed triangle. Spectrum sensing information from RF neighbors 2 and 3 may indicate the presence of an interfering signal while spectrum sensing information from RF neighbors 1, 4, 5 and 6 may not indicate the presence of interfering signal.

Identifying the presence of $w(n)$ from Equations (3.15) and (3.17) with regard to certain RF neighbors can allow the local node to fuse information per RF neighbor and estimate the direction of the emitter of the interfering signal even though the MANET nodes are using omnidirectional antennas. This is particularly important in military cognitive MANET that attempts to create and update an RF spatial map in real time. The outcome of the local decision fusion can mean any of the following actions:

1. Choose different local route tables to avoid routing over the direction where interfering is present[19]
2. Increase the transmission power in the direction of the interfering signal
3. Have the entire MANET switch to a different waveform type that can overcome the type of detected interference[20]
4. Have the entire MANET switch to a different frequency band to avoid interference.[21]

The key concept here is that local decision fusion for same-channel in-band spectrum sensing can be critical in identifying the direction of interference and can be critical in making some local routing decisions as well as assisting distributed and centralized decision fusion processes reach more optimal decisions. The probability of detection and the probability of false alarm of this local decision fusion process corresponding to Equations (3.15) and (3.16) can be expressed as:

$$P_D = \Pr(M_{Vn} > \lambda_{Evn}|\mathcal{H}_1), \tag{3.19}$$

$$P_F = \Pr(M_{Vn} > \lambda_{Evn}|\mathcal{H}_0), \tag{3.20}$$

where M_{Vn} is a vector normalization of the different vectors M from Equation (3.2) for the different RF neighbors when the transmitting signal is detected.[22]

The probability of detection and the probability of false alarm of this decision fusion process corresponding to Equations (3.17) and (3.18) can be expressed as:

$$P_D = \Pr(M_{Vn} > \lambda_{Evn}|\mathcal{H}_3), \tag{3.21}$$

$$P_F = \Pr(M_{Vn} > \lambda_{Evn}|\mathcal{H}_2), \tag{3.22}$$

---

19 This can increase the number of over-the-air relay hops but route around jammed areas.
20 Military waveforms that can be utilized in some cases may be using unlicensed spectrum, transmitting over ultra-wideband, using spread spectrum with high chip rate and low data rate and using other techniques such as fast frequency hopping to make the military signal seems as a background noise to other signals in the area or to an eavesdropping node. Military waveforms can also switch to an antijamming mode to overcome certain types of jammers.
21 Notice that in military communications, the action of switching to a different frequency band may not work when the enemy is using "follower jammers". These jammers can switch to the new frequency band and continue to jam the signal over the new frequency. The follower jammers control loop will always be faster than the DSA control loop. For that reason, once the military communications DSA technology identifies the interference signal as a jamming signal, relying on waveforms with antijamming capabilities is a better option than DSA. The mix of spread spectrum and frequency hopping is one common approach in creating antijamming capabilities in military communications.
22 Notice this important characteristic of DSA decision making versus symbol decoding decision making. First we projected the signal-in-space into one direction with positive values on one axis for energy detection. The detected energy is still considered a vector because we collect a large sample and we look for characteristics such deviation in the sample points and as such we still treat the sample points as a vectors. Now with adding a fusion process to estimate interference directions, we have created a different vector space affected by the RF neighbors' directions not the signal-in-space. We are using the term M_{Vn} because we want to preserve the fact that the energy detection projected into a vector in one dimension and then another spatial dimension is added.

where M_{Vn} is a vector normalization of the different vectors M from Equation (3.2) for the different RF neighbors when the transmitting signal is not detected.

Notice in Equations (3.19–3.22) that the term λ_{Evn} is used to indicate both vector normalization of the different RF neighbors decision thresholds but also indicates the adaptation of the decision threshold to the dynamics of the MANET where factors such as transmission power, geolocation of RF neighbors, terrain, rain, and fog are used by the machine learning process to adapt the decision threshold.

In a hybrid distributed cooperative MANET system that uses local fusion and distributed decisions, the MANET nodes would share information that includes the result of hypothesizing the presence of interference and the direction of the detected interference such that an accurate spectrum map can be made available to every node in the network. With hybrid heterogeneous network systems that use centralized spectrum managers, the centralized spectrum manager can create the most accurate spectrum map of the area of operation based on the fusions and decisions done at all the networks and also based on its own further decision fusion that can further fine-tune the direction and boundaries of interference based on how each network perceives interference directions.

3.3.1.2 Local Decision Fusion with Directional Energy Detection

While Section 3.3.1.1 showed how the same-channel in-band ROC model can grow from the two-threshold model to adding the RF neighbor dimension, this section shows that for the simple energy detection case illustrated in Figure 3.1 one can add the directionality dimension if the spectrum sensor is able to use a multisector antenna. Note that the single-threshold simple energy detection model, which can be utilized by an augmented sensor, has no consideration of an RF neighbor as the same-channel in-band case does. Directional energy detection can be done by a secondary user that has a directional antenna and can transmit directionally and thus would sense the primary user signal directionality as well as the primary user signal energy level.

Let us illustrate this case with the 12-sector antenna layout shown in Figure 3.8 where each sector is a 30° angle. Each sector senses energy independently from other sectors. The spectrum sensor would apply a single threshold ROC model per each sector and hypothesize the presence of a primary user per each sector independently.

First, the antenna sectors can be expressed as a matrix Ş as follows:

$$\text{Ş} = \begin{bmatrix} S_0 \\ S_1 \\ S_2 \\ . \\ . \\ . \\ S_{N-1} \end{bmatrix}. \tag{3.23}$$

If the ROC model of two neighboring sectors hypothesizes the presence of the primary user at a given time, there will be some spectrum sensing overlapping and we can express that

Figure 3.8 Directional sensing with multisector antenna.

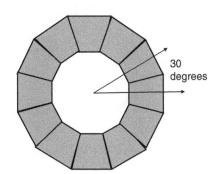

state as a union of the two sectors. For example, when sectors 0 and 1 hypothesize the presence of the primary user simultaneously, we have a union that can be expressed as $\cup\{S_0, S_1\}$. Similarly, when sectors 0, 1, and 2 hypothesize the presence of the primary user simultaneously, we create $\cup\{S_0, S_1, S_2\}$. The omni-state O when all the sectors hypothesize the presence of the primary user simultaneously can be expressed as:

$$O = \cup\{S_0, S_1, S_2, \ldots S_{N-1}\} \tag{3.24}$$

At any given time, the secondary user would want to decide not to emit spectrum at given sectors as it will interfere with the primary user. In order for the secondary user to reach this decision, it has to fuse the hypotheses of the different sectors. The outcome of the decision fusion process is a filter matrix that can eliminate the use of certain antenna sectors. This elimination matrix can be expressed as:

$$\underset{\sim}{E} = \begin{bmatrix} 1 \\ 1 \\ \varPhi \\ \cdot \\ \cdot \\ \cdot \\ \cdot \\ 1 \end{bmatrix}, \tag{3.25}$$

where 1 means the sector can emits spectrum to the destination secondary user and $\varPhi$ means the sector cannot emit spectrum to the destination secondary user.

Note that this decision fusion to reach Equation (3.25) is not straightforward and it requires the consideration of other factors such as the side lobes of the secondary user emitted spectrum. Figure 3.9 shows an example of the spectrum emitted by a single sector of the 12-sector antenna depicted in Figure 3.8 and how it is not purely directional. The side and back lobes and the beam width depend on the frequency range, among other factors. For that reason, the union expressions leading to Equation (3.24) are critical in deciding if a sector can be used or not.

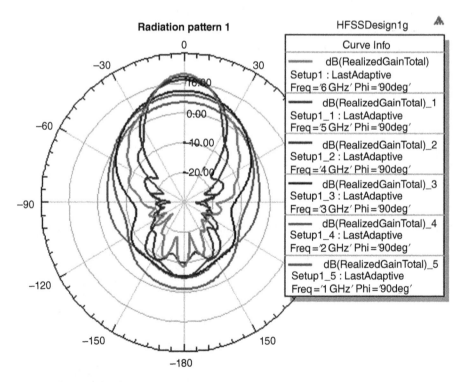

Figure 3.9 Example of a single-sector radiation pattern corresponding to the multi-sector antenna in Figure 3.8.

A simple decision approach would be to consider that the neighboring sectors and the sector at 180° may also emit spectrum in the elimination matrix in Equation (3.25). Another factor to consider is the power emission. If the secondary user has to always consider emitting a signal at relatively low energy to the primary user signal such that the side and back lobes impact in Figure 3.9 is minimal, consideration of neighboring sectors would differ and the number of elements with 1 in the matrix in Equation (3.25) would increase.

Now that we have shown cases for further local decision fusion concepts building on the ROC models, let us move to distributed and centralized decsion fusion that can make the decision fusion results more accurate.

3.3.2 Distributed and Centralized Decision Fusion

Depending on the DSA design, local decision fusion can be communicated to distributed peer nodes or to a centralized arbitrator. Distributed and centralized techniques can estimate a more comprehensive spectrum map of the area of operation. Spectrum awareness can be local (node-based), distributed (net-based), or centralized (area-of-operation-based).[23] Spectrum awareness can be expressed as a spatial map showing areas of interference per each frequency band used in the area of operation or can be potentially used.

Let us consider the local interference estimation exemplified in Figure 3.7 for a local node. Let us also consider that this type of directional interference estimation is shared between peer nodes in a distributed cooperative MANET. Each node can fuse the directional spectrum interference estimation communicated from multiple peer nodes to produce a cooperative spectrum map of the area of interference as shown in Figure 3.10. Notice how the estimated interference

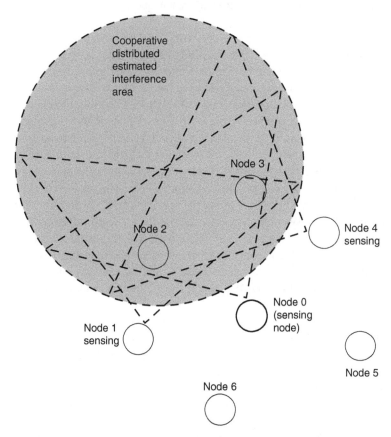

Figure 3.10 Cooperative distributed estimation of area of interference.

---

23 As Part 2 of this book presents, spectrum awareness can be more than these three cases. Distributed cooperative techniques between network gateways can be added as well as proxy of a centralized arbitrator by a gateway node.

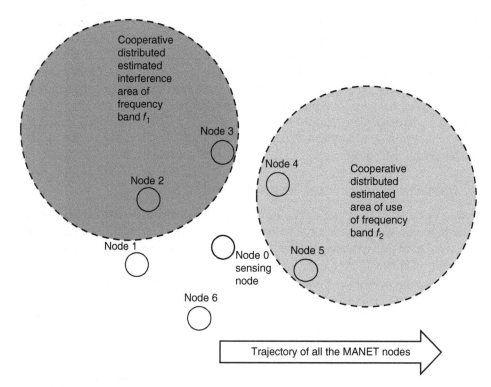

Figure 3.11 Cooperative estimation of overlay spatial use of different frequency bands.

area can be larger than the actual interference area as spectrum directional fusion may draw a circle around the estimated overlaid triangles.[24] If this sensed frequency is the same frequency used by the MANET, cooperative decisions can be made to continue to use the frequency band or switch to a different frequency band based on the results of the decision fusion.

Figure 3.11 shows a case where a distributed cooperative MANET DSA technique estimated the interference area of the frequency band in use by the MANET, f_1. The same DSA technique also estimated the (probed) frequency band f_2, which is a potential frequency band to be used by the MANET. If the MANET trajectory[25] is as shown in Figure 3.11, the distributed cooperative decision may continue to use f_1 as the MANET will move away from the area where interference of f_1 is detected and avoid using f_2. This decision will occur because probing f_2 as a potential frequency to use informed the DSA technique that the MANET will encounter interference if it is to switch to f_2. The key point here is for the distributed cooperative MANET to overlay estimated interference areas of the frequency band in use with the estimated spatial areas of use of potential frequency bands and consider other factors such as trajectory to make the best spectrum use decision, which may include avoiding f_2 and switching to a third frequency f_3 or staying with f_1. Recall that a good DSA design will attempt to avoid the rippling of DSA decisions.

A centralized arbitrator can start by using a pool of frequency bands to assign to different heterogeneous networks in a spatially separated manner, as shown in the previous chapter in Figure 2.15. Recall that the goal of the centralized arbitrator is to optimize the spatial use of spectrum resources. The challenge a centralized arbitrator faces with a case like Figure 2.15 is when a certain frequency is interfered with in a certain area, which may require tapping into a larger pool of frequency bands. Figure 3.12 shows how the frequency band f_3 can suffer from interference, as illustrated with the background large circle indicating the estimated inference area of frequency band f_3 forcing the central arbitrator to tap into an additional frequency band f_4, as indicated by the white circle.

---

24 Estimating the interference source geographical boundaries is an important aspect of decision fusion.
25 A trajectory is the general direction of the movements of all the nodes in the MANET.

Figure 3.12 A centralized arbitrator use of a larger frequency pool to overcome interference.

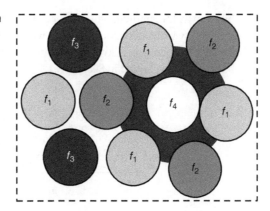

Decision fusion can occur at the local, distributed or centralized level. It all depends on the system under design. A good DSA design would create appropriate fusions at the appropriate level and share information in an optimal way to make the best use of spectrum resources. DSA is not a single solution to a single problem. Communication systems are complex and bounded by requirements, legacy systems interfacing with more up-to-date systems, and other dynamics that can influence the DSA design, information fusion, and decision making. The first three chapters of this book are intended to help the reader gain a broad understanding of DSA design challenges and how to approach DSA design for a given system. The next chapter covers examples of hybrid decision fusion cases and how decision fusion results can be leveraged for other cognitive capabilities such as reactive routing. Chapter 4 will give the reader an idea on how to design a hybrid DSA system while making the appropriate decision fusion local, distributed or centralized considering that DSA is part of the bigger goal of developing cognitive networks.

3.4 Concluding Remarks

The previous chapter covered the foundations of sensing techniques. This chapter builds on the previous chapter, covering ROC methodology and the foundations of DSA decision fusion techniques based on the ROC models. Two distinct ROC models were presented. The first is for sensing if a frequency band is occupied or not, which can be performed by an augmented sensor. This ROC model has its own challenges, including addressing the tradeoff between the probability of false alarm and the probability of misdetection. The second ROC model presented in this chapter is the same-channel in-band sensing model. The chapter examined how the ROC model can be best utilized to detect interference with the in-band signal taking advantages of certain signal characteristics such as a constant envelope. This chapter also covered building on the ROC model to generate local decision fusions for augmented sensing and for same-channel in-band sensing, and extending decision fusion to the spatial dimension and to distributed cooperative and centralized decision fusion. The next chapter covers creating a hybrid cognitive network decision fusion design, building on the foundations covered in this chapter and the previous chapters.

Exercises

1. Consider the binary antipodal signal detection model in the figure below where μ_0 is the mean of the PDF expressing the receipt of the signal y when the symbol S_0 is transmitted, and μ_1 is the mean of the PDF expressing the receipt of the signal y when the symbol S_1 is transmitted. The channel is assumed to introduce AWGN with variance σ and the two small shaded areas express the error probability on both sides of the decision threshold 0. The left-hand part of the shaded area expresses the error probability when decoding S_0

but S_1 was transmitted, while the right hand part of the shaded area expresses the error probability when decoding S_1 but S_0 was transmitted. State some of the parallels (similarities) and differences between this binary antipodal signal detection model and the ROC basic model explained in Section 3.1.

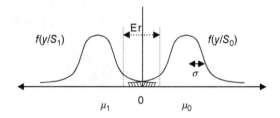

2. Using Equations (3.7) and (3.8) drive the equivalent to Equations (3.7) and (3.8) for the case of same-channel in-band local decision fusion building on Equations (3.19)–(3.22). Does the signal overlay model simplify this ROC model? Why?

3. A set of nodes in a cognitive MANET are moving in tandem while using directional antennas, as shown in the figure below. This could be a row of vehicles, illustrated by the oval shapes. Assume that this convoy is driving along a road in a straight line with the distance between each two vehicles being the same, D. Let us assume the road width is W. Let us also assume that the frequency used for communication between vehicles needs to be used again after a distance W away from the road, as shown in the figure, i.e. the spatial separation is $2W$. Consider a directional beam such that spectrum emission would reach the vehicle in front and the vehicle at the back and then fade away to the side of the road. Spectrum analysis of the directional beam power spectral density shows that $D1$ has to be $1.1 \times D$. The distance $D1$ defines the spatial separation regions such that when a node is communicating to the node ahead of it, the spectral density further away from $D1$ is negligible to allow for spectrum reusability.

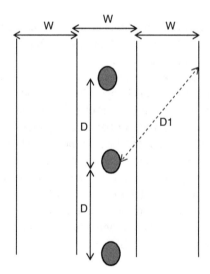

(a) If we use sectored antennas for the vehicles, find the minim number of sectors per antenna we need in order to meet the above spectral pattern design. Note that you would need to round up your calculation to an even number of antenna sectors to find a symmetrical division of the full circle. Note that this will require some geometric analysis.

   (b) Assume that we can apply power control on the antenna spectrum emission such that we can use a frequency after *n* hops. Find the number of frequencies we need for any number of nodes driving on this road.
   (c) Based on the nulling matrix explained in Section 3.3.1.2, do you recommend a higher number of sectors than estimated by (a) above? Why?

4. A cognitive military MANET is implementing a directional antenna technique with a sectored antenna with N equal sectors. Let us assume that there is no overlapping between the sectors. This directionality technique is evaluated based on its ability to avoid enemy eavesdropping nodes. Let us assume that we have node a that is communicating to node b. The maximum area of coverage (AoC) for node a is defined by a circle of diameter d where the radius R is the maximum distance for node a reachability to and beyond node b, as shown in the figure below. As the figure shows, at a certain point in time, the distance between nodes a and b was L.

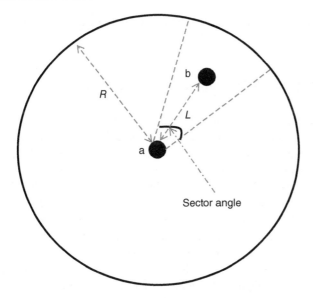

   (a) What is the sector angle of this sectored antenna as a function of N?
   (b) Let us assume that the enemy is able to put eavesdropping nodes randomly within the AoC of node a. What is the probability of the enemy succeeding in receiving the radio signal if the radio uses an omnidirectional antenna with the same AoC?
   (c) If the enemy succeeded in deploying a single eavesdropping node randomly in the AoC, what is the probability of the enemy succeeding in receiving the military signal if the military node used described the directional antenna with a single active sector?
   (d) If the enemy was able to deploy 1, 2, 3, and 4 eavesdropping nodes randomly within the AoC, create a table showing the probability of the enemy succeeding in receiving the radio signal for each of these four cases if the radio used the prescribed directional antenna with a single active sector.
   (e) If the number of sectors used by the military node $N = 12$, find the number of eavesdropping nodes the enemy would have to drop randomly so that the probability of the enemy succeeding in receiving the radio signal approach 0.5.

5. Refer to Appendix 3A. Let us assume we have two ROC curves as shown in the figure below. Are both curves in the "use" region? Based on these intersected curves, will you consider the area under the curve a good metric to assess a ROC curve? Why?

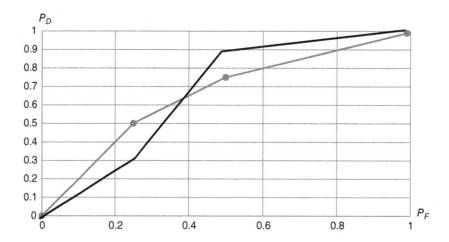

6. In military communications, is DSA alone sufficient to overcome an enemy's follower jammer? Why?

Appendix 3A: Basic Principles of the ROC Model

This appendix describes the ROC model in simple terms to give the reader who is not familiar with the different statistical decision concepts related to ROC models a basic understanding of the model characteristics. The ROC plots are well studied in multiple fields where two basic evaluation measures are needed. These evaluation measures are referred to as sensitivity and specificity in some fields. With DSA, sensitivity is known as the probability of detection of the sensed signal while specificity is known as the probability of false alarm indication of the sensed signal. Any DSA design has to consider a tradeoff between these two evaluation measures. Signal measurements in some cases can lend higher probability of detection at lower probability of false alarm, but the tradeoff always exist.

Let us consider a dataset where the values in the dataset can be classified as positive (P) or negative (N). As shown in Figure 3A.1, all the values in the dataset ideally can be classified as P or N.

An observer classifying the dataset into P or N may create four outcomes, as shown in Figure 3A.2. The four outcomes are true positive (TP), true negative (TN), false positive (FP), and false negative (FN). Notice that the observer hypothesizes the positive and negative value creating the FP and FN events.

The idea behind the ROC model is to create plots such that one axis specifies the false positive rate while the other axis specifies the false negative rate,[26] as shown in Figure 3A.3.

Let us assume that our ROC model plots the false positive (specificity) rate as the *x* axis and false negative (sensitivity) rate as the *y* axis. We refer to this ROC model as the ROC space and

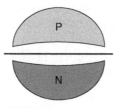

Figure 3A.1 Ideal labeling of a dataset.

---

26 The false negative rate is complementary to detection. The probability of TP or detection is one minus the probability of false negative. Thus, the false negative axis can lead to a probability of detection axis whereas the false positive axis leads to a probability of false alarm axis.

Figure 3A.2 A classifier outcome of the dataset.

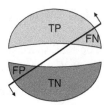

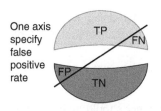

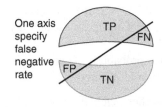

Figure 3A.3 Specifying FP and FN rates.

it is a two-dimensional space that allows us to create the tradeoff needed in DSA design. A DSA decision-making process is a classifier in the ROC space.

3A.1 The ROC Curve as Connecting Points

An ROC point is a point in the ROC space with x and y values where x is the probability of false alarm and y is the probability of detection. Each of the x and y axes spans from 0 to 1. Let us use an example of ROC curves in the ROC space where we simplify the curves by linearly connecting adjacent points. Let us assume that for an example dataset as explained above, we have four possibilities to classify the dataset:

1. Achieve a probability of detection equal to zero at a probability of false alarm equal to zero.
2. Achieve a probability of detection equal to 0.5 at a probability of false alarm equal to 0.25.
3. Achieve a probability of detection equal to 0.75 at a probability of false alarm equal to 0.5.
4. Achieve a probability of detection equal to 1 at a probability of false alarm equal to 1.

These four possibilities become four points in a ROC curve in the ROC space as shown in Figure 3A.4. Notice that each point constitutes a threshold that can be used to hypothesize the dataset value classification. In the ROC space, each point on a ROC curve can become a decision threshold. The key here is to decide an acceptable probability of false alarm and live with the associated probability of detection.

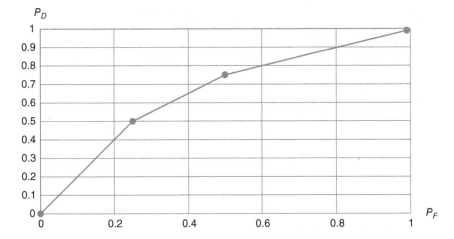

Figure 3A.4 An example of a ROC curve in the ROC space.

3A.2 The ROC Curve Classifications

The ROC space should be used only in certain areas. Figure 3A.5 shows different aspects of the ROC space as follows:

1. The poor performance area. This area should be avoided. The tradeoff can be replaced by a random process.
2. The random cutoff. This is the ROC curve associated with random decision making.
3. The use area where the tradeoff between false alarm and detection probabilities is acceptable.
4. The perfect curve where the probability of detection is always 1. Note that the vertical line should be the decision threshold line in this case.

With the ROC space, a classifier that yields acceptable performance should lie in the area between the random ROC curve and the perfect ROC curve. Figure 3A.6 shows multiple classifiers in the ROC space where the top classifier would yield better performance than the bottom classifier.[27]

Comparing Figure 3A.6 to Figure 3.2 one can see how the DSA ROC space can be conceptualized and how different SNIR ratios create different classifiers. Figure 3.2 shows that a higher

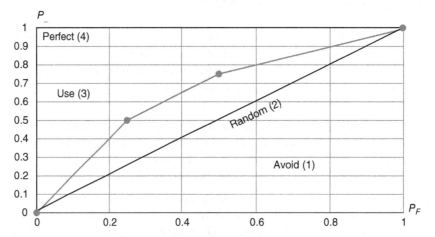

Figure 3A.5 ROC space working areas and thresholds.

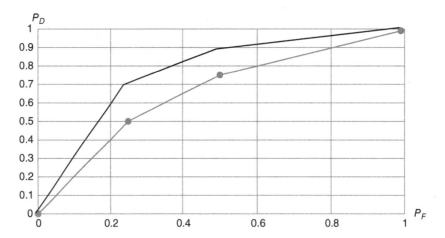

Figure 3A.6 Multiple classifier ROC cuves.

---

27 Notice that as the ROC curve approaches the perfect curve, the area under the curve approaches 1. The area under the curve maybe used to indicate if one ROC curve is better than another. Notice that the area under the random curve is 0.5.

SNIR brings the ROC curve closer to the perfect curve and how decision thresholds can be decided as vertical lines where at a given SNIR, a probability of false alarm should be identified as acceptable, leading to a probability of detection.

Notice the importance of decision fusion. A ROC based decision (e.g., signal detection) can be per an RF neighbor or per an antenna sector. While this single ROC decision can seem insufficient because of the presence of false alarm probability, decision fusion from all the RF neighbors or from all the antenna sectors can yield a more accurate signal detection outcome. Distributed cooperative DSA decisions can further increase the decision accuracy and have a centralized arbitrator with a bird's eye view of the area of operation, and a collection of local and distributed decisions can further increase the accuracy of DSA decision making.

Bibliography

Federal Communications Commission, Spectrum policy task force report. FCC 02-155, November, 2002.

Federal Communications Commission, Facilitating opportunities for flexible, efficient, and reliable spectrum use employing spectrum agile radio technologies. ET Docket No. 03-108, December 2003.

Federal Communications Commission, Unlicensed operation in the TV broadcast bands and additional spectrum for unlicensed devices below 900 MHz in the 3 GHz band. ET Docket No. 04-186, May 2004.

Federal Communications Commission, E911 requirements for IP-enabled service providers. ET Docket No. 05-196, May 2005.

Mitola, J., Cognitive radio: an integrated agent architecture for software defined radio. Ph.D. Dissertation, Royal Institute of Technology (KTH), Stockholm, Sweden, June 2000.

Qiu, R.C., Hu, Z., Li, H., and Wicks, M., *Cognitive Radio Communication and Networking: Principles and Practice*. Wiley,

Setoodeh, P. and Haykin, S., *Fundamentals of Cognitive Radios*. Wiley,

Elmasry, G.F., *Tactical Wireless Communications and Networks, Design Concepts and Challenges*. Wiley, October 2012. ISBN: 978-1-1199-5176-6.

Ganesan, G. and Li, Y., Cooperative spectrum sensing in cognitive radio networks. Proceedings of the IEEE DySPAN, November, 2005, pp. 137–143.

Stevens, J., Spatial reuse through dynamic power control and routing control in common-channel random-access packet radio networks. Ph.D. Thesis, University of Texas at Dallas, 1988.

Elmasry, G.F. et al., Augmenting OLSR with Priority Aware Dynamic Routing for Heterogeneous Networking. *Proceedings of Milcom* 2015, October 2015, pp. 401–406.

Olivieri, M.P., Barnett, G., Lackpour, A., Davis, A., and Ngo, P., A scalable dynamic spectrum allocation system with interference mitigation for teams of spectrally agile software defined radios. Proceedings of the IEEE DySPAN, November, 2005, pp. 170–179.

Shankar, S., Cordeiro, C., and Challapali, K., Spectrum agile radios: utilization and sensing architectures. Proceedings of the IEEE DySPAN, November, 2005, pp. 160–169.

Wild, B. and Ramchandran, K., Detecting primary receivers for cognitive radio applications. Proceedings of the IEEE DySPAN, November, 2005, pp. 124–130.

Haykin, S., Cognitive radio: Brain-empowered wireless communications. *IEEE Journal of Selective Areas of Communications*, vol. 23, no. 2, pp. 201–220, February 2005.

Cheng, B.-N., Block, F.J., Hamilton, B.R. et al., Design considerations for next-generation airborne tactical networks. *IEEE Communications Magazine*, May 2014.

Mishra, S.M., Sahai, A., and Brodersen, R., *Cooperative sensing among cognitive radios*. Available at: http://www.eecs.berkeley.edu/~sahai/Papers /ICC06 final.pdf, 2006.

DARPA Strategic Technology Office (STO), *Communications in Contested Environments (C2E)*. DARPA Broad Agency Announcement (BAA), no. DARPA-BAA-14-02, December 2013.

Wellenhoff, B.H., Lichtenegger, H., and Collins, J., *Global positioning system: theory and practice*, 4th edition. Springer Verlag, 1997.

Reisert, J.H., *Understanding and using antenna radiation patterns*. Available at: http://www.astronwireless.com/radiation patterns.html, 2007.

Elmasry, G.F., Aanderud, B., Kraus, W., and McCabe, R., Software-defined dynamic power-control and directional-reuse protocol for TDMA radios. *Proceedings of MILCOM* 2015, October 2015, pp.139–144.

Roos, T., Myllymaki, P., and Tirri, H., A statistical modeling approach to location estimation. *IEEE Transactions on Mobile Computing*, vol. 1(1), January/March 2002, pp. 59–69.

4

Designing a Hybrid DSA System

The previous chapter covered the ROC model and emphasized two distinct DSA ROC models. The simplest ROC model covered the cases similar to that of a secondary user hypothesizing the presence or absence of a primary user's signal in order to use a spectrum band opportunistically.[1] The second ROC model covered the same-channel in-band spectrum sensing which hypothesizes if the signal used for communications is suffering from interference by another signal or not. The previous chapter also showed how local decision fusion can add other dimensions to the spectrum sensing hypotheses such as the spatial dimension. The previous chapter also introduced some of the decisions that can be made locally and some of the decisions that can be made in a distributed cooperative or centralized manner.

As alluded to in the previous chapter, the DSA design may not stop at the local decision fusion and the solution may rely on cooperative distributed decision fusion or the use of a centralized spectrum arbitrator. Decision fusion can be made locally, in a distributed way and/or in a centralized arbitrator. This chapter covers the DSA design approaches that need to be thought of in cognitive networks, taking into consideration a variety of reasons to include the optimization of control traffic volume, the speed of making DSA decisions, the interdependency between DSA decisions and other cognitive networking processes, and the need for the different hierarchies of DSA decisions to work in harmony.

To make the best case for using a hybrid DSA design, this chapter uses examples from military communications systems where spectrum access needs to be more dynamic and the networks are heterogeneous and hierarchical. This book makes the case for considering hybrid DSA design in most applications. The second part of the book emphasizes approaches that can be common between different applications and areas where the DSA design approach may differ. In the second part of the book, Chapter 5 emphasizes the concept of developing DSA capabilities as a set of cloud services available at the different network hierarchical entities, which further emphasizes the hybrid DSA design consideration. Chapter 6 focuses on dynamic spectrum management for commercial cellular 5G systems. The cellular 5G dynamic spectrum management design is also a hybrid approach. Chapter 8 covers the inclusion of co-site interference mitigation as a subset of DSA cloud services, which also emphasizes the need for hybrid DSA design approaches.

4.1 Reasons for Using Hybrid DSA Design Approaches

The case of designing DSA solutions for a mix of heterogeneous hierarchical MANETs is the best example that can be used to illustrate the reasons for using a hybrid DSA design. Figure 4.1 shows examples of the trade space that can face the design of DSA capabilities. With heterogeneous hierarchical MANETs, there is a need for a centralized spectrum arbitrator and

---

1 Another case for using the simple ROC model is with MANET spatial frequency reuse. The spectrum sensor is sensing (probing) a frequency used by a peer node and the goal of spectrum fusion is to decide if the frequency can be spatially reused or not at a given geolocation and time without interfering with the other peer node.

Dynamic Spectrum Access Decisions: Local, Distributed, Centralized, and Hybrid Designs, First Edition. George F. Elmasry.
© 2021 John Wiley & Sons Ltd. Published 2021 by John Wiley & Sons Ltd.
Companion website: www.wiley.com/go/elmasry/dsad

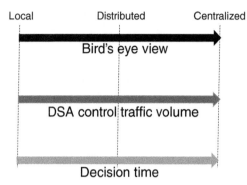

Figure 4.1 Trade space to be considered for hybrid DSA decision fusion.

there could be good design reasons for creating local, distributed cooperative and centralized DSA techniques. As Figure 4.1 shows, moving DSA decisions more towards distributed and centralized decision fusion can increase the bird's eye view (the overarching) understanding of the spectrum use map in the entire area of operation. The tradeoff here is the possibility of increasing the DSA control traffic volume and the increase in decision response time. Obviously, there is a trade space and one must be cognizant of the impact of this trade space depending on the specifics of the cognitive networking system under design.

Example

In a cognitive networking system we are faced with the following:

1. Control traffic volume is not an issue because we have an abundance of bandwidth.
2. DSA decision time can be long because the systems can create stable topology due to limited or no mobility.
3. The design is required to avoid processing at the lower hierarchy network nodes because of limited processing capabilities and power constraints.

In this specific example, the design may consider moving the DSA computational complexity more towards centralized decision making. However, in most systems, one will have to consider a hybrid approach in light of the trade space illustrated in Figure 4.1. Even when designing a distributed cooperative DSA system for a single network, there is a room to consider a mix of local and distributed cooperative decision fusion.

There is no magic bullet that suits every communications system when it comes to designing a hybrid DSA system. There is a trade space that can lead to designing different decision fusion techniques at the different hierarchical entities of the system. However, there are guidelines the design can follow. It is always better to increase the bird's eye view of the spectrum map, it is always better to reduce DSA traffic volume, and decision response times must be appropriate and must meet the system's requirements. The following section presents decision fusion cases that can be helpful to the reader to reach a good design approach for the different types of cognitive wireless networking systems.

DSA design has to create metrics that evaluate the performance of the system in real time and through post-processing, as covered in Chapter 5. Considering Figure 4.1, it is easy to conceptualize creating a metric for decision time trade space to be "response time" in microseconds, milliseconds or seconds depending on the system. It is also easy to conceptualize creating a metrics for DSA control traffic volume trade space to be "control traffic volume" in bps or packets per second. The design has to also create metrics to evaluate the bird's eye view trade space. The reader can refer back to Chapter 1, Section 1.2 for the comprehensive DSA design aspects and see how the bird's eye view considerations can lead to creating metrics that evaluate this trade space, which may include the following:

1. Measuring DSA decisions' rippling. This metric can measure how long a theater-based DSA decision can last before another decision has to be made. A good centralized spectrum arbitrator should make DSA decisions that last.

2. Adapting to traffic demand. If spectrum resources are allocated more where traffic demand is high, this will result in higher throughput efficiency of the heterogeneous networks and this higher throughput efficiency can be manifested in established network performance metrics such as packet or message completion ratio, packet delay, packet delay variation (jitter), and packet loss.

3. Avoiding hidden nodes. A bird's eye view capability should overcome the hidden node phenomena and reduce the likelihood of mistakenly using a primary user's spectrum. This metric can be the measuring of the number of complaints from a primary user.[2]

By the end of this chapter, the reader should realize that Figure 4.1 oversimplified the trade space areas in order to illustrate the most important aspects of this trade space. Other factors such as the used antennas' technologies, the use of licensed spectrum, unlicensed spectrum or a mix of both, and the interactions between the cognitive DSA process and other cognitive processes the system uses add more dimensions to this trade space.

4.2 Decision Fusion Cases

Let us consider the case of sensing the presence of a signal where the local decision fusion engine was able to collect a knowledge repository that can be expressed as depicted as in Figure 4.2. This repository is illustrated in Figure 4.2 with three different scenarios of the RSSI collected samples. Scenario 1 represents the case of low noise variance, scenario 2 represents the case of medium noise variance, and scenario 3 represents the case of high noise variance. The local knowledge repository and its associated cognitive engine were able to select a decision threshold based on the detected energy levels where the gray dots to the left of the decision threshold represent the energy detected in the absence of the sensed signal and the black dots to the right of the decision threshold represent the energy detected in the presence of the signal.

Notice that to obtain such knowledge repository, the sensor has to have some sort of pre-knowledge of the sensed signal characteristics. As explained earlier in this book, this can be achieved in both the case of same-channel in-band sensing and the case of an augmented sensor sensing the presence of a primary user signal with known cyclostationary characteristics. In the case of same-channel in-band sensing, signal marks allow for the differentiation between energy samples representing noise or noise plus interfering signal (left side of the decision threshold in Figure 4.2) and energy samples representing signal plus noise or signal plus noise plus interfering signal (right side of the decision threshold in Figure 4.2).[3] In the case of an augmented sensor probing a frequency band for potential use, the signal cyclostationary characteristics will allow the sensor to collect energy samples representing noise (left side of the decision threshold in Figure 4.2) and energy samples representing signal plus noise (right side of the decision threshold in Figure 4.2).

As scenarios 2 and 3 in Figure 4.2 show, when the noise power increases, the RSSI samples will tend to spread wider. Noise power can be due to pure AWGN or another secondary user overlaying its signal that has unknown characteristics. As scenario 3 in Figure 4.2 shows, a higher noise power will lead to the right side points and the left side points to encroach on each other. Note that higher noise power increases the standard deviation of the detected noise energy samples and the detected signal plus noise energy samples. With this case, the local decision fusion engine is able to hypothesize the presence or absence of a communications

---

2 Similar to all aspects of network design, DSA design will require simulation environment and laboratory testing during the system design and evaluation phases and before deployment. Hidden nodes can be simulated in scripted scenarios to evaluate the performance of this trade space

3 In Figure 4.2, for the same-channel in-band sensing case, noise and interfering signal are combined as noise.

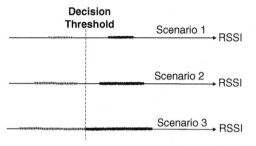

Figure 4.2 Local decision fusion based on single-dimensional knowledge base.[5]

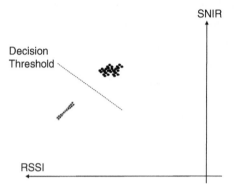

Figure 4.3 Decision fusion based on two-dimensional knowledge base.

signal but clearly noise power increase can lead to either a higher probability of false alarm or a higher probability of misdetection depending on where the decision threshold is chosen.

Now let us consider the case when the local decision fusion engine is able to create a more accurate energy detection fusion, as illustrated in Figure 4.3. With this case the knowledge repository can plot a two-dimensional curve of RSSI versus SNIR. The inclusion of SNIR in the knowledge repository adds a better depiction and more accurate hypothesizing as the increase in noise energy can cause spreading of the plotted points in two dimensions instead of one dimension. This will result in false alarm and misdetection hypotheses only in extreme cases such as the presence of very high noise power.

This example is used to illustrate the importance of coordinating between decision fusion hierarchies. If the local decision fusion follows the approach depicted in Figure 4.2 to reduce computational complexity, the distributed or centralized decision fusion may need to create knowledge repositories equivalent to Figure 4.3 to reduce false alarm and misdetection probabilities. On the other hand, if the local decision fusion engine was able to hypothesize based on Figure 4.3, distributed and centralized decision fusion engines can focus on other DSA aspects, such as spatial location of interference and the creation of a more accurate spectrum utilization map.

The trade space in Figure 4.1 illustrates some important aspects. In reality, there are more aspects in this trade space. For example, having a more detailed knowledge repository at the local node may not be achievable because of processing and power limitations.[4] At a centralized arbitrator, processing power may not be a limiting factor. On the other hand, sending more detailed spectrum sensing information to a centralized arbitrator can have its own drawbacks to include the use of more bandwidth for DSA control traffic.

Relying on the centralized arbitrator for more decision fusion can result in the loss of information when sensing information is fused locally. For example, local fusion of SNIR values may produce information about SNIR that includes the mean, average, and standard deviation out of a large sample of SNIR values. This processing can result in some information loss and may result in the centralized arbitrator failing to achieve the desired lower probability of

---

4 This can occur in sensor networks where the node has SWaP limitations.
5 Notice in Figure 4.2 that RSSI is a single-dimension positive value. The figure shows three different scenarios of RSSI reading and the vertical line is meant to show the decision threshold not a vertical axis.

false alarm or lower probability of misdetection. Local fusion using raw RSSI and SNIR values can have a more accurate cutoff threshold in Figure 4.3.

A designer of a hybrid DSA system has to consider the trade space taking into consideration all the factors that can affect the final solution to include bandwidth limitations, hardware limitation such as SWaP and processing power, speed and accuracy of decision making, and the system's requirements. One important factor that can affect hybrid design is the role of other cognitive processes in the cognitive wireless network nodes. These cognitive processes can influence the design of DSA decision fusion as detailed in the next section. Keep in mind that DSA capabilities are one of many aspects of wireless cognitive networking capabilities.

4.3 The Role of Other Cognitive Processes

Let us illustrate this critical factor with the case of a distributed cooperative directional MANET that is designed to route around jammed areas. The design of such a system has to feed the results of the DSA decision fusion to a cognitive routing process/engine and the cognitive routing engine has to use DSA information to create reactive directional routes relying on directional antennas. The cognitive routing engine has different objectives, including:

1. creating directional routes that allow communications around compromised areas
2. increasing spectrum reuse
3. reducing interference between the directional MANET nodes.

This type of cognitive routing relies on controlling the RF beam's direction and power to achieve these objectives while obtaining spectrum sensing information from a distributed cooperative spectrum decision fusion engine (see Figure 4.4). Some of the characteristics of a distributed cognitive routing engine that relate to a distributed cognitive DSA engine include the following:

- The cognitive routing engine may need to control transmitting power based on the receiving node location. This is needed to reduce the impact of the spectrum utilized by the directional MANET on the overall spectrum map. Power level is selected by the routing engine to be the minimum power needed to reach the next-hop node given the current interference level.
- Both routing and DSA engines need to know the antenna gain pattern associated with the utilized antenna.
- Nodes have to continually register with each other and disseminate their locations to each other using protocols such as terrestrial geolocation protocol (TGP). This is needed so that beams are formed optimally. TGP information would need to be shared between both the routing engine and the DSA engine.

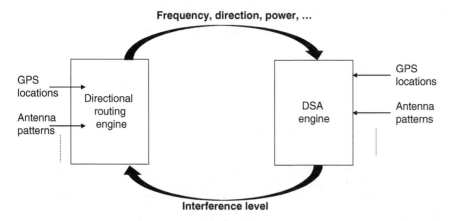

Figure 4.4 The interface between a cognitive routing engine and a cognitive DSA engine within a node.

- In the case of the routing engine considering a new route, this route can have multiple over-the-air hops. Each hop (the next-hop) is identified by source and destination nodes geolocations, a frequency band, direction of the formed link, and power level. This will require the cognitive routing engine to feed the DSA engine with its own potential links (spectrum use) such that the DSA engine draws a potential spectrum map that includes what spectrum would have been emitted had the new route considered by the routing engine been implemented. The estimated interference level affecting the new route can be calculated by the DSA engine and the interference levels can be sent to the routing engine.

Notice here the presence of another distributed cognitive engine for routing parallel to the DSA distributed decision fusion engine.[6] Notice also that each engine may impact the other engine directly through information exchange and indirectly through the implemented decisions. For example, routing decision made by the cognitive routing engine will impact the spectrum map estimation created by the DSA decision fusion engine and vice versa. Notice also that the implementation of such an engine can use heuristic approaches that make it hard to predict the impact of creating reactive decisions. These cognitive engines may also have to coordinate before executing a decision such that each engine can predict the impact of other engines' decisions. This coordination will create an integrated system that avoids bad decisions.[7]

Figure 4.4 illustrates examples of information exchange and indirect impacts between these two cognitive engines. The routing engine may send the DSA engine a set of frequency slots it wants to consider for transmission with each frequency associated with the direction of transmission (based on knowing the geolocation of the receiving node) and the considered transmission power. The routing engine may expect the DSA engine to send it a reply to this query with the estimated interference level per each frequency slot. Based on this estimated interference level, the routing engine may decide which slot to use for the targeted destination node. When the over-the-air links are established, the used frequencies will have an impact on the spectrum map estimated by the DSA engine.

One important aspect of the routing engine that impacts the spectrum map estimated by the DSA engine is the antennas' directional model used by the routing engine. This antennas' directional model directly affects the decision if a frequency slot can be used or not. Let us consider the use of an antenna directional model that shows the involvement of the routing engine with a spectrum map and how the routing engine influences the DSA engine. The transmitting node DSA engine would need to estimate the interference power at the receiving node per each candidate frequency slot so that the routing engine would have enough information to decide which frequency slot to use. Notice that the DSA engine estimation of interference power has to be comprehensive considering all the nodes in the geographical proximity that are using the same frequency, the directions this frequency is being used, and the transmission power level of each use in order to ensure that the estimated interference level is accurate enough for the routing engine. Assuming a free space path loss, the following formula can be used by the DSA engine to estimate the interference level had a candidate slot been used by the routing engine.

$$P_I = \sum_{j=0}^{N} P_j G_R(\delta_R) G_T(\delta_T) \alpha^2 / R^2, \tag{4.1}$$

where:

- I is the receiving node and P_I is the estimated interference

---

6 Some reference terminology is that a node has a single cognitive engine that contains multiple cognitive processes. Some other reference terminology is that a node has multiple cognitive engines. Regardless of using either terminology, a cognitive DSA engine/process has to work with other cognitive engines/processes.
7 This example makes a good case for developing DSA as a set of cloud services as further discussed in the second part of the book. One can see how in this example the routing engine is requesting a service (to establish a link) and the DSA engine is replying to the service request.

- P_j is transmit power from node j (all nodes j in geographical proximity are using this frequency slot at the same time)
- G_R is receiver antenna gain
- δ_R is the absolute value of the difference between the pointing angle of the receive antenna and the bearing to the transmit node
- G_T is transmitter antenna gain
- δ_T is the absolute value of the difference between the pointing angle of the transmit antenna and the bearing to the receive node
- R is distance between receiver and transmitter
- α is $(C/4\pi f)^2$
- link closure is a function of (δ_R, δ_T).

Notice that the DSA engine has to have knowledge of the geographical location of nodes in close vicinity and which frequencies are used at which direction for each node. This allows the DSA engine to estimate the interference power at the receiving node had the local node used the candidate frequency slot. Based on this interference level, the local node routing engine will determine if the interference level exceeds a given threshold or not before selecting the frequency slot to transmit at. Notice how such design can lead to relying more on distributed cooperative DSA decisions to coordinate between nodes.

The DSA engine has to estimate the pointing direction between two nodes in an accurate way. Figure 4.5 illustrates the different factors the DSA engine has to consider. The DSA engine may use an absolute north as a reference to how the different angles in this estimation are related. Figure 4.5 shows how "north" is used as a reference and how the angles are related.

In Figure 4.5, notice the following:

- γ is the antenna pointing direction, relative to north
- β is the bearing of the other node, relative to north
- δ is the angle between the pointing direction and the targeted node.

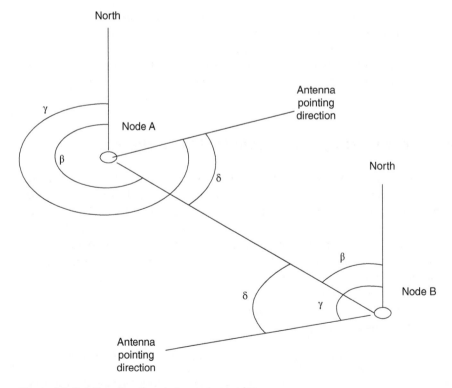

Figure 4.5 Pointing directions between two nodes.

Note in Figure 4.5 that β of either node is equal to $180° + \beta$ of the other node, which allows both δ angles to be determined at either end. This will require that the transmit power, antenna type, pointing angle, and location of the other nodes in the neighborhood using the same frequency slot must all be disseminated between the distributed DSA engines.

One can see the extensive distributed information that would need to be communicated between the different DSA engines in order for the DSA engine to be able to support a cognitive routing engine. In addition, the DSA engine is expected to perform extensive computations continually so that the routing engine has accurate estimations of the interference level of all potential frequency slots to use and all nodes in the neighborhood.[8]

While the disadvantages of using these two distributed cognitive engines include the increase of over-the-air resources use for DSA control traffic and the need for local computational power to perform comprehensive local decision fusion tasks, the advantages of such design include the following:

1. Efficient use of spectrum resources, which can lead to higher throughput.
2. The ability to reuse the same frequency slot relying on directionality and power control, which also can lead to higher throughput of the system.
3. The ability to route around jammers.
4. Making the transmitting nodes spectrum emission less prone to eavesdropping by reducing the spectrum footprint of the formed MANET in comparison to using omnidirectional antennas.
5. The ability to dynamically adapt spectrum resources use to the specifics of the area of deployment (e.g. terrain) and achieve connectivity in the presence of external benign and/or malicious interferences.

These advantages can be critical in the use of DSA in military cognitive MANETs. One can see that the trade space in Figure 4.1 is overly simplified and it can include other dimensions:

- The low probability of detection (LPD) trade space, which can be measured with a metric that calculates the ratio of the combined spectrum footprint of all directional links to the entire theater area of operation.
- The low probability of intercept (LPI) trade space, which can be measured with a metrics that calculates the probability of an eavesdropping node, randomly positioned in the theater's area of operation, to intercept a directional link.
- The probability of avoiding jammed areas trade space, which can be measured with a metric that counts the number of successes and failures to establish routes around jammed areas where jammed areas are randomly positioned in the theater's area of operation.

4.4 How Far can Distributed Cooperative DSA Design go?

In the previous section, we saw how a distributed DSA engine can interact with a distributed routing engine in order to create a low spectrum footprint MANET with antijamming, LPI/LPD, and dynamic spectrum reuse capabilities. In this section, we will show how this distributed cooperative routing engine concept, which is local to a single network, can be adapted to global heterogeneous routing, which is between the different types of hierarchical MANETs.[9] The goal here is to explore how far a system that is designed as a hybrid between local and distributed cooperative DSA decisions can go before we move to cases where a centralized DSA arbitrator is needed.

Making distributed DSA work in a global manner with heterogeneous hierarchical networks can be based on the following principals:

---

8 There are different propagation models the DSA engine can use, including a standard Gaussian model, a modified Gaussian model, a Lorentzian model or a modified Lorentzian model.
9 In heterogeneous hierarchical MANET, one can create two tears of routing: local routing within a network and global routing between networks.

- The creation of multitiered DSA engine architecture where a platform that is a node of more than one network (gateway node) has a distributed DSA engine for each network and a parent DSA engine (for global arbitration of DSA decisions).
- The creation of two distinct types of cognitive routing engines where one type is for routing local to the network and the other type is for gateway nodes that can create global routes.

For this multitiered architecture, let us refer to the lower tier DSA engine as the waveform DSA engine, the routing engine local to the network as the waveform routing engine, the upper tier DSA engine as the master DSA engine, and the upper tier routing engine as the master routing engine. Figure 4.6 illustrates how a node that is not a gateway node would contain only the two waveform engines while a gateway node would contain all four types of engines with the lower tier engines (waveform DSA engine and waveform routing engine) having one separate engine for each waveform type the gateway node has. The master DSA engine and the master routing engine will always have a single instantiation in each gateway node.

Notice that there are different approaches used to create dynamic heterogeneous global routing. One known approach creates an interface control definition (ICD) between each waveform type and the networking layer above the waveform. This interface is sometimes referred to as the router-to-radios (R2R) interface. In Figure 4.6, this interface may be replaced with the interface between the waveform routing engine and the master routing engine. The approach covered here relies on the master routing engine making routing decisions by choosing between the different available paths based on the condition of each wireless network in the path. This interface can override R2R protocols such as the point-to-point protocol over Ethernet (PPPOE), which have been shown to be insufficient in dynamic MANETs. The approach covered here leaves certain dynamic spectrum management local to each network through the waveform DSA engine and uses the master DSA engine for other DSA decisions. Here, we have a more comprehensive approach than that of Section 4.3 where the gateway node is part of the spectrum allocation negotiation between the different MANETs.

There are established methodologies that address getting statistics that convey the local MANET dynamics from the waveform (radio modem) to the routing layer above. These established methods assume that the wireless links below the routing layer have already been established. Protocols such as the dynamic link exchange protocol (DLEP) communicate metrics to include link quality, bandwidth, and neighbor discovery. With DLEP, the wireless links and communications with neighbors must be established before these metrics can be passed to a routing engine. As such, the master routing engine in the construct in Figure 4.6 will not be able to make routing decisions that include an underlying waveform network unless radio links are established. The multitiered approach illustrated in Figure 4.6 can generalize the DLEP a

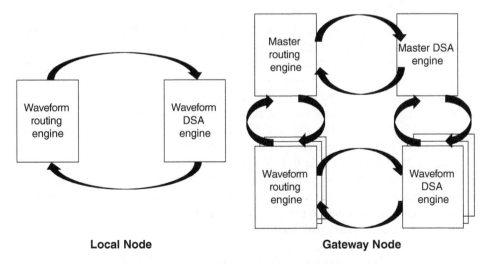

Figure 4.6 Local and gateway nodes cognitive engines.

step further by having the master DSA engine (higher tier engine) perform proactive sharing of the node's links capabilities over a "control plane" medium even when some of the available wireless links are not in use (i.e., in a large scale heterogeneous network, some MANET links have not been formed yet). In the gateway node, the master routing engine is able to receive (via the master DSA engine) location, status, capabilities, and spectrum resources utilization information from peer remote radios engines before all link establishment occurs and use this information to establish potential routes. The master DSA engine is able to use this information to perform spectrum resources deconfliction, link closure estimates, and bandwidth estimation before these links are established. The DSA master engine's ability to perform proactive sharing over different waveforms (mediums), through the master routing engine already established routes, gives the master routing engine the ability to ascertain potential data rates to neighbors on each network type prior to the establishment of all wireless links to these neighbors.

Note that with this type of heterogeneous waveforms formation, the interface between the waveform DSA engine and the master DSA engine has to use unified wireless link metrics regardless of the waveform type which the waveform DSA engine represents. One waveform type may express the health of its wireless links in a different way from another waveform type. This multitiered architecture requires some normalization of the link health metrics calculated by the different waveforms in order for a routing engine to create an optimum global routing table without being skewed to use one waveform over another due to the lack of uniformity of link health metrics. This normalization should also allow the different networks to use different routing approaches internal to the formed MANET independent of global routes. One of the most important essences of creating true seamless efficient heterogeneous networks is to allow each network to use the best protocols for its internal routing and link health representation while requiring the different types of waveform DSA engines to adhere to a unified representation of the network metrics when it comes to global aspects.

Let us illustrate how this hierarchical architecture can work with the case of a low bandwidth waveform that has global connectivity over a specific deployment. This low bandwidth waveform can be used to create a large RF footprint MANET that is used to establish the control plane over the deployed large-scale heterogeneous hierarchical MANETs. In addition to the low bandwidth global network, the theater deployment has different types of higher bandwidth waveforms that will establish different types of networks (e.g., mesh networks, omnidirectional multiple access networks, LPI/LPD directional networks, etc.) where spectrum resources can be allocated dynamically within these networks and utilized globally by the master routing engine. Let us refer to these higher bandwidth networks links as offload links. The goal of DSA in this construct is to use these offload links dynamically and on-demand considering the following:

1. When data traffic to a node over the low bandwidth global network exceeds a defined threshold, the master routing engine will ask the master DSA engine to allocate resources over a specific higher bandwidth network. The master DSA engine will allocate the required spectrum resources by asking the corresponding waveform DSA engine to create a *flow*[10] to that node. This action will create a new routing path for the master routing engine.
2. When data traffic levels exceed the currently allocated bandwidth, the master routing engine will ask the master DSA engine to increment the allocated resources. The master DSA engine will ask the corresponding waveform DSA engine to acquire more bandwidth over the established offload data link. The current waveform DSA agent might adjust allocations or the master DSA engine might switch to another offload link on another network as necessary to meet the traffic volume needs. In either case, the master DSA engine will inform the master routing engine if the created routing path has increased bandwidth or if a new routing path is created with the required bandwidth.

---

10 A *flow* can be dictated by source and destination address, and type-of-service (TOS), with the lowest amount of data rate allocation common between all waveforms.

3. If traffic levels go below the currently allocated levels, bandwidth over the offload datalinks will be reduced, releasing some of the spectrum resources. Corresponding message flow to the two cases above will ensure the release of resources is known to all involved engines.
4. If traffic levels go below a threshold, the flow over the offload data link will be torn down, releasing all resources back to the offload network.

Notice the following:

- The low bandwidth global network is a lifeline to all nodes and is the start of the control plan. The control plane may use dynamic resource allocation (offload links) if needed.
- All offload networks spectrum become shared spectrum resources pools for allocating spectrum resources per traffic demand.
- The master routing engine is independent from the master DSA engine and the master routing engine has no say on how spectrum resources are to be allocated. This allows the inclusion of heterogeneous waveforms that use different resources allocation protocols within this heterogeneous hierarchical architecture. Each waveform can have a specific waveform DSA agent that works independently of other waveforms' DSA engines. One key aspect here is normalization of waveform metrics such that the master DSA agent uses unified resource allocation terms (e.g., flow) and the routing engines sees created routing paths in terms of unified data rate units (e.g., flow) and source destination pair. For the routing engines, a flow may be associated with a routing cost to allow the routing engine to ascertain the best route (path of multiple over-the-air hops).

The use of such dynamic link establishment, link adaptation, and link teardown means the use of a dynamic reactive routing protocol creating route tables that get modified quickly. When there is a steady-state period, the cognitive engines will be monitoring the allocated flow's health and exchanging information. Information exchange during steady-state periods can include spectrum health awareness and mobile node geolocations. The shared information will allow the waveform DSA and routing engines to anticipate the need for changes[11] local to the network and react in a timely manner. Shared information will also allow the master DSA and routing engines to anticipate the need for global changes and react in a timely manner.

The master DSA engine is tightly coupled with the master routing engine in different ways. Consider the case when the master DSA engine is sharing spectrum awareness with its peer master DSA agents. The master DSA engine will rely on the dynamically created routes by the master routing engine and control traffic can become traffic demand on its own merit. If the global low-bandwidth network cannot accommodate all control traffic volume, the offload network's spectrum resources would be used for control traffic.

The available resources on an established data link can change due to mobility, link stability, and link quality. For the cases where the waveform DSA engine learns new information about the link resources, the waveform DSA engine can inform the master DSA engine about any changes. One case of dynamically increasing link bandwidth due to mobility would be when a waveform switches to a lower forward error correction (FEC) mode, increasing throughput at close range while using the same amount of spectrum resources.[12] On the other hand, a case of decreased bandwidth is when the waveform must switch to a higher FEC mode due to increased range or jamming. With dynamic MANET, situational awareness must be shared between all cognitive engines, which can result in constant change of routing tables, constant

---

11 One possible change that must not be ignored and needs to be considered first is power control. Node mobility may require increasing or decreasing power level while keeping all established links and routes unchanged. It will be left to the cognitive engines to decide if this change can be maintained or new links need to be established. Another similar change is directionality. The ability to keep the same topology with minimal changes to the beam direction should be explored and used before making changes that will affect route stability.
12 Many military communications waveforms use adaptive FEC. Some waveforms use concatenating codes where both the inner and outer codes are adaptive. The inner code can be turbo code while the outer codes can be RS code with erasure.

changes in spectrum resource allocation and teardown, and consequently increase in control traffic load.[13]

Let us summarize the advantages of this heterogeneous distributed cooperative DSA approach:

- Having a data plane that is independent of establishing offload links means the ability to exchange spectrum resources and link information before all networks are established.
- Waveform engines can be developed independent of each other as long as the exchange of waveform information is normalized between all types of waveforms.
- Waveforms can use their internal protocols for position awareness, antenna directionality, spectrum resources establishment, and teardown. Waveforms can decide how reactive routes local to the waveform network are established and torn down.
- A large-scale set of heterogeneous networks can be deployed together and these networks will morph to the unique scenario needs while the waveform cognitive engines manage the configuration of links internal to the waveforms network and the master engines manager global routes.
- Making global dynamic route tables independent of route tables local to a network makes it possible for heterogeneous networks to be connected and optimized seamlessly.

There are also disadvantages of this heterogeneous distributed cooperative DSA approach:

- The amount of DSA control traffic can grow exponentially as the number of nodes in a network increases and as the number of heterogeneous networks increases.
- Some nodes can go out-of-synch or temporarily lose connectivity and some information exchange control traffic can be lost. Retransmission of control traffic can further cause an already large volume of DSA control traffic to grow larger and can delay the adaptive route creation decisions.
- There are limits to global views of a large-scale set of heterogeneous networks through distributed protocols. The master DSA agent in each gateway node will create a spectrum map that is less accurate than the spectrum map that can be created by a centralized arbitrator.

To make different types of IP networks work seamlessly, Internet routing protocols faced similar challenges to what is described here for DSA challenges. This led to the creation of routing areas and gateway protocols such as the border gateway protocol (BGP). In some IP network design, certain routing configurations had to be left for a centralized network manager to decide. Similarly, the reliance on a distributed cooperative DSA for large-scale heterogeneous networks will always show some limitations and lead us to consider the need for a centralized spectrum arbitrator to curb control traffic volume and arbitrate fairly between heterogeneous networks in certain cases while allowing a human (network manager) to monitor the cognitive network's health and set policies and rule sets. The network manager can change policies when needed.

One simple way to curb control traffic volume is to reduce its rate. Figure 4.7 illustrates a conceptual view of how control traffic volume can depend on the updating intervals. As updating interval periods decrease (more frequent exchange of DSA messages occur), control traffic volume increases. On the other hand, as the updating interval periods increase (less frequent exchange of messages occur), control traffic volume decreases. There is a desired cutoff limit of the amount of DSA control traffic depending on the heterogeneous systems under design. The limitation of the amount of DSA control traffic can lead to curbing message exchange rate and the impact of this limitation can make resource allocation and dynamic routing tables less dynamic and less reactive to the deployment needs. This can be harmful, especially for fast-moving nodes in very dynamic MANETs.

---

13 The reader should keep in mind the different control loop time periods. Routing within the network, which can be waveform dependent, can have its own techniques that create and tear down reactive routes separate from global routes. Global routes need to be more stable than routes local to a network.

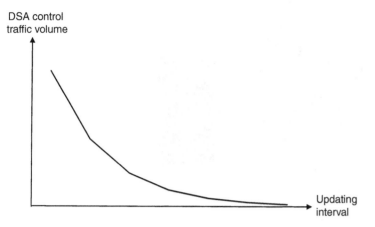

Figure 4.7 Conceptual view of how control traffic volume depends on updating rate.

Notice that within the heterogeneous MANET designs presented in this section, the distributed cooperative DSA[14] has a mix of local DSA decisions made by the waveform DSA engine and distributed DSA decisions made between the master DSA agents. Although these systems are referred to as distributed cooperative DSA systems, in essence they are hybrid between local DSA and distributed DSA decision systems with two different hierarchies of distributed cooperative decision fusion. Similarly, adding a centralized arbitrator to this construct creates a different type of hybrid DSA design that has more hierarchical DSA entities.

4.5 Using a Centralized DSA Arbitrator

Figure 4.8 illustrates an example of using a centralized DSA arbitrator with a simplified scenario of three heterogeneous MANETs. The large dashed circles represent a MANET current area of coverage. The small gray circles represent nongateway nodes while the small black circles represent gateway nodes. Gateway nodes are platforms with more than one waveform and are able to be nodes of more than one MANET. The black arrows represent the centralized DSA control plane where the centralized DSA arbitrator sends and receives control traffic to and from the gateway nodes. With this approach, the gateway nodes proxy their respective networks to the centralized DSA arbitrator such that local fusion in the gateway nodes is more comprehensive than fusion in the other network nodes.[15] The idea here is to have the gateway nodes send a network-based spectrum utilization and interference picture to the centralized DSA arbitrator. Although the centralized arbitrator would decide spectrum allocation dynamically between the networks, this allocation is a hybrid allocation that takes into consideration the role of distributed cooperative DSA decisions within each network.

The centralized DSA arbitrator would remove some of the heterogeneous spectrum allocation tasks from a distributed cooperative manner to a centralized manner. This will result in adding a centralized DSA control plane, as illustrated in Figure 4.8, but can drastically reduce the control traffic volume. Here, a gateway node is tasked to obtain the spectrum utilization map of its own network through its own fusion technique, send it to the centralized arbitrator, and proxy the centralized arbitrator's decision only when it cannot reach it. Note that messages sent over the centralized DSA control plane can be multicast messages that reaches the centralized arbitrator and a subset of peer gateway nodes with a single over-the-air transmission.

---

14 It is important to see that with this architecture there could be two distinct types of cooperative distributed DSA. One type is distributed cooperative DSA between nodes within a single network. The other type is the global distributed cooperative DSA between the gateways.

15 This proxy can go both ways. With a large-scale deployment of heterogeneous MANETs, some MANET gateways may not be able to reach the centralized arbitrator due to factors such as terrain. The gateways can proxy the centralized arbitrator decisions through the cooperative distributed approach described in the previous section.

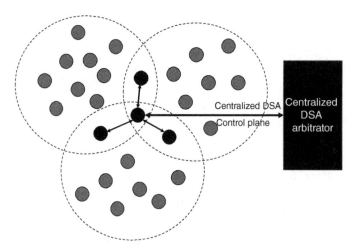

Figure 4.8 The use of a centralized DSA arbitrator in a hybrid DSA system of heterogeneous MANETs.

This will allow the gateway nodes to rely on the centralized arbitrator for global decision fusion in most cases. In rare cases, when the centralized arbitrator is not reachable for a decision, the gateway node will rely on the distributed cooperative protocol mentioned in the previous section[16] to act as a proxy for the centralized arbitrator.

Notice that the way a gateway can proxy a network spectrum utilization can depend on the gateway capabilities. This proxy approach can range from merely consolidating the network sensors' data into one message sent to the centralized arbitrator without fusion at the gateway node to performing extensive fusion at the gateway node and sending the centralized arbitrator a fused spectrum map. Factors such as SWaP limitations at the gateway node can guide the design of the gateway node fusion capabilities.

With this type of hybrid DSA system, we have a mix of centralized spectrum allocation and distributed cooperative spectrum allocation within each network. Each network is assigned a pool of frequency bands to use based on the network's own distributed cooperative protocols while the centralized arbitrator changes the allocation of this pool of frequencies (redistributes frequency pools between all the networks) when its own spectrum map shows the need for such change.

While the distributed cooperative hybrid system can work with a certain number of heterogeneous MANETs, as the number of networks in a large-scale deployment increases, the need for a centralized arbitrator will increase. Notice that if the design elects that a gateway node would proxy a network spectrum map using extensive fusion, this could reduce the amount of DSA control traffic drastically. The gateway node can abstract the network spectrum map, fusing spectrum sensing information into very compressed messages sent to the centralized arbitrator.

The global spectrum map within the centralized arbitrator has its own advantages. In addition to reducing DSA control traffic volume, the use of a centralized arbitrator can also further optimize spectrum resources allocation. Figure 4.9 illustrates how without a centralized arbitrator (left side of the figure), networks can negotiate the division of the global spectrum resources pool in a greedy way, causing spectrum utilization to be less dynamic.[17] With the use of a centralized arbitrator (right side of the figure), the spectrum pool can be divided into

---

16 The distributed cooperative protocol from the previous section is now partially implemented to reduce control traffic volume because it is only a fallback solution. Messages from a gateway node to the centralized arbitrator can use an unreliable multicast protocol. Dissemination of all these messages to all gateways is not needed. The fallback distributed cooperative solution can fuse the subset of information that reaches the gateway through multicast messages.
17 Gaming theory based approaches can be used in the absence of a centralized arbitrator and some networks master DSA engines negotiating spectrum resources allocation with peer master DSA engines can be "greedy" in negotiation (overestimating date rate) causing the gaming theory based technique to tend to evenly distribute spectrum resources between the networks.

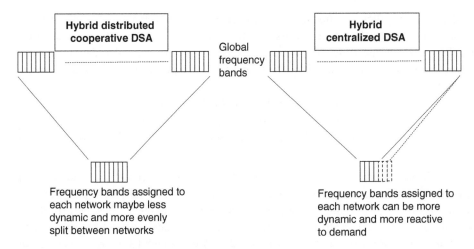

Figure 4.9 The use of a centralized DSA arbitrator can further optimize spectrum resource allocation between a large-scale set of heterogeneous networks.

smaller portions and each network can be assigned a smaller spectrum pool at the start. The spectrum assignment per network can be increased when the network local traffic demand increases and decreased when the network local traffic decreases. However, the centralized arbitrator can mitigate greed based on comparing traffic demand and actual traffic volume from the different networks while considering the global view of spectrum conflicts and interference levels.

One important advantage of using a centralized DSA arbitrator is the stability of spectrum assignments. The centralized arbitrator can assign each network a set of frequency bands as life lines and this set can be relatively small, as shown on the right of Figure 4.9. The centralized arbitrator can add more frequency band to a network based on traffic demand. The life-line frequency band assignment can be stable and only changes if mobility requires the redistribution of these life-line bands. This will create a form of stability in spectrum assignment that will result in creating more stable routes by the distributed cognitive routing engines. One of the challenges facing dynamic resource allocation techniques is rippling, as explained in the previous chapters. Rippling occurs when resources are assigned then quickly taken away to be reassigned.[18] This can lead to the establishment and teardown of link before routes are established. The use of a centralized DSA arbitrator offers the designer of these heterogeneous networks the means to reduce rippling.

The use of a centralized arbitrator can also help address energy-constrained nodes limitations. In heterogeneous MANET, some networks can be sensor networks with limited battery size, with only gateway nodes having enough energy to process high-volume traffic. The assignment of life-line frequency bands to energy-constrained nodes while assigning gateway nodes more frequency bands can result in links between gateway nodes having lower routing cost than that of the links that rely on energy-constrained nodes, which will decrease the likelihood that an energy-constrained node is considered for routing higher bandwidth traffic while still maintaining stable network connectivity.

The design of a hybrid DSA system with a centralized arbitrator has to take into consideration the following:

1. There is a place for local DSA decisions, distributed cooperative DSA decisions, and centralized DSA decisions.
2. The speed of these decisions has to be considered in the design where local DSA decisions are fast. Distributed cooperative DSA decisions should take considerably longer than local

---

18 As explained earlier, making global routes stable is more important than making routes local to one network stable.

decisions because they require some knowledge-based fusion and propagation of fused information between the nodes. Centralized DSA decisions are meant to create stability, including life-line spectrum allocation, and react to traffic demand increase in more optimum ways than using distributed cooperative decisions for demand-based resource allocation.

3. Distributed cooperative DSA decisions within a net will still use the spectrum resource pool assigned to the net by the centralized arbitrator dynamically to address temporary surges in traffic demand before attempting to request more spectrum resources from the centralized arbitrator.

The centralized DSA arbitrator can make predictive analysis based on node mobility, the spectrum map, and how it changes to create a new spectrum plan. This spectrum plan will have spectrum assignments with changes to most networks in the deployment. These new spectrum plans are meant to stabilize spectrum assignment for longer periods and result in topology and route stability. In other words, the design of a hybrid DSA system with a centralized arbitrator can consider different places for DSA decision making and different types of DSA decisions where some decisions are meant as quick reaction while other decisions are meant to create radical changes that result in more spectrum resources assignment stability and global route stability. In addition to creating stable topology and stable routes, DSA decision stability also reduces the volume of DSA control traffic as information exchange can be just "maintenance" information exchange during stable periods.

Another critical advantage of the use of the centralized DSA arbitrator is the creation of global route tables that avoid routing over the compromised areas where jammers or eavesdroppers are known to be present and avoid routing over energy-constrained nodes.

4.6 Concluding Remarks

With DSA design, there is no single design approach that can be used for all systems. Systems come with limitations, requirements, and vast differences in intended use. Regardless of whether the system under consideration is a civilian or a military system, local, distributed, and centralized DSA approaches have to be weighed in and the place of DSA cognitive engines has to be considered with other cognitive engines the system is using. This chapter is meant to help the reader analyze system limitations, requirements, especially the most critical requirements (e.g., avoid jamming, create low spectrum footprint map, avoid decision rippling, avoid routing over power constrained nodes, and adhere to the maximum bandwidth allocated for control traffic[19]), and create a design approach that makes the best of all DSA decision places in a hierarchical DSA system.

While this chapter considered mainly DSA systems for heterogeneous hierarchical MANETs, Chapter 6 addresses dynamic spectrum management in commercial 5G systems to draw parallels and contrasts between DSA design for military communications systems and DSA design for commercial systems eluding to the breadth of dynamic spectrum management needs and challenges. Chapter 7 presents some concepts for the use of 5G in military applications based on DSA concepts. It will be left to the reader to decide if 5G systems can be used for both civilian and military communications infrastructure with no modifications, or if some aspects of 5G can be borrowed for use in military communication systems. Can military networks, which are heterogeneous by definition, have a mix of military waveforms and 5G technologies such as mm-wave based links and networks?[20]

---

19 The system's requirements can define the maximum bandwidth that can be allocated to DSA control traffic.
20 The DARPA 100G program and the DARPA mobile hotspot program explored the adaptation of some 5G technologies for military communications use from two different aspects.

Exercises

1. Consider Figure 4.5, where we have the master DSA engine being able to establish flows with a TDMA based network through the TDMA waveform agent. Let us assume that the *flow* unit is defined as F bps, as shown in the figure below. The x axis is the time defined in unit T and the y axis is the flow defined in unit F. The curve defines the traffic demand between two peer nodes over the formed TDMA network. Let us assume the TDMA waveform DSA agent applies the following rules when creating flows (adding or subtracting flows):

   - A flow can be added or subtracted every $0.5T$ time. The protocols necessitate that flows cannot be added or subtracted in a rate faster than $0.5T$.
   - If a fraction of F is needed, the entire flow F is added.
   - If a flow is not needed, its resources are freed. Freeing resources cannot happen at a faster rate than $0.5T$.

     (a) Use the curve to draw the allocated flows versus time over the duration of this traffic demand.

     (b) Can you find time periods when the allocated bps are less than the traffic demand and time periods when the allocated pbs are more than the traffic demand?

     (c) State the reasons for encountering gaps between traffic demand and resources allocations.

     (d) Would you design such a system with TOS bit marking of traffic flows? Why?

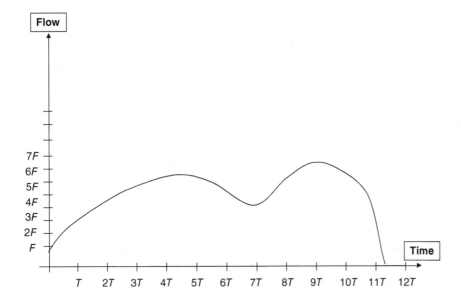

2. Consider a MANET with $N + 1$ nodes. One of these nodes is transmitting a DSA control traffic packet of size L bytes to all the other N nodes. The node under consideration can reach half of the network nodes through one over-the-air hop and the rest of the nodes through two over-the-air hops. Assume that we need spectrum allocation of 2 Hz per each transmitted bit.

   (a) If we assume 100% reliability, how much over-the-air resources in Hz does it take for the control traffic DSA packet to reach every node in the network, assuming unicast transmission?

   (b) Repeat (a) assuming broadcast transmission where every one-hop-away node that receives the packet will have to broadcast the packet in order to make sure it reaches the nodes that are two over-the-air hops away.

(c) Which method would you recommend for propagating DSA control traffic between nodes: unicast transmission or multicast transmission where the packet makes use of the multiple access capability of the waveform?

3. For the same MANET discussed in Problem 2 above, let us assume that the DSA control traffic packet is divided into smaller frames where over-the-air transmission is for one frame not the entire packet. Let us assume that the DSA control traffic packet is divided into 20 frames.

(a) Create a table that shows the probability of delivering the DSA control traffic packet with one transmission when the probability of delivering a frame is 1.0, 0.99, 0.98, 0.97, 0.96, and 0.95, respectively. Assume that losing frames are independent events and that you have to receive every frame in the packet to reconstruct the packet accurately. Make other appropriate approximations if needed.

(b) Let us consider the MANET case in problem 2 and consider the one over-the-air hop away and the two over-the-air hops away transmission cases. Let us consider the unicast transmission scenario. Create the same table as in part (a) if we have to use unicast and transmit the packet two over-the-air hops away to reach a node two over-the-air hops away from the transmitting node. Assume that packet segmentation to frames occurs with every over-the-air transmission.

4. Consider Figure 4.6 with a master routing engine, master DSA engine, waveform routing engine, and waveform DSA engine. What is your expectation of reactive routes stability when:

(a) master routing engine update time > waveform routing engine update time?
(b) master routing engine update time = waveform routing engine update time?
(c) master routing engine update time < waveform routing engine update time?

Bibliography

Charland, R., Veytser, L., and Cheng, B.-N., Integrating multiple radio-to-router interfaces to open source dynamic routers. In *IEEE Military Communications Conference, MILCOM 2012*, October 2012.

Cheng, J., Wheeler, J., and Veytser, L., Radio-to-router interface technology and its applicability at the tactical edge. *IEEE Communications Magazine*, vol. 50, no. 10, pages 70–77, October 2012.

Cheng, B., Charland, R., Christensen, P. et al., Evaluation of multihop airborne IP backbone with heterogeneous radio technologies. *IEEE Transactions on Mobile Computing*, vol. 13, no 2, February 2013.

Cheng, B., Coyle, A., and Wheeler, J., Characterizing routing with radio-to-router information in a heterogeneous airborne network. *IEEE Transactions on Wireless Communications*, vol. 12, issue 8, August 2013.

Chowdhury, K. and Akyildiz, I., A routing protocol for cognitive radio ad hoc networks. *IEEE Journal on Selected Areas in Communications*, vol. 29, no. 4, April 2011.

Elmasry, G., Aanderud, B., and Arganbright, T., A distributed multi-tiered network agent for heterogeneous MANET resource optimization. *Proceedings of Milcom* 2015.

Elmasry, G., Haan, B., and McCabe, R., Augmenting OLSR with a priority aware dynamic routing for hetergeneous netwoking. *Proceedings of Milcom* 2015.

Elmasry, G., Aanderud, B., Kraus, W., and McCabe, R., Software-defined dynamic power-control and directional-reuse protocol for TDMA radios. *Proceedings of Milcom* 2015.

Jawar, I. and Wu, J., Resource Allocation in Wireless Networks Using Directional Antennas. *Proceedings of PerCom* 2006.

Kanzaki, A., Uemukai, T., Hara, T., and Nishio, S., Dynamic TDMA slot assignment in ad hoc networks. In *Advanced Information Networking and Applications*, March 2003.

Patil, V.C., Biradar, R.V., Mudholkar, R.R., and Sawant, S., On-demand multipath routing protocols for mobile ad hoc networks issues and comparison. *International Journal of Wireless Communication and Simulation*, vol. 2(1), pp. 21–38, 2010.

Pearlman, M., Haas, Z., Sholander, P., and Tabrizi, S.S., Alternative path routing in mobile ad hoc networks. *Proceedings of Milcom* 2002.

Vergados, D.J., Koutsogiannaki, M., Vergados, D.D., and Loumos, V., Enhanced end-to-end TDMA for wireless ad-hoc networks. IEEE Communications Conference, Computers and Communications, July 2007.

Wang, J., Kong, L., and Wu, M.-Y., Capacity of wireless ad hoc networks using practical directional antennas. *Proceedings of WCNC* 2010, pp. 1–6.

Young, C.D., USAP: a unifying dynamic distributed multichannel TDMA slot assignment protocol. IEEE Military Communications Conference, MILCOM 1996, October 1996.

Zhang, J. and Liew, S.C., Capacity improvement of wireless ad hoc networks with directional antennae. *SIGMOBILE Mobile Computing and Communications Review*, vol. 10, no. 4, pp. 17–19, October 2006.

https://tools.ietf.org/html/draft-ietf-manet-dlep-02 .

https://tools.ietf.org/id/draft-dubois-r2cp-00.txt.

https://tools.ietf.org/id/draft-ietf-manet-dlep-16.txt .

https://tools.ietf.org/rfc/rfc5578.txt .

Part II

Case Studies

5

DSA as a Set of Cloud Services

The literature shows how different wireless communications systems study DSA from different angles. One can find many references on dynamic spectrum management for 5G cellular systems. The next chapter covers 5G dynamics spectrum management as one example of DSA systems. There are also cognitive radio references that look at DSA from the angle of opportunistic spectrum use and military-focused DSA references that are built on the military concepts of operations covered in Part 3 of this book. The previous chapters explained many DSA concepts from the cognitive MANET perspective. This chapter looks at DSA without having any specific system or application in mind. This chapter is a case study where the case is generic. DSA is presented as a collection of cloud services that can be accessed on any hierarchal entity[1] of a hierarchy of heterogeneous networks and DSA services are made available wherever and whenever they are needed.[2] As the reader moves to the next chapter, 5G dynamic spectrum management may be viewed as a specific case study that can be derived from the generic representation of DSA as a set of cloud services presented in this chapter.

The National Institute of Standards and Technology (NIST) defines cloud computing as "a model for enabling ubiquitous, convenient, on-demand network access to a shared pool of configurable computing resources – e.g., networks, servers, storage, applications, and services – that can be rapidly provisioned and released with minimal management effort or service provider interaction". The goal of this chapter is to present a case for DSA to be designed as a collection of cloud services that are ubiquitous, convenient, and on-demand for a shared pool of spectrum resources or frequency bands. Spectrum resources can be provisioned and released at different speeds depending on the hierarchy of heterogeneous networks with minimal management effort from a human in the loop. In addition to offering DSA services from provisioned spectrum resources, DSA cloud services can tap into opportunistic spectrum resources as well.

When interference is detected[3] by any entity, a service request for a new frequency band to operate on can be triggered. The client for DSA cloud services here is not an actual person. The client is a network entity that is suffering from interference. The service is to re-provision spectrum resources to accommodate all networks and nodes needed for seamless wireless communications.

NIST provides different service models for cloud computing that include infrastructure as a service (IaaS). DSA services can follow the IaaS model.

---

1 An entity is a network node, a gateway or the central arbitrator, as explained later in this chapter.
2 Most of the spectrum access systems (SASs) developed for the citizens broadband radio service (CBRS) in the USA, which operate around the 3.5 GHz band, offer DSA as a set of cloud services. CBRS allows enterprises to build their own LTE networks and share the band with the LTE cellular infrastructure.
3 Interference detection can be an outcome of decision fusion. As explained earlier, hypothesizing interference by a sensor at any time instant does not necessarily lead to interference detection. Fusion of many hypotheses that may come from different sensing sources is what produces a decision that interference is above a certain threshold, which requires new frequency assignment.

Dynamic Spectrum Access Decisions: Local, Distributed, Centralized, and Hybrid Designs, First Edition. George F. Elmasry.
© 2021 John Wiley & Sons Ltd. Published 2021 by John Wiley & Sons Ltd.
Companion website: www.wiley.com/go/elmasry/dsad

5.1 DSA Services in the Hierarchy of Heterogeneous Networks

Figure 5.1 shows a hierarchical heterogeneous network with a central network manager that is peered with a central spectrum arbitrator. The figure illustrates network gateways as another hierarchical layer where each network gateway has nodes within the network that are a lower hierarchical level. This hierarchy can grow to more layers[4] but for the sake of simplifying this generic concept we will adhere to these three layers. Within this construct, there are two distinct control planes, as shown by the solid and dotted lines. The solid line is a control plane between the peer gateway nodes and between the peer gateway nodes and the central spectrum arbitrator. The dotted line is the control plane within a network. Each gateway can be a gateway of a different network that has its own control plane.

This generic architecture construct has a goal: DSA services should be available as a set of cloud services for any entity that needs it regardless of the control planes' conditions. A system that offers DSA as a set of cloud services that optimize the use of spectrum resources will follow a hybrid approach that is a mix of local, distributed cooperative, and centralized DSA services. Notice that local DSA services can happen at any entity in this hierarchy. Distributed cooperative DSA can be between the peer nodes in the network or between the peer gateways. Each of these distributed cooperative techniques has a place in a hybrid design. Centralized services can happen at the central arbitrator or at the network gateway acting as a proxy of the central arbitrator's services. As Chapter 8 explains, co-site interference considerations can be another type of DSA cloud services that can be added to the collection of DSA services as a subset of services. The system's specifications and requirements can guide the designer on how to mix and match these DSA services in the system.

As the reader goes through the next chapter, the concept of 5G cellular dynamic spectrum management being a derivative of this generic construct should become clearer. The lowest hierarchical entity in 5G cellular that can seek DSA services is the end user device or user equipment (UE). The UE has access to different hierarchical cells (equivalent to gateways) and the higher tier DSA decision (e.g., from a macro-cell) can override the lower tier DSA decision (e.g., from a femtocell).

DSA decisions are often needed when wireless networks are suffering from interference and reduced connectivity. As a result, the design of DSA services must address the fact that reachability is not always guaranteed. For that reason, DSA as a set of cloud services should always work independently of the control plane status.

Using Figure 5.1, DSA decisions can be made in the following ways:

1. At the root DSA central arbitrator, which can make DSA decisions based on spectrum sensing information fused and propagated up the hierarchy to this central arbitrator. This central

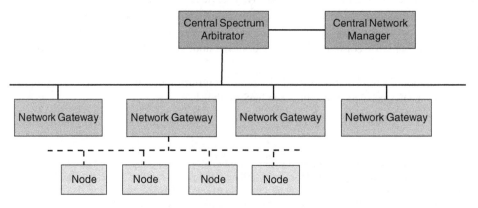

Figure 5.1 The construct of DSA as a set of cloud services in network hierarchy.

---

4 In military communications, all lower echelon networks are wireless MANETs. These wireless networks can be lower tier waveforms with small spectrum footprint and upper tier networks with large spectrum footprint.

arbitrator can also obtain information from external sources, such as what frequencies to avoid and which network can be assigned a static frequency. Rule sets for policy automation can also be entered at the central arbitrator.

2. At the network gateway. We assume that each network has a gateway and this gateway can make DSA decisions under different scenarios as follows:

   (a) Decisions local to this gateway node itself. The gateway node can make local decisions such as which frequency the network should attempt connectivity through first and the order of frequencies that can be used to attempt connectivity to other nodes in the network (the order of using backup frequency bands). The gateway can also decide how long it will attempt connectivity on one frequency band before trying the next frequency band in the backup frequency bands list.

   (b) Network distributed cooperative DSA decisions where the gateway node communicates with other nodes within a network for the network cooperative distributed DSA decisions.

   (c) Global distributed cooperative DSA decisions with peer gateway nodes. This scenario assumes that the gateway node may not be able to reach the central arbitrator but can reach peer gateway nodes through the solid line control plane in Figure 5.1. The gateway nodes can reach all or a subset of peer gateway nodes. Distributed cooperative DSA decisions under this scenario are taken with regard to a group of networks and can be an alternative (fallback solution) to the decisions made by the central arbitrator. When the central arbitrator is not reachable, this capability can make the gateway a proxy for the central arbitrator's services.

   (d) Deferred decision where the gateway node defers a DSA decision to the central arbitrator when it is reachable.

3. The third entity that can make DSA decision is the network node. Network nodes can make DSA decisions under different scenarios as follows:

   (a) Local decisions to the node. An example of such decision is the resorting to a backup list of frequency bands if connectivity to other peer nodes through the assigned frequency band fails.

   (b) Distributed cooperative DSA decisions with peer network nodes. This scenario assumes that the gateway node is not reachable.

   (c) Deferred decisions where the network node defers a decision to the network gateway node. Notice that the network gateway nodes can execute 2(c) or 2(d) above in order to make this decision.

In addition to offering DSA services seamlessly, a cloud architecture of DSA services should also allow for policy propagation such that policies set at the central arbitrator by the network manager can propagate seamlessly to the lower hierarchical entities. Policy automation is a core part of seamless DSA services. In addition to policies, some configuration parameters can be entered at the central arbitrator and processed to create configuration parameters for the networks. Network gateways can use the network configuration parameters it receives from the central arbitrator to create configuration parameters specific for each node in its network and send each node its configuration parameters. This hierarchical propagation of policies, rules, and configuration parameters can minimize the bandwidth used for DSA control traffic as it makes entities such as gateways process one network configuration message to produce many node configuration messages. The network gateway can use the network's waveform multiple access capabilities to send one message to more than one node using only one over-the-air transmission. The configuration message from the central arbitrator to the network gateway will have less impact on control traffic volume than having the central arbitrator attempt to configure each node in each network with a separate message.

Other considerations of this generic model include:

1. Reducing or eliminating the need for a human in the loop.
2. Creating harmony between the heterogeneous networks through consistent updates of policies and configuration parameters.

3. Utilizing the processing power of the upper hierarchy to process and disseminate global policies, rule sets, and configuration parameters.

4. In addition to reducing DSA control traffic volume going downward carrying policies, rules sets, and configuration parameters, this model reduces DSA control traffic going upward carrying spectrum sensing information where nodes and gateways fuse (and abstract) spectrum sensing information messages before forwarding them to upper networks entities.

5. Most importantly, make the best DSA decision available at any level of this hierarchy regardless of the control plane connectivity status.

5.2 The Generic DSA Cognitive Engine Skeleton

This section presents a generic construct of a DSA cognitive engine that can be followed at any of the hierarchical levels explained in the previous section. Although this book is not focused on cognitive engine design, it is appropriate to demonstrate that a common (skeleton) DSA cognitive engine concept can be followed at all entities of the hierarchy presented in this chapter in order to facilitate DSA as a set of cloud services. A DSA cognitive engine is responsible for many tasks, including spectrum sensing information fusion, offering DSA services and the propagation of policies, rule sets, and configuration parameters to the lower hierarchy entities.

Figure 5.2 shows this generic representation of the DSA cognitive engine. At the center of this DSA cognitive engine is an information repository. This information repository can differ from one entity to another based on its hierarchy and place in the hierarchical networks. The information repository can receive real-time input from local sensors, from peer DSA engines or from other hierarchic DSA engines. The information repository can also be affected by automation inputs such as policies, rules, and configurations. The information repository is coupled with an information fusion and evolver. The DSA engine also has a resource monitor that can create recommendations and warnings to external entities and can disseminate messages to other entities as needed. The resource monitor can trigger policy updates, rules updates or configuration parameter updates that propagate from higher DSA entities to lower DSA entities. The DSA decision maker is a separate process that receives DSA services requests and responds to these requests. A request can be a local request or a dissemination request as explained above. The resource monitor can create internally triggered decisions (as a result of spectrum sensing information fusion) and pass them to the decision maker for consideration.

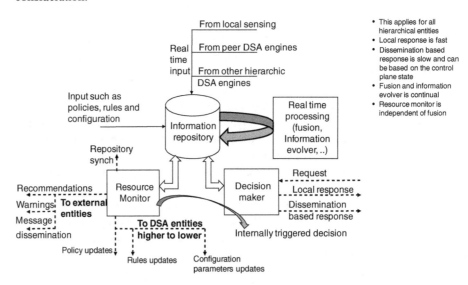

Figure 5.2 The generic cognitive engine representation that can be used at any network hierarchical entity offering DSA services.

As Chapter 8 explains, the decision maker can consider the impact of co-site interference before disseminating a response to a service request.

Notice that:

- this skeleton architecture applies for all hierarchical DSA decision making entities.
- local response should be instantaneous.
- dissemination based response should be slower and can depend on the control plane state.
- fusion and information evolver is a continual process that can update the information repository. The information repository can be updated based on external messages or based on local analysis. An example of this internal update is the ROC model hypothesizing threshold explained in Chapter 3. For example, local analysis at a gateway can update this threshold value and create a message that disseminates to the network nodes to update the value of this threshold in order to adjust the tradeoff between the probability of false alarm and the probability of misdetection.

This book is not about developing cognitive engines. However, some critical threads of this cognitive engine skeleton are worth covering to show some key aspects of the decision-making process.

5.2.1 The Main Thread in the Central Arbitrator DSA Cognitive Engine

This section shows how the generic cognitive engine construct used in the central arbitrator can execute the thread illustrated in Figure 5.3. The importance of this thread is that it is the initialization thread and subsequent threads can be considered a subset of this thread, as indicated by the curved arrow at the right-hand side of the figure. The curved arrow means that the update to the information repository, the fusion of information, and the monitor of resources will continue to occur as a loop. At any moment, this loop can trigger a decision to propagate a new policy, rule set or configuration parameters or propagate a new frequency assignments. Note that there is no need for a human in the loop. The central arbitrator decisions are autonomous. Configuration parameters can be updated and propagated autonomously.

Notice with Figure 5.3 that information repository updates can be:

- from external sources to include sensors, external database, and external tools.
- geolocation information updates propagated from lower hierarchy.
- a result of fusion that evolves the states of the DSA cognitive engine.

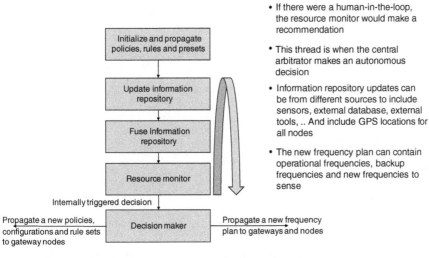

Figure 5.3 The main thread in the central arbitrator.

The decision maker can create a new frequency plan that includes:

- new frequency assignments to some or all networks (change of operational frequencies).
- new backup frequencies for each network (Note that backup frequencies, much like operational frequencies have to be a result of decision fusion so that they are spatially separated and when nodes use them for connectivity, nodes will not suffer from interference due to lack of spatial separation or from external systems in the same geographical location.).
- sensing frequency. (Note that in order to reduce the control traffic volume, the central arbitrator may select a subset of sensing capable nodes per each network. Not all sensors must be activated at all the times. This subset of sensors can be geographically dispersed to create a comprehensive spectrum map for the least amount of spectrum sensing control traffic volume.)

5.2.2 A Critical Thread in the Gateway DSA Cognitive Engine

One critical trigger that can reach the DSA cognitive engine in the gateway node is the receipt of a frequency plan from the central arbitrator. As mentioned earlier, the gateway DSA engine can defer frequency change in its network to the central arbitrator or the central arbitrator's own fusion can trigger a frequency change (frequency plan) that affects one or more networks. This frequency plan can contain frequency change for a network that the gateway is a node of, a list of frequencies to sense within the network identifying which sensor would sense which frequency, and a list of backup frequencies for the network to consider when network formation using the operational frequency fails. This is one of many possible paths that can go through the gateway DSA cognitive engine.

As Figure 5.4 shows, once the frequency plan is received by the gateway cognitive engine, it will result in an update to the gateway information repository and this update should trigger the decision maker to perform the following steps:

1. Disseminate frequency change to all nodes in the network with a frequency change time that is dependent on the waveform characteristics. For example, if the waveform uses TDMA with EPOCH time, the frequency change time defined by the gateway DSA engine can be at the start of the next EPOCH or it can be after N EPOCHs from the current time. It is critical to ensure that the frequency change message will reach all nodes before the actual change of frequency can occur and that all nodes synchronously switch to the new frequency.[5]

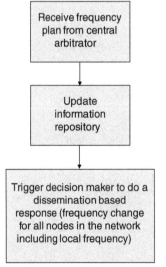

Figure 5.4 Gateway DSA cognitive engine thread for a frequency change.

---

5 The correct term to use here is frequency band not frequency. A frequency band is defined by a central frequency and bandwidth. The term frequency is used throughout this book loosely to mean a frequency band

2. Schedule for a frequency change locally to the gateway node so that the gateway and all the nodes in the network can continue the same network formation with the new frequency.
3. If the frequency change message is for a network that uses cooperative distributed DSA techniques, the frequency change message will contain a new pool of frequency bands for the network. The gateway DSA engine will have more work to do as it will have to disseminate the new pool and continue to perform cooperative distributed DSA within the network so that all the nodes in the network collectively try to optimize the use of the new pool of frequencies.

Note that Figure 5.4 is one possible thread the gateway DSA cognitive engine may execute. This is an example of a thread that propagates down the hierarchy. The next section gives an example of a thread that propagates up the hierarchy.

5.2.3 The Gateway Cognitive Engine Propagation of Fused Information to the Central Arbitrator Thread

This is an upward thread that can occur in the gateway DSA cognitive engine. With this thread, the gateway can be receiving spectrum sensing information from the following sources:

1. Local, if the gateway has augmented spectrum sensors or the gateway can be sensing the frequencies in use as explained in the previous chapters.
2. From peer gateway nodes. As explained earlier in this chapter, gateway nodes can be sharing their fused spectrum sensing information with peer gateway nodes.
3. From the network nodes. The gateway node can be a member node of one or more networks. These networks can have augmented sensors configured to sense (probe) certain frequency bands and can be forwarding spectrum sensing information of the used frequencies to the gateway.

Figure 5.5 shows the progress of this thread to the fusion part. Collected spectrum sensing information from the different sources are fused to produce actions such as the following:

1. Spectrum awareness map. This map shows what frequency bands in the area of operation of this network suffer from interference, what probed frequency bands can be used, and what

Figure 5.5 Gateway DSA engine upward propagation of spectrum sensing information.

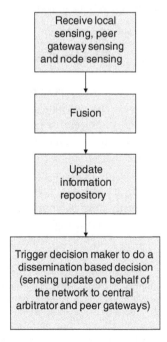

a network is operated with in order to differentiate an operational frequency band used by a network from the known defined frequency bands such as the HF band, the VHF band, and the UHF band.

probed frequency bands should be avoided because they are occupied. As explained earlier, this spectrum awareness map at the gateway is more accurate than the spectrum awareness map at the network nodes and the spectrum awareness map can be more accurate if peer gateways share spectrum sensing information with each other in a distributed cooperative manner.

2. Reprioritizing of backup frequencies. The probing of unused frequency within the network can lead to changes in the backup frequency order of use. Also for a network that uses cooperative distributed DSA internal to the net, spatial use of the assigned frequency pool can be altered with time.

As Figure 5.5 shows, fusion will result in updates in the gateway DSA engine information repository and this can trigger upward updates to the following entities:

1. The central arbitrator. Now the gateway DSA engine sees a different spectrum awareness map than the last time the gateway DSA engine updated the central arbitrator DSA engine. The gateway updates the central arbitrator with the new spectrum map.
2. The updates to the central DSA engine, which are sent over the same control plane, will reach the peer gateways.

Notice the asynchronous aspects of using cognitive engines where each cognitive engine works based on what information it collects and decisions occur as a result of fusion. Although these engines are not timely synchronized with each other, they work collectively, and sometimes using heuristics[6] algorithms, to optimize the use of spectrum resources. In the meantime, DSA is offered as a set of cloud services at any point of the heterogeneous networks hierarchies regardless of the status of the control plane. If some spectrum awareness propagation messages are lost, DSA services will always be available when a service is requested.

The thread in Figure 5.5 shows one possible gateway cognitive DSA engine flow. Fusion can always trigger an action at any entity and this action can trigger a message flow that leads to other cognitive DSA engine fusions and triggers. The DSA design of large-scale systems has to consider the amount of control traffic that can be generated from all of this information propagation upward, downward, and to peers. The design has to consider the tradeoffs mentioned in this book, which include the thresholds that trigger information dissemination. These thresholds must be selected and updated dynamically based on bandwidth availability.

5.3 DSA Cloud Services Metrics

A cloud service has to be evaluated by certain metrics per the NIST explanations of cloud services. The previous chapters have introduced metrics in areas such as the time between detecting interference (service request) and the time of overcoming interference and showed how this time as a metric can depend on the entity hierarchy and the specific mix and match of this hybrid approach to DSA. It is important to note that with IaaS, there may be direct metrics and indirect metrics. For example, measuring control traffic volume as a metric may not be used because measuring throughput efficiency or quality of service (QoS) metrics can indirectly consider control traffic volume impact.[7] Intuitively, the lower the control traffic volume impact, the higher the throughput efficiency achieved, and thus measuring throughput efficiency indirectly measures the impact of control traffic volume.

---

6 Many decision fusion techniques use heuristic algorithms because reaching an optimum solution for a large number of networks and a large number of nodes within each network can be too computationally extensive to produce timely DSA services. Reducing the order of computational complexity is beyond the scope of this book.
7 Notice that considering control traffic volume impact is not simple or straightforward. A DSA design approach can add more control traffic but in the meantime makes the use of spectrum resources much more efficient. Such design is better than lowering the control traffic volume at the expense of reducing the efficiency of spectrum resources use. Although one can use a metric to measure control traffic, this metric's impact comes secondary to the metrics that measure throughput efficiency.

5.3.1 DSA Cloud Services Metrics Model

In this section, we apply the cloud services model to DSA services. Figure 5.6 shows the standard cloud services model applied to DSA with three distinct stages:

1. The pre-run stage defines the service and the service agreement.
2. The runtime stage, represented by the gray rectangle, where the service is monitored and enforced to meet the service agreement.
3. Post processing, where service accountability is measured.

The first step in the Figure 5.6 model application to DSA services is simple. One of the DSA cloud services, such as response time, is selected. In typical cloud services, customer response time can be defined in the service agreement and the customer can know the response time before purchasing the service. DSA also needs a service agreement. The definition of a service agreement with DSA can be driven from system requirements and analysis. Response time can be the time between an entity reporting suffering from interference above a certain threshold (that can render connectivity to be lost or bandwidth to be below a certain value) and the time a new frequency band is assigned to overcome interference.

The gray rectangle in Figure 5.6 shows the runtime aspects of a cloud service where the service is monitored and policies, rule sets, and configuration parameters are adapted in order to force the service to adhere to the defined agreement. With DSA as a cloud service, policies, rule sets, and configuration parameter updates can be triggered by the DSA cognitive engine resource monitor, as shown in Figure 5.2, or by the decision maker, as shown in Figure 5.3. These actions are made in order to enforce the service to adhere to the service agreement during runtime. With DSA as a service, the design can create log files that can be analyzed as post processed in order to evaluate the DSA service accountability over a long period of time.

With DSA as a set of cloud services, while service agreements can be driven from system requirements, the design of DSA cloud services has to create metrics to help force the services to conform to the agreement and use metrics for measuring services accountability. The cognitive engine-based design would have to gain understanding of the properties of the metrics used to force the service to adhere to the service agreement and scripted scenarios must be used to assess service accountability before system deployment.

5.3.2 DSA Cloud Services Metrology

Metrology is the scientific method used to create measurements. With cloud services, we need to create measurements to:

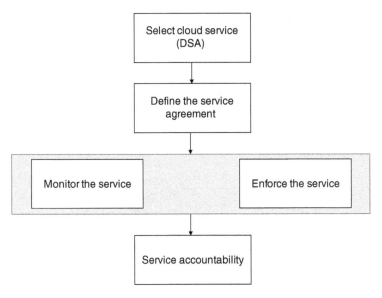

Figure 5.6 The DSA cloud services model.

1. quantify or measure properties of the DSA cloud services
2. obtain a common understanding of these properties.

A DSA cloud services metric provides knowledge about a DSA service property. Collected measurements of this metric help the DSA cognitive engine estimate the property of this metric during runtime. Post-processing analysis can provide further knowledge of the metric property.

It is important to look at DSA cloud services metrics not as software properties measurements. DSA cloud services metrology measures physical aspects not functional aspects of the services. The designer of DSA as a set of cloud services should be able to provide measurable metrics such that a service agreement can be created and evaluated during runtime and in post processing. Since the model used here is a hierarchical model, a metric used at different layers of the hierarchy is evaluated differently at each layer. For example, as explained in Chapter 1, response time when providing DSA as a local service should be less than response time when providing DSA as a distributed cooperative service, which is also less than when providing DSA as a centralized service.

With the concept of providing DSA as a set of cloud services, the design should be able to go through an iterative process before the model is deemed workable. The design should include the following steps:

1. Create an initial service agreement driven from requirements and design analysis.
2. Run scripted scenarios to evaluate how the agreement is met during runtime through created metrics.
3. Run post-processing analysis of these scripted scenarios to gain further knowledge of the properties of the selected metrics.
4. Refine the service agreement.

Figure 5.7 illustrates this iterative concept. The outcome of this processing is a defined service agreement with measurable metrics that a deployed system is expected to meet.

With standard cloud services, a customer should be able to compare two service agreements from two different providers and select the provider that best meets his needs. The provider of an IaaS attempts to optimize the infrastructure resources use dynamically in order to create an attractive service agreement. If the scripted scenarios in Figure 5.7 are selected to represent deployed scenarios accurately, and if the iterative process in Figure 5.7 is run sufficiently enough and with enough samples, the service agreement created should be met with the deployed system. However, there should still be room to refine the cognitive algorithms,

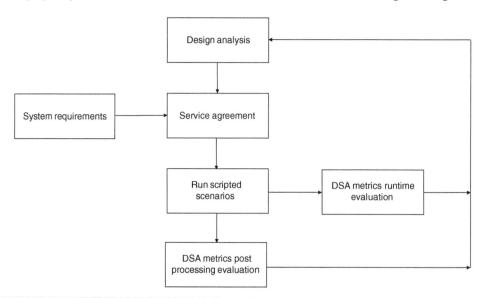

Figure 5.7 Iterative process to create a workable DSA service agreement.

policies, rule sets, and configuration parameters after deployment if post-processing analysis necessitates this change. A good system design should only require refining of policies, rule sets, and configuration parameters without the need for software modification. This system design should allow for the deployed cognitive engine to morph based on post-processing analysis results.

5.3.3 Examples of DSA Cloud Services Metrics

This section presents some examples of DSA cloud services metrics that can be considered in DSA design. Note that these are examples and the designer can choose to add more metrics depending on the system requirements and design analysis.

5.3.3.1 Response Time

Metric name: Response time.

Metric description: Response time between when an entity requests a DSA service and when the service is granted.

Metric measured property: Time.

Metric scale: Milliseconds.

Metric source: Depends on the hierarchy of the networks. The source is always a DSA cognitive engine but the response can be local, distributed cooperative, or centralized. The response can also be deferred to a higher hierarchy DSA cognitive engine.

Note: Response time can be more than one metric. Response time for a local decision is measured differently from response time from a gateway or a central arbitrator. The design can create more than one response time metric.

5.3.3.2 Hidden Node

Metric name: Hidden node detection/misdetection.

Metric description: Success or failure in detecting a hidden node.

Metric measured property: Success or failure.

Metric scale: Binary.

Metric source: An external entity, a primary user, files a complaint about using its spectrum by the designed system.

Note: Need scripted scenarios to evaluate this metric. It is evaluated by an external entity not the designed system.

5.3.3.3 Meeting Traffic Demand

Metric name: Global throughput.

Metric description: Traffic going through the system over time (global throughput efficiency).

Metric measured property: Averaged over time.

Metric scale: bps.

Metric source: Global measure of traffic going through the system. Successful use of spectrum resources dynamically should increase the wireless network's capacity to accommodate higher traffic in bps.

Note: This metric is system dependent. Some systems, such as cellular systems, link this traffic demand to revenue making. The metric is not only interested in getting insight into achieving higher throughput, but the higher number of users that increases revenues. Some users' rates can be lowered but the service continues in order to accommodate more users as long as the service agreement is met.

5.3.3.4 Rippling

Metric name: Rippling.

Metric description: The stability of the assigned spectrum.

Metric measured property: Time.

Metric scale: Minutes.

Metric source: The DSA cognitive engine can track the time between two consecutive frequency updates.

Note: Rippling can have a negative impact on the previous metric (meeting global through-put). It can reduce the network throughput. This metric can be measured at the node level, at the gateway level, and at the central arbitrator level. The rippling impact at higher levels (e.g., central arbitrator) can have much worse impact than rippling at the local node. Evaluation of this metric depends on where it was measured.

5.3.3.5 Co-site Interference Impact

Metric name: Co-site impact.

Metric description: The ability to reduce co-site impact on the assigned spectrum.

Metric measured property: SNIR.

Metric scale: dB.

Metric source: The DSA cognitive engines can track the SNIR from the collected spectrum sensing information and create an average for each waveform according to the waveform signal characteristics.

Note: Co-site impact may or may not be tolerable by a given waveform. This metric can show the average time SNIR exceeded a given threshold that is considered intolerable while co-site interference was known to exist and was accepted because of a policy or because of the limited availability of alternative spectrum bands that can be used.[8]

5.3.3.6 Other Metrics

The above metrics are just examples of what can be considered in DSA cloud services meteo-rology. The design can create categories of metrics. For example, the design can consider QoS category metrics to include packet delay, loss, and jitter. The design can consider security cate-gory metrics to include exposing a node to an eavesdropper or exposing a network to a jammer, the metric could measure the time of exposure to these security risks. The design can also con-sider the need for human intervention. A network administrator monitoring the use of DSA resources may intervene in cases of complete failure of the system to recover autonomously. The number of incidents that require human intervention can be turned into a metric. Some IaaS references name this metric the automatization degree.

5.3.3.7 Generalizing a Metric Description

One of the challenges that can face the designer of DSA as a set of cloud services for heteroge-neous networks is ensuring consistency. We discussed in the previous chapters how different waveforms can have different link status metrics and how a common definition for link health metrics between all the waveforms used in a heterogeneous system is needed such that services as reactive routing are optimized appropriately. The design of DSA as a set of cloud services needs to also ensure that a metric is measured consistently with regard to all types of wave-forms and all points where the metric is measured. Co-site interference impact as a DSA service metric is a good example of creating consistency. Different signals have different tolerance to co-site interference. Co-site interference can have different levels of spectral density in differ-ent frequency harmonics. The co-site impact as a DSA service metrics has to be normalized between all waveforms such that the metric evaluation is consistent and means the same for all waveforms.

5.4 Concluding Remarks

DSA can be designed to be a set of cloud services that are IaaS type. The design of DSA as a set of cloud services have some commonality with standard IaaS services and have some unique

---

8 Chapter 8 explains how co-site interference avoidance can be another DSA cloud service.

aspects that must be considered. This chapter offered a generic model in how to design DSA as a set of cloud services and some design concepts that can be followed. This chapter also covered some aspects of the metrology of DSA as a set of cloud services. As the reader goes through the next chapter, it will be seen how 5G cellular dynamic spectrum management is a special case of the generic model presented in this chapter. In 5G opportunistic and nonopportunistic spectrum bands are managed dynamically and spectrum resources are offered to users as services based on a service agreement between the user and the 5G service provider.

Exercises

1. Following the examples of DSA cognitive engine threads in Section 5.2, draft some DSA cognitive engine threads that can be added to the threads explained in Section 5.2.

2. Following the examples of DSA cloud metrics in Section 5.3, draft the details of the DSA metric of control traffic volume divided into two sub-metrics, one for upward control traffic and one for downward.

Bibliography

CSCC, Practical Guide to Service Level Agreements Version 1.0, April 2015. Available at: https://www.omg.org/cloud/deliverables/CSCC-Practical-Guide-to-Cloud-Service-Agreements.pdf.

Cloud Computing Service Metric Description, US Department of Commerce, Publication 500-307, NIST, April 2018.

Evaluation of Cloud Computing Services Based on NIST SP 800-145, US Department of Commerce, Publication 500-322, NIST, February 2018.

ISO/IEC 20926:2009, Software and systems engineering – Software measurement. IFPUG functional size measurement method, 2009.

Kushagra, K. and Dhingra, S., Modeling the Ranking of Evaluation Criteria for Cloud Services. *International Journal of Electronic Government Research*, vol. 14, pp. 64, 2018.

Polash, F., Abuhussein, A., and Shive, S., A survey of cloud computing taxonomies: Rationale and overview. *9th International Conference for Internet Technology and Secured Transactions 2014*, pp. 459–465, 2014.

Ramachandran, M. and Mahmood, Z., *Requirements Engineering for Service and Cloud Computing*. Springer, 2017. ISBN 978-3-319-51309-6.

Siegel, J. and Perdue, J., Cloud Services Measures for Global Use: The Service Measurement Index (SMI). Service Research & Innovation Institute (SRII) Global Conference, IEEE, San Jose, CA, July 24–27, 2012.

The NIST Definition of Cloud Computing, US Department of Commerce, Special Publication 800-145, NIST, September 2011.

6

Dynamic Spectrum Management for Cellular 5G Systems

Part 1 of this book presented the foundations of creating a hybrid DSA design while relying on examples of heterogeneous hierarchical MANETs that demonstrate the most critical aspects of DSA design needs. Chapter 5 showed how to design DSA as a set of cloud services with metrics to measure spectrum use optimization objectively. This chapter covers the use of DSA[1] with 5G cellular networks. There are commonalities and differences between hybrid DSA design for military cognitive MANET systems and DSA design for commercial cellular systems. With military MANET systems, all nodes in the network may be peers and all nodes may be mobile; there can be fast moving nodes such as fighter jets and slow-moving nodes such as tanks and much slower nodes such as dismounted soldiers. A deployment of heterogeneous networks can be provisioned spectrum blocks for use in a nonopportunistic way. Waveforms forming networks may switch to opportunistic spectrum use in some cases but for the majority of the cases, networks use provisioned spectrum dynamically. These networks may share the provisioned spectrum blocks for an entire deployment dynamically before attempting to resort to the use of unlicensed bands. The problem domains in 5G cellular systems and military MANET systems are different, but they share many of the basic DSA concepts such as relying on spatial and time based sensing, collecting a knowledge repository, fusing the knowledge repository to produce decisions, performing interference mitigation techniques, and utilizing metrics to quantify the performance of dynamic spectrum use. As Chapter 5 showed, DSA can be developed as a set of cloud services in a hierarchical manner regardless of the system under consideration.

There are differences in the way DSA concepts are applied in military communications and in commercial use. With commercial cellular 5G, there are standardization bodies, research and development findings from industry and academia, and a mix of licensed and unlicensed spectrum use. There is a wide range of spectrum bands considered for use with 5G, including:

- long-term evolution (LTE) enhancements to utilize the spectrum below 6 GHz; this band increases capacity to about 2.5 times of that of LTE
- segments of the 6–24 GHz band
- the millimeter wave (mm-wave) band between 24 and 86 GHz, which may bring the 20 Gbps theoretical speed to reality[2]
- bands above 86 GHz and up to 102.2 GHz for future considerations.

Notice that different countries have different ranges of spectrum allotted for the first trials of 5G. Table 6.1 gives an example of different spectrum allocations for below 6 GHz in different areas of the world. This trend of using different bands at different areas of the world is expected for all 5G bands as spectrum licensing agents such as the US FCC has plans for spectrum use for civil applications that are different from those of other countries.

---

1 5G literature often uses the term dynamic spectrum management (DSM) instead of dynamic spectrum access.
2 Technically mm-wave starts at 30 GHz, but it is common in 5G literature to consider the band above 24 GHz to be the mm-wave band.

Dynamic Spectrum Access Decisions: Local, Distributed, Centralized, and Hybrid Designs, First Edition. George F. Elmasry.
© 2021 John Wiley & Sons Ltd. Published 2021 by John Wiley & Sons Ltd.
Companion website: www.wiley.com/go/elmasry/dsad

Table 6.1 5G frequency bands allotted below the 6 GHz range.

Country	Range (GHz)
Europe	3.4–3.8
China	3.3–3.6
	4.4–4.5
	4.8–4.99
USA	3.1–3.55
	3.7–4.2
Japan	3.6–4.2
Korea	3.4–3.7

This chapter will reference below 6 GHz access as "lower frequency" access (LTE or enhanced LTE). Lower frequency access plays an important role in 5G deployment and 5G DSM.

Opportunistic spectrum use at the cell level is a core part of 5G dynamic spectrum management while points of presence of LTE are still part of 5G. Notice that 5G standardizations call for cells to be deployed dynamically, making dynamic spectrum access more challenging because mobility is not limited to the end user only. 5G will also operate in a wide range of frequencies, including the mm-wave range, which comes with its own challenges in designing advanced receiver techniques. 5G gives access to as much licensed spectrum as possible in addition to as much opportunistic use spectrum as possible, making it an umbrella for many commercial wireless capabilities outside of cellular technology. The mix of licensed spectrum and unlicensed spectrum makes the 5G dynamic spectrum management problem space as challenging as that of military communications systems, which do not have the concept of a fixed tower and all nodes are mobile.

There are different types of 5G cells, as shown in Table 6.2, where each size has an intended deployment. The ranges depicted in Table 6.2 can be increased with better antenna technologies and are included in this table to help explain the concept of spatial modeling with dense deployment of 5G access points detailed later in this chapter. Notice that the number of users (end users) varies as it is dependent on many factors, such as proximity to the access point (which can decide the chip code), requested data rate, and propagation models. Notice also that 5G makes use of fiber cable ground infrastructure for connectivity of 5G points of presence (cells and towers) to the core network but also relies on using existing infrastructure such as LTE access points to achieve complete wireless coverage in areas where fiber infrastructure is not abundant.

This chapter focuses on some of the 5G spectrum management techniques that are presented in research communities and standardization bodies. Note that 5G standardization gives a great deal of flexibility in dynamic spectrum management to the implementer, making the problem space of this area convoluted as there are different approaches to dynamic spectrum management and how it addresses challenges such as traffic allocation, network density, interoperability between the heterogeneous networks, and the use of orthogonal and nonorthogonal waveforms. The approach adopted in this chapter is to cover many DSM aspects

Table 6.2 5G cell types.

Cell type	Intended commercial deployment	Number of users	Power (W)	Range
Femto	Residential and enterprise	4–8	0.2–1	Few tens of meters
Pico	Public areas (airports, malls, etc.)	64–128	1–5	Tens of meters
Macro	Urban	128–256	5–10	Few 100s of meters
Metro	Urban for additional coverage	>256	10–20	100s of meters

without particularly diving into the details of building the cognitive algorithms that needs to be deployed in a service provider's network.

6.1 Basic Concepts of 5G

This section focuses on basic concepts for 5G that are needed to understand DSM. The reader is encouraged to refer to 5G references for more details. 5G can be conceptualized through the possible deployment scenarios of the 5G cells. Let us consider the following three possible deployments:

1. *Standalone mm-wave access*. This deployment scenario is illustrated in Figure 6.1. One can think of dense urban deployments for this scenario where a high-rise building would have most apartments using 5G cells connected to the core network through fiber links. This deployment scenario offers multiple access points to the end user and 5G calls for opportunistic serving of the end user where the cell that can offer the best service can be elected to serve the end user and hand over between cells can happen frequently.

   Opportunistic serving of an end user relies on DSM approaches. For example, the signal strength indicator from the different reachable cells can be used to decide which cell can be selected by the end-user device as the access point. Other factors, such as the availability of services (a cell may have no more resources to allocate to a new end user), are also considered. Handover between cells also relies on signal strength and the calculation of self-interference (SI) to decide which cell to switch to. This chapter elaborates more on how an end-user device arbitrates between different access points.

2. *Nonstandalone mm-wave access*. With this deployment scenario, which is illustrated in Figure 6.2, the end user can get services through either a 5G cell or an LTE (or enhanced LTE) tower. Opportunistic serving is also used with this deployment scenario where the end user can connect through the cell or tower access point, selecting the access point that would offer the best services. The tower DSM algorithms may override the end user selection. One can conceptualize such a deployment scenario in suburbs where cell access relying on fiber cables to houses may exist offering high density mm-wave access and high bandwidth reach to the core network through the fiber cables. This cell access does not cover all the areas, however, making the use of cellular towers using LTE or enhanced LTE necessary to create full wireless coverage.

   Notice in Figure 6.2 the difference between mm-wave access depicted by the dotted small circles and the cellular tower access using the below 6 GHz range depicted by the gray circles and wider areas of coverage.

3. *The mm-wave as an enabler*. With this deployment scenario, which is illustrated in Figure 6.3, the mm-wave is an enabler in the sense that the LTE or enhanced LTE access is already present and 5G cells can be added incrementally to the area. This scenario can be conceptualized at the start of early deployment of 5G in a given suburbs area as well

Figure 6.1 Standalone mm-wave 5G access.

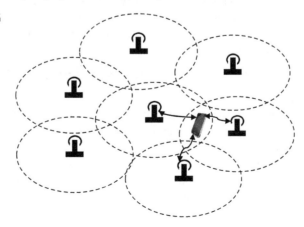

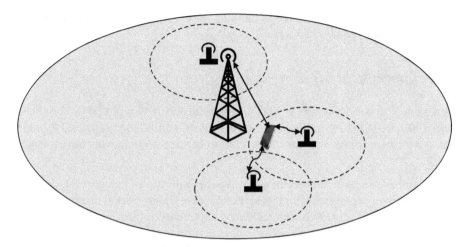

Figure 6.2 Nonstandalone mm-wave 5G access.

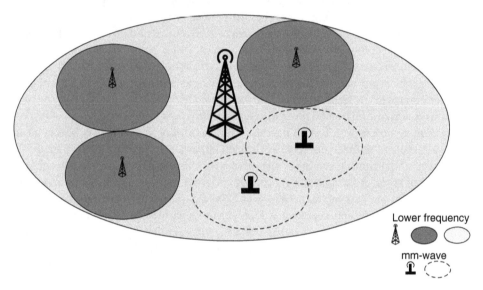

Figure 6.3 5G access with the mm-wave as an enabler.

as in rural areas where coverage can be sparse and fiber cables connectivity availability is limited. 5G cells can be added gradually to the area as user demand increases.

One important aspect of this deployment scenario is the mm-wave backhaul links where 5G cells can be deployed without the need for fiber cable connectivity. A 5G cell dropped anywhere can connect to the tower through a high capacity mm-wave point-to-point wireless link. This ability to just deploy a 5G cell in a busy area that will find the closest tower and establish a mm-wave point-to-point wireless link to the tower to flexibly offer 5G services is an important aspect of DSM. The concept of a self-organized network (SON) is a core 5G capability.

So far, this section has showed that 5G can use the following two features that are related to DSM:

- Opportunistic serving with which an end user can find the best suitable access point to use and frequent handover to a different access point at any time using metrics such as signal strength.
- 5G as an enabler where a 5G cell can be deployed in an area without needing fiber cable connectivity and the cell establishes a wireless mm-wave point-to-point link to the closest tower as part of the 5G SON. This mm-wave link should impact dynamic DSM decisions as explained later in this chapter.

Part 1 of this book alluded to military communications systems approaches to DSA with MANET nodes, which include:

1. using directional antennas
2. using adaptive power control.
3. spectrum allocation of planned spectrum and opportunistic use of unlicensed spectrum
4. the continual change of spectrum allocation due to mobility.

5G has similar concepts that influence DSM. For example, 5G considers full duplex (FD) wireless communication which enables the radio to directionally transmit and receive on the same frequency band simultaneously.[3] FD is considered because of the many advantages it brings, such as increasing transmission capacity and reducing end-to-end feedback delays while performing concurrent sensing.[4] FD implementation comes with many challenges, however, such as the need to mitigate SI. 5G implementers use different SI cancellation techniques such as analog and digital antenna cancellation. The rest of this chapter will present 5G-related DSA techniques, such as SI mitigation, in separate sections.

6.2 Spatial Modeling and the Impact of 5G Dense Cell Deployment

Commercial cellular technologies before 5G depended critically on spatial configuration and the separation of downlink (DL) and uplink (UL) frequencies. Frequency allocation to base stations ensured that interference from other base stations is minimal. In spread spectrum systems such as 3G and LTE, the base station assigns different spreading codes to the different end users to mitigate interference on the UL. Mobile end users can hand over from one base station to another while switching to a different frequency and after obtaining a different spreading code from the new base station. Figure 6.4 shows an oversimplified frequency planning illustration for a pre-5G cell tower deployment where seven frequency bands can be reused spatially to create full spectrum coverage. Factors such as terrain and weather can change this plan and cellular providers rely on comprehensive testing of the cellular infrastructure performance to change the frequency allocation of such a simple plan where areas that have more dense end users can use more base stations with more frequency bands to add capacity where demand is higher.[5]

Figure 6.4 Traditional cellular frequency spatial separation planning.

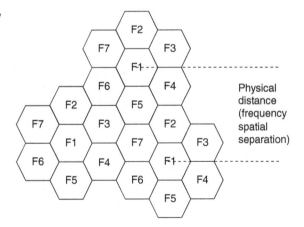

Physical distance (frequency spatial separation)

---

3 Many military communications waveforms use a mix of TDMA and FDMA, creating time and frequency slots for the allocated spectrum that can be shared between network nodes in omni-directional transmission. Part 1 of this book showed how military communications MANETs evolved to use directional sectored antennas to be able to transmit and receive on the same frequency simultaneously based on directionality. 5G introduces many antenna technologies that enhance this FD capability for commercial wireless systems.
4 FD and concurrent sensing can mitigate the hidden node problem explained in Chapter 2.
5 Figure 5.4 is an oversimplification of spatial separation. Base stations before 5G used sectored antennas allowing for frequency reusability, which is not illustrated in this figure.

With 5G and the dense deployment of different size cells, a received signal is impacted by the distance between many transmit and receive pairs using the same frequency. 5G includes overlay of different technologies and different access points, which have different areas of coverage that can overlap. Cellular 5G is heterogeneous in many aspects, including the following:

1. The deployment of different cell types, as shown in Table 6.1. Each cell type can have a different area of coverage and these areas can intersect and be overlaid on top of each other.
2. The mixed use of FD links, with directionalities to increase spectrum reuse, with LTE links that separate the uplink from the downlink channels.
3. The opportunistic use of available spectrum mixed with the use of provisioned spectrum.
4. The mix of unplanned deployment of 5G cells, which may or may not have fiber connectivity to the core network, with LTE fixed infrastructure.
5. The ability to operate in a very wide range of frequency bands spanning from below 6 GHz to 102.2 GHz.

In essence, cellular 5G provides high capacity access through randomly located nodes (end users and cells), irregular infrastructure, and dynamic spatial configurations. The cellular 5G paradigm is a major shift from previous cellular technologies that require the use of different spatial models.

The impact of the distance between the transmitter and the receiver on signal power has been studied with different propagation models. Wireless systems have long been designed based on link-budget analysis, fading margins, and the ability to tradeoff range for transmission rate. The 5G paradigm requires transmit and receive node pairs to continually consider the timely use of a frequency in light of spatial separation to avoid excessive interference. In the multidimensional spectrum sensing model presented in the previous chapters, space becomes the most challenging dimension to model with cellular 5G. While time and frequency separation is easier to model, space modeling encounters the leakage of undesired signals and the impact of co-site interference in addition to the continual change in the transmitting and the receiving nodes locations. 5G has limited practical options to reduce interference keeping in mind that reducing signal power would reduce the signal to interference ratio (SIR)[6] while increasing signal power will reduce the chance of spectrum reusability.

6.2.1 Spatial Modeling and SIR

Spatial modeling in 5G can use a set of metrics that can affect SIR. SIR becomes an instantaneous ratio of desired energy to all the additives of undesired interferences and noise. Thus SIR can be considered a random variable that depends on a set of factors that include the following:

1. The distance between the transmitting node and the receiving node. Much like traditional signals, this factor can be modeled by a path loss model. All path loss models follow an inverse-power law with an exponent trend. For example, in a free space model, signal power decrease with distance in a quadratic trend. Other path loss models can use different exponent values to model signal scattering and signal absorption.
2. The number of active transmitters in a given proximity. For a given receiver, and during the time of reception, there are different potential combinations of active transmitters where their signal will appear as interference. The sum of interference power from all the transmitting nodes, taking into consideration their distances from the given receiver, has to be modeled. In a dense deployment of 5G cells, this sum of interference power from other transmitters can exceed the noise power threshold that is selected to ensure reliable connectivity.

---

6 There are different terminologies in the literature that expresses the signal to noise ratio. SNR is one term that stands for signal to noise ratio; SNIR is another term that expresses signal to noise interference ratio. SINR is a third term that stands for signal to interference plus noise ratio. SIR is a fourth term for signal to interference ratio where interference can be additive noise and/or from another user using the same frequency. While in the previous chapters we used the term SNIR because military communications seek to distinguish noise from interference that can be malicious, in this chapter we will use the term SIR to emphasize the role of SI in 5G.

3. The ambient noise. SIR will be affected by noise power and this noise will depend on the received signal and the interference power. Notably, if a 5G deployment resorts to lower transmission power, the ambient noise impact on SIR can be the larger factor. On the other hand, if a 5G deployment resorts to a high transmission power, the sum of interference power in step 2 above becomes the larger factor.

4. Other factors. There are many other factors that can affect the SIR calculation such as fading and shadowing, transceiver design (e.g., the use of multiple antenna and interference cancelation techniques), and adaptive power control.

A simple representation of SIR at a typical receiving node can be expressed as:

$$SIR_o = \frac{h_{oo}\rho_o r^{-\alpha}}{N_o + \sum_{i\in\Phi}\rho_i h_{io}|X_i|^{-\alpha}} \tag{6.1}$$

where o represents the origin in the spatial model assuming the receiving node is at the origin of the spatial plan, h_{io} is the fading coefficient of the channel at the receiving node for the signal transmitted from node i, ρ_i is the transmit power of transmitter i, N_o is the noise power, Φ is the set of all interfering nodes,[7] and X_i expresses the distance between the ith interfering node and o.

Notice that Equation (6.1) performs the following:

1. It puts the receiving node at the origin of the spatial model o making the calculation with respect to o.
2. It creates a probabilistic spatial model where transmitting nodes can be randomly positioned with respect to the origin.
3. It consolidates path loss calculation of the subset of transmitting nodes presumed to cause interference with the receiving node at the origin.

Equation (6.1) can be used in a spatial model to calculate SIR, which can be used in a 5G dense deployment to calculate connectivity and coverage when assigning spectrum resources. In addition, SIR calculation can be used to estimate the capacity and throughput of a given deployment area. This calculation can further lead to collecting metrics that can measure the reliability of the 5G networks.[8]

6.2.2 SIR and Connectivity

Let us define the metric "connectivity" as the probability that a pair of nodes in a network will be able to exchange information at a specified rate R through a single over-the-air hop. There is a threshold β for SIR that can be used in a simple manner to express connectivity where connectivity can be assumed to occur if $\Pr[SIR > \beta]$. That is, if the estimated SIR is less than β, the node pair would decide that a link is not possible.[9] The desired rate R can be expressed in bits per second (bps) and can be estimated from β using the Shannon's equation as follows:

$$\beta = \Gamma(2^R - 1). \tag{6.2}$$

In Equation (6.2), $\Gamma \geq 1$, which represents the collective interference impact on data rate.[10] The node pair would calculate SIR from Equation (6.1) and calculate β from Equation (6.2) and decide if single over-the-air hop connectivity is possible or not.[11]

---

7 Notice that not all transmitting nodes cause interference. Φ is the subset of all transmitting nodes that can introduce interference at location o.

8 Metrics such as the probability of granting a connection when requested, the probability of keeping a connection during the duration of the session, and the probability of meeting the data rate defined in the service agreement can be used to quantify the reliability of the 5G network.

9 Notice that β can be a tunable parameter. Some literature uses the expression "link closure" to indicate that the condition $SIR > \beta$ is met.

10 Theoretically, one can always trade data rate for FEC to achieve link closure under low SIR. Practically, there is a computational power limitation for the use of FEC. A 5G node has limited modes of turbo code and can't keep trading data rate for lower SIR to achieve link closure. Also, a link is requested with a specific minimum data rate. The result of Equation (6.1) defines Γ in Equation (6.2).

11 Notice how β is related to the data rate. A link can always trade off bandwidth for higher dB gain using error control coding.

In 5G, connectivity has many use cases. Consider the following use cases:

1. If there are no other cells using the same frequency in the area (5G can be considered to be operating much like Figure 6.4) and there is a single active end user, the coverage area for a given rate is decided by the SIR estimation in Equation (6.1) and the threshold β in Equation (6.2) is calculated in a straightforward manner. If $SIR > \beta$, the end-user device has to simply refrain from transmitting when receiving to avoid SI in order to maintain the desired data rate.

2. If a source and destination pair have to communicate through one or more relay nodes, a path must be found where all of the over-the-air hops satisfy Equations (6.1) and (6.2). When there is a single flow, as shown in Figure 6.5, this estimation is also straightforward provided that interleaving of transmit and receive time is done accurately to prevent SI between all node pairs in the path.

3. When there are more than one flow, all flows have only one over-the-air hop, and there is one flow per cell, connectivity calculation will use the summation in Equation (6.1) to consider all transmitting and receiving node pairs. This case is illustrated in Figure 6.6. Here, accurate directionality, power control, and narrowing the beam as much as possible can reduce the effect of SI.

4. The general case when there is more than one flow and flows use relay nodes. With this case, there are different ways to describe network connectivity based on the single over-the-air

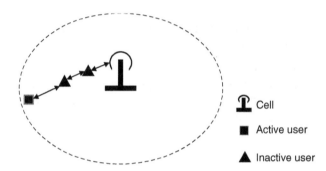

Figure 6.5 Single flow in a cell area of coverage.

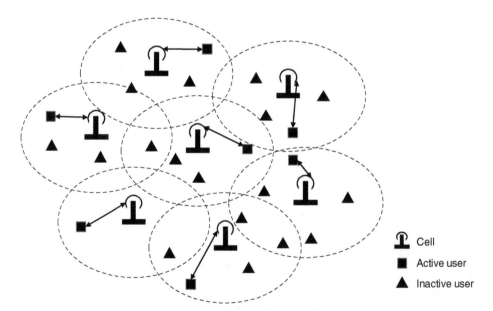

Figure 6.6 Multi-flow, each with a single hop to a different cell.

hop connectivity. Reaching full network connectivity means that all communicating pairs and all single over-the-air hops meet the conditions in Equations (6.1) and (6.2).[12] With this case, interleaving transmit and receive time per each node is not sufficient to avoid interference. Spatial separation becomes more complex and DSM is needed to ensure connectivity. This case is covered in more detail in Section 6.2.3.

6.2.3 General Case Connectivity and Coverage

Connectivity and coverage for the 5G general case are multifaceted. This section covers two important aspects of the general case: transmission capacity and cell overlay. There are more aspects of connectivity and coverage and how they relate to DSM in 5G but these two aspects should give the reader the important basis for 5G DSM.

6.2.3.1 Transmission Capacity

The main goal of DSM is to increase the network throughput or transmission capacity in a given area. Given that in 5G we have mobile users and cells that can also be mobile, we can conceptualize a 5G network as an ad hoc wireless network while putting aside the backhaul wired links. The goal here is to create a spatial metric in which throughput is measured as a function of both rate of transmission and the distance between the transmitting and receiving nodes. We not only put aside wired backhaul links, we also put aside the effect of implementation specific techniques such as physical layer algorithms and channel access protocol. The goal here is to create a spatial metric that is generic enough and points to some fundamental properties of the 5G network. The analysis below has the sole goal of pointing to spatial metrics as the problem domain can get complicated if we want to create a comprehensive model.

Let us consider the model in Figure 6.7, where we have transmitting and receiving pairs with the direction of transmission selected randomly. The transmitter and receiver locations can also be selected randomly within a given range such that the distance between a transmitter and the receiver, r, is bounded. If we do not consider interference due to background noise, we can create a formula that points to the most critical metric for the DSM technique to consider. If the DSM technique has a set target for SIR threshold as β, it can assume a Rayleigh fading model. With this model, the probability that SIR exceeds the threshold β is:

$$\Pr[SIR > \beta] = exp\left(-\lambda r^2 \beta^{\frac{2}{\alpha}} C(\alpha)\right), \tag{6.3}$$

where λ is the density of transmitters[13] and $C(\alpha)$ is a function of α used to simplify the formula by consolidating the Rayleigh α parameter impact.

For *SIR* to be less than β, a link closure must have failed. Thus, the transmission capacity in a given area can be related to an outage constraint ε, where the successful transmission in the given area (unit area) can be expressed as:

$$\tau_\varepsilon = \frac{\varepsilon - 1}{r^2 \beta^{\frac{2}{\alpha}} C(\alpha)} \log(1 - \varepsilon) \tag{6.4}$$

Equation (6.4), with its simplification approach of a complex problem domain, can point to the following critical aspects:

1. In a large ad hoc network, transmission capacity decreases as a function of r^2. This can lead to the concept of sphere packing where each successful transmission utilizes a ground area that depends on the distance between the transmitter and the receiver.

---

12 Note that different flows may be requesting different data rates.
13 Notice that the use of a Rayleigh model means that the number of transmitters is a multiplier in the exponential part of Equation (6.3).

2. The selected SIR threshold is critical. This is not a predetermined threshold. Cognitive techniques search for the SIR threshold that maximizes the area's spectral efficiency.
3. Transmit and receive node pairs can't be chosen in this model. Any node can be a transmitter or a receiver at any given time.
4. Although beam forming, spread spectrum, power control, and other DSA techniques are not included in this simple model, their impact can be reflected in the reduction of SIR, which leads to increasing spectral efficiency in Equations (6.3) and (6.4).
5. Other generalizations such as multihop transmission can be added to this model. Note that some cognitive techniques can search for the best multihop path utilizing the r^2 impact and SIR threshold shown in this model.

Equation (6.4) makes it possible to consider the transmission capacity problem starting from a simple model.

6.2.3.2 5G Cell Overlay

The model in the previous section and illustrated in Figure 6.7 considers transmission capacity in a given area in a general way. The model sheds light on the metrics a 5G DSM design would consider. However, 5G uses small cells that overlay on existing cellular networks to improve capacity and coverage. Indoor users specifically make use of small 5G cells such as femtocells. The best model for using 5G cells on top cellular networks is the two-tier model where one tier is a base station and the other is the cells. This overlay model is represented in Figure 6.8 where end users are randomly located. In this model, the density of the 5G cells can be expressed as l_C, while the density of the end users can be expressed as l_{EU}. Naturally, $l_{EU} \gg l_C$. One can extend the model in the previous section to consider other factors. For example, one can state the following:

1. The interference at an end user can be impacted by the interference from neighboring base stations and from the randomly placed small cells.
2. The interference at a small cell can be impacted by all the uplink connections from the end users within a certain vicinity of the small cell.
3. End users transmitting to a base station can use relatively higher power than end users transmitting to a small cell.
4. The aggregate calculated SIR may consider the impact of small cells, base stations, base station users, and cell users.

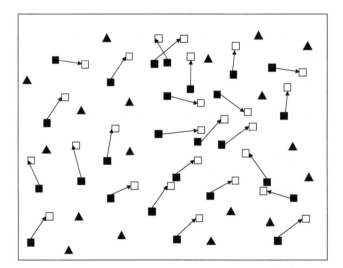

Figure 6.7 Transmission capacity general model. Triangles represent idle nodes, black squares represent transmitting nodes, and white squares represent receiving nodes.

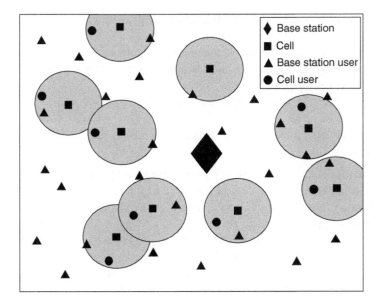

Figure 6.8 5G cell overlay over cellular base station.

Notice in Figure 6.8 that it is possible for an end user to be in the coverage of a 5G cell but use the base station not the cell because of the cell limited capacity or because of a required transmission rate and quality of service (QoS).

A DSM technique can consider the density of end users in a base station coverage area to create a metric for the transmission capacity needed and to point out the need for more cell deployment within the base station coverage area. Other factors that can be used are the transmission outage estimated from Equation (6.4) and actual measurements of events such as connection denial to an end user can be utilized to increase the cell density.[14]

Cellular infrastructure pre 5G is fixed. 5G has no fixed infrastructure since cells can be deployed anywhere where demand is needed. Spatial modeling, covered in Section 6.2, is essential to DSM for both the planning and runtime aspects. After deployment, the model can be used for finding out what new features and capabilities a 5G infrastructure needs in order to increase spectrum efficiency.

There is another layer of overlay that can complicate this spatial modeling. If and when 5G deploys nonterrestrial infrastructure, there will be a satellite or a high altitude platform (HAP) that will have a nonterrestrial base station overlaid on top of multiple ground base stations areas. This model, however, is far in the future and beyond the scope of DSM in this chapter.

6.3 Stages of 5G SI Cancellation

The previous sections focused on spatial modeling and evaluation metrics without considering some DSA techniques that can decrease the impact of SIR on the transmission capacity of 5G infrastructure. This section focuses on one important aspect of DSM in 5G, the maximization of SI cancellation. In order to increase the infrastructure capacity, 5G uses different stages of SI cancellation, as shown in Figure 6.9.

Let us consider the following aspects of SI cancellation with FD communications that makes DSM more efficient and possible with the higher frequency bands 5G is utilizing:

1. Directionality. 5G beam forming relying on MIMO antenna technology means the signal is as narrow as possible where the spectrum is concentrated to the receiving node, with

---

14 Note that such decisions to increase infrastructure resources can use the metrics runtime evaluation to get insight into this need but should use the more comprehensive post-processing evaluation explained in Chapter 5 to make the ultimate decision of adding infrastructure resources.

minimal spectrum leaks to other transmitting and receiving node pairs using the same frequency.

2. 5G MIMO antennas implement SI cancellation using multipath fading analysis stages that reach up to 20 dB gain at both the transmitting and receiving antenna. This is shown in Figure 6.9 as the MIMO antenna cancellation.
3. After using a low noise amplifier, the receiver implements further analog noise cancellation of the RF signal.
4. After the analog to digital converter, the 5G receiver further implements other digital signal noise cancellation techniques.

The 5G protocol stack and the SDR concepts of 5G allow the service provider more flexibility in implementation. Figure 6.10 shows the 5G protocol stack at a higher level. The use of 5G in specific industrial applications has the flexibility of adding software modules that can allow for adapting 5G to the industrial application needs. For example, a service provider can explore using a form of network coding in the open wireless architecture (OWA) layer for further dB gain that can be achieved when using 5G mm-wave links for long-range reachability. This technique will trade some bandwidth for additional error control coding redundancy.

The reader is encouraged to explore antenna design literature for more details about how 5G MIMO antenna interference cancellation is achieved in the mm-wave range, which is beyond the scope of this book. However, the next chapter introduces some MIMO techniques that can be considered for adapting 5G for military communications systems.

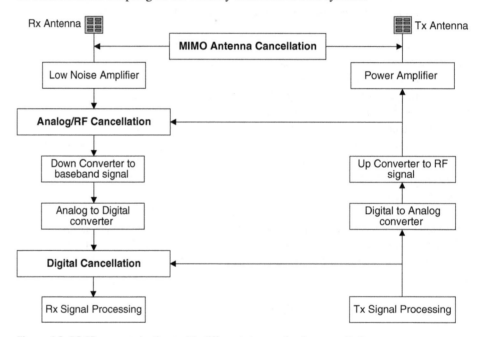

Figure 6.9 5G FD communications with different stages of noise cancellation.

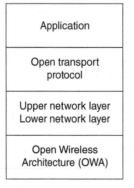

Figure 6.10 5G protocol stack.

6.4 5G and Cooperative Spectrum Sensing

Cooperative spectrum sensing was originally explored in the context of CR networks where the secondary user (SU) has the ability to detect spectrum utilization and collect spectrum sensing information either as a standalone radio terminal, cooperatively with peer nodes, or relying on an external spectrum sensing information. This was covered in Part 1 of this book, where energy detection and other sensing techniques that can be used by the SU are explained. Part 1 also covered DSA for heterogeneous MANET networks with some focus on military communications challenges and approaches to solving the challenges of DSA with heterogeneous MANETs using hybrid design that include cooperative distributed decision fusion. The 5Gs take on cooperative spectrum sensing has some different and unique angles such as:

- energy conservation of end-user devices
- leveraging the fact that 5G networks can be dense
- avoiding information redundancy
- optimizing resource utilization with heterogeneous networks overlay
- the consideration that end-use devices can connect to each other with device-to-device (DTD) communications protocols.

One of the core concepts in 5G is the use of a spectrum agent (SA) in order to offload some of the burdens of spectrum sensing and fusion away from the 5G end-user devices. Notice that 5G can use both licensed and unlicensed spectrum, making it important to detect what frequency bands other communication systems operating in the same geographical location are using.

The dense deployment of 5G small cells is leveraged to turn each cell into a SA and turn inactive[15] small cells into SAs. Figure 6.11 illustrates the compound overlay of active and inactive entities in a 5G deployment in an urban area where we can have a macrocell with the largest area of coverage shown in the figure with the large tower at its center, multiple picocells as illustrated by the small towers, and many femtocells as illustrated with the femtocell icons. This illustration uses inactive femtocells as SAs where an end-user device can obtain spectrum awareness information from an SA even though the device has not made the decision to select this SA as an access point or to use other access points.

Notice in Figure 6.11 that a macrocell user can be utilizing a licensed spectrum band while an end user obtaining spectrum awareness information from a SA can end up using unlicensed

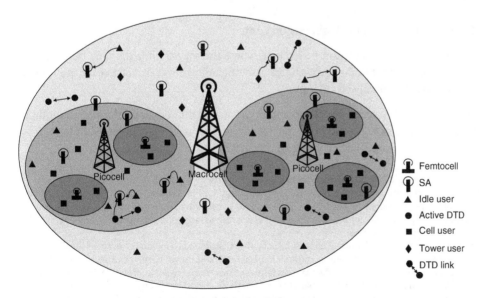

Figure 6.11 The compound overlay of 5G entities in a 5G deployment.

---

15 Inactive here means that there are no end-user devices utilizing the cell. The cell is always active as a SA.

spectrum band using the femtocell as the access point. With this approach, the end user does not have to perform the spectrum sensing functionality or the computationally extensive spectrum sensing information fusion. Instead, the end-user device obtains processed information with decision fusion results from the SA. One can see from this compound overlay that if and when an SA acts as a femtocell, it may have to resort to unlicensed spectrum because the macrocell is using the licensed spectrum bands. As illustrated be the dotted lines in Figure 6.11, an end user would send the SA a spectrum service request and the SA can then do any of the following:

1. Periodically detect spectrum utilization in the area and decide if an available licensed band can be used or if an unlicensed band must be used. This decision is based on the SA periodically detecting and analyzing spectrum utilization in the area. Here, the SA is acting as a spectrum fusion center, deciding spectrum availability and broadcasting it to the end users in the area.
2. Not perform fusion and rely on the macrocell as the fusion center where all the SAs, femtocells and picocells send the macrocell their own spectrum sensing information and the macrocell performs the fusion and send the fusion results to the SAs, femtocells, and picocells.[16]

In any case, the SA would respond to the spectrum access service request from the end user with a frequency band it can use regardless of whether that band is licensed or unlicensed, and regardless of whether the decision is made by the SA or the macrocell.

Note that 5G standards give the service provider flexibility in how to implement spectrum sensing, how to fuse sensing data, and how to create decision. A different hybrid form of local, cooperative distributed, and centralized[17] spectrum sensing technique can be used by the different 5G services providers and the utilized technique can always be enhanced with time. The 5G service provider would have prerequisites such as removing the burden of spectrum sensing from the end user, and the goal of achieving spectrum utilization efficiency. The use of a hybrid spectrum sensing approach where the macrocell has a bird's eye view of spectrum use in this example and the ability to turn idle femtocells into SAs can always improve spectrum efficiency. The macrocell would use computationally complex spectrum arbitration techniques that can better optimize spectrum resources use dynamically from a global perspective and these techniques can be enhanced with time with new software releases.[18]

6.4.1 The Macrocell as the Main Fusion Center

Figure 6.12 shows a 5G spectrum fusion case where the macrocell is the only fusion center for spectrum sensing information. With this case, the following events can occur (notice that these events are continually occurring):

1. The SA continually senses spectrum and collects spectrum sensing information (without any fusion). It consolidates spectrum sensing information for all frequency bands.
2. The SA sends the consolidated spectrum sensing information to the macrocell.

---

16 One can see the parallels and differences when comparing this 5G approach to the military communications approach of cooperative distributed and centralized fusion decisions. One can also see that regardless of the problem domain, good DSA design requires considering some form of local, distributed, and centralized spectrum fusion. System designers should be striving to create the appropriate hybrid approach for optimizing spectrum resources use.
17 Local can mean the SA or femtocell has fusion capability, distributed cooperative can mean the different cell types and SAs share spectrum sensing information and make distributed cooperative decisions, and centralized can mean the macrocell is the main performer of spectrum sensing information decision fusion.
18 From the software-defined network perspective, a centralized spectrum fusion and arbitration in a macrocell means an easier way to roll out new software releases as the software release is targeted to only the macrocell. One can debate if a 5G end-user device should or shouldn't be software defined but all 5G infrastructure from the femtocell to the largest tower are software-defined entities where new software releases can upgrade the infrastructure and enhance its capabilities without the need to deploy new hardware.

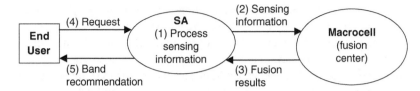

Figure 6.12 5G use of the macrocell as the fusion center.

3. Picocells and femtocells also send their consolidated spectrum sensing information to the macrocell.
4. The macrocell fusion center processes spectrum sensing information received from all SAs, all picocells, and all femtocells.
5. The macrocell fusion center creates fusion results and sends them to each SA, femtocell, and picocell.
6. An end user makes a spectrum service request from the SA.
7. The SA uses the fusion results from the macrocell and replies to the end user with a recommendation of which spectrum band(s) to use.

This case can simply be described as a hybrid of distributed sensing and centralized fusion. Notice that this case can be morphed to the case where the SA, femtocells, and picocells perform a form of spectrum sensing information fusion and send the macrocell fusion center the results of its own local fusion. In either case, the macrocell fusion results are broadcasted to all lower hierarchy entities and the SA can't rely on its own local fusion results to respond to frequency band service requests from end users. Notice that even with the macrocell being the only place for fusion, the SAs, femtocells, and picocells perform a form of cooperative spectrum sensing.

If some SAs have limited energy and must conserve energy, the macrocell can select a subset of the SAs to perform spectrum sensing and collect sensing information. The macrocell algorithm has to consider the hidden node challenge discussed earlier in this book and select SAs that are geographically distributed to optimize sensing impact.

6.4.2 Spectrum Agents Operate Autonomously

With this case, SAs can share spectrum sensing information in a distributed manner along with their geolocation information if they are not publically known to be static.[19] In this case, each SA is a fusion center of spectrum sensing information and the spectrum sensing information shared is abstracted as a spectrum map after fusion.

Figure 6.13 points to a potential disadvantage of this approach as the SA working autonomously will require a control plan to be created to exchange the spectrum map information, as Figure 6.13 shows. This control plan goes through the macrocell, which is supposed to have the most processing power and can act as the centralized fusion center.

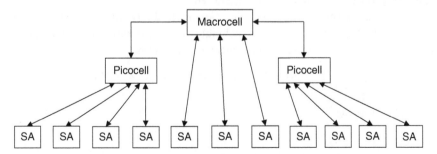

Figure 6.13 Autonomous SAs in a distributed cooperative fashion and their control place.

---

19 If SAs, femtocells, and picocells are static and their geolocation information is known publically, there is no need to amend spectrum sensing information messages with geolocation information.

There are reasons for ensuring the SAs can operate autonomously. Consider the case of deploying 5G communications assets in a disaster area where there is no macrocell to reach out to. There is a need to ensure that dynamically deployed small cells with SA capabilities can perform DSM decisions autonomously, in addition to being access points. The deployed small cells will have spectrum sensing fusion capabilities and the means to share spectrum sensing information with other small cells to make DSM decisions independently of a large tower (macrocell).

6.4.3 The End User as its Own Arbitrator

One can conceptualize this case with very dense urban deployment where the end-user device can obtain spectrum awareness information from multiple SAs and can use spectrum recommendations from multiple SAs to select the best spectrum band to operate on. Notice that if some QoS metrics are supplied to the end user as part of the recommendations, the end user can make the final decision on which band and which access point to select based on multiple factors, including least power consumption, highest data rate, and adhering to required QoS metrics thresholds.

Figure 6.14 illustrates this case where the end user obtains four different recommendations of spectrum use from four different SAs. The end user makes the final local DSA decision on which recommendation it will use.

It is important to note that this case with the end user making the final arbitration decision does not exclude designing a system that uses the macrocell as the fusion center or a system that makes SAs work autonomously. The arbitration here is in the context of arbitrating between different recommendations from different SA points and considering other metrics such as QoS and rate while making the end user reach the final decision on which recommendation to use.

A service provider building a 5G infrastructure can build all the above capabilities in the deployed infrastructure and can enable or disable certain capabilities based on network management decisions or some cognitive algorithms that can morph the deployed infrastructure functionalities based on sensing information. One always needs to distinguish between capabilities or assets and how to use them dynamically. There are pros and cons for some capabilities that make them worth enabling only under certain conditions. For example, the case in Figure 6.14 has the disadvantage of requiring the end-user device to perform an arbitration decision, which is an extra processing requirement on a device with limited battery. However, in very dense urban deployment with the close proximity of multiple cells, the device is already not consuming too much power in maintaining links, making it possible for the device to use some power for arbitration decisions. Also, in the case of autonomous SAs, we have the challenge of having a sparse deployment where an SA would have to rely on limited spectrum sensing information sources, which can lead to encountering a hidden node. However, one can see that in disaster areas, creating autonomous 5G access points is urgent enough to overcome the impact of higher probability of hidden node interference.

5G standardization offers the service provider a lot of flexibility and different service providers will use different spectrum arbitration techniques or have some capabilities enabled or disabled based on network monitoring and management decisions.

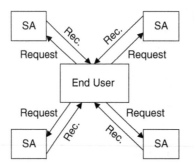

Figure 6.14 The end use as the final arbitrator.

6.5 Power Control, Orthogonality, and 5G Spectrum Utilization

So far, the models we used in Section 6.2 for spatial modeling and in Section 6.4 for cooperative spectrum sensing are at a relatively abstracted level and have not considered other factors such as power control. One can consider that one of the 5G optimization goals is to maximize data rate for the least power consumption and spectrum energy emission in addition to the maximization of throughput in terms of bits per Hertz. The dense deployment of 5G cells allows the end-user devices to achieve the desired data rate at low energy emission, but power control is still a critical factor in 5G DSM. In 5G, the policy given to a single end-user device regarding power control has to be dependent on the policies given to other end users. The DSM technique would have to consider an equilibrium (e.g., a gamming theory based equilibrium) while using constraints in order to approach optimality of spectrum assignment at any given point in time.

One important tool 5G DSM can use before considering power control is signal orthogonality. Spectrum resources are mutually orthogonal blocks[20] that can be cognitively utilized for data transmission to maximize data rate for the least power consumption and spectrum energy emission. Orthogonality is a critical tool in reducing SI.

With power control, Equations (6.3) and (6.4) are no longer enough to drive a model for SIR. Here, SIR encountered by a pair of transmitter and receiver nodes, k, has to consider the resource block n this pair is using. SIR can be expressed in the more general formula:

$$SIR_{k,n} = \frac{\alpha_{k,n}\, p_{k,n}}{\sigma_{k,n}^2 + \phi_{k,n} p_{k,n} + \sum_{j \neq k} \omega_{k,j,n}\, p_{j,n}}. \tag{6.5}$$

In Equation (6.5), notice the following:

- $p_{k,n}$ is the transmit power of the kth transmit/receive pair over the resource block n.
- $\alpha_{k,n}$, $\phi_{k,n}$, and $\omega_{k,j,n}$ are positive parameters that depend on the desired systems parameters and propagation model.
- The summation in the denominator is to consider orthogonal signal impact where $\omega_{k,j,n}$ depends on the impact of the other users' channels on the resource block n.
- $\alpha_{k,n}$ represents the impact of the kth pair on the signal dimension of the nth block, $\phi_{k,n}$.

Notice that for the special case of perfect orthogonality (where a signal in one dimension has no interference impact on signals in other dimensions), the summation in the denominator becomes zero and $\phi_{k,n}$ becomes zero except for the kth dimension, making Equation (6.5) render a simple signal to noise ratio. The 5G DSM model used here considers imperfect channel orthogonality.[21]

A comprehensive 5G DSM has to face many other challenges, including the following:

1. Power control has to lead to energy emission optimization in different scenarios, including when spectrum assignment decisions are made through a centralized arbitrator (i.e., the macrocell acting as main fusion center, as in Section 6.4.1) and when spectrum assignment decisions are made autonomously by cell access points, as in Section 6.4.2.
2. When spectrum assignment decisions are made by a centralized arbitrator, the 5G DSM technique should optimize energy emission globally. A metric for global energy efficiency, EE_G, can be used to gauge the efficiency of the decisions reached.
3. The optimization techniques used can be heuristic and may not render the best optimum way all the time. The DSM technique has to consider the tradeoff between accuracy, computational power, and speed of making decisions using heuristic algorithms. Techniques such as fractional programming and heuristic attempts to create sequential convex optimization in large-scale dense deployments can become challenging.[22]

---

20 A block is represented by a frequency band and signal orthogonality at a time slot that is part of a repeated epoch and can be utilized and released dynamically.
21 In practical MIMO antennas, orthogonal signals energy can be detected across different dimensions. This is different from intersymbol interference or co-site interference.
22 The development of fractional programming and heuristic technique is beyond the scope of this book.

4. When spectrum assignment decisions are made autonomously by a cell access point, the end-user device can act in a noncooperative manner (consider the case in Figure 6.14). The end-user device can be aiming to maximize its own energy efficiency and obtain a certain rate of transmission at the lowest power consumption regardless of the impact of its own arbitration decision on the global optimization of energy emission.
5. Resource blocks are employed for transmit/receive pairs dynamically and are continually assigned and freed, meaning the optimization problem is always trying to reach equilibrium.
6. There is a tradeoff between global performance in the centralized case and fairness. In an equilibrium problem settings, fairness has to be less subjective and be about not starving one end user of spectrum resources on the expense of another end user.

Now let us consider a wireless network with K transmit/receive pairs and N available resource blocks where SIR can be represented by Equation (6.5). Let us assume that for any given user K, energy efficiency can be measured in bit/Joule or the ratio of the achieved transmission rate (from the obtained resource block) to the dissipated power by the device. This device energy efficiency can be expressed as:

$$\eta_k \triangleq \frac{\sum_{n=1}^{N} \log_2(1 + SIR_{k,n})}{p_{c.k} + \mathbf{1}^T \boldsymbol{P}_K}. \tag{6.6}$$

Equation (6.6) can be synthesized as follows:

- $p_{c.k}$ is the power dissipated by the transmitter for the kth transmit/receive pair.
- $\boldsymbol{P}_K$ is the power allocation vector over the utilized N resource blocks, which is one of many possible vectors that can be allocated as denoted by $\boldsymbol{R}_+^N$ in the equation:

$$\boldsymbol{P}_K = [p_{k.1}, p_{k.2}, \ldots p_{k.N}]^T \in \boldsymbol{R}_+^N. \tag{6.7}$$

- The end-user device has local power constraints $\overline{p_k}$, which is the maximum power the device can dissipate. Thus:

$$\mathbf{1}^T \boldsymbol{P}_K \leq \overline{p_k}. \tag{6.8}$$

- In order to create fairness as explained above, we must assure that the end-user device will have a minimum transmission rate. However, this minimum transmission rate can be illusive if we want to also limit the minimum number of spectrum resources blocks we can assign to an end user. This will require creating a metric that measures the maximum rate that can be achieved for the minimum number of spectrum resource blocks. The term $SIR_{k,n}$ represents the maximum SIR we can expect when interference and noise are minimal.
- The end-user device can be assigned a target data transmission rate in order to achieve *fairness*. We have to consider that target data rate in terms of $SIR_{k,n}$. This means the nominator in Equation (6.6) has some boundary data rate. The data rate the end-user device can achieve, θ_k, is bounded as:

$$\theta_k \leq \sum_{n=1}^{N} \log_2(1 + SIR_{k,n}). \tag{6.9}$$

The goal of the DSM technique is to allocate to each end user a power vector as in Equation (6.7), where the power dissemination of the end-user device is bounded by Equation (6.8) and the data rate transmission is bounded by Equation (6.9). Globally, the DSM technique generates a *feasible*[23] set of power for all the transmitting nodes and all the resource blocks. The available combination of the global sets can be expressed as:

$$[\mathbf{p}_1^T, \mathbf{p}_2^T, \ldots, \mathbf{p}_K^T]^T \in \boldsymbol{R}_+^{KN}. \tag{6.10}$$

---

23 Here, the term "optimum" is avoided and the term "feasible" is used instead because the algorithms that can be used may have to be heuristic and have to consider the speed of making decisions and the computational power constraints making them reach a "feasible" decision.

The DSM technique generates the feasible set from Equation (6.10), which can be expressed as:

$$p \triangleq \prod_{k=1}^{K} p_k. \tag{6.11}$$

The selection of the feasible set considers many factors, as explained earlier, including minimizing SIR, maximizing date rate, minimizing power dissemination, and using sensing information to avoid interference with other systems.

Now let us look at a global metric a centralized DSM arbitrator may consider to achieve this multifaceted global energy efficiency, EE_G which can be expressed as:

$$EE_G \triangleq \frac{\sum_{k=1}^{K} \sum_{n=1}^{N} \log_2(1 + SIR_{k,n})}{\sum_{k=1}^{K} 1^T P_K}. \tag{6.12}$$

Notice that in Equation (6.12) the nominator is the global data rate while the denominator is the summation of assigned power. The search for an assigned power set to all transmit/receive pairs can be weighed based on maximizing this global metric (among other metrics). Notice that EE_G does not consider the device circuits' efficiency. The device circuits' efficiency is the ratio of dissipated power as spectrum emission to the assigned power vector. Also this optimization does not consider the presence of AWGN.

EE_G is one of many metrics the centralized arbitrator can use. Note that the more metrics one attempts to include in a spectrum fusion technique, the more complicated the technique can become. One has to look for the most effective modeling of the problem at hand and consider the most critical metrics that can help with the development of a practical cognitive spectrum fusion technique that can be evaluated in real time and their post processing can bring about meaningful policy and configuration parameters changes.

6.6 The Role of the Cell and End-User Devices in 5G DSM

The optimization problem described in the previous section may require the collection of information from the end-user device in order to ascertain if the central arbitrator resource blocks and power needs to be adjusted. Recall that Equation (6.12) did not consider the presence of AWGN or the relationship between allocated power and dissipated spectrum, which requires getting some performance metrics from the end-user device to adjust the allocation of resource blocks and power. The end-user device's lower protocol stack layer can collect different measurements such as block error rate, latency, jitter, and channel quality. This data can be utilized by the upper protocol stack layer algorithms for real-time adaptation of traffic load pushed to the lower stack layers and can also be abstracted and forwarded to the centralized arbitrator to fine-tune the global resource allocation modules.

It is important to note that the end-user devices and the cell access points also have some functions that help reduce the need for the central arbitrator to continually readjust resource allocations. These functions can include the following:

- A cycle adaptation function that is responsible for managing coexistence with other transmit/receiver pairs. This function adopts the time-domain utilization pattern.
- A listen-before-talk (LBT) function, which is tied to the coexistence function through detecting the state of the frequency resource prior to the data transmission. This function manages energy detection, preamble detection, and duration. This function can rely on tunable thresholds to compare the sensing results with.
- A frame format function, which is responsible for adapting the MAC frame according to the different active bearers (e.g., uplink versus downlink resource ratio, transmission time interval [TTI], or slot duration modification). This function can also react to the changes in the channel quality, enabling better coexistence without the need for intervention from higher hierarchy entity (e.g., cell).

- A contention coordination function, which manages any perceived contention on the utilized frequency bands. This function can adapt the random access schemes and block resources on licensed bands or can use a tunable contention access algorithms on shared spectrum bands.
- A multiple access (MA) function, which is responsible for the configuration and adaptation of the use of spectrum resources utilizing orthogonality and nonorthogonal multiple access schemes. Within the cell, this function can manage different active end users given the end-user locations and QoS requirements.
- A sensing function that coordinates between the different sensing mechanisms. This function can create sequential sensing patterns on the different bands while using configurable parameters such as sensing duration, minimum signal detection level, and sampling rate to create effective spectrum sensing information.

Within the cell to end-user radio access technology (RAT), the above functions make short-term decisions leaving longer-term decisions to the centralized function.

Notice that there are other aspects of DSM that have to be considered by the developed DSA technique, including the following:

- Traffic demand can be high in one area and low in another area. The centralized arbitrator allocation of spectrum resource blocks to the different RATs in different areas of the network it is managing can take into consideration traffic demand over time. The resource blocks (or the network infrastructure) are essentially a shared service between the different RATs.
- The small cell has limitations in handling traffic loads at a required QoS. Throwing more resource blocks to a small cell may be a waste of resources given the small cell limitations.
- Ultra-dense urban deployments can force the DSM central arbitrator to create layered architecture of spectrum sharing. The DSM technique would need to consider the tradeoff between spectrum and network density to optimize the network spectral efficiency of multiple RATs sharing a spectrum resources pool.

Other aspects include the goals of the service provider. Some service providers may market guaranteed QoS for higher prices to attract high paying customers while others may market lower prices with less QoS guarantees to create a mass market. These revenue-focused aspects will drive DSM implementation in the 5G infrastructure.

6.7 Concluding Remarks

Although the concept of local, distributed, and centralized spectrum sensing decision exists with 5G, one can see how it can differ from the military network examples used in Part 1 of this book. One can also see how the hybrid DSA design can be different in some aspects while others stay the same. Aspects that stay the same include sensing in geographically distributed locations, creating the framework for hierarchical fusions, and sharing of spectrum sensing information in a combined distributed and centralized manner when possible. The aspect of making DSA a cloud service with well-studied metrics that can be used to evaluate DSA services in real time and in post processing is also common between the many DSA cases.

5G gives the service provider many flexibilities in managing the infrastructure. One can expect a diverse set of DSM techniques to be present in different services providers' networks. However, there is a common theme in these optimization approaches where metrics are used and cognitive techniques are used to approach optimality of spectrum assignment. Relying on a good model for spectrum sharing and using expressive metrics to measure performance is always needed. Also, all service providers are likely to approach DSM as an IaaS set of cloud services, as explained in Chapter 5. One can expect that any service provider approach will have to be hybrid in one way or another and will have to morph based on deployment constraints and real-time measurements.

In some aspects, DSA for military networks may seem easier than commercial 5G, but the expected use of commercial technologies within military communications systems can complicate military network DSA decisions. The push for military networks to use unlicensed spectrum is another factor that can make military system DSA techniques more challenging.

This chapter presented a sample of DSM techniques for 5G. There is a wealth of literature sources for 5G DSM that further explain the developing of mathematical models for DSM, explain the implementation of these mathematical models, and present optimization techniques in the DSA cognitive engine implementation and MIMO antenna design that can aid DSM optimization techniques. This chapter focused on the core aspects of 5G DSM.

Exercises

1. There are many challenges with directional mm-wave links. Fog, rain, and other particles can negatively affect the signal. Beam forming, interference management, and using MIMO antennas are a few of the techniques used with mm-wave design to enhance over-the-air performance. The design of MIMO antennae can leverage multipath for further antenna gain. Consider the equation below used to calculate free space path loss (FSPL) in dB based on distance (d), frequency (f), transmitting antenna gain (G_t), and receiving antenna gain (G_r).

    $$FSPL = 20 \log_{10}(d) + 20 \log_{10}(f) + 20 \log_{10}\left(\frac{4\pi}{c}\right) - G_t - G_r$$

    (a) For the mm-wave frequency signal (chose any frequency in the mm-wave range), a distance of 14 km, and a 20 dB gain at both transmitting and receiving antennas, calculate how much free space path loss the signal faces.

    (b) State some of the options that could be considered to overcome this dB loss.

2. Microwave links are designed as point-to-point links between two stationary nodes. Some designs considered using a mechanical gimbal-based antenna to allow mobile nodes to track each other with point-to-point microwave links. For 5G mm-waves, would you consider using a mechanical gimbal-based antenna or a massive MIMO antenna? State your reasons for your choice.

3. Assume you are designing a voice call acceptance/rejection algorithm for a 5G cell. Your limitations include the air-interface (radio) resources limits and the backhaul link capacity limits.

    (a) Will you consider prioritizing voice over data for both air-interface and backhaul link resources? State your reasons for this prioritization.

    (b) Will you control the data rate access for users to allow more users to join in? State some use case(s) where you would consider data limits to increase the number of users versus other use case(s) where you consider limiting the number of users and guarantee access rates.

    (c) State two different practical scenarios where (i) air–interface resources are the limiting factor in accepting/rejecting a new calls and (ii) the backhauling link capacity is the limiting factor in accepting/rejecting new calls.

    (d) The Erlang distribution can be used to create a call admission control algorithm in a cellular base station such that there is room for a call handover between two cells and there is room for emergency calls. Without going into too much detail on the Erlang distribution, use Table 6.2 and add columns at the end to give an intuitive estimation of the number of voice calls you would allocate for new calls, the number of voice calls you would allocate for handover, and the number of voice calls you would allocate for emergency calls for each 5G cell type. In your estimation, consider the cell size, area of coverage, probability of having handover based on the area of coverage, and the role each cell regarding emergency calls. Use the upper limits of the number of users column.

4. Consider the relationship between SIR and link connectivity discussed in Section 6.2.2. Let us assume the simplest connectivity use case where we have no other cells using the same frequency in the area and there is a single active end user. If the end user desired data rate is 20 Mbps and the selected threshold in Equation (6.2) is $\Gamma = 2$, the technique was able to establish a connection for that end user. Will the technique be able to establish the connection if the end user's desired data rate was less? What if the end user's data rate was more?

5. Beam width plays a major role in 5G DSM. Let us assume the general case for the relationship between SIR and link connectivity discussed in Section 5.2 and assume that when all users use a beam of 10° width, the global interference power is 0.1 of the signal power and that global interference power doubles for every 10° increase in beam width.
   (a) Create a table showing a column for the beam angles incrementing from 10° to 50° in 10° incremental steps. Populate this table with a second column for power ratio as the beam width increases.
   (b) Add a column to the table showing the SIR in dB.
   (c) If link closure can only be achieved if SIR in dB > 0, show which entry in the table will cause the node pair to fail in creating connectivity.

6. Consider Figure 6.14 where an end-user device is given four different options for connectivity through four different access points as shown in the table below. The end-user device has a desired QoS for the traffic it is using and each access point option has a different QoS guarantee, as shown in the final column in the table.

Access point	Rate	Energy emission	QoS
1	R_1	E_1	Guaranteed for all rate R_1
2	$R_2 > R_1$	$E_2 > E_1$	Guaranteed for rate $R_x < R_1$
3	$R_3 > R_2$	$E_3 > E_2$	No QoS guarantees
4	$R_4 > R_3$	$E_4 > E_3$	No QoS guarantees

   (a) From the table, which access point should be eliminated from consideration as the first step?
   (b) If the end-user device required data rate is $R_r < R_1$, which access point should be considered?
   (c) If the end-user device requires the transmission of a total data rate $R_r > R_1$, but only a fraction of this data rate lower than R_1 requires the QoS guarantees, which access point should be considered?

7. A centralized spectrum arbitrator has a pool of eight orthogonal resource blocks and is allocating spectrum resources to four different simultaneous transmit/receive pairs. The algorithm output created the allocation of resource blocks per transmit/receive pair is shown in the table below, where 1 means the block is assigned to the pair and 0 means the block is not assigned to the pair. The spectrum arbitrator is attempting to generate the global energy efficiency metric EE_G based on this allocation using Equation (6.12). The EE_G calculation assumes that:
   (i) the power disseminated from an activate resource block (assigned 1 in the table below) is p and
   (ii) the power received at a block that is not assigned to the transmitter in any given transmit/receive pair is $0.143p$ where $0.143p$ is due to orthogonality leakage and other interferences.

	Block 1	Block 2	Block 3	Block 4	Block 5	Block 6	Block 7	Block 8
User 1	1	0	1	0	0	0	0	0
User 2	0	1	0	1	0	0	0	0
User 3	0	0	0	0	1	0	1	0
User 4	0	0	0	0	0	1	0	1

(a) Calculate the EE_G metric for this assignment in terms of p.

(b) If the power received at a block that is not assigned to a transmitter in a transmit/receive pair is $0.333p$, calculate EE_G metric.

(c) What is the impact of increased interference power on the EE_G metric?

(d) For the same system and the same number of users, let us assume that users 1 and 2 make a request for a high transmission rate while users 3 and 4 make a request for a low transmission rate. Can you redo the table above with different resource blocks assignments?

(e) Based on the new table created in (d), if you assume that each resource black can be assigned a transmission rate of r bps, how much rate increase do users 1 and 2 get and how much rate decrease do users 3 and 4 get?

(f) Based on 5G requirements for fairness between users, what is the minimum number of resource blocks you would assign to a given transmit/receive pair that makes a request for the lowest data rate?

(g) If the spectrum arbitrator recalculates the geographical separation of the four transmit/receive pairs and decides that users 1 and 2 are geographically separated from users 3 and 4 such that resource blocks can be reused, rework the table above with resource blocks reuse assignment based on geographical separation that maximizes the data rate for all four users.

(h) If the spectrum arbitrator calculation of geographical separation was not perfectly accurate and the assignment of the same block to two transmit/receive pairs caused the power received at a block that is not assigned to a transmitter in a given transmit/receive pair to increase from a fraction of p as in (a) and (b) above to become p, calculate the EE_G metric.

(i) Considering the three EE_G metric calculations in (a), (b), and (h), what conditions result in the worst EE_G metric and what conditions result in the best EE_G metric?

(j) How important is the role of the sensing function in the end-user device and how important it is to find a way to update a centralized spectrum arbitrator with the actual measurements of SIR?

(k) If you are designing the centralized spectrum arbitrator cognitive engine, what would you want this engine to do if it receives SIR from end-user devices and finds the EE_G metric is approaching the value you calculated in (h)?

8. Section 6.2.3.1 explained the optimization of 5G transmission capacity while relying on the spatial separation metric that uses SIR. Using the formats of the DSA cloud services metrics covered in Chapter 5, draft the details of this spatial separation metric.

Bibliography

3GPP standards. WWW.3GPP.org.

5G Network Transformation, 5G America. http://www.5gamericas.org/files/3815/1310/3919/5G_Network_Transformation_Final.pdf.

Andrews, J.G., Ganti, R.K., Haenggi, M. et al., A primer on spatial modeling and analysis in wireless networks. *IEEE Communications Magazine*, vol. 48, no. 11, pp. 156–163, November 2010.

Andrews, J.G., Buzzi, S., Choi, W. et al., What will 5G be? *IEEE Journal on Selected Areas of Communications*, vol. 32, no. 6, pp. 1065–1082, June 2014.

Aquilina, P., Cirik, A.C., and Ratnarajah, T., Weighted sum rate maximization in full-duplex multi-user multi-cell MIMO networks. *IEEE Transactions on Communications*, vol. 65, no. 4, pp. 1590–1608, April 2017.

Axell, E., Leus, G., and Larsson, E.G., Spectrum sensing for cognitive radio: State-of-the-art and recent advances. *IEEE Signal Processing Magazine*, vol. 29, no. 3, pp. 101–116, May 2012.

Belikaidis, I., Georgakopoulos, A., Demestichas, P. et al. Multi-RAT dynamic spectrum access for 5G heterogeneous networks: The speed-5G approach. *IEEE Wireless Communications*, vol. 24, no. 5, pp. 14–22, October 2017.

Chen, S. and Zhao, J., The requirements, challenges, and technologies for 5G of terrestrial mobile telecommunication. *IEEE Communications Magazine*, vol. 52, no. 5, pp. 36–43, May 2014.

Du, B., Pan, C., Zhang, W., and Chen, M., Distributed energy-efficient power optimization for CoMP systems with max-min fairness. *IEEE Communications Letters*, vol. 18, no. 6, pp. 999–1002, June 2014.

Elmasry, G., McClatchy, D., Heinrich, R., and Delaney, K., A software defined networking framework for future airborne connectivity. 2017 Integrated Communications, Navigation and Surveillance Conference (ICNS), Herndon, VA, 2017, pp. 1–19.

Haider, F. and Gao, X., Cellular architecture and key technologies for 5G wireless communication networks. *IEEE Communications Magazine*, vol. 52, no. 2, pp. 122–130, May 2014.

He, S., Huang, Y., Jin, S., and Yang, L., Coordinated beamforming for energy efficient transmission in multi-cell multi-user systems. *IEEE Transactions on Communications*, vol. 61, no. 12, pp. 4961–4971, December 2013.

Hong, X., Wang, J., and Wang, C.X., Cognitive radio in 5G: A perspective on energy-spectral efficiency tradeoff. *IEEE Communications Magazine*, vol. 52, no. 7, pp. 46–53, July 2014.

Ismail, M. and Zhunag, W., A distributed multi-service resource allocation algorithm in heterogeneous wireless access medium. *IEEE Journal on Selected Areas in Communications*, vol. 30, no. 2, pp. 425–432, February 2012.

Koudouridis, G. and Soldati, P., spectrum and network density management in 5G ultra-dense networks. *IEEE Wireless Communications*, pp. 30–37, October 2017.

Mahmood, N., Sarret, M.G., Berardinelli, G., and Mogensen, P., Full duplex communications in 5G Small CELLS. In 2017 13th International Wireless Communications and Mobile Computing Conference (IWCMC), Valencia, 2017, pp. 1665-1670.

Ng, D., Lo, E., and Schober, R., Energy-efficient resource allocation in multi-cell OFDMA systems with limited backhaul capacity. *IEEE Transactions on Wireless Communications*, vol. 11, no. 10, pp. 3618–3631, October 2012.

Osseiran, A., Boccardi, F., and Braun, V., Scenarios for 5G mobile and wireless communications: The vision of the METIS project. *IEEE Communications Magazine*, vol. 52, no. 5, pp. 26–35, May 2014.

Rodriguez, J. (ed.), *Fundamentals of 5G Mobile Networks*, Wiley, 2015. ISBN 9781118867525.

SPEED-5G Public Deliverable, D3.2: SPEED-5G enhanced functional and system architecture, scenarios and performance evaluation metrics. ICT-671705, H2020-ICT-2014-2, June 2016.

Venturino, L., Zappone, A., Risi, C., and Buzzi, S., Energy-efficient scheduling and power allocation in downlink OFDMA networks with base station coordination. *IEEE Transaction on Wireless Communications*, vol. 14, no. 1, pp. 1–14, 2015.

Weber, S., Yang, X., Andrews, J.G., and de Veciana, G., Transmission capacity of wireless ad hoc networks with outage constraints. *IEEE Transactions on Information Theory*, vol. 51, no. 12, pp. 4091–4102, December 2005.

Yang, C., Li, J., Guizani, M. et al., Advanced spectrum sharing in 5G cognitive heterogeneous networks. *IEEE Wireless Communications*, vol. 23, no. 2, pp. 94–101, April 2016.

Zhang, J. and Andrews, J.G., Distributed antenna systems with randomness. *IEEE Transaction on Wireless Communications*, vol. 7, no. 9, pp. 3636–3646, September 2008.

Zhang, Z, Zhang, W., Zeadally, S, Wang, Y., and Liu, Y., Cognitive radio spectrum sensing framework based on multi-agent architecture for 5G networks. *IEEE Wireless Communications*, vol. 22, no. 6, pp. 34–39, December 2015.

7

DSA and 5G Adaptation to Military Communications

5G applications are not limited to cellular. Many industries are attempting to leverage the bandwidth released for 5G and the capabilities developed for cellular applications to build enterprise wireless systems that are based on 5G spectrum, protocol stack, and open source protocols to build new high capacity private wireless systems. One of these industries is the military communications field. Most concerns regarding the use of 5G technologies with military applications are related to security and vulnerabilities. Thus, the most important aspect in bringing 5G capabilities to military communications is addressing security and vulnerabilities concerns.

There are many security challenges that face adapting 5G to military communications that can be addressed with DSA techniques. Security challenges can be addressed in multiple fronts but the DSA front is very critical. When military communications technologies were in their infant phase at the start to the middle of the twentieth century, transmission security (transec) was an important capability to consider. The designers of military radios invested in research and development of powerful transec techniques. Considerable efforts were given to building hardware, such as high end oscillators that can allow for fast frequency hopping to make military radios overcome jammers. When it comes to considering 5G for military communications, the parallel concept to these infant phases of developing early military communications technologies is to look for the equivalent to transec. That is, to look for the air-interface enhancement techniques that increase 5G security. This chapter briefly discusses areas where 5G can address security vulnerabilities but focuses mostly on DSA-related security enhancements over the air-interface.

7.1 Multilayer Security Enhancements of 5G

Figure 7.1 illustrates how the commercial 5G protocol stack, on the left-hand side of the figure, can be adapted to the military wireless communications shown on the right-hand side of the figure. Figure 7.1 shows locations for different security enhancement capabilities that can be added at the different protocol stack layers. The best approach to make 5G secure enough for military communications is to leverage every opportunity 5G offers to increase security and resilience capabilities.

While Figure 7.1 shows different security enhancement areas such as the use of communication security (COMSEC) encryption and application layer security, it is the air-interface of the 5G open wireless architecture (OWA) that offers the greatest opportunities to enhance security. The air-interface enhancements can ensure that the use of the same spectrum, freed for commercial use, in military applications is secure in many critical aspects to include low probability of interception and low probability of detection (LPI/LPD). A military communications system design can consider any or all of the upper protocol stack techniques covered in this section but the air-interface security enhancements are unique since they are related to DSA and may require special hardware development.

Dynamic Spectrum Access Decisions: Local, Distributed, Centralized, and Hybrid Designs, First Edition. George F. Elmasry.
© 2021 John Wiley & Sons Ltd. Published 2021 by John Wiley & Sons Ltd.
Companion website: www.wiley.com/go/elmasry/dsad

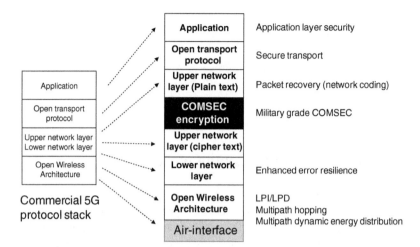

Figure 7.1 Adapting the 5G protocol stack into a military secure stack with multilayers of security enhancements.

Notice the split of the 5G upper network layer on the left-hand side of Figure 7.1 to the plain text upper network layer and the cipher text upper network layer. This split is needed to add the COMSEC capabilities mandated in military communications.

Following Figure 7.1, making 5G secure enough for military applications can include the following areas of security enhancements:

1. *Add strong application layer security.* As defense communications radios moved from proprietary protocols to IP-based waveforms and networks, the US National Security Agent (NSA) introduced COMSEC type 1 encryption, known as high assurance internet protocol encryption (HAIPE). The use of 5G within military networks could be coupled with the development of an application layer security that rivals the capabilities of IP-based type I encryption where the application layer encryption code is virtually impenetrable. This adds a security layer to the applications layer that increases resilience against eavesdropping. Even if an eavesdropper was able to reconstruct an IP packet through mimicking all the lower protocol stack layers, the packet will not be decodable to give useful information.
2. *Add a strong secure transport protocol.* The military industry developed its own secure transport layers with its own security capabilities. One example of secure transport protocols is the nack oriented reliable multicast (NORM) developed by the US Navy. NORM or any other secure transport layer can be used within the adapted protocol stack shown in Figure 7.1.
3. *Increase the plain text layer packet recovery capabilities.* One of the drawbacks of using IP-based type 1 encryption is the high percentage of packet loss. Any cipher text packet decrypted that does not pass the type 1 encryption parity check decoding is considered compromised and is not forwarded to the plain text IP layer. 5G offers an abundance of bandwidth that can be utilized to increase the security and resilience of military applications traffic. The adaptation of 5G to military communications can include the use of network coding techniques that can recover packets lost by type I decryption. Here, we trade bandwidth for enhanced resilience. This will essentially help the adapted 5G technology to be more resistant to jammers. Because some ratio of the lost decrypted packets can be recovered and forwarded to the transport layer at the receiving node, jammers have to increase the impact of the jamming signal on the 5G air-interface signal to increase the plain text packet loss ratio
4. *Fit the 5G protocol stack with a military grade type I COMSEC encryption.* This will ensure that all the security capabilities that come with type 1 encryption are leveraged with the 5G protocol stack. As eluded to in Section 7.4.2, adapting 5G to military communications should be coupled with the use of software-defined radios where the 5G-based waveform is one of many waveforms a platform can use. These military platforms are built with a plane

text side general purpose processor (GPP) and a cipher text side GPP. These two GPPs are hardware separated by a COMSEC chip. A 5G-based waveform should be portable to these military communications platforms.

5. *Increase the lower network layer error resilience.* The 5G open protocol stack offers another opportunity to increase error resilience through software modules at the lower network layer. It is important to overcome the impact of jamming[1] or the interference with commercial use by increasing error resilience. The lower network layer of 5G offers the ability to add techniques such as Reed-Solomon (RS) coding with erasure such that an IP packet can be reconstructed from the OWA frames even when there is a high frame loss rate. Frame loss happens when error patterns cannot be corrected by the OWA layer turbo codes error correction capability.

6. *Develop special MU MIMO hardware.* The rest of this chapter focuses on how to modify the 5G MIMO module to make it a military grade MIMO module with capabilities such as LPI/LPD, multipath hopping, and multipath dynamic energy distribution. MIMO design for 5G is the most critical part of bringing 5G capabilities to military communications. While the upper protocol stack layers in 5G are software modules, adapting them for military communications can be less challenging and adding new software modules for security and resilience needs can be done as explained above without the need to modify the communications platforms hardware. It is the air-interface design that needs the most consideration and the MIMO antenna design that requires the most difficult modifications since it may require adding new hardware to the communications platform. This antenna module may require developing new and unique firmware capabilities. The gained air-interface security from this antenna module (e.g., beamforming, multipath hopping, and power distribution and adaptation) will meet the unique requirements of military communications that cannot be simply met by adding a new software module to any of the upper protocol stack layers. The following section lists MIMO design considerations for a military grade MIMO antenna module.

7.2 MIMO Design Considerations

There are multiple areas of MIMO design that can enhance the security of 5G over-the-air use in military applications. While commercial 5G DSA techniques are focused on increasing throughput and optimizing the use of spectrum to maximize data rate and monetary revenues, military applications have no concept of monetary revenues and have to consider DSA techniques that enhance OTA security capabilities such as LPI/LPD.

7.2.1 The Use of MU MIMO

With the 3GPP standards moving from 802.11ac to 802.11ax, a critical capability was added that military communications can utilize. 802.11ax uses MU MIMO at both the base station and the end user. In commercial networks, the use of MU MIMO at all nodes is leveraged for capabilities such as making a node relay another node's traffic to the base station while optimizing spectrum resource use through beamforming and power adaptation, as explained in Chapter 6. This MU MIMO capability in all nodes can be leveraged to create a 5G-based MANET network instead of the cellular mesh network with the cell or tower as the focal point in commercial wireless. There are many reasons for military communications systems to use MANET instead of mesh networks. These reasons include the need to avoid having a

---

1 There are many types of jammers. The air-interface resilience techniques explained in this chapter should make the job of jammers harder and hence one should expect jammers to cause high bit error rate, frame error rate or packet error rate without being able to render the system totally unusable (e.g., not being able to make the probability of bit error equal 0.5). The more we introduce error recovery techniques that trade off some of the 5G abundant bandwidth for error resilience, the more we make 5G suitable for military communications.

single point of failure. A military network using a 5G-based waveform may not have a base station as its focal point. The use of MU MIMO at all nodes can result in all nodes in the military network having similar capability (all nodes can be peer nodes). This will allow MANET capabilities with 5G protocol stack adaptation to be created. The 5G upper protocol stack layers software may need modifications for MANET functionalities. However, MANET capabilities will require an antenna module at each node that uses specifically designed MU MIMO antennas.

7.2.2 The Use of MIMO Channel Training Symbols for LPD/LPI

Using 5G technologies with military networks requires achieving LPD/LPI. MIMO antenna design can be utilized for LPD/LPI. Training symbols that have been developed for use in commercial 5G cellular systems to optimize the MIMO channel mutual information, can be repurposed in military communications to discover which transmit and receive antenna pair (which over-the-air path) has the best performance. They can also be used to discover RF paths suffering from the most interference. This discovery of paths with the highest interference and paths with the best performance can be utilized to make the transmitter distribute the power spectral emission among the transmit antenna's paths accordingly. This power distribution can minimize the MIMO channel signal RF footprint and enhance LPD and LPI capabilities.[2]

7.2.3 The Use of MIMO Channel Feedback Mechanism for LPD/LPI

As explained above, LPD/LPI capabilities require that the transmitter distributes the power unevenly among the different MIMO antenna elements. A feedback mechanism would be needed such that the receiver estimation of which antenna transmit and receive pairs have the least faded path and which antenna transmit and receive pairs have best performance can be conveyed to the transmitter.

It is important that the multichannel fading be tracked accurately. This requires that MIMO channel feedback is frequent and must face the least instantaneous delay.[3] Notice that between two nodes a and b the transmit and receiver pair with the least fading when a transmits to b may not necessarily be the same transmit and receive pair with the least fading when b transmits to a. In some commercial systems, the design can avoid the use of a feedback mechanism and assumes channel symmetry such that node a distributes the power among the transmit antennas based on its local decision fusion on which antenna transmit and receive pair has the least fading. With military communications, this assumption should be avoided because jammers can target a specific node transmission creating the need to avoid this symmetry assumption.

7.2.4 The Use of MU MIMO for Multipath Hopping

With a single input single output (SISO) channel, time and frequency domain hoping is used with military communications to make it harder for a jammer to follow the transmit receiver pattern. Hopping jammers and communications systems play a cat-and-mouse game where the hopping pattern is changed in a pseudorandom manner by the communications system while the jammer is attempting to decipher the hopping pattern to follow the signal so that it can jam it more precisely. Military communications adaptation of 5G technology has an opportunity to create another security layer where the MIMO module would dynamically change the power distribution among the transmit antenna elements, creating the equivalent of multipath hopping. This multipath power variation can use a code that is based on a pseudorandom algorithm to change power distribution among the transmit antennas such that the jammer will not

---

2 As will become clear later in this chapter, this power distribution should not null the power on any transmit and receive pair of MIMO antennas, and may not concentrate all the power within one pair. This power distribution will create variation in power distribution that can achieve the goal of LPD/LPI among other goals.
3 The channel feedback mechanism should rely on the antenna module firmware, not an upper protocol stack layer, to be instantaneous.

be able to follow the random change in power distribution between the different over-the-air paths. This capability will require a new generation of jammers that try to follow this multipath hopping pattern in order to precisely jam the target RF signal. It is obvious that the larger the number of antenna elements in the MIMO design, the more powerful this multipath hopping becomes and the more resilience to jammers the design achieves.

5G adaptation in military communications comes with opportunities and vulnerabilities. Creating another domain for hopping (multipath hopping or multipath power variations) can be considered an opportunity to increase signal resilience to jammers. Notice that it is harder for a directional jammer to track a MU MIMO channel than to track a SISO channel because of the multipath nature of the MIMO signal. Also, the jammer would have to consider additional factors to change direction based on node mobility. Increasing the multipath channel dynamics will increase the complexity of the jammer design. The increase in the number of antenna elements will further increase the complexity of the jammer design.

7.2.5   The Use of MU MIMO to Avoid Eavesdroppers

Military communications has a history of developing security techniques that adapt to newer technologies. Different versions of transec have been developed for different military waveforms. When IP networks made their way to military communications, strong COMSEC capabilities were developed for the military communications without precluding the use of transec techniques. HAIPE is an added security layer to IP-based military waveforms that still use transec techniques.[4] The use of MU MIMO with a 5G-based military waveform offers the opportunity to develop a new transec techniques that leverage the MU MIMO characteristics to create the equivalent of transec keys that would be hard to break by an eavesdropper. The new transec capabilities make it harder for an eavesdropper to find the signal and to track it reliably enough to reconstruct OTA frames or IP packets. MU MIMO have characteristics that can be leveraged to hide the signature of the signal over the air in ways that are much stronger than all current transec techniques that are based on SISO antennas.[5]

The MU MIMO module of a 5G-based military waveform offers the military communications system designer another opportunity to increase 5G-based waveform security by adding to the MU MIMO module a transec key for the power spectral variations between the antenna's transmit and receive pairs. Only the peer MU MIMO receiver knows the pseudorandom key used to control power spectral variations and only the peer receiver can follow the signal power spectral variations while the eavesdropper fails to follow this multipath signal power spectral variations to reconstruct meaningful information from the signal.

7.2.6   The Use of MU MIMO to Discover Jammers

Two peer nodes communicating on a MANET can discover and send feedback to each other with information regarding which multipath channel suffers from the most interference. If this information is communicated to a centralized spectrum arbitrator, it can be fused to discover jammers' locations provided that the centralized arbitrator can analyze the multipath signal fading based on terrain information to discover unexpected fading patterns that are persistent between certain nodes. These patterns can be indicative of the presence of jammers that are targeting certain locations with certain directional beams.

Certainly this capability will come at the cost of developing complex fusion engines. A centralized arbitrator performing analysis for SISO systems using terrain information can pinpoint the direction of a jammer based on a single path and terrain information using fading models. With MU MIMO, this type of fusion complexity will increase multiple times given that the

---

4 HAIPE encryption surpasses commercial IPSec techniques, making it possible to route classified IP packets over secure tunnels through commercial satellite vendors' networks.
5 The addition of a new transec layer of security for MU MIMO may not preclude the use of other transec techniques currently used by different military systems.

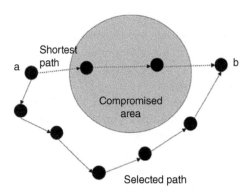

Figure 7.2 Selecting a path to avoid a compromised area through beamforming.

MU MIMO antennas create multiple transmit and receive pairs, and given the need to calculate fading for each multipath channel independently.

7.2.7 Beamforming and LPI/LPD

Beamforming is a standard 5G feature used for the global optimization of spectrum use as explained in the previous chapter. Military communications can use beamforming for increased LPI/LPD. Military systems can opt to decrease power emission and narrow the beam in order to increase the number of hops between the transmitting and receiving nodes and avoid hoping through a compromised area. Figure 7.2 shows an example of a detected compromised area where jammers or eavesdroppers can exist, as indicated by the gray circle. A 5G system designed for military communications will decide routing based on this information where a path with more hops than the shortest path is selected to avoid the compromised area. One can see how important beamforming can be with the military communications use of 5G capabilities to address LPI/LPD needs while not losing sight of the rule of beamforming in minimizing SI and optimizing the use of spectrum resources.

It is critical for military communications applications to distinguish between optimizing for the MIMO channel capacity, optimizing for global spectrum resources use, and optimizing to avoid compromised areas and increase the signal resilience. The next section provides more details on this multifaceted tradeoff.

7.3 Multifaceted Optimization of MU MIMO Channels in Military Applications

Let us get an insight into this area through a mix of basic known channel capacity expressions and intuitive analysis to find ways to adapt MU MIMO channel optimization for military applications.

If we use training symbols with feedback mechanisms, then we can assume that the channel gain matrix H is known at both the transmitter and the receiver. The capacity of such a channel can be expressed as:

$$C = \sum_{i=1}^{n} log_2 \left(1 + \frac{\lambda_i P_i}{\sigma_n^2} \right), \tag{7.1}$$

where λ_i is the ith eigenvalue of the matrix HH^*, n is the number of transmit and receive antenna pairs, and $P_1, P_2, \ldots P_n$ is the power distribution among the transmit antennas such that $\sum_{i=1}^{n} P_i = 1$.

Equation (7.1) is simply an adaptation of the SISO channel capacity where there is only one antenna, there are no summations, and the last term in the logarithm bracket is the signal energy over the noise variance.

The eigenvalues $\lambda_1, \lambda_2, \ldots \lambda_n$ can be changed by changing the directionality of each transmit antenna. If the goal is to maximize the capacity of the MIMO channel, then each transmit antenna i would seek an ideal line-of-sight beam such that changes in the antenna location do not change λ_i. With this ideal situation, $\sum_i \lambda_i = \sum_{ij} |h_{ij}|^2$. Thus, the constraints on the eigenvalues are simply dependent on the channel gain. In order to maximize the capacity of this channel, one would want to find $\lambda_1, \lambda_2, \ldots \lambda_n$ and $P_1, P_2, \ldots P_n$ such that $\sum_i P_i$ is bounded by the total transmit signal energy, E_s. If we express $\sum_i \lambda_i = L$ for a given signal energy, then we are seeking to determine how to distribute the transmit signal energy among the different transmit antennas and how to influence directionality such that we create variation in $\lambda_1, \lambda_2, \ldots \lambda_n$. Mathematical references use the Lagrange multipliers to solve this problem mathematically. However, let us consider a practical solution that correlates to the Lagrange multiplier solution.

We can influence the eigenvalues such that $\lambda_1 \geq \lambda_2 \geq \ldots \geq \lambda_n$ without losing any generality. This inequality is simply varying the eigenvalues regardless of their order. Consider the following cases:

1. When SNR is low (we are emitting low power and the target node is nearby), the Lagrange multiplier solution will yield to making P_1 equal to the total signal energy and making $\lambda_1 = L$. All other eigenvalues and power will be zero, that is, $\lambda_j = P_j = 0$ for all $j > 1$, in other words we use the MIMO antenna to create a focused beam. The channel model is similar to that of a SISO and Equation (7.1) becomes similar to that of a SISO channel. One can reach the same conclusion intuitively. If the two nodes are very close to each other, then the transmitting node would focus the beam and lower the power to barely reach the receiving node in a focused beam and thus we achieve LPD/LPI.

2. When SNIR is sufficiently large, that is, we are trying to reach a far node, we increase the power to overcome the distance. The optimal solution will call for $\lambda_1 = \lambda_2 = \ldots = \lambda_n = \frac{L}{n}$ and $P_1 = P_2 = \ldots = P_n = \frac{E_s}{n}$, where E_s is the total signal energy. This corresponds to creating n parallel channels and distributing the signal energy among them evenly while attempting to control directionality such that all eigenvalues are equal. Recall that the goal of equal distribution of power here is to maximize channel capacity.

3. Now let us consider the case when an adversary affects $\lambda_1, \lambda_2, \ldots \lambda_n$. This is the presence of a jammer case. Recall that we either use one transmit and receive antenna's pair for low power emission to reach a nearby node or we have crossed a certain power threshold to reach a faraway node and we are distributing power evenly among transmit antenna and attempting to control directionality to achieve equal eigenvalues and high capacity of transmission. At any given time, the receiver may detect the presence of a jammer that reduces the channel capacity considerably. The presence of the jammer will be detected at the receiver as $\sum_i P_i < E_s$ and $\sum_i \lambda_i > L$. The feedback mechanism explained earlier would communicate what the receiver detected to the transmitter.

In the presence of a jammer, maximizing the channel capacity means:

$$\max C = \begin{array}{c} \min \\ \lambda_i: \\ \sum_i \lambda_i > L \end{array} \begin{array}{c} \max \\ P_i: \\ \sum_i P_i < E_s \end{array} \sum_{i=1}^{n} \log_2\left(1 + \frac{\lambda_i P_i}{\sigma_n^2}\right). \tag{7.2}$$

Equation (7.2) is intuitive. It says that the transmitter should attempt to redistribute the power among the transmit antenna and adjust directionality to overcome the jammer. Using the Lagrange multiplier solution, it will yield that all the eigenvalues need to be nonzero, that is, the antenna directionality should be selected such that the antenna directionality should be selected while ensuring that the receiver reports to the transmitter eigenvalues that are nonzero for all paths.[6] This is intuitive in one sense in that we do not want to put all our eggs in one basket but we have to utilize all channels unevenly. One needs to consider that a MU MIMO channel has multipaths and adjusting directionality is not that simple. The antenna module would need a machine learning technique that accounts for changing terrains and learns how to control directionality to evade the jammer.

---

6 The worst-case scenario occurs when all eigenvalues are zeros. This will prevent the receiver from discovering the impact of the jammer.

The Lagrange multiplier solution will also yield that all channels should have power transmission. The transmitter should not evade the jammer by nulling any transmit and receive pair of antennas that results in zero eigenvalues for that path; rather adjust this pair's directionality.

The most important step for the transmitter to do is with regard to power distribution. If a transmit and receive antenna pair consistently yields eigenvalues that are close to zero, the transmitter can reduce the power over that transmit antenna and add that reduced power to the transmitters in the transmit and receive pairs that yield nonzero eigenvalues.[7] This redistribution of power makes the MIMO module rely more on uncompromised paths and less on compromised paths without affecting the channel capacity.

One can see the need to develop a special MU MIMO module for military communications adaptation of 5G technology. Special firmware maybe needed to overlay all the techniques that influence the adaptation of power spectral distribution among the antenna elements at the transmitter. Extensive processing is also needed at the receiver to fuse the received signal and generate the proper feedback to the transmitter.

7.4 Other Security Approaches

Adapting 5G for military applications can consider other security approaches. The two subsections below introduce some examples of security approaches to be considered when gradually bringing 5G capabilities to the tactical theater.

7.4.1 Bottom-up Deployment Approach

With this approach, 5G is first deployed in small enclaves surrounded by existing security trusted military networks. Small footprint 5G networks can grow with time to include directional mm-wave links for reach-back between different networks. Directional mm-wave links can be used first in secure areas where it is important not to emit spectrum in an area occupied or accessible by the enemy. This approach will allow for the introduction of 5G technology in military communications while adding the security capabilities explained in this chapter gradually. The DARPA mobile hotspots program studied this approach.

A military waveform has to go through many certifications before it is deployed. The larger the RF footprint of the waveform, the harder it is to get the waveform certified. The 5G-based small RF footprint MANET waveform maybe the first 5G technology to be certified for military use. This bottom-up approach will allow for the gradual deployment of 5G capabilities in a military theater.

7.4.2 Switching a Network to an Antijamming Waveform

One of the advantages of using SDR in military communications is the ability to make a platform use different waveforms at different times based on deployment needs. When 5G is used and no security compromises are detected, 5G can continue to be used with its high bandwidth advantage. When security breaches are suspected, the same network formed using a 5G waveform can be switched to a different waveform that is considered an AJ waveform. In other words, the same network using 5G technology will switch to an AJ waveform to form the same exact network. The network will lose the high bandwidth advantage of 5G but it will be secure. This calls for the use of 5G in military applications as a software defined waveform.

---

7 If this technique is implemented accurately with low latency feedback, there will always be a group of transmit and receive antenna pairs that lend eigenvalues that are close to being evenly distributed.

7.5 Concluding Remarks

There are many considerations when it comes to adapting 5G capabilities to military communications applications. In this chapter, we focused mainly on the DSA aspects of the use of 5G in military applications where the adapted 5G over-the-air signal security is enhanced to meet military communications needs.

The chapter showed that 5G use in military communications can be in the form of a SDR waveform that a military communications platform can use. The chapter also showed that developing a special MU MIMO antenna module for military applications may be needed to create MANET capabilities and enhance over-the-air security.

There are many technical and nontechnical points of views on adapting 5G for military communications. At the time of writing this book, the cases for adapting 5G in military communications were mere demonstrations for mobile hotspots and mm-wave point-to-point links range extension while tracking fast movers. For these reasons, this chapter is about breadth not depth. The chapter is intended to open the reader's eyes to the many areas that can be considered to allow 5G-based waveforms to be deployed in military applications.

Exercises

1. When building a MU MIMO for 5G use in military communications we should consider increasing the number of antenna elements. Is this statement true or false?

2. What is the drawback in making the number of antenna elements very large (use of massive MIMO) in a MANET that uses 5G technologies?

3. The larger the RF footprint of a 5G-based MANET, the greater the need for security enhancements. Is this statement true or false?

4. Two MU MIMO antennas communicate through N transmit/receive pairs. What level of complexity as a function of N does a designer of an eavesdropper face to be able to track the signal?

5. The adaptation of 5G for military applications may need to develop a military grade MU MIMO antenna module. Is this statement true or false?

Bibliography

Grushevsky, Y.L. and Elmasry, G.F., Adaptive RS codes for message delivery over encrypted mobile networks. *IET Proceedings – Communications*, vol. 3, no. 6, pp. 1041–1049, June 2009.

Hassibi, B. and Hochwald, B., How much training is needed in multiple antenna wireless links? *IEEE Transactions on Information Theory*, vol. 49, no. 4, pp. 951–963, April 2003.

He, S., Huang, Y., Jin, S., and Yang, L., Coordinated beamforming for energy efficient transmission in multi-cell multi-user systems. *IEEE Transactions on Communications*, vol. 61, no. 12, pp. 4961–4971, December 2013.

Ismail, M. and Zhunag, W., A distributed multi-service resource allocation algorithm in heterogeneous wireless access medium. *IEEE Journal on Selected Areas in Communications*, vol. 30, no. 2, pp. 425–432, February 2012.

Ponnaluri, S., Soltani, S., Shi, Y., and Sagduyu, Y., Spectrum efficient communications with multiuser MIMO, multiuser detection and interference alignment. MILCOM 2015, 2015 IEEE Military Communications Conference, Tampa, Florida, 2015, pp. 1473–1478.

Soysal, A. and Ulukus, S., Joint channel estimation and resource allocation for MIMO systems. Part II: Multi-user and numerical analysis. *IEEE Transactions on Wireless Communications*, vol. 9, no. 2, pp. 632–640, February 2010.

Telatar, E., Capacity of multi-antenna Gaussian channels. *European Transactions on Telecommunication*, vol. 10, no. 6, pp. 585–596, November 1999.

Venturino, L., Zappone, A., Risi, C., and Buzzi, S., Energy-efficient scheduling and power allocation in downlink OFDMA networks with base station coordination. *IEEE Transaction on Wireless Communications*, vol. 14, no. 1, pp. 1–14, 2015.

Wicker, S., *Error Control Systems for Digital Communication and Storage*, Prentice Hall, 1995.

Wicker, S. and Bhargava, V., *Reed Solomon Codes and Their Applications*, IEEE Press, New York, 1994.

Yoo, T. and Goldsmith, A., Capacity and power allocation for fading MIMO channels with channel estimation error. *IEEE Transactions on Information Theory*, vol. 52, no. 5, pp. 220–2214, May 2006.

Zhang, J. and Andrews, J., Distributed Antenna Systems with Randomness. *IEEE Transactions on Wireless Communications*, vol. 7, no. 9, pp. 3636–3646, September 2008.

https://www.darpa.mil/program/mobile-hotspots.

https://www.darpa.mil/program/100-gb-s-rf-backbone.

8

DSA and Co-site Interference Mitigation

The concept of *blanking signal* was mentioned at the conclusion of Chapter 2. With this concept, a platform that has both communications waveforms and a spectrum sensor has to consider the impact of communications activities on sensing accuracy. The sensor should not be sensing all the times. It will be important for the communications waveforms to inform the spectrum sensor of transmission time intervals on the operational frequency bands. During these intervals, the spectrum sensor would refrain from performing spectrum sensing because spectrum sensing will be impacted (blanked) by the communications signal emitters at the same platform. Even if sensing is at a different frequency band from the transmission band, frequency-domain lobes (co-sites interference) can impact the sensing accuracy.

Blanking signal is a technique used to ensure that spectrum sensing is not impacted by a known communications signal on the same platform. In addition to ensuring sensing accuracy, DSA design faces the critical aspect of making sure new frequency assignments will not suffer from co-site interference. DSA as a set of cloud services should include a subset of services for co-site interference mitigation that takes into consideration the different sources of co-site interference. Co-site interference can impact new frequency assignments, causing the rippling mentioned earlier in this book. Co-site interference mitigation as a set of DSA services should minimize or eliminate the impact of co-site interference on frequency assignments taking into consideration all the sources that can cause co-site interference.

On a communications platform that uses a single waveform, that is, the platform uses only one operational frequency at a time, co-site interference within the platform may not be a concern. It is when a communications platform has more than one waveform with more than one antenna module operating simultaneously that the challenge of co-site interference over the same platform has to be addressed.[1] Spectrum emission over one antenna at a given frequency f_1 can emit power at the first, second, and third frequency bands adjacent to f_1. This spectrum leakage can be in the same range as the frequency f_2 over which another antenna on the same platform is operating. Co-site interference on the same platform is one of many sources of co-site interference. DSA design has to be coupled with understanding of the relationship between a signal spectrum leakage and spectrum interference with other signals.

There are other sources of co-site interference such as external systems communicating over a frequency band that can create co-site interference to the system under design in given geographical areas. This chapter addresses different possible sources of co-site interference and shows how the cognitive engine-based DSA cloud services approach presented in Chapter 5 can include co-site interference mitigation as a subset of cloud services without an excessive increase in computational complexity.

This chapter does not cover baseband circuitry, switching power supplies, system clocks, and other noise generating sources. It also does not cover minimizing local radiation using shields, RF chokes, or RF absorbing material. It is assumed that the platform is already designed to

---

1 Many military communications platforms have more than one hardware channel loaded with more than one waveform. All upper echelon nodes are multichannel platforms and are nodes of more than one network. Even at the lower echelon, light vehicles can have more than one channel. A platoon commander can carry a multichannel radio (platform) where the platoon soldiers may carry a single-channel radio.

Dynamic Spectrum Access Decisions: Local, Distributed, Centralized, and Hybrid Designs, First Edition. George F. Elmasry.
© 2021 John Wiley & Sons Ltd. Published 2021 by John Wiley & Sons Ltd.
Companion website: www.wiley.com/go/elmasry/dsad

minimize these interfering sources. The focus here is on co-site interference mitigation as it relates to DSA. Co-site interference should not exceed a given energy level that can affect the use of a given frequency band. If DSA does not consider co-site interference, dynamic frequency assignments can ripple and negatively affect the throughput efficiency of the managed networks.

This chapter lays the foundations for understanding and using the IEEE Standard 1900.2 – The IEEE Recommended Practice for the Analysis of In-Band and Adjacent Band Interference and Coexistence Between Radio Systems, which is included in Part 4 of this book. For engineers building a DSA system, it is recommended that they comprehend and adhere to IEEE Standard 1900.x in their development.

8.1 Power Spectral Density Lobes

Co-site interference is a sort of a necessary evil. Some signal design intentionally creates higher power leakage outside of the baseband. Theoretically, a signal is not bounded to a limited frequency bandwidth. The Fourier transform of the signal has infinite extent in the frequency domain. Practically, a signal has a range of frequencies where most of its power spectrum is located. The designer of a communications signal faces a tradeoff with regard to spectral design. A signal that has low spectral density at the side lobes (less leakage to adjacent bands), which reduces its interference to other signals, may have achieved this at the expense of making the main loop wider. A wide main lobe can suffer more distortion during channel filtering.[2]

Figure 8.1 shows the power spectral density of a typical signal at a carrier f_c and bandwidth W. The main lobe of this signal, which has most of the signal energy, is between $f_c - \frac{W}{2}$ and $f_c + \frac{W}{2}$. The side lobes, which have a considerable portion of the signal power spectral density, are over other frequency bands outside of the signal's assigned bandwidth, as shown in Figure 8.1, where:

- The first leakage band is in the side lobe between $f_c - \frac{W}{2}$ and $f_c - \frac{3W}{2}$ to the left of main lobe, and $f_c + \frac{W}{2}$ and $f_c + \frac{3W}{2}$ to the right of the main lobe.
- The second leakage is in the side lobe between $f_c - 3\frac{W}{2}$ and $f_c - \frac{5W}{2}$ to the left of main lobe, and $f_c + \frac{3W}{2}$ and $f_c + \frac{5W}{2}$ to the right of the main lobe.
- ....

When an antenna module is receiving over a frequency that is neighboring to the transmitting frequency, the transmit and receive signals can interfere with each other due to the overlap with leaked spectrum from other signals. As Figure 8.1 shows, the first leakage loop

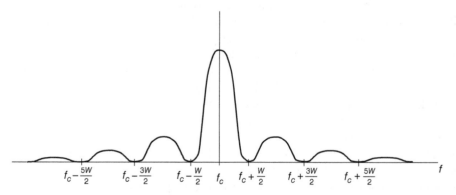

Figure 8.1 The power spectral density of a typical signal versus frequency.[3]

---

2 Note that a signal power spectral density describes the signal in probabilistic terms. Regardless of what information a signal carries, its power spectral density expresses power distribution over frequency.
3 This figure shows only the positive side of the power spectral density.

of a signal has more power spectral density than the second lobe, which has more power spectral density than the third lobe and so on. One can see that the first order of business in a DSA design that mitigates co-site interference is to avoid making simultaneous transmit and receive frequencies[4] over the same node neighboring frequencies.

A waveform that uses TDMA and schedule transmit timing for each node over a single operational frequency band will not face this type of interference since the transmit and receive bands are the same but the node is either transmitting or receiving at a given time.

8.2 Co-site Interference between Frequencies in Different Bands

In the previous section, we concluded that the first order of business for DSA in avoiding co-site interference is to avoid making transmit and receive frequencies neighboring to each other. The second order of business is to avoid working in a single frequency band without considering the neighboring bands.

Figure 8.2 shows a representation of the IEEE frequency band chart. A pitfall in DSA design can result from working in one band, e.g., the very high frequency (VHF) band, addressing co-site interference within this band and ignoring interference from other bands. In the VHF band, ignoring power spectral leakage from the high frequency (HF) band or the ultra-high frequency (UHF) band can negatively affect a frequency in the VHF band. Spectrum is continual and making a waveform operate in a certain band has to do with regulators and spectrum licensing. Co-site interference mitigation as a set of cloud services has to look at spectrum as continual regardless which band a frequency assignment belongs to.

8.3 Co-site Interference From Unlicensed Frequency Blocks

Notice that the frequency band chart in Figure 8.2 has a mix of licensed frequency blocks and unlicensed frequency blocks. There is a tendency (not a rule) that transmission over a licensed frequency block would have more power than transmission over an unlicensed frequency block. Transmission over unlicensed frequencies tends to have a small RF footprint to allow for spatial reuse. In addition, transmission over unlicensed blocks tends to use spread spectrum modulation to allow a large number of users to share the same spectrum resources. This may make it harder for a spectrum sensor to detect unlicensed frequency use. On the upside, DSA has more than one technique to indirectly address co-site interference from these small RF footprint signals using unlicensed frequency blocks. Consider the following:

1. These unlicensed frequency blocks tend to be used by wideband signals. Power decay with spectral leakage lobes is faster with wideband signals than narrow band signals. This diminishes the co-site interference impact of these unlicensed frequency blocks.
2. Because of the use of spread spectrum modulation coupled with a small RF footprint, the co-site impact of these signals will further diminish. The co-site inference impact may be considered as background noise or may be dealt with as an underlay signal that can appear as noise.

Figure 8.2 The IEEE frequency band chart.

---

4 There is inconsistency in using the term frequency band in the literature. This book uses the term frequency to mean a carrier frequency with a defined bandwidth. The book follows the literature sources that tend to divide and classify the spectrum into known bands, as in Figure 8.2. Thus, a frequency band in this chapter is a publicly defined band such as the HF band, the VHF band, the UHF band, etc. A frequency in this chapter is defined by a carrier frequency f and a bandwidth W Hz, which can be within any of the bands of Figure 8.2.

3. The same-channel in-band sensing covered in Part 1 of this book shows that the communications node can hypothesize the presence of an interfering signal but can also hypothesize that the noise power has crossed a certain threshold. Same-channel in-band sensing will be able to lead the DSA cloud services to decide the continuation of the use of a given unlicensed frequency or to avoid the use of that frequency because of any type of unknown interference, which includes co-site interference.

A practical implementation of DSA as a set of cloud services may ignore calculating co-site interference for unlicensed frequency blocks and rely on same-channel in-band sensing to hypothesize if this co-site impact measured as a background noise can be tolerated or not.

8.4 Adapting the Platform's Co-site Interference Analysis Process for DSA Services

This section introduces a methodological approach to include co-site interference impact in DSA decisions. This approach is practical to implement. DSA, as a set of cloud services, needs considerable computational complexity for information fusion and decision making. Taking into consideration co-site analysis during the decision-making process can only be practical if the design relied on preprocessing to generate lookup tables. These lookup tables should allow the DSA decision-making process to mitigate co-site interference or minimize its impact without the need for extensive computational power or delaying the DSA services response time. When assigning frequencies dynamically, any extra delay from the computation of co-site analysis in real time can degrade the service by increasing the service's response time considerably. The approach presented here is heuristic and relies on creating approximation models. These models help the DSA decision-making process to consider the critical impacts of co-site interference as part of a multifactor decision-making approach. For example, if the spectrum resources pool is limited at any given time, DSA services can allow interference from higher order leakage lobes of another signal to coexist with a new frequency assignment as it is likely to be seen as background noise. The limited spectrum resources pool will limit the options to use other frequencies, forcing the DSA decision-making process to tolerate some level of co-site interference.

Figure 8.3 illustrates an adaptation of a standard co-site interference analysis process. The standard co-site process gathers data regarding the platform such as the antenna locations, emission pattern from each antenna, signals characteristics, receivers' susceptibility, and antenna coupling.[5] Once the data have been collected, they are entered into a simulation platform such as Matlab and the results are analyzed. This process is iterative and can be performed multiple times until a conclusion is reached on how co-site interference occurs over the simulated platform. With DSA, this standard process is adapted to generate a set of lookup tables that can be used during the DSA decision making. These lookup tables are generated for each platform and for all the waveforms the platform may use, that is, Figure 8.3 is repeated for each platform separately and the co-site impact between all possible combinations of waveforms is analyzed.[6] The goal of these lookup tables is to allow the DSA decision-making process to mitigate the co-site interference impact without having to perform any real-time analysis for all possible combinations of waveform types that can operate simultaneously on the platform.

---

5 Every antenna on the platform has to be considered. Sometimes platform-based antennas that are not tied to a waveform are overlooked. For example, if the platform has a GPS antenna or a radar warning receiver, these antennas have to be part of the platform co-site interference analysis.

6 There are scenarios that may be overlooked when performing platform analysis. For example, consider a platform that can carry two separate waveforms, that is, the platform can be a node in two separate networks at the same time. If the waveform used for both networks is the same waveform type where each waveform has to operate on a separate frequency, co-site analysis here has to be considered although it is the same waveform type.

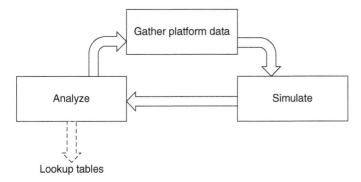

Figure 8.3 Adapting a standard platform co-site interference analysis process for DSA.[7]

With a large platform that uses directional antennas that are far away from each other,[8] co-site interference can be minimal. For other platforms that are small in size, use omni-directional antennas, and are ground platforms,[9] co-site interference can have a large impact on signal performance.

The nature of the generated lookup tables in Figure 8.3 will become clear in the subsequent sections of this chapter.

8.5 Adapting the External System's Co-site Interference Analysis for DSA

In addition to the platform-based co-site interference analysis explained in the previous section, DSA co-site interference consideration has to include the impact of external systems. The designed system that is offering DSA as a set of cloud services has to be able to interface to external systems and obtain a database of external licensed users that includes their locations and their signal characteristics.

Similar to platform-based co-site analysis, a form of preprocessing is needed for the consideration of the impact of external systems on the system under design. Before assigning a new frequency to a network or a node, DSA services should consider the co-site interference impact of external systems without performing real-time co-site analysis.

Figure 8.4 illustrates a similar process to Figure 8.3 where the external systems data are collected and a different form of simulation than Figure 8.3 is performed in order to create

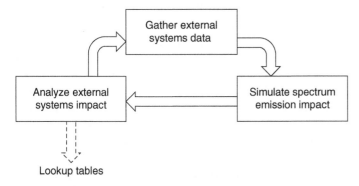

Figure 8.4 Creating DSA co-site interference lookup tables for external systems.

---

7 Notice that the arrow from the "analyze" step to the "gather platform data" step is needed in case the analysis shows that some platform data need to be fine-tuned or added to the simulation.
8 An example of such platform is an aircraft carrier.
9 Airborne platforms are less problematic than ground platforms because some antennas can be placed on top of the platform and other antennas can be placed on the belly of the airborne platform, creating better separation.

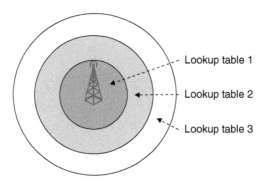

Figure 8.5 Creating different co-site lookup tables for an external system for DSA services based on the distance from the external emitter.

Lookup table 1

Lookup table 2

Lookup table 3

lookup tables for an external system for a specific location. For example, if the external system is a cellular tower with a defined RF footprint, a lookup table within a certain distance of the cellular tower location can be created. Different lookup tables may be created based on the distance from the tower, as shown in Figure 8.5.

Notice that analyzing the impact of the external system through simulation may not be very accurate. The runtime may require fine-tuning of how the co-site interference impact is considered. A good design of DSA services with co-site mitigation considered should include a set of metrics that evaluate the performance co-site interference mitigation. These metrics should allow the DSA cognitive engine to change the lookup tables used in a certain scenario without the need to perform the preprocessing steps after deploying a system. The DSA services design should include the use of lookup tables for all possible scenarios and should have policies, rule sets, and configuration parameters to change the behavior of the cognitive engines regarding co-site interference mitigation without the need for software upgrades to the system.

8.6 Considering the Intersystem Co-site Interference Impact

Dynamic frequency assignment relies on spatial separation to reuse the same frequency within the managed system such that two networks or two node pairs using the same frequency will not interfere with each other. If we consider Figure 8.1, the dynamic frequency assignment process spatially separates the geographical areas where the main loop in Figure 8.1 is emitted for the same frequency. Let us refer to this spatial separation as the first order of spatial separation.

The intersystem co-site interference impact requires considering a second order of spatial separation. As explained in the previous two sections, DSA cloud services can include the platform co-site interference impact and the external systems co-site interference impact. Second-order spatial separation can take into consideration minimizing or eliminating co-site interferences internal to the system. One can consider platform co-site interference mitigation as just one type of this second order of spatial separation where one signal power spectral leakage impact on another signal main lobe over the same platform (same geographical location) is considered. DSA services can also consider this second order of spatial separation where the managed networks are assigned spatially separated frequencies that prevent interference between the main power spectral lobes and also minimize the impact co-site interference.

This second order of spatial separation can be calculated after the first order of spatial separation is concluded, producing a set of new frequency assignments. The second order of spatial separation can confirm that the new frequency assignments have no or minimal intersystem co-site impact. If flags are raised by the second order of spatial separation that there are concerns about co-site impact, fine-tuning of frequency assignments can be performed or a decision can be reached to accept some level of intersystem co-site interference. Accepting some level of co-site interference can occur because there are not enough spectrum resources to overcome this level of interference. Note that the first order of spatial separation has to take precedence over the second order of spatial separation because co-site impact dwindles quickly with spatial separation. The cases that will mandate that this second order of spatial separation is rigorous include:

- the overlay of different networks within the same geographical area
- when narrow band signals are used extensively within a small area of operation: narrowband signals have considerable power spectral density.

8.7 Using Lookup Tables as Weighted Metrics

As Figure 8.1 shows, power spectral densities for the higher order leakage lobes are less than power spectral densities for the lower order leakage lobes. All the above-mentioned techniques should boil down to approximating co-site impact as a weighted matrix where the power of each lobe is mapped to a weight. Table 8.1 shows an example of mapping the impact co-site interference of a certain signal. The weighted matrix abstracts co-site impact in numbers where higher numbers mean higher impact and lower numbers mean lower impact. A weighted matrix can be used as a lookup table for the co-site impact of an external signal, for the co-site impact on a platform, or for the intersystem co-site impact of the signals that would have been created had new frequency assignments been executed.

Let us take an example where the DSA service is going to assign a certain frequency f to a node on a platform. The DSA services may perform the following steps:

1. If the same frequency is used on the same platform, the co-site impact is ∞, which means the frequency should not be assigned. It is an overlap of the main lobes in Figure 8.1. This is part of the first order of spatial separation.
2. If there is a frequency f_1 used on the platform that has a first spectral leakage lobe that shares a frequency range with f, the impact is weighted as 16. This weight can be correlated to a policy that may only allow this assignment to occur if there are no other frequencies available in the spectrum resources pool to overcome this co-site interference. There could be cases where the only option available is to live with the co-site interference resulting from the first spectral leakage lobe of f. If there are enough frequency resources to reconsider a different spectrum assignment, the policy will allow the DSA technique to explore the impact of the different assignment.
3. If there is a frequency f_1 used on the platform that has a second spectral leakage lobe that shares a frequency range with f, the impact is weighted as 4. This weight can be correlated to a policy that allows this assignment to occur if there are no other frequencies available in the resource pool to overcome the co-site interference or if the frequency pool resources are diminishing and what is left is considered a reserve only to be used to address emergency situations.
4. Steps similar to steps 2 and 3 can be performed with policies for all higher spectral leakage lobes. Note that a certain co-site interference mitigation may or may not be tolerated based on policies and rule sets, which consider factors such as spectrum resources availabilities.
5. Steps 2 through 4 can be repeated for external systems impact as well. External systems impact is with respect to a geographical location.
6. A common method to create these weights is to translate the co-site interference to a weight that expresses the dB loss impact on the receiver of the signal under consideration. If the

Table 8.1 Example weighted matrix as a lookup table.

Main lobe	First leakage lobe	Second leakage lobe	Third leakage lobe	Fourth leakage lobe	Fifth leakage lobe
∞	16	4	2	1	0

design of DSA as a set of cloud services strives to achieve higher fidelity, it will create a one-to-one relationship between the signal we are calculating co-site impact for and all the signals that would be created if a new frequency assignment took place. Different signals react to the same level of co-site interference differently. The loss in dB can have a higher impact on one signal than another and can be used as the ultimate reference, that is, dB loss can be used to normalize the impact of co-site interference between all possible signals. With this normalization, the weights in Table 8.1 would represent dB loss.

The zero weight in Table 8.1 for the fifth spectral leakage lobe means that the impacts of the fifth lobe and all higher lobes are negligible.

8.8 Co-site Interference Incorporation in Decision Fusion and Fine-Tuning of Co-site Impact

A system with a set of DSA cloud services will have a large collection of weighted matrices similar to the one in Table 8.1. In addition to the (16, 4, 2, 1, 0) example in Table 8.1, the system can use other lookup tables to include:

- (9, 3, 1, 0)
- (4, 2, 1, 0)
- (3, 1, 0)
- (2, 1, 0)
- (1, 0)

Note that these weighted matrices are examples of the outcome of co-site interference analysis. As mentioned above, these weights can be dB loss values. There are common signal characteristics that can lead to creating the matrices as listed above, including the following:

1. Power spectral density decays exponentially or close to exponentially with higher spectral leakage lobes. Different signals have different decay patterns.
2. Some signals are designed for this spectral leakage lobe power decay to be fast while other signals are designed for the spectral leakage lobe power decay to be slow.[10] Fast decaying signals can be mapped to a matrix such as (1, 0) while slow decaying signals can be mapped to a matrix such as (16, 4, 2, 1, 0).
3. The signal power emission (in W) has to be considered as part of creating the weights in the matrices. The integer units of the above matrices are functions of the signal power. Two signals that have the same decay pattern but one emits 5 W while the other emits 1 W do not have the same impact. The co-site interference impact of the 5 W signal can be weighed as multiple times the impact of the 1 W signal.[11] The integer values in these matrices have to be normalized to the co-site impact as a loss in dB.
4. With external systems, distances can be mapped to different matrices, as illustrated in Figure 8.5. The impact of an external system co-site interference on the DSA managed system can differ based on where the DSA managed system is operating with respect to the external system geographical location (spatial distance).
5. With intersystem co-site interference, the second order of spatial separation may use short matrices such as the (1, 0) matrix in most cases because the power spectral density of leakage lobes decays quickly with distance. The (1, 0) matrix will lead the second order of spatial separation to consider intersystem co-site impacts only for the first spectral leakage lobe and when there is close proximity. If the intersystem co-site interference is being considered for two signals in the same geographical location but for different networks and over different platforms, the lookup tables will be similar to platform co-site interference lookup tables.

---

10 Narrowband signals tend to have slower power decay while wideband signals tend to have faster decay because of signal processing needs. Spread spectrum wideband signals may have the least co-site interference impact.
11 The power in a signal leakage lobe as a function of the emitted wattage can be signal dependent.

6. The weights have to always be tied to policies. DSA as a set of cloud services is dynamic and has to consider many factors. Co-site analysis should be a guide to frequency assignments based on policies. A rigid design that can stop frequency assignments because of co-site interference from higher spectral leakage lobes can lead to inefficient use of spectrum resources.

8.9   DSA System co-site Interference Impact on External Systems

When considering co-site interference based on a platform, the DSA service naturally considers reciprocity (mutual impact). If a platform is part of two networks A and B, the DSA service will ensure that frequency assignment to A does not suffer from co-site interference from B when assigning a new frequency to A. Similarly, the DSA service will ensure that frequency assignment to B does not suffer from co-site interference from A when assigning a new frequency to B. When assigning a new frequency to a network in a certain area and co-site interference from an external system is considered through weighted matrices, the DSA service has to also consider the impact of any frequency assignment, within the system under design, on the external system. This consideration is most important when the external system is a primary user. The DSA service has to consider the impact of co-site interference of frequencies assigned in the area of operational external primary users so that it does not violate the primary user's rules. Weighted matrices can be varied based on the distance from the center of the external system spectrum emission. Higher spectral leakage lobes may be allowed to exist in the peripherals of the external system area of coverage under conditions such as scarcity of the available spectrum resources pool.

8.10   The Locations Where Co-site Interference Lookup Tables and Metrics are Utilized

In a hybrid DSA system, the design may consider the location where the weighted matrices are utilized. Because higher hierarchy decisions can override lower hierarchy decisions, one may think that co-site interference impact should be considered with the centralized arbitrator only. This design approach may not be sufficient. A comprehensive design for DSA services as a set of cloud services should include co-site interference consideration as a hybrid design and matrices lookup should be included in any entity that makes a DSA decision.

Let us consider DSA decision making at the node level. The node may have weighted matrices specific for its platform to check against when a new frequency assignment reaches it from a higher hierarchy. The node may send the higher hierarchy entity that performed the frequency assignment a feedback message indicating how tolerable this frequency assignment is considered to be based on the node's own assessment.

Needless to say, checking against these weighted matrices locally can also occur if the decision to use a frequency is reached locally, that is, the node is an entity that offers DSA services.

A hybrid DSA design should have these lookup tables (matrices) at any entity even if it is expected that a higher hierarchy entity will override the lower hierarchy entity DSA decision. This local node lookup can be used for decision feedback. Decision feedback can be a code that expresses the amount of tolerance to co-site interference when accepting a decision from a higher hierarchy. This decision feedback can help a cognitive-based system of the higher hierarchy to fine-tune its parameters. For example, the centralized arbitrator can reach a conclusion through its machine learning techniques and using decision feedbacks that a certain platform has lower tolerance to co-site interference than anticipated. This decision can be reached because of the high ratio of frequency assignments that do not last when co-site interference is forced to be tolerated for some reason or another and as expressed by the platform decision feedback. The machine learning technique can use the feedback from the lower hierarchical entity and its own performance metrics to adjust its policy. The use of weighted matrices will allow the centralized arbitrator machine learning technique to decide to switch to a different matrix for this platform co-site impact. For example, if the centralized arbitrator was relying on the matrix (3, 1, 0) for co-site impact, it can switch to the matrix (4, 2, 1, 0) to increase the weight of the co-site interference impact for this platform and hence adapt its policy towards this platform.

This book presented the case for hybrid DSA design in the previous chapters. This hybrid design concept also applies to co-site interference mitigation.

8.11 Concluding Remarks

Including co-site interference mitigation in a DSA design as a set of cloud services cannot use real-time analysis every time a frequency assignment decision is made. This can cause delay in the response time to a request to change of frequency and may require extensive processing power. The utilization of preprocessing can produce lookup tables that can be used by the designed system. DSA services that rely on policies and rules sets that incorporate co-site interference considerations into the frequency assignment decision making adaptively may be better.

This chapter pointed to the basic principles DSA co-site interference should consider. The chapter also covered the use of lookup tables to offer a set of co-site avoidance DSA services that include:

1. platform co-site interference avoidance
2. external systems co-site interference avoidance
3. intersystem co-site interference avoidance
4. avoid creating co-site interference to external systems.

Co-site interference consideration should be part of the hybrid design approach of DSA where any entity that makes a DSA decision, whether a local node, a gateway or a central arbitrator, should perform some form of co-site interference mitigation using lookup tables. Although higher hierarchy entities can override the DSA decisions made by lower hierarchy entities, lower hierarchy entities can still use co-site interference lookup tables to generate decision feedback or for local DSA services. Decision feedback can be helpful in the scenarios where the lower hierarchy entity has to accept the frequency assignment decisions made by the higher hierarchy entity, although it anticipated a form of co-site interference impact. Decision feedback will help fine-tune future frequency assignment decisions performed by the upper hierarchy entity. Decision fusion considering the use of co-site interference lookup tables can become an integral part of providing DSA as a set of cloud services with a subset of these services related to co-site interference mitigation.

Note how the co-site interference mitigation techniques presented in this chapter can be incrementally added to an existing DSA set of cloud services later, that is, a system can be designed and deployed for a set of DSA cloud services that do not include co-site interference mitigation services and a later release of the DSA capabilities can superimpose co-site interference impact as a new capability without the need to modify the existing DSA services. Co-site interference mitigation can become a subset of DSA cloud services that can be added in incremental steps. A system design may opt to add co-site interference impact mitigation in incremental steps. For example, a system design for DSA capabilities of a geographically dispersed set of heterogeneous MANETs in an area that is not considered crowded with spectrum users can add co-site interference mitigation capabilities in the following order:

1. Consider the impact on external systems so that primary user rules are not violated.
2. Consider the platform co-site interference impact since it is likely to have the most co-site impact on frequency assignments.

3. Consider external systems co-site interference impact. This consideration has lower priority because the area of operation is not considered to be crowded with spectrum users.
4. Consider inter-system co-site interference impact. This consideration comes last because the system is a set of geographically dispersed networks.

Exercises

1. What type of signal may lead a DSA cloud services design to create the co-site interference lookup matrix (32, 16, 4, 2, 1, 0)?

2. You are given two platforms that have the same waveform modems and the same types of waveforms running on them. One of these platforms is a large ship and the other is a person carrying communications equipment. Are the co-site analyses for these two platforms the same? What other factors can make the co-site interference analyses differ?

3. What type of signal may lead a DSA cloud services design to create the co-site interference lookup matrix (1, 0)?

4. You are given a power spectral emission of a signal measured at the transmitter, as shown in the figure below. Given the pattern of signal power decay with leakage lobes, what type of signal this could be? Give an example of a lookup matrix that you may use for this signal when you are considering co-site interference on the same platform.

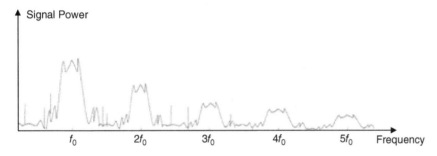

5. The goal of considering co-site interference analysis with DSA as cloud services is to eliminate co-site interference. Is this statement true or false?

6. Signals have positive and negative frequency representation, that is, they have real and imaginary components. When considering co-site interference impact, do you need to consider the complex representation of signals or is the one-sided representation shown in Figure 8.1 sufficient? Why?

Bibliography

Allsebrook, K. and Ribble, C., VHF co-site interference challenges and solutions for the United States Marine Corps' Expeditionary Fighting Vehicle Program. IEEE MILCOM, Military Communications Conference 2004, Monterey, California, 2004, pp. 548–554.

Balanis, C., *Antenna Theory – Analysis and. Design*. John Wiley & Sons, Inc., pp. 575–580, 2005.

Elmasry, G., *Tactical Wireless Communications and Networks, Design Concepts and Challenges*. Wiley, 2012. ISBN 978-1-1199-5176-6.

German, F., Young, M., and Miller, M.C., A multi-fidelity modeling approach for co-site interference analysis. 2009 IEEE International Symposium on Electromagnetic Compatibility, Austin, Texas, August 2009, pp. 195–200.

Olivier, R., Marchand, P., and Chenu, S., On the use of electromagnetic simulation in front door radio frequency interference. 2017 International Conference on Military Communications and Information Systems (ICMCIS), Oulu, 2017, pp. 1–5.

Smith, W. and Palafox, G., Electromagnetic interference among cables and antennas on military ground vehicles. 2016 IEEE/ACES International Conference on Wireless Information Technology and Systems (ICWITS) and Applied Computational Electromagnetics (ACES), Honolulu, Hawaii, 2016, pp. 1–2.

Sreenivasa Rao, M., EMI/EMC effects on EW receiver systems of military aircraft. *Proceedings of the 10th International Conference on Electromagnetic Interference & Compatibility*, pp. 63–67, 2008.

Weinmann, F., Knott, P., and Vaupel, T., EM simulation of installed antenna performance on land, aerial and maritime vehicles. 2013 IEEE Antennas and Propagation Society International Symposium (APSURSI), Orlando, Florida, 2013, pp. 2179–2180.

https://www.mathworks.com/products/matlab.html.

Part III

United States Army's Techniques for Spectrum Management Operations

9

Overview

This chapter introduces the frequency spectrum, provides an overview of spectrum management operations, and describes the core functions related to spectrum management operations within the context of Army operations. This chapter also provides an overview of spectrum management operations task.

9.1 Electromagnetic Spectrum

9.1. The electromagnetic spectrum is a continuum of all electromagnetic waves arranged according to frequency and wavelength. Multiple radiated signals coexist in the same physical space and selectively detected using the appropriate equipment and channel. The spectrum extends from below the frequencies used for radio (at the long-wavelength end) through gamma radiation (at the short-wavelength end). Divided into alphabetically designated bands for specific wavelengths and frequency ranges, the spectrum encompasses wavelengths from thousands of kilometers to a fraction of an atom. Radio signals are able to coexist in the same physical space. Radio frequency spectrum is the continuum of frequencies of electromagnetic radiation from 3,000 Hertz (Hz) or 3 kilohertz (kHz) to 300 gigahertz (GHz). Isolation of multiple users of spectrum is possible by allocating different bands of this continuum to them.

Note. See Appendix C for an overview of spectrum physics.

9.1.1 Constrained Environment

9.2. Gaining and maintaining control of the electromagnetic spectrum is a critical requirement for the commander. From communications, to intelligence collection, to electronic warfare, all forces, and supporting agencies depend on the electromagnetic spectrum to execute operations in the air, land, maritime, space, and cyberspace domains. Within the electromagnetic spectrum, joint forces contend with civil agencies, commercial entities, allied forces, and adversaries for use of a common electromagnetic spectrum resource. This demand for electromagnetic spectrum use results in a constrained, congested, and contested environment that affects operations across all domains and functions. This contention and competition produces a constrained environment regarding how, when, and where to use electromagnetic spectrum resources.

9.3. Congestion in the electromagnetic spectrum results when multiple users attempt to use the same portions of the spectrum simultaneously. This competition and congestion can potentially lead to the operational failure of systems during critical missions due to electromagnetic interference. Adversaries can exploit modern technologies to develop sophisticated electronic attack capabilities, contesting the ability of all military assets to access and use the electromagnetic spectrum.

9.4. Army spectrum managers' tasks include planning, managing, coordinating, and providing policies and regulations for the use of the electromagnetic spectrum. The Army shares spectrum related resources with other Services, civilian counterparts, and friendly forces. Due

Dynamic Spectrum Access Decisions: Local, Distributed, Centralized, and Hybrid Designs, First Edition. George F. Elmasry.
© 2021 John Wiley & Sons Ltd. Published 2021 by John Wiley & Sons Ltd.
Companion website: www.wiley.com/go/elmasry/dsad

to the large quantity of devices and forces using the spectrum, portions maybe unavailable. Environmental factors such as solar activity and weather can adversely affect SMO. The Army spectrum manager uses knowledge and spectrum management tools to determine how to best support a mission with limited spectrum resources. Solutions can be as simple as having a unit switch to a different frequency or as complex as adjusting the entire spectrum plan. There are times when the commander prioritizes spectrum use to conduct operations.

9.1.2 Spectrum Dependent Devices

9.5. The spectrum manager is the commander's resident expert who provides course of action (COA) recommendations based on software modeling and simulation to mitigate spectrum use problems. The spectrum manager is vital to ensuring all spectrum dependent devices (SDD) operate as intended without suffering or causing harmful interference. Devices that utilize the electromagnetic spectrum to emit, receive, monitor frequencies are referred to as SDDs. SDDs include any conceptual, experimental, developmental, operational transmitter, receiver, or device that uses any portion or part of the electromagnetic spectrum.

9.6. SDD systems include, but are not limited to transmitter, receivers, command and control systems and platforms, electronic warfare assets, sensors, beacons, navigational aids, radio and radio systems, radar systems, remote controlled robotic equipment, manned and unmanned aircraft systems. Electromagnetic interference is any electromagnetic disturbance that interrupts, obstructs, or otherwise degrades or limits the effective performance of

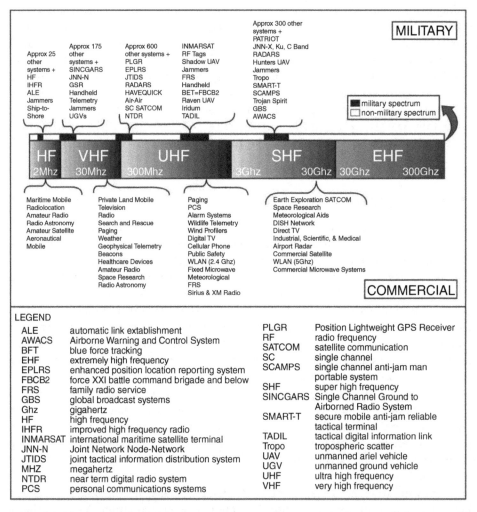

Figure 9.1 Electromagnetic spectrum competition. From Techniques for Spectrum Management Operations, Dec 2015. Headquarters, Department of the Army.

electronics or electrical equipment. It can be induced intentionally, as in some forms of electronic warfare, or unintentionally, as a result of spurious emissions and responses and intermodulation products.

   Note. See CJCSM 3320.02F for further information on electromagnetic interference.

   9.7.   Spectrum users should understand that it is not a replaceable resource like fuel or ammunition. Once the allotted spectrum to support a specific capability or system is in use, it is no longer available for use depending upon system and environmental variables. The commander may need to operationally assess the impact of sacrificing other potentially critical capabilities to ensure the use of another spectrum dependent user. Spectrum management operations are the oversight of all characteristics of electromagnetic radiation. The goal is to protect systems from harmful interference while allowing the optimum use of the spectrum. The process is complex since the characteristics of electromagnetic radiation vary with time, space, and frequency.

   9.8.   Figure 9.1, on page 9.3, displays a portion of the spectrum used by various systems and devices both commercial and military that compete for these bands (acronyms in graphic are not essential to understanding the text).

9.2   Definition

9.9.   SMO are the interrelated functions of spectrum management, frequency assignment, host nation coordination, and policy that together enable the planning, management, and

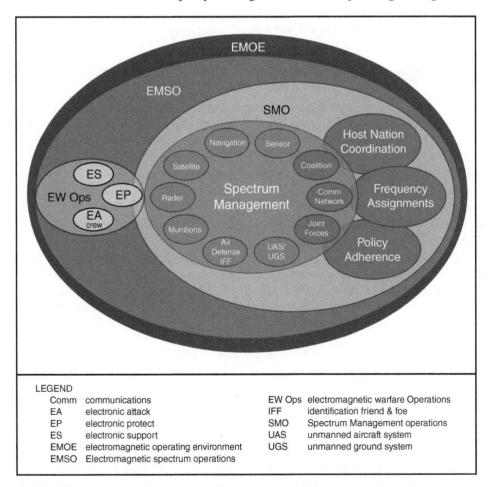

Figure 9.2 The electromagnetic operational environment (EMOE). From Techniques for Spectrum Management Operations, Dec 2015. Headquarters, Department of the Army.

execution of operations within the electromagnetic operational environment (EMOE), during all phases of military operations. The portions of the electromagnetic spectrum, experienced and influenced by military operations is the EMOE. Electromagnetic spectrum operations include electronic warfare for which the Cyber Center of Excellence is the proponent. Electromagnetic spectrum operations includes all activities in military operations to control the spectrum. SMO is the management portion of electromagnetic spectrum operations. Figure 9.2, on page 9.4, depicts the various areas of responsibility as they pertain to EMOE.

9.10.　Army spectrum managers coordinate and collaborate with spectrum managers working in joint environments. Collaboration with joint personnel is common and necessary for the Army spectrum manager while using the highly saturated and limited spectrum available. The primary goal of joint electromagnetic spectrum operations is to enable SDD to perform their functions in the intended environment without causing or suffering unacceptable interference. Joint electromagnetic spectrum operations are those activities consisting of electronic warfare and joint electromagnetic spectrum management operations used to exploit, attack, protect, and manage the electromagnetic operational environment to achieve the commander's objectives.

9.11.　SMO is a supporting function or enabler for many of the Army unified land operations. SMO is a primary component of CEMA, which consists of cyberspace operations, electronic warfare, and spectrum management operations. CEMA enables the management of the electromagnetic spectrum in support of mission command. These activities employ the same technologies, capabilities, and enablers to accomplish assigned tasks resulting in the commander's integration and synchronization across all command echelons and warfighting functions as part of the operations process.

Note. Refer to ADRP 3-0 for further information.

9.3　Objective

1-12.　The objective of Army SMO is to ensure access to the electromagnetic spectrum in support of users conducting the Army's operational missions. SMO enables the allotment of the vital, but limited, natural resource that directly supports operational forces throughout the world. The Army is dependent upon the use of the radio frequency spectrum at all levels of unified land operations. An effective spectrum management operations program enables electronic systems to perform their functions in the intended environment without causing or suffering unacceptable performance.

9.13.　Commanders must have the ability to see the use of their assigned spectrum resources so they can apply systematic management controls in the logistics and mission command arenas. The electromagnetic spectrum is a vital warfighting resource that requires the same planning and management as other critical resources such as fuel, water, and ammunition. Spectrum managers, with the appropriate expertise and tools, ensure that commanders have adequate knowledge of the utilization of the frequency spectrum to make decisions that positively influence accomplishment of their missions.

9.4　Core Functions

9.14.　The SMO core functions determine the tasks and requirements of the Army spectrum manager. These four functions are—

- **Spectrum Management:** Spectrum management is the planning, coordinating, and managing of joint use of the electromagnetic spectrum through operational, engineering, and administrative procedures. Spectrum management consists of evaluating and mitigating electromagnetic environmental effects, managing frequency records and databases, de-conflicting frequencies, frequency interference mitigation, allotting frequencies, spectrum supportability assessments, and electronic warfare coordination to ensure SDD operate as intended.

- **Frequency Assignment:** The request and issuance of authorizations to use frequencies for specific equipment such as combat net radio and Army common user systems is a task of frequency assignment. This also includes the planning necessary for combat net radio, Army common user systems, and associated systems. Examples of frequency assignment are assigning the frequencies necessary to generate single-channel ground and airborne radio system (SINCGARS) hopsets, providing frequencies for unmanned aerial systems and line of sight networks, or assigning frequencies for the Warfighter Information Network-Tactical (WIN-T) network.

- **Host Nation (HN) Coordination:** Each nation has sovereignty over its electromagnetic spectrum within its geographic area and negotiates the use of the spectrum on a case-by-case basis. A representative of the sovereign country evaluates each Department of Defense (DOD) request for the use of spectrum based on the perceived potential for electromagnetic interference (EMI) to local receivers. Use of military or commercial spectrum systems in host nations requires coordination and negotiation that result in formal approvals and certifications.

- **Policy Adherence:** The commanders ability to access and maneuver within the electromagnetic spectrum is dependent on policy. Policy are those authoritative instruments from the national strategic through the tactical level that nest and shape the spectrum management, frequency assignment, and host nation coordination process. Countries coordinate global international spectrum use through the International Telecommunications Union and the World Radio Communication Conference. At the U.S. national level under U.S. Code Title 47, the division of spectrum management responsibility rests with the National Telecommunications and Information Administration (NTIA) for federal frequencies and the Federal Communications Commission for non-federal frequencies. The Military Communications-Electronics Board (MCEB) is the main coordinating body for spectrum matters among DOD components. Overseas, the U.S. mission, working with DOD strategic partners, negotiates treaties and agreements when stationed or training U.S. forces are within a host nation. These agreements establish lines of communications between the host-nation and senior military commands to negotiate spectrum usage in support of training and operations. Examples of policy instruments include International Telecommunications Union and World Radio Communication Conference agreements, status of forces agreements, host-nation agreements, operational orders, U.S. Code Title 47, and operations plans.

9.5 Army Spectrum Management Operations Process

9.15.   The Army SMO process comprises three interacting and continuous activities: planning, coordinating, and operating (see figure 9.3, on page 9.6). The Army SMO process is a means of planning that continues through all phases of the mission. During the execution of unified land operations, these functions occur concurrently.

9.5.1   Planning

9.16.   SMO planning includes the identification of spectrum requirements for training, pre-deployment, deployment, and reconstitution of Army forces, both in and outside the continental U.S. SMO planning is an on-going process that must be deliberate as well as dynamic to support unified land operations. It requires the collection, storage, and protection of critical spectrum data, and assured access to this data by spectrum planners on a global scale. Additionally, planning for the establishment of lines of communications for coordination of

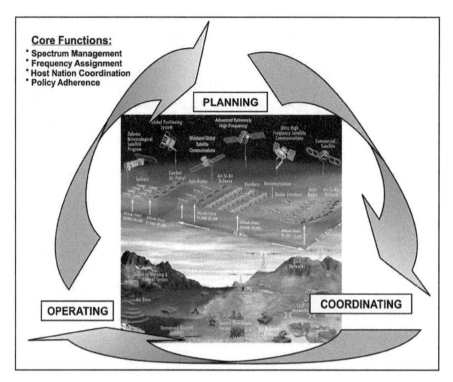

Figure 9.3 Army spectrum management operations process. From Techniques for Spectrum Management Operations, Dec 2015. Headquarters, Department of the Army.

spectrum use with national and international government and non-government agencies is critical to the spectrum planning process.

9.17.　The CEMA element and the CEMA working group have an assigned spectrum manager that provides expertise in planning and coordinating horizontally and vertically to support unified land operations. The spectrum manager's primary role is to assist with de-conflicting detection and delivery assets through the planning and targeting processes.

9.5.2　Coordinating

9.18.　Coordination ensures initial spectrum availability and supportability for operations. Lines of communication for coordinating spectrum allocation at the national and international level are primarily a matter of policy established in the planning process. Enemy nations or their military do not receive U.S. host nation coordination. Spectrum managers coordinate adjacent countries spectrum, particularly if forces stage, train, or operate within these countries, to include airspace, sovereign waters, and frequencies for satellites. Coordination at the operational Army level requires prior coordination as well as a dynamic, instantaneous collaboration tool.

9.19.　**Staff Coordination**-Spectrum managers coordinate with various staff sections to ensure effective SMO. Commanders engage spectrum managers early in the planning process when forecasting for the use of spectrum dependent devices. Staff coordination, electronic warfare (EW) coordination, communications security coordination, satellite coordination, frequency deconfliction, frequency interference resolution, and joint restricted frequency lists are SMO tasks that support spectrum functions.

9.20.　Spectrum managers work with many systems that are not exclusively communications systems. They must interact with other staff members to provide guidance, assistance, and advice to the commander regarding the use and prioritization of the spectrum. Systems such as unmanned aerial systems, common user jammers, radars, navigational aids, and sensors all use the spectrum for operation. Their extensive use and unique operating characteristics necessitate special planning and coordination to mitigate frequency fratricide.

9.21. **Unified Action Partners**-Coordinating spectrum use is the process of collaborating with unified action partners. Unified action partners are those military forces, governmental and nongovernmental organizations, and elements of the private sector with which Army forces plan, coordinate, synchronize, and integrate during the conduct of operations. This function ensures initial spectrum availability and supportability for operations. Lines of communication for coordinating spectrum allocation at the national and international levels are primarily a matter of policy established in the planning process.

9.22. **Host Nation Coordination**-Use of military or commercial spectrum systems in host nations requires coordination and negotiation that result in formal approvals and certifications. Coordination for use of the spectrum in host nations is required if forces stage, train, or operate within these countries to include airspace, sovereign waters, and frequencies for satellites. Prior coordination as well as dynamic, immediate collaboration tools results in a seamless use of the spectrum. Failure to request frequency usage in a timely manner results in the inability to operate communications equipment in the host nation. Each nation has sovereignty over its spectrum within its geographic area and negotiates the use of spectrum on a case-by-case basis. A representative of the sovereign country evaluates each DOD request for the use of spectrum based on the perceived potential for EMI to local receivers.

9.23. The host nation spectrum worldwide database online (HNSWDO) is a tool, used by military service department spectrum management offices, to track DOD host nation spectrum supportability request to determine equipment supportability. Host nation access request are added to HNSWDO by the sponsoring service spectrum management agency. Requests are sent to the respective combatant command's joint frequency management office (JFMO) to annotate comments in HNSWDO for visibility. Tactical spectrum managers coordinate frequency assignments through established spectrum coordination channels. Spectrum management offices assuming the role of the Joint Spectrum Management Element (JSME) may be delegated by the combatant command JFMO to perform person-to-person host nation coordination in support of joint task force operations.

9.24. **Electronic Warfare Coordination**-The spectrum manager should be an integral part of all EW planning to provide awareness of spectrum conflicts initiated by friendly systems for personnel protection, enemy exploitation, or enemy denial. The advent of common user "jammers" has made this awareness and planning critical for the spectrum manager. In addition to jammers, commanders and staffs must consider non-lethal weapons that use electromagnetic radiation. EW coordination normally takes place in the CEMA working group. It may take place in the EW Cell if it is operating under a joint construct or operating at a special echelon.

9.25. **Communications Security Coordination**-Spectrum managers work closely with communications security personnel to ensure the proper keying material for the appropriate frequency resource of SINCGARS loadsets. Spectrum managers only manage and process communications security for SINCGARS by way of loadsets. They do not manage communications security for other emitters.

9.26. **Satellite Coordination**-Spectrum managers coordinate with satellite managers to maintain awareness of channels (frequencies) used by satellite communications systems. The satellite manager generates and processes satellite access requests. Spectrum managers receive and verify the information provided in the satellite access request for all satellite communications. Once approved, the spectrum manager enters the frequencies into the spectrum database for frequency deconfliction with all other emitters in the area of operations.

9.27. **Frequency Deconfliction**-Frequency deconfliction is a systematic management procedure to coordinate the use of the electromagnetic spectrum for operations, communications, and intelligence functions. Frequency deconfliction is one element of electromagnetic spectrum management and applies practices to minimize or prevent spectrum dependent devices from suffering or causing interference while being used as intended. It is easy to confuse EMI mitigation with frequency de-confliction. The main difference is that frequency deconfliction occurs during the planning phase of a mission while EMI mitigation occurs during mission execution.

9.28.   **Joint Restricted Frequency List (JRFL)**-The JRFL is a concise list of highly critical protected frequencies and nets categorized as Taboo, Protected, and Guarded. Commanders and planners prohibit jamming or attacking frequencies listed on the JRFL. The JRFL includes command channels of senior commanders and safety-of-life frequencies used by local civilian noncombatants. Usually listed in the JRFL are international distress, safety, and controller frequencies.

9.29.   High priority nets, bands, and frequencies are protected from friendly electronic attack (EA) when possible however, the concern of the spectrum manager is to ensure that all friendly systems have the ability to operate unimpaired. This can be accomplished by simply adding the offending jammer to a database and using spectrum management techniques (such as changing frequencies, assignments, or moving to an unaffected area) to accomplish the mission. The spectrum manager has tools that can identify potential frequency fratricide if properly utilized, ultimately saving lives. Refer to paragraph 5-36 and appendix A for further information on the JRFL.

> *Note.* Use of the JRFL will not deconflict all frequency issues. The JRFL does not provide communications planners with frequencies EA systems transmit or the technical information needed to deconflict EA from friendly operations including lower echelon maneuver forces. Efficient utilization of spectrum management tools identifies potential interference and frequency conflicts during mission planning reducing frequency fratricide.

9.30.   **Interference Resolution**-The spectrum manager performs interference resolution at the echelon receiving the interference. Interference is the radiation, emission, or indication of electromagnetic energy; either intentionally or unintentionally causing degradation, disruption, or complete obstruction of the designated function of the electronic equipment affected. The spectrum manager should utilize available near-real time monitoring and analysis capabilities to aid in the interference resolution. The reporting end user is responsible for assisting the spectrum manager in tracking, evaluating, and resolving interference. Appendix D contains further information on frequency interference resolution and reporting.

9.5.3   Operating

9.31.   The operating activity for SMO enables and sustains the functions of planning and coordinating. It includes the process to plan, conduct, coordinate, and sustain spectrum operations. SMO ensures the efficient use of allocated spectrum and associated frequencies in a given area of operations. Spectrum managers use the operating function to enable dynamic, near instantaneous frequency assignment, re-assignment, interference mitigation, and frequency deconfliction across all users in an area of operations. The architecture provides for interoperability with U.S. national, local government and non-government agencies as well as unified action partners.

10

Tactical Staff Organization and Planning

SMO is dynamic and requires continuous coordination among all echelons and warfighting functions both laterally and horizontally to mitigate harmful interference. This chapter describes SMO functions for staff organizations at the corps and below level, and provides an overview of division, brigade and battalion spectrum operations. This chapter also describes how SMO is incorporated within the military decisionmaking process and shows how the spectrum manager supports the common operational picture.

10.1 Spectrum Management Operations for Corps and Below

10.1. The goal of tactical SMO is to protect and provide access to the spectrum so that it serves the needs of friendly forces. Spectrum operations at the tactical level can be a very complicated and time-consuming process.

10.2. In the past, the bulk of spectrum management was concerned with networked communications emitters and combat net radio networks. Today, the tactical environment includes a vast number of SDD operating in all regions of the spectrum across the battlefield. The key to sound spectrum management is having an understanding of all emitters and receivers in the operational area while being able to deconflict these systems. As stated earlier, the commander must be aware that the spectrum is a limited resource and that efficient spectrum use is critical to enabling the warfighting functions.

> **Note.** SMO is bottom driven for requirements while top fed for resources. The brigade combat teams represent the pointy end of the spear and it is critical that the staff at each echelon captures all requirements to ensure commanders receive the proper resources. Maximizing the use of the spectrum requires coordination between EW, network operations, intelligence staffs, and other known users.

10.3. Figure 10.1, on page 10.2, illustrates the competing systems that cause challenges throughout the spectrum. The assistant chief of staff for communications, signal staff officer (G-6) or the battalion or brigade signal staff officer (S-6) is responsible for coordination with all spectrum users within a given operational area, to identify all requirements for spectrum access, and to conduct frequency deconflictionThey also maintain a database of all known emitters and receivers in the operational area to identify and prioritize competing systems for frequency assignments.

10.1.1 Corps Spectrum Operations

10.4. There are three spectrum managers within the G-6 staff and one located in the assistant chief of staff, operations section (G-3). The spectrum management chief is the principle advisor to the commander for spectrum management related matters and is the Army spectrum authority in a corps operational area. Two other spectrum managers assist the spectrum

Dynamic Spectrum Access Decisions: Local, Distributed, Centralized, and Hybrid Designs, First Edition. George F. Elmasry.
© 2021 John Wiley & Sons Ltd. Published 2021 by John Wiley & Sons Ltd.
Companion website: www.wiley.com/go/elmasry/dsad

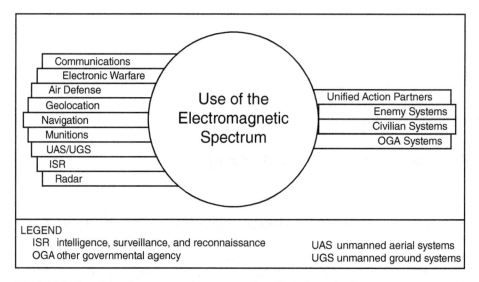

Figure 10.1 Use of the electromagnetic spectrum. From Techniques for Spectrum Management Operations, Dec 2015. Headquarters, Department of the Army.

management chief in performing corps spectrum management duties. Normally, one assists the network planners and the other manages the signal operating instructions (SOI) and other functions that fall outside of the network. The spectrum manager assigned to the G-3 conducts electronic warfare deconfliction, coordination and planning and advises the Electronic Warfare Officer (EWO) on potential spectrum conflicts and issues. The corps spectrum management chief may be designated as the joint task force spectrum manager if the Army is the lead service in a joint operation.

10.5.   At the corps, spectrum operations place more emphasis on host nation coordination, establishing policy and procedure to assure the necessary spectrum is available for operations, and ensuring subordinate units efficiently use spectrum resources. The spectrum manager accomplishes this through the development of standard operating procedures based on joint and service regulations, instructions, policies, and doctrine.

10.6.   The corps spectrum manager's responsibilities include the following—

- Assist in the development and publishing of communications annexes and appendices.
- Develop, produce, and disseminate spectrum operations standard operating procedures.
- Create and send a data call message (see appendix A for a description of data call message).
- Determine if the unit's devices have spectrum supportability.
- Coordinate host nation spectrum use.
- Develop, create, and distribute the SOI.
- Coordinate and participate with other staff sections and cells.
- Coordinate with unified action partners.
- Perform prescribed reviews for frequency use and requirements.
- Provide frequency-engineering support for communications network design and operations.
- Conduct EMI identification, analysis, mitigation, and reporting.
- JRFL production and promulgation.
- Advise the EWO on potential spectrum conflicts and issues.
- On order, assume the roles and responsibilities of the joint task force joint spectrum management element (JSME).

Note. See CJCSM 3320.01 for JSME responsibilities.

10.2   Division, Brigade and Battalion Spectrum Operations

10.7.   The roles of division and brigade spectrum managers are similar at their respective levels. The brigade spectrum manager gathers, validates, and forwards requirements for all spectrum support to the division. In turn, the division forwards its requirements to the next higher authority.

10.8.   There are two spectrum managers at the G-6 and one in the G-3 per division. Normally, one spectrum manager is responsible for the network frequency assignments to include satellite access authorization deconfliction. The network planners design the network that determines the spectrum requirement and the spectrum manager uses this design to request the spectrum requirements necessary for the communications network. Another spectrum manager is responsible for combat net radio, radar, and other systems requirements. The G-3 spectrum manager provides support for fires, EW deconfliction, and coordination and planning.

10.9.   Brigade spectrum managers are located in the S-6 and S-3 to maintain visibility of all spectrum related matters in the brigade. The brigade and in some instances the battalion (selected maneuver units) is currently the lowest echelon to have a spectrum manager. Echelons lower than brigades or battalions coordinate their spectrum requirements and concerns through the brigade spectrum manager.

10.10.   The division, brigade or battalion spectrum manager's responsibilities include the following—

- Advise the commander in spectrum prioritization and implementation.
- Build and distribute SINCGARS loadsets.
- Request, obtain, and distribute frequencies for all devices.
- Perform spectrum network analysis to engineer line of sight radio links and assign frequencies.
- Advise network planners in matters concerning spectrum management.
- Maintain and update spectrum related databases.
- Advise and coordinate with EW personnel for frequency planning and use.
- Perform spectrum analysis and frequency deconfliction.
- Coordinate satellite frequency deconfliction.
- Conduct electromagnetic interference identification, analysis, mitigation, and reporting.
- Coordinate with Army Aviation Units to determine and mitigate interference.
- Perform situational awareness and analysis using a spectrum analyzer or monitoring receiver.
- Perform propagation analysis for high frequency and tropospheric scatter systems.
- Assist in JRFL production and promulgation.
- Assist in spectrum supportability determinations.
- Develop and maintain the EMOE picture by capturing and recording all unit SDDs with appropriate tools and databases.
- On order, the Division SMO assumes the roles and responsibilities of the joint task force J-6 JSME as required.

   Note. See CJCSM 3320.01 for JSME responsibilities.

10.3   Spectrum Managers Assigned to Cyber Electromagnetic Activity Working Group

10.11.   When established, the CEMA working group is accountable for integrating CEMA and related actions into the concept of operations. CEMA working groups do not add additional structure to an existing organization. The CEMA working group is a collaborative staff meeting led by the EWO to analyze, coordinate, and provide recommendations for a particular purpose, event, or function.

10.12.   A spectrum manager's inherent duties include many affiliations and activities based on their assignment. Spectrum managers participate in CEMA working groups or CEMA elements as required by the command. As a member of these groups, they provide the specialized technical knowledge to enable the working group or element to provide the commander with expert knowledge on spectrum related activities.

10.13.   The CEMA working group is responsible for coordinating horizontally and vertically to support unified land operations and primarily deconflict detection and delivery assets through the planning and targeting processes. (FM 3-38) Staff representation within the CEMA working group may include the G-2 (S-2), information operations officer, battalion or brigade civil affairs operations staff officer assistant chief of staff, civil affairs operations, fire support officer, space support element, judge advocate general representative (or appropriate legal advisor), and a joint terminal attack controller when assigned. Based on requirements capabilities, the CEMA working group staff may delete or modify members. The CEMA working group augments the function of the permanently established CEMA element. When scheduled, the CEMA working group is a critical planning event integrated into the staff's battle rhythm.

10.14.   The CEMA working group requires a spectrum manager positioned within the working group to deconflict spectrum, identify conflicts, and mitigate possible frequency fratricide during the planning phase of all forms of fire. Frequency fratricide is the unintentional interruption of friendly frequencies. Frequency fratricide can cause many problems for operations and prevention is the key. Spectrum managers provide the working group with frequency options and advice that follows internal and external policies that minimize frequency fratricide.

10.15.   To accomplish the task of integration of CEMA into all unit operations, the EWO leads the CEMA working group, which determines EW requirements and integrates these requirements into the unit's planning and targeting processes. One role of the EW team in CEMA is to coordinate the operational targeting of effects in cyberspace.

10.4   Cyber Electromagnetic Activities Element

10.16.   The CEMA element consists of personnel that plan, prepare, and synchronize cyberspace operations, EW, and SMO. The element, led by the electronic warfare officer (EWO), provides staffs expertise for the planning, integration, and synchronization of cyberspace operations, EW, and SMO. When the mission dictates, the CEMA element can leverage other additional skill sets of the CEMA working group. When operating in a joint, multinational, or intergovernmental environment, commanders may reorganize their staffs to better align with higher headquarters. The CEMA element is an organic organization in brigade, division, corps, and theater Army staffs (FM 3-38).

10.17.   The key personnel involved in planning and coordination in the CEMA element are the—

- EW staff.
- Spectrum manager.
- Assistant chief of staff, intelligence (G-2) or battalion or brigade intelligence staff officer (S-2).
- Assistant chief of staff, signal (G-6 [S-6]) staff.

10.4.1   Electronic Warfare Staff

10.18.   The EWO functions as the commander's designated staff officer for the planning, integration, synchronization, and assessment of CEMA and uses other members of the staff to integrate CEMA into the commander's concept of operations. The EWO is responsible for understanding all applicable classified and unclassified policy relating to cyberspace, EW, and electromagnetic spectrum to properly inform the commander on the proper planning, coordination, and synchronization of CO, EW and SMO. The EWO is the commander's subject matter expert on CREW.

10.4.2 Spectrum Manager

10.19.   As a key member of the CEMA element, the spectrum manager coordinates spectrum use for a wide variety of communications and electronic resources. Some of the primary functions the spectrum manager provides include—

- Coordinates the preparation of the JRFL and issuance of emissions control guidance.
- Coordinates frequency allotment, assignment, and use.
- Coordinates electromagnetic deception plans and operations in which assigned communications resources participate.
- Coordinates measures to eliminate, moderate, or mitigate electromagnetic interference.
- Coordinates with higher echelon spectrum managers for electromagnetic interference resolution they cannot resolve internally.
- Assists the EWO in issuing guidance to the unit (including subordinate elements) regarding deconfliction and resolution of interference problems between EW systems and other friendly systems.
- Participates in the CEMA working group to deconflict friendly spectrum requirements with planned EW, CO, and intelligence collection.
- Synchronizes frequency allotment and assignment use with the G-6 or J-6 spectrum manager.

10.20.   The working groups may include, but are not limited to, key members of operations, intelligence, communications, training, air liaison officer, fires, special technical operations, and liaisons from supported units. Figure 10.2. CEMA depicts the working group organizational framework.

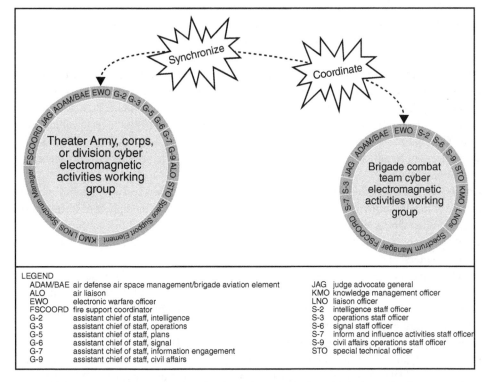

LEGEND

ADAM/BAE	air defense air space management/brigade aviation element	JAG	judge advocate general
ALO	air liaison	KMO	knowledge management officer
EWO	electronic warfare officer	LNO	liaison officer
FSCOORD	fire support coordinator	S-2	intelligence staff officer
G-2	assistant chief of staff, intelligence	S-3	operations staff officer
G-3	assistant chief of staff, operations	S-6	signal staff officer
G-5	assistant chief of staff, plans	S-7	inform and influence activities staff officer
G-6	assistant chief of staff, signal	S-9	civil affairs operations staff officer
G-7	assistant chief of staff, information engagement	STO	special technical officer
G-9	assistant chief of staff, civil affairs		

Figure 10.2 CEMA working group organizational framework. From Techniques for Spectrum Management Operations, Dec 2015. Headquarters, Department of the Army.

10.5   Tips for Spectrum Managers

10.21.   The following tips provide guidance to spectrum managers for further understanding of the types and number of emitter devices in their unit. This information leads to a firm understanding of the spectrum requirements and mitigates the need to request resources multiple times.

- Obtain a detailed in-briefing from the outgoing spectrum manager.
- Meet and establish a rapport with other staff entities, such as electronic warfare, intelligence, communications, operations, cyber, and logistics. This builds avenues for coordination during mission planning and execution.
- Identify all devices by analyzing the modified table of organization and equipment and other documents that contain equipment lists.
- Understand the requirements of the unit's devices by understanding their mission.
- Provide recommendations to the commander regarding unauthorized frequency use by devices causing interference to certified SDD and suspending or modifying spectrum use.
- Visit all entities within the unit and ask what types of emitter devices they have. This can often identify devices that are not in current databases.
- All emission devices must have a completed DD Form 1494 to operate in the operational area of the unit. Completion of the DD Form 1494 is the responsibility of the material developer. The material developer must provide a collection of technical data about the device to begin the planning process by placing the technical data in SMO tool databases. Report the characteristics to higher echelon spectrum management agencies to receive authorization for using the device in the operational area.
- Meet with unit commanders to discuss spectrum management options. This provides the commander with valuable information to incorporate in the military decisionmaking process (MDMP) and provides a point of contact for spectrum concerns.
- Maintain a library, either paper or electronic, of spectrum related manuals, to include national, international, and governmental regulations and policies. An excellent start to this library includes the manuals listed in the reference section of this text.
- Become familiar with the operational area for the unit. In particular, know the agencies, national and international, that regulate spectrum use and obtain their contact information.
- Become familiar with unit SMO tools and develop databases for them. Build force templates that include spectrum devices and spectrum requirements to aid in mission planning. Obtain map files, such as digital terrain elevation data for the operational area of the unit.
- Begin an informal handbook relating to spectrum management functions specific to the unit. The goal of the handbook is to serve as a reminder for completing the same tasks in the future. The handbook provides for an excellent start to the in briefing to future incoming spectrum managers.
- Collaborate with unit the G-6 signal officer and unit commanders to discuss policy impacts on use of locally procured equipment and the ability to operate in support of training and operational environment.
- Understand command and policy relationships between combatant commands and the joint force providers.
- Development or acquisition of systems that meet operational requirements, but fail to obtain formal spectrum supportability are not allowed operation in the U.S. or in host nations. These systems create the potential for severe mutual interference between the system and other spectrum users, to squander resources and delay fielding war fighting capabilities to units. An approval number, documented on the emitters DD Form 1494, indicates approval to coordinate for spectrum resources. The joint spectrum center database web server contains approved DD Form 1494s. Completion and submission of an emitter's DD Form 1494 is conducted thru the sponsoring military department, further coordination is the responsibility of the material developer.

Note. For more information on the Army spectrum supportability process, see AR 5-12.

- Home station operations provide unique challenges and opportunities for tactical spectrum managers. The challenges come in the form of equipment returning from overseas theaters that does not have spectrum supportability within U.S. territories.
- Another challenge is that the spectrum manager on a post, camp, or station now normally deals with a civilian counterpart for spectrum support who may not be as familiar with tactical requirements. Within these challenges lie opportunities for training with other units, agencies, and directorates on the installation. Informal luncheons or meetings are a good way to share information and lessons learned from previous experiences.

  Note. According to the MCEB Publication 8, complete the spectrum supportability process for all devices (emitters and receivers) as early as possible to prevent the use or acquisition of a device that is unsupportable, cause interference, or not authorized for use within the operational area of a mission.

10.6   The Military Decisionmaking Process

10.22.   The MDMP is an iterative planning methodology to understand the situation and mission, develop a course of action, and produce an operation plan or order. The MDMP is the Army's analytical approach to problem solving. The MDMP is a tool that assists the commander and staff in developing estimates and a plan.

10.23.   SMO has inputs to each step in the MDMP. The MDMP produces the greatest integration, coordination, and synchronization for an operation and minimizes the risk of overlooking a critical aspect of the operation. The complete MDMP results in a detailed operation order or operation plan. The disadvantage of using the complete MDMP is that it is a time-consuming process. For further information concerning the MDMP please review FM 6-0, chapter 9.

10.24.   Key inputs for the MDMP are actions, processes or information spectrum managers provide to the MDMP. SMO key outputs for MDMP are the completed SOI, reports, frequency proposals or data call messages. Figure 10.3, on page 10-8, depicts the key SMO inputs and outputs for each step of the MDMP.

10.7   Support to the MDMP Steps

10.25.   SMO supports the MDMP through each step of the process. The SMO planning process incorporates each step of the MDMP in to support the commander. The following are some examples for each step—

- **Step 1: Receipt of Mission—**
  - ➢ Conducting a data call provides a list of SDD and the requirements that those devices need to perform the mission.
  - ➢ Compiling force structure templates allows the commander to determine the amount and type of SDD available for the mission.
  - ➢ Modeling the operational area, using SMO tools, with digital topography and electromagnetic environmental effects information to analyze spectrum supportability.
  - ➢ Determine from governmental and host nation spectrum allocation tables, which frequencies maybe assigned in a given operational area.
  - ➢ Compiling restrictions or constraints on spectrum use that prevent planning and use of protected, taboo, and restricted frequencies in the operational area. (see CJCSM 3320-01C, appendix I, enclosure C for a listing of the worldwide taboo frequencies.
  - ➢ Defining the EMOE provides a common source for spectrum use information, particularly all available blue (friendly), red (enemy), and grey (neutral and civil) spectrum occupancy.
- **Step 2: Mission Analysis—**
  - ➢ Analyze the EMOE, highlighting unified action partners' spectrum users, and aid the commander in determining spectrum priority.

Key SMO Inputs	Steps	Key SMO Outputs
• Updated EMS databases • Unit force structure • Library of EMS documents • HN allocation tables • Gather SMO tools	Step 1: Receipt of Mission	• Defined EMOE • Data call message • Identify EMS constraints • JFRL guidance
• Identified EMS capabilities pertaining to combat power • List of unit's spectrum dependent device • Frequency requests • JRFL requests	Step 2: Mission Analysis	• Prioritized EMS use • Completed JRFL • Frequency reuse plans • Initial EMS risk assessment
• Commander's intent • Frequency allotments • Initial frequency assignments • SSA/DD-1494 for unit's spectrum dependent device	Step 3: Course of Action (COA) Development	• M&S of EMS to develop multiple COAs • EMI/EW deconfliction • Initial Spectrum Plan • EMS COP
• Initial Spectrum Plan • Mitigating factors to decrease EMS risk	Step 4: COA Analysis (War Game)	• M&S shows EMS advantages / disadvantages for each COA • Continues analysis of EMS risk assessment
• Optional unit movement routes for planning COTM • Refines EMS COAs	Step 5: COA Comparison	• M&S depicts EMS use to compare COAs • Recommended EMS COAs
• Recommended EMS COA • Coordinated frequency conflicts	Step 6: COA Approval	• Commander selected EMS COA and any modifications • Frequency proposals
• Frequency assignments/allotments from higher echelon ESM • HN frequency clearance • CREW loadsets	Step 7: Orders Production, Dissemination and Transition	• The Spectrum Plan • SOIJCE01 • Annex H of OPORD • Distribute frequency assignments to requestors • CNR loadsets

LEGEND

CNR	combat net radio	HN	host nation
COP	common operational picture	JCEOI	joint communications-electronics operating instructions
COTM	communications on the move		
CREW	counter radio-controlled improvised explosive device electronic warfare	JRFL	joint restricted frequency list
		M&S	modeling and simulation
EMI	electromagnetic interference	OPORD	operational order
EMOE	electromagnetic operating environment	SMO	Spectrum Management operations
EMS	electromagnetic spectrum	SOI	signal operating instructions
EW	electronic warfare		

Figure 10.3 Key SMO inputs to the MDMP. From Techniques for Spectrum Management Operations, Dec 2015. Headquarters, Department of the Army.

➢ Conducting an initial spectrum risk assessment identifies the spectral impact of a mission on unified action partners in the operational area. This process also identifies frequency usage conflicts such as EMI and frequency fratricide.

➢ Generating frequency reuse plans provides for spectrum optimization and increased spectrum capabilities.

➢ Identifying spectrum constraints where certain frequencies are taboo, such as those not allocated for use by the host nation.

➢ Determining spectrum capabilities pertaining to combat power, such as EW and counter radio controlled improvised explosive device electronic warfare systems.

- **Step 3: Course of Action Development—**
  - ➢ Modeling the unit's boundaries and movement formations, using SMO tools, to develop COA recommendations.
  - ➢ Performing EMI and EW frequency deconfliction, using SMO tools, for COA development and spectrum supportability.
  - ➢ Generating frequency allotment and allocation tables for subordinate units.
  - ➢ Identifying the unit's spectral impact on civilian spectrum users.
  - ➢ Identifying primary, alternate, contingency, and emergency communications for each COA based on unit capabilities, software simulation, and spectrum supportability.
- **Step 4: Course of Action Analysis (War Game) —**
  - ➢ Depicting the spectrum advantages and disadvantages for each COA.
  - ➢ Identifying mitigating factors for the spectrum risk assessment to reduce or eliminate risks.
  - ➢ Recommending modifications to the COA based on spectrum supportability during the war game.
- **Step 5: Course of Action Comparison—**
  - ➢ Comparing spectrum use over multiple COAs, using SMO tools, to allow the commander to determine which COA provides the best flexibility during execution while minimizing risks.
  - ➢ Analyzing routes over a movement of forces determines which route provides the best options for the commander.
- **Step 6: Course of Action Approval—**
  - ➢ Allows the unit to submit frequency proposals and receive frequency assignments.
  - ➢ Modifying COAs in accordance with commander's decision.
  - ➢ Coordinating frequency conflicts through higher echelons for mitigation assistance.
- **Step 7: Orders Production Dissemination, and Transition—**
  - ➢ Producing the SOI and joint communications-electronics operating instructions (JCEOI) and disseminate as needed to units.
  - ➢ Providing input to Annex H (Signal) of the Operations order (OPORD) that addresses all signal concerns, to include spectrum use information.

  Note. Refer to FM 6-0 for additional information on Annex H of the OPORD.

10.8   The Common Operational Picture

10.26.   The common operational picture is a single display of relevant information within a commander's area of interest tailored to user requirements and based on common data and information shared by more than one command. The SMO core functions and common tasks support completion of the COP for the commander. SMO tools, when used in conjunction with Intelligence and EW Cell information, allow the spectrum manager to collect spectrum related information tailored to the commander's operational area. These tools provide a visually depiction of force structure and geographical locations in a three-dimensional picture that personnel can understand quickly and easily. The following are some examples of SMO support to the COP—

- Live spectrum analysis of a given area of operations: Using SMO planning tools to receive analysis of the signal, allows the spectrum manager to perform EMI mitigation. A spectrum analyzer or monitoring receiver, a direction-finding antenna, and analysis software show persistent unplanned signals that interfere with assigned frequencies. These tools provide a three-dimensional picture to the commander with a graphical depiction of the spectral footprint of the signal, along with recommendations for frequency reassignment to maintain communications in the area and the impacted units. The commander, based on mission priority, deems it necessary to obtain new frequencies in order to accomplish the mission

- **Force Movement to a New Location:** The commander orders movement to a new location. The spectrum manager creates the proposed movement route with the SMO planning tool along with adjacent units' communications systems, sensors, and receivers. The SMO planning tool performs a simulation and provides courses of action to determine if the mission command system remains operational during the movement. The tool calculates that a direct path will cause counter radio-controlled improvised explosive device EW (CREW) interference on friendly communications along the route. The tool then presents a report with actionable information such as sources, victims, levels, and duration of interference. This provides the commander with supplementary information to make knowledgeable decisions.

11

Support to the Warfighting Functions

SMO enables and supports the Army's warfighting functions described in ADP 3-0,
Unified Land Operations: mission command, intelligence, fires, movement and
maneuver, protection, and sustainment. A warfighting function *is a group of tasks*
and systems (people, organizations, information, and processes) united by a common
purpose that commanders use to accomplish missions and training objectives. This
chapter links Army SMO to the warfighting functions; it also describes how SMO
supports and enables commander's efforts as they exercise mission command.

11.1   Movement and Maneuver

11.1.   The movement and maneuver warfighting function are the related tasks and systems
that move forces to achieve a position of advantage in relation to the enemy. SMO enables
movement and maneuver by maintaining freedom of action within the electromagnetic
spectrum. Commanders are able to leverage the spectrum manager's advice to provide lethal
and non-lethal effects against enemy combat capability, protection from adversary use of the
spectrum. SMO supports movement and maneuver through—

- Spectrum resource planning, analysis, and simulation to determine spectrum supportability
  over a projected movement of forces.
- Analysis, location, and direction finding of unknown and unplanned signals.
- Planning and simulating spectral use over the operational area.
- Frequency deconfliction planning over a movement.

11.2   Intelligence

11.2.   The intelligence warfighting function is the related tasks and systems that facilitate
understanding of the enemy, terrain, and civil considerations. It includes tasks associated
with information collection. SMO supports intelligence through the provision of spectrum
situational understanding and the ability to gain a greater understanding of the EMOE.
This understanding occurs through the successful frequency deconfliction of SDD, greater
fidelity in threat recognition, and support to the denial and destruction of adversary counter-
intelligence, counter-surveillance, and counter-reconnaissance systems. SMO supports
intelligence through—

- Measurement, analysis, and assessment of spectrum situational awareness.
- JRFL production and promulgation to protect intelligence operations.
- Centralized databases facilitate planning requirements and assessing collection through
  subordinate and adjacent units.

Dynamic Spectrum Access Decisions: Local, Distributed, Centralized, and Hybrid Designs, First Edition. George F. Elmasry.
© 2021 John Wiley & Sons Ltd. Published 2021 by John Wiley & Sons Ltd.
Companion website: www.wiley.com/go/elmasry/dsad

11.3   Fires

11.3.   The fires warfighting function is the related tasks and systems that provide collective and coordinated use of Army indirect fires, air and missile defense, and joint fires through the targeting process (ADRP 3-0). It includes tasks associated with integrating enemy counter mission command activities. SMO provides crucial support to the fires warfighting function through the ability to discriminate friendly forces from adversary targets, increased spectrum awareness, and direct support to EW.

11.4.   Electromagnetic environmental effects influence the operational capability of military forces, equipment, systems, and platforms. Spectrum managers support the fires warfighting function by mitigation of interference and ensuring systems are compatible. Hazards of electromagnetic radiation to personnel (HERP), hazards of electromagnetic radiation to ordnance (HERO), and hazards of electromagnetic radiation to fuels (HERF), are examples of electromagnetic environmental effects.

11.5.   A **hazard of electromagnetic radiation to personnel** is the potential hazard that exists when personnel maybe exposed to a radiation field of sufficient intensity to heat the human body. Radar, communication systems, and EW systems that use high-power transmitters and high-gain antennas represent a hazard to personnel working on, or near these systems. Leaders should ensure areas are clearly marked off to avoid injury to personnel.

11.6.   A **hazard of electromagnetic radiation to ordnance** is the danger of accidental activation of electroexplosive devices or otherwise electrically activating ordnance because of the radio frequency fields. This unintended actuation could cause premature firing of ordnance.

11.7.   **A hazard of electromagnetic radiation to fuels is** the potential hazard that exists when volatile combustibles, such as fuel, exposed to radiation fields of sufficient energy may cause ignition. The hazard is likely to occur when refueling operations are taking place. Leaders must adhere to proper grounding and static discharge procedures. Cease or minimize transmissions during refueling operations to prevent the potential hazard and exposure to radiation fields.

11.8.   SMO supports fires through—

- Coordination of the EMOE to prevent EMI to firing devices, sensors and data links that use the spectrum.
- Coordination with the CEMA element that allows effective use of spectrum resources and EW.
- Integration and synchronization of CEMA by assignment and allocation of spectrum use in joint environments.

> *Note.* Coordinated execution of joint electromagnetic spectrum operations with other lethal and nonlethal operations that enable freedom of action in the electromagnetic operational environment comprises electromagnetic spectrum control. (JP 3-13.1)

11.4   Sustainment

11.9.   The sustainment warfighting function is the related tasks and systems that provide support and services to ensure freedom of action, extend operational reach, and prolong endurance. SMO ensures that all spectrum dependent activities necessary for sustainment function properly and with minimal interference. Further, through coordination with EW, SMO contributes to overall sustainment in a hostile EMOE. SMO supports sustainment through—

- Design and development, acquisition, and distribution of advanced tools that manage the spectrum use.
- Protection of sustainment forces from friendly and adversary use of spectrum in static or mobile environments.

- Obtaining frequency clearance for all devices for the duration of the mission.
- Frequency deconfliction and emissions control procedures in support of sustainment mission command.
- Provides deconfliction within the spectrum to mitigate negative impacts to aircraft survivability.

11.5 Mission Command

11.10.   The mission command warfighting function develops and integrates those activities enabling a commander to balance the art of command and the science of control. Mission command emphasizes the centrality of the commander. Commanders exercise mission command through the conduct of the operations process, knowledge management and information management, synchronize information related capabilities, and through the conduct of CEMA, which includes SMO. SMO enhances mission command in light of other spectrum dependent activities (such as jamming and passage of intelligence) through effective spectrum management. In a contested, congested, and competitive EMOE, the mission command function must remain effective. SMO plays a key part in planning and battle management process and enables situational awareness of the EMOE. Spectrum managers are assigned to aviation units to support mission command. Aviation units require support to flight dispatch elements, and airfield services elements with robust communications requirements. Figure 11.1 shows the relationship between two SMO tools (described in Chapter 13) that support mission command. Many

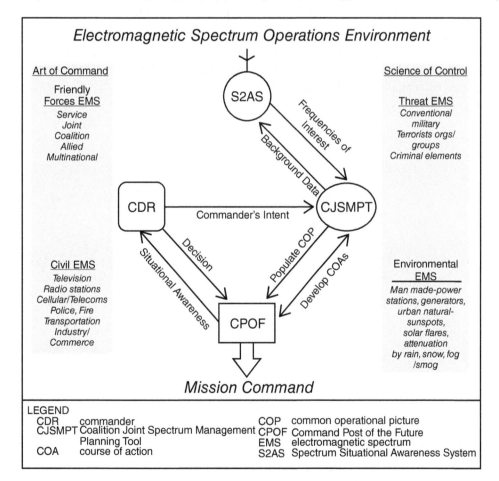

Figure 11.1 Spectrum situational awareness system and CJSMPT support to mission command. From *Techniques for Spectrum Management Operations*, Dec 2015. Headquarters, Department of the Army.

SMO tools can be substituted for the two SMO tools depicted in the graphic. These tools support the commander using the Command Post of the Future.

11.11.   SMO supports mission command through—

- Planning and preparing the spectrum in response of a mission.
- Assessment of the EMOE in response to commander's intent.
- Preparation and maintenance of the EMOE database.
- Understanding the impact of a mission on friendly, neutral, adversary, enemy, joint, interagency, intergovernmental, and multinational entities.
- Collecting spectrum information and visualizing this information in quick and easy to understand formats for completion of the COP.
- Control of the spectrum through force tracking and visualization, frequency deconfliction, reprogramming of SDD, and registration of all spectrum users (such as emitters, sensors, and receivers) with the spectrum manager.
- Development of SMO planning and management tools that support the net-centric environment (NCE) and become interoperable with Army and joint task force spectrum users.

11.6   Protection

11.12.   The protection warfighting function is the related tasks and systems that preserve the force so the commander can apply maximum combat power. SMO supports the protection warfighting function through the conduct of frequency deconfliction, interference mitigation, and support to EW defensive actions. SMO supports protection through—

- Network and frequency fratricide avoidance, detection, and mitigation.
- Development of the JRFL to prevent frequency fratricide and mission degradation.
- Coordination with CEMA Element to protect against blue force EMI during EW operations, such as counter radio-controlled improvised explosive device EW use.

12

Joint Task Force Considerations

Modern warfare is inherently a joint operation. Joint operations require precise coordination and establishment of procedures for effective spectrum use. This chapter describes the information and products for planning, coordination, and control of the spectrum at the joint task force level.

12.1 Inputs and Products of Joint Task Force Spectrum Managers

12.1. Spectrum manager assignments within a joint task force include multiple organizations: JFMO, joint spectrum management element (JSME), CEMA Element and the G-6. These agencies have a wide variety of inputs, collaboration, and products. Figure 12.1 shows a visual description of the spectrum management workflow between organizations in a joint task force environment. The following paragraphs contain an in depth look at the workflow between organizations in a joint task force environment.

12.2 Joint Frequency Management Office

12.2. The JFMO is a permanent organization within the operational area of a combatant command. The JFMO Chief is a DA civilian. Various personnel focused on region-specific spectrum requirements, staff the JFMO. The JFMO staff size varies, and is dependent on regional requirements. Figure 12.2, on page 12.2, shows the structure of the JFMO.

12.3. Table 12.1 shows the inputs to the JFMO by agency. The table provides the agency, the action conducted and the input the agency provides.

12.4. Table 12.2 shows a sample of the products of the JFMO and includes the agency and action taken.

12.3 Joint Spectrum Management Element

12.5. The JSME is a temporary organization that activates only for the duration of a specific joint task force mission. The JSME is an element within the J-6 and is sometimes an entity within a joint network operations control center of a joint task force. The senior spectrum manager may lead the JSME from the lead service of the joint task force. If designated as the joint task force lead, the Army Service component command headquarters, corps headquarters, or division headquarters G-6 spectrum management office assumes the role and responsibilities of the JSME.

Dynamic Spectrum Access Decisions: Local, Distributed, Centralized, and Hybrid Designs, First Edition. George F. Elmasry.
© 2021 John Wiley & Sons Ltd. Published 2021 by John Wiley & Sons Ltd.
Companion website: www.wiley.com/go/elmasry/dsad

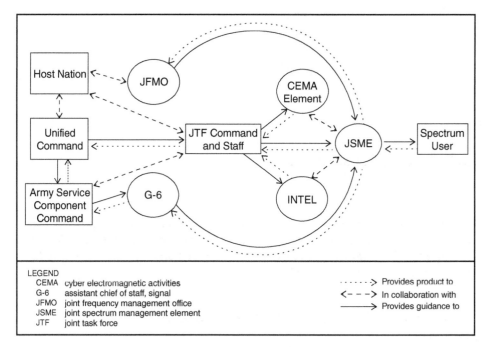

Figure 12.1 Interagency workflow in a joint task force environment. From Techniques for Spectrum Management Operations, Dec 2015. Headquarters, Department of the Army.

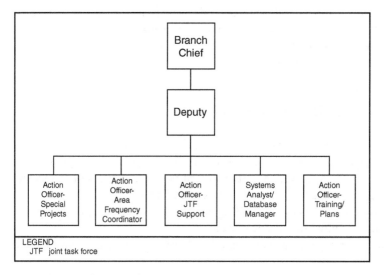

Figure 12.2 JFMO structure. From Techniques for Spectrum Management Operations, Dec 2015. Headquarters, Department of the Army.

12.6.   Spectrum management personnel from coalition forces and the sister services augment organizations designated as JSME during deliberate planning using the global force management process. Initial augmentation during the crisis action-planning phase of operations may come from joint enablers such as the Joint Spectrum Center (JSC), the Joint Electronic Warfare Center, or directly from the combatant command, J-6 JFMO. Depending on the size of the force structure supported, JSME end strength ranges from three to ten military service members.

12.7.   The JSME has a primary function to ensure authorized assigned joint task force military forces receive sufficient use of the spectrum to execute their designated missions. The JSME satisfies spectrum needs and ensure frequency deconfliction, prior to assignment or allotment, of all SDD including systems used by joint task force and component forces, such

Table 12.1 Agency inputs to the JFMO.

Agency	Action	JFMO Inputs
Combatant Command	Provides guidance and direction	JCEOI
JTF staff, JSME	Provides	Inputs to develop a JTF JRFL
JSME	Receives	Frequency proposals
Joint staff, civil affairs (J-5), HN	Provide input and responsible for	Host nation frequency authorizations
Spectrum users experiencing EMI	Submit	JSIR report

Legend	
EMI	electromagnetic interference
JCEOI	joint communications-electronics operating instructions
JFMO	joint frequency management office
JRFL	joint restricted frequency list
JSIR	joint spectrum interference resolution
JSME	joint spectrum management element
JTF	joint task force

From Techniques for Spectrum Management Operations, Dec 2015. Headquarters, Department of the Army.

Table 12.2 Products of the JFMO.

Agency	Action	JFMO Products
Spectrum users	Provides guidance and direction	Administrative and technical support for spectrum use
JTF staff, JSME, spectrum users	Provides guidance and direction	JCEOI
JTF staff, JSME	Provides guidance and direction	The Spectrum Plan, to include frequency use, reuse, and sharing schemes
JFMO	Provides guidance and direction	• Frequency assignments and allotments • Production and management of common spectrum use databases • The JRFL, upon approval from Joint Staff, Operations (J-3) • Mitigation assistance for EMI suffered

Legend	
EMI	electromagnetic interference
JCEOI	joint communications-electronics operating instructions
JFMO	joint frequency management office
JRFL	joint restricted frequency list
JSIR	joint spectrum interference resolution
JSME	joint spectrum management element
JTF	joint task force

From Techniques for Spectrum Management Operations, Dec 2015. Headquarters, Department of the Army.

as the United Nations, the North Atlantic Treaty Organization, and coalition forces. The JSME provides additional support based on strategic agreements between DOD, the U.S. mission, and the host nation. Table 12.3, on page 12.4, shows the agency inputs to the JSME.

> **Note.** Collaboration and coordination with varying agencies, especially host nations, occurs through a variety of processes. These processes are generally very formal and setup through the fostering of mutual trust and rapport between the JFMO and host nation. The spectrum

Table 12.3 Agency inputs to the JSME.

Agency	Action	JSME Inputs
JTF commander, JFMO	Provide guidance / direction	JCEOI Guidance
JTF STAFF	Responsible for	Inter service considerations, such as data formats, tools in use, frequency request procedures.
J-6	Provide Guidance / Direction	Nets to be included on the JCEOI
Component Commanders	Receive Guidance / Direction from	Friendly Force Spectrum use requirements and call words for inclusion on the JCEOI
Joint Staff, Intelligence (J-2)	Responsible for	Priority of intelligence gathering requirements
Spectrum Users	Provide Input	JRFL requirements
Spectrum Users Experiencing EMI	Provide Input	JSIR report
JTF Staff and various databases	Provide Guidance / Direction for	Spectrum use information on all friendly military and civilian, available enemy, and neutral forces
Spectrum Users	Responsible for	Requests for frequency authorization, modification, and deletion
JFMO	Provide Guidance / Direction for	Frequency allocations
CEMA ELEMENT	Responsible for	Instances of hostile EW
J-3	Provide Guidance / Direction for	Spectrum user priority
Joint staff, civil affairs (J-5), Host Nation	Provide	Host Nation spectrum authorization

Legend

CEMA	cyber electromagnetic activities
EMI	electromagnetic interference
EW	electronic warfare
J-2	intelligence directorate of a joint staff; intelligence staff section
J-3	operations directorate of a joint staff
J-6	communications system directorate of a joint staff
JCEOI	joint communications-electronics operating instructions
JFMO	joint frequency management office
JRFL	joint restricted frequency list
JSIR	joint spectrum interference resolution
JSME	joint spectrum management element
JTF	joint task force

manager must keep in mind customs and cultures, tact and courtesy, and the concerns of other agencies while still attempting to obtain the amount of spectrum resources necessary for the mission. The spectrum manager must also maintain accurate records of all dialogue and agreements made with the host nation.

From Techniques for Spectrum Management Operations, Dec 2015. Headquarters, Department of the Army.

12.9. Table 12.4 shows some of the products produced by or for the JSME. These products include the data call message, JCEOI, and the spectrum plan.

12.10. Figure 12.3 shows the spectrum manager input to the JSME.

Table 12.4 Products of the JSME.

JSME Products	Action	Agency
The JRFL, upon approval from J-3 and JFMO	Provides Guidance / Direction	Spectrum Users
Data Call Message	Provides Guidance / Direction	Spectrum Users
JSIR assistance	Provides Guidance / Direction	Spectrum Users
Annex K of the OPORD, upon JTF commander approval	Provides Guidance / Direction	Spectrum Users
JSIR report	Disseminates Product	JFMO
Frequency proposals	Disseminates Product	JFMO
Frequency assignments and allotments for stationary units and those on-the-move or at-the-quick-halt	Provides Guidance / Direction	Spectrum Users
The Spectrum Plan, to include frequency use and reuse and sharing schemes	In Collaboration With	JTF Staff, Spectrum Users
JCEOI	Provides Guidance / Direction	Spectrum Users
Frequency usage conflict identification, risk assessment, COA recommendations, and deconfliction	Identified in Collaboration with	CEMA Element
Definition of EMOE	Identified in Collaboration with	Joint Staff
Live spectrum monitoring	Disseminated Product	JTF Commander

Legend

CEMA	cyber electromagnetic activities
COA	course of action
EMOE	electromagnetic operational environment
J-3	operations directorate of a joint staff
JCEOI	joint communications-electronics operating instructions
JFMO	joint frequency management office
JRFL	joint restricted frequency list
JSIR	joint spectrum interference resolution
JSME	joint spectrum management element
JTF	joint task force
OPORD	operation order

From Techniques for Spectrum Management Operations, Dec 2015. Headquarters, Department of the Army.

12.4 Spectrum Management Support to Defense Support of Civil Authorities

12.11.   Army Defense Support of Civil Authorities (DSCA) encompasses all support provided by the components of the Army to civil authorities within the U.S. and its possessions and territories. This includes support provided by the Regular Army, Army Reserve, and Army National Guard when in Title 10 or Title 32 status. United States Code Title 10, Armed Forces, enables the Army to lawfully organize, train, equip, and conduct operations in coordination with other military services, federal departments, and agencies. United States Code Title 32, National Guard, consist of National Guard forces conducting DSCA while under authority of the specific State. Army forces conduct DSCA in response to requests from Federal, state, local, and tribal authorities for domestic incidents, emergencies, disasters, designated law enforcement support, and other domestic activities (ADRP 3-28).

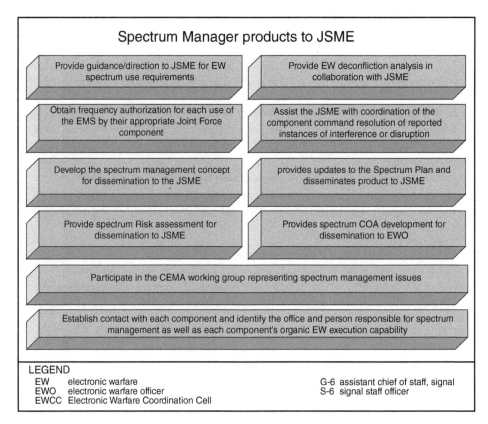

Figure 12.3 Spectrum manager inputs for a JSME. From Techniques for Spectrum Management Operations, Dec 2015. Headquarters, Department of the Army.

12.12.   Spectrum management in support of domestic operations requires liaison with state, municipal, local, and tribal authorities as well as first responders. Spectrum management for domestic operations within the U.S. must comply with a complex legal, regulatory and policy environment. United States Northern Command (USNORTHCOM) and United States Pacific Command (USPACOM) are the principal planning agents for DSCA and have the responsibility to provide joint planning and execution directives for peacetime assistance rendered by DOD within their assigned areas of responsibility. The other combatant commands provide capabilities to USNORTHCOM and USPACOM for DSCA as directed by the Secretary of Defense.

12.13.   Various resources may provide spectrum management support when Army forces are operating within the homeland. The separate joint forces headquarters for each state retains responsibility for forces operating within that state or territory. The National Guard (NG) J-6 spectrum management branch is the office responsible for coordinating and planning spectrum management for NG forces and provides support to the State's spectrum managers for domestic operations. The NG J-6 spectrum management branch provides coordination between the State's spectrum managers and all federal agencies. NG J-6 authorizes state spectrum managers direct liaison with the Army frequency management office or the Army spectrum management office. Joint forces headquarters state spectrum managers form a JSME in support of operational task forces under state active duty, dual status (Title 32 or Title 10) domestic operations.

12.14.   Activated forces, after acquiring frequency assignments may operate both civil and military systems within a domestic area of operations to achieve interoperability with other Federal agencies and civil authorities. NG forces may request frequency assignments through the NG, or through Title 10 military channels depending on their duty status for a given operation. The National Telecommunications and Information Administration control the spectrum within the homeland. They certify and license civilian usage of the electromagnetic spectrum.

NG forces and U.S. Coast Guard may operate both civil and military systems within a domestic area of operations, as well as numerous states, local and federal agencies.

12.15. After receiving orders to conduct movement for a domestic operation, each operational element initially contacts their local state spectrum manager or JSME for a JCEOI extract detailing the frequencies and procedures to use for communications. The local spectrum manager or the JSME submit a standard frequency action format (SFAF) request for frequencies on behalf of the end user. A state's qualified spectrum manager provides spectrum management for a given geographical state to the greatest extent possible. Spectrum managers coordinate for interstate operations, and for spectrum deconfliction for operations adjacent to another spectrum manager's area of responsibility. Local spectrum managers form the JSME and work directly for the incident commander (or the local state joint force headquarters prior to the appointment of an incident commander).

12.16. Domestic operations lessons learned have demonstrated that both unity of effort and coordinated spectrum management are critical to the success of the operation. Congress and the DOD, through the implementation of a dual status commander (commander of both Title 32 NG forces, and Title 10 NG and active Army forces for a domestic operation) have addressed unity of effort. Unity of command is not applicable between Federal military forces and the state NG, but unity of effort can be achieved if the President and the Governor formally agree to appoint a dual-status commander. Federal authorities have been established that allow a designated dual-status commander to serve in a hybrid Federal and state status. A dual-status commander will usually be a National Guard officer who is given simultaneous but separate

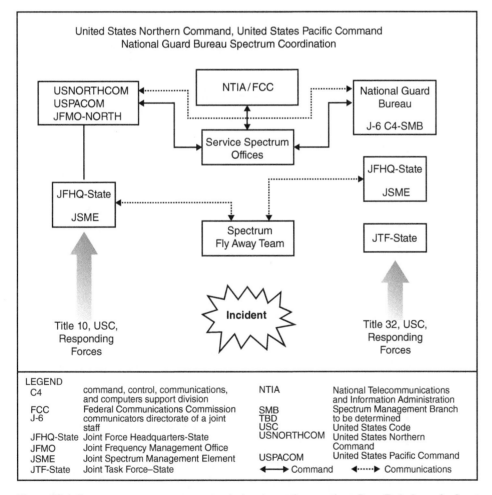

Figure 12.4 Spectrum management support during domestic operations. From Techniques for Spectrum Management Operations, Dec 2015. Headquarters, Department of the Army.

authorities over Federal and state military forces. Spectrum management for a domestic operation requires consolidation under one JSME, (or at a minimum coordinated Title 10 and Title 32 JSMEs led by the same commander) to minimize confusion and provide seamless support to tactical communications.

12.17.   The JSME initially assigned to an incident continues to work for the incident commander as the operation transitions to a different duty status or legal authority, to ensure continuity of spectrum management. Typically, a state Governor or Adjutant General creates a standing joint task force including a JSME, or establish a joint task force with a JSME in response to an incident. Optionally, each state builds its JSME from qualified spectrum managers, and equipment from within the state National Guard's tables of distribution and allowance allotment. Should a state not have qualified spectrum managers, the Adjutant General and Governor may request qualified spectrum managers prepare to deploy from other states to form a JSME under a formalized emergency management assistance compact. The JSME may request a spectrum flyaway team from the NG bureau J-6 and USNORTHCOM to supplement the element.

12.18.   Figure 12.4, provides a graphic of the collaboration process during domestic operations.

Note. See JP 3-28 and JP 6-01 for more information regarding domestic operations.

13

Spectrum Management Operations Tools

Spectrum managers have access to a wide variety of tools to aid in effective and efficient spectrum planning and management. This chapter provides a technical description of several tools used to facilitate spectrum management operations. Included in this chapter are hardware and infrastructure requirements, software used, and capabilities of spectrum management tools.

13.1 Tool Considerations

13.1.   There are a variety of spectrum related tools used to plan and manage communications networks and SDDs. Many of these tools do specific functions of limited scope precluding the sharing of relevant information among these functions creating inefficiencies. This can lead to erroneous planning and assignments that can cause frequency interference. It is essential for the benefit of all spectrum stakeholders, tools should share data in a consistent manner to improve efficiencies. As an example, EW operators should use the same tool that the spectrum manager uses in order to allow the spectrum manager to mitigate harmful interference to friendly systems possibly caused by EW systems.

13.2.   Gathering and managing spectrum data requires considerable time in order to ensure accuracy. Tools that support the automation of spectrum management functions can drastically reduce this time constraint. Tools that promote the flow of information between spectrum stakeholders reduce the planning cycle leading to quicker decisions. Spectrum managers are able to perform the core SMO functions much more efficiently when tools comply with the net-centric environment.

13.3.   The NCE, a common shared virtual space used within and among differing authenticated units and organizations, has facilitated numerous advantages for spectrum managers of all levels. Central access to multiple databases reduces or eliminates the need to visit agencies to obtain a list of devices used in the area of operations. Having central databases requires SMO tools to have interoperable and compatible formats in order to function. The net-centric environment is very effective in joint task force operations.

13.4.   There exist many data file standards regarding frequency proposals. Standard Frequency Action Format (SFAF) is a line oriented text format used by DOD, and by U.S. allies and unified action partners who use Spectrum XXI. SFAF is the standard format for frequency proposals, assignments, modifications, renewals, reviews, and deletions.

13.1.1 Spectrum Situational Awareness System

13.5.   The Spectrum Situational Awareness System (S2AS) is theater provided equipment used to assist in maintaining effective use of the spectrum. The S2AS provides fixed site and portable spectrum-monitoring receiver that performs instantaneous analysis of captured spectrum data. The S2AS consists of government off-the-shelf software referred to as multi spectral ambient noise collection and analysis tool (MANCAT) and commercial off-the-shelf hardware

Dynamic Spectrum Access Decisions: Local, Distributed, Centralized, and Hybrid Designs, First Edition. George F. Elmasry.
© 2021 John Wiley & Sons Ltd. Published 2021 by John Wiley & Sons Ltd.
Companion website: www.wiley.com/go/elmasry/dsad

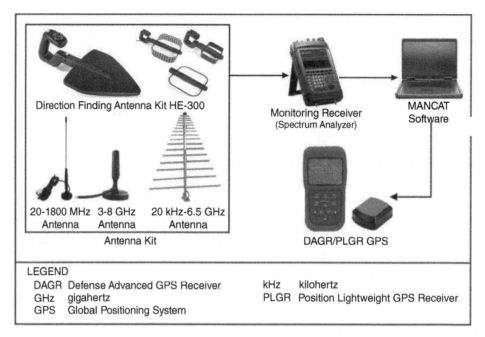

Figure 13.1 S2AS and supporting equipment. From Techniques for Spectrum Management Operations, Dec 2015. Headquarters, Department of the Army.

Rohde & Schwarz PR100 monitoring receiver and ancillary antennas. The S2AS requires the unit to provide a computer to run the software and a global position location device. The system comes with a ruggedized carrying case that protects the PR100 and HE-300 antenna while in transit. Figure 13.1, on page 13.2, shows the components that comprises S2AS.

> **Note.** An upgrade to the HE-300 antenna is available that includes a built in global positioning system location device and electronic compass. This capability, in conjunction with an available software update for the PR100, allows for rudimentary direction finding of signals and triangulation between multiple S2AS users and measurements. The S2AS provides direction-finding antenna and triangulation capability intended for post, camp, or forward operation base locations. System use of S2AS is for a relatively small area. This upgrade also eliminates the need for a unit provided global positioning system receiver.

13.6. The advantages of a monitoring receiver over a traditional spectrum analyzer are its rapid precision measurements and analysis of spectrum over a wide bandwidth. The monitoring receiver provides the measured spectrum to the MANCAT software for automated analysis. The PR100 can operate on 100 to 240 volts alternating current and comes with a wide variety of connectors to allow for connecting to differing voltage sources around the world. The PR100 also has an easily removable and rechargeable battery with an approximate lifespan of 3.5 hours.

13.7. When used with a location device, S2AS allows for mobile measurements of the spectrum. For example, the spectrum manager can take S2AS on a convoy route to measure persistent signals that are in use in the area. The location device senses the grid coordinates of the route for mapping captured signals by other tools (Global Electromagnetic Spectrum Information System [GEMSIS] and MANCAT). When imported, Google maps provide the software overlays with a color-coded spectrum map over a threedimensional digitally mapped terrain to support the COP. Figure 13.2, on page 13.3, depicts Soldiers operating the direction-finding antenna and the PR-100 handheld receiver.

13.8. The S2AS provides a fast panoramic scan across the frequency range of 9 kHz to 7.5 GHz. This enables Soldiers to quickly access the spectrum and begin to incorporate data into the required database. The display on the device provides a spectrum and spectrogram display which users of the spectrum analyzer may be familiar with on a portable 6.5″ color

Figure 13.2 S2AS in use by Soldiers. From Techniques for Spectrum Management Operations, Dec 2015. Headquarters, Department of the Army.

screen. The unit provides storage of measurement data to the receiver's built-in storage card. The PR-100 design is ergonomic and rugged for portable use and low weight. The device has a setting for manual location of spectrum emissions using the active directional antenna or automatic location of spectrum emissions with direction finding algorithms.

13.9.   The operator of the S2AS can save spectrum measurements in comma-separated value format and spectrum screenshots in portable network graphics format. MANCAT software exports reports in PDF, HTML, JPEG, or TIFF formats.

13.10.   The S2AS includes a continuous band antenna that is vehicle mounted or fixed to a tripod stand for measurements of spectrum from 30 to 6000 MHz. The system includes a hand-held HE-300 antenna with three interchangeable modules for the 20 MHz to 7.5 GHz ranges and the HE-300HF module for high frequency ranges from 9 kHz to 20 MHz. Antenna HE-300 provides the capability to perform direction finding of unplanned signals in specific frequency ranges. S2AS is compatible with some antennas that units may already have, including the SINCGARS vehicle whip and the OE-254 antenna.

13.11.   The S2AS standard package can measure and analyze signals from 20 MHz to 7.5 GHz with the included continuous band antenna and HE-300 antennas. This range encompasses the majority of spectrum conflicts the spectrum manager encounters. An upgrade package, that expands the range to 18 GHz, is available for users that need to analyze higher frequencies, but cost is an issue for users that do not regularly analyze those frequencies.

13.12.   S2AS uses standard data formats compatible with CJSMPT, GEMSIS and Spectrum XXI. Once the captured signals are available in the CJSMPT or Spectrum XXI tools, the spectrum manager can update known databases. The MANCAT software allows the user to import frequencies of interest from planned databases, such as Spectrum XXI, and provide a visual display to the operator of planned or known signals.

13.13.   S2AS visually differentiates signals that are not in any planned database so that the spectrum manager can further investigate the source of the signal. Figure 13.3 shows the functional relationship between the S2AS key capabilities. The figure provides a graphical depiction of the sense characteristic. Sensing and monitoring frequencies that are available to the user is an initial operational function of the S2AS. The system then analyzes the information captured and shares the data with the listed database.

13.1.2   Global Electromagnetic Spectrum Information System

13.14.   Global Electromagnetic Spectrum Information System (GEMSIS) is a joint program of record that provides access to several spectrum management tools. Spectrum managers

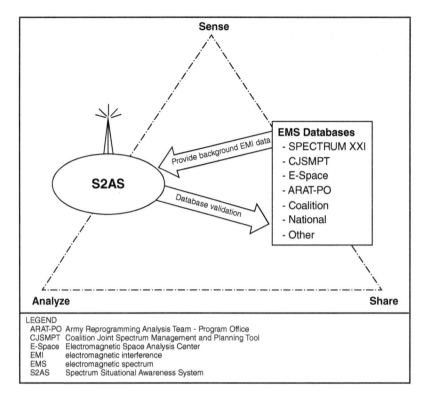

Figure 13.3 S2A2 functional relationships. From Techniques for Spectrum Management Operations, Dec 2015. Headquarters, Department of the Army.

access GEMSIS on the internet via NIPRNET and SIPRNET connections. GEMSIS increases the effectiveness of the COP, accelerates spectrum access, increases interoperability, and support to NCE. GEMSIS increment 2 incorporates other SMO tools, such as CJSMPT, Spectrum XXI-Online, systems planning, engineering, and evaluation device (SPEED), and Afloat Electromagnetic Spectrum Operations Program (AESOP) as an effort to further transition spectrum management to a NCE compliant and provide all the needed capabilities to the spectrum manager in one central tool.

13.15. GEMSIS provides worldwide visibility of host nation supportability of SDD equipment. The system automates distribution of host nation coordination requests and Combatant command submission of host nation supportability comments. This enables spectrum managers to determine the historical supportability of other systems in the same frequency band.

13.1.3 Coalition Joint Spectrum Management Planning Tool

13.16. Coalition Joint Spectrum Management Planning Tool (CJSMPT) is a capability delivered by the GEMSIS program. CJSMPT developed as a joint capability technology demonstration that integrates spectrum management, modeling, simulation, and planning tools that enables spectrum managers at all levels (joint task force and below) to perform spectrum planning and frequency deconfliction for mission planning and combat operations.

13.17. CJSMPT provides the capability to predict interference as units move across a simulated EMOE. CJSMPT uses this simulation to perform deconfliction analysis that is compatible with EW operations and future rapid maneuvering forces. This means that the spectrum manager can simulate and visualize a unit's movement, perform spectrum interference analysis and frequency deconfliction, and provide recommendations to the commander for complete spectrum use during the movement.

13.18. The CJSMPT database is the spectrum knowledge repository. CJSMPT is compatible with S2AS and Spectrum XXI using common data formats. The map manager functional area of the software allows the user to import any national geospatial-intelligence agency map resource. The spectrum data repository provides users with a single authoritative data source of known databases, such as joint, equipment, tactical, and space.

13.19.   CJSMPT performs spectrum optimization and conflict mitigation using environmental factors, operational priorities, frequency allocation and assignments, and international spectrum management policies and regulations. The main visualizer panel within CJSMPT displays spectrum use in a color-coded two and three dimensional picture that is available throughout the mission's duration. CJSMPT enables the spectrum manager to provide the commander with accurate spectrum information for civil (grey), hostile (red), friendly (blue) and counter radio-controlled improvised explosive device EW, referred to as CREW, and Intelligence operations on blue force SDDs.

13.20.   CJSMPT allows the operator to submit frequency proposals to the Spectrum XXI system using the SFAF and standard spectrum resource format compliant formats. Upon approval by Spectrum XXI, CJSMPT can import frequency assignments into the spectrum knowledge repository. CJSMPT can automatically format a satellite access authorization into the appropriate format (SFAF) saving the spectrum manager time.

13.21.   The spectrum requirements advisor utility within CJSMPT automatically generates spectrum reuse plans and calculates the minimum spectrum requirements for an interference free operation over a given movement of forces. This allows for rapid force movement while minimizing the spectral impact of a mission. CJSMPT can generate formatted reports, such as the joint spectrum interference resolution (JSIR) report, based on the communications effects simulator utility. The operator can save detailed reports in extensible markup language (XML), HTML, or comma separated values formats. Spectrum planning within CJSMPT can account for bandwidth, frequency locking, guard bands, and frequency allocation tables.

13.22.   CJSMPT allows the spectrum manager to develop scenarios quickly by placing forces into the database using force templates. Force templates within CJSMPT include force structure, SDDs characteristics, and spectrum usage information for devices that have passed the spectrum certification process. CJSMPT allows the user to place device characteristics into the database for devices that have not received spectrum certification. Without this capability, the spectrum manager must manually place force structure and SDD characteristics into a variety of locations, such as XML spreadsheets. Manual input can cause data format inconsistencies, possible human error, and time delays.

13.23.   CJSMPT functions in a NCE by granting network access through SIPRNET. Users perform peer-to-peer collaboration and retrieve information from the master spectrum knowledge repository while connected to the SIPRNET. CJSMPT also provides support to joint task force environments by providing features targeted to key joint task force agencies. JFMO and JSME agencies develop and maintain the JRFL and JCEOI, which resides over the spectrum knowledge repository database for the area of responsibility. Spectrum managers assigned to the JFMO or JSME control and update the spectrum knowledge repository with specific locally operated equipment and identify the effects of EW on emitter devices.

13.24.   The CJSMPT administrator, serves as the overall oversight of the CJSMPT database by maintaining force structure and equipment, control of the master spectrum knowledge repository, and updating a detailed list of known SDD worldwide. CJSMPT functions in a standalone environment to operate while not connected to the SIPRNET.

13.1.4   Systems Planning, Engineering, and Evaluation Device

13.25.   The systems planning, engineering, and evaluation device (SPEED), developed as United States Marine Corps government off the shelf software. SPEED is a modular software application that provides modules that target user specific needs. SPEED allows the spectrum manager to complete and edit SFAF forms while using an equipment database that includes tactical platforms, equipment, and antennas. The Asset Manager module within SPEED provides the capability to import, export, build vehicle manifests, personnel rosters, and equipment deployment lists. SPEED is free to all federal agencies but primarily used by the United States Marine Corp communications and spectrum managers. Army spectrum managers may interface with SPEED in joint operations.

13.26.   SPEED provides both two and three-dimensional views of the operational area to support the COP. The Advanced Prophet and Terrain Integrated Rough Earth Model and National Geospatial Intelligence Agency provide map data to the system. SPEED provides the user with a color-coded display of spectrum use over the operational area. SPEED can generate JRFL input in the correct format to the next higher echelon. The system allows the user to manually input, store, and view information for a tactical satellite network defined in a satellite access authorization, but cannot automatically format the authorization into the SFAF or standard spectrum resource format (SSRF). SPEED is a software package that is distributed with the automated communications engineering software or joint automated CEOI system image on the unit provided AN/GYK-33 computer.

13.1.5   Afloat Electromagnetic Spectrum Operations Program

13.27.   AESOP is an integrated operational radar, combat system, and communications frequency-planning tool primarily used by U.S. Navy and U.S. Coast Guard spectrum managers. This tool calculates optimal frequency use and distance separation that considers all strike group SDDs. AESOP minimizes electromagnetic interference in accordance with national and international frequency regulations. The strike group staff or designated frequency coordinator can select frequencies and separation distances for the group's ships to ensure that the radars operate with a minimum of electromagnetic interference.

13.28.   In addition to the ships of the U.S. Navy, AESOP contains data from fleets of over 60 countries. AESOP supports radar and communication analysis and spectrum planning for joint warfare operations on platforms for the following—

- Ships.
- Submarines.
- Aircraft.
- Military and civilian ground sites.

13.29.   AESOP periodic updates have improved the performance of communications networks in the presence of counter radio-controlled improvised explosive device electronic warfare and other EW. The AESOP master database is shore based and is only available via connection to the SIPRNET. For users with no access to SIPRNET or having limited bandwidth, AESOP is available in a standalone mode with a local database.

13.30.   AESOP is compliant with the SFAF and SSRF. Measurements taken by AESOP provide input and development of the DD Form 1494. AESOP provides spectrum visualization after analysis of spectrum use. AESOP can import and export XML files. The system can provide the Navy's input to the JRFL during joint task force operations. Army spectrum managers may interface with AESOP when coordinating spectrum use in operational areas collocated with Navy missions.

13.1.6   Spectrum XXI

13.31.   Commanders have several configuration options within Spectrum XXI. Spectrum XXI is a client and server, Windows-based software system that provides spectrum managers with a single information system that addresses spectrum management automation requirements. The JSC manages Spectrum XXI. Spectrum XXI supports operational planning as well as near instantaneous management of the electromagnetic spectrum with an emphasis on assigning compatible frequencies and performing spectrum-engineering tasks. Spectrum XXI client version is a software package that requires a unit funded computer.

13.32.   The joint spectrum center central repository for Spectrum XXI provides the DOD with a central database that contains spectrum certification for compliant systems, topography and electromagnetic environmental effects data, and all DOD spectrum proposals and assignments. The repository also serves as the mechanism to transfer data between the DOD and NTIA for permanent frequency assignments in the U.S. and its possessions. Spectrum XXI users may access the government master file through the central repository as needed.

13.33. Spectrum XXI users can connect to one of the three regional servers through local area network access, SIPRNET access, or secure telephone for dial-up access. The Spectrum XXI database uses the Oracle database management system based on structured query language that requires licenses and training for the regional servers. The client version that Army spectrum managers use does not require an Oracle license or training. The client can function in standalone mode using the local database with limited functionality when network connectivity is unavailable. Spectrum XXI contains a table of International Telecommunications Union allocations by region to aid the spectrum manager in international spectrum planning compliance. The Spectrum XXI database also contains geographical boundaries and utilities. The system can plot SDD based on frequency records.

13.34. Spectrum XXI allows the user to create and maintain permanent, temporary, proposed assignments, including background on frequency assignments. Spectrum XXI analyzes frequency assignments for operating conditions, interference, intermodulation, allocation and allotment tables, and compliance with technical and administration standards. A simulated spectrum analyzer provides a display of current spectrum occupancy and projected spectrum use at user-defined sites.

13.35. Spectrum XXI allows for the creation of the JSIR to aid in the mitigation of EMI. The system also creates and manages input to the JRFL. Spectrum XXI can analyze the impact of EW on spectrum users. Spectrum XXI is compliant with the SFAF as outlined in the MCEB Publication 7 format.

13.1.6.1 Spectrum XXI Key Components

13.36. This section provides readers with information on the various components of the Spectrum XXI software. The modules described in this section are a small sample of the capabilities of what Spectrum XXI provides to commanders and leaders.

13.37. **Interference Analysis Module** analyzes existing frequency assignments for potential interference. This analysis, is normally performed when the holder of a frequency assignment reports interference from an unknown station. An interference analysis maybe accomplished to determine whether a transmitter on a single frequency would potentially cause interference to an existing environmental receiver represented by a frequency record in the database. The interference analysis module performs analysis to determine if a receiver potentially receives interference from an existing environmental transmitter represented by a frequency record in the database.

13.38. **Interference Report Module** generates interference reports that describe interference problems and provides information to resolve the problem. Interference reports can also document a history of problems, and thus identify possible causes for subsequent interference. If interference problems exist, the first step as a spectrum manager is to verify that the person reporting the interference has authorization to use that frequency. Spectrum managers attempt to resolve interference problems at the lowest level possible. If this is not possible, create a report for distribution to higher authorities. If a resolution is not found, the interference information is reported to the unified or specified command (usually the combatant commander or the service representative) who then may call upon the JSIR team (as part of the JSIR program located at the JSC) to investigate.

13.39. **EW Deconfliction Module** assesses the impact of a planned electronic attack and jamming on existing receivers during contingency operations and exercises. The joint staff, operations must know the operational situation to make intelligent decisions when using this module. The EW Deconfliction Module, used in conjunction with the JRFL Module, documents a list of frequencies protected from jamming. In addition, the module analyzes the impact a frequency jammer has on environmental receivers using a range of azimuths. Analysis results comprise three types of conflicts—

- Frequency assignment conflicts.
- JRFL conflicts.
- Communications-Electronics Operating Instructions conflicts.

13.40.   The Joint Restricted Frequency List Module is a management tool used by various operational and support elements to identify the level of protection they desire, applied to specific spectrum, to preclude these assets from being "jammed" by friendly forces conducting electronic warfare activities. The JRFL identification and building process begins at the unit level, works upward through the military services' chain of command, then consolidated within the combatant command or joint task force staff. The module allows users to select frequency assignment from Spectrum XXI or JCEOI nets. Selecting these frequency assignments is done by importing the generated CEOI in the JRFL module.

13.41.   The **Engineering Tools Module** is a collection of utilities used to perform several types of analyses—

- **Coordinate Conversion:** This utility provides a graphical representation of the conversion between latitude, longitude, and military grid coordinates.
- **Co-site Analysis:** Used to perform co-site analysis on a list of frequencies and emissions.
- **Coverage Plots:** Used to create terrain elevation plots, line-of-sight plots, and signal strength plots. This function provides the commander analytics help determine the best placement of sensors.
- **Geomagnetic Conversion:** The Geomagnetic Conversion utility converts magnetic azimuths to true azimuths.
- **High Frequency Skywave Analysis:** The High Frequency Skywave Analysis utility in calculates the high frequency skywave, propagation prediction values for the maximum usable frequency, the frequency of optimum transmission, and the lowest usable frequency based on the time of day between a transmitting and receiving location.
- **Point-to-Point Link Analysis:** The Point-to-Point utility displays the terrain profile and aids in determination of the viability of radio links between transmitting and receiving locations.
- **Satellite Look Angles for Multiple Earth Stations and Multiple Satellites:** Used to calculate the azimuth and take-off (elevation) angle from earth stations to geostationary satellites.
- Spectrum Occupancy: Used to display a graphical representation of the calculated received signal power at a specified location indicated in the frequency records of the assignment database (this is similar to the view seen on a spectrum analyzer).

13.42.   The **Topographic Manager Module** is an automated capability that reformats the level-1and level-2 digital terrain elevation data obtained directly from the National Geospatial-Intelligence Agency on compact disk-read only memory disks. The Topographic Manager can register and manage reformatted topographical data files.

13.43.   The **Frequency Assignment Module** automates the processing of requests for the use of frequency resources from spectrum managers in support of authorized users. The process includes the preparation of frequency assignment proposals, validation of those proposals, determination of possible interference with the background environment, distribution and status tracking of proposals. The Frequency Assignment module also provides processes for frequency assignment database updates and retrievals.

13.44.   The **Allotment Plan Generator Module** creates a list of frequencies commonly referred to as Allotment Plans, Channelization Plans, Spectrum Use Plans, or Radio Frequency Authorizations. These plans are a frequency resource for nominating proposals using the Frequency Assignment module. In some cases, allotment plans disseminate authorized temporary frequencies used for training or tactical exercises.

13.45.   The **Compliance Module** checks the format and content of frequency records saved to a file and are not in the proposal editor. Three types of compliance checks maybe performed: allocation table checks, Canadian and Mexican coordination checks, or field validation checks. You also have the option to perform all checks. The record source for your records determines which validation checks are performed. International users should use the Validation option only.

13.46.   **Spectrum Certification System** is an automated system used to prepare a DD Form 1494, Application for Equipment Frequency Allocation, at frequency management offices that support materiel acquisition.

13.47.   The **Data Exchange Module** electronically exchanges data between servers and client computers. The Data Exchange Module manages the server accounts, job accounts, and domains used for data exchanges. A stand-alone client (not network connected) cannot use the functions of Data Exchange. When first installed, Spectrum XXI is a stand-alone client until the initial connection to a server. When connected it becomes a data-exchanging client.

13.1.7   Host Nation Spectrum Worldwide Database Online

13.48.   The Host Nation Spectrum Worldwide Database Online (HNSWDO) is a web application that facilitates warfighter deployment and communications by providing worldwide visibility of host supportability of SDD. The HNSWDO automates the distribution of host nation coordination requests allowing combatant command submission for host nation supportability, reducing time requirements for managing the host nation spectrum authorization process. The design of the database provides informed decision making concerning frequency bands. This mitigates the risk of acquiring potentially unsupportable systems. HNSWDO provides the user with near instantaneous updates and dramatic reductions in process lag (from years to months). HNSWDO requires an approved account and NIPRNET access using a unit provided laptop computer. The Defense Spectrum Organization processes account request.

> *Note.* Host nation allocation tables and SDD certification does not constitute the authorization to assign frequencies within the host nation. Send all formal frequency requests to the host nation to obtain frequency authorization. See JP 6-01 for more information on host nation coordination.

13.1.8   Automated Communications Engineering Software and Joint Automated Communications Engineering Software

13.49.   Automated Communications Engineering Software (ACES) and Joint Automated Communications Engineering Software (JACS) are part of the Army key management system that automates the management of communications security (COMSEC) keys, electronic protection (EP) data, and SOI. These multipurpose programs reside on a laptop computer. Key features of the software for SMO purposes are SOI generation, viewing and printing, EP identifiers, transmission security keys, data generation, creating loadsets for SINC-GARS and SINCGARS compatible radios and electronic distribution of the joint automated communications-electronics operation instructions system.

13.50.   ACES and JACS integrate secure network planning, EP distribution, and SOI generation and management. The workstation functions in conjunction with the data transfer device, hosting tier 3 software, to automate cryptographic control operation for networks with electronically keyed COMSEC equipment.

13.51.   The resident software components on the ACES or JACS workstation include the following—

- General purpose module.
- Core module.
- Area common user system module.
- Resource manager module.
- Master net list module.
- Signal operating instructions (SOI) module.

- Combat net radio module.
- ARC-220 Module
- Satellite Communications Module

13.1.8.1 General Purpose Module

13.52. The general-purpose module provides the information and operations necessary to satisfy the planning requirements for cryptonets that operate independently of area common user systems and combat net radio networks. It allows the planning capability for manual key assignments for compatible COMSEC equipment in an operator-designed cryptonet configuration. It allows for the importing of the Black Key packages from the local COMSEC management system.

13.1.8.2 Area Common User System Module

13.53. The area common user system module lists information that encompasses both Joint Network Node equipment and echelons above corps networks. The module contains procedures for creating and drawing an area common user system backbone network and creating and drawing network extensions. It also describes the procedures required to validate and generate area common user system networks, and modify area common user system member properties.

13.1.8.3 Resource Manager Module

13.54. The resource manager module contains the list of available frequency resources and allows creation, editing, merging, deleting, and printing of these resources. Each frequency resource is described by attributes that specify the authorized use and location of all the frequencies contained within the resource. The resource module also provides planners the capability to import and export resources in JACS, Integrated System Control, and SFAF formats.

13.1.8.4 Master Net List Module

13.55. The master net list module provides a communications list containing the net name or description, net identification, organizational code, restrictions, frequency type, power, reuse class, reuse zone, and call word or color word requirements. The master net list is developed for an operations plan. The master net list module provides the capability to create, edit, merge, delete, and print nets. The master net list module incorporates a number of SFAF-compatible fields to facilitate the transfer of data to and from other frequency management systems such as Spectrum XXI, as well as service unique systems. The database capabilities of the workstation allow the data in the master net list to create the initial SFAF frequency proposal and the SOI.

13.1.8.5 SOI Module

13.56. The SOI module contains call signs, call words, frequency assignments, signs and countersigns. The SOI module also contains pyrotechnic and smoke signals, dictionaries, groups, quick reference, and title pages. Generation of pyrotechnics and smoke signal components may be separate or randomly selected. SOI also provides the capability to create the Master Call Sign packets, as well as separate extract packets, while maintaining a database link to nets in the master net list.

13.1.8.6 Combat Net Radio Module

13.57. The combat net radio module provides the necessary functions to create, modify, and generate hopsets or loadsets for SINCGARS transmission security keys. It also provides the capability to plan combat net radio nets in all bands. Combat net radio network planning provides integration with the master net list module. Loadsets are packages of frequency hopping data and COMSEC keys required to load up to six channels of the SINCGARS radios. One loadset consists of COMSEC keys tags, hopsets, lockouts, transmission security keys and net identifiers. Hopsets consists of a set or sets of resources converted into SINCGARS useable frequency hopping data. The complexity of the hopset may be directly related to the amount of memory needed in the receiver-transmitter. Hopset resources maybe constructed with minimal pattern interruption, as the radio is frequency hopping at a rate of 100 channels per second. Lockouts are digitized hopset data generated and stored in the combat net radio. Lockouts electronically map all available frequencies by relaying to the radio's memory frequencies it cannot use. This deliberately disables unused frequency channels, avoiding interference with another service.

13.1.8.7   ARC-220 Module

13.58.   The ARC–220 module allows platforms and equipment assignment to ARC-220 nets. ARC–220 is a radio network that supports long-range communications between military aircraft and ground stations. This network type provides support for the AN/ARC-220 (aircraft version) and AN/VRC-100 (ground version) radios. These radios operate in three different modes: single channel (Basic Preset or Manual), automatic link establishment, and electronic counter-counter measure. The net validate function ensures that the platforms intended to communicate with each other can in fact do so with the equipment they have been allocated. The net generated function automatically creates COMSEC key tag assignments to secure the network.

13.1.8.8   Satellite Communications Module

13.59.   The satellite communications module allows the operator the capability to support the crypto planning for two of the Army's satellite communications terminals. These terminals are the Single Channel Anti-Jam Man Portable Terminal and the Secure Mobile Anti-Jam reliable Tactical Terminal. These satellite systems operate at radio frequencies in the extremely high frequency range.

13.60.   The network planning functionality of ACES or JACS incorporates cryptonet planning, key management, and key tag generation. The planning concept relates to the development of network structures supporting missions and plans. The data for a given plan includes individual nets, and assigned individual net members. Net members are associated with a specific platform and equipment. Once designation of all variable information (net members, platforms, and equipment), specific equipment fill locations defined, keys are associated with the equipment locations. The equipment records, which include platform data, net data, and key tags, maybe downloaded to the data terminal device, and subsequently associated with the required key. Similarly, the EP data and SOI generated by the JACS workstation operator maybe downloaded to the data terminal device.

13.2   Joint Spectrum Interference Resolution Online

13.61.   Joint spectrum interference resolution online (JSIRO) collaboration portal is the preferred tool for reporting EMI occurrences. JSIRO is a Web-based, centralized application containing data and correspondence for reported EMI, intrusion, and jamming incidents dating back to 1970. JSIRO is the repository for the results of analyses, collected data, and supporting documentation for EMI resolution to support both trend and future interference resolution analysis. To access the JSIRO tool utilize a SIPRNET link. JSC provides management and control of the JSIRO. The tool is hosted through Intelink and the joint worldwide intelligence communication system. The spectrum manager may access the tool through a SIPRNET connected computer without the need for loading software onto the computer. JSIRO allows the user to upload files that may be instrumental in mitigating the EMI; such as spectrum analyzer traces, recorded audio, or comma separated values files. Use the manual JSIR report format when SIPRNET access is not available.

13.62.   When reporting online, JSIRO prompts the user for required information using a fill-in-the-blank form. Checkboxes and dropdown menus supply choices where possible. The JSIRO provides free text input space for input of directly into the report. Text from e-mail and other documents maybe copied and included into the JSIRO report or added as attachments. Submitted reports maybe updated as further information becomes available.

13.3   Joint Spectrum Data Repository

13.63.   The Defense Spectrum Organization collects, standardizes, and distributes spectrum-related data. The Defense Spectrum Organization provides direct on-line data access to the joint spectrum data repository (JSDR) and provides customized reports. The JSDR contains DOD, national, and international spectrumrelated information up to the secret level and can be accessed via the joint spectrum center data access web server (JDAWS) tool. JDAWS provides user access to the database components of JSDR.

13.64.   The JSDR provides access to a collection of over 100 area studies. Area studies are Defense Spectrum Organization produced country-specific telecommunication profiles hosted on Intelink. Area studies found within the JDAWS provide a hyper-link access to the Intelink site.

13.65.   The JSDR contains various resources in a variety of formats. The following are the primary features of the JSDR—

- **Joint Equipment, Tactical, and Space (JETS) Database**: The JETS segment of JSDR is a Defense Spectrum Organization created and maintained resource that includes: Parametric data for DOD; commercial and multinational equipment; platform data, including equipment complements; U.S. military unit names, locations and hierarchy; U.S. military unit equipment and platform complements; and space satellite parametric and orbital data.
- **Host Nation Spectrum Worldwide Database Online (HNSWDO)** is a web-based application for processing DOD Host Nation Coordination Requests and responses.
- **Spectrum Certification System (known as SCS) Database** is the central archive repository for all DOD spectrum certification system data, including information from the joint force 12 (known as the J/F-12), Application for Equipment Frequency Allocation. J/F-12 is the unique tracking number assigned by the Army Spectrum Management Office.
- **Background Environmental Information (known as BEI) Database**: To accurately represent the electromagnetic environment, the Defense Spectrum Organization collects additional non-U.S. Federal and international frequency assignments, which are stored in the BEI database. The BEI currently includes International Telecommunication Union, Federal Communications Commission, Canadian, and Radio Astronomy assignments.
- **Government Master File (known as GMF) Database**: The GMF is a data source containing records of the frequency assigned to all U.S. Federal Government agencies in the U.S. and its possessions. Data is obtained from NTIA.
- **Frequency Resource Record System (known as FRRS) Database:** FRRS contains information on DOD frequency assignments used throughout the world that is controlled by the Commanders of the Unified Commands and the Military Departments.
- **Electronic Order of Battle (known as EOB) database**: The JSDR contains nearly 25,000 Defense Intelligence Agency EOB foreign equipment locations.

Appendix A

Spectrum Management Task List

This appendix describes the current spectrum manager task list to include each task and their supporting sub-tasks. This appendix also contains flow charts that describe the collaboration process between spectrum managers and the EW Cell.

A.1 Tasks

A.1. Tasks assigned or associated with spectrum management are based on unit specificity. Spectrum management encompasses a wide range of military activities and missions. Each unit will have standard operating procedures to enable spectrum management operations.

A.2. Each unit provides the spectrum manager with a unique set of circumstances. As an example, assignments to Aviation units differ from Special Operations units in the deployment and use of SDD and related systems.

A.1.1 Plan the Use of the Electromagnetic Spectrum for all Spectrum Dependent Devices

A.3. Planning for spectrum use requires information from a variety of sources. The spectrum manager uses force structure templates to plan missions. Forces submit spectrum requirements for all devices used for the mission to the spectrum manager. The spectrum manager submits frequency proposals to appropriate agencies in the correct format (SFAF or SSRF). Frequency record creation in the appropriate database prevents other units from requesting the same spectrum resources. Supporting sub-tasks for plan the use of the electromagnetic spectrum for all spectrum dependent devices include—

- Conduct a data call.
- Generate frequency proposal.
- Process frequency proposal from subordinate units.
- Analyze spectrum resource allocations and partition them into allotment plans and assignments.
- Nominate assignments against allotments (spectrum resources).
- Create and edit a frequency record.
- Provide input to the production of Annex H (OPORD).

A.1.2 Conduct Electromagnetic Interference Analysis

A.4. The spectrum manager conducts analysis of the spectrum's impact on the mission. Identification of EMI caused by a mission occurs during the initial planning process using SMO tools. This allows course of action (COA) development to eliminate or mitigate the interference. Spectrum users and spectrum managers identify EMI during mission execution through various ways, such as reports of degraded communications, inoperable sensors, or malfunctioning equipment. The spectrum manager analyzes the EMI to identify the cause of the EMI.

Dynamic Spectrum Access Decisions: Local, Distributed, Centralized, and Hybrid Designs, First Edition. George F. Elmasry.
© 2021 John Wiley & Sons Ltd. Published 2021 by John Wiley & Sons Ltd.
Companion website: www.wiley.com/go/elmasry/dsad

EMI happens for various reasons, such as operator programming errors, or blue, grey, or red force jamming (intentional or otherwise). The primary resources that the spectrum manager has for EMI mitigation is spectrum monitoring and direction finding devices used in conjunction with the JSIR process and interagency collaboration. As outlined in the JSIR procedures, mitigate EMI at the lowest echelon possible. EMI reporting, to higher echelons, occurs for all EMI occurrences. Reporting EMI occurs regardless of a resolution for the interference.

Note. See CJCSM 3320.02D for more information on JSIR.

The following sub-tasks support the task conduct electromagnetic interference analysis—

- Identify EMI.
- Provide recommendation to eliminate and or mitigate interference.
- Prevent frequency substitution by locking nets, and assignments.
- Provide recommended frequency modification or substitution by user assigned priority.
- Import and validate JSIR input from subordinates.
- Export JSIR to higher headquarters.

Note. The JSC serves as the center for EMI mitigation and monitors the JSIRO collaboration portal. JSIRO is accessible through the SIPRNET link provided in the reference portion of CJCSM 3320.02D. JSIRO is currently the preferred method of reporting EMI occurrences.

A.1.3 Assign Frequencies Within the Operational Parameters of SDD and Available Resources

A.5.   The use of SMO tools provides the spectrum manager with operational characteristics of all SDDs validated by the DD Form 1494 process. The spectrum manager performs analysis of the operational requirements of a mission based on the characteristics of each device. Host nation comments and agreement allows the spectrum manager to construct allocation tables for the operational area. The spectrum manager assigns frequencies based on these allocations to requesting units for use during the mission. The following sub-tasks support this task—

- Conduct data call.
- Determine if SDD is supportable in area of interest.
- Coordinate for spectrum usage with host nation.
- Create and edit a frequency record.

A.1.4 Obtain Requests and Provide Electromagnetic Spectrum Resources to Requesting Unit

A.6.   Subordinate units submit frequency requests, in the correct format (SFAF or SSRF), to the spectrum manager after a unit receives a mission and determines spectrum requirements to support that mission. The following sub-tasks support this task—

- Conduct data call.
- Determine if SDD is supportable in area of interest.
- Coordinate for spectrum usage with host nation (HNSWDO).
- Create and edit a frequency record.

A.1.5 Provide Electromagnetic Operational Environment Information in Either a Networked or Stand-Alone Mode

A.7.   Sharing of information within and between agencies is critical for accurate and efficient spectrum management. As SMO tools become more NCE compliant, sharing of critical information among agencies becomes easier. As the spectrum manager may not always have access to the network, SMO tools must remain functional in a stand-alone mode. The following sub-tasks support this task—

- Derive specific mission requirements from operational plan.
- Maintain situational awareness of the EMOE.
- Conduct EMOE information data exchange with peer-to-peer, subordinate to higher and higher to subordinate users.
- Delete, modify, and export user selected background data.
- Conduct analysis.

A.1.6 Perform Modeling and Simulation of the emoe Via User Selected Data Fields of the Impact of the emoe on Projected Spectrum Plans

A.8. Modeling and simulation of the EMOE using SMO tools allow for mitigating the effects of SDD on unintended bystanders. It also allows for development of various COAs during the MDMP upon receipt of an OPORD or fragmentary order. It is critical for the spectrum manager to monitor the spectrum continually in order to detect EMI or EW during mission performance. The following sub-tasks support this task—

- Conduct data call.
- Maintain situational awareness of the EMOE.
- Derive specific mission requirements from the operation plan (OPLAN) or OPORD.
- Conduct analysis.

A.1.7 Monitor and Use Spectrum Common Operational Picture Information in Support of Unified Land Operations

A.9. The COP provides commanders with an easy to understand picture of all relevant information that pertains to a mission. This requires an accurate and up-to-date depiction of spectrum use within the operational area. For instance, the spectrum manager uses a spectrum analyzer or monitoring receiver to identify signals in the operational area and overlay the results with a color-coded display on a two or threedimensional picture of the area. The following sub-tasks support this task—

- Maintain situational awareness of the EMOE.
- Export the Spectrum Plan in a format compatible for import by mission command systems.
- Provide spectrum situational awareness to the common operational picture.

A.1.8 Prioritize Spectrum Use Based on Commanders Guidance

A.10. When the requirement for spectrum exceeds the supply, spectrum use priority becomes established. The commander, normally with input from the G-6 or S-6 spectrum manager, institutes prioritization. Priorities placed into various SMO tools for planning missions makes prioritization very efficient. Prioritization of spectrum users allows interference mitigation in accordance with the commander's intent. The following sub-tasks support this task—

- Maintain situational awareness of the EMOE.
- Identify conflicts.
- Perform spectrum course of action analysis.

A.1.9 Utilize Electronic Warfare Reprogramming During the Nomination, Assignment, and Deconfliction Processes

A.11. Blue force electronic warfare can easily disturb other spectrum users within the EMOE. Coordination between the CEMA element spectrum manager and the G-6 or S-6 spectrum manager can mitigate many of these disturbances. SMO tools allow the spectrum manager to analyze the effects of EW and provide frequency deconfliction recommendations to return spectrum users to operational status (if possible). There are no sub-tasks associated with this task.

A.1.10 Import Satellite Access Authorization

A.12.   The Defense Information Systems Agency regional satellite support center responsible for the area of operations for the mission disseminates satellite access authorizations to all required agencies concerned with satellite resources, to include the brigade satellite communications noncommissioned officer. Spectrum managers import the authorizations for all satellite users within the unit and transfer it to the proper MCEB format. This allows for a more complete picture of the spectrum for all spectrum management agencies. The supporting task is modify satellite access authorization record to ensure required data fields comply with the MCEB standard for assignment.

> ***Note.*** The satellite access authorization authorizes frequencies for use on satellite systems. The satellite access authorization does not provide area frequency clearance in the operational area. Spectrum managers must obtain frequency clearance from the host nation using guidelines for the respective geographic commander prior to allowing units to transmit on the assigned uplink frequency. Deconflict these frequencies from other ground-based emitters to prevent interference during mission execution.

A.1.11   Generate and Distribute SOI and JCEOI

A.13.   The SOI and JCEOI provide the Army and joint units with detailed regulations concerning spectrum use for the duration of a mission. The spectrum manager must use SMO tools with SOI or JCEOI generation and distribution capabilities to provide units with this regulation. The following sub-tasks support this task—

- Conduct data call.
- Build and test base SOI or JCEOI.

A.1.12   Create, Import, Export, Edit, Delete, Display, and Distribute the Joint Restricted Frequency List

A.14.   The JRFL is a management tool used by various operational, intelligence, and support elements to identify the level of protection desired for a critical function utilized within the electromagnetic spectrum. EW planners utilize the JRFL to conduct mission planning and to mitigate the effects of friendly offensive and defense electronic attack when possible. The JRFL does not provide protection from other spectrum users. Planners limit JRFL entries to the minimum number of radio frequencies and intelligence equities necessary for friendly forces to accomplish mission objectives. The JRFL entry contains at a minimum—

- Tactical/operational point of contact for frequency usage.
- Center channel of the frequency assignment.
- Emission designator.
- Name of receiver location.
- Geolocation of receiver.
- Protection radius of receiver.
- Justification for protection.
- JRFL code (protected, taboo, guarded).
- Serial number of Spectrum XXI frequency record for transmitters only. Receivers or sensors do not have a record.

A.15.   The spectrum manager receives requests from subordinate units to place friendly force spectrum users into the JRFL. Spectrum managers validate organizational and subordinate JRFL requests and forward them to higher echelons for approval. The command with responsibility for developing and promulgating the JRFL validates subordinate unit input. Upon completion of the JRFL, the spectrum manager disseminates the JRFL to subordinate users. This task is supported by the following sub-tasks—

- Gather and compile JRFL input.
- Validate JRFL input (codes: taboo, guarded, and protected).
- Export JRFL input to higher headquarters.
- Import completed JRFL from higher headquarters.
- Export completed JRFL to subordinates.

A.1.13 Access and Use Spectrum Operations Technical Data

A.16. Every SDD has operational characteristics that allow it to perform the intended functions. The spectrum manager accesses these characteristics through various spectrum databases and uses them during the frequency assignment process to ensure that spectrum resources support the proper operation of the device. Some of these characteristics include waveforms, number of frequencies used, transmit and receive power, and frequency bands. The following sub-tasks support this task—

- Delete, modify, and export user selected background data.
- Determine if spectrum dependent device is supportable in the operational area.

A.1.14 Manage, Store, and Archive Spectrum Use Data (Frequency Management Work History) and Utilize Host Nation Comments in the Spectrum Nomination and Assignment Process

A.17. The spectrum manager uses SMO tools to file spectrum use data and utilize host nation comments during the spectrum nomination and assignment process. This process not only aids current mission planning but also planning for future missions. The following sub-tasks support this task —

- File data according to regulatory records.
- Coordinate for spectrum usage with host nation (HNSWDO).

A.2 Sub-task List

A.18. The following list provides a description of the sub-tasks as they pertain to the functions of the Army spectrum manager—

- **Conduct Data Call:** The spectrum requirements data call message provides guidance to staff elements, components, and supporting agencies on how to request spectrum support for SDD systems that operate under their control within the area of operations. This multi-part message should cover the following subjects—
  - Spectrum management policy and guidance.
  - Security classification guidance.
  - Frequency and communications-electronic operating instructions.
  - Master net list request procedures.
  - Guidance for identifying nets and frequencies to be included on the JRFL.

  Note. For a sample of the data call format, see CJCSM 3320.01C, Annex A, appendix A, enclosure C.

- **Process frequency proposals from subordinate units:** The spectrum manager receives frequency requests from subordinate units in the format described in the data call message. This allows the spectrum manager to place the required information into the planning software and analyze the impact of the request on the spectrum. Also of concern is receipt of agency approval, host nation supportability operations using host nation comments, receiving and updating spectrum related databases, and input from the area frequency coordinator. Once the spectrum requirements exist within databases, the spectrum manager determines spectrum supportability of the request.

- **Generate frequency proposal:** Once the frequency proposal processing is complete, the spectrum manager submits the proposal in the correct format (SFAF or SSRF) to obtain frequency assignment. Use of SMO tools allows the manager to accurately generate and submit frequency proposals to the appropriate agencies.

- **Analyze spectrum resource allocations and partition them into allotment plans and assignments:** spectrum managers receive a range of frequency allocations in a given area for SDD. The spectrum manager can use SMO tools to analyze force spectrum requirements and submit frequency proposals based on the analysis.

- **Nominate assignments against allotments (spectrum resources):** If provided allotments for use within given bands of the spectrum, the spectrum manager assigns frequencies to spectrum users. The SMO tool in use during the planning process, determines possible frequency assignments and if they are supportable.

- **Create and edit a frequency record**: A frequency record includes all information pertaining to spectrum use of a specific unit or force (blue, red, or grey). Frequency records include characteristics, capabilities, frequency proposal and assignment, frequency clearance, and the force structure supporting the frequency use. Frequency records consolidation occurs during the normal procedures for obtaining frequency assignment with SMO tools. Location of the frequency records are in various databases.

- **Provide input to the production of Annex H (OPORD):** Annex H of the OPORD concerns signals. The spectrum manager places key spectrum information in Annex H of the OPORD. This allows the commander and subordinate units to have a clear picture of the operational environment.

- **Identify EMI:** EMI can present itself in various ways. For instance, a communications terminal may contact the brigade or battalion headquarters concerning difficulty receiving a signal from another communications terminal. The primary tool used to identify immediate EMI is S2AS. The S2AS can scan the specific frequency range that the terminal is operating within for jamming, intermodulation, and noise, and eventually locate (through direction finding) and assist the spectrum manager in determining the cause of the EMI (frequency fratricide or enemy EW).

- **Provide recommendation to eliminate and or mitigate interference:** The SMO tools in use can perform mitigation or frequency deconfliction of EMI occurrences. Recommendations provided to the commander from the spectrum manager enhance decision-making. The commander may decide to continue with limited spectrum use or obtain frequency reassignment.

- **Prevent frequency substitution by locking nets and assignments:** Based on mission priority and commander's discretion, the JRFL lists frequencies and networks that require protection from friendly force spectrum users. A variety of SMO tools allow for automatically locking nets and assignments during the mission planning process.

- **Provide recommended frequency modification or substitution by user:** Frequency modification or substitution occurs to obtain new frequencies for users that experience unresolved EMI. The commander may deem frequency modification necessary based on user priority during EW operations.

- **Conduct analysis:** The spectrum manager conducts analysis when using SMO tools to plan spectrum use. Tools determine the impact of spectrum use in the operational area by calculated EMI, spectrum requirements, and force structure. The analysis results determine if the spectrum can support a given COA.

- **Export Spectrum Plan in a format compatible for import by mission command systems:** The SMO tools currently in use are capable of exporting the correct format for use by various command systems. The spectrum manager verifies accuracy and completeness of the spectrum plan prior to exporting it in the correct format to various mission command systems.

- **Provide spectrum situational awareness to the COP:** This occurs during mission performance by using spectrum analyzers or receivers. The spectrum manager can use these tools while stationary to detect unknown or unplanned signals. Mobile packages or antennas allow for direction finding and locating these signals to determine spectrum COA analysis.

- **Identify conflicts:** Spectrum awareness identifies when spectrum conflicts occur. These conflicts may be blue, grey, or red forces. Use the JSIR procedures and spectrum awareness tools to locate, characterize, and determine critical information concerning the signal(s) in question.

- **Perform spectrum COA analysis:** Differing SMO tools develop COAs during the planning phase of a mission. This allows the commander to choose the best COA. During mission execution, EMI occurrence requires the development of COAs. The nature of the EMI (blue, red or grey force caused EMI) determines the development of COAs. The spectrum manager may possibly require new frequencies for users. Another COA, based on the impact of the EMI and mission priority, may be simply to do nothing. The JSIR procedures include directions and reporting procedures to mitigate EMI.

- **Modify Satellite Access Authorization records to ensure required data fields comply with the MCEB standard for assignment:** Spectrum users that depend on satellite resources require a satellite access authorization from the regional satellite communications support center responsible for the location of the user. The regional satellite communications support center disseminates satellite access authorizations to the brigade satellite communications operations noncommissioned officer that requested the satellite resources. The spectrum

  manager must receive the authorization and transfers the information into the correct SFAF or SSRF (MCEB Publication 7 or 8) format prior to obtaining frequency clearance in the area. SMO tools automatically complete this process after importing the authorization.

  Note. The regional satellite communications support center generally interfaces with brigade satellite operations. In some cases, the brigade spectrum manager is also the satellite operations NCO.

- **Import and validate JSIR input from subordinates:** Report EMI at the lowest level recognized. The spectrum manager attempts to mitigate the EMI at the lowest level possible using the JSIR procedures (CJCSM 3320.02D). If that level cannot rectify the situation, it escalates to the next higher level until EMI resolution. Spectrum users and managers of all levels report EMI occurrences to the next higher echelon, regardless of severity or cause. SMO tools allow the spectrum manager to import a JSIR report and determine the validity of the information. If SIPRNET access is available, use the JSIRO collaboration portal for EMI reporting.

- **Export JSIR to higher headquarters**: Once imported and validated the next higher headquarters takes action. If SIPRNET access is available, use the JSIRO collaboration portal for EMI reporting. If not, various SMO tools allow for exporting the JSIR to higher headquarters.

- **Derive specific mission requirements from OPLAN or OPORD:** The OPLAN or OPORD contains a variety of information that spectrum managers may use to perform key tasks, such as generating the SOI or performing a data call.

- **Maintain Spectrum Analysis of the EMOE:** This task is an ongoing task for the duration of a mission. Ideally, the spectrum manager performs live spectrum analysis even before the mission becomes active to determine whether the planned frequencies have interference once active. Live spectrum monitoring plays a critical role in identifying, analyzing, and mitigating EMI.

- **Conduct EMOE information data exchange with peer-to-peer, subordinate to higher and higher to subordinate users:** Spectrum managers update a variety of databases, especially in a joint environment, to remain effective in spectrum use. SMO tools currently in use allow for easy data exchange through common formats and central databases.

- **Delete, modify, and export user selected background data:** User selected background data involves obtaining detailed SDD data and characteristics. Background data characteristics are located in spectrum related databases. The spectrum manager must update the selected background data periodically to ensure that the databases reflect accurate information.

- **Build and test base SOI or JCEOI**: The spectrum manager uses SMO tools to develop the SOI or JCEOI based on mission requirements and commander's intent. The SOI or JCEOI gives the spectrum user guidelines for operating within the spectrum and instructions for reporting spectrum issues.

- **Determine if spectrum dependent device is supportable:** Completion of the DD Form 1494 is critical in determining the area of interest supportability. Also of use are the various spectrum databases, such as Spectrum XXI, CJSMPT's spectrum knowledge repository, and HNSWDO.

  Note. The user of the SDD is responsible for DD Form 1494 processing and completion.

- **File data in accordance with regulatory records:** Data compliance with SFAF or SSRF, Federal Communications Commission, NTIA, International Telecommunications Union, and host nation formatting to file data correctly. Use of various SMO tools automates the process of formatting during the frequency acquisition process.

- **Gather and compile JRFL input:** Depending on mission priority and commander's discretion, some (but not all) spectrum users may be on the JRFL protected list.

- **Validate JRFL input (codes: taboo, guarded, and protected):** Many users request placement on the JRFL. However, JRFL code selection requires validation of mission priority and commander's discretion.

- **Export JRFL input to higher headquarters:** Once the JRFL validation is complete, the spectrum manager exports it to higher headquarters to place the user's SDD on the central JRFL. SMO tools allow the user to export JRFL information in the correct format.

- **Import completed JRFL from higher headquarters:** The higher headquarters completes and compiles the JRFL based on subordinate unit's inputs. The spectrum manager then imports the JRFL from the higher echelon and prepares to disseminate it to subordinate units.

- **Export completed JRFL to subordinates:** The spectrum manager disseminates the approved JRFL to subordinate units to place the JRFL into effect. Various SMO tools allow for the easy distribution of the completed JRFL.

- **Coordinate for spectrum usage with host nation (HNSWDO):** When operating outside the U.S. and its possessions, it is critical to coordinate spectrum use within the area of operations with the host nations. Use of the spectrum within a host nation without authorization from that nation causes international consequences, such as fines, imprisonment, or loss of life. HNSWDO is the primary means for the spectrum manager to determine host nation spectrum supportability for SDD.

- **Perform person-to-person host nation coordination:** When delegated under combatant command authority the Joint Task Force J6 JSME may be required to conduct host nation coordination in support of Joint Task Force spectrum access within the joint operational area.

- **Distribute JRFL electronically or by printed text:** The spectrum manager disseminates the completed JRFL to the units that require it. SMO tools currently in use allow for easily disseminating the JRFL to required agencies electronically or by printed text.

A.3   SMO to EW Flow Charts

A.19.   The following flow charts describe the collaboration process between the G-6 or S-6 spectrum manager and the EW Cell. Figure A.1 shows an overview of the entire process. Descriptions of tasks shown in these figures that relate to spectrum manager are located beneath the chart. For more information concerning EW Cell tasks, review ATP 3-36 and JP 3-13.1.

A.20.   Figure A.2 shows a detailed description of the SMO tasks that support the collaboration process.

A.21.   **Generate Tactical Spectrum Plan and Develop COAs:** The spectrum managers use various SMO tools to plan a mission (CJSMPT, Spectrum XXI, and SPEED). The spectrum manager generates a data call message to all subordinate units. The data call message directs the units to—

- Identify all SDD.
- Define spectrum policy.
- Defines the procedures included on the JCEOI, and defines JRFL Guidance.

A.22.   **Receive Frequency Request:** As the units answer the message, the spectrum manager receives frequency requests in accordance with mandatory formats (SFAF, SSRF, NTIA or International Telecommunications Union required items). Spectrum managers review the spectrum dependent device characteristics and determine if each has passed the spectrum certification process (DD Form 1494), and is supportable in the operational area by reviewing host nation comments (HNSWDO). The spectrum manager validates frequency requests by checking for inflated requests (such as the unit requests more frequency than needed). The spectrum manager validates JRFL requests with the G-2 and G-3; this ensures warranted protection requests. The spectrum manager also prioritizes spectrum users, with the G-3, to aid in planning and prioritizing frequency requests.

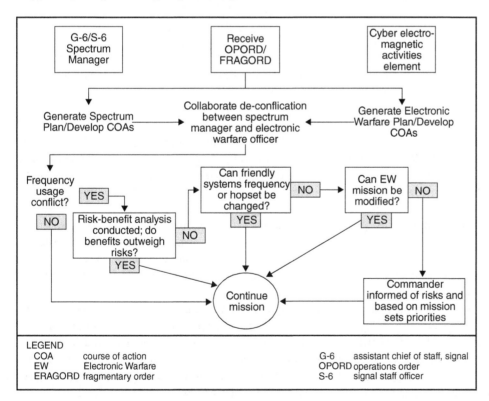

Figure A.1 The SMO to EW collaboration process. From Techniques for Spectrum Management Operations, Dec 2015. Headquarters, Department of the Army.

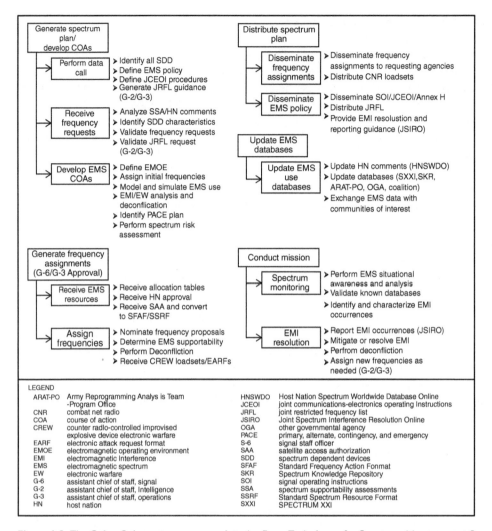

Figure A.2 The G-6 or S-6 spectrum manager's tasks. From Techniques for Spectrum Management Operations, Dec 2015. Headquarters, Department of the Army.

A.23. **Develop COA:** Spectrum managers develop COAs for the mission and issue initial frequency assignments to perform modeling and simulation for the spectrum. Various SMO tools identify EMI caused by various sources and provide deconfliction recommendations. Spectrum managers also perform a spectrum risk assessment to determine the effects of the SDD in the area of responsibility. The commander may choose the COA according to all of the identified spectrum issues and risks.

A.24. **Generate frequency assignments:** The spectrum manager receives spectrum resources, in the form of allocation tables and permissions, from higher echelon (such as Army Spectrum Management Office, JFMO, and host nation coordination). The spectrum manager uses SMO tools to transfer the information into SFAF or SSRF format and place it into spectrum use databases.

A.25. Spectrum managers assign frequency and nominate frequency proposals to the approving authority using SMO tools. Upon receiving approved frequency assignments, the spectrum manager determines spectrum supportability of any new or revised frequencies. Spectrum managers receive counter radio-controlled improvised explosive device EW loadsets and electronic attack request format frequencies from the EW Cell. The spectrum manager performs deconfliction to mitigate EMI caused by EW efforts.

A.26. **Distribute Tactical Spectrum Plan:** The spectrum manager disseminates the plan to all required agencies (JFMO, JSME, and CEMA element) and provides spectrum data to communities of interest such as unified action partners. The spectrum manager generates and distributes combat network radio loadsets. The spectrum manager disseminates approved policies for spectrum use, to include the SOI or JCEOI Annex H (OPORD), the completed JRFL, and EMI resolution guidance (CJCSM 3320.02D).

A.27.   **Update Spectrum Databases:** The spectrum manager updates various spectrum related databases HNSWDO, Spectrum XXI.

A.28.   **Conduct Mission:** Spectrum managers conduct spectrum monitoring prior to conducting the mission to validate spectrum databases and identify differences between planned authorized frequencies and spurious or unauthorized frequencies in use. Spectrum managers use spectrum analyzers and spectrum analysis software to monitor frequencies. Spurious and unauthorized frequencies may be found through direction finding and triangulation. Spectrum monitoring during the mission identifies and characterizes EMI occurrences. Upon EMI occurrence, the spectrum manager performs EMI resolution mitigation and reporting procedures. The spectrum manager uses the characterized data to submit a JSIR report. The spectrum manager then follows the steps in the CJCSM 3320.02D, to attempt to resolve and mitigate the EMI at the lowest echelon possible. If resolution is not possible, the spectrum manager provides spectrum users with new frequency assignments.

A.29.   Figure A.3 provides a graphic depiction of EW Cell tasks that support the SMO collaboration process.

A.30.   The CEMA element, with guidance from the S-6, G-6, G-2, and G-3, generates the EW Plan that includes EA, electronic warfare support, and EP planning. This plan results in

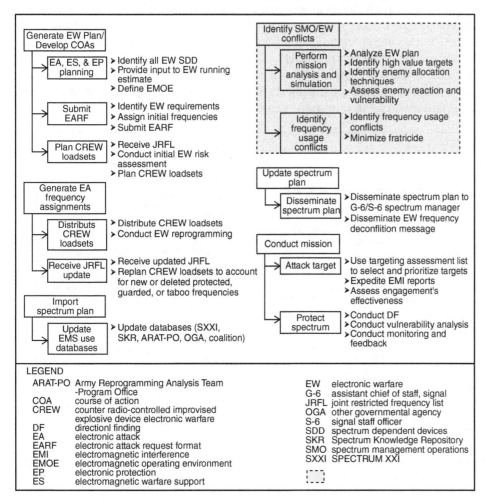

Figure A.3 The CEMA element tasks. From Techniques for Spectrum Management Operations, Dec 2015. Headquarters, Department of the Army.

development of the electronic attack request format and includes spectrum use requirements. The CEMA element plans counter radiocontrolled improvised explosive device EW loadsets.

A.31. The CEMA element plans and distributes counter radio-controlled improvised explosive device EW loadsets to required users. The JRFL, once imported, aids in planning counter radio-controlled improvised explosive device EW loadsets.

A.32. The CEMA element receives the tactical spectrum plan from the G-6 and S-6 spectrum manager. The CEMA element also uses the tactical spectrum plan and SMO tools to identify conflicts caused by the EW plan.

A.33. The CEMA element, upon determining that the EW plan causes no conflicts, updates the tactical spectrum plan and disseminates it to the G-6 and S-6 spectrum manager. This allows the G-6 and S-6 spectrum manager to update spectrum databases and prevent frequency assignments that conflict with the EW Plan. Battalion and brigade staff elements receive the EW plan and approve the validated plan. Upon approval, put the EW plan into action.

A.34. Figure A.4 shows the collaboration between the spectrum manager, the G-6, and the G-3.

A.35. Collaboration and deconfliction with spectrum manager occurs when the CEMA element identifies frequency conflicts. The collaboration determines if friendly systems can change frequencies, if not, consider possible modification of EW mission, determine if friendly forces can use a different system. If these steps resolve the conflict, continue to conduct the mission. The G-3 determines which services or missions to end or alter. Services or mission termination are done by priority of the EW mission or the service. Based on G-3 guidance, the spectrum manager performs an assessment on the new, altered mission. Determine the spectrum supportability for the mission. Develop COAs for the mission change. With direction from the G-3, select and enact the appropriate COA. Refer to G-6 or S-6 spectrum manager and CEMA element to conduct mission blocks.

A.36. Upon mission completion, each agency conducts after mission actions. These include submitting frequency assignments for deletion, updating spectrum databases, and updating host nation comments to aid in future mission planning.

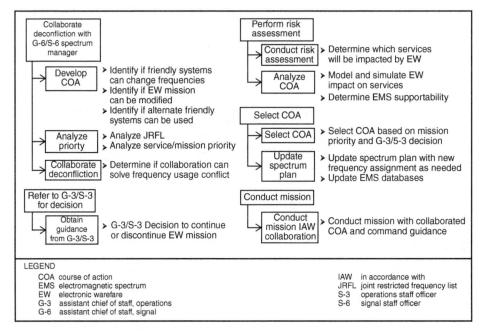

Figure A.4 SMO collaboration tasks. From Techniques for Spectrum Management Operations, Dec 2015. Headquarters, Department of the Army.

Appendix B

Capabilities and Compatibility between Tools

This appendix provides an overview of the capabilities and compatibilities of various SMO tools. There are many tools spectrum managers use to perform their duties. Due to the many tools available, compatibility understanding is of great importance.

B.1 Capabilities and Compatibility

B.1. The spectrum manager should have an understanding of SMO tool capabilities and compatibility in order to complete required tasks. Table B.1 shows the current compatibilities among tools with a description of known compatibility fixes. Table B.2 on page B.2 shows the capabilities by tool with a description of each capability.

From Techniques for Spectrum Management Operations, Dec 2015. Headquarters, Department of the Army.

B.1.1 Compatibility between SMO Tools

B.2. For the purpose of this ATP, SMO tools have format compatibility if they can import and export spectrum related files from other SMO tools without the need to modify the format. Format compatibility between tools complies with the NCE concept and reduces time constraints for spectrum manager in a joint environment. Limited format compatibility means that the tool has the capability of importing or exporting spectrum related files between another tool but requires the spectrum manager to manipulate format inconsistencies. For instance, legacy Spectrum XXI records modify or delete records in Spectrum XXI Online, but cannot create new records.

B.3. Table B.2, on page B.2, compares SMO tool and shows the capability of each tool. A description of each capability follows the table.

From Techniques for Spectrum Management Operations, Dec 2015. Headquarters, Department of the Army.

B.1.2 SMO Tool Capabilities

B.4. SMO tool capabilities are the attributes of a system or tool utilized to perform spectrum management operations. The following list is not all inclusive and many of the SMO tools are constantly updated.

- **Spectrum measurement and direction finding:** This tool takes measurements (live) of the spectrum and provides direction finding of unknown, unplanned, EW, or EMI signals.
- **Live spectrum analysis**-This tool receives spectrum measurements and provides analysis of the measured signals over time for the purpose of frequency records, trend analysis, and EW interference detection.

Dynamic Spectrum Access Decisions: Local, Distributed, Centralized, and Hybrid Designs, First Edition. George F. Elmasry.
© 2021 John Wiley & Sons Ltd. Published 2021 by John Wiley & Sons Ltd.
Companion website: www.wiley.com/go/elmasry/dsad

Table B.1 Compatibility between SMO tools.

TOOL	Spectrum XXI	Spectrum XXIO	CJSMPT	S2AS	AESOP	SPEED	ACES/JACS	HNSWDO
Spectrum XXI		*	X	X	X	X	X	
Spectrum XXIO	*		X	X	X	X	X	X
CJSMPT	X	X		X		X	X	X
S2AS	X	X	X					
AESOP	X	X						X
SPEED	X	X	X					X
ACES/JACS	X	X	X					X
HNSWDO		X	X		X	X	X	
Legend:	**X** = formatting compatible * = limited format compatibility							
	blank = not currently compatible							
ACES	automated communications engineering software							
AESOP	Afloat Electromagnetic Spectrum Operations Program							
CJSMPT	Coalition Joint Spectrum Management Planning Tool							
HNSWDO	Host Nation Spectrum Worldwide Database Online							
JACS	joint automated communications-electronics instructions system							
S2AS	Spectrum Situational Awareness System							
SPEED	systems planning, engineering, and evaluation device							

- **EW and EMI analysis and frequency deconfliction (fixed location):** This tool analyses EW and EMI effects (actual or planned) on spectrum use and provides recommendations and COAs for deconfliction of the EW and EMI for stationary SDD. This also includes the analysis of second or third order harmonics, intermodulation and electromagnetic environmental effects in the areas impacted by spectrum use.
- **EW and EMI analysis and frequency deconfliction (on-the-move):** Same as above, with the exception that the tool analyzes and deconflicts SDDs while conducting communications on-the-move.
- **Two or three-dimensional simulation and modeling of EMOE:** This tool provides both a two and three-dimensional model of the EMOE to include topography, electromagnetic environmental effects, and color-coded spectrum footprints.
- **Plan spectrum reuse and minimize requirements:** This tool minimizes the impact of a mission on the spectrum through the reuse of frequencies in different locations and planning for the minimum requirements for spectrum users and provides for more flexible and available spectrum resources for all users.
- **Import Satellite Access Authorization and convert to SFAF or SSRF:** This tool imports an authorization and automatically converts it to the SFAF or SSRF approved format. This capability provides a more complete spectrum database and aids in the mitigation of EMI caused by or affecting space based SDD.
- **Force structure templates:** This tool has the capability of creating or accessing force structures and placing them quickly and easily on the three-dimensional map of the battlefield using drag and drop, with associated SDD and the general spectrum requirements for those devices.
- **Assign frequencies:** The tool can assign frequencies to users that have submitted a frequency proposal in the correct format (SFAF or SSRF). A check mark means that the tool is capable and authorized to assign frequencies. An asterisk means that the tool may assign and plan for projected frequencies approved by another tool prior to use of the projected frequency.

Table B.2 SMO tool capabilities.

SMO Capability	Spectrum XXI	Spectrum XXIO	CJSMPT	HNSWDO	S2AS	AESOP	SPEED	ACES/JACS
Spectrum measurement and direction finding					X			
Live Spectrum Analysis					X			
EW/EMI analysis and frequency deconfliction (fixed location)	*	X	X			X	X	
EW/EMI analysis and frequency deconfliction (on-the -move)		X	X			X	X	
2D/3D Simulation and Modeling of EMOE	*	X	X		X	*	X	
Plan spectrum reuse and minimize requirements			*				*	X
Import satellite access authorization and convert to SFAF/SSRF			X				X	
"Drag and Drop" Force structure templates		*	X			X	X	
Assign Frequencies	X	X	*			*	*	*
Access JETS database		X	X					
Access 2D/3D Digital Terrain databases	*	X	X		X	X	X	
Access CREW loadsets			X			X		
Access HN comments				X				
XML Format		X	X		X	X	X	
HTML Format			X		X			
CSV Format			X		X		X	
Generate SOI/JCEOI						X		X
Provide input to SOI/JCEOI			X					
Generate JSIR report	X	X	X					
Provide JRFL input to higher echelon	X	X	X			X	X	
SFAF Format	X	*	X		X	X	X	X
SSRF Format		X	*			X	X	
Provide input to COP							X	
COA Development			X			X	*	
NCE Compliant	X	X	X			*	*	
Standalone operations	X		X	X	X	X	X	X

Legend: **X** = formatting compatible ***** = limited format compatibility
blank = not currently compatible

(continued)

- **Access joint spectrum center equipment, tactical, and space (JETS) database:** This tool may query the JETS database for information and receive updates (refreshes) from the database.
- **Access three-dimensional Digital Terrain databases:** This tool imports detailed three-dimensional digital terrain data from a variety of sources, such as the terrain integrated

Table B.2 (Continued)

SMO - spectrum management operations ACES - automated communications engineering software AESOP - Afloat Electromagnetic Spectrum Operations Program CJSMPT - Coalition Joint Spectrum Management Planning Tool COA - course of action COP - common operational picture CSV - comma separated values EMI - electromagnetic interference EMOE - electromagnetic operational environment	EW - electronic warfare HN - host nation HNSWDO - Host Nation Spectrum Worldwide Database Online HTML - hypertext markup language JACS - joint automated communications-electronics operation instructions system JCEOI - joint communications-electronics operating instructions JRFL- joint restricted frequency list JSIR - joint spectrum interference resolution NCE - network-centric environment	S2AS - Spectrum Situational Awareness System SFAF - standard frequency action format SOI - signal operating instructions SPEED - systems planning, engineering, and evaluation device SSRF - standard spectrum resource format XML - extensible markup language

rough earth model, or Google Maps for the purpose of planning, managing, and visualizing the EMOE. If marked with an asterisk, the tool has limited capabilities, such as two-dimensional maps only.

- **Access counter radio-controlled improvised explosive device EW loadsets:** This tool can access counter radio-controlled improvised explosive device EW loadsets for frequency deconfliction planning.
- **Access Host Nation comments:** This tool has access to host nation allocation tables that aid in planning spectrum use in various host nation locations. If marked with an asterisk, the tool has limited access, such as relying on importing data from another tool.
- **XML Format-Extensible Markup Language (XML)** is a file format that is rapidly becoming the standard for compatibility between software packages. If selected, the tool may import or export various files (records, reports, and data) in XML format.
- **HTML Format:** Hypertext Markup Language (HTML) is the main markup language for creating web pages and other information displayed in a web browser. HTML elements form the building blocks of all websites. HTML allows embedding images and objects within web pages to create interactive forms. It provides a means to create structured documents by denoting structural semantics for text such as headings, paragraphs, lists, links, quotes and other items.
- **Comma-separated values (CSV) Format:** CSV is a common, relatively simple file format that is widely supported by consumer, business, and scientific applications. Among its most common uses is moving tabular data between programs that natively operate on incompatible (often proprietary or undocumented) formats. This works because so many programs support some variation of CSV at least as an alternative import or export format. If checked, the tool can import or export various files in CSV format.
- **Generate SOI or JCEOI:** SOI or JCEOI provide for policies and regulations to subordinate units. If selected, the tool may generate and disseminate SOI or JCEOI in the correct format to subordinate units.
- **Provide input to SOI or JCEOI:** This tool provides input to the SOI or JCEOI in compatible formats, but cannot generate the SOI or JCEOI.
- **Generate JSIR report:** Use the JSIR report format to report EMI occurrence to the next higher echelon in a joint task force operating environment. This tool has the capability of generating the JSIR report in the correct format and exporting the report to the next higher echelon.
- **Provide JRFL input to higher echelon:** The JRFL is a time and geographically oriented listing of functions, nets, and frequencies requiring protection from friendly EW. This tool

allows the operator to process JRFL input from subordinate forces, provide the input to higher echelons in the correct format, and distribute the JRFL to concerned units upon approval from higher echelon.

- **SFAF Format:** This tool can propose, assign, modify, renew, review, and delete radio frequencies in the SFAF (MCEB Pub 7) approved format. If marked with an asterisk, the tool has limited capabilities, such as it can only import and read a SFAF frequency record.
- **SSRF Format:** This tool can propose, assign, modify, renew, review, and delete radio frequencies in the SSRF (MCEB Pub 8) approved format. If marked with an asterisk, the tool has limited capabilities, such as the tool requires manual workarounds to generate proposals in the SSRF approved format.
- **Provide input to COP:** The COP is a single identical display of relevant information shared by more than one command. This tool can interface and provide relevant spectrum information to the COP.
- **COA Development:** This tool can analyze spectrum resources impacted by a mission and develop many COAs to determine how to best support the mission and mitigate spectrum conflicts.
- **NCE Compliant:** This tool is compliant with the NCE concept by providing central locations for access of information (through SIPRNET or NIPRNET) to authorized users both within an agency (vertically) and between agencies (horizontally) in joint environments. If marked with an asterisk, the tool is only partially NCE compliant, such as not providing information between agencies.
- **Operate in standalone environment:** This tool can operate while disconnected from outside agencies or central databases.

Appendix C

Spectrum Physics

This appendix describes the physics of radio frequency (RF) spectrum. A basic understanding of the underlying principles of RF energy is necessary to the execution of spectrum management operations.

C.1 Radio Frequency

C.1.   RF communications, based on the laws of physics, describes the behavior of electromagnetic energy waves. RF communication works by creating electromagnetic waves at a source and being able to receive those electromagnetic waves at a particular destination. These electromagnetic waves travel through the air at the speed of light. The wavelength of an electromagnetic signal is inversely proportional to the frequency; the higher the frequency, the shorter the wavelength.

C.2.   Frequency measurements are in Hz (cycles per second) and radio frequency measurements are in kHz (thousands of cycles per second), MHz (or millions of cycles per second) and GHz (or billions of cycles per second). The wavelength for a device utilizing a frequency in the MHz range is longer than frequency in a GHz range. In general, signals with longer wavelengths travel a greater distance and penetrate through, and around objects better than signals with shorter wavelengths.

C.3.   Waveforms are patterns of electrical energy over time. A Sine wave is the fundamental building block of electricity and other energies. A Sine wave mathematically defines a natural action describing a harmonic alternating event.

C.4.   Figure C.1 provides a graphic depiction of a simple waveform. Displacement is the crest (high point) and trough (low point) of a wave. The wavelength is the distance from one crest to another or trough to another. Amplitude is the height of a crest or trough.

C.1.1 Harmonics and Intermodulation Products

C.5.   Frequencies are associated with different standing wave patterns that produce wave patterns known as harmonics. Figure C.2 on page C-2, displays the relationship between the wave that produces the pattern and the length of the medium in which the pattern is displayed. The pattern for the first harmonic reveals a half wavelength where each point on the line represents nodes and the arching middle represents antinodes. The second harmonic displays a complete wavelength; this pattern described as starting at the rest position, rising upward to a peak displacement, returning down to a rest position, then descending to a peak downward displacement and finally returning back to the rest position.

C.6.   One complete wave in a standing wave pattern consists of two *loops*. Thus, one loop is equivalent to one-half of a wavelength. The third harmonic pattern consists of three anti-nodes. Thus, there are three loops within the length of the wave. Since each loop is equivalent to one-half a wavelength, the length of the wave is equal to three-halves of a wavelength. The table has a pattern when inspecting standing wave patterns and the length-wavelength

Dynamic Spectrum Access Decisions: Local, Distributed, Centralized, and Hybrid Designs, First Edition. George F. Elmasry.
© 2021 John Wiley & Sons Ltd. Published 2021 by John Wiley & Sons Ltd.
Companion website: www.wiley.com/go/elmasry/dsad

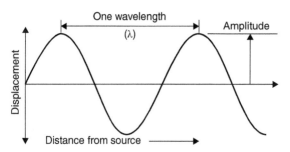

Figure C.1 Waveform characteristics. From Techniques for Spectrum Management Operations, Dec 2015. Headquarters, Department of the Army.

Harmonic	# of Nodes	# of Antinodes	Pattern	Length-Wavelength(λ) Relationship
1st	2	1		$L = 1/2 \cdot \lambda$
2nd	3	2		$L = 2/2 \cdot \lambda$
3rd	4	3		$L = 3/2 \cdot \lambda$
4th	5	4		$L = 4/2 \cdot \lambda$
5th	6	5		$L = 5/2 \cdot \lambda$
6th	7	6		$L = 6/2 \cdot \lambda$

Figure C.2 Wavelength relationship. From Techniques for Spectrum Management Operations, Dec 2015. Headquarters, Department of the Army.

relationships for the first three harmonics. The number of antinodes in the pattern is equal to the **harmonic number** of that pattern. The first harmonic has one antinode; the second harmonic has two antinodes; and the third harmonic has three antinodes. The mathematical relationship simply emerges from the pattern and the understanding that each loop in the pattern is equivalent to one-half of a wavelength. The general equation that describes this length-wavelength relationship for any harmonic is on the right side column of Figure C.2.

C.7. Harmonics develop into currents and voltages with frequencies that are multiples of the fundamental frequency. Harmonic signals that fall within the pass band of a nearby receiver and the signal level are of sufficient amplitude can degrade the performance of the receiver. Receivers live under constant bombardment of signals which enter through the antenna port. Some of these signals quickly attenuate due to front-end filtering, which is often referred to as pre-selection.

C.8. Intermodulation generation occurs when multiple signals reach a non-linear element such as a detector, mixer, or amplifier and are mixed. Whenever signals are mixed, two additional signals are introduced as the sum and difference of the original frequencies. This process is often intentional as in the case of mixing a frequency with the intermediate frequency in a system to produce the desired operating signal. Harmonics of the original two frequencies are still present but most occur well outside the pass band of the RF and intermediate frequency filters and cause no problems. The harmonics that tend to cause the most problems are the odd-order products. For example, if a 50 MHz mixing frequency is combined with a 98 MHz intermediate frequency to produce a desired transmission signal of 148 MHz, this is very close to the 3rd order harmonic of 50 MHz (150 MHz) and may cause interference at the desired frequency. Channelized communications systems tend to suffer more from these issues due to the uniform spacing of the channels.

C.1.2 Transmission, Propagation and Reception

C.9.   A radio transmits a signal by driving a current on an antenna where the current amplitude is the changing quantity of the signal. This changing current, in turn, induces an electromagnetic field about itself, with a field strength that corresponds to the current amplitude. This electromagnetic field propagates away from the antenna as a wave at the speed of light. As the signal propagates, it attenuates. At a distant receiver, the electromagnetic wave passes across the receiver's antenna and induces a current. Figure C.3 shows transmit waves and propagation.

C.10.   Electromagnetic radiation in the area passes across the receiving antenna. To detect and receive the correct signal, the receiving antenna must be able to isolate the desired signal from all others. If the receiver is in range of two transmitters using the same frequency band it is attempting to receive, then the receiver may not properly capture the desired signal for demodulation. The receive signal captured may be unintelligible. The spectrum management process attempts to prevent this situation from occurring. The goal is not to prevent transmitters from using the same frequencies but to ensure that receivers are capable of receiving and distinguishing the desired signals. There may be more than one transmitter using the same carrier frequency as long as the receivers are able to distinguish the desired signal over the others.

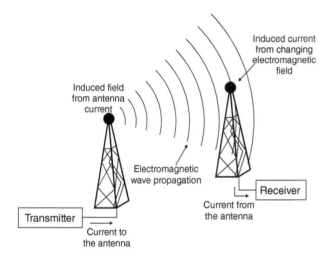

Figure C.3   Transmission and propagation of electromagnetic waves. From Techniques for Spectrum Management Operations, Dec 2015. Headquarters, Department of the Army.

Appendix D

Spectrum Management Lifecycle

The Army spectrum management lifecycle serves as a guide to follow in establishing a functional and efficient spectrum management program. The lifecycle encompasses the complete process of providing spectrum management support to the commander and is applicable to all spectrum managers regardless of duty location. The Army spectrum management lifecycle mirrors the joint task force lifecycle.

D.1 Spectrum Management Lifecycle

D.1. The spectrum management lifecycle consist of 12 activities that enhance SMO. It is not imperative to conduct the lifecycle activities in order as presented. Many spectrum managers conduct activities in the lifecycle simultaneously. Listed below are steps of the spectrum management lifecycle—

1. Define command specific policy and guidance.
2. Gather requirements.
3. Develop electromagnetic spectrum requirements summary.
4. Define the EMOE.
5. Obtain spectrum resource.
6. Develop spectrum management plan.
7. Nominate and assign frequencies.
8. Generate the joint communication-electronic operating instructions.
9. Develop the JRFL.
10. Perform electronic warfare deconfliction.
11. Resolve interference.
12. Report interference.

D.1.1 Step 1. Define Command Specific Policy and Guidance

D.2. Unit standard operating procedures establish specific guidance for managing, requesting, coordinating, and assigning spectrum use, the JRFL process, communications electronic operating instructions, and other processes. Policy and guidance information should be available in the commander's spectrum guidance, spectrum management manual, command regulations, instructions, or existing plans.

D.3. Spectrum managers require many resources. The unit standard operating procedures would establish the basic spectrum management resources needed to establish a spectrum management element in support of operations anywhere within the commander's operational area. Such resources should include digitized terrain data, background electromagnetic environment records, country area studies, copies of agreements for spectrum use or sharing with involved or adjacent host nations, and historical spectrum use records involving the operational area. This step generates two deliverable products: the spectrum concept and the spectrum requirements data call message.

Dynamic Spectrum Access Decisions: Local, Distributed, Centralized, and Hybrid Designs, First Edition. George F. Elmasry.
© 2021 John Wiley & Sons Ltd. Published 2021 by John Wiley & Sons Ltd.
Companion website: www.wiley.com/go/elmasry/dsad

D.4.   The spectrum management concept is the vision of spectrum management operations best practices performed to support the mission. The spectrum management concept comprises assumptions, considerations, and restrictions that, when analyzed together, can illustrate the best approach to managing the EMOE.

D.5.   The spectrum requirements data call message provides guidance to staff elements, components, and supporting agencies on how to request spectrum support for spectrum dependent systems that operate under their control within area of influence. This multipart message should cover the following subjects: spectrum management policy and guidance, security classification guidance, frequency and communications electronic operating instructions master net list request procedures, as well as provide guidance for identifying nets and frequencies to be included on the JRFL.

D.1.2 Step 2. Gather Requirements

D.6.   Gathering requirements can begin as soon as spectrum management receives guidance and coordination channels are defined. Spectrum managers must also obtain the requirements of spectrum users, primarily the staff elements. These requirements must address both communications and non-communications such as radar and weapons systems and stated in terms of spectrum requirements to support the command. This step involves undocumented requirements from sources external to the spectrum management coordination chain.

D.1.3 Step 3. Develop the Spectrum Requirements Summary

D.7.   This summary quantifies the amount of spectrum necessary to support the command, determine the necessity of using frequency sharing and reuse plans, and help in the development of allotment or channeling plans. This process requires compiling and analyzing the data previously generated. The spectrum manager analyzes the summary and determines the amount of spectrum required. In addition, the spectrum requirements summary determines the number of different radio services competing for spectrum in the same frequency band, determines the different emissions utilizing a particular band, and supports development of a plan for frequency sharing.

D.8.   The spectrum requirements summary generated is a compilation of the requirements identified in response to the spectrum requirements data call message. This product is for the sole use of the spectrum manager and provides a tool to base future decisions about efficient spectrum-use and initial requirements definition. This product may assist the spectrum manager in requesting spectrum from a host nation, or to better allocate portions of the spectrum, to support emitters utilizing varying bandwidths.

D.1.4 Step 4. Define the EMOE

D.9.   Military operations require a common, single, authoritative source for spectrum use information for all friendly, enemy (to the extent available), neutral, and civil emitters and receivers to achieve and manage successful joint spectrum use. This common source of spectrum use information found within the EMOE must be current, accurate, and accessible to authorized users. The spectrum manager is responsible for building and managing this common source of information.

D.10.   The EMOE database contains spectrum use information on all friendly military and civilian, available enemy, and neutral forces. Defining the EMOE not only creates a database of frequency assignments, but also identifies factors that affect signal propagation such as environmental characteristics and terrain. This activity starts with defining your operational area and its environmental characteristics, locating necessary terrain data and then locating the data for and creating a database of the known spectrum use information. Defining the EMOE is an ongoing activity. The information produced by this activity provides a baseline database digitally depicting the EMOE and the basis for all spectrum interaction analyses.

D.1.5   Step 5. Obtain Spectrum Resources

D.11.   Obtain spectrum resources needed to support the command. The spectrum manager coordinates military spectrum use with the spectrum management authority of the host nation or coalition forces involved. The host nation can request spectrum resources for exercises or most military operations other than war. Operations that preclude prior coordination with a host nation, such as forced entry, require the spectrum manager to determine the spectrum resource; evaluation of the background and history of the electromagnetic environment provides support to the spectrum manager. If required, an evaluation of the background environment is essential to establish well-defined spectrum requirements and for the EMOE to remain as up-to date as possible.

D.12.   The spectrum requirements summary can help quantify the amount of spectrum needed and identify the different radio services and emissions that may be operating within each frequency band.

D.1.6   Step 6. Develop the Spectrum Management Plan

D.13.   Unit standard operating procedures establish specific guidance for managing, requesting, coordinating, and assigning spectrum use and procedures for JRFL and JCEOI processing. Additionally, the spectrum manager is the focal point for inclusion of spectrum use considerations in the Annex H development and provides administrative and technical support for military spectrum use. This process uses the spectrum management concept, developed in the first activity, along with existing policy and guidance. Other sources of information are lessons learned from previous operations and exercises, the JSC and other spectrum managers.

D.14.   The spectrum manager devises a plan to use spectrum resources available. This plan depends upon the products of all the previous activities. Spectrum managers evaluate spectrum management plans for possible improvement on a regular basis. The spectrum management plan is typically included as an appendix to Annex H of an OPLAN or OPORD and evolves from guidance as the operation or exercise transitions from the planning to execution phase.

D.15.   The spectrum management plan provides guidance for all spectrum management functions, including information exchange, expected coordination channels, format for deliverable products, interference and reporting resolution procedures, and suggested resolution steps.

D.1.7   Step 7. Nominate and Assign Frequencies

D.16.   Nominate and assign frequencies is the actual implementation of the spectrum management plan. Authority, delegated to components, to issue frequency assignments or allotments provides the maximum latitude and flexibility in support of combat operations. This activity involves the initial assigning of frequencies. The spectrum manager may assign frequencies or delegate (decentralize) assignment authority using frequency pools (allotment plans) provided to functional and service component spectrum managers allowing them to assign frequencies.

D.17.   The frequency assignment database, which conforms to and is created based on the table of frequency allocations, radio regulations, and channel plans, is the most important resource the spectrum manager has available and forms the basis for nominating interference free assignments, providing impact analyses of EW operations, and identifying and resolving interference issues.

D.1.8   Step 8. Generate a Communications-Electronic Operating Instructions

D.18.   The CEOI is a two-part document. Part 1 is a directory of radio nets or units and their associated frequencies, call signs, call words, and network identification listed by time period.

Part 2 contains supplemental procedures for electronic, visual, and verbal interactions, such as sign or countersigns, obscurants or pyrotechnics and suffix or expanders. CEOI development and distribution is an S-6 or G-6 responsibility and delegated to the spectrum manager.

D.19.   The JCEOI provides communications details and information for joint forces, service-specific elements and units including—

- Daily changing and non-changing frequency assignments.
- SINCGARS cue, manual and net identification assignments.
- Call sign assignments.
- Call words assignments.
- Daily changing code words.
- Running passwords.

D.20.   Information found in the JCEOI includes document-handling instructions, controlling authority data, effective dates and reproduction instructions. Due to the sensitive information contained in the JCEOI, classification should be at the same level. When jointly used, the Army CEOI becomes the Joint CEOI or JCEOI.

> *Note.* The JCEOI is the most widely used communication control document in any given operational area.

D.21.   Overarching regulatory guidance for JCEOI management, to include call signs and call words, is contained within the CJCSM 3320.02, JCEOI publication. Additional guidance may apply based on command relationships with other unified action partners.

D.22.   During operations, the combatant commander is the authority for the JCEOI. The combatant commander may delegate this authority, to the ground component commander or the respective joint task force commander. There is a distinction between the air and ground component JCEOI. The relevant air component issues a Special Instructions document that is the air operations equivalent of the JCEOI.

D.23.   Within garrison, the Army Command, Army Service component command, or direct reporting unit commander has responsibility for CEOI production and distribution in support of training requirements. U.S. Forces Command may delegate authority to a corps, division or remain centralized to meet installation-training objectives. Regardless of echelon, the commander is responsible for the JCEOI. The J-6 or G-6 develops and promulgates the JCEOI. The J-3 or G-3 validates master net list requirements and resolves conflicts.

D.1.8.1   Distribution and Development

D.24.   The COMSEC facility provides distribution of the final ACES or JACS produced JCEOI product. Doing so ensures all units receive the latest JCEOI with the distribution of COMSEC. Communications cards are derivative products of the JCEOI and are METT-TC driven. Document and protect communications cards in the manner appropriate for their security classification level. These products inherit the classification level of the source JCEOI material.

D.25.   When operationally required to maintain administrative tempo with the pace of operations it may be necessary to go without management of the SINCGARS compatible loadset via ACES or JACS and maintain the list of SINCGARS networks using a spreadsheet. ACES or JACS generates the loadset.

D.1.8.2   Call Words, Call Signs, Suffixes and Expanders

D.26.   Call signs and call words establish and maintain communications. They identify the radio stations of command authorities, activities, facilities, units, elements, or individual positions. Call signs do not identify people. Tactical call sign systems meet specific military requirements under an exemption to the International Telecommunications Union radio regulations.

- **Call signs** are a combination of alphanumeric or phonetically pronounceable characters that identifies a communication facility, command, authority, activity, or unit; used primarily for establishing and maintaining communications.
- **Call words** identify units when communicating within a secured communications net. The generation of call words differ based on service component command guidance or directed based off command authority.
- **Suffixes and expanders** further assist in identifying a radio station's position or function. Care in the management of call words ensures that each station sounds phonetically different over voice transport (for example MAD DAWG 6 or MAD DOG 6). Call word usage is for secure networks only.

D.27.   Table D.1 provides an explanation of call sign, call word, suffix and expander.

From Techniques for Spectrum Management Operations, Dec 2015. Headquarters, Department of the Army.

D.1.8.3 Security Classification

D.28.   The content of the JCEOI, master net list, and communications card extracts determine the classification levels. The level of COMSEC key tag information entered into the system for ACES terminal and corresponding generated loadsets determine the classification. Similar to a classified presentation, the overall classification of a specific product would be the highest level of classification it contains.

Note. See AR 380-5 for security classification markings.

D.1.9   Step 9. Develop Joint Restricted Frequency List

D.29.   The JRFL is a time and geographically oriented listing of functions, nets, and frequencies requiring protection from friendly spectrum users. Developing the JRFL requires the spectrum manager to prepare and combine G-2, G-3, G-6, and component inputs to develop a JRFL for approval by the G-3, and when required, periodically update and distribute the JRFL.

D.30.   The JRFL is a G-3 product; it protects communications nets, from enemy communications nets exploitation, and safety of life frequencies used by the command and local civil noncombatants. The development, distribution, and maintenance of the JRFL is a task of the S-6 or G-6 and normally accomplished by the spectrum manager. Creation of the JRFL is for the CEMA element and based on guidance established by the commander, EWO and the CEMA working group.

D.31.   Leaders should become familiar with the types of protection status codes that exist for the JRFL. Knowing these status codes allows the EWO to plan jamming operations on the unrestricted frequencies for training and during operations. Three types of protection status codes apply to frequency assets identified for inclusion in a JRFL. Sample JRFL restriction status codes include—

- **Taboo frequencies:** Taboo frequencies are any friendly frequencies of such importance that they must never be deliberately jammed, interfered with by friendly forces. Normally, these include international distress, safety, stop buzzer, and controller frequencies. These frequencies include international distress, safety, and controller frequencies. They are generally long-standing as well as time-oriented. (JP 3-13.1)
- **Protected frequencies:** Those friendly frequencies used for a particular operation, identified and protected to prevent them from inadvertent jamming by friendly forces while engaged in active EW operations against hostile forces. These frequencies are of such critical importance that jamming should be restricted unless necessary or until coordination with the using unit is made. These frequencies are generally time-oriented, may change with the tactical situation, and updated periodically. Protected frequencies are friendly frequencies used for a particular operation. An example of a protected frequency would be the command net of a maneuver force engaged in the fight. (JP 3-13.1)

Table D.1   Call signs, call words, suffix and expander.

Item Name	Example	Explanation
Call Sign	X6Y24E B9K60H	The term "call sign" refers to the letter-number-letter combination that typically designates a unit element in the CEOI or JCEOI. CJCSI 3320.03A specifies that the call sign should remain daily, changing in the event that the ability to communicate securely is lost. Examples: X6Y= 1/A/1-25 INF (1st Platoon, A Co., 1-25 INF Bn)
Call Word	Bulldog24E Fury60H	Pronounceable words that identify a communications facility, command, authority, activity or unit; serves the same functionality as the call sign. The Army does not have set call words per unit. Call word deconfliction is typically handled by the highest level spectrum manager. EXAMPLES: Bulldog = 1/a/1-25 INF (1st Platoon, A Co., 1-125 INF Bn)
Suffix	X6Y24E B9K60H Bulldog24E Fury60H	The term "suffix" refers to the two digits assigned to a particular position, mission or function within a unit or element. EXAMPLES: 24 = AVN Officer/NCO 60 = G-6 or S-6
Expander	X6Y24E B9K60H Bulldog24E Fury60H	A single letter code (A through Z) used in conjunction with a suffix and call sign to identify a sub-element of the position, mission, or function. EXAMPLES: E = NCOIC H = Officer in charge
Legend	ACES	automated communications engineering software
AVN	aviation	
Bn	battalion	
CEOI	communications-electronics operating instruction	
CJCSI	Chairman of the Joint Chiefs of Staff Instruction	
Co	company	
INF	infantry	
JACS	joint automated communications-electronics operation instructions system	
JANAP	joint Army, Navy, Air Force publication	
JCEOI	joint communications-electronics operating instructions	
NCO	noncommissioned officer	
NCOIC	noncommissioned officer in charge	
SOI	signal operating instructions	

- **Guarded frequencies:** Guarded frequencies are those enemy frequencies that maybe currently exploited for combat information and intelligence. Guarded frequencies are time-oriented in that the list changes as the enemy assumes different combat postures. These frequencies may be jammed after the commander has weighed the potential operational gains against the loss of the technical information gained. (JP 3-13.1)

D.1.10   Step 10. Perform Electronic Warfare Deconfliction

D.32.   The S-3 or G-3 EW spectrum manager participates in the CEMA element representing spectrum management issues. This includes providing EW deconfliction analysis. The EWO identifies planned EA missions and request the spectrum manager perform an analysis on the impact of these missions to operations. This process requires information from the JRFL, communications electronic operating instructions, and EMOE. The analysis determines what impact the EA mission has on communication nets, systems, enemy communications nets exploitation, and possible safety of life situations.

D.33.   This product provides the CEMA element with an analysis of the potential impact of friendly EW operations on friendly forces. The CEMA element then decides if the benefits of the jamming mission outweigh the dangers of the potential fratricide. This product is time sensitive and produced on an as needed basis.

D.1.11   Step 11. Resolve Interference

D.34.   Resolving interference is a daily activity once forces have deployed and is part of the planning process. This activity encompasses the reporting and attempting to resolve EMI. Interference maybe created by various factors such as unauthorized users, faulty nomination criteria, lack of timely data exchanges, or equipment problems. Victims of interference should ensure every effort to resolve frequency interference locally. Multiple interference problems may indicate adversary EW operations, unintentional impact of blue or grey EW operations or errors in the spectrum management plan. The spectrum manager should define and analyze the EMOE to help determine the cause of an EMI problem.

D.1.12   Step 12. Report Interference

D.35.   Spectrum congestion and the nature of military operations make some level of EMI likely. Interference reporting and tracking provides the spectrum manager with a valuable historical reference for resolving future EMI problems. After performing interference analysis, always create an interference report to document the results.

D.36.   Keep these reports in a database used as a history of interference problems. The purpose of the interference report database is to provide the spectrum manager with a repository for previous interference incidents and steps taken to resolve them. This database provides a wealth of information on unit discipline, training deficiencies, and a starting place for the spectrum manager to begin resolving interference issues. Spectrum managers share this database with all. To the extent, unexplained interference persists or recurs coincident with either red, blue or grey operations, notification to the CEMA element occurs.

D.37.   Spectrum managers must be involved at the onset of interference. Spectrum managers are responsible for resolving and reporting of interference within their responsible area. This includes setting alerts in JSIRO for interference affecting the units operations. They receive notifications by secure email of interference reports submitted for action and situational understanding of the interference. Spectrum managers assist and mitigate spectrum interference at the lowest level possible and should be knowledgeable of all forms of jamming, deception, and interference. Users experiencing EMI may change frequencies only when the spectrum manager coordinates authorized replacement frequencies. Guidance for the JSIR program is contained within the CJCSM 3320.02 series manuals and instructions. Additional procedural guidance in support of the JSIR program may apply based off command relationships such as military departments, Army commands, and combatant commands.

D.1.12.1   Joint Spectrum Interference Report

D.38.   Victims of interference report EMI using JSIRO. JSIRO is a web-based, centralized application containing data and correspondence for reported EMI, intrusion, and jamming incidents. It is the repository for the results of analyses, collected data, and supporting documentation for EMI resolution to support both trend and future interference resolution analysis. JSIRO and CJCSM 3320.02 provide an operator checklist for local investigations.

D.39. EMI is any electromagnetic disturbance that interrupts, obstructs, degrades, or limits the effective performance of electronics and electrical equipment. EMI can be induced intentionally, as in some forms of electronic warfare, or unintentionally, as a result of spurious emissions, responses or intermodulation products.

D.40. EMI mitigation begins with operator-level troubleshooting and reporting. It is imperative that affected users attempt to resolve EMI incidents at the lowest possible level. Troubleshooting may identify the source of the interference as truly EMI or, as in most cases, an equipment or operator failure. Reporting facilitates situational understanding and supports the development of solutions. Report and investigate all prohibitive EMI through the JSIR program. Not all EMI incidents are prohibitive however; prohibitive EMI has an operational impact. Trained equipment operators should identify the difference between prohibitive EMI, equipment failure, and purposeful interference by the adversary. The JSIRO report is submitted through intelligence channels by the appropriate authority if the interfering signal is determined to be from a hostile source,

D.41. The spectrum manager or the victim of interference is responsible for reporting interference using the JSIR format with the information described in table D.2.

From Techniques for Spectrum Management Operations, Dec 2015. Headquarters, Department of the Army.

Table D.2   Data input for JSIR offline reporting.

Item Number	Data Input
1	Frequencies affected by the interference.
2	Locations of systems experiencing the interference.
3	The affected system name, nomenclature, manufacturer (with model number), or other system description. If available, include the equipment characteristics of the victim receiver, such as bandwidth, antenna type, and antenna size.
4	The operating mode of the affected system. If applicable, include the following: frequency agile, pulse Doppler, search, and upper and lower sidebands.
5	The characteristics of the interference (noise, pulsed, continuous, intermittent, frequency, or bandwidth).
6	The description of the interference effects on victim performance (reduced range, false targets, reduced intelligibility, or data errors).
7	Enter the dates and times the interference occurred. Indicate whether the duration of the interference is continuous or intermittent, the approximate repetition rate of the interference, and whether the amplitude of the interference is varying or constant. Indicate if the interference is occurring at a regular or irregular time of day, and if the occurrence of the interference coincides with any ongoing local activity.
8	The location of possible interference sources (coordinates or line of bearing, if known; otherwise, state as unknown).
9	A listing of other units affected by the interference (if known) and their location or distance, and bearing from the reporting site.
10	Clear and concise narrative summary information about the interference, and any local actions taken to resolve the problem. The operator is encouraged to provide any other information, based on observation or estimation that is pertinent in the technical or operational analysis of the incident. Identify whether the information furnished is an actual observation, measurement or estimate. Avoid the use of Army or program jargon and acronyms.
11	Reference message traffic related to the interference problem reported. Include the message date-time group, originator, action addressees, and subject line.
12	Indicate whether identification or resolution of the problem is completed.
13	Indicate if joint spectrum interference resolution (JSIR) technical assistance is desired or anticipated.
14	Point of contact information, including name, unit, and contact phone numbers.

D.42.   The spectrum manager or victim of interference, reports the types of interference signals, the actions used to overcome the interference, the suspected cause and other comments related to the interference signal. Send this report online or forward offline as soon as feasible, based on situation.

D.1.12.2   Types of Jamming Signals

D.43.   Jamming is an effective way for the adversary to disrupt mission command. All the adversary needs to jam is a transmitter tuned to our frequency with enough power to override friendly signals at our receivers. There are two modes of jamming. Spot jamming is concentrated power directed toward one channel or frequency. Barrage jamming is power spread over several frequencies or channels at the same time.

D.44.   Jamming can be difficult, and sometimes impossible to detect. Users of spectrum devices have the potential of being jammed and should be able to recognize jamming. The two types of jamming most commonly encountered are obvious and subtle jamming. Obvious jamming is normally very simple to detect. When experiencing a jamming incident, it is more important to recognize and overcome the incident than to identify it formally. The spectrum manager or victim reports the type of jamming signal during the JSIR process. The more commonly used jamming signals of this type are—

- **Random noise.** This is synthetic radio noise. It is random in amplitude and frequency. It is similar to normal background noise and can degrade all types of signals. Operators often mistake it for receiver or atmospheric noise and fail to take appropriate actions.
- **Stepped tones.** These are tones transmitted in increasing and decreasing pitch. They resemble the sound of bagpipes. Single-channel voice circuits are normally the victims of stepped tones.
- **Spark.** The spark signal produces the most easily and effective type of jamming. Bursts are of short duration and high intensity. Spark jamming signals, repeated at a rapid rate, is effective in disrupting all types of radio communications.
- **Gulls.** The gull signal is a quick rise and slow fall of a variable radio frequency and is similar to the cry of a sea gull. It produces a nuisance effect and is very effective against voice radio communications.
- **Random pulse.** In this type of interference, pulses of varying amplitude, duration, and rate are generated and transmitted. They disrupt teletypewriter, radar, and various types of data transmission systems.
- **Wobbler.** The wobbler signal is a single frequency modulated by a low and slowly varying tone. The result is a howling sound that causes a nuisance effect on voice radio communications.
- **Recorded sounds.** Recorded sounds are any audible sound, especially of a variable nature, to distract radio operators and disrupt communications. Music, screams, applause, whistles, machinery noise, and laughter are examples of recorded sounds jamming.
- **Preamble jamming.** This type of jamming occurs when a broadcast resembling the synchronization preamble speech of security equipment over the operating frequency of secure radio sets. Preamble jamming results in all radios being locked in the receive mode. It is especially effective when employed against radio nets using speech security devices.
- **Subtle jamming.** Subtle jamming is not obvious; no sound from the receiver radio. The radio cannot receive the intended incoming signal, even though everything appears normal to the radio operator. In effect, the threat jammers block out these radios' ability to receive a friendly transmission without the operator being aware it is happening. This is squelch capture and is a subtle jamming technique. The radio operator can readily detect jamming in all other function control modes. Often, we assume that our radios are malfunctioning instead of recognizing subtle jamming for what it is.

D.1.12.3 Recognizing Jamming

D.45.   Equipment operators must be able to recognize jamming. Threat jammers may employ obvious or subtle jamming techniques. In addition, interference caused by sources having nothing to do with adversary jamming may be the source. Jammers affect receivers and do not affect transmitters.

D.46.   Prohibitive EMI may be caused by the following—

- Unintentionally by other radios (friendly and enemy).
- Other electronic or electric or electromechanical equipment.
- Atmospheric conditions.
- Malfunction of the radio.
- Improper operation of the radio.
- Combination of any of the above.

D.47.   **Internal or external interference.** The two sources of interference are internal and external. If the interference or suspected jamming remains after grounding or disconnecting the antenna, the disturbance is most likely internal and caused by a malfunction of the radio. Contact maintenance personnel to assist in troubleshooting. Further examinations could reveal external interference from adversary jamming or unintentional interference.

D.48.   **Jamming or unintentional interference.** Causes of unintentional interference include other radios, some other type of electronic or electromechanical equipment, or atmospheric conditions. The battlefield is so crowded with radios and other electronic equipment that some unintentional interference is virtually unavoidable. Static electricity produced by atmospheric conditions can negatively affect radio communications. Unintentional interference normally travels only a short distance and a search of the immediate area may reveal the source of this type of interference. Moving the receiving antenna for short distances may cause noticeable variations in the strength of the interfering signal. These variations normally indicate unintentional interference. Conversely, little or no variation may indicate inadvertent friendly or adversarial jamming. Regardless of the source, take actions to reduce the effect of interference on our communications.

D.49.   In all cases, report suspected adversary jamming and any unidentified or unintentional interference that disrupts our ability to communicate. This applies even if the radio operator is able to overcome the effects of the jamming or interference. Information provided to higher headquarters in the JSIR report mitigates the adversary jamming efforts.

D.50.   The adversary can use two types of jamming signals: powerful un-modulated or noise-modulated signals. Un-modulated jamming signals lack any noise and noise modulated jamming signals have obvious interference noises.

D.1.12.4 Overcoming Jamming

D.51.   The adversary constantly strives to perfect and use new and more confusing forms of jamming. Our equipment operators must be increasingly alert to the possibility of jamming. Training and experience are the most important tools operators have to determine when a particular signal is a jamming signal. Exposure to the effects of jamming in training or actual situations is invaluable. The ability to recognize jamming is important, because jamming is a problem that requires action.

D.52.   Continue to operate if jamming does occur. Usually, adversarial jamming involves a period of jamming followed by a brief listening period. The adversary is attempting to determine how effective jamming has been. What the victim is doing during this short period when listening, tells the jammer how effective jamming has been. If the operation is continuing in a normal manner, as it was before the jamming began, the enemy assumes that jamming has not been particularly effective. If the adversary finds users discussing the jamming over the radio or shut down our operation entirely, the adversary may very well assume that jamming has been effective. Because the enemy jammer is monitoring operations, unless otherwise ordered, never shut down operations or in any other way disclose to the enemy that you may be adversely affected. Normal operations should continue even when degraded by jamming.

D.1.12.5 Improve the Signal-to-Jamming Ratio

D.53. The signal-to-jamming ratio is the relative strength of the desired signal to the jamming signal at the receiver. Signal refers to the frequency users are attempting to receive. Jamming refers to the hostile or unidentified interference received. A signal-to-jamming ratio in which the desired signal is stronger than the jamming signal cannot significantly degrade the desired signal.

D.54. Users experiencing jamming may take a variety of steps to improve the signal-to-jamming ratio. Adjust the receiver and ensure frequency tuning is as precise as possible to the desired incoming signal to improve the signal-to-jamming ratio. Additional techniques to improve signal-to-jamming ratio include—

- Adjusting the radio frequency bandwidth.
- Adjusting the gain or volume control.
- Fine-tuning the frequency.
- Increasing the transmitter power output.

D.55. Increasing the power output of the transmitter emitting the desired signal improves the signal-to-jamming ratio. To increase the power output at the time of jamming, the transmitter must be set on something less than full power when jamming begins. Using low power as a preventive technique depends on the adversary not being able to detect radio transmissions. Once the adversary begins jamming radios, the threat of being detected increases.

D.56. Users experiencing jamming should ensure antennas are optimally adjusted to receive the desired incoming signal. Additional techniques to improve receive signal strength regarding antenna include—

- Reorienting the antenna.
- Changing the antenna polarization. (Perform this action at all stations.)
- Installing an antenna with a longer range (higher gain).

D.57. **Relocate the antenna.** Frequently, the signal-to-jamming ratio maybe improved by relocating the antenna and associated radio set affected by the jamming or unidentified interference. This may mean moving a few meters or several hundred meters. It is best to relocate the antenna and associated radio set to an area that has a terrain feature between the user and any suspected enemy jamming location.

D.58. **Establish a retransmission station.** A retransmission station can increase the range and power of a signal between two or more radio stations. Depending on the available resources and the situation, this may be a viable method to improve the signal-to-jamming ratio.

D.59. **Use an alternate route for communications.** In some instances, enemy jamming prevents us from communicating with a radio station with which we must communicate. If radio communications have degraded between two radio stations that must communicate, there may be another radio station or route of communications that can communicate with both of the radio stations. Use the alternate radio station or route as a relay between the two other radio stations.

D.60. **Change frequencies.** If a communications net cannot overcome adversarial jamming using the above measures, the commander (or designated representative) may direct the net to switch to an alternate or spare frequency coordinated through your spectrum manager. If practical, dummy stations can continue to operate on the frequency being jammed to mask the change to an alternate frequency. Frequency changes that are preplanned result in minimal communications loss. During adversarial jamming, it is very difficult to coordinate a change of frequency.

Appendix E

Military Time Zone Designators

Spectrum managers provide support for the EMOE in support of unified action partners. Spectrum managers communicate with users across different time zones and provide commanders with operational times established during missions. Support to airborne operations provides the widest use of military time zones during flight missions. Knowledge of global military time zone differences for spectrum managers is a vital skill. Spectrum managers may be called upon to prepare briefings, conduct and assist planning or provide input to critical documents, accurate time zone information is essential.

E.1 Overview

E.1.   Military time uses the 24-hour clock beginning at midnight (0000 hours) and ending at 2359 hours. Military time format eliminates the need for using A.M. and P.M. designations as regular time uses numbers 1 to 12 to identify the hours in a day. In Military time 12 P.M. is 1200 hours, 1 P.M. is 1300 hours up until 11 P.M. where it is 2300 hours. The military uses this standard as it leaves less room for confusion than standard time. The world is divided into 24 military time zones and each military zone has a letter designation and the military phonetic alphabet word.

E.2.   The time zone for Greenwich, England is the letter "Z" and the military phonetic word is "Zulu". Since many U.S. military operations must be coordinated across times zones, the military uses Coordinated Universal Time (formerly Greenwich Mean Time) as the standard time. The U.S. Military refers to this as Zulu (Z) time and attaches the suffix to ensure the referred time zone is clear.

E.3.   When referring to specific military time zones, speak the letter or word attached. As an example, if a military exercise began at 3:00 P.M. Zulu time; or "fifteen hundred hours Zulu time" and written as 1500Z.

E.4.   The time zones from the U.S. are Romeo, Sierra, Tango, Uniform, Victor, Whiskey, X-Ray. Local time uses the letter J or Juliet. Written format for 9 A.M. local time is 0900J and spoken as "Zero 900 hours Juliet time." The Lima time zone designator does not equate to local time. See Time Zone chart for the location of Lima time zone.

E.5.   Some countries have a 15, 30, or 45-minute offset from the designated time zone designator. Although located in the Delta time zone, Juliet time in Afghanistan is Coordinated Universal Time +4 hours 30 minutes. An asterisk behind the affected time zone designator denotes that a Juliet offset is in effect. See Time Zone Chart for affected regions. This offset is recorded as 0430D* or 190430RDEC13* within a Date Time Group.

E.1.1 Military Time Zone Considerations

E.6.   While conducting military operations, spectrum managers must consider all rules for determining a specific time. The following are some consideration to be aware of—

Dynamic Spectrum Access Decisions: Local, Distributed, Centralized, and Hybrid Designs, First Edition. George F. Elmasry.
© 2021 John Wiley & Sons Ltd. Published 2021 by John Wiley & Sons Ltd.
Companion website: www.wiley.com/go/elmasry/dsad

- The military observes daylight savings time when recognized by the state or country.
- The 12 time zones west of the Zulu time zone (coordinated universal time), starting from the International Date Line and ending in the Pacific Ocean are November through Yankee.
- The 12 time zones east of the Zulu time zone (coordinated universal time), starting at the International Date Line and ending in the Pacific Ocean are Alpha through Mike.
- 12 A.M. can be both 0000 and 2400 hours. However, clocks that display military time always display it as 0000.

E.7.   Many countries use military time as their main time format. European, African, Asian, and Latin American countries commonly use military time as their main time format. In some countries, both the 12 and 24-hour clock are used. Figure E.1 shows the world map and military zone designators for each zone.

E.8.   Table E.1 outlines each time zone around the world and provides its relationship to Zulu time.

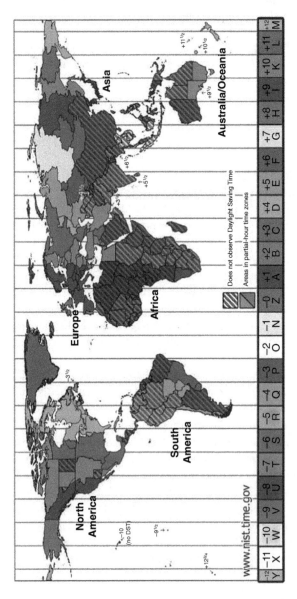

Figure E.1   World military time zone designator chart. From Techniques for Spectrum Management Operations, Dec 2015. Headquarters, Department of the Army.

Table E.1 Example of world time zone conversion (standard time).

Military Time Zone Designators

Y	X	W	V	U	T	S	R	Q	P	O	N	Z	A	B	C	D	E	F	G	H	I	K	L	M

Civilian Time Zones

IDLW	NT	HST	ASDT	PST	MST	CST	EST	AST	NST	AT	WAT	UTC	CET	EET	BT	ZP4	ZP5	ZP6	WAST	CCT	JST	GST	SBT	IDLE
1200	1300	1400	1500	1600	1700	1800	1900 b)	2000	2100	2200	2300	2400	0100	0200	0300	0400	0500	0600	0700	0800	0900	1000	1100	1200 a)

Standard Time = Universal Time + Value from Table

Z 0	E +5	K +10	P −3	U −8
A +1	F +6	L +11	Q −4	V −9
B +2	G +7	M +12	R −5	W −10
C +3	H +8	N −1	S −6	X −11
D +4	I +9	O −2	T −7	Y −12

a) = Today
b) = Yesterday

Legend
ASDT-Alaska Standard Time
AST-Atlantic Standard Time
AT-Azores Time
AWST-Australian Western Standard Time
BT-Baghdad
CCT-China Coast Time
CET-Central European Time
CST-Central Standard Time
E ET-Eastern European Time

EST-Eastern Standard time
GST-Guam Standard Time
HST-Hawaii Standard Time
IDLE-International Date Line East
IDLW-International Date Line West
JST-Japan Standard Time
MST-Mountain Standard Time
NST-Newfoundland Standard Time
NT-Nome Time

PST-Pacific Standard Time
SBT-Solomon Island Time
UTC-Coordinated Universal Time
WAST-West Africa Time Zone
WAT-West Africa Time
ZP-4 Azerbaijan, Oman, Mauritius
ZP-5 Maldives, Pakistan, Tajikistan
ZP-6 Bangladesh, Kazakhstan

From Techniques for Spectrum Management Operations, Dec 2015. Headquarters, Department of the Army.

References

Required Publications

Most joint publications are available online: www.dtic.mil/doctrine/new_pubs/jointpub.htm.
These documents must be available to intended users of this publication.
ADRP 1-02. *Terms and Military Symbols*. 7 December 2015.
JP 1-02. *Department of Defense Dictionary of Military and Associated Terms*. 8 November 2010.

Related Publications

These documents contain relevant supplemental information.

Joint Publications

Most joint publications are available online: http://www.dtic.mil/doctrine.
CJCSM 3320.01C. *Joint Electromagnetic Spectrum Management Operations in the Electromagnetic Operational Environment*. 14 December 2012.
CJCSM 3320.02D. *Joint Spectrum Interference Resolution (JSIR) Procedures*. 3 June 2013.
JP 3-13.1. *Electronic Warfare*. 8 February 2012.
JP 3-28. *Defense Support of Civil Authorities*. 31 July 2013
JP 6-01. *Joint Electromagnetic Spectrum Management Operations*. 20 March 2012.
MCEB Pub 7. *Standard Frequency Action Format (SFAF)*. 20 November 2012.
MCEB Pub 8. *Standard Spectrum Resource Format (SSRF)*. 4 April 2012.

Army Publications

Most Army doctrinal publications and regulations are available online: http://www.apd.army.mil.
ADP 3-0. *Unified Land Operation*. 10 October 2011.
ADP 5-0. *The Operations Process*. 17 May 2012.
ADRP 3-0. *Unified Land Operations*. 16 May 2012.
ADRP 3-28. *Defense Support of Civil Authorities*. 14 June 2013.
AR 5-12. *Army Use of the Electromagnetic Spectrum*. 15 February 2013.
AR 380-5. *Department of the Army Information Security Program*. 29 September 2000.
ATP 3-36. *Electronic Warfare*. 16 December 2014.
FM 3-38. *Cyber Electromagnetic Activities*. 12 February 2014.
FM 6-0. *Commander and Staff Organization and Operations*. 05 May 2014.
FM 6-02. *Signal Support to Operations*. 22 January 2014.
FM 27-10. *The Law of Land Warfare*. 18 July 1956.

Dynamic Spectrum Access Decisions: Local, Distributed, Centralized, and Hybrid Designs, First Edition. George F. Elmasry.
© 2021 John Wiley & Sons Ltd. Published 2021 by John Wiley & Sons Ltd.
Companion website: www.wiley.com/go/elmasry/dsad

Other Publications

National Telecommunications and Information Administration Redbook. *Manual of Regulations and Procedures for Federal Radio Frequency Management.* May 2013 found online at the following; http://www.ntia.doc.gov/osmhome/redbook/redbook.html.

Title 10 of the United States Code. *Armed Forces.*

Title 32 of the United States Code. *National Defense.*

Title 47 of the United States Code. *Telecommunication.*

http://uscode.house.gov/.

International Telecommunication Union Radio Regulations. Edition of 2008.

Prescribed Forms

None.

Referenced Forms

Unless otherwise indicated, DA Forms are available on the Army Publishing Directorate (ADP) website: www.apd.army.mil.

DA Form 2028. *Recommended Changes to Publications and Blank Forms.*

DD Forms are available on the Office of the Secretary of Defense (OSD) web site: www.dtic.mil/whs/directives/infomgt/forms/formsprogram.htm.

DD Form 1494. *Application for Equipment Frequency Allocation.*

Websites

Acquisition Community Connection.

https://acc.dau.mil/CommunityBrowser.aspx?id=18002&lang=en-US.

Spectrum XXI overview. http://www.disa.mil/jsc/pdf/SPECTRUMXXI_JSC.pdf.

Part IV

The IEEE Standards 1900x – 2019 – Dynamic Spectrum Access Networks Standards Committee (DySPAN-SC)

14

IEEE Standard for Definitions and Concepts for Dynamic Spectrum Access: Terminology Relating to Emerging Wireless Networks, System Functionality, and Spectrum Management

14.1 Overview

14.1.1 Scope

This standard provides definitions and explanations of key concepts in the fields of spectrum management, spectrum trading, cognitive radio, dynamic spectrum access, policy-based radio systems, software-defined radio, and related advanced radio system technologies. The document goes beyond simple, short definitions by providing amplifying text that explains these terms in the context of the technologies that use them. The document also describes how these technologies interrelate and create new capabilities while at the same time providing mechanisms supportive of new spectrum management paradigms.

This revision to IEEE Std 1900.1™-2008 adds additional definitions, modifies existing definitions, and removes outdated definitions; it updates the auxiliary text and informative annexes to reflect new concepts and developments in advanced radio systems; introduces a taxonomy of terms which depicts relationships between definitions and concepts, and updates the document structure to align revised definitions, concepts, and relationships between terms and definitions.

14.1.2 Purpose

New concepts and technologies are rapidly emerging in the fields of spectrum management, spectrum trading, cognitive radio, dynamic spectrum access, policy-based radio systems, software-defined radio, and related advanced radio system technologies. Many of the terms used do not have precise definitions or have multiple definitions. This document facilitates the development of these technologies by clarifying the terminology and how these technologies relate to each other.

14.2 Acronyms and Abbreviations

For the purposes of this document, the following terms and definitions apply. The *IEEE Standards Dictionary Online* should be consulted for terms not defined in this clause.[1]

ACS	adaptive channel selection
A/D	analog/digital
AI	artificial intelligence
B	bandwidth
BER	bit error rate; bit error ratio
CCA	clear channel assessment
CDMA	code division multiple access

---

1 *IEEE Standards Dictionary Online* is available at: http://dictionary.ieee.org.

Dynamic Spectrum Access Decisions: Local, Distributed, Centralized, and Hybrid Designs, First Edition. George F. Elmasry.
© 2021 John Wiley & Sons Ltd. Published 2021 by John Wiley & Sons Ltd.
Companion website: www.wiley.com/go/elmasry/dsad

cdma2000	code division multiple access 2000
CR	cognitive radio
CRN	cognitive radio network
CSMA/CA	carrier sense multiple access with collision avoidance
DCA	dynamic channel assessment
DECT	Digital Enhanced Cordless Telecommunications
DFS	dynamic frequency selection
DSM	dynamic spectrum management
DSA	dynamic spectrum access
DSP	digital signal processor
EDGE	Enhanced Data Rates for GSM Evolution
FCC	Federal Communications Commission
FPGA	field programmable gate array
GERAN	GSM EDGE Radio Access Network
GNSS	Global Navigation Satellite System
GPS	Global Positioning System
GSM	global system for mobile communications
IMT-2000	International Mobile Telecommunications-2000
I/O	input/output
ITU	International Telecommunication Union
ITU-R	International Telecommunication Union Radiocommunication Sector
J2EE	Java 2 Enterprise Edition
J2ME	Java 2 Micro Edition
LAN	local area network
LMSC	Local and Metropolitan Area Network Standards Committee
MAC	media access control
MEMS	microelectromechanical systems
MExE	mobile executable environment
MMS	multimedia messaging service
OSI	Open Systems Interconnection
PBAR	policy-based adaptive radio
PHY	physical layer of the Open Systems Interconnection (OSI) model
PPDR	public protection and disaster relief
QoS	quality of service
RAT	radio access technology
RAN	radio access network
RF	radio frequency
RR	reconfigurable radio
RTLS	real-time location system
SCR	software-controlled radio
SDR	software-defined radio
SLA	service level agreement
SMS	short messaging service
SO	spectral opportunity
SPTF	Spectrum Policy Task Force
T/R	transmit/receive
UMTS	Universal Mobile Telecommunications System
UTRA	UMTS Terrestrial Radio Access
Weff	wireless efficiency

Wi-Fi	registered trademark of the Wi-Fi Alliance
WiMAX	Worldwide Interoperability for Microwave Access
WM	wireless medium
WPAN	wireless personal area network

14.3   Definitions of Advanced Radio System Concepts

14.3.1   Adaptive Radio

A type of radio system that has a technical means of varying its operational parameters in response to stimuli originating internally or externally to the radio environment.

> NOTE—An adaptive radio may change its operational parameters in order to improve its own performance or that of the network in which it operates, typically by a closed-loop action.[2]

14.3.2   Cognitive Radio

Cognitive radio[3, 4]is:

a) A radio that utilizes cognitive processes to control its behavior. *See also:* **cognitive process**.
b) A radio that exploits cognition to control its behavior. *See also:* **cognition**.
c) A type of radio in which communication systems are aware of their environment and internal state and can make decisions about their radio-operating behavior based on that information and predefined objectives. These systems learn from past experience regarding these actions and adapt their decisions based on that knowledge.
d) A radio, as defined in item a) that uses software-defined radio, adaptive radio, and other technologies to adjust its behavior or operations to achieve desired objectives.

> NOTE 1—The environmental information may or may not include location information related to communication systems.

> NOTE 2—The term *cognitive radio* is often used to describe systems, as defined in item a), that do not incorporate learning functionality.

> NOTE 3—A cognitive radio commonly includes a cognitive control mechanism. *See also:* **cognitive control mechanism**.

> NOTE 4—The desired objectives may exist in the scope of a certain use case, and may adapt based on a change in that use case.

14.3.3   Hardware-defined Radio

See: **Hardware radio**.

> NOTE—Hardware-defined radio is often used synonymously with hardware radio.

14.3.4   Hardware Radio

A type of radio that implements communications functions entirely through hardware such that changes in communications capabilities can only be achieved through hardware changes. *Contrast:* **software-defined radio**.

---

2 Notes in text, tables, and figures of a standard are given for information only and do not contain requirements needed to implement this standard.

3 The IEEE recognizes that the terminology commonly used is *cognitive radio*. However, generally the cognitive functionality may be outside the boundary normally associated with a radio (e.g., environment sensing is a cognitive function that is not normally part of a radio).

4 The IEEE notes that the terms *dumb radio*, *aware radio,* and *smart radio* are used in the technical literature, but the IEEE does not define these terms at this time. They are additional descriptive terms that are sometimes applied to radios.

NOTE 1—This term represents an idealized abstraction that is useful in designating categories of radio devices (e.g., hardware radio, software-defined radio, and cognitive radio) to which certain regulatory provisions or functional capabilities may apply. The term is also useful in describing the general evolution in the software reconfigurability of radio devices with hardware radio not being software reconfigurable and software-defined radio being software reconfigurable.

NOTE 2—Replacing a hardware component with an identical component containing different stored data or executable instructions is considered to be a physical modification of the device. Furthermore, a device having regulated parameters that can be changed without physical modification is not considered a hardware radio, even if such change requires specialized equipment or proprietary procedures.

NOTE 3—For consistency with regulatory definitions of software-defined radio, regulators often consider a hardware radio to be a type of radio, regardless of the communications function implementation technique, in which regulated emission or reception parameters cannot be changed in the field, post manufacture, without physically modifying the device.

14.3.5 Intelligent Radio

A type of cognitive radio that is capable of machine learning. *See also:* **machine learning**.

NOTE—Intelligent radios are considered to be a subset of software-controlled radios. This dependency occurs because the definition for machine learning in 8.6 requires an intelligent radio to adapt the decision-making process, which implies dynamic reprogramming. The reprogramming of radio control would not be possible unless the radio were also software controlled. However, such a radio need not be software defined since it would be possible for an intelligent radio to use software control to adapt the decision-making process controlling how or when a fixed set of hardware-defined physical layer implementations are used. A software-defined, software-controlled intelligent radio would have the additional ability to change the physical layer implementation dynamically under software control.

14.3.6 Policy-based Radio

A type of radio in which the behavior of communications systems is governed by a policy-based control mechanism. *See also:* **policy-based control mechanism**.

NOTE 1—Policies may restrict behaviors (e.g., policies constraining time, power, or frequency use) associated with a specific set of radio functions, but they do not necessarily change the functional capability of a radio. Because policies often do not change basic radio functionality, a policy-based radio need not also be a reconfigurable radio.

NOTE 2—Because the definition for the term policy-based control mechanism in 4.9 considers radio policy to be a type of radio-control software, policy-based radios are considered to be a subset of software-controlled radios.

14.3.7 Reconfigurable Radio

A type of radio whose technology allows the modification of communications functions by software.

14.3.8 Software-controlled Radio

A type of radio where some or all of its radio interface functions and parameters can be set or managed by software. *See also:* **hardware radio**.

NOTE 1—Software-controlled radio implies more dynamic operational flexibility from radio interfaces, in contrast with software-defined radio which might in some cases be seen as a static implementation of the radio interface by software.

NOTE 2—In some cases software control of the radio interface may be understood to be a function of a software-defined radio.

14.3.9 Software-defined Radio

A type of radio in which some or all of the physical layer functions are software defined. *See also:* **waveform processing**. *Contrast:* **hardware radio**.

> NOTE 1—Radios in which the communications functions are implemented in software are considered hardware radios for regulatory purposes if the regulated emission or reception parameters cannot be changed in the field, post manufacture, without physically modifying the device. However, a device having regulated parameters that can be changed without physical modification is considered a software-defined radio, even if such a change requires specialized equipment or proprietary procedures.

> NOTE 2—This term represents an idealized abstraction that is useful in designating categories of radio devices (e.g., hardware radio, software-defined radio, and cognitive radio) to which certain regulatory provisions or functional capabilities may apply. The term is also useful in describing the general evolution in the software reconfigurability of radio devices with hardware radio not being software reconfigurable and software-defined radio being software reconfigurable. Software-defined radios include software-reconfigurable hardware such as microprocessors, digital signal processors, and field-programmable gate arrays that are used with software to implement communications functions. The degree of software reconfigurability will depend on the radio implementation.

> NOTE 3—U.S. Federal Communications Commission definition: A radio that includes a transmitter in which the operating parameters of frequency range, modulation type, or maximum output power (either radiated or conducted), or the circumstances under which the transmitter operates in accordance with Commission rules, can be altered by making a change in software without making any changes to hardware components that affect the radio frequency emissions (FCC, FCC05-57 [B13][5]).

> NOTE 4—ITU-R definition (ITU-R M.2063 [B37] and ITU-R M.2064 [B38]) including the text in the dashed list:—A radio in which radio frequency (RF) operating parameters including but not limited to frequency range, modulation type, or output power can be set or altered by software, or the technique by which this is achieved.—Excludes changes to operating parameters that occur during the normal preinstalled and predetermined operation of a radio according to a system specification or standard.—SDR is an implementation technique applicable to many radio technologies and standards.—Within the mobile service, SDR techniques are applicable to both transmitters and receivers.

> NOTE 5—For the purposes of this standard, the term *SDR* may be applied to radios consisting of one or more transmitters, receivers, or combinations of both. It is noted that the FCC definition of SDR is applicable only to the transmitter portion of the radio. The ITU-R and IEEE definitions of software-defined radio are applicable to both the transmitter and the receiver.

> NOTE 6—For the purpose of this standard, SDR applies to a radio in which major parts of the physical layer are implemented by software.

14.4 Definitions of Radio System Functional Capabilities

14.4.1 Adaptive Modulation

A radio system function that changes the modulation type in response to a stimulus.

> NOTE—Typically, these adjustments are in response to link and other external conditions and are used to achieve desired performance (e.g., bit error ratio, data rate, robustness, and range).

14.4.2 Cognition

The ability to acquire and apply knowledge and to reason based on that knowledge, within a given scope, to reach a certain goal or objective.

---

5 The numbers in brackets correspond to those of the bibliography in Annex 14D.

14.4.3 Cognitive Control Mechanism

Cognitive control mechanism is an instantiation of cognitive processes that enable a radio to acquire knowledge about its environment and internal state, and to apply that knowledge to adapt its behavior based on predefined objectives.

14.4.4 Cognitive Process

A cognitive process is a series of actions which, as a whole, build an essential constituent of cognition.

14.4.5 Cognitive Radio System

A cognitive radio system:

a) Is a radio system that exploits cognition to control its behavior.
b) Is a radio system that utilizes cognitive processes to control its behavior.
c) Is a radio system employing technology that allows the system to obtain knowledge of its operational and geographical environment, established policies, and its internal state; to dynamically and autonomously adjust its operational parameters and protocols according to its obtained knowledge in order to achieve predefined objectives; and to learn from the results obtained (adopted from ITU-R SM.2152 [B39]). *See also:* **cognitive radio network**; **intelligent radio**.

> NOTE—In some contexts, a cognitive radio system is seen as a collection of communicating entities each implementing cognitive radio technologies.

14.4.6 Frequency Agility

The ability of a radio to change its operating frequency automatically; typically these changes are rapid or span a wide frequency range or have both of these change characteristics.

14.4.7 Geolocation Capability

The ability to locate one's position in respect to the surface of the Earth.

> NOTE—An adaptive radio having a geolocation capability may use awareness of its position as an input that influences its behavior. Accurate location awareness could improve the performance of dynamic spectrum access and support the implementation of radio-based policies having geographic constraints. Geolocation awareness may be in terms of absolute coordinates (e.g., global navigation satellite system [GNSS]), by indoor localization (e.g., real-time location system [RTLS]), or relative to other radios (e.g., multilateration using precise signal time of arrival measurements).

14.4.8 Location Awareness

The ability of a radio to determine its location.

> NOTE—Location may refer to geographical coordinates (longitude, latitude, and elevation) or to other information such as what country or city the radio is in.

14.4.9 Policy-based Control Mechanism

A mechanism that governs radio behavior by sets of rules, expressed in a machine-readable format, that are independent of the radio implementation regardless of whether the radio implementation is in hardware or software.

> NOTE 1—The definition and associated modification of radio functionality occur: a) During manufacture or reconfiguration; b) During configuration of a device by the user or service provider; c) During over-the-air provisioning; d) By over-the-air or other real-time control.

NOTE 2—As implied by the scope of this standard, the control of radio dynamic spectrum access behavior is expected to be a typical application of a policy-based control mechanism. However, the concepts of policy-based control could be applied to network management policies as well. Policy sources include spectrum regulators, manufacturers, and network operators.

14.4.10 Policy Conformance Reasoner

The system component of a policy-based radio system that evaluates the policy compliance of transmission requests (adopted from IEEE Std 1900.5™-2011 [B26]). *See also:* **policy-based radio**.

NOTE—The policy conformance reasoner is capable of making logical inferences from a set of asserted facts and rules (i.e., policies). It is able to formally prove or disprove a hypothesis (e.g., that a transmission request is policy compliant), and is capable of inferring additional knowledge (e.g., identifying transmission opportunities for unbound transmission requests).

14.4.11 Policy Enforcer

The realization of the policy enforcement point in a policy-based radio system (adopted from IEEE Std 1900.5-2011 [B26]). *See also:* **policy-based radio**; **policy conformance reasoner**; **system strategy reasoning capability**.

NOTE 1—The policy enforcer examines transmission control commands received from the system strategy reasoning capability to ensure the policy conformance reasoner has found them to be policy compliant and outputs only policycompliant commands to the transmitter for execution.

NOTE 2—The policy enforcer might maintain a record of previous system strategy reasoning capability transmission decision requests, policy conformance reasoner transmission decision replies, and validity periods to support its policy enforcement function.

NOTE 3—The policy enforcer might be co-located with the radio system.

14.4.12 Radio Awareness

An attribute or characteristic incorporated in a radio to internally maintain and detect changes in information about its own location, spectrum environment, and internal state.

NOTE—Radio awareness is required to support the cognitive control mechanisms.

14.4.13 Software Controlled

The use of software processing within the radio system or device to select the parameters of operation.

14.4.14 Software Defined

The use of software processing within the radio system or device to implement operating (but not control) functions.

14.4.15 System Strategy Reasoning Capability

The system functional capability of a policy-based radio system that generates transmission requests and proposes the parameters of each transmission (adopted from IEEE Std 1900.5-2011 [B26]). *See also:* **policy-based radio**.

NOTE—The system strategy reasoning capability gathers data to optimize radio operation, formulates communications strategies, and coordinates these strategies with the policy conformance reasoner to evaluate compliance with the active policy set.

14.4.16 Transmit Power Control

Transmit power control is a mechanism to adjust automatically, or in response to a received command, the transmission power of a radio.

14.5 Definitions of Decision-making and Control Concepts that Support Advanced Radio System Technologies

14.5.1 Coexistence Policy

Policy specifying coexistence constraints and parameters (adopted from IEEE Std 1900.5-2011 [B26]). *See also:* **policy**.

> NOTE 1—Coexistence policy may be specified by the regulator as a subset of the regulatory policy, or might be specified by the spectrum manager/planner or the system administrator.

> NOTE 2—For example, a coexistence policy might specify a listen-before-talk coexistence mechanism and might specify the sensor detection threshold (e.g., -90 dBm in a 10 kHz bandwidth).

14.5.2 DSA Policy Language

Dynamic spectrum access (DSA) policy language is a specialization of a policy language, taking policy attributes from a knowledge domain pertaining to dynamic spectrum access. *See also:* **policy language**.

14.5.3 Formal Policy

A set of formulas in the logic associated with the policy language that specifies how a resource (e.g., radio spectrum) may be used (adopted from IEEE Std 1900.5-2011 [B26]). *See also:* **policy language**.

14.5.4 Meta-policy

One or more assertions in the policy language that state relationships between policies (adopted from IEEE Std 1900.5-2011 [B26]). *See also:* **policy language**.

14.5.5 Model-theoretic Computational Semantics

Defines logical consequents by relating statements in the language to entities in a given structure, the so-called "model" (adopted from IEEE Std 1900.5-2011 [B26]).

> NOTE—The importance of the model-theoretic semantics lies in its simplicity that allows us to understand the effect of a policy without the need to understand the reasoning process.

14.5.6 Policy Language

A formal language designed for the special purpose of expressing policies. *See also:* **DSA policy language**.

> NOTE—A policy language syntactically defines valid combinations of contextual attributes and renders applicable rules or actions as semantic output.

14.5.7 Reasoner

The decision-making entity which uses a logical system to infer formal conclusions from logical assertions. It is able to formally prove or disprove a hypothesis and is capable of inferring additional knowledge (adopted from IEEE Std 1900.5-2011 [B26]).

14.6 Definitions of Network Technologies that Support Advanced Radio System Technologies

14.6.1 Cognitive Radio Network

A type of radio network in which the behavior of each radio is controlled by a cognitive control mechanism to adapt to changes in topology, operating conditions, or user needs.

NOTE—Nodes in a cognitive radio network do not have to be cognitive radios. Rather a cognitive network is a network of radio nodes in which the nodes are subjected to cognitive control mechanisms. Each node may have cognitive capabilities, or it may receive instructions from another node with such capabilities. The cognitive capabilities potentially include awareness of the network environment, network state and topology, and shared awareness obtained by exchanging information with other nodes (typically neighboring nodes) or other network-accessible information sources. Cognitive decision-making considers this collective information, and this decision-making may be performed in coordination or collaboration with other nodes.

14.6.2 Dynamic Spectrum Access Networks

Wireless networks that employ dynamic spectrum access functionality. *See also:* **dynamic spectrum access**.

14.6.3 Reconfigurable Networks

Networks that have a capability to be reconfigured dynamically (e.g., change in network topology) or a capability to change communication protocols dynamically under the control of network management services.

NOTE 1—Reconfigurable networks do not require reconfigurable radios.

NOTE 2—The term *protocols* is defined in *IEEE Standards Dictionary Online.*[6]

NOTE 3—Software-defined networks (SDN) are considered as special cases of reconfigurable networks.

14.7 Spectrum Management Definitions

14.7.1 Allocation

The *allocation* entry in the Table of Frequency Allocations of a given frequency band for the purpose of its use by one or more terrestrial or space radiocommunications services or the radio astronomy service under specified conditions. This term shall also be applied to the frequency band concerned (adopted from ITU-R, The Radio Regulations [B32]).

NOTE—In the above definition, the *Table of Frequency Allocations* refers to the ITU's Table of Frequency Allocations or any other appropriate regional or national allocation table. An example of using the term to refer to the allocation of a frequency band is as follows: the 2450 MHz to 2483.5 MHz band allocation to the mobile service.

14.7.2 Clear Channel Assessment Function

A function for ascertaining via RF sensing that the communications channel is not in use prior to initiating a transmission.

NOTE 1—The RF measurement may be made by a radio at a single point or involve the collaborative sharing of measurements made by multiple radios at multiple locations and times.

---

6 *IEEE Standards Dictionary Online* is available at: http://ieeexplore.ieee.org/.

NOTE 2—For IEEE 802.11 WLANs (IEEE Std 802.11™-2018 [B20]), the term is defined as follows: "That logical function in the physical layer (PHY) that determines the current state of use of the wireless medium (WM)."

14.7.3 Coexistence

Coexistence is the state of two or more radio devices or networks existing at the same time and at the same place in a shared spectrum space.

NOTE—Coexistence may be achieved by numerous methods such as coordinating time usage (e.g., time sharing), geographic separation, frequency separation, directive antennas, orthogonal modulations, and so on. In the past, the employment of these mechanisms has typically been on a static, preplanned basis. In advanced radio systems, the employment and configuration of these features is increasingly dynamic and may be implemented in real time by the radio device or network in response to changing conditions and objectives.

14.7.4 Coexistence Mechanism

A technique for supporting or facilitating coexistence in the operation of two or more radio devices or networks, that may be implemented at any layer of the protocol stack.

NOTE—IEEE 802.15 WPAN™ Task Group 2 (IEEE Std 802.15.2™-2003 [B22]) has defined this term as follows: "A method for reducing the interference of one system, which is performing a task, on another different wireless system, that is performing its task."

14.7.5 Cognitive Interference Avoidance

Any cognitive process by which a radio system adapts its operating parameters to avoid interference with remote spectrum-dependent devices. *See also:* **cognitive process**.

NOTE—There are several techniques for cognitive interference avoidance. For example, a cognitive radio may provide a new orthogonal modulation waveform with respect to the interferer or search for available spectrum.

14.7.6 Collaboration

Working together involving active communication and sharing of information to achieve a common goal. *See also:* **cooperation**.

NOTE—Collaboration involves direct communication between radio nodes, perhaps over a control channel, so that the information used for the basis of spectrum access decisions is shared among several nodes.

14.7.7 Collaborative Decoding

A collaboration for collecting multiple signal observations at disparate locations to improve the chances of decoding the signal.

14.7.8 Cooperation

Performing a task as requested by other entities. *See also:* **collaboration**.

NOTE—Cooperation does not necessarily involve direct communication between radio nodes while performing a task. Cooperation is usually orchestrated in a way that all participants do their assigned parts separately and share their results with each other. Entities requesting collaboration may be technical or jurisdictional entities, for example regulators, standards bodies, or similar.

14.7.9 Data Archive

A logical entity in distributed spectrum sensing storing systematically sensing-related information (adopted from IEEE Std 1900.6™-2011 [B27]).

14.7.10 Distributed Radio Resource Usage Optimization

The distributed optimization of radio resource usage by a composite wireless network to satisfy global network objectives and by terminals to satisfy local devices and user objectives.

> NOTE—This definition is specific to a class of network and device dynamic reconfiguration scenarios that enable coordinated network-device distributed decision-making, including spectrum access control in heterogeneous wireless access networks as described in IEEE Std 1900.4™-2009 [B24].

14.7.11 Distributed Sensing

The process of sensing where sensors are distributed in acquisition time, space, frequency, and function (adopted from IEEE Std 1900.6™-2011 [B27]). *See also:* **data archive**; **sensing control information**; **sensing information**; **sensor**; **spectrum sensing**.

> NOTE—Sensing information obtained from distributed sensors is qualified either implicitly or explicitly by time and location of the acquisition of measurement data as well as by the attributes or characteristics of the acquisition method utilized.

14.7.12 Dynamic Channel Assignment

The process of assigning different channels in real time to various entities or devices, by making use of the available data regarding the operating environment.

> NOTE—Dynamic channel assignment may be performed by external parties that do not take part in the communication process (*see also:* **spectrum broker**) or by the system/network itself where system/network has one or more logical entity responsible for transient radio frequency channel assignment.

14.7.13 Dynamic Frequency Selection

Dynamic frequency selection is:

a) The ability to dynamically select different RF frequencies in accordance with spectrum etiquette.

b) The ability of a system to switch to different physical RF channels based on channel measurement criteria (adapted from IEEE Std 802.16™-2009 [B23]).

> NOTE—The term *dynamic frequency selection* was originally created to describe a method for protecting 5 GHz radars from IEEE 802.11a WLANs (ITU-R M.1652 [B35]; IEEE Std 802.11h™-2003 [B21]), to conform to particular regulatory requirements. More recently, the term is also being used to describe any dynamic frequency coordination in which a user acquires information about the presence of other users (through, for example, listening to the RF channels, or a remote database, among other solutions), in real time, and automatically avoids transmitting on frequencies in use by other users. This is commonly deployed in the context of secondary users avoiding interference with primary users, although it might be deployed for other purposes such as horizontal spectrum sharing.

14.7.14 Dynamic Frequency Sharing

The implementation of frequency-sharing techniques on a changing basis, possibly in real time, in response to changing circumstances and objectives. Dynamic frequency-sharing techniques are a subset of techniques for implementing dynamic spectrum access.

> NOTE—The term *dynamic spectrum sharing* has superseded *dynamic frequency sharing*. Hence, the term *dynamic frequency sharing* may be removed in a future revision of this standard.

14.7.15 Dynamic Spectrum Access

The real-time adjustment of spectrum utilization in response to changing circumstances and objectives.

NOTE 1—Changing circumstances and objectives include (and are not limited to) energy-conservation, changes of the radio's state (operational mode, battery life, location, etc.), interference-avoidance (either suffered or inflicted), changes in environmental/external constraints (spectrum, propagation, operational policies, etc.), spectrum-usage efficiency targets, quality of service (QoS), graceful degradation guidelines, and improvement of radio lifetime.

NOTE 2—The term *utilization* in this case refers to the parameters of the physical access of the spectrum by a radio or group of radios. These parameters might include, for example, the frequency and frequency range that is accessed, and the characteristics of the transmission spectrum mask.

14.7.16 Dynamic Spectrum Assignment

Dynamic spectrum assignment is

a) The dynamic assignment of frequency bands to radio access networks within a composite wireless network operating in a given region and time to optimize spectrum usage.
b) The process of making continuously changing spectrum assignments. *See also:* **dynamic channel assignment**.

NOTE 1—The definition in item b) is specific to a class of network and device dynamic reconfiguration scenarios that enable coordinated network-device distributed decision-making, including spectrum access control in heterogeneous wireless access networks as described in IEEE Std 1900.4-2009 [B24].

NOTE 2—This may be done by the owner of the associated spectrum, or by a regulatory authority.

NOTE 3—The term *dynamic spectrum assignment* is considered a generalization of dynamic channel assignment. In cases where channelization is applied to the spectrum, either defined on the regulatory or standardization level, the term *dynamic channel assignment* is more commonly used.

14.7.17 Dynamic Spectrum Management

A system of spectrum management that dynamically adapts the use and access of spectrum in response to changing circumstances and objectives.

NOTE 1—Dynamic spectrum management helps address the inherent inflexibility of static band allocations, and the ability of future networks to carry traffic simultaneously that corresponds to multiple radiocommunications services.

NOTE 2—The circumstances and objectives might be influenced by entities under control of the same managed system that is undertaking the dynamic spectrum management, or by other entities that are not operating under the scope of that managed system.

14.7.18 Electromagnetic Compatibility

The condition that prevails when devices and networks are performing their individually designed function in a common electromagnetic environment without causing or suffering intolerable or unacceptable degradation from unintentional electromagnetic interference to or from other equipment in the same environment (adapted from the NTIA Manual [B43]).

14.7.19 Frequency Hopping

A technique in which the instantaneous carrier frequency of a signal is periodically changed, according to a predetermined code, to other positions within a frequency spectrum that is much wider than that required for normal message transmission (adopted from IEEE Std 802.15.2-2003(R2009) [B22]).

NOTE—In the scope of DSA, the term *frequency hopping* includes the case where the carrier frequency changes according to predetermined patterns and timings which themselves might be opportunistically or dynamically changed.

14.7.20 Frequency Sharing

Frequency sharing is:

a) The common use of the same portion of the radio frequency spectrum by two or more users where a probability of interference exists.
b) The common use of a portion of the radio frequency spectrum by two or more users.
c) The concurrent use of a frequency spectrum by at least two radio systems at the same location or which are within transmission range of at least one of those radio systems within mutual reach of their transmissions.
d) The concurrent use of a radio frequency spectrum by at least two radio systems, all within the interference range of at least one of those radio systems.
e) *See:* **spectrum sharing**.

 NOTE 1—Definition (a) is from the NTIA Manual [B43].

 NOTE 2—Frequency sharing is often used synonymously with spectrum sharing.

14.7.21 Hierarchical Spectrum Access

A type of spectrum access in which a hierarchy of radio users or radio applications determines which radios have precedence.

 NOTE 1—The most common hierarchy proposed today is one that distinguishes between primary users and secondary users. In this hierarchy, secondary users may only access spectrum when primary users are not occupying it. However, other hierarchies are possible, including the existence of tertiary users or hierarchies based on the type or criticality of the communication.

 NOTE 2—The hierarchy may be determined by a central authority, such as a regulator, a spectrum broker, or through active collaboration among affected systems. The hierarchy may be static or it may be established dynamically based on the current environment. Practical implementations of hierarchical spectrum access can also be found in tier-based models.

 NOTE 3—In the context of hierarchical spectrum access, a notification may be issued to radio systems currently accessing spectrum that a higher precedence radio system intends to use that spectrum. Such a mechanism may be the means by which higher precedence systems reclaim spectrum from lower precedence systems.

14.7.22 Horizontal Spectrum Sharing

Spectrum sharing between users that have equal precedence in spectrum access.

 NOTE—Spectrum sharing in an unlicensed spectrum is an example of horizontal spectrum sharing.

14.7.23 Interference

Interference is:

a) In a communication system, interference is the extraneous power entering or induced in a channel from natural or man-made sources that might interfere with reception of desired signals or the disturbance caused by the undesired power (adapted from IEEE Std 802.15.2-2003(R2009) [B22]).
b) Radio-frequency interference (adapted from ITU-R *The Radio Regulations* [B32]):

1) *Interference*: The effect of unwanted energy due to one or a combination of emissions, radiations, or inductions upon reception in a radiocommunication system, manifested by any performance degradation, misinterpretation, or loss of information that could be extracted in the absence of such unwanted energy.

2) *Harmful interference*: Interference that endangers the functioning of a radionavigation service or of other safety services or seriously degrades, obstructs, or repeatedly interrupts a radiocommunication service operating in accordance with these regulations.

NOTE—The term *regulations* in the definition of harmful interference refers to the ITU radio regulations.

14.7.24 Opportunistic Spectrum Access

The real-time adjustment of spectrum utilization in order to exploit spectral opportunities. *See also:* **spectral opportunity**.

14.7.25 Opportunistic Spectrum Management

A system of spectrum management that dynamically adapts the use of spectral opportunities in response to changing circumstances and objectives. *See also:* **dynamic spectrum management; spectral opportunity**.

14.7.26 Policy Authority

An entity that has jurisdiction over spectrum usage and is authorized to create policy for that jurisdiction (adopted from IEEE Std 1900.5-2011 [B26]).

NOTE—An authority may be, for example, a regulatory agency or a primary user who is authorized to lease its spectrum to other users.

14.7.27 Policy Traceability

The ability to provide evidence of the source of a policy that cannot be repudiated.

NOTE—Policy includes machine-understandable policy.

14.7.28 Radio Environment Map

An integrated space-time-frequency database consisting of multi-domain information, such as geographical features, available services, spectral regulations, locations and activities of radios, relevant policies, and experiences that characterizes the radio environment for DSA applications (adopted from Zhao et al. [B50]).

14.7.29 RF Environment Map

A subset of a radio environment map pertaining only to information on the RF environment. *See also:* **radio environment map; system strategy reasoning capability**.

14.7.30 Sensing Control Information

Information describing the status and configuration of the sensor, cognitive engine, and data archive, as well as information controlling and configuring the acquisition and processing of sensing information (adopted from IEEE Std 1900.6-2011 [B27]). *See also:* **data archive; distributed sensing; sensing information; sensor; spectrum sensing**.

14.7.31 Sensing Information

Any information acquired by and obtained from sensors, including related spatiotemporal state information such as position, time, and confidence of acquisition (adopted from IEEE Std 1900.6-2011 [B27]). *See also:* **data archive**; **distributed sensing**; **sensing control information**; **sensor**; **spectrum sensing**.

14.7.32 Sensor

A logical entity that performs sensing within a radio system. Sensors may also act as clients of other sensors (adopted from IEEE Std 1900.6-2011 [B27]). *See also:* **data archive**; **distributed sensing**; **sensing control information**; **sensing information**; **spectrum sensing**.

14.7.33 Spectral Opportunity

The existence of a frequency band segment satisfying an availability criterion that dynamic spectrum access or opportunistic spectrum access devices can exploit for their communications purposes.

> NOTE—Non-interference with primary spectrum users is an example of an availability criterion for the use of spectral opportunities for opportunistic secondary spectrum access. Possible interference to avoid can occur in the temporal, spatial, or code domain.

14.7.34 Spectrum Access

Spectrum access is defined as the:

a) Transmission or reception on the radio spectrum.
b) Ability to obtain or make use of the radio spectrum.

> NOTE—Spectrum access includes the attributes of frequency, location, time, power (spectral flux density), and angle of arrival. These attributes also may be characterized by additional parameters. For example, modulation further characterizes the frequency attribute of spectrum access. *See also:* **spectrum utilization**.

14.7.35 Spectrum Broker

An entity, device, or device capability responsible for dynamic assignment of spectrum access rights.

> NOTE 1—This term complements the regulatory and technical perspectives of the term *spectrum database*. It is to be considered an economically-oriented term with some relation to the technical or regulatory perspectives. The term *spectrum assignment* complements spectrum brokerage under a regulatory perspective, while the term *spectrum management* complements under a technical perspective.

> NOTE 2—Spectrum brokerage is a concept which might have various realization options including technical realization as a manager or realization as a mediator between stakeholders.

14.7.36 Spectrum Efficiency

A general measure of how well a spectrum segment of interest is being used that is determined from the ratio of the benefits derived from the spectrum usage to the resource costs of providing those benefits. *See also:* **spectrum utilization**; **spectrum utilization efficiency**.

> NOTE 1—A common usage of spectrum efficiency is as a metric representing the bit rate that is achieved per unit of spectrum bandwidth (i.e., b/s/Hz).

> NOTE 2—It is observed that the terminology related to spectrum efficiency has expanded in the context of spectrum sharing systems, creating a number of related terms depending on the field and scope of usage. Additional explanatory information is provided in 14B.8.

14.7.37 Spectrum Etiquette

A consistent set of rules, policies, procedures, and protocols, or a subset thereof, that govern spectrum-sharing behavior.

14.7.38 Spectrum Leasing

The process of obtaining or providing a grant to use spectrum based on a contract between a license holder or its agent and a second party.

> NOTE—The method of lease is commonly defined by the regulator as a sharing of license between multiple stakeholders, transfer of license between stakeholders having exclusive usage rights, or by giving permission to a license holder to share the spectrum usage rights with third-party stakeholders. The lease time and geographical extend are limited to the duration and geographical extent of the spectrum license or any fraction thereof.

14.7.39 Spectrum Management

Spectrum management is defined as:

a) Implementation, realization, or coordination of measures to use the radio spectrum in accordance with the objectives of stakeholders.
b) Controlling and making decisions on the use of spectrum resources in accordance with the objectives of stakeholders.

> NOTE 1—These two complementary definitions have been made to cover the different interests of various types of stakeholders present in the current and evolving spectrum regulatory scenarios. Definition a) refers to spectrum management as a regulatory process, while definition b) addresses the perspective of spectrum management to utilize spectrum as a resource. This distinction reflects the current evolution of spectrum management in the light of international spectrum harmonization and increased stakeholder involvement in the process.

> NOTE 2—According to definition a), spectrum management involves rulemaking and implementing regulatory frameworks including stakeholders' objectives that address socio-economic benefits in the wider sense. Examples are activities in the scope of rule-making of regulatory authorities, such as public consultations, spectrum allocations, or spectrum harmonization. According to definition b), spectrum management might focus on efficient utilization of spectrum as a resource by stakeholders with the focus on technologies and implementations of business cases in the scope of their networking business. Examples are technology development, decision-making, implementation, and deployment by operators, spectrum brokers, and spectrum users.

14.7.40 Spectrum Overlay

Dynamic spectrum access by secondary spectrum users that exploits spectral opportunities in a noninterfering manner. *See also:* **dynamic spectrum access**; **spectral opportunity**. *Contrast:* **spectrum underlay**.

> NOTE—In practice, spectrum-sharing deployments that are considered as spectrum overlay often operate under the assumption of an interference limit to incumbent services, and therefore are realized as spectrum underlay schemes.

14.7.41 Spectrum Owner

An individual, group, corporation, organization, or governmental body that has the sole responsibility and authority over a band of frequencies for a time determinate or indeterminate period.

> NOTE 1—Spectrum ownership rights are typically determined by the government having authority in the particular geographical area or by international treaties.

NOTE 2—In the context of dynamic spectrum access, the preferred term is *incumbent* rather than *owner*.

14.7.42 Spectrum Pooling

Spectrum pooling is a management strategy for merging spectral bands assigned to different operators into a common pool for mutual use.

NOTE 1—Pooling describes the common use of one resource, but not necessarily at the same time. Use of common resource at same time, means "sharing." *See also:* **spectrum sharing**.

NOTE 2—The goal of spectrum pooling is to enhance spectral efficiency and is commonly implemented by spectrum overlay techniques. *See also:* **spectrum overlay**.

14.7.43 Spectrum Sensing

Spectrum sensing is:

a) The act of categorizing and evaluating radio signals for the purpose of obtaining information.

NOTE—A spectrum-sensing device according to this definition is expected to obtain information from the sensed signals applying or not a-priori knowledge that enables analysis or decision-making on the spectrum occupancy.

b) The act of measuring information indicative of spectrum occupancy (information may include frequency ranges, signal power levels, bandwidth, location information, etc.). Spectrum sensing may include determining how the sensed spectrum is used (adopted from IEEE Std 1900.4a™-2011 [B25]).

c) The act of measuring information indicative of spectrum occupancy (information may include frequency ranges, signal power levels, bandwidth, location information, etc.) in the context of radio frequency spectrum. Sensing may include determining how the sensed spectrum is used (adopted from IEEE Std 1900.6-2011 [B27]).

NOTE 1—Spectrum sensing as defined in a) pertains to the presence of signals, whereas spectrum sensing as defined in b) and c) pertains to sensing the presence or absence of a signal or its features for the purpose of identifying spectrum opportunities.

NOTE 2—Definitions b) and c) recognize that in some contexts the term *sensing* is assumed to refer to spectrum sensing.

See also: **data archive**; **distributed sensing**; **sensing control information**; **sensing information**; **sensor**.

14.7.44 Cooperative Spectrum Sensing

The act of spectrum sensing performed by a group of cooperating sensors *See also:* **cooperation**.

14.7.45 Collaborative Spectrum Sensing

The act of spectrum sensing performed by a group of collaborating sensors *See also:* **collaboration**.

14.7.46 Spectrum Sharing

The application of technical methods and operational procedures to permit multiple users to coexist in a shared spectrum space. *See also:* **coexistence**.

NOTE—The notion of spectrum space sharing is introduced to set it apart from the concept of shared spectrum utilization. This is done to emphasize the fact that spectrum is only one of the dimensions of spectrum space.

14.7.47 Spectrum Underlay

Dynamic spectrum access by secondary spectrum users that exploit spectral opportunities transmitting below an interference threshold, not causing harmful or even disruptive interference to the incumbent services *See also:* **dynamic spectrum access**; **spectral opportunity**. *Contrast:* **spectrum overlay**.

> NOTE—In dynamic spectrum access practice, spectrum sharing deployments are considered as spectrum underlay when the interference generated to an incumbent service might degrade the system but does not disrupt its capability to communicate.

14.7.48 Spectrum Utilization

Spectrum utilization is defined as:

a) The spectrum space denied to other potential users.

> NOTE—Transmitters and receivers both use spectrum space. Transmitters use spectrum space by denying the use of that space to certain receivers (other than the intended receiver) that would receive interference from the transmitter. This space is called *transmitter-denied space* or simply *transmitter space*. Receivers use spectrum space by denying the use of nearby space to additional transmitters (assuming that the receiver is entitled to protection from interference). A transmitter operating in that space would cause interference to the receiver's intended operation. This space is called *receiver-denied space* or simply *receiver space*.

b) The product of the frequency bandwidth, the geometrical (geographic) space, and the time denied to other potential users (adapted from *Handbook: National Spectrum Management* [B29]):

$$U = B \times S \times T$$

where
U   is the amount of spectrum space used ($Hz \times m^3 \times s$)
B   is the frequency bandwidth
S   is the geometric space (desired and denied)
T   is time

> NOTE—The determination of the amount of bandwidth, space, and time occupied will be a function of the characteristics of other systems desiring to use or share the same spectrum and may involve numerous assumptions, such as the level of protection to be provided or the propagation model used to determine signal loss. Consequently, the comparison of spectrum utilization values may only be meaningful between like systems where the assumed conditions are similar.

c) A metric for the amount of spectrum collectively used by communicating parties in a geometrical (geographic) space and for a time under consideration.

> NOTE—While definition a) can be deduced from definition b), definition c) is a generalization capturing the potential for shared use of the spectrum space.

14.7.49 Spectrum Utilization Efficiency

The spectrum utilization efficiency (SUE) is defined as the ratio of information transferred to the amount of spectrum utilization (adapted from *Handbook: National Spectrum Management* [B29]):

$$SUE = \frac{M}{U} = \frac{M}{B \times S \times T}$$

where
M is the amount of information transferred
U is the amount of spectrum utilization

NOTE 1—Because the computation of SUE is primarily of interest in comparing the efficiency of similar types of systems, the quantity M should take the form most meaningful and convenient for the systems being compared. M could be in terms of bits/sec, Erlangs, analog channels, radar channels, and so on.

NOTE 2—Found to be used synonymously to the term *spectrum efficiency*.

14.7.50 Vertical Spectrum Sharing

Spectrum sharing between users that do not have equal precedence in spectrum access.

NOTE—Spectrum sharing between primary and secondary users is an example of vertical spectrum sharing.

14.7.51 White Space

Part(s) of spectrum allocated to a particular radio system (primary radio system) in particular location(s) that may be temporarily unused by this primary radio system in some location(s) and thus allowed by radio regulations to be used by another radio system(s) (secondary radio system) on a temporary secondary basis without causing harmful interference to the primary radio system, where harmful interference and protection mechanisms are defined in the radio regulations (adopted from IEEE Std 1900.4a-2011 [B25]). *See also:* **white space database**; **white space frequency band**.

NOTE—Depending on the specific context, several definitions are commonly used for this term, reflecting different stakeholders' perspectives. These might be the perspectives of spectrum regulators, technology providers, or the spectrum user.[7]

14.7.52 White Space Database

A database that provides information on currently available white space resources in particular location(s) (adopted from IEEE Std 1900.4a-2011 [B25]).

14.7.53 White Space Frequency Band

Any frequency band where white space may exist.

14.7.54 White Space Spectrum Band *See:* White Space Frequency Band.

NOTE—White space spectrum band is often used synonymously with white space frequency band.

14.8 Glossary of Ancillary Terminology

14.8.1 Air Interface

The subset of waveform functions (i.e., coding/decoding, modulation/demodulation, transmission and receiving) designed to establish communication between two radio terminals.

NOTE—In the ISO-OSI reference model, this is the waveform equivalent of the wireless physical layer and the wireless data link layer.

---

7 No attempt is made here to generalize the use of this term in other groups, within the IEEE or elsewhere. Specifically, it is recognized that this term is often used in reference to TV frequency bands. Contrary to that approach, the IEEE 1900.1 working group prefers to recognize the term *TV white space* as "white space observable in TV frequency bands," and does not limit that perspective to certain technologies or to the case of unused frequency bands. It is noted that occupied frequency bands might be opportunistically used if unacceptable interference is not caused to the incumbent.

14.8.2 Digital Policy

A machine actionable policy that is semantically and consistently interpreted by system(s). *See also:* **machineunderstandable policies**; **policy**.

14.8.3 Domain

An area of knowledge or activity characterized by a set of concepts and terminology understood by practitioners in that area (adopted from IEEE Std 1900.5-2011 [B26]).

14.8.4 Interference Temperature

A metric used to determine the interference threshold to be applied to transmissions from dynamic spectrum access systems. The metric is computed from the dynamic spectrum access system's signal bandwidth as follows:

$$P = kTB$$

where

P is the power

k is Boltzmann's constant (1.38×10^{-23} W/Hz/K)

B is the signal bandwidth

T is the interference temperature (K)

NOTE—Due to the complexity of calculating the interference temperature, it is mostly referred to in the context of an interference model rather than used as a metric (interference temperature model).

14.8.5 Interoperability

In the context of dynamic spectrum access, the capability of different radio systems to exchange information with each other and use that information.

NOTE 1—Interoperability may have operational and policy aspects as well as technical aspects.

NOTE 2—Systems that are not interoperable might be able to interoperate through an interface that translates exchanged information between incompatible formats.

14.8.6 Machine Learning

A field of study that explores the design of algorithms that facilitate learning from observed data.

NOTE—In the field of dynamic spectrum access, machine learning is one aspect process in the development of algorithms that are aiming to enhance cognitive processes by observing data to build knowledge from it. *See also:* **cognitive process**.

14.8.7 Machine-understandable Policies

Policies expressed in a form that allows a machine to read and interpret them automatically and consequently to apply them without requiring human intervention.

14.8.8 Ontology

Ontology is:

a) The common words and concepts used to describe and represent an area of knowledge (adapted from Obrst et al. [B44]). *See also:* **domain**; **policy language**.
b) A specification of a conceptualization (adopted from a draft version of IEEE 1900.5.1).

NOTE 1—An ontology models the vocabulary and meaning of domains of interest: the objects in domains; the relationships among those objects; the properties, functions, and processes involving those objects; and constraints on and rules about those objects.

NOTE 2—In the scope of policy languages, the ontology consists of definitions that associate the names of entities and concepts in a problem domain (e.g., objects, classes, relations, functions) with text describing what the names mean, and axioms (expressed in a formal language) that constrain the interpretation of these entities and concepts (adopted from IEEE Std 1900.5-2011 [B26]).

14.8.9 Policy

A set of rules governing the behavior of a system.

NOTE 1—More specifically, the Internet Engineering Task Force (IETF) defines policy as a definite goal, course, or method of action to guide and determine present and future decisions being implemented or executed within a particular context (such as policies defined within a business unit) (Westerinen et al. [B48]).

NOTE 2—Policies may originate from regulators, manufacturers, developers, network and system operators, and system users. A policy may define, for example, allowed frequency bands, waveforms, power levels, and secondary user protocols.

NOTE 3—Policies are normally applied after manufacturing the radio as a configuration to a specific service application.

NOTE 4—In some contexts, the term *policy* is assumed to refer to machine-understandable policies. *See also:* **machine-understandable policies**.

14.8.10 Quality of Service

Totality of characteristics of a telecommunications service that bear on its ability to satisfy stated and implied needs of the user of the service (adopted from ITU-R, Recommendation E.800 [B33]).

14.8.11 Radio

Radio is defined as:

a) A technology for wirelessly transmitting or receiving electromagnetic radiation to facilitate transfer of information.
b) A system or device incorporating technology as defined in item a).
c) A general term applied to the use of radio waves (adapted from ITU-R *The Radio Regulations* [B32]).

NOTE 1—Users of the term *radio* should understand that it has different meanings depending on its context. For example, it can refer to a general type or class of technology (e.g., AM radio, land mobile radio, or amateur radio) and to instantiation of the technology (e.g., a radio or the radio). Next-generation radio concepts introduce additional ambiguity in the use of the term. For instance, next-generation radio components can be virtual, implemented in software, and geographically distributed using computer networks, all of which can make the boundaries of radio difficult to ascertain. Further confounding the definitional problems is that radio functions are often tightly integrated with a variety of non-radio applications. What should and should not be considered part of a radio may be particularly difficult to determine in the presence of cognitive radio functionality, which can reside on computers physically and logically separated from radio hardware.

NOTE 2—More precise terms than radio are often available to reduce ambiguity and should be used whenever possible. For example, when discussing the source of electromagnetic radiation, "the location of the antenna" is preferable to "the location of the radio" because what is called "the radio" may be multiple devices in several locations. Similarly, when discussing collaborative decision-making, "communication between cognitive control mechanisms" is preferable to "communication between radios" because the cognitive control mechanism may be independent from the system that wirelessly transmits information and may, in some cases, occur over a wired medium.

NOTE 3—To avoid confusion between a device supporting multiple functions and its radio subsystem, the term *radio* should be limited to components providing radio functionality in the form of wireless connectivity with radio-frequency electromagnetic transmission. For example, a smart phone is not a radio, but it contains a radio.

14.8.12 Radio Node

A radio device being part of a network.

14.8.13 Radio Spectrum

The radio-frequency portion of the electromagnetic spectrum.

NOTE—The radio spectrum is divided into several frequency bands (see *IEEE Standards Dictionary Online* and ITU-R, *The Radio Regulations* [B31]).

14.8.14 Receiver

In the context of dynamic spectrum access, a device that accepts radio signals for the purpose of radiocommunication.

14.8.15 Software

Modifiable instructions executed by a programmable processing device.

NOTE 1—For the purpose of this standard, software is considered as that part of the system consisting of instructions and data.

NOTE 2—Instructions and data that are embedded in a hardware device or that define the operation of a programmable logic device are a specialized type of software called firmware.

14.8.16 Transmitter

In the context of dynamic spectrum access, a device that produces radio-frequency energy for the purpose of radiocommunication.

14.8.17 Waveform

In the scope of software radios, the time-domain or frequency-domain representation of an RF signal obtained from a transformation of information into a suitable channel symbol.

NOTE—This is usually achieved by line and channel coding. *See also:* **waveform processing**.

14.8.18 Waveform Processing

In the scope of software radios, the set of transformations and protocols to convert between information to be communicated and channel symbols.

NOTE—This usually includes signal-processing functions for waveform generation, waveform shaping, and waveform analysis.

Annex 14A

(informative)

Implications of Advanced Radio System Technologies for Spectrum

A growing number of regulatory agencies around the world believe that there is a need for a new approach to spectrum management, spectrum allocation, and spectrum utilization. The new spectrum paradigm is driven, in part, by the increasingly keen competition for spectrum—a problem common to many parts of the world and to all segments of the communications industry: government, commercial wireless, public safety, and so on. This annex describes how the advanced radio system technologies defined in this standard potentially have spectrum management and regulatory implications that may lead to a more effective utilization of spectrum worldwide.

14A.1 Regulatory Issues to Which Advanced Radio System Technologies and New Spectrum-sharing Concepts are Applicable

The following list outlines some current regulatory issues that are being considered by various national spectrum regulatory agencies:

— Increasing demands for access to more spectrum
— Requirement for more efficient use of the spectrum
— Spectrum trading
— Dynamic spectrum access
— Dynamic frequency selection
— Dynamic spectrum management
— Relative mixture of different types of spectrum management and transition from one to another
  1) Command and control: Inflexible frequency assignments
  2) Market mechanisms: The market manages the spectrum within the constraints of the licenses
  3) License-exempt use: Nobody controls who uses the spectrum; power constraints or other mechanisms restrict usage to reduce interference
— Interrelationship of developments in technology, market, and regulatory practices.
— Pace of technology development; regulation has to keep up
— International coordination
— Security (ensure that disruption to communication services cannot occur as a result of inadvertent or malicious changes to software in advanced communications devices and systems)
— Interference (ensure that users can use the spectrum assigned to them without disruption)
— Interference and noise temperatures
— Certification and conformity issues
— Circulation issues

The terminology for advanced radio system technologies defined in this standard is applicable to many of these issues. For the bands designated by national spectrum regulatory agencies for use for unlicensed services, each regulatory agency is responsible for establishing the rules associated with these bands. Manufacturers are most interested in the use of advanced radio system technologies because of the potential for decreased costs and quicker time to market. Regulatory agencies have demonstrated interest in these technologies because of the potential of dynamic spectrum access to improve the efficiency with which the total spectrum is used.

14A.2 New Spectrum Management Concepts

Research and studies conducted under the purview of some national spectrum regulatory agencies have concluded that spectrum management should increasingly depend on the marketplace rather than on administrative systems (Australian ACA [B1]; Cave [B4]; Japan MPHPT [B42]). Several questions are associated with the development of a new spectrum management paradigm using advanced radio system technologies, such as cognitive radio systems and policy-based radio systems. The questions include:

a) To what extent can the technologies delineated in this document be applied to spectrum management in the unlicensed bands? How much spectrum should be set aside for the unlicensed bands? These are questions that regulatory agencies may choose to address.
b) For licensed bands, should national regulators permit cognitive radio access if they wish, but not mandate the use of cognitive radio in the band on a secondary-use basis? This may be an issue of import to the ITU-R World Radio Conference as well as to national regulatory agencies.
c) To what extent can the market be relied on as a major part of the new spectrum management concepts?
d) What, if any, broad framework of international rules are needed to promote new spectrum management concepts and the use of advanced radio system technologies to enhance efficient use of the spectrum and provide other benefits?

14A.3 Frequency Band Consideration in the Application of Advanced Radio System Technologies

In general, the advanced radio system technologies defined herein are applicable to all bands. However, there may be practical limitations such as power, size, weight, and cost as well as legacy considerations that may restrict the use of these advanced radio system technologies. Some national spectrum regulatory agencies have already started investigating the possibility of increased unlicensed spectrum in which advanced radio system technologies could be used.

14A.4 Radio Network Control Considerations in the Application of Advanced Radio System Technologies

There is currently much research and investigation by many industrial organizations and national spectrum regulatory agencies on the closely related topics of dynamic spectrum management, flexible spectrum management, advanced spectrum management, dynamic spectrum allocation, flexible spectrum use, dynamic channel assignment, and opportunistic spectrum management. Advanced radio system technologies of policy-based radio, cognitive radio, software-controlled radio, and reconfigurable radio are enabling technologies to implement these new spectrum management and usage paradigms. These concepts are equally applicable to a wide variety of mobile communications systems, including public protection and disaster relief (PPDR), government, and commercial wireless.

As noted elsewhere in this document, more efficient use of the spectrum is one benefit associated with software-defined radios and the closely related technologies described herein. To be able to achieve this benefit, it is necessary for the software-defined radio (SDR)/policy-based adaptive radio (PBAR)/cognitive radio (CR)/reconfigurable radio (RR) to be controlled in such a way that underused portions of the spectrum can be used more efficiently. This has been called opportunistic spectrum management. For many scenarios, the method of control needed to achieve opportunistic spectrum management through the use of advanced radio system technologies is a network issue as well as a radio issue. Network control of these advanced radios includes control of the configuration of the radio and the radio frequency (RF) operating parameters. Regulatory policies that govern the allowable behavior, i.e., RF operating parameters, are part of this network control. The control policies may, for some scenarios, also include network operator and user policies.

In general, there are two control models for opportunistic spectrum access or flexible spectrum usage, namely the centralized control model and the distributed control model. The centralized control model is one in which the management of spectrum opportunities is controlled by a single entity or node that has been referred to as the spectrum broker. The spectrum broker is responsible for deciding which spectrum opportunities can be used and by which radios in the network. A central broker may use sensors from the distributed nodes or may use other means for sensing and spectrum awareness. One application of centralized control is real-time spectrum markets.

The second opportunistic spectrum access or flexible spectrum usage control model is the distributed control model. In this model the interaction is "peer-to-peer." In other words, the advanced radio nodes in the network are collectively responsible for identifying and negotiating use of underused spectrum. For some scenarios, the distributed control may be between cooperative radio access networks.

14A.5 Progressing Toward Regulatory Harmonization

The potential for spectrum sharing should be improved as it generally supports innovation and market entry. Opening spectrum resources increases competition and business-drive to innovate and improve services. Spectrum sharing increases supply of services leading to decreased costs. Removing or decreasing obstacles in the spectrum use opens an opportunity to a larger number of potential investors and allows other systems to exploit under-utilized spectrum resources, thus increasing spectral efficiency. Moreover, resales rights of spectrum resources enable dynamic spectrum market. Conversely, multiple incompatible systems using the same resource waste more than one optimized system, leading to increased costs.

Indeed, the success of the license-exempt (LE) bands has been the most surprising and consequential regulatory action in the previous 15 years of spectrum management. The future of spectrum usage will be marked by a growing diversity of uses—especially due to the emergence of the Internet of Things—with demand growing most strongly on LE networks. In such a scenario, the opportunity costs of dedicating exclusive spectrum rights for each new use are likely to become progressively more difficult to justify—especially as advances in technology permit greater and more reliable spectrum sharing. License-based DSA harks back to an era of few users of spectrum and few networks. In addition, it risks repeating the mistakes of exclusive allocation that have led to today's artificial spectrum shortage. Instead, grasping the possibility offered by policy-based DSA permits to create modes of spectrum access that respond quickly to market conditions, allowing for continuous technology upgrades and enabling networks with finely tailored speeds, capacities, and qualities of service (QoS).

The full economic and social benefits of this digital transformation will only be achieved if national and international standardization bodies and agencies can ensure widespread deployment and take-up of very high capacity networks, in rural as well as in urban areas. Since the telecoms sector today is an enabler for the entire digital economy and society, nations need to act quickly to secure their future global competitiveness and prosperity.

Globally, the ITU-R produced its final recommendations for possible deployment of DSA-based technologies in TV bands, although spectrum regulations are done on a national basis (ITU, "Provisional final acts" [B30]). On a national level, the USA's FCC and the National Telecommunications and Information Administration (NTIA) have continued to spearhead the adoption of much more efficient and flexible radio spectrum regulations. The two bodies have proposed and developed several guidelines and regulations on how DSA technologies can be used to exploit both government and commercial/civilian spectrum in co-primary sharing and primary/secondary users sharing scenarios. The new regulations include allowing unlicensed operations in the 5 GHz band using dynamic frequency selection (DFS) techniques and allowing shared licensed operations in the 3.65 GhZ to 3.7 GHz bands using contention-based protocols. Moreover, in a FCC ruling (FCC "Third memorandum" [B16]), the requirement to use spectrum sensing was relaxed for white spaces devices (WSDs) that use a combination of geo-location and database lookup functions. Furthermore, the FCC has been instrumental in encouraging RF spectrum to be traded in secondary markets using various dimensions such as time, space, and frequency (FCC "Promoting efficient" [B12]). In the wake of its decision, the FCC has passed a landmark regulation to allow innovative sharing of TV broadcast channels among broadcasters (FCC "Innovation" [B11]). A complementary regulation to allow incentive auctions (perceived as a legal instrument to encourage TV broadcasters to trade their unused spectrum) has also been proposed (FCC "Statement from FCC chairman" [B15]). The FCC's rules permit unlicensed radio devices to transmit on white space in the spectrum bands used by the broadcast television service, i.e., 54 MHz to 72 MHz, 76 MHz to 88 MHz, 174 MHz to 216 MHz, 470 MHz to 608 MHz, and 614 MHz to 698 MHz. More recently, the FCC established the 3.5 GHz band as an innovation band, where it is breaking down age-old regulatory barriers to create a space for a wide variety of users to coexist by sharing spectrum (FCC "Amendment" [B10]). As a result of technological innovations and a focus on spectrum sharing, the FCC opened up 100 MHz of spectrum previously unavailable for commercial uses, which it added to existing commercial spectrum to make a 150 MHz contiguous band.

In Europe, the European Commission has proposed a series of initiatives designed to establish the right conditions for the necessary investments to take place, primarily to be achieved by the market (EC [B8]). These consist of a major reform of the regulatory framework for electronic communications, in the form of the accompanying legislative proposal for a European Electronic Communications Code and BEREC Regulation, an Action Plan on 5G Connectivity for Europe. The proposed code establishes key principles for spectrum assignment in the European Union, new EU-level instruments to establish assignment deadlines and license periods, and a peer review among national regulators to ensure consistent assignment practices. Moreover, it promotes a consistent approach to coverage obligations, to small-cell deployment, and to network sharing, thereby stimulating 5G deployment and rural connectivity, hence facilitating spectrum sharing in 5G networks.

Finally, most of the pioneering DSA regulatory and standardization activities in the Asia-Pacific region come from the Info-communications Media Development Authority (IMDA), Singapore's spectrum regulator (IMDA [B28]). IMDA adopts a license-exempt policy for the use of radio frequencies in the designated spectrum band to deploy TV white spaces (TVWS) technology to facilitate adoption and to lower cost barriers for the industry. The regulations stipulate the TVWS equipment requirements, spectrum channels to be made available for TVWS use, and how such equipment should communicate with geo-location databases in order to identify the available spectrum channel to use, among others.

It is worth noting that most countries in the developing world including Africa are still lagging behind on the regulatory front pertaining to possible deployment approaches for DSA-CR technologies, with the exception of the Information and Communication Technologies Authority (ICTA) of Mauritius (ICTA [B19]; Ministry of Information [B41]).

In addition to faster processes to designate spectrum for electronic communications, with clear deadlines for when the spectrum is to be made available to the market, investors in the next generation of wireless broadband need more predictability and consistency regarding future licensing models and the key conditions for assigning or renewing national spectrum rights. These include a minimum license duration to ensure returns on investment, greater scope for spectrum trading and leasing, and consistency and objectivity in market-shaping regulatory measures (reserve prices, auction design, spectrum blocks and caps, exceptional spectrum reservations, or wholesale access obligations). On the other hand, operators should commit to use the spectrum assigned to them effectively. The shared use of spectrum, either on the basis of general authorization or individual rights of use, can enable more efficient and intensive exploitation of this scarce resource, and this is particularly relevant for the new, very short-range ("millimetre") spectrum bands foreseen for 5G communications.

Annex 14B

(informative)

Explanatory Notes on Advanced Radio System Technologies and Advanced Spectrum Management Concepts

14B.1 Relationship of Terms

A common understanding of the terminology used to describe various interrelated advanced radio system concepts, including **software-controlled radio** (SCR), **software-defined radio** (SDR), **cognitive radio** (CR), and **policy-based radio** (PBR) is critical to the furtherance of new spectrum management concepts. This annex is provided as informative explanatory material to be used in conjunction with the normative definitions provided in Clause 3.

Figure 14B.1 depicts several advanced radio system and radio-control concepts. The figure is intended to be notional rather than a complete architectural block diagram. As such, it does not include functionality such as protocol processing and applications.

For many years, there have been conflicting views as to what constitutes a **software-defined radio** particularly in regard to the radio signal processing via software and in regard to software control of the radio. Figure 14B.1 is intended to amplify on the definitions contained herein, particularly in regard to software-defined radio functionality and software-controlled functionality as explained below.

Figure 14B.1 illustrates the differentiation of the radio signal-processing functionality and the radio-control functionality. Radio signal-processing functionality includes all of those operations between the input to the radio and the transmission of the radio signal (upper portion of the figure going from left to right). The radiocontrol functionality is depicted in the lower portion of the figure.

A radio could be considered to be a software-defined radio if:

a) Some or all of the physical layer functions are accomplished through the use of digital signal-processing software, or field programmable gate array (FPGA) firmware, or by a combination of software and FPGA firmware.

b) This software or firmware or both can be modified after deployment.

Figure 14B.1 illustrates the possibility of multiple signal-processing configurations that could be switched under software control. In the past, such a radio has been considered by some to be a software-defined radio even if the signal processing is accomplished by hardware. According to the IEEE definitions contained herein, such a configuration would be considered to be a **software-controlled radio** but not a **software-defined radio**. For many years, there have been conflicting views as to what constitutes a **software-defined radio**.

A **software-controlled radio** has the property that the radio-control functionality is accomplished through software. Many of the benefits typically ascribed to software-defined radios, such as improved utilization of the spectrum, can be achieved only if the SDR is under software control. Software-controlled adaptive, cognitive, or intelligent radios possessing either **dynamic frequency selection** or **dynamic spectrum access** capability have the potential for improved efficiency in the utilization of spectrum.

A cognitive radio that is also a policy-based radio is able to sense its operational environment and, on this basis, makes appropriate adjustments to radio-operating parameters, while remaining within the constraints defined by the policy. In a **cognitive radio** that is also a **software-controlled radio**, the cognitive control mechanism is implemented through

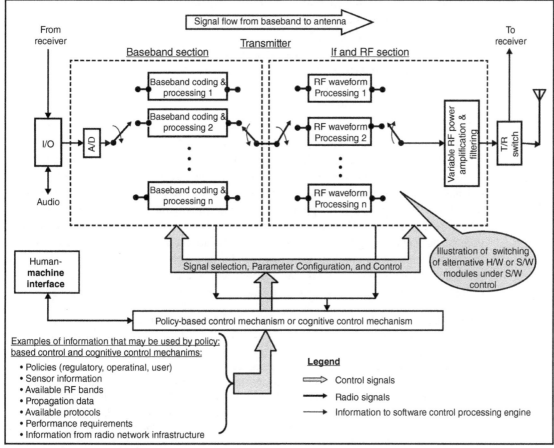

Note—Receiver functions not shown

Figure 14B.1 —Illustration of advanced radio system concepts for radio transmitter signal processing and radio functionality control. From IEEE Standard for Definitions and Concepts for Dynamic Spectrum Access: Terminology Relating to Emerging Wireless Networks, System Functionality, and Spectrum Management, Feb 2019. IEEEq.

software processing. However, the radio-specific human–machine interface for controlling frequency, waveforms, and protocols is replaced by a set of algorithms that enable the software control to adjust dynamically and automatically its radio-operating parameters (both at the transmitter and at the receiver side). In both an **adaptive radio** employing a **policy-based control mechanism** and in a **cognitive radio** employing a **policy-based control mechanism**, **policy** provides constraints. The difference is that in a **cognitive radio**, the cognitive control function identifies opportunities to optimize within these constraints. Optimization requires a broader set of sensing capabilities and optimization algorithms. A **cognitive radio**, employing a **policy-based control mechanism**, usually performs both policy logic and cognitive reasoning within the host processor.

A **reconfigurable radio** has functionality that can be changed either through manual reconfiguration of hardware-defined or software-defined radio modules or through software-controlled reconfiguration of these radio modules. The following summarizes key aspects of these radio technologies:

a) If any of the physical-layer functions are implemented using software, the radio is a **software-defined radio**.

b) If the radio uses software for control of any functionality of the radio, it is a **software-controlled radio**. For example, if the radio provides alternative **waveform** processing functionalities that are switched under software control, it is a **software-controlled radio**. However, it is not a **software-defined radio** unless some of the **waveform** processing is done in software. A radio is both a SCR and an SDR if both of the following conditions hold:

1) Some portion of the **waveform** processing is done in software.
2) Control of the **waveform** processing (e.g., switching to a different **waveform** processing module) is done via software.

c) A **reconfigurable radio** is also a **software-controlled radio** if the waveform processing is selected under software control.

d) **Policy-based control** and **cognitive control mechanisms** can be used as part of a software-controlled radio system.

To aid in understanding the relationships among the terms *adaptive radio, cognitive radio, hardware radio, intelligent radio, policy-based radio, reconfigurable radio, software-controlled radio,* and *software-defined radio* that are defined in Clause 3 the set of radio relationship rules listed in Table 14B.1 were developed. These rules follow directly from, or are implied by, the normative definitions and indicate the combinations of radio types that are consistent with the definitions. These rules can be applied to determine whether a specific combination of radio types—for example intelligent, cognitive, adaptive, policy-based, software-controlled, software defined, but not reconfigurable radio—is consistent with the definitions. Since there are eight terms, 256 combinations of these terms are possible, but according to the rules in Table 14B.1, some of the combinations may be inconsistent with the relationships implied by the definitions. A "truth table" enumerating all possible combinations is presented in Table 14B.2.

Table 14B.1 —Rules defining radio terminology relationships interference.

Rule #	Rule
1	Every radio must either be a **software-defined radio** or a **hardware radio,** but not both. This rule is derived from the definitions of hardware radio and software-defined radio.
2	A **reconfigurable radio** may be either a software-defined radio or a hardware radio. This rule is derived from the definition for reconfigurable radio and the notes following the definition.
3	The definitions do not limit the relationship between **adaptive radio** and hardware radio, software-defined radio, reconfigurable radio, policy-based radio, and software-controlled radio. Consequently, a hardware radio, software-defined radio, reconfigurable radio, policy-based radio, or software-controlled radio may or may not also be an adaptive radio.
4	Except as specified in rule #2, the definitions do not define the relationship between **reconfigurable radio** and hardware radio, software-defined radio, adaptive radio, policy-based radio, and software-controlled radio. Consequently, a hardware radio, software-defined radio, adaptive radio, policy-based radio, or software-controlled radio may or may not also be a reconfigurable radio.
5	The definitions do not define the relationship between **software-controlled radio** and hardware radio, software-defined radio, and adaptive radio. Consequently, a hardware radio, software-defined radio, or adaptive radio may or may not also be a software-controlled radio.
6	The definitions do not define the relationship between **policy-based radio** and hardware radio, software-defined radio, reconfigurable radio, and adaptive radio. Consequently, a hardware radio, software-defined radio, reconfigurable radio, or adaptive radio may or may not also be a policy-based radio.
7	A **policy-based radio** is a subset of a software-controlled radio since the definition of a policy-based radio requires that the policy is modifiable. Some scenarios for modification of the policy would require software control. Also, the definitions for policy-based control mechanism and software imply that radio-readable policy is a form of software.
8	**Intelligent radio** is a subset of cognitive radio. This rule is implied by the definition of intelligent radio.
9	**Cognitive radio** is a subset of adaptive radio. This relationship is implied by the definitions of adaptive radio and cognitive radio.
10	Definition b) of **cognitive radio** implies that **cognitive radio** must intersect software-defined radio, adaptive radio, and other technologies that enable a radio to adjust automatically its behavior or operations to achieve desired objectives. These other technologies may include reconfigurable radio, policy-based radio, and software-controlled radio.
11	An **intelligent radio** must be a software-controlled radio. This relationship is implied by the definition of machine learning.

Table 14B.2 — Permitted combinations of radio types.

	Radio types						
	Hardware/SDR		Software	Policy-based			
1	Hardware	Yes	Yes	Yes	Yes	Yes	Yes	Yes
2	Hardware	Yes	Yes	Yes	Yes	Yes	No	Yes
3	Hardware	Yes	Yes	Yes	Yes	No	Yes	No
4	Hardware	Yes	Yes	Yes	Yes	No	No	Yes
5	Hardware	Yes	Yes	No	Yes	Yes	Yes	Yes
6	Hardware	Yes	Yes	No	Yes	Yes	No	Yes
7	Hardware	Yes	Yes	No	Yes	No	Yes	No
8	Hardware	Yes	Yes	No	Yes	No	No	Yes
9	Hardware	Yes	Yes	Yes	No	Yes	Yes	No
10	Hardware	Yes	Yes	Yes	No	Yes	No	No
11	Hardware	Yes	Yes	Yes	No	No	Yes	No
12	Hardware	Yes	Yes	Yes	No	No	No	Yes
13	Hardware	Yes	Yes	No	No	Yes	Yes	No
14	Hardware	Yes	Yes	No	No	Yes	No	No
15	Hardware	Yes	Yes	No	No	No	Yes	No
16	Hardware	Yes	Yes	No	No	No	No	Yes
17	Hardware	Yes	No	Yes	Yes	Yes	Yes	No
18	Hardware	Yes	No	Yes	Yes	Yes	No	No
19	Hardware	Yes	No	Yes	Yes	No	Yes	No
20	Hardware	Yes	No	Yes	Yes	No	No	No
21	Hardware	Yes	No	No	Yes	Yes	Yes	No
22	Hardware	Yes	No	No	Yes	Yes	No	Yes
23	Hardware	Yes	No	No	Yes	No	Yes	No
24	Hardware	Yes	No	No	Yes	No	No	Yes
25	Hardware	Yes	No	Yes	No	Yes	Yes	No
26	Hardware	Yes	No	Yes	No	Yes	No	No
27	Hardware	Yes	No	Yes	No	No	Yes	No
28	Hardware	Yes	No	Yes	No	No	No	No
29	Hardware	Yes	No	No	No	Yes	Yes	No
30	Hardware	Yes	No	No	No	Yes	No	No
31	Hardware	Yes	No	No	No	No	Yes	No
32	Hardware	Yes	No	No	No	No	No	Yes
33	Hardware	No	Yes	Yes	Yes	Yes	Yes	Yes
34	Hardware	No	Yes	Yes	Yes	Yes	No	Yes
35	Hardware	No	Yes	Yes	Yes	No	Yes	No
36	Hardware	No	Yes	Yes	Yes	No	No	Yes
37	Hardware	No	Yes	No	Yes	Yes	Yes	Yes
38	Hardware	No	Yes	No	Yes	Yes	No	Yes
39	Hardware	No	Yes	No	Yes	No	Yes	No
40	Hardware	No	Yes	No	Yes	No	No	Yes
41	Hardware	No	Yes	Yes	No	Yes	Yes	No
42	Hardware	No	Yes	Yes	No	Yes	No	No
43	Hardware	No	Yes	Yes	No	No	Yes	No
44	Hardware	No	Yes	Yes	No	No	No	Yes
45	Hardware	No	Yes	No	No	Yes	Yes	No

(continued)

Table 14B.2 (Continued)

	Radio types						
	Hardware/SDR		Software	Policy-based			
46	Hardware	No	Yes	No	No	Yes	No	No
47	Hardware	No	Yes	No	No	No	No	No
48	Hardware	No	Yes	No	No	No	No	Yes
49	Hardware	No	No	Yes	Yes	Yes	Yes	No
50	Hardware	No	No	Yes	Yes	Yes	No	No
51	Hardware	No	No	Yes	Yes	No	Yes	No
52	Hardware	No	No	Yes	Yes	No	No	No
53	Hardware	No	No	No	Yes	Yes	Yes	No
54	Hardware	No	No	No	Yes	Yes	No	No
55	Hardware	No	No	No	Yes	No	Yes	No
56	Hardware	No	No	No	Yes	No	No	Yes
57	Hardware	No	No	Yes	No	Yes	Yes	No
58	Hardware	No	No	Yes	No	Yes	No	No
59	Hardware	No	No	Yes	No	No	Yes	No
60	Hardware	No	No	Yes	No	No	No	No
61	Hardware	No	No	No	No	Yes	Yes	No
62	Hardware	No	No	No	No	Yes	No	No
63	Hardware	No	No	No	No	No	Yes	No
64	Hardware	No	No	No	No	No	No	Yes
65	SDR	Yes	Yes	Yes	Yes	Yes	Yes	Yes
66	SDR	Yes	Yes	Yes	Yes	Yes	No	Yes
67	SDR	Yes	Yes	Yes	Yes	No	Yes	No
68	SDR	Yes	Yes	Yes	Yes	No	No	Yes
69	SDR	Yes	Yes	No	Yes	Yes	Yes	Yes
70	SDR	Yes	Yes	No	Yes	Yes	No	Yes
71	SDR	Yes	Yes	No	Yes	No	Yes	No
72	SDR	Yes	Yes	No	Yes	No	No	Yes
73	SDR	Yes	Yes	Yes	No	Yes	Yes	No
74	SDR	Yes	Yes	Yes	No	Yes	No	No
75	SDR	Yes	Yes	Yes	No	No	Yes	No
76	SDR	Yes	Yes	Yes	No	No	No	Yes
77	SDR	Yes	Yes	No	No	Yes	Yes	No
78	SDR	Yes	Yes	No	No	Yes	No	No
79	SDR	Yes	Yes	No	No	No	Yes	No
80	SDR	Yes	Yes	No	No	No	No	Yes
81	SDR	Yes	No	Yes	Yes	Yes	Yes	No
82	SDR	Yes	No	Yes	Yes	Yes	No	No
83	SDR	Yes	No	Yes	Yes	No	Yes	No
84	SDR	Yes	No	Yes	Yes	No	No	No
85	SDR	Yes	No	No	Yes	Yes	Yes	No
86	SDR	Yes	No	No	Yes	Yes	No	Yes
87	SDR	Yes	No	No	Yes	No	Yes	No
88	SDR	Yes	No	No	Yes	No	No	Yes
89	SDR	Yes	No	Yes	No	Yes	Yes	No
90	SDR	Yes	No	Yes	No	Yes	No	No

(*continued*)

Table 14B.2 (Continued)

	Radio types							
	Hardware/SDR		Software	Policy-based				
91	SDR	Yes	No	Yes	No	No	Yes	No
92	SDR	Yes	No	Yes	No	No	No	No
93	SDR	Yes	No	No	No	Yes	Yes	No
94	SDR	Yes	No	No	No	Yes	No	No
95	SDR	Yes	No	No	No	No	Yes	No
96	SDR	Yes	No	No	No	No	No	Yes
97	SDR	No	Yes	Yes	Yes	Yes	Yes	Yes
98	SDR	No	Yes	Yes	Yes	Yes	No	Yes
99	SDR	No	Yes	Yes	Yes	No	Yes	No
100	SDR	No	Yes	Yes	Yes	No	No	Yes
101	SDR	No	Yes	No	Yes	Yes	Yes	Yes
102	SDR	No	Yes	No	Yes	Yes	No	Yes
103	SDR	No	Yes	No	Yes	No	Yes	No
104	SDR	No	Yes	No	Yes	No	No	Yes
105	SDR	No	Yes	Yes	No	Yes	Yes	No
106	SDR	No	Yes	Yes	No	Yes	No	No
107	SDR	No	Yes	Yes	No	No	Yes	No
108	SDR	No	Yes	Yes	No	No	No	Yes
109	SDR	No	Yes	No	No	Yes	Yes	No
110	SDR	No	Yes	No	No	Yes	No	No
111	SDR	No	Yes	Yes	No	No	Yes	No
112	SDR	No	Yes	No	No	No	No	Yes
113	SDR	No	No	Yes	Yes	Yes	Yes	No
114	SDR	No	No	Yes	Yes	Yes	No	No
115	SDR	No	No	Yes	Yes	No	Yes	No
116	SDR	No	No	Yes	Yes	No	No	No
117	SDR	No	No	No	Yes	Yes	Yes	No
118	SDR	No	No	No	Yes	Yes	No	Yes
119	SDR	No	No	No	Yes	No	Yes	No
120	SDR	No	No	No	Yes	No	No	Yes
121	SDR	No	No	Yes	No	Yes	Yes	No
122	SDR	No	No	Yes	No	Yes	No	No
123	SDR	No	No	Yes	No	No	Yes	No
124	SDR	No	No	Yes	No	No	No	No
125	SDR	No	No	No	No	Yes	Yes	No
126	SDR	No	No	No	No	Yes	No	No
127	SDR	No	No	No	No	No	Yes	No
128	SDR	No	No	No	No	No	No	Yes

Because of rule #1 in Table 14B.1, which requires a radio to be either a hardware radio or a software-defined radio but not both, the combination of hardware radio and software-defined radio can be represented by a single column, which reduces the number of possible combinations to 128. The radio types are the column headings in the table, and at the intersection of each column and row is an entry of "Yes" or "No" to indicate whether the radio type is present for that combination. The entries below the column heading for hardware radio/SDR are an exception; for this column, the entry is either "hardware" or "SDR." In addition, a right-most column is provided to indicate, by an entry of "Yes" or "No," if the particular combination is permitted by the rules, and a left-most column is provided to number the combinations uniquely from 1 to 128.

Rows containing shaded cells correspond to combinations of radio features that violate one or more of the rules listed in Table 14B.1. The shading below the radio features columns indicates the specific feature combinations that are not permitted. Specifically note that rule #7 requires a policy-based radio to also be a software-controlled radio; that rule #8 requires an intelligent radio to also be a cognitive radio; that rule #9 requires a cognitive radio to also be an adaptive radio; and that rule #11 requires an intelligent radio to also be a software-controlled radio. These four rules account for all 84 combinations that are not permitted. In preparing this table, no consideration was given as to the practicality, utility, or commercial value of any combination.

A series of Venn diagrams are presented in Figure 14B.2 to assist in visualizing the relationships and dependencies of the radio types. Figure 14B.2 depicts the relationship of software-defined radio, hardware radio, software-controlled radio, intelligent radio, and policy-based radio as required by rule #1, rule #5, rule #7, and rule #11 in Table 14B.1. In Figure 14B.3, cognitive radio and adaptive radio have been added and the relationship between these radio types and with other radio types is depicted. This figure illustrates the effect of radio relationship rule #3, rule #6, rule #8, rule #9, and rule #10. The addition of reconfigurable radio in Figure 14B.4 depicts the relationship among all the radio types. It highlights radio relationship rule #2 and rule #4 pertaining to reconfigurable radio. To illustrate the consistency between this figure and the radio relationship truth table in Table 14B.2, the corresponding regions in the figure have been annotated with the row numbers from the table as shown in Figure 14B.5. The size of the "radio regions" and overlapping radio regions depicted in Figure 14B.2 are not drawn to scale and are not intended to indicate the prevalence, popularity, or future likelihood of any radio type or combination of radio types.

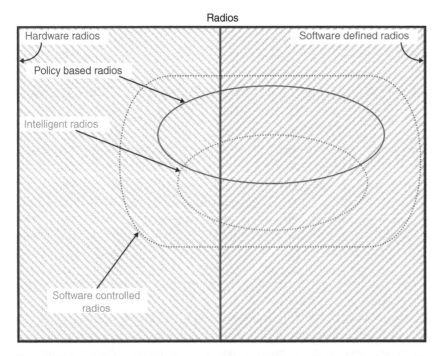

Figure 14B.2 —Relationship of radio types: Software-defined radio, hardware radio, software-controlled radio, intelligent radio, and policy-based radio as required by rule #1, rule #5, rule #7, and rule #11. From IEEE Standard for Definitions and Concepts for Dynamic Spectrum Access: Terminology Relating to Emerging Wireless Networks, System Functionality, and Spectrum Management, Feb 2019. IEEEq.

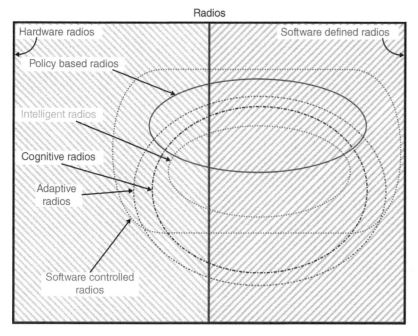

Figure 14B.3 —Relationship of radio types: Addition of cognitive radio and adaptive radio and relationship with other radio types as required by rule #3, rule #6, rule #8, rule #9, and rule #10. From IEEE Standard for Definitions and Concepts for Dynamic Spectrum Access: Terminology Relating to Emerging Wireless Networks, System Functionality, and Spectrum Management, Feb 2019. IEEEq.

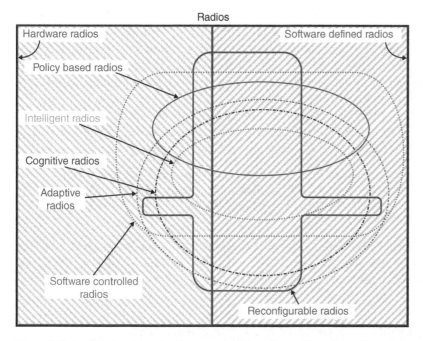

Figure 14B.4 —Relationship of radio types: Addition of reconfigurable radio and the relationship with other radio types as required by rule #2 and rule #4. From IEEE Standard for Definitions and Concepts for Dynamic Spectrum Access: Terminology Relating to Emerging Wireless Networks, System Functionality, and Spectrum Management, Feb 2019. IEEEq.

The primary purpose of Table 14B.1, Table 14B.2, and Figure 14B.2 is to illustrate the types of radio relationships, combinations, and dependencies implied by the normative definitions and to assist the reader in understanding the definitions and in correctly applying the radio terminology to specific situations. Each advanced radio type has certain features and functionality associated with it, and the materials in this annex may be useful in understanding

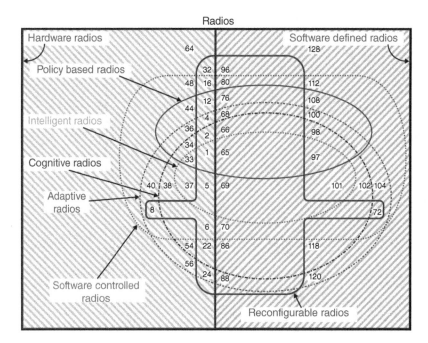

Figure 14B.5 —Relationship of radio types: Cross-reference of permitted radio combination shown in Table 14B.2 to the corresponding regions shown in Figure 14B.4. From IEEE Standard for Definitions and Concepts for Dynamic Spectrum Access: Terminology Relating to Emerging Wireless Networks, System Functionality, and Spectrum Management, Feb 2019. IEEEq.

the permitted feature sets associated with a specific radio term or combinations of terms. Some terms are tightly coupled and imply a specific hierarchy of features or functionality. This is evident in the progression from adaptive radio to cognitive radio to intelligent radio, and implies a sequential timeline in the evolution of these radio technologies. The dependency of intelligent radio and policy-based radio on software-controlled radio illustrates another type of implied progression in technology development.

From IEEE Standard for Definitions and Concepts for Dynamic Spectrum Access: Terminology Relating to Emerging Wireless Networks, System Functionality, and Spectrum Management, Feb 2019. IEEEq.

> NOTE—Rows containing shaded cells correspond to combinations of radio types that violate one or more of the rules listed in Table 14B.1. These rules derive directly or indirectly from the normative definitions. The shading below the radio features columns indicates the specific feature combinations that are not permitted.

From IEEE Standard for Definitions and Concepts for Dynamic Spectrum Access: Terminology Relating to Emerging Wireless Networks, System Functionality, and Spectrum Management, Feb 2019. IEEEq.

14B.2   Explanatory Note on Software-defined Radio

The software-defined radio (SDR) technology has been on the research agenda for more than 20 years. During this period both the original intended use, as well as the associated implications, have changed. The basic idea of the SDR technology was to reduce the number of communication platforms required without introducing new system functionality. Now, SDR allows multimode, multi-band, and/or multi-functional wireless devices to be enhanced using software upgrades, hence providing an efficient and relatively inexpensive solution for supporting multiple waveform standards. SDR provides a path toward the realization of concepts such as reconfigurability, run-time reconfiguration, and cognitive radio.

A popular misconception about SDR is that extremely wide RF coverage (e.g., 2 MHz to 2 GHz) is mandatory. Such wide RF coverage is not necessary for many applications, especially in the consumer and commercial industries that operate within limited RF spectrum. Another popular misconception about SDR is that software reconfigurability is needed all the way to the antenna terminals. A radio can be an SDR as long as the characteristics of the transmitted signal can be modified after manufacture through software and/or firmware downloads. Several different types of software may be part of a radio device, including:

a) Application software
b) Ancillary software
   1) Operating system
   2) Middleware
   3) Drivers

Application software directly affects the radio functionality (e.g., frequency, power, and modulation). This software is tightly coupled with the radio hardware to derive the overall radio functionality. Ancillary software refers to radio software that affects the use of the device such as input/output driver and user interfaces, and does not affect the radio functionality.

Radio devices may have many types of software, but they are not considered to be a software-defined radio device unless they conform to the definition provided in 3.9. In fact, SDR should be seen as a concept or enabler rather than a system implementation. There will be system implementations based on SDR technology (for example cognitive radios) as well as equipment enhanced with SDR capabilities. To be able to distinguish the different implementation types, the SDR Forum (now the Wireless Innovation Forum) introduced a classification scheme for radios. The scheme identifies four levels of reconfigurability according to the class type, as noted in Table 14B.3.

From IEEE Standard for Definitions and Concepts for Dynamic Spectrum Access: Terminology Relating to Emerging Wireless Networks, System Functionality, and Spectrum Management, Feb 2019. IEEEq.

The SDR technology itself has evolved a lot in the last decade and it is still progressing. It is likely that in the near future the SDR technology will be available to implement even the most demanding radio configurations in software. For example, the flexibility provided by SDR gives more and better options for virtualization. Virtualization in SDR allows a single physical radio device to transmit and receive traffic for multiple wireless standards, consuming less additional resources without requiring a fully functional virtual machine or any additional hardware. Virtualization using SDR offers a new solution to the coordination of various wireless standards operating close to each other in the frequency spectrum. Since a single access point or base station is able to cope with different standards via software manipulation, the coordination or scheduling work can be done in a centralized manner. This will save a lot of wireless traffic used to detect, estimate, and negotiate. But, this will be highly dependent on external

Table 14B.3 — Classification of radios according to the SDR Forum (now the Wireless Innovation Forum).

Class type	Reconfigurability
Software-controlled radio (SCR)	Reconfigurations through control functions in software, limited to pre-defined set of configurations
Software-defined radio (SDR)	Software control and reconfigurability of a variety of modulation techniques
Ideal software radio (ISR)	Analog conversion takes place at antenna, speaker, and microphone, everything else is digitized and software configurable
Ultimate software radio (USR)	Understands all traffic and control information and supports (most) applications and radio air interfaces

incentives, such as a regulatory regime that allows the use of reconfigurable radio technology. Finally, it is expected that the major foreseeable usage areas for SDR include dynamic spectrum allocation and management scenarios.

14B.3 Explanatory Note on Cognitive Control and Cognitive Functionality

The understanding of the term *cognitive radio* has developed toward focusing on systems of radios rather than on single devices. More recently, the term *cognitive* is generally understood as a capability of rendering a system more efficient by applying various levels of adaptation capacity. Now it is a definition that is applicable to distributed systems and potentially to virtualized systems. Note that a *cognitive radio system* does not need to be composed by cognitive radios alone, rather it is a radio system that exploits cognition to control its entire behavior. This IEEE standard defines *cognitive radio* and *cognitive radio system* by utilizing the definitions of *cognition* as a concept and of *cognitive process* as a realization of that concept.

In this standard, the definition of cognitive radio is drawn from technical definitions from the artificial intelligence and computation science disciplines along with specific considerations of the radio communications system domain. One of the central objectives of AI research is the study of intelligence in reasoning systems. Currently, these systems fall into two broad categories: 1) systems that appear to think and act like humans, and 2) systems that appear to think and act in a purely rational (e.g., "logical") manner (Russell and Norvig [B45]).

In the AI literature, the term *cognitive* is often applied to systems that exhibit human-like qualities in their externally visible behavior. Cognitive science is concerned with modeling reasoning processes in accordance with those exhibited by humans aiming to cast these models into paradigms that may be algorithmically processed. The human-like processes are not limited to purely rational ones, but can encompass processes that are inconsistent with strict rationality when reasoning, problem solving, planning, and learning. Furthermore, the inputs, outputs, and internal behaviors of a cognitive system are targeted to be consistent with human behavior in both conduct and timing. A system is said to be rational (rather than cognitive) if it does the "right thing," given what it knows. In this sense, a rational system obeys the well-defined laws of inference and logic in processing information. It may use a combination of deductive and inductive processes for reasoning, problem solving, planning, and learning (Obrst, Smith, and Daconta [B44]).

Clearly, a radio that is useful and supportable must be deterministic insofar that it will always obey a set of rules or policies that govern its behaviors. These rules may be regulatory in nature (e.g., ensure the radio is not harmful to other radios) or optimizing (e.g., improve or reduce a certain aspect of the radio's operation). Within that deterministic bound, the radio may be free to adapt by whatever processes are deemed appropriate. These processes may be purely rational and deterministic or may incorporate non-rational reasoning for learning and adaptation.

Cognition, as defined in the *Webster's New Collegiate Dictionary* [B47], is "the act or process of knowing including both awareness and judgment." However, this IEEE standard takes a more limited view of the term *cognition* than that of either the AI community as described above or in the classic dictionary definition of *cognition*. This IEEE standard considers the following attributes as the fundamental blocks of the definition of cognition: i) memory and reasoning; ii) learning; iii) goal driven; iv) adaptiveness. The definition of cognitive radio has a strong focus on the contemporary view. It is likely that the current understanding of terminology might change in the near future. It is expected that the terms *cognitive* and *cognitive radio* will continue to undergo changes incorporating evolving concepts such as cloud and virtualized radios; software-defined networks; separation of radio, decision-making, and cognitive functions.

14B.4 Explanatory Note on Adaptive Radios that Employ a Policy-based Control Mechanism

The rapid growth in the demand for wireless broadband applications has led to an ever-increasing need for radio spectrum. As confirmed by several measurements about spectrum occupancy, the current policy adopted by many governments and regulatory agencies concerning the assignment of the spectrum results in an under-utilization of this valuable resource, and limits access to the available (i.e., vacant) frequency bands which can be used to deploy new communication services or to enhance existing ones. Consequently, the continuous development of new technologies requires a more flexible and efficient management of the spectrum to satisfy the goals of the European Digital Agenda (EC [B8]) and the future market demands for mobile and broadband services.

Large portions of allotted spectrum are unused when considered on a temporal or geographical basis (FCC "Spectrum policy" [B14]). Portions of the assigned spectrum are used only in certain geographical areas, and some portions of the assigned spectrum are used only for brief periods of time. Studies have shown that reuse of such "wasted" spectrum can provide an order of magnitude improvement in the available communication capacity (Wireless World Research Forum [B49]). The currently common regulatory framework is based on static allocation and assignment of spectrum, which is generally referred to as the command and control mechanism. Accordingly, the existing radios are mostly policy-based radios in which the policy forbids a dynamic access to the spectrum.

Alternatively, adaptive radios with more-flexible policies allowing a dynamic access to the spectrum can be used to achieve better spectrum utilization, dynamic spectrum management, and higher flexibility in spectrum use. These goals, which are encapsulated by the term *dynamic spectrum access (DSA) technology*, are facilitated by the measurements and control of a policy-based management software that reside in adaptive policy-based radios. Such radios are software-controlled radios in which the control information includes:

— Policies (regulatory, operational, user)
— Sensory information
— Data characterizing the radio propagation environment
— Available protocols
— Performance requirements
— Information about the radio network

The use of policy-based radio for dynamic spectrum access is an approach wherein static assignment of spectrum is complemented by the opportunistic use of unused spectrum in a manner that limits interference to primary users. This approach is called *opportunistic spectrum access*. The basic parts of this approach are:

— Sense: Sense the spectrum to discover spectral opportunities (spectrum holes).
— Adapt: Adapt the radio parameters according to the spectral opportunities discovered.
— Communicate: Communicate by using the discovered opportunities without making interference on users with a right to access the spectrum.

Figure 14B.6 illustrates the components of an adaptive radio that employs a policy-based control mechanism (adapted from IEEE Std 1900.5-2011 [B26]).

The concept of policy-based control mechanism employed in adaptive radios aims at addressing the spectrum scarcity problem and improving the spectrum efficiency through a dynamic allocation of vacant (or under-utilized) frequency bands. For licensed bands, it potentially provides the license holders with a tool to enhance the utilization of the spectrum covered by their license. In addition, this mechanism allows for diversity of policy sources from different regulatory bodies and enables adopting policies that change by time and geographical location. The concept will facilitate regulatory traceability provided the computer-coded policies trace to the original regulatory documents.

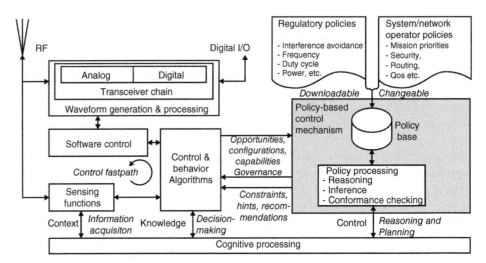

Figure 14B.6 —Components of an adaptive radio that employs a policy-based control mechanism. From IEEE Standard for Definitions and Concepts for Dynamic Spectrum Access: Terminology Relating to Emerging Wireless Networks, System Functionality, and Spectrum Management, Feb 2019. IEEEq.

The key technologies enabling policy-based control mechanisms in adaptive radios include:

— Real-time (preferably wideband) spectrum monitoring capability achieved at low-power consumption
— The capability to automatically and dynamically synthesize waveforms with different temporal, spectral, and spatial characteristics
— The ability to perform network reconfiguration operations
— Policy-based meta-language which:
  1) Translates policy rules into radio system behavior controls
  2) Includes radio control operating rules based on policies and situations
  3) Decouples the radio technology from the regulatory process

Policy-based control mechanism employed in adaptive radios promises several benefits to improve the spectrum utilization and to implement reconfigurable and cost-effective architectures for wireless devices. But, it also poses several challenges that are not to be overlooked in the design and implementation of a cooperative environment, including:

— Wideband sensing
— Opportunity identification
— Network aspects of spectrum coordination when using opportunistic spectrum access
— Traceability so that sources can be identified in the event that interference does occur

14B.5 Explanatory Note on Dynamic Spectrum Access

Wireless networks that implement dynamic spectrum access through utilizing cognitive processes, policy-defined control mechanisms, or spectrum management protocols aim to improve spectrum efficiency and the utility thereof in a shared spectrum space that is governed by appropriate regulatory rules. Coexistence among participants in a shared spectrum space ideally is guided by fairness principles and sharing with others on a non-interfering basis. For coexistence with "grandfathered" systems that are not aware of spectrum-sharing technologies, opportunistic spectrum access along with further protective measures may apply.

Figure 14B.7 illustrates a possible configuration of technologies to accomplish dynamic spectrum access using cognitive and policy-based control patterns under consideration of supportive or governing infrastructure-based components, such as spectrum authorization systems. The distinct processes outlined herein perform simultaneously, although implementations

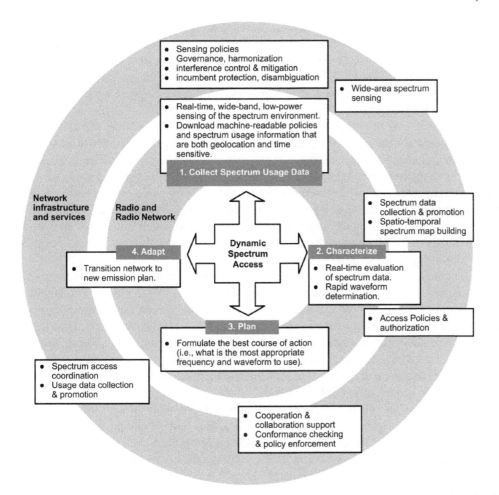

Figure 14B.7 —Key technology components of representative policy-based radio implementation of dynamic spectrum access including supportive infrastructure actions. From IEEE Standard for Definitions and Concepts for Dynamic Spectrum Access: Terminology Relating to Emerging Wireless Networks, System Functionality, and Spectrum Management, Feb 2019. IEEEq.

might be required to iterate repeatedly through the underlying functions for computational or performance reasons as indicated by the numbering of actions in the figure. As can be seen further in Figure 14B.7, infrastructure-based processes might perform concurrently with regards to the radio network and might be governing or responding to the corresponding radio network process.

An essential prerequisite for implementing such control patterns is in the acquisition of information about the operational environment, in knowledge structuring, and in sharing among processes. Systems that realize dynamic spectrum access as illustrated in Figure 14B.7 and based on concepts as defined in Clause 7 might further require external knowledge sources, decision-making, and control processes to meet particular coexistence requirements, for example.

Consequently, dynamic spectrum access networks in contrast to wireless networks implementing dynamic frequency selection algorithms constitute a complex system. A suitable design requires much more consideration of operational environment and co-existing spectrum users, in particular under changing regulatory and operational constraints. Such updates of regulatory constraints and resulting access policies might be required in the course of introducing new spectrum-sharing technologies, or in the course of worldwide spectrum harmonization as currently going on in the 5G spectrum, for example.

The functionality requirements in block 1 of Figure 14B.7 are critical technology drivers. Dynamic spectrum access requires the ability for a real-time, wide-band sensing of the spectral environment. This is the process of sampling shared spectrum space to determine occupancy. It should be noted that no consensus has been reached in regard to a definition of when a channel is occupied; several factors are involved, including receiver sensitivity, the sampling time and sampling interval, thresholds for discriminating wide-band noise from signals, and so on. It should also be noted that along with dynamic spectrum sensing and access policies there comes a need for policy agility. Policy agility thus allows the behavior of a radio or spectrum sensor to change over time and location following operational or regulatory demands. Such policy changes could be downloaded from local or remote sources through the network infrastructure in a machine-readable format. The timescale for such changes is potentially much different from the frequency agility timescale needed for dynamic spectrum access.

Recent technologies mostly focus on significant infrastructure-based support such as spectrum databases, spectrum-sensing policy enforcement, and access authorization to protect coexisting spectrum users, such as broadcasting or radar systems, or to control disclosure of sensible location information that might have been acquired through an uncontrolled spectrum-sensing process.

The functionality requirements in block 2 of Figure 14B.7 include an analysis of the data to determine whether the particular spectrum space bears opportunities for usage. This identification process includes the characterization of the data and uses this information to determine whether the spectrum can be used by another communications service or system. The identification process also includes communication with some subset of its neighbors or the infrastructure because what may seem to be a clear channel at one end of the link may not be a clear channel at the other end of the link. For some mobile radio subsystems, this communication may require a low data-rate control channel.

Recent technologies might collect related coverage information through infrastructure-based spectrum sensing, maintaining spectrum databases and promoting the information available upon request or through the use of dynamic access policies.

Block 3 of Figure 14B.7 is the synthesis of the specific dynamic waveform and frequency that are appropriate for use at this time and this location. This leads to the need for the network to adapt (block 4).

The formulation of appropriate access methods has many aspects that can have a severe impact on the performance of all communication systems co-existing at this time and location, in particular in the presence of defective or even malicious equipment. Recent technologies therefore often rely on the infrastructure-based processes as a corrective. For the current state of technology maturity, a dynamic spectrum access radio or radio system rarely implements a new emission plan solely based on local knowledge and decisions. In consequence, that mandates the presence of infrastructure-based control and coordination as well as governance and enforcement processes.

14B.6 Explanatory Note on Collaborative Spectrum Management

Coordination of shared spectrum use by independent radios or radio networks, potentially including further infrastructure-based coordination functions, has been considered for some time. Collaboration in shared spectrum management in contrast to conventional, coordinated frequency band assignment based on spatio-temporal diversion of resources used, or on binding with dedicated technologies, extends these concepts by means of active collaboration among shared spectrum users. Collaboration can be achieved "among equals" or across "tiers," and underlying mechanisms can be centralized or distributed. Current approaches often rely on more than a single mechanism at a time, in particular in cases of networked or cloud-based, service-oriented infrastructures that may involve multiple stakeholders (e.g., regulators, resource owners, operators, virtual network operators, and manufacturers).

A coarse classification of scenarios of practical relevance can be made based on the type and level of coordination and collaboration between non-cognitive legacy radios, cognitive radios, and the cognitive functions part of some supportive infrastructure. Ranging from eviction, coexistence, and coordination to active collaboration, these scenarios mainly aim on increasing spectrum efficiency, incumbent protection, and interference avoidance.

With upcoming next-generation wireless systems, there is also a vital interest for spectrum efficiency and interference avoidance in dense local area use cases (e.g., in license-assisted access for small cells or narrow-band communication in massive IoT), incumbent protection and interference avoidance in wide area use cases, and in particular in the coexistence between local area and wide area use cases and their spectrum and energy efficiency.

In particular scenarios, such as those involving wireless network virtualization, collaborative spectrum management can become a joint task in the wider scope of resource and

Table 14B.4 —Examples of collaborative scenarios.

Scenario	Coordination type	Coordination method and spectrum management information exchange
Homogeneous—Non-cognitive legacy spectrum users	Configuration	1. Spectrum access is manually coordinated
		2. Radio networks exchange no information
	Database	1. Spectrum access is coordinated and potentially controlled by a spectrum database
		2. Radio networks exchange no information directly or through an infrastructure
Heterogeneous—Non-cognitive legacy and cognitive shared spectrum users	Sensing-based	1. Cognitive radios independently sense and avoid legacy radio spectrum
		2. Radio networks exchange no information
	Clear channel assessment	1. Legacy radios signal their presence (e.g., signature beacon or by detection of known waveforms) to cognitive radios
		2. Radio networks exchange no other information
	Multi-tier	1. Legacy spectrum access is manually coordinated
		2. Shared spectrum access is coordinated and potentially controlled by a spectrum database
		3. Cognitive radios apply coordination methods to avoid conflicts with legacy spectrum and interference with other cognitive radios
		4. Radio networks exchange no other information
Homogeneous—Cognitive spectrum users	Sensing-based	1. Cognitive radios independently sense and avoid spectrum conflicts with each other (e.g., in frequency, time, modulation, or space)
		2. Radio networks exchange no information
	Common control channel	1. Cognitive radios coordinate over a common channel using a common protocol to avoid spectrum conflicts
		2. Various levels of information sharing and collaboration through the common control channel are possible
		3. Radio networks exchange no other information
	Network-based	1. Cognitive radios coordinate over a common channel using a common protocol to form a common network that avoids spectrum conflicts
		2. Radios form a mesh network and act as relays for the common control channel
	Virtualized	1. Cognitive radios form a cloud-based infrastructure and coordinate through a control function on top of a virtualization platform
		2. Radios form a configurable network exchanging information directly or through a shared infrastructure

interference management. Collaborative spectrum management then must consider an additional "self-interference" dimension that results from the demands of isolating virtual network slices that utilize at a certain point in time and space the same spectrum resource that is shared among many dynamic spectrum users.

Based on this classification, Table 14B.4 summarizes common scenarios of collaborative spectrum management considering the various configurations of radio networks and infrastructure involved.

From IEEE Standard for Definitions and Concepts for Dynamic Spectrum Access: Terminology Relating to Emerging Wireless Networks, System Functionality, and Spectrum Management, Feb 2019. IEEEq.

14B.7   Explanatory Note on Spectrum Efficiency

The definition of the term *spectrum efficiency* given in this standard in 7.36 provides a conceptual basis for computing spectrum efficiency. Specification of the benefits derived from spectrum use and of the resource costs will depend on the specific situation. Measures of benefits may relate to the economic or societal value of a communications system or to the value of the information transferred. Information transfer could be valued in terms of its quantity, quality, or importance. Similarly, numerous measures of resource cost can be computed. These measures include the economic cost of acquiring spectrum or the quantity of spectrum used and denied to others.

Accordingly, government agencies have developed categories of systems to which they have applied different efficiency metrics. For example, the Federal Communications Commission Technological Advisory Council (FCC TAC) has used six challenges and considerations in defining spectrum efficiency categories: two for satellite systems and four for terrestrial systems. These categories include satellite broadcast systems, point-to-point satellite systems, terrestrial broadcast systems, terrestrial personal communication systems, terrestrial point-to-point systems, and terrestrial hybrid systems (FCC Technological Advisory Council [B17]). The ITU [B29] has defined an important type of spectrum efficiency measure called spectrum utilization efficiency, in which the benefit is the information transferred and the cost is the amount of spectrum used. The Spectrum Efficiency Working Group of the FCC's Spectrum Policy Task Force (SPTF) [B14] identified additional measures such as Technical Efficiency and Economic Efficiency. Then, the IEEE 802 Local and Metropolitan Area Network Standards Committee (LMSC) [B6] suggested a new metric called Wireless Efficiency (Weff).

More recently densified topologies with two main approaches, i.e., massive multiple-input multiple output (M-MIMO) and small-cell densification, have been considered promising candidate technologies for fifth generation (5G) wireless systems (ITU-R "IMT Vision" [B34]). M-MIMO schemes are able to achieve dramatic gain in spectrum efficiency, increasing the data rate to a large extent because of the large number of antennas. Conversely, small cells scenarios allow to improve the spectrum efficiency by increasing the cell density, hence employing more cells for a specific area. In M-MIMO, each base station employs a large-scale antenna array in linear, cylindrical, or other shapes, thus providing more diversity to the transmission. In contrast, small-cell networks consist of many micro/femto cells with limited number of antennas, typically one or two, at each access point (AP). Small-cell networks have smaller path loss and co-channel interferences while requiring lower power consumption, thus improving both spectral and energy efficiencies.

Finally, it is likely that the request of higher data rate communications might continuously increase in the future, as well as the number of technologies to improve spectrum efficiency.

14B.8   Definition of Informative Terms

14B.8.1   Cognitive Engine

A cognitive engine is defined as:

a) A component of a cognitive radio that assesses inputs, such as environment and internal state, and can make decisions about the radio-operating behavior based on that information and predefined objectives (adopted from IEEE Std 1900.5-2011 [B26]).
b) The portion of the cognitive radio system containing the policy and rules, decision database, and decision-making capability that constantly evaluates the inputs from the sensing mechanism and directs the spectrum access of the reconfigurable radio platform. In Figure 14B.3, the typical elements of a cognitive engine are identified.

14B.8.2   Primary Users

Users with higher priority or legacy rights on the usage of a particular spectrum frequency band.

> NOTE 1—The term is originating from earlier spectrum regulations denoting organizations or technologies qualified for exclusive use of a frequency band, but is ambiguous for recent dynamic spectrum management applications where multiple users can exist at a given time with equivalent or less spectrum usage rights.

> NOTE 2—The term *incumbent user* is sometimes used interchangeably with the term *primary user*, even if an incumbent user is the holder of the licensee.

14B.8.3   Secondary Users

Users with lower priority or less access rights taking second place in the presence of primary users.

> NOTE—The term is originating from earlier spectrum regulations denoting organizations or technologies qualified to operate in a frequency band in case no primary user is present. The term is ambiguous for recent dynamic spectrum management applications where multiple users can exist at a given time each having its own distinct spectrum usage rights and interference protection guarantees.

14B.8.4   Knowledge

In the context of dynamic spectrum access, the term *knowledge* refers to processed data that can be obtained through a cognitive control mechanism, and can be used for decision-making.

> NOTE—In artificial intelligence, the term *knowledge* refers to information that is required for reasoning and decisionmaking, and to solve complex tasks in cognitive control mechanisms, for example. Knowledge representation is a particular field in artificial intelligence aiming to provide such knowledge in a machine-understandable form.

14B.8.5   Radio Access Technology (RAT)

Radio access technology is defined as:

a) A set of techniques realizing wireless access to a communication network.
b) A set of techniques enabling wireless access to a radio network of a particular radio technology.

> NOTE—Examples include: UTRA (Universal Mobile Telecommunications System [UMTS] Terrestrial Radio Access), code division multiple access (CDMA), Digital Enhanced Cordless Telecommunications (DECT), and GERAN (GSM EDGE Radio Access Network), LTE (Long-Term Evolution), and WLAN (wireless local area network).

14B.8.6 Radio Access Network (RAN)

A radio access network is defined as:

a) The network that connects radio base stations to the core network. The RAN provides and maintains radio specific functions, which may be unique to a given radio access technology, that allow users to access the core network (adopted from ITU-T Recommendation Q.1742.3 [B40]).

b) A radio network using one or more radio access technologies to provide wireless access to a core network.

NOTE—Definition b) emphasizes on the use of flexible or re-configurable radio technologies, which is state-of-the-art technology in current multi-standard radio access networks.

14B.8.7 Reasoning

In the context of dynamic spectrum access, the process of decision-making using the available knowledge or outcomes of the learning process.

NOTE—In artificial intelligence, the process of inferencing facts from other facts and knowledge represented in machine-understandable form. Reasoning is an essential pre-requisite for decision-making and learning in cognitive processes, for example.

14B.8.8 Waveform Specification

A waveform specification contains the description of a waveform that includes information about the wireless medium, and transformations and protocols for at least layer 1 (Physical) of the Open Systems Interconnection (OSI) reference model.

14B.8.9 Wireless Sensor Networks

A wireless sensor network (WSN) is a wireless network consisting of spatially distributed autonomous devices using sensors to monitor physical or environmental conditions.

NOTE—WSNs are frequently deployed in the form of mesh networks where node sensors are programmed to monitor information in the surrounding area and to deliver detailed data about the physical environment to neighboring nodes. Sensor nodes and their embedded radio interfaces are strictly constrained in power, processing capability, and memory capacity.

14B.9 Explanatory Notes on Wireless Virtualization

14B.9.1 Introduction

In the information and communications technology (ICT) sector, *virtualization* has become a popular concept in different areas, e.g., virtual memory, virtual machines, virtual storage access network, and virtual data centers.

Virtualization involves abstraction and sharing of resources among different parties. With virtualization, the overall cost of equipment and management can be significantly reduced due to the increased hardware utilization. As well, virtualization eases the decoupling of functionalities from infrastructure, enabling easier migration to newer services and products. It is also poised to facilitate a more flexible management and scalability of resources.

This concept is applicable not only to computing platforms, but to telecommunication and networking application alike (Chowdhury and Boutaba [B5]). For example, time division multiplexing (TDM) access on a line may be viewed as a form of virtualization of a dedicated line to a respective user, as may be tagging of 802.3 MAC frames to provide a virtual LAN.

The omnipresent duality in tradeoff of cost versus functionality drives the virtualization of wireless networks even further.

Wireless network virtualization can have a very broad scope ranging from spectrum sharing, infrastructure virtualization, to air interface virtualization. Similar to wired network virtualization, in which physical infrastructure owned by one or more providers can be shared among multiple service providers, wireless network virtualization needs the physical wireless infrastructure and radio resources to be abstracted and isolated to a number of virtual resources, which then can be offered to different service providers. In other words, virtualization, regardless of wired or wireless networks, can be considered as a process splitting the entire network system.

However, the distinctive properties of the wireless environment, in terms of time-various channels, attenuation, mobility, broadcast, etc., make the problem more complicated. Furthermore, wireless network virtualization depends on specific access technologies, and wireless networks make use of many more access technologies compared to wired network virtualization. And as each access technology has its particular characteristics, this makes convergence, sharing, and abstraction truly difficult to achieve. Therefore, it might even be inaccurate to consider wireless network virtualization as a subset of network virtualization.

For starters, consider two different set-ups (A and B), each with its own definitions of wireless links and communication patterns spanning the actual network.

If these two set-ups (device/devices/cluster, etc.) are to co-exist on the same hardware, communication activities from one set-up should not affect any reception behavior on the second experiment, and vice versa. This observation translates to two important requirements:

— Coherence: When a transmitter of one set-up is active, all of the corresponding receivers and potential sources of interference as defined by the experiment should be simultaneously active on their appropriate channels of operation.
— Isolation: When a node belonging to one set-up is receiving some signal pertinent to the set-up, no transmitter of a different set-up within the communication range of the receiver should be active in the same or a partially-overlapping channel. To enforce such requirements, careful scheduling and selection of transmission activities across different set-ups is required.

Isolation is the basic issue in virtualization that enables abstraction and sharing of resources among different parties. Any configuration, customization, or topology change of any virtual network should not affect and interfere with other coexisting parties. Hence, isolation in wireless networks is far more challenging than in on-wire networks. Unlike wired networks, where bandwidth resource abstraction and isolation can be done on a hardware (e.g., port and link) basis, radio resource abstraction and isolation are not straight-forward, due to the inherent broadcast nature of wireless communications and fluctuations of wireless channel quality observed in time and spatial dimensions. For example, in wireless networks, especially cellular networks, any change in one cell may introduce high interference to neighbor cells.

Not only wireless links, but furthermore, network functions themselves are to be virtualized to reduce the amount of bespoke hardware utilities, particularly with the next significant wave of telecommunications infrastructure deployment: 5G. While what *5G* exactly encompasses is still under discussion in the wider industry, 5G is being defined in relevant Standards Development Organizations (the reason we do not seek to define it here). Re-use of existing wireless infrastructure, e.g., LTE, is anticipated and important for early deployments. The evolved 5G system will very likely be characterized by an agile resilient converged fixed/ mobile core network based on network functions virtualization (NFV) and software-defined networking (SDN) technologies and capable of supporting network functions and applications from different domains.

14B.9.2 Defining Entities in Wireless Virtualization

In wireless network virtualization, virtual resources are being utilized and marketed by parties that may differ from stakeholders owning the actual physical resources.

Today's business models basically define two different roles. Generally, after wireless network virtualization, these are mobile network operators (MNO) and service providers (SP).

All of the infrastructures and radio resources of physical substrate wireless networks, including the licensed spectrum, radio access networks (RANs), backhaul, transmission networks (TNs), and core networks (CNs), are owned and operated by MNOs. MNOs execute the virtualization of the physical substrate networks into some virtual wireless network resources. For brevity, *virtual wireless network resources* are also termed *virtual resources*.

With MNOs providing the virtual resources, SPs lease these virtual resources, operate and program them to offer end-to-end services to end users. In some scenarios, the MNO is identified as infrastructure or resource provider (InP, ReP), which is only responsible for owning and leasing wireless network resources to SPs. SPs will create and deploy the virtual resource by themselves based on the leased and allocated resource to satisfy the requirements of end-to-end services. Further decoupling leads to more specialized roles, including InP, mobile virtual network provider (MVNP), mobile virtual network operator (MVNO), and SP. The functions of them are describing as follows:

a) InP/ReP: Owns the infrastructure and wireless network resources. In some cases, the spectrum resources may or may not be owned by InP/ReP.

b) MVNP: Leases the network resources and creates virtual resources. Some MVNPs may have some licensed spectrum such that they do not need request spectrum resources from InP/ReP. Sometimes an MVNP is called a mobile virtual network enabler (MVNE).

c) MVNO: Operates and assigns the virtual resources to SPs. Meanwhile, in some approaches, MVNOs consists of the roles of both MVNOs and MVNPs. Actually, this model is fit for the emerging concept of so-called "anything as-a-service" (XaaS) in cloud computing. In this model, InPs provide infrastructure-as-a-service (IaaS) while MVNOs provide network-as-a-service (NaaS).

d) SP: Concentrates on providing services to its subscribers based on the virtual resources provided by MVNOs. In other words, virtual resources are requested by SPs, managed by MVNOs, created by MVNPs, and running at InPs/RePs physically. Obviously, this four-level model can create more opportunities in the market and simplify the functions of each role intuitively. Nevertheless, more coordination mechanisms and interfaces should be used, which may increase the complexity and latency significantly.

14B.9.3 NFV Features and Topics that Arise in Communication Including Wireless Networks

In the sequel some features and topics are listed that network operators in NFV standards development organizations (SDOs) as well as the wider industry are addressing to realize use cases. The explanatory remarks below are attempting a neutral or bipartisan point of view:

a) Network slicing: From a network operator viewpoint, network slicing is a service-oriented network construct providing network-on-demand to concurrent applications. As such it may be categorized as a business model. In other words, network slicing can be seen as an implementation of the network-as-a-service paradigm, where a common network is able to provide and expose concurrent, partitioned, and self-contained "slices" to support different services in an efficient way and provide the required quality of service (QoS). When viewed from a standards definition viewpoint, however, network slicing may be barrier to convergence of requirements, potentially leading to increased complexity to implement network slicing in a converged (i.e., common) core infrastructure supporting 5G and other evolved network services.

b) Cloud-native design principles: In contrast to virtualized network functions that basically are re-purposed software coming from existing physical network functions/appliances (PNFs), cloud-native network functions are network functions implemented using generic IT cloud techniques beyond virtualization (e.g., functions composed from re-usable components rather than monolithic implementations of functions). Their goal is to improve efficient use of resources, including non-palpable resources from operations or maintenance, through a finer-grained multiplexing on the infrastructure, and to be amenable to advanced cloud orchestration techniques as used in IT environments while still offering deterministic performance deriving from NFV requirements.

c) Cloudification of the radio access network (RAN): NFV should be applied to communication networks end-to-end, i.e., including both core networks and radio access networks (RANs), as well as network management systems. Network virtualization to date has focused on the core network to underpin growth of existing systems. Virtualization is now also moving into the RAN area. Challenges particular to radio functions virtualization have been identified as follows:

1) RAN function split and the functional definition of RAN virtual network functions (VNFs). Even though not all RAN functions may be amenable to virtualization and run on general-purpose processors or centralized in data centers, it is crucial to assess the most suitable targets for virtualization and define a RAN architecture with clear identification of both VNFs and PNFs, and the related standard interfaces.

2) Potential new requirements imposed by RAN-VNFs on the NFV infrastructure due to unique features of RAN function processing need to be considered.

3) Potential RAN requirements for acceleration, in particular interface definition and management, need to be addressed.

4) RAN-related management and orchestration functions, including self-organizing networks (SONs), should evolve to interwork with the NFV-management and orchestration (MANO) framework, where orchestration can be defined as the automated arrangement, coordination, and management of ICT systems, middleware, and services.

5) Detrimental energy-efficiency effects due to the use of generic computing hardware instead of objective-tailored proprietary hardware.

d) Scalability: Using 5G as an example, market deployment will drive a need to make the NFV architecture massively scalable. 5G networks will be composed of thousands, perhaps millions or even billions, of distributed compute nodes. For example, it is entirely feasible that each home could host a compute node running 5G VNFs and applications. This degree of scalability and distribution is not possible with today's virtual infrastructure managers (VIMs). A key challenge in this context may emerge as to develop a VIM architecture and implementations that are massively scalable while at the same time open and flexible enough to address future unknown 5G applications. While SDOs can help define the scalability requirements, it is most likely that VIMs will be developed through open source initiatives elsewhere.

Annex 14C

(informative)

List of Deleted Terms from the Previous Versions of IEEE Std 1901.1

This annex contains terms that have been considered for deletion in the current revision of the standard.

14C.1 Composite Network

A type of wireless communications network that consists of multiple radio access technologies under single or multiple network management control to support efficient communication.

NOTE—In a composite network, cognitive radio techniques may be applied to enable the radio to select the best available option for communication. Composite networks differ from heterogeneous networks in that the focus of composite networks is on sharing of resources, particularly spectrum resources; this is not the focus of heterogeneous networks. An example of composite networks is Wi-Fi, WiMAX, digital audio broadcasting, digital video broadcasting, and second-, third-, and fourth-generation commercial wireless radio access technologies controlled by dynamic resources management (Bourse et al. [B2]).

14C.2 Collaborative Spectrum Usage

The process of two or more radio nodes combining their capabilities and spectrum-usage resources via negotiated or predetermined policies and agreements to improve the expected utility of the network.

14C.3 Communications Mode

An operational configuration of a radio device, comprising selections of particular behaviors at all layers of the protocol stack from among the options supported by the device, specified by the radio access technology (RAT) or supported by the radio access network (RAN) used by the device.

14C.4 Prioritized Spectrum Access

Spectrum access in which the precedence of radio transmissions is determined by message criticality rather than by user class or type.

NOTE—In prioritized spectrum access, a user that might typically be termed secondary might assert precedence in an emergency. For example, public safety users might be secondary users in a frequency band for routine communications, but assert precedence to assure spectrum availability during major incidents.

14C.5 Interference Event

A circumstance in which a quantified threshold level of interference has been exceeded. Interference events are specified in terms of the relevant variables for a specific scenario (e.g., time, frequency, amplitude, and performance metrics) (adopted from IEEE Std 1900.2™-2008).

NOTE—A single variable or more than one variable may be used in quantifying the threshold level for which an interference event is registered.

14C.6 Negotiated Spectrum Access

A spectrum access protocol under which multiple radios or networks agree to mutual use of a common spectrum band via negotiated agreements.

NOTE—Negotiated spectrum access agreements may be prearranged or automatically established on an ad hoc or realtime basis without the need for prior agreements between all parties.

14C.7 Noncollaborative Coexistence Mechanism

A coexistence mechanism not based on the cooperative exchange of information among systems attempting to coexist.

14C.8 Performance Metric

A parameter, which can be measured or estimated, or a function of multiple parameters that quantifies the performance of a system in some way.

NOTE—Examples of performance metrics for spectrum-dependent devices or networks include:

— Throughput
— Latency
— Packet error rate
— Time for a dynamic frequency selection (DFS) system to change channels
— Probability of detection and probability of false alarm in identifying which channels are occupied as part of a DFS system.

14C.9 Precedence Assertion

In the context of hierarchical spectrum access, a notification to radio systems currently accessing spectrum that a higher precedence radio system intends to use that spectrum.

NOTE—Precedence assertion is the mechanism by which higher precedence systems reclaim spectrum from lower precedence systems. Potential precedence assertion mechanisms include control messages or information sent through spectrum availability beacons, control pilot channels, or predefined embedded signaling properties. In some cases, a radio may assert precedence merely by transmitting regardless of current usage.

14C.10 Protocol Agility

The ability to change protocols as different networks, or protocols within one network, become available and are used. These protocols include the complete communications and transport layers, from layers 1 through 4, as well as the application layer protocols.

NOTE—As multimode or multinetwork software-defined radio (SDR) functionality expands, the dynamic discovery and optimization of available protocols becomes possible. Instead of the current technique of specifying a preferred network, its set of available protocols, and backup networks in the event the preferred one is not available (e.g., digital to analog), a protocol agile radio may interrogate its network or scan the available network(s) and service(s) for possible protocols. Such a radio may pick a best protocol for one service and then jump to another. In such a scenario, protocol agile radios will be able to select the protocols that provide the best link, and the best application, in line with the user needs. The radio will need to be aware of possible protocols and how and when to change. Examples of protocol agility that are less cognitive include the awareness necessary for (transmitted signal) adaptive power control, and data rate and modulation agility in Wi-Fi and other protocols where the modulation technique is changed as a function of signal-to-noise ratio (SNR) to maintain a certain bit error rate (BER). In the cognitive approach, multiple parts of the protocols may be changed, possibly at a protocol component level, or at a higher level.

14C.11   Quality-of-service Management

Management of the network to conform with the QoS requirements for each application as agreed on between the service provider and the end user.

NOTE 1—Agreement between the service provider and the user may be implicit, such as when the provider and the user are related entities, or it may be explicitly defined in a service level agreement.

NOTE 2—Network management systems accommodate different QoS requirements for distinct applications by prioritizing network traffic, as well as by monitoring and maintaining the network as a whole.

14C.12   Radio Quiet Zone

A radio quiet zone is:

a) A frequency band and corresponding geographic region in which no radio energy emission shall occur.
b) A zone in which spectrum usage must be explicitly coordinated or authorized.

NOTE—This may protect sensitive deep space radio receiver systems, radio astronomy systems, or areas where explosive blasting equipment is in use. Typically, appropriate policies loaded in a policy-based radio would protect these sensitive spectrum uses.

14C.13   Restricted Dynamic Spectrum Access Etiquette

A type of spectrum sharing etiquette that shares spectrum access only with a known, prespecified set of other systems.

14C.14   Spatial Awareness

Spatial awareness is defined as:

a) Awareness by a device of its relative orientation and position.

NOTE—Radios may use this knowledge to improve network performance and to control the dynamic spectrum access process. For example, a radio may be able to use spatial awareness information to control the operation of an adaptive antenna and thereby to reject undesired signals and enhance reception of desired signals.

b) Capability to geolocate a system or device through the use of mechanisms involving RF signals or related information, or network information.

> NOTE—The cognitive application includes the selection and optimization of the techniques, signals, and networks to use for geolocation. This may involve the use of some GPS signals. However, the cognitive approach is more appropriate if GPS signals are totally or partially unavailable and if signals in the environment, specifically those being used for communications purposes (networks or other specific users), are available. It may also include information available, processed or unprocessed, from the network(s) to which the system or device has access.

14C.15 Spectrum Access Behavior

The spectrum access actions and responses of a radio device, system, or network.

14C.16 Spectrum Availability Beacon

Auxiliary transmitters that primary users of the spectrum may deploy to control access of secondary users to specified channels or frequency bands.

> NOTE—Beacons can be unmodulated continuous-wave transmitters or more elaborate depending on whether they also transmit information about the availability of spectrum. Beacons may be used to signal either the availability or the unavailability of spectrum.

14C.17 Unrestricted Dynamic Spectrum Access Etiquette

A type of spectrum-sharing etiquette that shares spectrum access with other systems that were unknown at the time the system was designed.

14C.18 Wireless Network Efficiency

Expressed by the following equation (adapted from the Comments of the IEEE 802 Local and Metropolitan Area Network Standards Committee [B6]):

$$\text{Weff} = \frac{(C \times N_s)}{(B \times A)}$$

where

C   is the capacity of the system in delivery of information bits per second, after decoding, demodulation, and including the vagaries of the network protocol and duty cycle

N_S is the number of logical connections or users in the network, within the coverage area and using the allocated bandwidth B

B   is the allocated bandwidth to the network in Hertz

A   is the area covered (in units of square meters) by the radio system over which the bandwidth B is uniquely associated

14C.19 Firmware

Firmware is defined as:

a) Software that is embedded in a hardware device that allows reading and executing the software, but does not allow modification, for example, writing or deleting data by an end user (adapted from FEDSTD-1037C [B18]).

b) Modifiable or unmodifiable binary instruction and configuration data that is loaded into a programmable logic device to define its operation.

NOTE—Firmware is a specialized type of software. *See also:* **software**.

14C.20 Quality of Spectral Detection

A metric of the quality of spectrum-sharing opportunities.

NOTE—Quality of spectral detection implies the specification of a specific level of interference to the primary users of the channel, which will not limit secondary users from using the spectrum when not used by the primary user.

14C.21 Vision and Roadmap for Application of Advanced Radio System Technologies

Figure 14C.1 provides a visual description of how the technologies referenced in the document are related and indicates the relative time sequence of their development. The figure is based on a qualitative estimation of the foreseen developments in the framework of radio devices, and thus, it should not be considered as an exact timeline.

Within the figure, the advanced radio terminology has been divided into four categories. The first category indicates that the technique for implementing the physical layer of a waveform may either be hardware defined or software defined. Initially, the physical layer of all waveforms was implemented using nonprogrammable hardware. Although hardware-defined implementations will continue, the overall trend over time is increasing use of software-defined radio implementation techniques. This is part of a larger trend of expanding usage of software-reconfigurable devices within radios that extends beyond physical layer

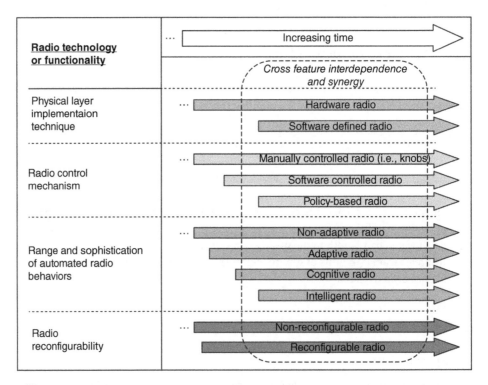

Figure 14C.1 —Conceptual timeline for advanced radio system technologies. From IEEE Standard for Definitions and Concepts for Dynamic Spectrum Access: Terminology Relating to Emerging Wireless Networks, System Functionality, and Spectrum Management, Feb 2019. IEEEq.

signal processing and affects radio control and spectrum access behavior as denoted in the second and third radio feature categories presented in the figure. In simple radio implementations, the operator controls the operation of the radio through a user interface such as control knobs. As radios have become more complex, control is increasingly being performed by software-reconfigurable devices. Software control of radio functions may be implemented in both hardware radio and software-defined radio. Often the control software is proprietary to a specific manufacturer and radio device, and consequently, only the manufacturer may be able to reconfigure this software. In advanced radios, such as adaptive, cognitive, or intelligent radio that engage in complex behaviors such as dynamic spectrum access, there may be a need for others such as spectrum regulators to be able to inspect or modify radio-control software or to be able to reuse control software across a range of dissimilar hardware platforms. This has given rise to the development of policy-based control mechanisms that use machine readable rules that are in a format independent of the radio implementation. Such an approach could support the development of standardized policy languages and authoring tools that may be applied to broad categories of advanced radio devices. Ideally, standardization would free policy authors from the need to have in-depth knowledge of the inner workings of each radio device and allow one set of policies to function correctly in the same category of advanced radio devices. Consequently, over time, increased use of software-controlled radio and policy-based radio is expected.

Radios with a wider range and sophistication of spectrum access behaviors are being developed to support the increasing usage of wireless networks and dynamic spectrum access. Wireless LAN devices incorporating a form of dynamic spectrum access called *dynamic frequency selection* are an early example of this trend. These types of radios generally fall within the broad category of adaptive radio because they need to be able to monitor their own performance automatically and to vary their parameters of operation to improve their performance. Cognitive radio is a more capable subset of adaptive radio that also has the capability to be aware of its environment and internal state and can make decisions about its operating behavior based on that information and predefined objectives. Intelligent radio improves on cognitive radio by also incorporating a machine learning capability. The use of adaptive, cognitive, and intelligent radio is expected to grow in parallel with the growth of wireless networks and dynamic spectrum access. Other types of radio technologies such as policy-based radio, software-controlled radio, and software-defined radio provide capabilities that contribute to and are in some cases necessary for the development and implementation of adaptive, cognitive, and intelligent radio.

The last radio feature listed in Figure 14C.1 is radio reconfigurability. Although radio reconfigurability is not new and it is possible to reconfigure hardware modules in a hardware radio, the advent of software-defined radio and software-controlled radio have introduced new aspects and widened the potential scope of reconfiguration. The expanding use and functional capability of software-reconfigurable devices within radios will make reconfigurable radio easier to implement and provide added value to manufacturers, users, operators, and regulators. In the past, reconfiguration was often limited to manufacturers or to individuals with special training and tools. Increasingly, reconfiguration will be performed by the user, network operator, network, or spectrum regulators. The growth of reconfigurable radio is expected to increase over time in parallel with the deployment of radios incorporating software-reconfigurable devices.

14C.22 Spectrally Aware Networking

The network layer (of an Open Systems Interconnection [OSI] stack) having the intelligence regarding the availability and occupancy as well as any other details of the spectrum that are being used by its lower layers (PHY). With this knowledge, the network configuration or auto-configuration is managed so as to increase the efficiency.

Annex 14D

(informative)

Bibliography

Bibliographical references are resources that provide additional or helpful material but do not need to be understood or used to implement this standard. Reference to these resources is made for informational use only.

B1 Australian Communications Authority (ACA), Vision 20/20—Future Scenarios for the Communications Industry—Implications for Regulation, May-Jun. 2004.

B2 Bourse, D. et al., "Business perspectives of end-to-end reconfigurability," IEEE Wireless Communications, vol. 13, no. 3, pp. 44–57, June 2006. [Special issue on European perspective on Composite Reconfigurable Networks.] http://dx.doi.org/10.1109/MWC .2006.1700071.[8,9]

B3 Buddhikot, M. M. and K. Ryan, "Spectrum management in coordinated dynamic spectrum access based cellular networks," Proceedings of IEEE DySPAN, Baltimore, MD, pp. 299-307, 8-11 Nov. 2005, http://dx.doi.org/10.1109/DYSPAN.2005.1542646.

B4 Cave, M., "Radio spectrum management review; an independent review for the DTI and HM treasury," final report, March 2002.[10]

B5 Chowdhury, N. and R. Boutaba, "Network virtualization: State of the art and research challenges," *IEEE Communications Magazine*, vol. 47, no. 7, pp. 20–26, July 2009, http://dx.doi.org/10.1109/MCOM.2009 .5183468.

B6 Comments of the IEEE 802 Local and Metropolitan Area Network Standards Committee in the matter of: Spectrum Policy Task Force Seeks Public Comment on Issues Related to Commission's Spectrum Policies, ET Docket No. 02-135, DA 02-1311, 13 Jul. 2002.

B7 European Commission Directorate-General for Communication Citizens information, Digital agenda for Europe. Brussels, Belgium; 2014, pp. 1-8, 10.2775/41229.[11]

B8 European Commission (EC), "Communication from the Commission to the European Parliament, the Council, the European Economic and Social Committee and the Committee of the Regions. 5G for Europe: An action plan," COM (2016) 588 final, *Brussels, Belgium*, 2016.[12]

B9 eXtensible Markup Language (XML) 1.0 (Fourth Edition), World Wide Web Consortium Recommendation, 16 Aug. 2006.[13]

B10 Federal Communications Commission (FCC), "Amendment of the commission's rules with regard to commercial operations in the 3550–3650 MHz band," FCC-16-55, May 2016.

B11 Federal Communications Commission (FCC), "Innovation in the broadcast television bands: allocations, channel sharing and improvements," WT Docket No. 10-235, Apr. 2012.

8 The IEEE standards or products referred to in Annex 14D are trademarks owned by The Institute of Electrical and Electronics Engineers, Incorporated.

9 IEEE publications are available from The Institute of Electrical and Electronics Engineers (http:// standards.ieee.org/).

10 Available: http://web1.see.asso.fr/ICTSR1Newsletter/No004/RS%20Management%20-%202_title-42.pdf.

11 Available: https://www.google.it/url?sa=t&rct=j&q=&esrc=s&source=web&cd=1&cad=rja&uact=8 &ved=2ahUKEwiK36nQo_LgAhVhMOwKHcgfDuwQFjAAegQICRAC&url=https%3A%2F%2Feuropa.eu% 2Feuropean-union%2Ffile%2F1497%2Fdownload_en%3Ftoken%3DKzfSz-CR&usg=AOvVaw3RadlBjrZ19Tc VzB4VsXOc.

12 Available: http://ec.europa.eu/newsroom/dae/document.cfm?doc_id=17131.

13 Available: http://www.w3.org/TR/2006/REC-xml-20060816.

B12 Federal Communications Commission (FCC), "Promoting efficient use of spectrum through elimination of barriers to the development of secondary markets, second report and order," WT Docket No. 00-230, Sep. 2004.

B13 Federal Communications Commission (FCC), "Report and order in the matter of facilitating opportunities for flexible, efficient, and reliable spectrum use employing cognitive radio technologies," FCC05-57, ET Docket No. 03-108, 11 Mar. 2005.

B14 Federal Communications Commission (FCC), "Spectrum policy task force report," ET Docket No. 02-135, 15 Nov. 2002.[14]

B15 Federal Communications Commission (FCC), "Statement from FCC chairman Julius Genachowski on incentive auction legislation," Feb. 2012.[15]

B16 Federal Communications Commission (FCC), "Third memorandum opinion and order in the matter of unlicensed operation in TV broadcast bands (ET Docket No 04186), additional spectrum for unlicensed devices below 900 MHz and 3 GHz band," ET Docket No 02-380, Apr. 2012.

B17 Federal Communications Commission (FCC) Technological Advisory Council, Sharing Work Group, "Spectrum efficiency metrics," Sep. 2011.

B18 FED-STD-1037C, Glossary of Telecommunication Terms.

B19 Information and Communication Technologies Authority (ICTA), "Annual report on the development of the information and communication industry in Mauritius: 2009," Jan. 2010.[16]

B20 IEEE Std 802.11™-2018, IEEE Standard for Information Technology—Telecommunications and Information Exchange Between Systems—Local and Metropolitan Area Networks—Specific Requirements—Part 11: Wireless LAN Medium Access Control (MAC) and Physical Layer (PHY) Specifications.

B21 IEEE Std 802.11h™-2003, IEEE Standard for Information Technology—Telecommunications and Information Exchange Between Systems—Local and Metropolitan Area Networks—Specific Requirements—Part 11: Wireless LAN Medium Access Control (MAC) and Physical Layer (PHY) Specifications, Amendment 5: Spectrum and Transmit Power Management Extensions in the 5 GHz Band in Europe.

B22 IEEE Std 802.15.2™-2003(R2009), IEEE Recommended Practice for Information Technology—Telecommunications and Information Exchange Between Systems—Local and Metropolitan Area Networks—Specific Requirements—Part 15.2: Coexistence of Wireless Personal Area Networks with Other Wireless Devices Operating in Unlicensed Frequency Bands.[17]

B23 IEEE Std 802.16™-2009, IEEE Standard for Local and Metropolitan Area Networks—Part 16: Air Interface for Fixed Broadband Wireless Access Systems.

B24 IEEE Std 1900.4™-2009, *IEEE Standard for Architectural Building Blocks Enabling Network-Device Distributed Decision Making for Optimized Radio Resource Usage in Heterogeneous Wireless Access Networks.*

B25 IEEE Std 1900.4a™-2011, IEEE Standard for Architectural Building Blocks Enabling Network-Device Distributed Decision Making for Optimized Radio Resource Usage in Heterogeneous Wireless Access Networks Amendment 1: Architecture and Interfaces for Dynamic Spectrum Access Networks in White Space Frequency Bands.

B26 IEEE Std 1900.5™-2011, IEEE Standard for Policy Language Requirements and System Architectures for Dynamic Spectrum Access Systems.

B27 IEEE Std 1900.6™-2011, IEEE Standard for Spectrum Sensing Interfaces and Data Structures for Dynamic Spectrum Access and Other Advanced Radio Communication Systems.

B28 Infocomm Media Development Authority (IMDA), "Technical specification: television white space devices," Oct. 2016.

---

14 Available: http://hraunfoss.fcc.gov/edocs_public/attachmatch/DOC-228542A1.pdf.

15 Available: https://www.fcc.gov/document/chairmans-statement-incentive-auction-legislation.

16 Available: http://www.icta.mu/documents/publications/ict_report09.pdf.

17 IEEE Std 802.15.2-2003 has been withdrawn; however, copies can be obtained from The Institute of Electrical and Electronics Engineers (http://standards.ieee.org/).

B29 ITU Radio Communications Bureau, Handbook: National Spectrum Management, v. 1.01, Geneva, Switzerland, ITU Radio Communications Bureau, 2005.[18]

B30 ITU-R, "Provisional final acts: World Radiocommunication Conference (WRC-12)," Feb. 2012.[19]

B31 ITU-R, Article 2, Vol. 1. Geneva, Switzerland: ITU Radio Communications Bureau, 2008, The Radio Regulations, Edition of 2008.

B32 ITU-R, *The Radio Regulations, Edition of* 2012, "Article 1, terms and definitions," Geneva, Switzerland, ITU Radio Communications Bureau, 2012.

B33 ITU-R Recommendation E.800, Series E: Overall Network Operation, Telephone Service, Service Operation and Human Factors: Quality of Telecommunication Services: Concepts, Models, Objectives and Dependability Planning—Terms and Definitions Related to the Quality of Telecommunication Services.[20]

B34 ITU-R Recommendation ITU-R M.2083–0, IMT Vision—Framework and Overall Objectives of the Future Development of IMT for 2020 and Beyond.

B35 ITU-R Report M.1652, Dynamic Frequency Selection (DFS) in Wireless Access Systems Including Radio Local Area Networks for the Purpose of Protecting the Radio-determination Service in the 5 GHz Band.

B36 ITU-R Report M.2038, Technology Trends.

B37 ITU-R Report M.2063, The Impact of Software Defined Radio on IMT-2000, the Future Development of IMT-2000 and Systems Beyond IMT-2000.

B38 ITU-R Report M.2064, Software-Defined Radio in the Land Mobile Service.

B39 ITU-R Report SM.2152, Definitions of Software Defined Radio (SDR) and Cognitive Radio System (CRS).

B40 ITU-T Recommendation Q.1742.3, IMT-2000 References (Approved as of 30 June 2003) to ANSI-41 Evolved Core Network with cdma2000 Access Network.

B41 Ministry of Information and Communication Technology of Mauritius, "National Broadband Plan Policy: 2012–2020," Jan. 2012.[21]

B42 Ministry of Public Management, Home Affairs, Posts and Telecommunications, Japan (MPHPT), "Outline of report 'radio policy vision'."

B43 National Telecommunications and Information Administration (NTIA), Manual of Regulations and Procedures for Radio Frequency Management, Washington, DC: U.S. Department of Commerce.[22]

B44 Obrst, L., K. T. Smith, and M. C. Daconta, The Semantic Web: A Guide to the Future of XML, Web Services, and Knowledge Man. Topeka, KS: Topeka Bindery, 2003.

B45 Russell, S. J. and P. Norvig, Artificial Intelligence: A Modern Approach. Englewood Cliffs, NJ: Prentice Hall, 1995.

B46 Accessible at, http://web1.see.asso.fr/ICTSR1Newsletter/No004/RS %20Management%20-%202_title-42.pdf

B47 *Webster's New Collegiate Dictionary.* Springfield, MA: Merriam-Webster, Inc.

B48 Westerinen, A., J. Schnizlein, J. Strassner, M. Scherling, B. Quinn, S. Herzog, A. Huynh, M. Carlson, J. Perry, S. Waldbusser, RFC 3198, Terminology for Policy-Based Management, RFC Editor/The Internet Society, 2001.[23]

B49 Wireless World Research Forum, "Cognitive radio and management of spectrum and radio resources in reconfigurable networks," *Working Group 6 White Paper*, pp. 1–28, 2005.

B50 Zhao, Y., S. Mao, J. Neel, and J. H. Reed, "Network support: the radio environment," in Cognitive Radio Technology, Fette, B. A., ed. Oxford, UK: Elsevier, Apr. 2009, pp. 337–363, http://dx.doi.org/10.1016/ B978-0-12-374535-4.00011-4.

---

18 ITU-T publications are available from the International Telecommunications Union (http://www.itu.int/).

19 Available: http://www.itu.int/md/R12-WRC12-R-0001/en

20 Available: http://www.itu.int/rec/T-REC-E.800-200809-I

21 Available: http://www.icta.mu/documents/nationalbroadbandpolicy2012.pdf

22 Available: http://www.ntia.doc.gov/osmhome/redbook/redbook.html.

23 Internet Requests for Comments (RFCs) are available on the World Wide Web at the following ftp site: venera.isi.edu; logon: anonymous; password: user's e-mail address; directory: in-notes.

15

IEEE Recommended Practice for the Analysis of In-Band and Adjacent Band Interference and Coexistence Between Radio Systems

> *IMPORTANT NOTICE: This recommended practice is not intended to assure safety, security, health, or environmental protection in all circumstances. Implementers of the standard are responsible for determining appropriate safety, security, environmental, and health practices or regulatory requirements.*
>
> *This IEEE document is made available for use subject to important notices and legal disclaimers. These notices and disclaimers appear in all publications containing this document and may be found under the heading "Important Notice" or "Important Notices and Disclaimers Concerning IEEE Documents." They can also be obtained on request from IEEE or viewed at http://standards.ieee.org/IPR/disclaimers.html.*

15.1   Overview

15.1.1   Relationship to Traditional Spectrum Management

The introduction to the 2005 edition of the ITU-R National Spectrum Management Handbook states:

> *Society's increasing use of radio-based technologies, and the tremendous opportunities for social development that these technologies provide, highlight the importance of radio-frequency spectrum and national spectrum management processes. Technological progress has continually opened doors to a variety of new spectrum applications that have spurred greater interest in, and demand for, the limited spectrum resource. Increased demand requires that spectrum be used efficiently and that effective spectrum management processes be implemented. In this framework, modern data handling capabilities and engineering analysis are important to accommodate the variety of potential users seeking access to the spectrum.*
>
> *Radio communications is heavily used in a growing number of services such as national defense, public safety, broadcasting, business and industrial communications, aeronautical and maritime radio communications, navigation, and personal communications. Radio communication links, as opposed to wireless telecommunications, are necessary in a dynamic or mobile environment, where wire-line telecommunication may not be available, or where telecommunications have been disrupted, such as in emergency or natural disaster situations. Radio communication systems may operate from satellites or from terrestrial platforms.*
>
> *If the spectrum is to be used efficiently, its use must be coordinated and regulated through both national regulations and the Radio Regulations of the International Telecommunication Union (ITU). The ability of each country to take full advantage of the spectrum resource depends heavily on spectrum management activities that facilitate the implementation of radio systems and ensure minimum interference. To this end, administrations should, as appropriate, make use of computerized spectrum management systems.*

The demand for spectrum continues to increase dramatically, and with it, the importance of efficient spectrum management is increasing. Fixed, universal rules to coordinate spectrum are not efficient. The most efficient spectrum allocation is not fixed to a single set of rules for all locations and situations nor is it static over time. Optimally spectrum management would be adaptive to specific situations and variable to the specific conditions at a given time and place. Dynamic and adaptive management of spectrum is now possible. The development of computer science, radio engineering, and related disciplines has reached the point where spectrum usage can now be controlled automatically by coexisting radio systems in real time. This replaces the manual method in which human radio-frequency (RF) planners used static frequency assignments.

15.1.2   Introduction to this Recommended Practice

Interference between systems occurs when operation of one system affects the performance of another system.[1] Two or more radio systems successfully coexist when the level of performance of all systems is judged to be acceptable. There may be some interference, but it is generally judged to be within acceptable limits.

Emerging technologies such as cognitive radio, ad hoc wireless networks, as well as dynamic and adaptive systems complicate the interference and coexistence analysis. For example, if a network can reroute data so loss of a particular link has no impact on end-to-end user performance, does loss of that link constitute interference? If it is judged to be interference, does it constitute harmful interference? If so, what is the appropriate remedy?

This recommended practice provides a structure for interference and coexistence analysis. The purpose of providing such a structure is to guide the analyst in considering all relevant issues in a systematic way. Furthermore, a uniform structure makes the comparison of different analyses easier. When different analyses utilize a common structure and method of analysis the reasons for similarities or differences in conclusions are more quickly identified. Thus, focus may be brought to the critical elements, where more data or further analysis may either confirm results or aid in the understanding of differing conclusions.

15.1.3   Scope

15.1.3.1   Formal Scope[2]

This recommended practice will provide technical guidelines for analyzing the potential for coexistence or in contrast interference between radio systems operating in the same frequency band or between different frequency bands.

15.1.3.2   Discussion of Scope

This is a recommended practice for analyzing the interference and coexistence among radio systems. It establishes a structured framework for analyzing interference and coexistence. It also establishes a structured document format for presenting the analysis to others.

This recommended practice does not establish interference criteria. It does not propose judgments on what constitutes interference or on what constitutes harmful interference. It does, however, indicate a number of effects and conditions that should be considered to provide a complete interference and coexistence analysis.

Subclause 15.5.3 of this recommended practice outlines the report section and subsection[3] headings for a report of a recommended analysis of interference and coexistence. This may be used as the framework for a report complying with this recommended practice.

Annex 15D through Annex 15G provide sample analyses, in accordance with the structure and method of this recommended practice.

---

1 Although the focus of this standard is interference and coexistence of radio systems, other kinds of systems utilize spectrum and are dependent on spectrum for their operation. The analytical methods proposed in this standard may be used when analyzing scenarios, including spectrum-dependent systems other than radio systems.
2 The formal scope is the scope approved in the Project Authorization Request for this project.
3 In this document, clause and subclause refer to portions of this document. Section and subsection refer to portions of an analysis report, prepared in compliance with this recommended practice.

15.1.4 Purpose

15.1.4.1 Formal Purpose[4]

New concepts and technologies are rapidly emerging in the fields of spectrum management, policy-defined radio, adaptive radio, and software-defined radio. A primary goal of these initiatives is to improve spectral efficiency. This recommended practice will provide guidance for the analysis of coexistence and interference between various radio services.

15.1.4.2 Discussion of Purpose

By standardizing the framework and documentation structure for interference and coexistence analysis, this recommended practice enables side-by-side comparisons of the analyses presented by various parties. It thereby may streamline the process of determining whether interference and harmful interference have or will occur. Furthermore, this document provides a framework within which differing analysis can be understood, considered, and resolved.

In the case of any particular standardized radio systems that may coexist or interact with others dynamically, it may be desirable to develop standard interference and coexistence criteria. The structured framework described in this document should be used by the authors of such standards. However, this recommended practice does not state or attempt to define any particular interference or coexistence criteria.

15.1.5 Rationale

A reader may ask why this recommended practice was created. This recommended practice has been developed in response to the emergence of new radio technologies and changes in spectrum policy and regulation. Changes in regulations have the potential to enable flexibility in services, radio access technology, and spectrum assignments. Together these changes have the potential to significantly improve spectral efficiency and communications performance through exploiting:

—Flexibility
— Dynamic, autonomous, and adaptive behavior
— Awareness, cognition, and intelligence
— Networking
— Coexistence and sharing of radio resources

However, present spectrum management practices do not provide effective techniques for identifying interference among these types of advanced radio devices or for managing their access to the radio spectrum dynamically. This may create barriers to the introduction of these new radio technologies. At the same time, premature standardization of fixed interference criteria could also stifle innovation, since the sophisticated, complex behaviors of future environmentally aware and cooperative radio systems and services will continue to develop over time.

The application of the techniques outlined in this recommended practice will facilitate the continuing deployment of advanced radio technologies and thereby improvement in spectral efficiency and communications system performance.

Questions of interference may arise regarding actual or proposed radio systems. Normally there is a two-step process to understand and answer these questions. First, there is a technical debate to determine whether interference has occurred or will occur. Second, there is a broader, more general debate over what constitutes an acceptable level of interference, which includes issues such as regulatory history, public policy, and economic tradeoffs. The second debate often seeks to determine whether the interference, determined in the first debate, rises

---

4 The formal purpose is the purpose approved in the Project Authorization Request for this project.

to the level of harmful interference or if system effects on coexisting systems require corrective action.The process can be highly contentious. In the technical debate over interference, it often emerges that the differing conclusions of disputing parties derive from differing assumptions rather than from faulty analysis. When controversy arises, analysts will typically differ in one or more of the following areas:

— What scenarios and use cases should be assumed?
— What should be considered an interference event?
— What constitutes harmful interference?
— Who owns the margin?[5]

When differing viewpoints are anticipated, the analyst should take care to treat these questions with particular care. In the broader debate over whether interference is harmful, it often emerges that the disputing parties have different and incompatible definitions of harm. Merely determining the points on which the parties agree and disagree can require substantial investigation. Therefore, using a common structure and clearly identifying assumptions greatly assists in guiding discussion to the critical issues.A well-prepared analysis answers the following questions:

— Why is this analysis being done?
— What are the questions being addressed?
— What need is being served?

If the possibility of significant misunderstanding exists about the scope of an analysis, the analysis report should clearly define areas that are being considered and, also, those areas beyond its scope.

15.2 Normative References

The following referenced documents are indispensable for the application of this document (i.e., they must be understood and used, so each referenced document is cited in text and its relationship to this document is explained). For dated references, only the edition cited applies. For undated references, the latest edition of the referenced document (including any amendments or corrigenda) applies.

ANSI C63.14-1998, American National Standard Dictionary for Electromagnetic Compatibility (EMC), Electromagnetic Pulse (EMP), and Electrostatic Discharge (ESD) (Dictionary of EMC/EMP/ESD Terms and Definitions).[6]

ITU-R Handbook on National Spectrum Management, 2005 edition.[7]

NIST Technical Note 1297, Guidelines for Evaluating and Expressing the Uncertainty of NIST Measurement Results, Sept. 1994.[8]

---

5 Margin is used here generically to refer to any of a number of buffers that exist to allow some protection to the desired level of operation. Various margins are maintained to allow products to vary, for example due to manufacturing variation, the influence of temperature, or other environmental factors. Providing some margin or buffer is necessary, if a desired level of service or reliability is to be provided, over a foreseeable but uncontrollable set of variables. Some examples of the types of margin that are discussed at various places in this recommended practice are design margin, operating margin, interference margin, safety margin, system margin, protection margin, fading margin, and reliability margin. Although each of these margins defines a different quantity, all of them serve to allow variation to not unduly influence some desired requirement.
6 ANSI publications are available from the Sales Department, American National Standards Institute, 25 West 43rd Street, 4th Floor, New York, NY 10036, USA (http://www.ansi.org/).
7 ITU-R publications are available from the International Telecommunications Union, Place des Nations, 1211 Geneva 20, Switzerland (http://www.itu.in/).
8 For information on how to purchase FCC and NIST publications, contact the Superintendent of Documents, US Government Printing Office, Washington, DC 20402 USA.

15.3 Definitions, Acronyms, And Abbreviations

15.3.1 Definitions

For the purposes of this document, the following terms and definitions apply. The glossary in Annex 15H and *The Authoritative Dictionary of IEEE Standards Terms* [B12] should be referenced for terms not defined in this clause.

15.3.1.1 Adaptive Radio

A type of radio in which communications systems have a means of monitoring their own performance and a means of varying their own parameters by closed-loop action to improve their performance.

15.3.1.2 Co-channel

The condition that exists when one transmitter operates in the same channel as a second transmitter.[9]

15.3.1.3 Coexistence

The ability of two or more spectrum-dependent devices or networks to operate without harmful interference.

15.3.1.4 Cognitive Radio/Cognitive Radio Node

(A) A type of radio in which communication systems are aware of their environment and internal state and can make decisions about their radio operating behavior based on that information and predefined objectives. The environmental information may or may not include location information related to communication systems. **(B)** Cognitive radio [as defined in **(A)**] that uses software-defined radio, adaptive radio, and other technologies to automatically adjust its behavior or operations to achieve desired objectives

15.3.1.5 Cooperative (Device or System)

Devices or systems that can coordinate their use of spectrum.

15.3.1.6 Dynamic Spectrum Management

A system of spectrum management that dynamically adapts the use of spectrum in response to information about the use of that spectrum by its own nodes and other spectrum-dependent systems.

15.3.1.7 Electromagnetic Compatibility

The condition that prevails when devices or networks perform their individually designed function in a common electromagnetic environment without causing or suffering unacceptable degradation due to electromagnetic interference to or from other equipment in the same environment.

15.3.1.8 Interference Event

A circumstance in which a quantified threshold level of interference has been exceeded. Interference events are specified in terms of the relevant variables for a specific scenario, e.g., time, frequency, amplitude, and performance metrics.

---

9 In frequency bands that do not have fixed channel allocations, a determination must be made as to how much overlap constitutes a cochannel situation and alternately when two signals are operating in adjacent channels.

15.3.1.9 Noncooperative (Device or System)

Devices or systems that operate independently of a source device or system with regard to the use of spectrum.

15.3.1.10 Recipient Device

A device being analyzed as to its potential for receiving interference from a source device or system.

> NOTE—An analysis seeks to determine the level of impact on these devices. The protection need not be a legal requirement but may arise from ethical or social concerns.

15.3.1.11 Recipient System

A system being analyzed as to its potential for receiving interference from a source device or system.

> NOTE—An analysis of a recipient system will often study the potential for system impact and may build on an analysis of the potential for interference to individual devices in the system.

15.3.1.12 Software-controlled Radio

A type of radio where some or all of the physical layer functions are software controlled.

15.3.1.13 Software-defined Radio (SDR)

(A) (Normative) A type of radio in which some or all of the physical layer functions are software defined. **(B)** (Informative) U.S. Federal Communications Commission Definition [1]: A radio that includes a transmitter in which the operating parameters of frequency range, modulation type or maximum output power (either radiated or conducted), or the circumstances under which the transmitter operates in accordance with Commission rules, can be altered by making a change in software without making any changes to hardware components that affect the radio frequency emissions. **(C)** (Informative) ITU-R Definition (including the notes) [2] [3]: A radio in which RF operating parameters including but not limited to frequency range, modulation type, or output power can be set or altered by software, and/or the technique by which this is achieved.

> NOTE 1—Excludes changes to operating parameters that occur during the normal pre-installed and predetermined operation of a radio according to a system specification or standard.

> NOTE 2—SDR is an implementation technique applicable to many radio technologies and standards.

> NOTE 3—Within the mobile service, SDR techniques are applicable to both transmitters and receivers.

15.3.1.14 Source Device

A device being analyzed as to its potential for interference to recipient devices or systems.

> NOTE—In some previous documents in the field, these devices were called secondary devices.

15.3.1.15 Source System

A system being analyzed as to its potential for interference to recipient devices or systems.

> NOTE—An analysis of a source system will often focus on the combinatorial effects of the source devices that compose the system and will build on an analysis of the potential for interference from the individual source devices in the system.

15.3.1.16 Spectrum Utilization

The spectral space denied to other potential users.

15.3.2 Acronyms and Abbreviations

AM	amplitude modulation
ANSI	American National Standards Institute
AREPS	Advanced Refractive Effects Prediction System
ATSC	Advanced Television Systems Committee
CDF	cumulative distribution function
CDMA	code division multiple access
CFR	Code of Federal Regulations
C/I	carrier-to-interference
CW	carrier wave
dB	decibel
dB SPL	decibels referenced to a sound pressure level of 20 µPa
dBm0	power level in dBm relative to zero transmission level point
DSA	dynamic spectrum access
DTV	digital television
DTX	discontinuous transmission
E3	electromagnetic environmental effects
EMC	electromagnetic compatibility
ETSI	European Telecommunications Standards Institute
f	frequency
FDD	frequency division duplex
FDP	fractional degradation of performance
GSM	Global System for Mobile
HFA	High-frequency average
iDEN	Integrated Digital Enhanced Network
IEC	International Electrotechnical Commission
IEEE	Institute of Electrical and Electronics Engineers
ISO	International Organization for Standardization
JTC	Joint Technical Committee
LOS	line of sight
MIMO	multiple in multiple out
N	noise
NADC	North American Digital Cellular
OLOS	obstructed line of sight
OSI	open systems interconnection
P	transmit power or forward power (Watts)
PAR	peak-to-average ratio
PC	personal computer
PCS	personal communications services
PEP	peak envelope power
P-MP	point to multipoint
Q	quality factor
QAM	quadrature amplitude modulation
RBW	resolution bandwidth
RF	radio frequency

RMS	root mean square
RSE	relative spectrum efficiency
Rx	receiver
SDR	software-defined radio
S/N	signal-to-noise ratio
SPE	standard parabolic equation
SPL	sound pressure level
SUE	spectrum utilization efficiency
TDMA	time division multiple access
TDD	time division duplex
TV	television
Tx	transmitter
UARFCN	UTRA absolute radio frequency channel
UKAS	United Kingdom Accreditation Service
UMTS	Universal Mobile Telecommunications System
UTRA	Universal Terrestrial Radio Access
VoIP	Voice over Internet Protocol
WAPE	wide-angle parabolic equation

15.4 Key Concepts

15.4.1 Interference and Coexistence Analysis

This document provides a method for performing an interference and coexistence analysis. An interference analysis and coexistence analysis are complementary and related concepts.

An interference analysis starts with energy introduced by a source system and studies the effect of that energy from the most basic physical layer through higher levels until the impact of that energy on recipient systems is understood. A coexistence analysis starts at the system level, and studies the ability of a recipient system to communicate or accomplish its intended useful effect in the presence of interference introduced by a source system. A coexistence analysis is focused on system level performance issues, seeking a comprehensive understanding of benefits and impacts of coexistence, including an understanding of mitigations of interference and the user behavior.

The differing approach of an interference analysis, in contrast to a coexistence analysis, is useful in assuring that a case is completely understood. Starting from the energy introduced by the source is most helpful in identifying all the potential impacts. A focus on system performance and moving down toward the impact of energy is often more helpful in giving a meaningful context to the impact of that energy.

15.4.2 Measurement Event

A measurement event defines the basic quantities measured in order to analyze the interference or coexistence. A measurement event is a single instance of a sample to be used to characterize the state of interference. It is a data sample, measured or calculated, using a specified set of parameters. In some cases, the measurement event may capture multiple variables or the same variable from multiple devices at the same time and contain their values. Therefore, a measurement event may contain multiple values. The measurement sample is selected to provide a context for interference events. A measurement event may be as simple as the average power received from a source over an interval of time. However, no matter how simple or complex, it must be well defined by radio system variables, such as the frequency bandwidth, averaging time, or other system variables, such as a time period during which a specified quality of service will be measured.

15.4.3 Interference Event

An interference event is a measurement event in which a source device or system has a quantifiable performance effect on the recipient device or system or for the user of a recipient device or system. The concept of an interference event is used in the analysis for determining the amount and severity of interference. Interference events are therefore a subset of the measurement events. An interference event is scenario dependent. Depending on the service, performance degradation can be manifested in many ways. For example, as lower data throughput, lower voice quality, video distortions, decreased battery life, increased incidence of blocked or dropped links, delay, reduced system capacity, reduced interference margin, or reduced capability of the recipient system to adapt to new conditions or advance with new technology. An individual interference event may not in itself be deemed harmful. The definition of when interference events degrade performance to an unacceptable level is termed, harmful interference, and is dealt with when defining the criteria for harmful interference for an analysis.

15.4.4 Harmful Interference

Harmful interference is the level at which the analysis deems interference events have created unacceptable interference. The level shall be defined in terms of interference events across time and/or users or systems that cause an unacceptable degradation of the recipient system's performance, in the judgment of the analyst. This threshold will be used when determining whether harmful interference has occurred. The analysis of a system may involve more than one threshold. The analysis shall state the reasons for selecting the harmful interference criteria used in the analysis.

Determining when interference events become excessive is a difficult value judgment. It may be that different analyses are in general agreement except for specific thresholds defined in the criteria. The viewpoint of those connected with the recipient system often is different than those connected to the source system. When this is the case and the differing analyses use the structure proposed in this recommended practice, then the debate may be quickly focused on the critical points of difference. This facilitates the comparison of analyses, understanding differences in conclusions and recommendations and, ultimately, the decision-making process.

15.4.5 Physical and Logical Domains

Communication systems have often been described in terms of a physical layer and several higher layers. This document refers instead to the *physical domain* and the *logical domain*. The physical domain covers the transmission and reception of the radio signals, including power, antenna effects, noise, signal modulation, coding, interference, sensitivity, intended and unintended emissions, and radio performance. The logical domain covers all the behaviors of the higher OSI (Open Systems Interconnection) layers.[10] Examples of issues that may attract significant interest are medium access control, packetization, retransmission, routing, scheduling, services, and system performance. System performance may include, for example, capacity, blocking, power consumption, link/system reliability, quality of service, and excess delay. The logical domain considers of other device and system behaviors that interact with the communication system, such as battery management algorithms and location-dependent control algorithms.

---

10 ISO/IEC 7498-1:1994 [B17].

Interference analyses have traditionally focused on the physical domain. Such studies consider how the RF energy emitted by one system affects other systems. One major new aspect of interference and coexistence analysis raised by the new radio technologies is whether to broaden the scope of the traditional physical domain to also include the logical domain.

A canonical example of such a question was stated earlier in this introduction (see 15.1.2): *[I]f a network can reroute data so loss of a particular link has no impact on end-to-end user performance, does loss of that link constitute interference?* In one view, no interference has occurred because no user experiences reduced performance. In another view, the rerouting capacity of the network represents a designed-in safety margin for tolerating a variety of effects such as link congestion and node mobility as well as against external interference. This safety margin is part of the user's expected performance. Degradation of that margin due to operation of another radio system reduces the statistically expected performance of the network going forward, and therefore, interference has occurred.

Such questions must be considered as part of an interference and coexistence analysis. However, these questions must often be answered on a case-by-case basis. This recommended practice is designed to permit physical and logical domain behaviors to be properly considered in analyzing an interference and coexistence scenario. The structured framework described in this recommended practice should be applied to both physical domain effects, to logical domain effects, or to both, considered together, as appropriate in a particular situation.

15.5 Structure of Analysis and Report

15.5.1 Structure for Analysis

An interference and coexistence analysis considers the effects of one radio system (the *source device or system*) on the operation of one or more other radio systems (the *recipient device or system*).

The analytical framework adopted in this recommended practice consists of four cascaded, interdependent steps: scenario definition, establishment of interference and coexistence criteria for the recipient system, definition of variables or behaviors used in modeling, and the modeling, measurement or analysis itself.

Figure 15.1 depicts the analytical process and structure recommended by this recommended practice. The structured approach depicted in the figure is recommended to serve as both a logical, step-by-step procedure to be followed in conducting the analysis and an organization template for the documentation of the entire analysis. Clause 15.5 through Clause 15.10 provide guidance on the performance and documentation of each of these steps.

15.5.1.1 Scenario Definition

The deployment, usage and architectures contemplated for the recipient and source systems are described in the scenario definition step. For example, the recipient system may be a broadcast service, for which there is a known, fixed transmitter configuration requiring protection of receive-only, independently-operated nodes, distributed across wide geographical areas around the transmitter with little or no capability to predict or measure precise receiver locations. The source may be a mobile ad-hoc network that has GPS to provide location information for each node.

15.5.1.2 Establishment of Interference and Coexistence Criteria

Interference and coexistence criteria for the subject system are established and described in this step. The goal is to provide realistic criteria for the defined scenario(s).

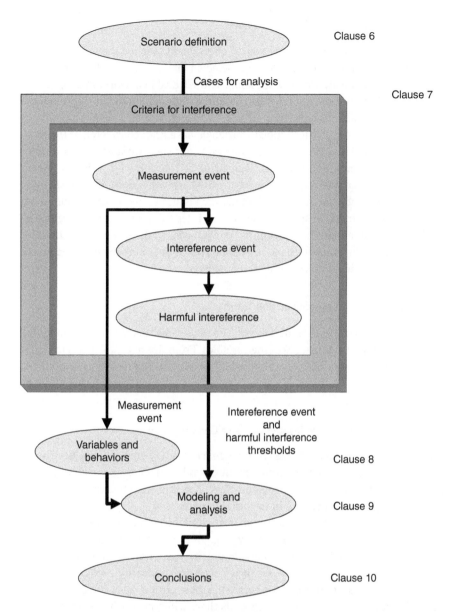

Clause 6

Clause 7

Clause 8

Clause 9

Clause 10

Figure 15.1 Overview of the interference and coexistence analysis process. From IEEE Recommended Practice for the Analysis of In-Band and Adjacent Band Interference and Coexistence Between Radio Systems, Mar 2018. IEEE.

In some instances, purely physical domain, discrete criteria may be appropriate, such as the maximum interference power in a single link. For the broadcast service example, as just described in 15.5.1.1, the appropriate criteria may involve statistical analysis of a signal-to-interference ratio over a region. In other instances, more subtle criteria invoking logical domain variables may be established. These criteria may include effects such as dropped calls in a cellular network, reduced service quality, or reduction of overall throughput in an ad hoc digital communications network.

In this step, issues such as erosion of system margin or diminished system capacity are addressed. This recommended practice provides standard definitions and a categorization of these logical domain effects to assist in making clear and supportable arguments for what constitutes harmful interference.

15.5.1.3 Selection of Relevant Variables or Behaviors

In this step, the variables or behaviors of the source or recipient systems that are relevant to the particular scenarios and interference criteria are selected for use in the analysis, modeling, measurement, or testing of the interference situation. The selected behaviors are defined precisely, such as by using pseudo-code or other algorithmic representation, to fully represent the systems' interactions and reactions to interference.

In many past analyses, purely physical domain variables have been considered in the models, simulations, or calculations. Such analyses only consider such variables as transmitter power, location, and frequency, as these are assumed to be the only key variables. Newer analyses, addressing new adaptive systems and spectrum-sharing concepts, also consider the logical domain behaviors. These analyses may include such issues as frequency abandonment algorithms, networked spectrum sensing information, or integration protocols. These logical domain variables may be dominant in determining the behavior of the systems and the coexistence outcome. For cases in which logical domain behavior is operating in either the source or the recipient systems, such variables should be considered.

15.5.1.4 Modeling, Analysis, Measurement, and Testing

In this step of the analysis, the effects of the source(s) on the recipient(s) is modeled, analyzed, measured, or tested for appropriate ranges of the variables. From this analysis, a declaration is made based on the established criteria regarding whether harmful interference has occurred or will occur.

15.5.2 Process Flow—divergence, Reduction, and Convergence

The analysis process is initially characterized by divergence. A study question may have many possible use cases. Each use case has multiple frequency relationships, and each frequency relationship potentially has many system relationships. The range of choices introduced at each step leads to a combinatorial growth in the number of cases.

Periodic reduction steps help bound the number of cases being advanced for treatment. A matrix reduction step may be introduced at any stage of the analytical process. When the important cases for analysis can be identified early in the process, this may allow several steps in the analysis to be combined. In other cases, it is necessary to identify explicitly multiple cases and to perform some part of the analysis before the more critical cases can be identified. It is critical that the important cases are identified, as there will typically be too many possibilities for an analysis to deal with all possible cases. An exhaustive analysis would require an extraordinary effort to accomplish.

At the end of an analysis, the cases are converged to a summary and conclusion of the findings. The convergence may be done in several steps or in a single, final step at the end of the analysis.

Figure 15.2 illustrates an initial divergent characteristic of an interference and coexistence analysis followed by recursive reduction steps and concluding with convergence to a summary and conclusion.

All analyses begin with a study question. Many study questions might be applied in multiple scenarios, and so there is a divergence at this first step in the process. Each scenario or use model has multiple frequency relationships (e.g., in-band, low-frequency adjacent band, or high-frequency adjacent band). Each frequency relationship may have many system relationships. For example, many different types of incumbent systems may be currently operating in the proposed frequency band or in each adjacent band. It will be common to have a great many potential cases for analysis at the end of the scenario definition process. Matrix reduction steps during or at the end of this phase of the analysis will be a common occurrence to reduce the number of cases to a manageable level.

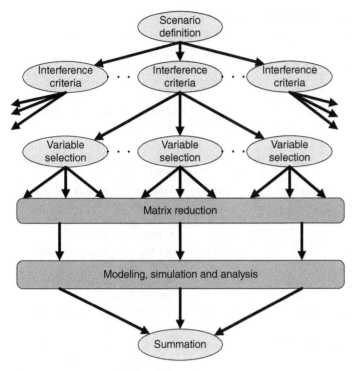

Figure 15.2 An interference and coexistence analysis initially is characterized by successive stages of divergences with recursive reduction steps and concludes with converging and summarizing the findings. From IEEE Recommended Practice for the Analysis of In-Band and Adjacent Band Interference and Coexistence Between Radio Systems, Mar 2018. IEEE.

Looking further it will become clear that each system relationship may have multiple types of interference (e.g., audio interference, operational interruption, reduction of operating margin, and increased latency or jitter). Each type of analysis may be analyzed on more than one level (e.g., user perceptible interference, increased bit errors, carrier to interference ratio, and quality of service).

A full treatment of all possibilities will seldom be possible. Therefore, the matrix of possible cases may be reduced to a manageable number of critical cases for detailed analysis. The cases for detailed analysis are selected using some guiding logical process. In some cases, the defining or "corner cases" are selected, where it can be argued that when these are understood, all other cases are also understood. In other situations, the most critical cases may be selected, with the argument that other cases are of relatively minor concern. Other selection logic may be used but should be clearly stated and defensible if challenged.

At the end of the process, the findings are converged, sometimes in multiple steps, until a single summary and conclusion(s) can be proposed.

15.5.3 Report Structure

15.5.3.1 Quick Reference Guide

Table 15.1 provides a "quick reference" of the report structure for an analysis, in tabular form. It was developed as a tool to assist analysts as they prepare their reports, following this recommended practice. A complete discussion of what should be contained in each section of the report is provided in the main body of this recommended practice.

> NOTE—In this recommended practice, clause and subclause are used to refer to this document, whereas section and subsection are used to refer to portions of the report of an analysis developed following this recommended practice.

Table 15.1 Structure of an interference and coexistence analysis report.

Report section	Clause	Description
I. Executive summary	15.5.3.2	Give a short overview of the analysis.
II. Findings	15.5.3.2	What is the conclusion of the analysis?
III. Scenario definition	15.6	
A. **Study question**	15.6.2	What is the question or proposal being analyzed?
B. **Benefits and impacts of proposal**	15.6.3	What are the benefits and impacts of this proposal?
C. **Scenario(s) and usage model**	15.6.4	Where and how might this proposal be applied?
1. Frequency relationships	15.6.4.1	What is the frequency context being considered?
2. Usage model	15.6.4.2	In what ways will the subject being studied be used?
3. Characteristics of usage model	15.6.4.3	What are the characteristics of the usage model(s)?
a. Spatial and power characteristics	15.6.4.3.1	How closely will devices be used in relationship to each other? What transmit power is allowed? What are the characteristics of the transmission waveform?
b. Temporal characteristics	15.6.4.3.2	How frequently will each device transmit? When transmitting, how long does it transmit? Is the transmission governed by a fixed frame period?
c. Frequency characteristics	15.6.4.3.3	What frequency band and bandwidth will be used?
d. Other orthogonal variables	15.6.4.3.4	Are there other variables that will separate transmissions, like coding, MIMO, or antenna polarization?
4. System relationships	15.6.4.4	
a. Systems considered	15.6.4.4.1	What systems share the same band? What systems operate in adjacent bands? What other equipment might be affected by the transmission?
b. Protection distance	15.6.4.4.2	What protection distance is required or assumed for this analysis?
c. Geographic area for analysis	15.6.4.4.3	What geographic area is being analyzed?
d. Impact of interference	15.6.4.4.4	If there is interference, what are the consequences?
e. Interference mitigation	15.6.4.4.5	If there is interference, what can be done about it?
f. Baseline	15.6.4.4.6	What interference already exists?
D. **Case(s) for analysis**	15.6.5	What specific case will be moved forward for further analysis?

Table 15.1 (Continued)

Report section	Clause	Description
IV. Criteria for interference	15.7	
A. **Interference characteristics**	15.7.2	What are the characteristics of the interference? What kind of interference is of concern?
1. Impacted level	15.7.2.1	At what level will the interference be measured, C/I, bit errors, user perception?
B. **Measurement event**	15.7.3	What parameters will define a measurement event?
C. **Interference event**	15.7.4	When does a measurement event become an interference event?
D. **Harmful interference criteria**	15.7.5	At what point do interference events become harmful interference?
V. Variables	15.8	
VI. Analysis: modeling, simulation, measurement and testing	15.9	Perform the modeling, simulation, or analysis.
A. **Selection of the analysis approach**	15.9.2	Provide an explanation for the approach selected.
B. **Matrix reduction**	15.9.3	Which cases need to be modeled, simulated, or analyzed?
C. **Performing the analysis**	15.9.4	This subsection contains the models, simulations, or description of the analysis applied, including models, simulations measurements, or testing for the cases selected.
D. **Quantification of benefits and interference**	15.9.5	Quantify the benefits and interference.
E. **Analysis of mitigation options**	15.9.6	What are the mitigation options?
F. **Analysis uncertainty**	15.9.7	What is the uncertainty of the analysis?
VII. Conclusion and summary	15.10	What is the conclusion of this analysis? Present the findings and summarize the results.
A. **Benefits and impacts**	15.10.1	Summarize the benefits and impacts.
B. **Summation**	15.10.2	Summarize the analysis.

15.5.3.2 Introductory Material

An executive summary section should be placed as the first section or near the first section of a report. The executive summary should provide a brief summary of the problem and the findings of the analysis. It should give the reader an understanding of the scope and purpose of the analysis performed, the major assumptions and decisions, the analysis, measurements or modeling performed, and a summary of the findings.

After the executive summary should be a section on findings, providing a top-level summary of findings. This provides a presentation of the core outcome(s) of the analysis. The full details and supporting data are placed in the body and the closing summary section of the analysis.

15.6 Scenario Definition

15.6.1 General

This recommended practice recommends that an interference and coexistence analysis begin by describing its context. Why is this analysis being performed? What is being proposed that needs analysis? What are the underlying assumptions? What are the assumptions and focus areas being analyzed? By answering these and similar questions in the early part of an analysis, other analysts and the readers may quickly understand both what will be done and if the analysis addresses their issues of interest.

As depicted in Figure 15.3, the scenario definition, which described in this clause, delivers one or more, at times many, cases for analysis. Each case will look at the potential for interference from the source device(s) or system(s) in the proposal in a specific recipient situation. Each case for analysis is then delivered to succeeding steps in the analysis for

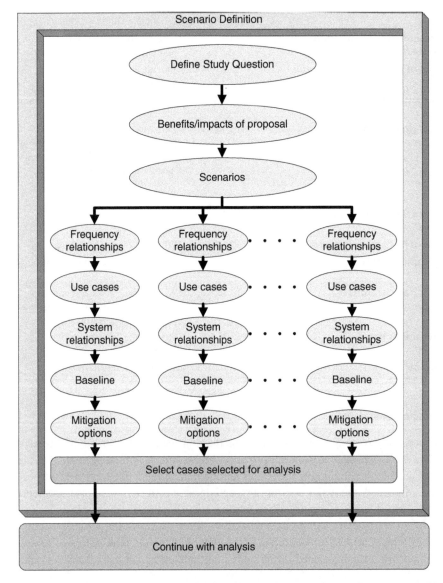

Figure 15.3 Process diagram of the scenario definition clause. From IEEE Recommended Practice for the Analysis of In-Band and Adjacent Band Interference and Coexistence Between Radio Systems, Mar 2018. IEEE.

determination of interference event(s), harmful interference criteria, variable identification, and finally modeling and analysis.

From IEEE Recommended Practice for the Analysis of In-Band and Adjacent Band Interference and Coexistence Between Radio Systems, Mar 2018. IEEE.

15.6.2 Study Question

The process starts with a clear statement of the question to be studied. This study question will often be in the form of a proposal. What new system, regulation, or operating policy is being proposed for analysis?

15.6.3 Benefits and Impacts of Proposal

If the analysis is being done to explore a proposed innovation, regulation, policy,[11] or new technology, a description of the anticipated benefits and impacts should be placed in this section of the report.

A description of the anticipated benefits and impacts from the innovation, regulation, policy, or technology under study should be included in this section of the report. The anticipated benefits and impacts may affect many different parties and affect them in different ways. The parties benefited and the parties impacted should be identified. This discussion is particularly important when the parties benefiting are different from those that are impacted.

The description in this subsection should summarize the more detailed discussion of the benefit and interference impact developed in 15.10.1. This report section should make as explicit as possible the tradeoff between benefits and impacts and provide the analyst the opportunity to state their assessment of the value and cost of the proposal.

15.6.4 Scenario(s) and Usage Model

The scenario(s) being analyzed are described in this report section. A scenario description shall include a description of the source device(s) and system(s); the recipient device(s) and system(s); how they will be used; as well as the system users and the environment in which they will operate. The usage model should be described in a general way. This description may address, for example, questions such as follows: Does the scenario assume a home, office, public, or industrial environment? What kinds of equipment are most commonly used together in the scenario being addressed? What services will be provided, e.g., voice communications, entertainment, personal communications, data, or mission critical or life critical real-time data? Characteristics of the services such as reliability, range, delay, and interaction with other services or networks should also be described.

Each scenario will contain one or more usage models. A well-prepared usage model answers at least the following questions:

— What frequency bands are involved?
— What kind of use environment is being considered, e.g., home, office, industrial, or hospital?
— What device population and population density is anticipated?
— How will these devices be used?
— What is the distribution of the systems' transmitters and receivers?

The answers to these questions will allow the analyst to identify the important variables necessary for the analysis.

The scenario provides the understanding from which the technical variables of the analysis are derived. Once the analyst has defined the scenario, it is possible to identify the radio

---

11 In this document, the term "policy" is used generally to mean a radio policy for use in a policy-defined radio. Other uses of the term "policy," such as "policy makers," are clear by their context. See IEEE Std 1900.1-2008 [B15] for a specific definition of this use of policy.

services involved, their characteristics, and the potential interactions. The scenario definition sets the stage for the following steps of the analysis:

— Defining the interference criteria
— Identifying variables and behaviors
— Performing the modeling and analysis

15.6.4.1 Frequency Relationships

The following frequency regions, depicted in Figure 15.4, should be considered for each analysis:

— Co-channel
— Adjacent channel (directly adjacent channels)
— In-band, nonadjacent channels (noncontiguous, in-band channels)
— Band edge (near-band edge and out-of-band)
— Out-of-band (far out-of-band)

The analysis should quantitatively specify the frequency boundaries defining each region. For each frequency region, a review is necessary of the types of systems operating in that region. In 15.6.4.4, the various types of equipment operating in each frequency region and their characteristics, which are relevant to the question being studied, should be identified.

15.6.4.2 Usage Model

This subsection of the report defines what usage models are to be considered in the analysis. When beginning an analysis, a great many usage models can be envisioned. Typically, after some consideration, a few usage models are found to be sufficient to define the question being studied. These defining models may be the most significant or difficult or they may define the boundary for the analysis. The configuration and operation of both the source and recipient systems should be explicitly defined.

If there are usage models that are worse than those selected for study, the analysis will describe why those are rejected. Some possible reasons for rejecting a use model or an entire scenario may be that it occurs too infrequently, it is beyond the purpose of a particular analysis for other reasons, or the consequences of interference are of less practical significance than the cases selected.

The material provided in this subsection of the report provides the foundation for the interference event determination in 15.7.4.

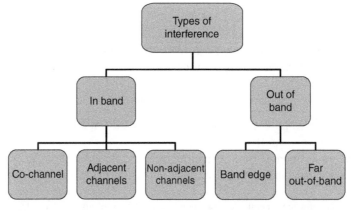

Figure 15.4 Frequency relationships. From IEEE Recommended Practice for the Analysis of In-Band and Adjacent Band Interference and Coexistence Between Radio Systems, Mar 2018. IEEE.

15.6.4.3 Characteristics of Usage Model

After explaining in 15.6.4.2 how users will use the system, the technical characteristics of that usage model should be identified. This sub-section of the report states how the use model interacts with the technology being considered.

Both the source and recipient system characteristics should be explicitly defined.

> NOTE—A technology likely to be used in implementing DySPAN is software-defined radio (SDR) technology. SDRs are reconfigurable, which means many parameters documented in the analysis could be subject to change in the field, either inadvertently or by hackers. Describing the reconfiguration controls and security measures used to protect the devices and networks using them will be very important for a full treatment and understanding of a scenario, but SDR security is beyond the scope of this document.

Spatial and Power Characteristics This subsection shall contain a description of the following characteristics:

— Spatial distribution of transmitters and receivers
— Mobility of devices
— Transmission power and power control
  1) Power limits
  2) Dynamic power control (as applicable)
— Directionality of antenna and control of directionality
— System range and coverage
— Operating margins expected or used in system planning

In cases where an adaptive user will coexist with a recipient device or system, the variables that allow for spatial reuse of spectrum are very important. For example, in a situation where the recipient device or system transmits on a relatively static basis, such as a broadcast service, the dynamic user is required to operate with the spatial and power limits required by the recipient device or system's transmission and reception characteristics and spatial distribution.

Both the source and recipient system spatial characteristics should be explicitly defined.

Temporal Characteristics This subsection of the report shall contain a description of the following characteristics:

— Transmitter and receiver temporal characteristics (continuous, packetized, time division, carrier sensed) and temporal distributions
— Acquisition and abandonment time of dynamic systems (if applicable)
— TDMA parameters (if applicable)
— Time division or frequency division duplexing (if applicable)
— Timescale of cooperation
— Time-dependent traffic parameters.

Source and recipient devices may have distinct traffic characteristics over time. The source may have peak traffic usage at a period of the day when the recipient is generally off or operating below capacity. When an adaptive user will be operating with a recipient device or system that transmits periodically rather than continuously, the opportunity may exist for frequency channels to be used during quiet periods between the recipient device's or the system's transmission. In such a situation, the temporal characteristics of the recipient device or the system's transmission, and the expectation of the receivers during non-transmission intervals, becomes significant. The time required to detect a recipient device's or a system's signal and to abandon the channel is a major concern. In addition, source system transmissions during intervals expected to be quiet by the recipient receiver may decrease the recipient battery life due to processing of the unexpected signals. The source transmissions also may block a recipient from receiving signals from other transmitters or block the recipient from replying to a signal due to unexpected channel occupancy.

Sources and recipients may cooperate to avoid interference over very different timescales. Longer timescales allow manual frequency and siting coordination. Shorter timescales may allow coordination through automated sense and avoid techniques. In some cases, it may be possible to interleave transmissions on an ongoing basis. When this occurs, multiple systems can operate in an ongoing fashion, separated by the temporal characteristics of their transmission.

Both the source and recipient system temporal characteristics should be explicitly defined.

Frequency Characteristics This subsection of the report shall contain a description of the following characteristics:

— Channel plans
— Regulatory frequency limits
— Spurious and out-of-band emissions
— Intended frequency range of operation
— Device frequency capability or possible frequency range of operation
— Other frequency characteristics

Both the source and recipient system frequency characteristics should be explicitly defined.

Other Orthogonal Variables This subsection of the report shall contain a description of other relevant orthogonal variables that may affect coexistence:

— Polarization
— MIMO
— CDMA techniques
— Diversity
— Dynamic routing behaviors
— Adaptive network management algorithms
— Adaptive error correction
— Other variables used to separate radio signals

Although power, frequency, space, and time are the primary variables for separating transmissions, there are other variables for signal orthogonally.[12] These variables have more subtle behavior characteristics than the traditional variables of power, frequency, distance, and time. However, these other variables are growing in importance as experience and technology develop. A more complete set of the variables that may be relevant is offered in Clause 15.8.

15.6.4.4 System Relationships

Systems Considered Once the usage model(s) are identified, the spectrum characteristics of the baseline should be described. This is the basis to be used in 15.6.4.4.6 when analyzing the interference baseline for the scenario. To analyze the system relationships, first the systems that are potentially involved should be identified. Table 15.2 gives an example system interaction matrix for scenarios with a number of possible interacting systems. Each intersection of a column and row identifies a coexistence scenario, which should be studied.[13]

From IEEE Recommended Practice for the Analysis of In-Band and Adjacent Band Interference and Coexistence Between Radio Systems, Mar 2018. IEEE.

---

12 Other parameters used in modern radio systems to discriminate among signals include frequency hopping, spreading codes, pulse compression, and channel propagation characteristics (such as used in MIMO and spatial-coded transmissions).

13 As dynamic spectrum access comes into more common use, such analyses will become significantly more complex. The flexibility of these systems will make a full analysis of their mutual interaction increasingly difficult.

Table 15.2 Sample system interaction matrix.

		Incumbent System				
		1	2	3	4	5
		Broadcast	Cellular Mobile	Public Safety	Data Telemetry	RLAN
New System(s)						
1	Fixed Microwave					
2	In-home					
3	P-MP Last Mile					
4	Public Safety					

In the merged data region of the table: **Explore coexistence following recommendations**

- **Criteria for Interference**
- **Variables and Behaviors**
- **Modeling and Analysis**
- **Conclusions**

Each scenario in 15.6.4 will have its own channel plan, frequency, and system relationships. After identifying these system relationships, Clause 15.7 through Clause 15.9 of this recommended practice identify and quantify the variables to be used in the analysis and the impact of interference on the recipient systems.

Categories of System Cooperation There are a number of categories of intrasystem spectrum cooperation. These categories are illustrated in Figure 15.5. The systems may be totally noncooperative, making no attempt to coordinate their use of spectrum. Alternatively, if there is spectrum adaptation, it may be one sided, with one system adapting its behavior in the presence of others, or mutual, with all systems adapting. In either case, the spectrum adaptation may be passive or active. In passive adaptation, there is no communications between the systems, and the adaptation is achieved through techniques such as sensing the presence of the others and coordinating through a set of rules or spectrum etiquette. In active adaptation, systems may actively communicate among themselves to coordinate their use of spectrum. The five categories of spectrum cooperation are illustrated in Figure 15.5.

It is important to note that there may be multiple systems sharing a spectrum assignment. Some systems may be adaptive, whereas some may not be. Some (nonadaptive) systems may already be sharing with others through design or deployment techniques (e.g., sharing of bands between space service uplinks and terrestrial fixed links). In other sharing scenarios, there may be multiple, independent adaptive systems sharing with each other and a (nonadaptive) primary user. The scenario use cases should include consideration of behavior with multiple (disparate) systems coexisting together.

The analysis should specify the level and type of system cooperation expected.

Non-radio Equipment and Non-antenna Coupling When considering system relationships, the potential to impact equipment other than radio systems and non-antenna-coupled effects

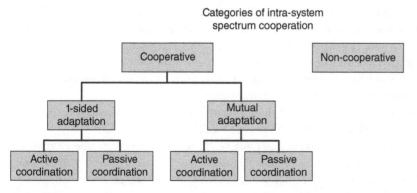

Figure 15.5 Categories of spectrum cooperation between systems. From IEEE Recommended Practice for the Analysis of In-Band and Adjacent Band Interference and Coexistence Between Radio Systems, Mar 2018. IEEE.

should be considered. The potential for the source radio signal to be demodulated and to create audio interference in hearing aids, conference call equipment, speakers, microphones, or other equipment handling voice should be considered. The potential for audio interference is highly dependent on how much of the demodulated energy falls inside of the audio band. If a signal does not demodulate into the audio band, there is limited potential for audio interference.

The impact on medical equipment shall also be considered. The possibility of source radio signals to appear to be biologic signatures when demodulating into medical equipment should be considered.

Similarly, the analysis should consider the effect of electromagnetic interference from source system transmissions to recipient radio systems through coupling mechanisms other than through the recipient antenna terminal. Such "case penetration" effects can produce interference when systems are in close proximity.

Protection Distance A fundamental parameter for an analysis is the protection distance. The protection distance is the smallest separation distance over which protection is to be provided. This subsection of the report should state the protection distance to be used in the analysis and give the rationale for it.

The calculation of the protection distance is typically based on the characteristics of the source transmitter, the propagation channel and the recipient receiver. The establishment of a protection distance is, for example, a well-recognized technique for spectrum assignment and deployment planning in radio systems with fixed locations for pairs of transmitters and receivers. In the analysis of radio systems with mobile transmitters and receivers, a protection distance is more difficult to establish as the receivers and transmitters are not in fixed locations. In these cases, the protection distance may be set by a decision that within some proximity, the radio systems can be reasonably expected to be under the control of a single user, who can separate them.

Geographic Area for Analysis The geographic area considered for analysis is significant. If the area is too small, the results may suffer from "edge effects" and from the possibility of being statistically nonrepresentative of a larger area. If too large an area is analyzed, the ability to perform a thorough analysis may become burdensome or even impossible. Furthermore, if the area is too large, the presence of isolated but significant pockets of interference may be hidden from view. It is, therefore, important that realistic assumptions be made to guide the selection of the area for analysis.

Impact of Interference Each usage model will define a situation that determines the impact of interference. The impact of interference should be specifically described.[14]

A critical component of the impact of interference is its significance. For example, if interference occurs, would the result be a major disaster or an annoyance? In some cases, any momentary interference may be of major concern. In other scenarios, if interference can be dealt with in a reasonable time or can reasonably be dealt with by the user, it may be acceptable. Some interference may be invisible to the user, such as a blocked call or lost message. Such invisible interference should be considered, when appropriate.

The concepts described in Clause 15.7 should be used to help categorize and describe the impact of interference related to a particular usage model.

If multiple types or severities of impact are possible, the analysis should provide estimates of the probability of each type occurring. The probabilities should be based on empirical evidence when practical but can involve expert judgment, particularly when data are not available.

---

14 It should be understood that the impact in some cases will have an objective criteria, but in others, it will be subjective, e.g., inability to respond to critical medical emergencies.

Interference Mitigation The significance of interference is partially dependent on the remedies available for mitigating interference. If interference can be dealt with quickly and with little effort or impact on the recipient system, then it may be judged that some probability of interference is acceptable.

For example, when discussing possible remedies for interference, the response time for those remedies and the level of effort required when applying them should be described in this subsection of the report. The response time within the scenario(s) being analyzed should be described as they relate to interference. Thus, analysis of mitigation must account for the cost, complexity, and timeliness of mitigation approaches. In some cases, if unacceptable levels of interference occur, the only remedy is to upgrade, replace, or prohibit use of the equipment involved. In these cases, knowing the average equipment replacement time may provide an avenue for mitigation. With software-defined radio systems, equipment upgrades have the potential to be performed very quickly and efficiently. If a class of equipment is routinely upgraded within a short time, then the ability to deal with interference may be significantly better than equipment that is typically deployed and left in use for a period of a decade or longer. If spectrum-dependent systems are reconfigurable through software or other means, the analysis should note how quickly systems can be reconfigured and whether such reconfiguration will involve a significant cost or administrative burden. It should also note what controls are in place to prevent unauthorized reconfiguration of systems.

For other kinds of interference, the mitigation approach would not be to upgrade the equipment but to use some field management mechanism to reduce the interference to an acceptable level. In these scenarios, knowing the response time of the field management technique is of major interest.

Similarly, it is important to know the level of effort required to remedy interference.[15]

Baseline A scenario definition shall include a description of the baseline condition to which the analysis compares the new system or changes being considered.

In this subsection of the report, the baseline needs to address the following:

— What interference already exists?
— What are the points of difference between the baseline state and that being analyzed?

Common examples of these sources of electromagnetic noise and interference are as follows:

— Thermal noise
— Natural environmental noise
— Device internal noise
— System internal noise
— Existing interference from the ambient environment

When a new device is introduced, it adds its own effect into the existing context of operating conditions for incumbent, recipient systems. This subsection of the report describes and quantifies that existing context.

The context, combined with the scenario, is important information in determining the interference event and harmful interference criteria.

---

15 An example occurred in the early days of the study of the mobile-phone-to-hearing-aid interference issue. Analyses were prepared of the potential for one user's mobile phone to interfere with another person's hearing aid. The consensus conclusion was that with most hearing aids of that era, the mobile phone would need to be closer than 0.61 m (2 ft) from the hearing aid to cause unacceptable levels of interference. One remedy would be for the hearing aid user to move away, to a distance of more than 0.61 m (2 ft) or to ask the mobile phone user to move. This was generally agreed to be a sufficiently available remedy so as to allow the judgment that in general this bystander interference was not an issue. It is significant that it was noted that use scenarios existed where this remedy might not be available; an example might be two passengers sitting side-by-side on a crowded passenger train. It was also recognized that some hearing aids experienced interference at substantially larger distances. It was judged that the percentage of such hearing aids was relatively low. Furthermore, in those cases, the hearing aid wearer would already be receiving interference from several other sources, such as some types of florescent lighting.

When evaluating harmful interference caused by a new innovation or system, it shall be in comparison with the baseline. A harmful interference criteria should not be set to be significantly more stringent than the current environment. This criteria should be established with an understanding of the limiting conditions of the recipient systems' current and planned baseline context.

Context is important because it gives a basis for the judgments made in an analysis as to what level of interference events rise to the level of harmful interference. Without a context, it is very difficult to decide whether a given level of interference or protection is acceptable or unacceptable.

15.6.5 Case(s) for Analysis

This subsection of the report defines and delivers a set of cases for analysis. Each case is considered for analysis and treated according to the processes described in succeeding subclauses.

When a proposal results in a large number of cases, a reduction of the complete set to a more manageable set, for modeling or analysis, may be desirable. When a reduction in the number of cases for analysis is performed, the rationale for prioritizing some cases over others should be explicitly stated. This reduction may be performed at any point in the analysis or in 15.9.3.

The total set of cases should be described along with the subset of cases that will receive more careful analysis.

When selecting the cases for analysis, a rationale should be stated. For some analyses, the rationale is that the cases selected are the most difficult, and if these are satisfied, then all remaining cases are also satisfied. In other situations, the cases that are most critical to the objective may be selected, eliminating those that are less important to the desired result. The analyst should be able to defend the thesis that when the cases selected are analyzed, the potential for interference will be understood.

15.7   Criteria for Interference

15.7.1   General

The preceding clauses defined the process for identifying and describing the scenarios to be analyzed and usage models to be considered. The distinct analysis case(s) derived from the preceding step define the unique system, setting, and situations that define a coexistence scenario. The next step in preparing for the interference and coexistence analysis is to define the levels of interference that will be considered as criteria for concern in the analysis. Realistic criteria are essential to support a realistic and reasonable evaluation of coexistence.

Three items shall be quantified to support an interference and coexistence analysis, as follows:

— What will be the quantity used for measurement, the measurement event?
— When does interference rise to the level of an interference event?
— When do interference events rise to the level of harmful interference?

Determining what an interference event is and when harmful interference has occurred are complex and case-dependent questions. This section of an analysis shall provide a justification for the rationale used in determining these criteria.

This clause begins by looking at the characteristics of the particular interference under consideration. These characteristics are used to structure the interference analysis. The structure used is illustrated in Figure 15.6. First, they are used to quantify the measurement event correctly and in deciding what thresholds or conditions will be considered an interference event. Second, they are used to determine what quantity of interference events will constitute harmful interference.

Figure 15.6 Process diagram for criteria for interference clause. From IEEE Recommended Practice for the Analysis of In-Band and Adjacent Band Interference and Coexistence Between Radio Systems, Mar 2018. IEEE.

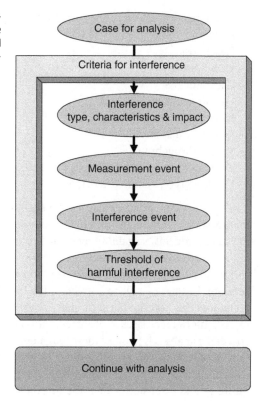

The analysis can be structured in a wide variety of ways depending on the scenario under consideration. For some cases, interference events are defined in terms of a single device. A single interference event in a single device may be generally agreed to constitute harmful interference. Alternatively, it may be agreed that interference events must either occur with some frequency in a single device or be experienced by a significant percentage of users to rise to the level of harmful interference. In some other cases, interference events are defined in terms of system behavior such as network delay, quality of service, or throughput. As with a single device, a single interference event may constitute harmful interference, or interference events must occur with sufficient frequency or by sufficiently many systems to rise to the level of harmful interference. Subclause 15.7.5 provides a structure and guidance for performing a harmful interference analysis.

15.7.2 Interference Characteristics

The process for determining the criteria for interference begins by extracting the specific characteristics of the interference for the case being considered. These characteristics are examined and reformulated into variables that are conducive to treatment as measurement and interference events.

15.7.2.1 Impacted Level

Interference may be analyzed for its impact on many levels.[16] The characteristics of the interference should be used to determine the level to be used in an analysis. Among the levels of impact that may be analyzed are as follows:

— Carrier-to-Interference C/I and C/I + N (sometimes referred to as Desired to Undesired D/U)

---

16 The seven-layer OSI communication model or other similar models may provide a guide in defining at what level(s) the impact is being analyzed. In particular, such a model emphasizes that the interference can lie at levels above pure physical layer measurements.

— Fractional degradation of performance (FDP)
— Bit loss
— Packet loss
— Packet latency
— User data loss
— Link loss or device impact
— Electromagnetic environmental effects (E3) or system impact[17]
— System availability[18]
— System throughput
— Equipment resource utilization, e.g., battery, processor, or memory utilization
— Limitations on system extensions

A source system may degrade a recipient system's carrier to interference ratio but be below the threshold of producing user data losses. However, many systems have a protection margin to account for link variability over time or across locations. In this case, the question is as follows: "Who owns the protection margin?" Has interference occurred if the recipient system suffers no user data loss and continues to operate normally? Similarly, a recipient system may suffer bit loss but be able to correct for the lost data and continue without data loss. At what point does this become harmful interference? Ultimately the answer to the question of what level of interference rises to become harmful interference is a complex value judgment.[19] The goal of the analysis is to make these judgments explicit. For instance, if the margin is important to minimize retransmission in order to meet end-to-end delay goals, then measurement events should include delay as a metric. If the margin has some intrinsic value (e.g., for enabling future system extensions), then the thresholds for measurement events that are classified as interference events can be set to reflect this needed margin. If the margin is important for protecting against time or location variability, then the pattern of interference events that are classified as harmful interference should include criteria that considers across time or across users. A complete analysis will provide enough understanding so that measurement events, interference events, and harmful interference can be defined based on an understanding of the impact of the interference from degradation of signal quality to system network impact. The analysis should provide information that justifies these definitions based on an understanding of the impact of the interference on a recipient system due to degradation of signal quality and assumptions about the shared responsibility of the source and recipient system.

15.7.3 Measurement Event

A measurement event is a specific instance of a sample to be used to characterize the state of interference. A measurement event is defined according to the service that is being considered. For instance, it could consist of one or more measurements over a defined interval of time or over a single connection. A broadcast service might define a 1 min measurement interval. A mobile telephone service might measure over a single call attempt. During the event, various variables are measured such as signal-to-interference ratios, delays, whether the broadcast was received correctly, or whether the call attempt was successful. In some cases, the measurement event may capture multiple variables or the same variable from multiple devices at the same time and contain their values. Therefore, a measurement event may contain multiple values.

---

17 This category is relatively broad, including erosion of capability caused by an increase in the noise floor for CDMA systems, resulting in a system loss of capacity, network impact on a mesh network when nodes are removed by interference, and other effects.
18 This is especially important for interference that is only periodic, for example, radar sweeps that occasionally cause interference when they occur during a specific operating state of a protected system.
19 ITU Radio Regulations are useful for definitions and examples of link budgets or fade margin that have been prepared for various purposes. These are usually divided up and attributed to various factors (e.g., rain, multi-path, or other interferers). A particularly troublesome issue occurs if there are many interferers. When multiple sources of interference contribute, which one causes a certain threshold to be violated—the first interferer or the most recent?

A measurement event should be defined to enable quantifiable analysis. The variables of the measurement event shall be defined in the analysis. The analysis should identify how the variables are measured, where in the device or system they are measured, and at what level in the protocol stack they are measured. The analysis should identify measurement windows, whether the measurements are peak or average, and measurement units. In short, the specification should be complete enough so that different analysts would be able to interpret or reproduce the data meaningfully. Measurement events should be assessed as to whether there is performance degradation. Decision criteria should be defined, to be used when deciding whether performance degradation has occurred during a measurement event. If a performance degradation threshold is reached, then the measurement event has risen to the level of an interference event.

15.7.4 Interference Event

An interference event is a measurement event where the source device or system causes significant, quantifiable performance degradation in the recipient device or system or for the user or operator of a recipient device or system. The criteria for when this degradation is significant may be unique for each service. However, in some cases, acceptable ranges have been established in formal standards or de facto by industry practice for classes or services, such as voice, video, or best-effort data communications. In principle, interference should only be considered significant if it has a measurable impact on performance or is observable by an operator or end users. For instance, a source of interference may cause more errors in a digital signal, but if the end user cannot differentiate the performance with and without the interferer, then it may be considered negligible. A robust communication system may be able to compensate for many sources of interference. At the physical layer, power control can increase power if necessary to overcome an interfering signal. At the link layer, error correcting codes can correct bit errors. At the network layer, communicating devices may route around areas of high interference. The transport layer can implement an end-to-end retry mechanism. Applications can adapt by using lower rate audio or more buffering when data are being lost. These mechanisms may collectively provide an acceptable communication performance for the end user in the presence of interference. However, performance degradation may still be present in the form of greater battery use, excess delay, blockage, reduced system capacity, reduced coverage, or other noncommunication degradations. In addition, degradation of margin can occur even if immediate impact on communications quality is not evident in a particular interference event. When establishing a performance degradation threshold, the analysis should identify the level in the protocol stack measured and what criteria are used.

15.7.5 Harmful Interference Criteria

This subclause provides a structure and framework for determining when interference will rise to the level of harmful interference. The central notion is that an interference event alone is not necessarily harmful interference. The framework is a set of interference concepts or categories that the analyst can choose from when defining the harmful interference criteria.

Several concepts of interference may be used in an analysis. These concepts identify and explicitly define the differing types of interference. For some cases, only one or a few of these concepts will be of interest. For others, all of the concepts of interference will be of interest.

> NOTE—A dark square indicates that the measurement event is classified as an interference event. The widespread graph sums interference events across users. The excessive graph sums interference events across measurement events.

These concepts can be explained in terms of the graphs in Figure 15.7. The user/ measurement event graph in the lower left plots which measurement events are classified as interference events for different users (or systems as appropriate). The excessive graph on the right is the percentage of measurement events that are classified as interference events for each user. The widespread graph on the top is the percentage of users in a measurement event that are classified as having an interference event. The widespread graph can be applied when the events by different users are synchronized.

For these concepts to have meaning, the population of users and measurement event sample for evaluation must be defined. The population of users might include an entire country, a metropolitan area, or a specific recipient system. The measurement sample might be measurement events over the previous year, the previous week, or the previous hour; or it may be determined by a specific number of measurement events. These choices are defined by the characteristics of the scenario under analysis. These characteristics may put limits on what classes of devices, users, or allowed uses are considered in the analysis. Practical aspects of this process must be considered. In a simulated system or laboratory analysis, it may be possible to evaluate and collect every measurement event at every simulated device. In a real system, only some devices may be capable of evaluating and reporting on measurement events. Other devices may only be able to report sporadically. Special monitoring devices may be used that make occasional site surveys or are located at only a few locations. These considerations will shape the final definition of the population of users and measurement sample. With these considerations defined, we can consider the following harmful interference concepts:

a) Conceivable interference event: Conceivable interference exists when a configuration of a recipient system or device and another source device can be postulated, in which an interference event can theoretically occur. Conceivable interference is usually explored on a device to device basis but may also be explored on a multiple device or system basis.[20] In Figure 15.7, conceivable interference has occurred if an interference event is possible in the user/measurement event graph.

b) Observed interference event: Observed interference exists when there are experimental or field reports of an interference event in a recipient device or system due to the operation of the source device under typical usage of the recipient and source device or system. In Figure 15.7, an observed interference occurs if any interference event occurs in the User/Measurement graph.

c) Excessive interference event: Excessive interference exists if a recipient system or device under typical usage has more than a specified fraction of the measurement events classified as interference events; e.g., no recipient device can experience an outage more than X minutes per year. In Figure 15.7, excessive interference occurs if any user exceeds the threshold x_e in the Excess graph.

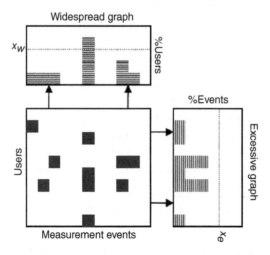

Figure 15.7 User/measurement event graph. From IEEE Recommended Practice for the Analysis of In-Band and Adjacent Band Interference and Coexistence Between Radio Systems, Mar 2018. IEEE.

---

20 Conceivable interference is unique among these categories in that it is based on only the possibility of interference and also that it does not have a threshold associated with it.

d) Widespread interference event: Widespread interference exists if more than a specified fraction of recipient devices or systems experience an interference event at any time; e.g., not more than $Y\%$ of receivers experience an outage at any time. In Figure 15.7, widespread interference occurs if at any time the threshold x_w is exceeded in the Widespread graph.

e) Multiple widespread interference event: Multiple widespread interference exists if more than a specified fraction of the time, more than a specified fraction of recipient devices or systems experience interference, e.g., interference events exceeding X minutes per year affecting more than $Y\%$ of devices. In Figure 15.7, multiple widespread interference occurs if the threshold x_w is exceeded in the Widespread graph more than the fraction x_{mw} of measurement events.

f) Multiple excessive interference event: Multiple excessive interference exists if more than a specified fraction of recipient devices or systems experience interference events for more than a specified fraction of the time; e.g. interference events affecting more than $Y\%$ of devices or systems cannot exceed X minutes per year, or the recipient system performance, e.g., system capacity, is below acceptable levels for in excess of X minutes per year. In Figure 15.7, multiple excessive interference occurs if the threshold x_e is exceeded in the Excessive graph by more than the fraction x_{me} of users.

g) Probability of interference event: The probability of interference summarizes the fraction of recipient devices or systems for a specific scenario/usage model experiencing an interference event, averaged over time; e.g., no more than $Z\%$ of devices experience an outage on average. In Figure 15.7, the probability of interference occurs if more than the fraction x_{pi} of all measurement events in the User/Measurement graph are interference events.

The first three recipient device models are on a per-user or per-device basis. Conceivable interference implies that an interference event could appear in the user/measurement event graph. This can be determined either through analysis, component measurement, or specialized laboratory experiment. In all remaining models, an actual interference event must appear in the user/measurement event graph. In observed interference, it is enough that at least one interference event appears anywhere in the graph across the measurement sample and the recipient system population to claim interference.

Excessive interference is defined in terms of the excessive graph. A threshold x_e is defined as the maximum percentage of measurement events that can be interference events for any one user or system. If any one user or system exceeds this threshold, then there is harmful interference.

The next four models are aggregate standards defined for some set of recipient devices comprising a recipient system. Aggregate here refers to the total effect across many recipient devices or the system. It is not related to the issue that a receiver may suffer an interference event as a result of the sum of multiple source device signals. Individual recipient devices may experience many interference events without being declared harmful interference as long as the recipient system does not satisfy the aggregate harmful interference criteria. Widespread interference is defined in terms of the widespread graph. A threshold x_w is defined as the maximum percentage of users that can simultaneously suffer an interference event. If at any measurement event this threshold is exceeded, then there is harmful interference.

The next two models combine the excessive and widespread concepts. Multiple widespread interference is when there is widespread interference for more than the fraction x_{mw} of the measurement events. Multiple excessive interference is when there is excessive interference for more than the fraction x_{me} of the recipients. These models are not the same. Multiple widespread interference is based on the fraction of measurements that are above threshold in the widespread graph, whereas multiple excessive is based on the fraction of recipients that are above the threshold in the excessive graph. These fractions can be different. In Figure 15.7, for example, no recipient exceeds the threshold in the excessive graph, whereas at least one measurement exceeds the threshold in the widespread graph.

As can be seen in Figure 15.7, the excessive and widespread models may or may not deem a use/measurement event graph as harmful depending not only on the number of interference events but also on their distribution. The final model, probability of interference, considers the fraction of interference events across recipients and measurement events. It sets a threshold x_{pi} on the maximum fraction allowed.

The relationships between the models are shown graphically in Figure 15.8. As indicated by the arrows, a recipient device model lower on the graph can be used to satisfy a model higher on the graph. For instance, if no interference events are ever observed (no observed interference event), then in all of the higher models, there will be no harmful interference. Similarly, if no user experiences a rate of interference more than x_e (no excessive interference), then there will be no multiple excessive interference; and furthermore, there will be no probability of interference if $x_{pi} > x_e$. The relationship between parameters is shown in Table 15.3.

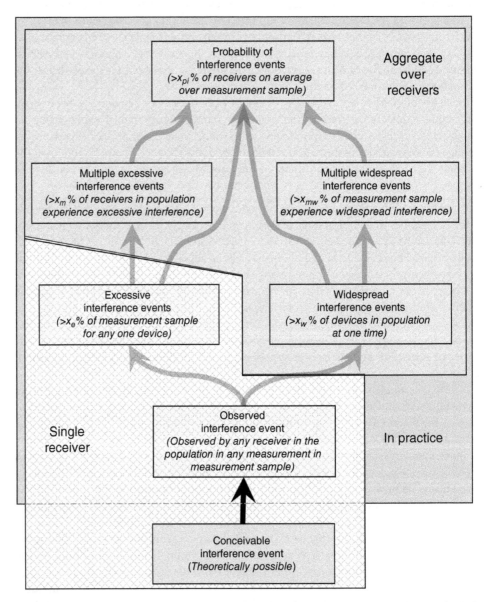

Figure 15.8 Harmful interference categories. From IEEE Recommended Practice for the Analysis of In-Band and Adjacent Band Interference and Coexistence Between Radio Systems, Mar 2018. IEEE.

Table 15.3 Relationship between model parameters: If Model A has no harmful interference, then Model B has no harmful interference with the parameters defined.

Model A	Model B	Parameter[a]
Excessive	Multiple Excessive	$(x_e)_B \geq (x_e)_A$
Widespread	Multiple widespread	$(x_w)_B \geq (x_w)_A$
Excessive	Probability of Interference	$(x_{pi})_B \geq (x_e)_A$
Widespread	Probability of Interference	$(x_{pi})_B \geq (x_w)_A$
Multiple widespread	Probability of Interference	$(x_{pi})_B \geq (x_w + x_{mw} - x_w\,x_{mw})_A$
Multiple Excessive	Probability of Interference	$(x_{pi})_B \geq (x_e + x_{me} - x_e\,x_{me})_A$

a) The subscript A or B indicates to which model the parameters refer.

For a given analysis, the harmful interference category or categories selected as the criteria for determining harmful interference should be identified and a rationale should be given for the selection. For example, in some situations, if a single device experiences the defined event, that may be deemed harmful interference. In other situations, where the consequences are less severe, it may be that a certain percentage of devices must experience the defined event for a percentage of time.

From IEEE Recommended Practice for the Analysis of In-Band and Adjacent Band Interference and Coexistence Between Radio Systems, Mar 2018. IEEE.

The analysis may include multiple dimensions at any level. For instance, the measurement events may include multiple criteria such as C/I and packet losses and a measurement event may be classified as an interference event if either criteria is sufficiently degraded. Alternatively, two separate types of interference events might be generated with a different harmful interference criteria applied to each.

Each analyst is free to choose the category they feel is appropriate for the case they are dealing with and to propose the threshold they believe to be appropriate. Analysts may disagree on what category is appropriate and what level rises to the level of harmful interference. The purpose of this analysis is to bring clarity to the discussion and to facilitate debate and perhaps a consensus among all stakeholders.

15.8 Variables

15.8.1 General

The preceding clauses define the process for identifying and describing the scenarios to be analyzed and establish the relevant interference criteria to be considered. The next step in preparing for the interference and coexistence analysis is to define the variables that are to be included in the analysis to quantify a realistic and reasonable evaluation of coexistence in the defined scenarios for the relevant interference mechanisms.

The analysis should state the variables being considered and the rationale used in selecting them. The relative impact of variables differs for each case being analyzed. Some variables are critical to the case being analyzed, whereas others have little impact. Some variables may be extremely important but are fixed and unchanging; from the viewpoint of a specific analysis, they will receive less treatment. Subclause 15.8.1.1 and subclause 15.8.1.2 provide guidance on the selection of variables for analysis.

15.8.1.1 Relationship with Advanced Radio Technologies

Equipment and systems employing advanced radio technologies[21] are being proposed that will provide users, network operators, and regulators with a multitude of new opportunities and challenges. Some general features of these emerging radio systems are as follows:

— Flexibility
— Dynamic, autonomous, and adaptive behavior
— Awareness, cognition, and intelligence
— Networking for collaboration and interaction management
— Highly efficient use of radio resources
— Very high-quality performance and reliability of communications

These features create many new avenues for improving use and access to the radio spectrum and for managing the effects of interference and coexistence. Specifically, they provide additional degrees of freedom that allow systems and devices to choose the best method of spectrum access for a particular situation, adapting the method of access to respond to changing conditions. A common thread in the migration to emerging spectrum management and access structures is the dependence on such technologically advanced radio systems to automate and distribute the frequency/channel management process to make judicious use of the spectrum and to mitigate the effect of interference.

For example, dynamic spectrum access (DSA) systems provide a fundamental, technology-enabled technique to make effective and efficient use of scarce available spectrum. In principle the DSA concept empowers the radio devices with the local authority and responsibility to administer spectrum management, avoid interference, and manage coexistence within their domains.

Traditional spectrum management practices do not provide effective techniques for identifying interference among these advanced radio systems or for managing their access to the radio spectrum. To protect other users from interference, advanced radio technology devices must meet certain standards for reasonable behavior. Some of the more sophisticated, complex behaviors envisioned for these systems involve environmentally awareness and self-coordinated, cooperative behavior. These types of scenarios suggest the need for including in an analysis the complete and comprehensive set of variables and behaviors that affect coexistence performance.

For an interference and coexistence analysis, a consequence of this dynamic, adaptive behavior is that the operation of these advance radio technologies is far more complex. Consequently, the analysis of their potential for interference is far more complex. Historically, radio system variables were fixed and could be analyzed with the assumption that variables were single dimensioned and established over the life of the device. Adaptive, dynamic devices can be designed to modify their behavior based on a variety of parameters. A device may set its transmission power level based on its internal battery capacity and onn the spectrum environment in which it is operating. Variables are no longer single dimensioned and fixed. Now variables may include multiple inputs and complex algorithms to determine the devices operating behavior. A thorough interference and coexistence analysis should include the more traditional types of variables, when they are relevant, but also include complex, multi-input behaviors and the algorithms that determine a devices behavior.

---

21 The advancements made in modern communications is directly traceable to digital signaling and processing techniques unattainable with legacy analog radios. The Nyquist theorem relating digital sampling rate f_s to signal bandwidth B ($f_s > 2B$), and the Shannon theorem for channel capacity (i.e., bps) relating signal bandwidth and noise energy ($C = B \log_2(1 + S/N)$ provide the framework for maximum utilization of spectrum for information transfer in noisy or interference environment. The introduction of error detection and error correction codes for communications by Hamming, Viterbi, Reed-Solomon, and others is fostered by the increase in radio-frequency spectrum (in GHz) wide-band signals(in MHz) and high-speed, low-power digital processors. Finally, the use of orthogonal signaling techniques in time and frequency permits multimode operation (e.g., data, voice, or security) within a single channel without loss of quality.

15.8.1.2 Analytical Framework for Selecting Relevant Variables

This recommended practice provides an analytic framework that is based on previous interference analysis frameworks but has been expanded to include analysis of variables and resources in both the physical domain, e.g., power, and the logical domain, e.g., routing protocols. This expanded framework is needed to address new and emerging radio and wireless network technology that employs advanced features. Advanced features provide many new mechanisms to improve use of the radio spectrum and to manage interference and coexistence. The analytical framework demands the use of variables that characterize additional degrees of freedom found in advanced radio networks that allow systems and devices to adapt the methods of spectrum access in response to changing conditions. The framework deals with signal resources composed of the various physical and logical resources employed in the radio system both to achieve a communications objective and to also coexist with others. Such a framework helps to identify, characterize, and categorize the variables and associated behaviors of radio technologies that are relevant to interference and coexistence analysis. This analytical framework is used to define the set of variables needed for an interference and coexistence analysis of advanced systems, and to provide guidance on their applicability.

Figure 15.1 (see Clause 15.5) depicts the overview of the analytical framework and the role of the variable within that framework. As shown in the figure, definition of the scenario and interference mechanisms for both the source system(s) and the recipient system(s) will determine the nature of the analysis and the available models for harmful interference. Some considerations include geographical deployment densities and distributions, time domain aspects of usage, and a description of the functional mission and expected use of the systems. Once the relevant measurement events for the scenarios are determined, the variables necessary to characterize and quantify the measurement events can be identified.

As an example of using the framework for a scenario and usage model considering a new fixed-site, singlefrequency transmitter with an adaptive antenna that is proposed to be located in close proximity to an extremely sensitive, fixed, receiver. The analysis should provide traditional carrier-to-interference ratio analysis addressing all of the physical domain variables (e.g., propagation loss, antenna gains, modulation, and spectrum spreading). However, it is not necessary to analyze widespread effects, aggregate effects, or any other nonapplicable effects for this scenario. Logical domain attributes are not needed to define the relevant measurement events in this scenario.

Another example of variable selection is considering the introduction of a mobile, ad hoc network system into an environment already populated with other such systems sharing a common frequency resource. In this scenario, the harmful interference model is likely to be statistically based. Analysis will focus on widespread interference and excessive interference models, and the list of variables should include such things as cognitive functions (for example, networked spectrum-state awareness), routing protocol efficiency/effectiveness (to reroute around interference events), and other relevant variables. The objective is to identify and analyze the variables that are significant to adaptive, dynamic behaviors, and impact the interference and system coexistence. In this scenario, logical domain variables play a significant role in defining the measurement events for the interference and coexistence analysis.

15.8.2 Variable Selection

Some of the adaptive behaviors envisioned in 15.8.1.1 involve environmentally aware and cooperative systems. An interference and coexistence analysis should give some consideration of those behaviors. The variables selected for analysis should reflect the complete slate of variables and commensurate with the complexity of the system and the intended level of fidelity or reliability of the analysis. In the examples described previously, a very conservative analysis identifying "conceivable" interference to a sensitive, fixed deep-space receiver may be adequate and appropriate. Alternatively, a far more complex and statistically based analysis of "widespread" and "excessive" interference would be warranted to justify the coexistence of two different mobile ad hoc DSA networks in a high-density environment. Each situation demands attention to a very different set of variables to characterize interference interactions.

Table 15.4 provides guidance for selecting the variables that should be considered relevant players in a particular situation. Many of the variables are interrelated and interdependent. That relationship changes as new technologies are introduced, especially in the interplay between logical domain and physical domain variables. Variables in the logical and physical domains interact and affect each other, in sometimes very complicated ways. New technologies change these dynamics. The way they interact and affect each other can only be accurately known in the context of implementation. For example, in a system employing "frequency selection and agility" algorithms (logical domain), those algorithms clearly affect and determine the tuned frequency in the physical domain. The "logic" within those

Table 15.4 Major groupings of variables affecting RF coexistence operations and behaviors.

RF physical domain	Logical domain
Frequency-related variables	**System management-related variables**
• Frequency(ies)	• Error correction
• Frequency range	• Compression technologies
• Frequency division duplexing	• Packet structure
• Emission bandwidth	• Application software effects
• Unintentional emissions	• Delay
• Modulation management	• System capacity (peak, average, user)
• Channel plan	• Battery lifetime
• Frequency reuse plan	• Reliability margin
Power-related variables	• Spectrum utilization efficiency
• Link power budget	• System growth and extensions
• Power control	• System testing and verification
• Demodulated waveform characteristics peak-to-average (PAR)	• Performance monitoring
• Peak or average power?	**Network-related variables**
• Tx spectral density	• Transport control
• Receiver sensitivity	• Frequency selection and agility
• Receiver spurious response	• Network routing
• Dynamic range	• Directional routing
• Signal distortion	• Access priority
• Intermodulation	• Quality of service
Time-related variables	• Services-system integration
• Time division duplexing and transmission	• Operational protocols to manage interference
• Transmission duty cycle	• Collision detection and management
• Multipath tolerance	• Interference mitigation or tolerance
• Time interval management	**Cognitive functions-related variables**
• Dynamic spectrum access reaction and response time	• Geographic awareness
Spatial-related variables	• Environmental awareness
• Directionality of signal	• Spectrum sensing
• Propagation	• Signal identification capability
• Adaptive antennas	• Information sharing/cooperation
• Polarization	**Policy management-related variables**
• Location and terrain features (shadowing, indoor vs. outdoor)	• Sensing sensitivity
• Mobility	• Sensing discrimination
• Directionality of antenna	• Ability to measure harmonics
• Range	• Geographic location information
• System coverage	• Group information availability and quality
	• Policy update capability
	• Use restrictions
	• User priority

algorithms has a significant effect on interference, and therefore, it must be represented in characterizing a measurement event. In turn the physical characteristics of the RF electronics may behave differently based on the precise way the commands are applied, and so the same logical command may produce somewhat different results based on the context in which it is issued. It is recommended that each analysis provide a discussion of the variable considered and the subset selected as relevant to the case being analyzed. Many variables may affect an interference and coexistence analysis. For each case, for analysis, a subset of these variables is important. The report should provide justification for the use of the chosen variable for each analysis and should also justify why other variables were not considered relevant for the analysis. Table 15.4 categorizes the major variables in groups belonging to the two major domains defined earlier: the physical domain and the logical domain.

When an analysis is comparing two situations, for example, a baseline and a new situation, some variables are different between the cases and many others are not. In many cases, only the variables that are affected by the differences being considered need to be analyzed. These are "contrasting variables." However, in some cases, it will be important to include in the selection noncontrasting variables that are constant among scenarios. Although the question being analyzed is primarily intended to compare two situations, both should be sufficiently complete, in and of themselves, to be acceptable. In such cases, it will be important to determine not only which situation is superior but that it is also sufficient for the intended purpose of analysis.

The variables chosen and identified should be those that will allow a fair and complete analysis to be prepared. The choice of which variables should be considered in the various selected cases should be justified. The selection of variables and identification of relevant behaviors sets the stage for the process of analysis. The analysis may involve selecting and applying analytical approaches along with appropriate tools, including modeling, simulation, measurement, and tests. By defining the variables that will be considered relevant to the cases and measurement events selected for interference and coexistence analysis, the analyst can balance and choose the best approaches for quantifying the ranges of the variables without enduring undue complexity and cost for the analysis.

15.9 Analysis—modeling, Simulation, Measurement, and Testing

15.9.1 General

The preceding clauses addressed the problem to be analyzed and identified the relevant variables and criteria to be considered. The heart of the analysis is to reduce the array of possible cases to be analyzed to a manageable number of the most important cases and to perform the analysis by applying modeling, simulation, measurements, and testing, as appropriate. The objective is to determine the impact of the source(s) being analyzed on the recipient(s) (see Clause 15.6).

There are a number of considerations in performing, interpreting, and documenting the results of the modeling, simulation, and analysis, as discussed in 15.9.2 through 15.9.7.

From IEEE Recommended Practice for the Analysis of In-Band and Adjacent Band Interference and Coexistence Between Radio Systems, Mar 2018. IEEE.

15.9.2 Selection of the Analysis Approach, Tools, and Techniques

The analyst is set to address the final process step, i.e., performing the analysis, by properly identifying the scenarios/usage models, interference criteria, and variables to be considered in the analysis. At this point, the choice of analysis approach, and the tools to support that approach, can be defined and documented.

The analysis approach can range from a very simple, single closed-form calculation that convincingly demonstrates the absence of any conceivable interference, to a very involved analysis, measurement, and testing program that may be needed to address the full scope of a complex scenario. For example, a modeling, measurement, and testing program may be needed to quantify a statistical representation of widespread interference for large systems affecting hundreds of thousands of users.

The analyst should define the approach by working through the following series of questions to define that approach and the appropriate tools and methods for each case being considered:

a) Can the relevant cases be analyzed by methods less complex than models or simulations (to establish either the absence or the existence of harmful interference) that would convincingly settle the issue? If so, then this may be all the analysis that is required. If the answer to question a) is no, then can/should the analysis apply modeling and simulation tools to calculate/predict interference events? If so, then the tools should be selected to address the full complement of system scenarios and variables needed to quantify the interference events and the occurrence of harmful interference. Selection of modeling and simulation tools[22] should consider:

   1) What propagation calculation model/technique is appropriate?
   2) What signal processing model is appropriate?
   3) What type of antenna modeling is needed?
   4) What logical domain variables must be addressed?
   5) Does the approach demand a network-level simulation, a device-level simulation, or both?
   6) Is it a one-on-one interaction, or one-on-many, or many-on-many?

b) Could the analysis benefit[23] from measured data and testing? If so, then the measurements and testing should address the critical areas of the systems needed for analysis. Selection of the test and measurements should consider:

   — Are measurements or test results already available? If not, what measurement and testing programs are appropriate to achieve the level of confidence and fidelity necessary to support the analysis?
   — Are the needed systems and deployment scenarios available for practical testing and measurements?
   — Can the testing be done using deployed "live" systems without compromising existing users and system performance?

15.9.3 Matrix Reduction

The resulting set of variables, frequency regions, systems, and other aspects of the analysis will often result in a matrix of cases for analysis that exceeds the available resources. Therefore, it is necessary to select the spectrum regions and systems operating in those regions that are most critical to the outcome of the analysis. In this subsection of the analysis, the rationale for these selections will be discussed and reported.

It is permissible, for example, to have some prioritization within the analysis. So although several frequency regions and systems may be selected for analysis, some may be identified as

---

22 Note that many sources of modeling and simulation capabilities and services can address these requirements. These capabilities range from separate models to address specific parameters and phenomena (such as propagation; see Annex 15A) to fully integrated many-on-many interference analysis software applications. Many available techniques are summarized in public sources (for example, see the ITU-R Handbook on National Spectrum Management).

23 Note that theoretical analysis using modeling and simulation alone does not typically invoke the same level of confidence as real measured data; yet test and measurement programs are often expensive and time-consuming. The analyst should assess the tradeoffs between the required level of confidence and the cost of measurement and test support.

having clearly greater impact on the question being examined. These higher impact cases may be highlighted for more detailed analysis. Other, lower impact cases, while still being analyzed, may receive more cursory attention. It is often necessary to select from the full matrix a sparse matrix for analysis and perhaps even a smaller matrix for in-depth analysis. In this subsection, the analyst will describe the rationale and justification for these selections.

15.9.4 Performing the Analysis

Having reduced, defined, and justified the cases for analysis, as well as the approach and tools to be applied, the analyst performs the interference and coexistence analysis.

15.9.5 Quantification of Benefits and Interference

Measures appropriate for the scenario may be used to quantify the benefits of the proposed scenario. The value of the proposal is important to understand when assigning significance to the potential interference created by the scenario being considered.

Quantification of interference and coexistence is based on the concepts of harmful interference described in Clause 15.6. The interference and coexistence analysis builds on the previous subclauses to reach a conclusion as to the probability of interference effects and potential for harmful interference. Analysis of interference and determination of interference events should follow accepted procedures for traditional established systems.[24] However, other types of systems have more complex behavior and traditional measures such as the impact to total information delivered may not be the dominant concern, In some cases, priority, delay, reliability, or capacity of information transferred may be the more important aspects. In these cases, the effect on these aspects of performance should be included in the quantification of the interference and coexistence scenario.

15.9.6 Analysis of Mitigation Options

The scenarios should include identification of options that can mitigate the impact of interference. At a minimum, the analysis should include an assessment of the extent to which such options can realistically address potential harmful interference. The level of depth of analysis of mitigation options will depend on the scenarios and on the purpose of the analysis activity. This can range from simple informal analysis to detailed simulation to reflect alternative cases. If informal analyses are performed in place of theoretical computations, empirical testing, or computer simulations, then key assertions should be accompanied by detailed explanations of the basis for the assertions along with references to other sources for their validity.

15.9.7 Analysis Uncertainty

All analyses and measurements have an associated uncertainty. This subsection of the report requires the analyst to state explicitly the uncertainty of their findings.

15.9.7.1 Uncertainty Distribution

When developing an analysis emphasis may be given to minimizing the possibility of a false positive finding, a false negative finding or balancing the two. The analysis should state the emphasis used and the rational for it. There are legitimate reasons why an analysis will choose one emphasis or the other. The effect will be, where there is uncertainty, to deal with it to achieve the desired goal. If one analysis intends to assure with 95% confidence that interference will not be underestimated while a second seeks to assure with the same confidence that inference will not be overestimated then the results may be quite different. By stating the intent of the analysis the reason for divergent results can be more readily identified.

---

24 One example for land mobile radios coexistence analysis guidance is provided in the TSB-88-B-1 bulletin published by the Telecommunications Industry Association [B27].

15.9.7.2 Reduction of Uncertainty

Uncertainty of an analysis can be reduced by several methods. One method is to identify the variables contributing to uncertainty, select those making the greatest contribution, and minimize them. By minimizing the uncertainty of the variables with the greatest uncertainty, the total uncertainty of the analysis is reduced.

A second method is to compliment an analysis with measurements. Results that are corroborated by analysis and measured results have less uncertainty than results supported by only analysis or measurements.

Results supported by analysis performed using multiple approaches and differing methods also gain greater credibility. Therefore, a method of reducing uncertainty is to perform an analysis using multiple methods and approaches.

Results supported by analyses performed by independent analysts also have greater credibility than those performed by a single analyst. However, the analyst must be truly working independently. If multiple analysts work using similar source material and models, systemic errors can affect all of the analyses.

When measurements are made, uncertainty can be reduced by having the results reproduced in independent laboratories. Having the same laboratory repeat a measurement can also be helpful to identify measurement variability.

15.10 Conclusions and Summary

15.10.1 Benefits and Impacts

In this section of the report, analysts provide their assessment of this balance between the potential benefits and the negative impacts, particularly the possibility for interference. The parties benefited and those impacted should be identified.

15.10.2 Summation

The summary subsection provides an opportunity for the analyst to discuss the findings of the analysis and conclusions reached.

If there is linkage between variables, those should be explicitly identified and the interaction between the variables described. In particular, if an analysis has been performed on several variables and reported independently on each variable, an interaction between these variables should be reported so that the summation accurately conveys the composite result.

The summary provides the opportunity for a staged presentation of the analysis, including a concise review of:

— The interference event and harmful interference criteria
— Analysis of variables / attributes
— Summation of groups of variables
— Summation by category and use model
— Description of analysis approach, tools, and techniques applied
— System impacts
— Conclusions

Annex 15A

(informative)

Propagation Modeling

15A.1 General

Propagation modeling is an integral component of estimating the interaction of differing radio systems on each other. Unfortunately, propagation is a complex function of many variables making it one of the most controversial areas when analyzing radio system coexistence. This annex gives an introduction to propagation modeling and some guidance on using the propagation modeling tools most applicable to a particular coexistence case.

The received power depends on the transmit power and the channel gain, denoted path loss L_p. The path loss depends on four high-level variables, as follows:

— Scale: What are the communication distances relative to the environment and to the wavelength?
— Terrain and environment: Is this indoors or outdoors? Is it an urban, suburban, rural, mountainous, marine, aeronautical, or space environment?
— Signal: What is the frequency? Is the signal narrowband or wideband?
— Mobility: How fast does the signal vary because of transmitter, receiver, or environmental mobility?

This annex describes each of these variables, their role in propagation modeling, and some commonly used propagation models.

15A.2 Scale

The appropriate model depends on the communication distances. Indoor models are used when the use case is an indoor environment, involving walls and other common features of buildings. Outdoor models are used for communications over several kilometers. Most propagation models specify ranges over which they are appropriate. Communication in the near field is defined as distances less than approximately $\frac{2D^2}{\lambda}$, where D is the largest dimension of the antenna (not including the antenna mounting) and λ is the wavelength. The near field is typically within one wavelength of the transmitter. However, it may be important for some applications such as RFID readers where the wavelength is long and communication distances are short. The rate of change varies widely in the near field based on the characteristics of the transmitting antenna and other variables. Beyond the near field, signals generally propagate using free-space propagation whereby the received signal is proportional to the inverse of the distance squared.

The free-space region is where the propagation does not have significant interaction with the ground or surrounding objects. Consider the line connecting a transmitter and receiver of length d. The first Fresnel zone is the ellipse with foci at the transmitter and receivers such that the distance from the transmitter to any point on the ellipse and on to the receiver is $d + \lambda / 2$. As long as objects do not intersect this ellipse, the attenuation can be considered as line of sight and attenuating as in free space.[25] For example, assuming two antennas over a flat surface, the ellipse will touch the ground when:

$$d > d_f = \frac{4h_{tx}h_{rx}}{\lambda} \tag{15A.1}$$

where h_{tx} and h_{rx} are the height of the transmitter and the receiver above the ground. If the ground is not flat, then a careful analysis would need to show whether any portion of the ground intersects the first Fresnel ellipse. Beyond d_f, the path loss is typically much worse than free space. If the line-of-site path from transmitter antenna to receiver is obstructed, then other variables come into play, depending on the obstructions.

15A.3 Terrain and Environment

The channel loss (or gain) is the path loss (or gain) between a transmitter and a receiver. In general, the path loss can be decomposed into subvariables $L_p(d) = \overline{L_p}(d)X_f X_s$, where $\overline{L_p}(d)$ is the mean path loss, X_f is a small-scale (aka fast) fading component, and X_s is a large-scale (aka slow) fading component.

As a receiver moves away from its transmitter, the received signal becomes weaker because of the growing propagation attenuation with increasing distance. $\overline{L_p}(d)$ denotes the mean path loss, which is a function of the distance d separating the transmitter and receiver, carrier frequency f_c, antenna heights and gains, and other variables.

The fast fading is dominated by the multipath propagation and is independent of the distance between the transmitter and receiver. It can be characterized by a Rayleigh, Rician, or Nakagami distribution.[26] Fast fading varies over distance scales less than a wavelength. Many applications are only interested in the local average, which measures the signal strength at several points within a few wavelengths and computes the average. In such models, $X_f = 1$. The slow fading (sometimes denoted shadowing) represents the variability around the mean path loss because of the specific environment in the vicinity of the transmitter or receiver, such as whether a receiver is on, inside, or behind a building. It is often characterized with a log-normal distribution. The first-order statistics of log-normal shadowing are characterized by the standard deviation σ_w (in decibels), which can be obtained from measurements. For example, 8 dB is the typical value of σ_w for the Okumura-Hata Model, 5 dB is the value for JTC Model, and 3 dB is the value for terrain-based models.

The fast- and slow-fading components emphasize that many models are statistical in nature. In principle, these stochastic components could be eliminated if a propagation model could have the exact placement and construction of every object in the environment and then perform ray tracing from transmitter to receiver accounting for every object interaction. Software packages can perform this kind of analysis for both indoor and outdoor propagation. Other software packages allow detailed finite-element modeling on and around antennas on the same object (e.g., two antennas on the same airplane). However, these packages always make approximations at some level, and so the role of these stochastic components is never completely eliminated.

Ray tracing or finite-element packages are appropriate when the interference and coexistence analysis only considers one or a few specific instances. Otherwise, more general models are required. The appropriate channel gain model depends on the propagation environment for the specific case being analyzed. Various models have been proposed, mainly based on field measurements. In this clause, we describe four popular path loss models: Terrain based models, Advanced Refractive Effects Prediction System (AREPS), Okumura Hata, and Joint Technical Committee (JTC) models.

---

25 The assumption of having a line-of-sight (LOS) channel if the first Fressnel zone is not obstructed is only true for antenna systems having a circular aperture. If one uses omnidirectional antennas, a (reflecting) object right behind the transmitter may cause very deep fades. However, this is not typical for LOS environments.
26 See the *User's Manual for Advanced Refractive Effects Prediction System*, Version 3.0 [B28].

15A.3.1 Terrain-based Models

Terrain-based models are suited for open terrain without significant foliage and for long link distances (1 km to 50 km). These models use, as input, a terrain profile described by a set of discrete terrain elevation points. Also required is information on the transmitter (frequency, antenna height, and antenna polarization), the receiver (antenna height and polarization), atmospheric constants (surface refractivity and humidity), and ground constants (permittivity and conductivity). These variables are not directly incorporated into a closed-form solution. Rather, the intervening terrain between transmitter and receiver are considered, and for instance, if terrain intersects the direct interference path, then diffraction losses over the obstacle are computed.

15A.3.2 Advanced Refractive Effects Prediction System (AREPS)

Long-range radar signals from airplanes in the troposphere can have deep fades, sometimes losing the signal completely. The parabolic equation (PE)[27] is a key to the modeling of this behavior. The wide-angle parabolic equation (WAPE) and its narrow-angle version, the standard parabolic equation (SPE), are used to model wave propagation in the troposphere for near-horizontal propagation, i.e., the direction with significant refractive effects. The PE is also used in strong fluctuations caused by atmospheric turbulence.[28]

AREPS[29] is freely available PC software with a support service to compute path loss and other quantities of relevance to tropospheric wave propagation. AREPS is based on the advanced propagation model (APM) coded in FORTRAN. APM is a hybrid model to enhance speed, effectively merging the terrain propagation equation model (TPEM)[30] with a radio physical optics (RPO) model. The capabilities include range-dependent refractivity profiles, variable terrain height, and the selection of various antenna types with horizontal h or vertical v polarization. The primary analytical technique used in APM is the split-step Fourier PE algorithm, developed by Hardin and Tappert [B11]. The split-step PE is a small-angle approximation to the Helmholtz elliptical wave equation and is accurate, robust, and more efficient than equally sophisticated methods. The novel approach of using fast Fourier transforms (FFTs) to "march" or propagate the field solution in range at many different receiver heights simultaneously provides a relatively fast technique to produce coverage diagrams of field strength for multiple receiver heights and ranges. Some examples are shown in Figure 15A.1 and Figure 15A.2. These figures depict propagation loss (dB) over a range of 0 km to 400 km and height of 0 km to 10 km calculated using the AREPS Research Project computations at 3 GHz with a Gaussian transmitter antenna at 2000 m altitude with horizontal axis, 3 dB beamwidth of 3°.

15A.3.3 Okumura-Hata Propagation Model

The Okumura-Hata model was developed by curve-fitting urban propagation measurement data collected in Tokyo, Japan, using large height, base station antennas and handheld radios. The path loss $\overline{L_p}(d)$ is represented as a function of[31,32]

— Carrier frequency $f_c \in [150,1000]$ MHz
— Antenna heights of transmitter, $H_T \in [30,200]$ m, and receiver $H_R \in [1,10]$ m

The shadowing standard deviation is between 8 dB and 10 dB.

---

27 See Mireille Levy [B21].
28 Akira Ishimaru [B16].
29 *User's Manual for Advanced Refractive Effects Prediction System*, Version 3.0 [B28].
30 Amalia E. Barrios [B4].
31 Mark and Zhuang [B23].
32 Rappaport [B24].

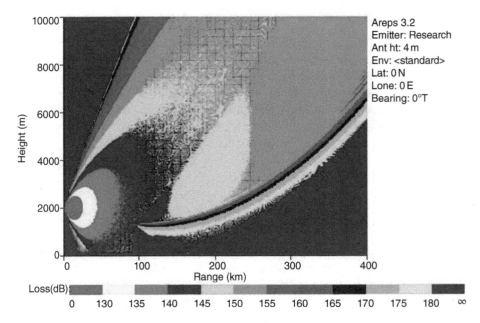

Figure 15A.1 Standard atmosphere with knife-edge shape peak at 1400 m. From IEEE Recommended Practice for the Analysis of In-Band and Adjacent Band Interference and Coexistence Between Radio Systems, Mar 2018. IEEE.

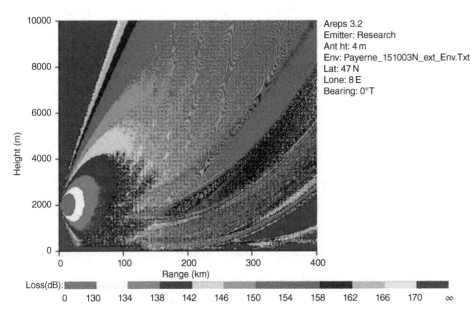

Figure 15A.2 Atmosphere defined by Payerne radiosonde profile of Oct. 15, 2003. From IEEE Recommended Practice for the Analysis of In-Band and Adjacent Band Interference and Coexistence Between Radio Systems, Mar 2018. IEEE.

15A.3.4 JTC Path Loss Model

The JTC path loss model is designed for microcells. It assumes that the distance between the base and the mobile stations is less than 1 km and that the base station antenna height is below the roof-top level. It is recommended by a technical working group of the TIA/ANSI Joint Technical Committee (JTC) on the Air-Interface Specification for PCS.

The model divides the distances into two line-of-sight (LOS) and one obstructed line-of-sight (OLOS) region. The first LOS region is the Fresnel zone defined by the break-point distance d_{bp}:

$$d_{bp} = \frac{4h_b h_m}{\lambda} \tag{15A.2}$$

where h_b and h_m are the heights of the base and mobile station antennas, respectively, and λ is the wavelength of the carrier frequency. In this region, the power received from the LOS path dominates the total power of the other paths and the propagation loss is the same as for free-space propagation. The second LOS region starts at d_{bp} and continues to d_{cor} (distance to corner) where the mobile unit turns a corner and loses the LOS path. In this region, the gradient is assumed to be proportional to $40 \log_{10}\left(\frac{d}{d_{bp}}\right)$, to include the direct LOS path as well as the path reflected from the ground. The third region starts from d_{cor} where the mobile loses the LOS path. The gradient in this region is assumed to be proportional to $50 \log_{10}\left(\frac{d}{d_{cor}}\right)$, and an additional path loss of L_{cor} is added to compensate for the immediate power drop after turning the corner.

The formula for calculating the path loss is then given by:

$$\overline{L}_p = 38.1 + \begin{cases} 20 \log_{10} d, & d < d_{bp} \\ 20 \log_{10} d_{bp} + 40 \log_{10} \dfrac{d}{d_{bp}}, & d_{bp} < d < d_{cor} \\ L_{cor} + 20 \log_{10} d_{bp} + 40 \log_{10} \dfrac{d}{d_{bp}} + 50 \log_{10} \dfrac{d}{d_{cor}} & d > d_{cor} \end{cases} \tag{15A.3}$$

The standard deviation of the shadowing is between 3 dB and 4 dB.

15A.4 Relationship of Outage Probability and Radio Coverage

This subclause describes the relationship of outage probability and the radio coverage distance. An outage event occurs when the received signal strength falls below a minimum acceptable level of intelligible communication. The minimum acceptable level depends on the Noise Figure (NF) and on E_b/N_0 of the receiver, where E_b is the bit energy and N_0 is the noise level. It is assumed that the noise is a white Gaussian noise with zero mean and two-sided power spectral density. The unit for N_0 is watts/Hertz.

Let $R(d)$ denote the received signal level at a node located at a distance d from the transmitter. Then the outage event at distance d is given by:

$$O(d) = \{R(d) \leq -174 + NF + E_b/N_0 + 10 \log_{10}(B)\}. \tag{15A.4}$$

Let the outage probability at distance d denoted by $P_{out}(d)$ and be the probability of the outage event from Equation (15A.5):

$$P_{out}(d) = \Pr\{O(d)\} = \Pr\{R(d) \leq -174 + NF + E_b/N_0 + 10 \log_{10}(B)\} \tag{15A.5}$$

The received signal level $R(d)$ is:

$$R(d) = P - (\overline{L}_p(d) + w) \tag{15A.6}$$

where P is the transmit power, $\overline{L}_p(d)$ is the mean propagation loss that occurs over the distance d, and w is random fluctuation of local mean loss around the median loss.

From Equation (15A.5) and Equation (15A.6), the outage probability can be further derived as follows:

$$P_{out}(d) = \Pr\{\overline{L}_p(d) + w \geq P + 174 - NF - E_b/N_0 - 10 \log_{10}(B)\}$$
$$= 1 - F_L(P + 174 - NF - E_b/N_0 - 10 \log_{10}(B), d). \tag{15A.7}$$

where $F_L(x,d)$ is the cumulative distribution function (CDF) of the propagation loss given distance parameter d between the transmitter and the receiver. In this study, the CDF, $F_L(x, d)$, is considered for four propagation prediction models: terrain-based models, AREPS, Okumura-Hata, and JTC.

Now, for a given outage probability p_{out} from Equation (15A.8), we can determine the radio coverage radius d_{cov} as follows:

$$F_L^{-1}(1 - p_{out}, d_{cov}) = P - (NF + E_b/N_0) + 174 - 10 \log_{10}(B) \qquad (15A.8)$$

In particular, for Okumura-Hata and JTC models, the Equation (15A.9) becomes

$$\overline{L}_p(d_{cov}) = P - (NF + E_b/N_0) + 174 - 10 \log_{10}(B) - \sigma_w Q^{-1}(p_{out}) \qquad (15A.9)$$

where $Q(x) = \frac{1}{\sqrt{2\pi}} \int_x^\infty \exp\left(-\frac{t^2}{2}\right) dt$.

15A.5 Signal

The transmitted signal has two key characteristics that influence propagation, frequency, and bandwidth.

15A.5.1 Frequency

The frequency affects every interaction with the environment. The models in the previous subclause include frequency dependence, but in some cases, it may be useful to consider independently the frequency effects. These effects are best understood in terms of wavelength λ.

— **Reflection:** When a surface has dimensions greater than λ and its surface has variations that are small, relative to λ, then it will act as a reflector. Power incident on the surface will be reflected in the usual way. A building, mountain, automobile, and the ground are examples. When an object is small relative to a wavelength, the reflection mechanism is called volume scattering, which is discussed later in this subclause. When an object is large, relative to λ and also has variations that are large relative to λ, the reflective mechanisms are called surface scattering, which is also discussed later in this subclause.
— **Penetration:** Unless the object is a conductor some power will simply go through the object. The penetrating signal will be attenuated. Typical losses from building materials are 15–20 dB.[33]
— **Absorption:** Certain materials simply absorb the signal power at certain frequencies and convert it to another form (e.g., heat). This is the principle of the microwave ovens (usually operating at 2.45 GHz where the absorption of water is particularly high). This is a simple loss expressed in decibels/kilometer. Absorption is frequency (wavelength) selective. Occasionally scattering is called absorption because it has a similar effect although the mechanism is different.
— **Diffraction:** When an edge is longer than λ, then it will bend the signal around the edge. A building corner and mountain top are examples. However, to be effective, the edge must be sharp, meaning that the curvature is much smaller than λ.
— **Volume Scattering:** When an object has dimensions much smaller than λ, but there are many of them, it has the effect of scattering the signal power. Foliage on trees and particles in the air are examples. This has two effects. The first example is that the signal scatters to otherwise blocked locations (e.g., light still reaches "shadows"), and the second example is that power scatters from the propagation path so that the signal is attenuated as a function of the distance traveled.

---

33 See Chapter 3 in Rappaport [B24].

— **Surface Scattering:** When reflector is much larger than λ, but has surface variations that are large relative to the wavelength, then the signal will reflect in many directions. The effect is to diffuse the power in all directions in front of the reflector.

In general an object that is small relative to the wavelength can be ignored in an interference and coexistence analysis. Objects large relative to the wavelength must be considered at some level.

15A.5.2 Bandwidth

A characteristic of a propagation channel is its coherence bandwidth B_c. Signals that have bandwidth smaller than B_c are denoted narrowband and will suffer path loss uniformly across the signal bandwidth. Thus, although the signal suffers attenuation, it suffers no frequency selective distortion. As a result, the effect of one signal on another does not require separate propagation models for different parts of the signal bandwidth.

Signals that have bandwidth greater than B_c are denoted wideband and will suffer path loss that varies across the signal bandwidth. In this case, the interference and coexistence analysis has to consider whether these variations are significant. It may be possible that significant interference requires high power simultaneously in two different parts of the band. With frequency selective attenuation, this may be more or less likely.

15A.6 Mobility

Mobility introduces two variables: exposure and coherence time. Mobility may limit the exposure of any one device to another. Thus, in a mobile environment, a signal source may only affect any given receiver a small fraction of the time. Propagation models would have to be rerun periodically to determine the time varying effect.

Exposure represents a large-scale effect. Channel characteristics also vary over short timescales. The period over which the channel is considered stable is the coherence time T_c. The coherence time is related to fast fading, and as was noted earlier, many types of analysis simply average out this effect. However, if detailed propagation modeling is required, the coherence time determines how often a channel would need to be remodeled.

Annex 15B

(informative)

Audio interference

15B.1 General

For real-time voice services, interference to the audio transmission may be the primary interference of interest when specifying the interference event. This annex provides guidance on determining when audio interference rises to the level of an interference event. The material in this annex is based on the psycho-acoustic impact of interference on the ability of humans to recognize speech. Speech intelligibility is a well-studied phenomenon that can be measured and quantified by experts in psycho-acoustics. Annoyance is a more subjective determination that is highly variable with individual listeners and the listening environment. Interference that would be completely unacceptable to a user listening to music in a quiet environment may be negligible when listening to a conversation in an acoustically complex environment, e.g., talking on a cell phone in an airport. There is some research data to suggest that annoyance, on average, occurs at about 5 dB below the level at which word recognition begins to degrade significantly. If these data come to be supported by future research, it may be that word recognition levels may also provide guidance for setting annoyance levels as well.

Another complicating factor is the variation of hearing function within the user population. Hearing loss is one of the most common disabilities, and the type and degree of loss vary widely. Hearing loss runs a range from very mild to complete deafness. There are many types of hearing loss. The levels of speech to interference recommended in this document rely heavily on the levels developed in ANSI C63.19 [B3]. The speech-to-interference values used by the ANSI C63.19 committee were accepted as providing reasonable performance for people with hearing loss. As such, they are proposed in this document as providing acceptable and even conservative levels for the general population.

15B.2 Relative Amplitude of Audio RF Interference

ANSI C63.19 [B3] identifies a 20 dB signal-to-interference level as the threshold for acceptable interference to a voice signal intended for regular use, e.g., a personal cell phone used routinely. The data supporting this conclusion vary by ±5 dB. A defensible argument can be made for requiring a speech-to-interference level anywhere in the range of 15 dB to 25 dB. The ANSI C63.19 committee agreed to use a value of 20 dB.

15B.3 Absolute Amplitude of Audio RF Interference

Assuming the recommended relative signal-to-interference level of 20 dB is accepted, determining the absolute level of interference depends on several variables. The first variable is the average speech level provided. A typical value for human speech in a quiet environment is 65 dB SPL. Mobile phones often deliver an average of 85 dB SPL because there is significant loss between the phone output and the user's ear. Holding a mobile phone tightly pressed to the ear, so as to form a good acoustic seal, will illustrate how much loss there is in the normal use position. Furthermore, mobile phones are often used in acoustically noisy environments. In the first case, if interference rises above 45 dB SPL, it would exceed the desired level of 20 dB speech-to-interference. In the second case, a speech level of 85 dB and an audio interference level of up to 65 dB SPL would be acceptable.

To complicate the issue even more, when measured, the values measured are highly dependent on the coupling method and ear simulator used to make the measurement. Hence, to assure the desired outcome, it is important that the same assumptions are made throughout as to how the values of speech and interference are to be measured.

Volume control is another significant variable. Should the determination be made based on the nominal volume control level, at its minimum or at its maximum? Different answers are appropriate for different cases. It is important when selecting a level that the assumptions be stated to assure that the comparison is maintained on an equivalent basis.

15B.4 Model of Audio RF Interference

A signal-to-interference level only has precise meaning in the context of a defined model for audio interference with parameters for the measurement of such interference.

The following discussion, adopted from ANSI C63.19 [B3], is given to describe clearly and succinctly the physical quantity to be measured when analyzing or measuring the signal-to-interference level for a voice service.

Conceptually, the user perception of interference is characterized by the following attributes:

— The full RF signal bandwidth is presented to a wideband detector. This means that the elements in the recipient device that receive and demodulate the RF signal into an audio band signal have a bandwidth ≥ the emission bandwidth of the RF signal.
— The detection elements are assumed to function as a square law detector.
— The post-detection, recovered audio signal is limited to the audio band.[34]
— The typical response of human hearing is applied to the detected signal before determining the final signal-to-interference level. A-weighting has been shown to be a good predictor of human perception for steady-state interference but is not necessarily valid for interference that has substantial variation over time. Research is being actively pursued as to the impact of interference that has substantial temporal variation. Until that research reaches generally recognized conclusions, A-weighting of the interference signal provides the best, generally agreed model for impact to human perception of an interference signal.

Figure 15B.1 identifies the audio interference process, starting with coupling of the RF energy into a circuit through some receiving element, acting as an antenna. The RF is demodulated across a nonlinear junction, typically responding as a square law detector. The demodulated signal is then passed through an audio band pass filter, which is often the combination of the frequency characteristics of the path between the point of demodulation and the user's hearing ability.

---

34 If the recommendation of A-weighting is adopted, it should be noted that A-weighting provides a frequency weighting that essentially obviates the importance of the following discussion of the audio frequency band. However, it is important to understand the reasons why differing values for the audio band may be chosen. The audio band is defined in *The Authoritative Dictionary of IEEE Standards Terms* [B12] as the frequency range from 15 Hz to 20 000 Hz. However, for the purposes of RF interference and in this document, the range is defined more narrowly as 100 Hz to 10 000 Hz. In general, this is the frequency range that is passed by many hearing aids. It is wider than the typical frequency range used in telephony of 300 Hz to 3300 Hz. A wider range is selected because newer research has demonstrated that important speech cues exist to 7000 Hz. Higher fidelity voice communications is trending toward this wider bandwidth for voice communications. Another factor is that although speech communication is contained in a narrower audio band, noise outside of this band may be disruptive to speech intelligibility.

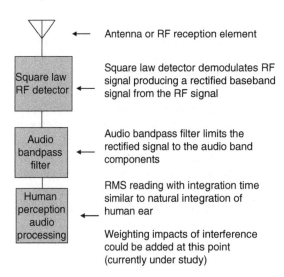

Figure 15B.1 Conceptual model of RF interference level. From IEEE Recommended Practice for the Analysis of In-Band and Adjacent Band Interference and Coexistence Between Radio Systems, Mar 2018. IEEE.

Antenna or RF reception element

Square law detector demodulates RF signal producing a rectified baseband signal from the RF signal

Audio bandpass filter limits the rectified signal to the audio band components

RMS reading with integration time similar to natural integration of human ear

Weighting impacts of interference could be added at this point (currently under study)

15B.5   Other Considerations for Audio Interference

In some cases, the interference of audio may be measured at a point before the production of the final audio signal. These points may be more convenient or more consistent places to measure. Analog signals can measure directly C/I at the receiver input. Digital signals often organize the data stream into packets. Groups of packets are then organized into frames. In some cases, frame error rates become the relevant metric. Note that some services such as IS-95 CDMA mobile cellular dynamically control the power level so as to maintain the frame error rate within a certain range.

It may be necessary to consider the source device behavior that is in the interference measure. A packet data radio will send data in bursts that may have low average C/I or low average impact on frame error rates. However, they may or may not have significant impact on the perception of the audio or video. Occasional transmission errors are expected and compensated for in digital audio communication. A source device that produced sporadic but concentrated bursts could be considered acceptable. The same sporadic bursts in an analog system might manifest themselves as annoying pops. Thus, the definition of interference may need to consider more than average interference levels.

Annex 15C

(informative)

Spectrum Utilization Efficiency

15C.1 Spectrum Utilization Efficiency

Spectrum utilization efficiency (SUE) is the ratio between the useful effect obtained with the aid of the communication system in operation and the spectrum used. The useful effect intended is different for different types of systems. For many communications systems, the amount of user information transferred is the most important measure of useful effect. However, other types of systems have different purposes and, therefore, different measures of their useful effect.

Different types of systems have different purposes, and therefore, the measurement of their useful effect is different. The differences in how useful effect is measured means the SUE for different types of systems is based on a different reference. For example, the computation of SUE for point-to-point, satellite, and land mobile systems will be based on a different reference. This means that comparisons between systems of the same type are relatively straightforward; however, comparisons of different kinds of systems are very difficult or impossible. The reason the comparison of SUE for different types of systems is difficult or even sometimes meaningless because each type of system uses a different frame of reference. Furthermore, the implications of SUE are different for different types of systems. Therefore, when any comparison is attempted, it should be done with careful attention to the reference being used.

ITU-R Recommendation SM-1046-2 [B19] on spectrum management presents two different methods for calculating SUE.[35] The methods are based on characterizing the bandwidth, space, and time "volume" denied to other users of spectrum because of the transmissions from the system being considered. These are presented in this subclause along with additional material, relevant when analyzing dynamic and adaptive spectrum management systems.

Transmitters and receivers both use spectrum space. Transmitters deny the use of their spectrum space to receivers other than their intended receiver. This is called "transmitter-denied space." Transmitters generally do not deny spectrum space to other transmitters. However, denial is possible if a band requires "listen-before-talk" etiquettes, or if a cognitive radio is required to defer to certain types of transmitters.

Receivers use spectrum space by requiring that certain spectrum space be protected from use by transmitters other than the transmitter for that receiver. A different transmitter operating in the protected spectrum space would interfere with the receiver's operation. This space is called "receiver-denied space." Receivers generally do not deny spectrum space to other receivers. But, many devices function as transceivers and, therefore, have requirements and potential impacts for their receiver and transmitter functions. As one example, many communication protocols require the receiver to acknowledge the successful receipt of a message with a transmission back to the sender.

When comparing two proposed systems, it is often sufficient to calculate the relative spectrum efficiency (RSE). The advantage of calculating the RSE is that it is often less complex to

---

35 ITU-R Recommendation SM-1046-2 [B19] provides the first definition and calculation method, and ITU-R Recommendation SM-1599 [B20] provides the second. These ITU-R recommendations may be referenced for a more detailed treatment of each method.

calculate and can be stated with greater certainty. An RSE calculation will often drop many variables from a full SUE calculation because they are identical for both proposed systems.

15C.1.1 ITU-R Method 1 for Calculation of SUE

ITU-R method 1 for calculation that defines spectrum utilization is determined by the amount of frequency, geometric space, and time that is used and may be calculated using Equation (15C.1).

$$U = B \times S \times T \tag{15C.1}$$

where

U is the amount of spectrum space used ($Hz \times m^3 \times s$)
B is the frequency bandwidth
S is the geometric spaced (desired and denied)
T is the time

Equation (15C.1) identifies three dimensions of spectrum utilization: frequency, space, and time. These dimensions suggest the variables that affect spectrum utilization. However, the three dimensions are not exclusive. Other variables may be introduced to make transmissions orthogonal or otherwise allow them to interoperate. When new variables are introduced, Equation (15C.1) should be revised to reflect the impact of those variables.

The frequency variable includes the effect of the RF and IF bandpass filters, transmitter modulation, with any orthogonality included, transmitter-occupied bandwidth, off-frequency rejection characteristics, signal processing, coding, and allowable signal-to-interference ratios. Harmonic and other spurious responses are also included. In summary, all variables that affect the frequency-related interaction of radio systems should be included in this variable.

The space variable includes the factors that affect the geometric space used by a transmission. This includes the pointing angles, antenna patterns of both the transmitting and the receiving antennas, as well as factors related to terrain and signal propagation. Although space is inherently three-dimensional, at times less than three dimensions are of interest. The shape of the space of interest may not be spherical; it may be hemispherical or, as is often the case with directional antennas, cone shaped.

The time variable will include the duty cycle for TDMA or packet-based systems and may include items such as the rotation of a radar antenna.

Spectrum utilization efficiency is the ratio of the useful effect to the amount of spectrum utilized. The SUE is then calculated from:

$$\text{SUE} = \boldsymbol{M}/\boldsymbol{U} = \boldsymbol{M}/(\boldsymbol{B} \times \boldsymbol{S} \times \boldsymbol{T}) \tag{15C.2}$$

where

M is the useful effect of the system[36]
U is the amount of spectrum utilization

M is the quantified value of the system's useful effect. The useful effect for some systems is quantified as the amount of user information transferred. For other types of systems, useful effect is quantified in different ways, relevant for the purpose of the service. For instance, a garage door opener would be best characterized by the number of open or close commands successfully sent and not the number of bits used to transmit these commands. In voice systems, the traffic measure of interest is the Erlangs of traffic. Or, M could represent a more general concept such as economic or social value. For instance, we can compare the revenue potential of two services per unit of spectrum utilization.

---

36 Note that ITU-R Recommendation SM-1046-2 [B19] provides the same equation for SUE as the ITU-R National Spectrum Handbook. However, ITU-R Recommendation SM-1046-2 defines the parameters "M" and "U" as follows:
M is the useful effect obtained with the aid of the communication system in question
U is the spectrum utilization factor for that system

15C.1.2   ITU-R Method 2 for Calculation of SUE

The second ITU-R method is based on a special procedure for redesignating the frequencies of operating radio stations and uses Equation (15C.3) to calculate the spectrum utilization. This method is very different from the first ITU-R method because it is attempting to provide an approach that is independent of frequency band or the type of service.

$$Z = \Delta F / DF_0 \tag{15C.3}$$

where

ΔF is the minimum necessary frequency band to permit the functioning of the operational facilities of interest.

ΔF_0 is the frequency band being analyzed.

The lower limit for spectrum utilization is achieved by determining the ΔF of the optimum or near-optimum frequency use algorithm. Actual values of ΔF_0 are determined by analysis of the actual frequency assignment data.

This method allows comparison of different frequency bands, even those used for different services. This method does not require identifying a standard service or resource to serve as a basis of comparison.

The SUE is then calculated using Equation (15C.2) which incorporates the measure of the useful effect, selected for use.

15C.2   Relative Spectrum Efficiency

SUE calculations may be performed for different systems of the same type and compared to obtain the relative spectrum efficiency (RSE) of the systems. Often, the RSE is much easier to calculate than the SUE because the ratio of two SUE equations will remove the identical and unchanging variables leaving a simplified form of the equation to be computed.

Alternatively, RSE may be computed based on a standard system. All candidate systems are then compared with the standard system, as shown in Equation (15C.4).

$$RSE = SUE / SUE_{std} \tag{15C.4}$$

where

SUE   is the spectrum utilization efficiency of the source system

SUE_{std} is the spectrum utilization efficiency of the "standard" system

Commonly chosen candidates for the standard system are as follows:

— The most efficient system that can be built, on a practical basis
— A system that is easily defined and understood
— A system in wide use or a system defined by a real or *de facto* industry standard

Care should be taken when using RSE to include all the relevant variables. An example of how other variables may be omitted can be seen when comparing systems using different modulations. A criterion is based on the comparison of the frequency bandwidth (F_c) required to transmit a given volume of information in a real radio system and an optimum frequency bandwidth (F_{opt}) of an ideal radio system. The RSE might be reduced to a comparison based on this single variable, as shown in Equation (15C.5).

$$RSE = F_c / F_{opt} \tag{15C.5}$$

If a calculation were performed for a 256-QAM system and compared with a 16-QAM system, the conclusion would be that the 256-QAM system is about two times as efficient as the 16-QAM system. However, when the signal-to-noise ratio (S/N) requirements of the two systems are considered, it is revealed that the 256-QAM system requires a substantially higher S/N. The requirement for higher S/N and more freedom from interference of the 256-QAM system may cancel any benefit received from its more efficient modulation. In some environments, the 16-QAM system may well be the more efficient system. This example illustrates how oversimplification and focusing exclusively on too small a set of variables may lead to misleading results.

In cellular this concept is captured using a Erlangs/(MHz km^2) as the measure. This captures the traffic carried per unit area per unit bandwidth. It explicitly captures the tradeoff among frequency reuse, cell sizes, and modulation efficiency. For instance, CDMA systems use inefficient modulation that is more than compensated by their ability to reuse channels in every sector of every base station. Measures of efficiency, such as this one for cellular spectrum efficiency, demonstrate that more variables than just frequency may be required to get an accurate result.

Annex 15D

(informative)

Sample Analysis—selection of Listen-before-talk Threshold

15D.1   Executive Summary

[NOTE—The reason for this example:

This sample analysis demonstrates the pivotal importance an early assumption can make in an analysis. In this example the analysis is seeking to determine the optimum value for a threshold in a listen-before-talk, least-interfered channel protocol. The frequently, and often unstated, assumption is that range is the parameter to be optimized. However, density of devices in some use cases is a more important variable than range. The issue becomes, is range or density of devices more important? If range is selected as the principle value, a threshold of 20 dB less will be selected, than that which would be selected if density of devices were given the highest value.

This example also shows management of interference vs total avoidance of interference. In this case a far greater density of devices can be supported if a higher threshold is allowed. However, the densely populated devices would lose range during times of high use, necessitating additional base stations to support the entire population. If it can generally be assumed that densely located devices will normally be under the control of a single organization, then the organization that is receiving the benefit of the densely populated devices will also bear the cost of additional base stations. Thus the tradeoff becomes a network administrative issue and should be left to be optimized by the network administrator rather than be given a fixed value. If the installation of additional base stations is deemed a reasonable cost for the value of being able to support a high density of devices, then the higher threshold would be a reasonable choice. However, that conclusion depends on the validity of the assumption that in the vast majority of cases densely populated devices will be controlled by the same organization.

Alternately, a regulating authority may require some further protections that a single organization control exists before allowing a relaxed threshold. In a policy defined implementation, the regulator could allow the network administrator to adjust the threshold under certain conditions could be met. For example, the network administrator or in an automated implementation, the device or network, could determine if all the densely populated devices were under the control of the same network and only in that case allow the higher population density. So if one organization wanted to put a device in every cubical and the control logic could confirm this, then a more relaxed threshold could be allowed. If the local devices were under the control of differing organizations, then the logic could be required to utilize a lower threshold.]

In systems that use a listen-before-talk, least-interfered-channel protocol, there is a defined monitoring threshold above which transmission is not allowed. Before a unit is allowed to transmit, it is required to listen to its desired transmission channel. If the unit senses energy above the defined threshold, it is required to either wait for that channel to clear or move to a different transmission channel.

An example of such a protocol is given in the FCC rules for the Unlicensed Personal Communications Services band. The relevant sentence of 47CFR15.323(c)(5) [B8] reads as follows:

(5) If access to spectrum is not available as determined by the above, and a minimum of 40 duplex system access channels are defined for the system, the time and spectrum windows with the lowest

power level below a monitoring threshold of 50 dB above the thermal noise power determined for the emission bandwidth may be accessed.

The issue is as follows:

What is the optimum value for the monitoring threshold, set in this FCC rule at 50 dB above the thermal noise power? Stated alternately, what is the best compromise between range and density of devices?

This analysis will show that the FCC selection of a threshold at 50 dB above the thermal noise is a credible choice when optimizing for range. However, to optimize for density of devices, a threshold of as much as 70 dB above the thermal noise yields better results.

15D.2   Findings

This analysis will demonstrate that although a monitoring threshold of 50 dB above thermal noise is valid, assuming range is the primary issue. A value of 70 dB is justified so as to also support dense installations. The analysis argues that in almost all cases densely populated devices will be under the control of a single organization, which should be afforded the option of optimizing range versus density of base stations to meet the needs of its network.

15D.3   Scenario Definition

15D.3.1   Study Question

This analysis is being performed to support the setting of a monitoring threshold in a listen-before-talk, least-interfered-channel etiquette. The analysis is seeking to find a means for supporting densely populated devices.

15D.3.2   Benefits and Impacts of Proposal

Many organizations have workers located in densely packed cubicles. When it is desirable to equip those workers with similar wireless devices, e.g., cordless phones or wireless headsets, the density of devices can exceed the available channels, creating an access problem during high usage periods. By raising the monitoring threshold, a greater density of devices can be supported. The cost is that devices further from a base station will receive interference from other devices and lose effective range. However, if the monitoring threshold allows the greater density of devices, the access issue can be addressed by installing additional base stations. Therefore, the situation becomes a network planning issue. The primary benefit of this analysis is that it preserves for an organization the option of installing a denser population of devices.

15D.3.3   Scenario(s)

The scenario being analyzed is that created by a listen-before-talk, least-interference-channel spectrum etiquette.

15D.3.3.1   Frequency Relationships

Of the frequency relationships, only the in-band, co-channel relationship is considered. The threshold being analyzed only affects the frequency reuse decision for co-channel devices. It is assumed that other requirements give adequate protection for adjacent channel and out-of-band devices.

NOTE—In this sub-section, a matrix reduction step is combined with the analysis of frequency relationships. The analyst demonstrates awareness of other frequency relationships, such as adjacent channel and out-of-band devices but states a conclusion that these are not relevant for this scenario).

15D.3.3.2 Usage Model

Three usage models will be considered in this analysis. The first and baseline case is a single lightly populated installation where maximum range is desired. This baseline case will be compared with a second case where a single densely populated system is operating under the control of a single entity. A third case is that of multiple-entity operating systems in close proximity, such as in an office or apartment building.

Although some use scenarios should be optimized for distance, in other use scenarios, it is preferable to subordinate range for density of devices. In some use models, it is preferable that several devices are able to operate in close proximity, and density of devices is preferable to range.

There are situations where it is desirable to have a number of devices operating in close proximity. An example of such an operating environment would be a cubicle (partitions between offices that do not fully extend to the ceiling of the building) office environment where every cubicle might have a wireless device in it. In such a scenario, each device would lose range due to the density of spectral use. However, in such dense systems, it is common practice to install a system in which devices may operate a short distance from the nearest base station, and in this way, the loss of range has little if any effect.

15D.3.3.3 Characteristics of Usage Models

The three use cases are assumed to share the following characteristics:

a) There is a listen-before-talk, least-interfered-channel requirement for all the devices.
b) There is a 10 MHz wide frequency band and devices with emissions bandwidth of slightly less than 2 MHz. Hence, there are 5 available transmit frequencies.
c) The devices operate under a protocol similar to DECT. The DECT transmission protocol uses TDMA techniques with symmetrical TX and RX timeslots on a 24 timeslot frame, 12 TX slots and 12 RX slots in each frame.

The equipment being used is assumed to be typical home or office devices, primarily telephones or wireless headsets. The usage is assumed to be to support typical office or home telecommunications services.

Spatial and Power Limits The devices are assumed to be operating under relatively low power requirements, between 10 mW and perhaps 300 mW. Devices may be located arbitrarily, and the devices are mobile.

Temporal Limits It is assumed for this analysis that the band uses a 10 mS frame. It is also assumed that a device is required to monitor a channel for 1 frame period, 10 mS, and find it clear before it can use the channel.

It is then assumed that at maximum loading, the devices are in use 70% of the time; that is that the probability of any one device being used at a given time is 70%. It is also assumed that while in use, each device transmits data in every time slot available to it.

In most cases it is sufficient if remedies can be provided on a temporary basis in minutes and on a permanent basis in a few days.

Frequency Characteristics As stated in 15D.3.3.3, the systems are assumed to operate in a 10 MHz wide frequency band and each of the devices has an emissions bandwidth of slightly less than 2 MHz. There are, therefore, 5 available transmit frequencies.

Other Orthogonal Variables No other orthogonal variables are critical to this analysis.

15D.3.3.4 System Relationships

Systems Considered The systems are assumed to be systems operating according to the requirements of the band. For this analysis, other characteristics do not impact the analysis.

Protection Distance A protection distance of 0.4 m is selected for this analysis. The basis for this protection distance are use cases, such as a tight cubical environment or users on public transportation. In such environments, users may be only 0.6 m apart. Therefore, an individual user should be able to separate two devices at least 0.4 m apart. Even where the devices are being used by different, adjacent users, normally a 0.4 m separation may be arranged with relative convenience.

Geographic Area for Analysis This analysis is assuming relatively low power devices, operating between 10 mW and 300 mW. The geographic area for analysis is line of sight and obstructed line of sight as determined by the operating power. The maximum operating range considered is 1000 m with a focus on operation under 500 m.

Impact of Interference For the cases under study, the only impact of interference is that a channel is not available to another system. Because all devices are operating in a listen-before-talk protocol, they will not impact each other once a device has gained the right to transmit on a channel. However, the threshold levels do affect which channels are available for use by other systems.

Interference Mitigation No interference with voice transmission or dropped calls is ever desirable. However, this analysis assumes that if there are temporary remedies available to the user within minutes and permanent remedies available, at reasonable cost, within days, then these are acceptable.

When a user experiences interference with a voice transmission, the following remedies are immediately available:

a) Move away from the interfering device.
b) Request that the interfering device be moved or its use discontinued.
c) Reinitiate the call; at which time, the system will probably locate a different frequency and time slot.

Beyond these temporary remedies, a permanent remedy is to install a higher density of base stations.

Baseline The comparative context is the established 50 dB over thermal noise limit established with range as the primary objective. The analysis will study the impact of higher limits on range, density, and interference that exists for this baseline case.

15D.3.4 Case(s) for Analysis

There are then two cases for analysis, the in-band, co-channel interference for the baseline case and the case with an elevated threshold.

15D.4 Criteria for Interference

15D.4.1 Interference Characteristics

Interference that rises to the level of disrupting transmission is of primary interest. At that level of interference, data may be lost resulting in a number of effects. Although best-effort data service will require retransmission of the lost data, the largest impact will be on real-time services, e.g., a telephone call. Under the use scenario described, worst-case interference could create interference with voice calls due to lost packets. In the extreme worst case, calls could be dropped.

15D.4.1.1 Impacted Level

The impacted level for this case will be the received signal strength of the interfering signal.

15D.4.2 Measurement Event

The measurement event will be defined as one channel for one frame period. The frequency of each measurement event will be centered on a frequency channel and have a bandwidth equal to the transmission bandwidth, approximately 2 MHz. Each measurement event will last for one frame period, 10 mS.

15D.4.3 Interference Event

For the analysis, an interference event will be defined as a device being denied use of a channel it is monitoring for transmission.

15D.4.4 Harmful Interference Criteria

It is proposed that harmful interference is deemed to occur if, under the worst-case loading of 70%, with devices spaced at 0.4 m, a device is unable to find an available channel or if a device would continuously transmit on a channel at a level that would cause audible interference for the user of another device.

15D.5 Variables

The relevant variables are the monitoring threshold, distance, power, and time.

The contrasting variable is the monitoring threshold. All other variables are assumed to be identical.

15D.6 Analysis—modeling, Simulation, Measurement, and Testing

15D.6.1 Selection of the Analysis Approach, Tools, and Techniques

This analysis is performed using fundamental calculations. It is believed that a more complex analysis is not required to explore the concept being examined. A number of variables, such as using a more complex propagation model or device usage model, are not included. However, for the purposes of this study, more complex analysis is not deemed necessary.

15D.6.2 Matrix Reduction

To simplify the analysis, the matrix of possible use scenarios will be reduced to a cubical environment with heavy voice usage of identical wireless devices, such as cordless phones or cordless headsets.

15D.6.3 Performing the Analysis

Simulations can be developed for high traffic density open areas (e.g., large office landscapes and exhibition halls with close to free space propagation). These simulations can show the impact of different monitoring thresholds on device density. Figure 15D.1 is a simulation of a system covering a three-floor 100 × 100 m building. There are 25 equally spaced base stations on each floor (20 m base station separation). The system has 60 duplex access channels (5

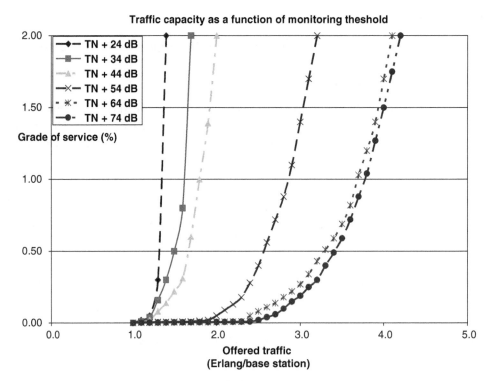

Figure 15D.1 Capacity as a function of the monitoring threshold limit (free space model of 120-system access channels). From IEEE Recommended Practice for the Analysis of In-Band and Adjacent Band Interference and Coexistence Between Radio Systems, Mar 2018. IEEE.

carriers with 12 duplex channels each) on a 10 MHz spectrum allocation. Moving portables, intracell and intercell handover are included in the simulation.

Figure 15D.1 shows that for this specific simulation, the system capacity (1% grade of service limit) would increase by at least 60% if the monitoring threshold is changed from TN + 50 dB to TN + 65 dB.

For cases where access channels are relatively limited, it is even more important that an appropriate monitoring threshold be used. In such scenarios, devices have relatively few channels to which they can escape. In the example cited, the device has 5 frequencies available and 60 access channels. If the monitoring threshold is too low, it will restrict use of channels that are perfectly useful for communication. In dense usage environments, there would be a loss of range. However, range in such environments is not the critical component and is typically compensated for by providing additional base stations to service the area.

15D.6.4 Quantification of Benefits and Interference

This analysis has been prepared to look and explore the impact the underlying assumption will have on establishing a listen-before-talk threshold. The insight gained may lead to more flexible protocols. If use environments can be identified by devices with sufficient certainty, then more appropriate thresholds might be allowed. In an environment where range is the primary concern, then a lower threshold would be used. However, in an environment where density of devices is the primary concern, a high threshold would be selected, trading density for range. The ultimate benefit would be to allow operation that is more appropriate for a specific use environment.

15D.6.5 Analysis of Mitigation Options

The need for mitigation in an environment that focuses on range is primarily on the user of the device. When range is of primary interest, a lower threshold will be selected. This increases the possibility that few or no channels will be available to a device. The user of the device will then be presented from using the device, or its operation may be somewhat erratice, with it at times finding a qualifying channel in which to operate and at other times not finding one.

When the environment qualifies for density of devices, one entity has control of all the devices. This is a critical component of what qualifies an environment for density of devices. The entity, perhaps a company, can mitigate the loss of range by installing more base stations. It would not do this unless it found the benefit of having a number of devices operating in close proximity to justify the added expense.

15D.6.6 Analysis Uncertainty

This analysis is analytical, not experimental, and therefore, there is not measurement uncertainty. The primary uncertainty with an analytical analysis is that simplifying assumptions may not be valid. This is not believed to be the case in this analysis, but it could be true.

Additional uncertainty is introduced by not using a more realistic and complex use model. In reality some transmissions will not overlap in time and others will only partially overlap. On the boundaries, some transmissions will commence just as a device finishes its monitoring and prepares to transmit on a channel it believes to meet the threshold criteria.

15D.7 Conclusion and Summary

Thus, having an upper threshold of "thermal noise floor + 65 dBm" would considerably increase the utilization (+60 %) and decrease infrastructure costs for high-capacity installations.

15D.7.1 Benefits and Impacts

The primary benefit of implementing a higher threshold is that the number of simultaneous users would be increased greatly. The impact, beyond a loss or range, would primarily be in situations that were qualified to be optimized for density of users but in reality should not have been. An example would be a small office environment with different tenants operating separate systems in close proximity. One user may lose range due to the operation of a neighbor's system.

15D.7.2 Summation

This analysis has demonstrated that if range is assumed to be the primary variable to be optimized when establishing a listen-before-talk threshold, the threshold may be set at a level that is lower than necessary. A more flexible approach will consider both range and density and may allow the use of multiple thresholds, if a means can be found to qualify environments for appropriate use of a threshold.

Annex 15E

(informative)

Sample Analysis—effect of Out-of-band Emissions on a LBT Band

[NOTE—This scenario is included as an example of how different spectrum management principles can affect each other, even across a band boundary.

This analysis is looking at the potential for interference in a very specific condition. How commonly this condition will exist is not examined. The structure of the analysis makes explicit how many limiting assumptions are being made, which is the value of the structure.]

15E.1 Executive Summary

The issue addressed in this analysis is the effect of an out-of-band emission limit on a nearby band using a listen-before-talk protocol. It demonstrates that if not carefully crafted, rules for different but adjacent bands can profoundly influence each other. In the case being considered, one band has a relatively typical out-of-band emission limit and an adjacent band is using a listen-before-talk protocol. The question is as follows: "How often could out-of-band emissions from one band block a device in the adjacent band from transmitting because its out-of-band emissions are above the threshold?"

15E.2 Findings

What is found is that the out-of-band emissions from a single device has the potential for blocking the use of the first megahertz of the adjacent band for over 10 m and can block an entire 10 MHz band for over 3 m.

15E.3 Scenario Definition

15E.3.1 Study Question

The question being asked is as follows: "What is the effect of an out-of-band emission limit in one band on an adjacent band, using a listen-before-talk protocol?"

15E.3.2 Benefits and Impacts of Proposal

The benefit of this analysis is to guide the more judicious crafting of rules so as to avoid interference to adjacent bands. The consequences could be to limit the spectrum management principles recommended for use adjacent to each other, to have more restrictive out-of-band emission limits, or raise the transmission threshold in a listen-before-talk band.

15E.3.3 Scenario and Usage Model

Assumptions of this scenario are as follows:

a) Devices in both bands are assumed to be consumer products that can be expected to be used in close proximity to each other.
b) Typical use environments for these devices are offices, factories, and homes with typical separation distances of 1 m to 5 m.
c) The out-of-band requirement measures the allowed out-of-band emission using 1% of the transmit emission bandwidth filter in the first megahertz beyond the band edge and a 1 MHz bandwidth filter for frequencies beyond 1 MHz from the band edge.

15E.3.3.1 Frequency Relationships

This analysis looks at a single frequency relationship, the effect across a band boundary with one device transmitting and a device in the adjacent band, using a listen-before-talk protocol monitoring a channel in an attempt to transmit.

15E.3.3.2 Usage Model

A single usage model will be considered, two portable consumer devices, commonly used in close proximity to each other, such as in a home or office environment. An example would be a mobile phone and a cordless phone.

15E.3.3.3 Characteristics of Usage Model

This usage model is characterized by two devices that are both portable. Both are assumed to be transmitting voice and so have real-time connectivity requirements.

Spatial and Power Characteristics It is assumed that the devices may be used closer than 1 m from each other, such as in a cubical environment. How frequently two devices would be used in close proximity is not determined. It is assumed that this would be a reasonably common condition.

The two devices are assumed to have an unrestricted, line-of-sight condition to each other. It is recognized that architectural barriers may attenuate the signal of one device to the other. For this analysis, only the line-of-sight condition is treated.

> NOTE—A matrix reduction step has been incorporated, limiting the analysis to only considering the condition when the two devices are used in close proximity and in a line-of-sight condition.

The transmitter, whose out-of-band emissions are being considered, is assumed to be a mobile phone with a transmit power of up to 2 W.

The analysis assumes an out-of-band transmit power of −13 dBm/MHz[37] would be permitted.

The interference level can be expressed as the equivalent level above Thermal Noise floor, TN. TN is −114 dBm for 1 MHz bandwidth.

Using the thermal noise floor as a reference, the assumed out-of-band emission limit of −13 dBm/MHz can be expressed as TN + 101 dB.

Within the first megahertz of the LBT band, the allowed out-of-band transmit power from a device in the neighboring band is −13 dBm/1% of B, where B is the bandwidth of the device transmission. If B = 1.25 MHz (as for CDMA 2000), the allowed interference becomes −13 dBm/12.5 kHz. TN is −133 for 12.5 kHz. Thus, −13 dBm/12.5 kHz can be expressed as TN + 120 dB.

In summary, the out-of-band emissions can be TN + 120 dB in the first megahertz and TN + 101 dB in the remainder of the band.

---

37 −13 dBm/MHz is a value used in some frequency bands.

Temporal Characteristics This analysis is restricted to the condition where one device is transmitting and the recipient device is monitoring a channel in an attempt to transmit. How frequently this condition exists is not considered.

> NOTE—A matrix reduction step has been incorporated, limiting the analysis to only considering the condition when one device is transmitting and the other is monitoring a channel in an attempt to transmit.

Frequency Relationships The transmitting device is assumed to be operating near the band edge. The listen-before-talk band is assumed to be 10 MHz wide, and the impact of the out-of-band emissions is considered on the entire band.

The purpose of this analysis is to evaluate the impact of an out-of-band emission limit, and so it is assumed that a transmitting device may be putting energy into the entire adjacent band at the level set by the limit.

> NOTE—Two matrix reduction steps have been incorporated, limiting the analysis to only considering the condition when the transmitting device is near the band edge and further assuming that its out-of-band emissions are at the allowed limit over the entire adjacent band.

Other Orthogonal Variables No other orthogonal variables are being considered.

> NOTE—Yet another matrix reduction step is incorporated. Other variables are not considered, which may affect this situation, such as the degree to which the antennas on the two devices are cross polarized and how frequently intervening barriers will shield their transmissions from each other.

15E.3.3.4 System Relationships

Systems Considered The only system relationship being considered is that which exists between a typical mobile phone operating at 2 W across a band edge to a device monitoring a channel in an attempt to transmit. The characteristics of the devices are not relevant as this analysis is looking at the impact of the out-of-band emissions limit, which is common to all devices in one band, on the monitoring threshold, which is common to all devices in the adjacent band.

Protection Distance It is assumed that the two devices should be able to operate within 0.3 m of each other without impact.

Geographic Area for Analysis The geographic area being analyzed is relatively close, within 10 m to 20 m. The devices being analyzed are low-power devices with limited transmission range. The source device is a higher power device with a potential range of a few kilometers.

Impact of Interference The impact of the interference is to deny the use of one or more channels to the listen-before-talk device.

Interference Mitigation In this scenario, the listen-before-talk device will monitor a different channel after finding one channel blocked. If all channels are blocked, it will not be able to transmit.

The interference could be remedied by moving the devices away from each other or waiting until one transmitting device stops transmitting.

Baseline No baseline interference is assumed. This analysis is only looking at the additional impact from the single variable considered.

15E.3.4 Case(s) for Analysis

A single case is proposed for analysis. A mobile phone is assumed to be transmitting at 2 W in the channel nearest the band edge. In the adjacent band, a device is monitoring a channel, using a listen-before-talk (LBT) protocol.

15E.4   Criteria for Interference

15E.4.1   Interference Characteristics

Interference is characterized as energy that is gathered in to the recipient device while monitoring before transmission.

15E.4.1.1   Impacted Level

The impacted level is energy at the receiver input.

15E.4.2   Measurement Event

The measurement event is 1 channel bandwidth wide and 1 monitoring period in duration. That is, it is the monitoring period leading to a decision to transmit.

15E.4.3   Interference Event

An interference event is any measurement in which the threshold is exceeded and transmission denied. The monitoring threshold is assumed to be TN + 50 dB. Anytime the out-of-band emissions are above TN + 50 dB during a monitoring period will be considered an interference event.

15E.4.4   Harmful Interference Criteria

If more than 10% of the band, 100 kHz, is blocked from use, it is proposed that harmful interference has occurred.

15E.5   Variables

The relevant variables are as follows:

— Transmit power
— Out-of-band emissions limit
— Monitoring threshold
— Frequency separation
— Spatial separation

The only contrasting variable is spatial separation.

15E.6   Analysis—modeling, Simulation, Measurement, and Testing

15E.6.1   Selection of the Analysis Approach, Tools and Techniques

This analysis uses fundamental calculations to explore the question being addressed.

15E.6.2   Matrix Reduction

Of the possible cases for analysis, a single case is being considered. In this case, a higher power device, capable of transmitting up to 2 W, is operating near its band edge. The adjacent band is operating under a listen-before-talk, least-interfered-channel protocol. It is postulated that this scenario is worst case for the question addressed.

15E.6.3   Performing the Analysis

Assuming free-space propagation, the attenuation at 1 m, 3.2 m, and 10 m is about 38 dB, 48 dB, and 58 dB, respectively, for the 2 GHz frequency range. Table E.1 gives the interference levels experienced in the LBT band.

The interference power is expressed as the equivalent level above Thermal Noise floor, TN, for a transmitter with an out-of-band emission power of −13 dBm/12.5kHz in the first megahertz beyond the band edge and 13 dBm/MHz in frequencies more than 1 MHz from the band edge. This equates to an out-of-band emissions limit of TN + 120 dB in the first megahertz and TN + 101 dB in the remainder of the band.

From IEEE Recommended Practice for the Analysis of In-Band and Adjacent Band Interference and Coexistence Between Radio Systems, Mar 2018. IEEE.

15E.6.4   Quantification of Benefits and Interference

This analysis is performed to investigate a single aspect of interaction between dissimilar spectrum management methodologies. What is shown is that dissimilar methodologies may interact in undesirable ways even though each methodology is entirely acceptable in isolation.

15E.6.5   Analysis of Mitigation Options

Few mitigation options are available for uses in the scenario being explored. The mitigation options are in the hands of spectrum managers and regulators, who have the ability to place compatible systems adjacent to each other.

15E.6.6   Analysis Uncertainty

Since this is an analytical analysis, there is no measurement uncertainty. There is considerable uncertainty as to the preliminary conclusions due to use of simple propogation and use models. The potential for interference, even very significant interference, has been identified. More complex analysis would be required to determine how frequently that interference would exist in more realistic use environments. This analysis proves that there is conceivable interference. Further exploration would be necessary to verify that this interference exists at higher levels of analysis.

15E.7   Conclusion and Summary

15E.7.1   Benefits and Impacts

Reviewing the interference levels of Table 15E.1, we find:

a) The first megahertz at the band edge is not usable with a monitoring threshold of less than 63 dB above thermal noise and loses a great deal of utility if the monitoring threshold is less than 72 dB.
b) If the monitoring threshold is set lower than 53 dB above thermal noise, a single transmitter potentially can block an entire neighboring band for a distance of 3 m to 10 m.

As can be seen in Table 15E.1, out-of-band emissions requirements can have a significant impact on bands utilizing a listen-before-talk protocol. Unless the values of the monitoring threshold and the out-of-band emissions are carefully coordinated, there can be severe impacts on the utility of LBT bands.

Table 15E.1 Interfering power at different separation distances.

Portion of the LBT band	Separation distance between TX devices and LBT band equipment		
	1 m	3.2 m	10 m
First megahertz	TN + 82 dB	TN + 72 dB	TN + 62 dB
>1 MHz from the band edge	TN + 63 dB	TN + 53 dB	TN + 43 dB

15E.7.2 Summation

This analysis has shown that significant interference is conceivable for the case examined. Where this possibility would become a reality requires further analysis using more realistic propagation and use model. Also of significance would be the possibility of identifying mitigations that may be available or reasonably made available to the users operating in such an environment.

Annex 15F

(informative)

Sample Analysis—low-power Radios Operating in the TV Band

15F.1   Executive Summary

This analysis examines whether low-power cognitive radios can operate in the television (TV) bands without causing *harmful interference*. A single cognitive radio, if *properly designed*, is unlikely to have a significant impact to TVs across a broadcast coverage area. However, as the *cognitive radio density* increases, the impact can increase to the point of harmful interference. In this analysis, we define harmful interference in terms of the probability of a TV suffering a reception degradation. We show that a proper design implies a transmitter design (power, modulation, etc.) that enables a cognitive radio to operate at a distance $r_{\min}$ to a TV on the same channel without causing signal degradation; and a cognitive radio TV band occupancy detector that has detection probability $C_{\det}$ that is sufficiently accurate. The source device density D that is supported is a function of these parameters and under the assumptions in this analysis can be quite high.

15F.2   Findings

The analysis indicates that a high cognitive radio device density is supported even with a small fraction of TV receivers affected. Specifically, this fraction is $F = \alpha D$, where D is the density of source devices and based on the analysis: $\alpha = 10^{-7}\frac{fraction\ impacted}{sources/km^2}$. For instance, if $D = 1000$ source devices per km^2 then $F = 1/10\ 000$ TV receivers will suffer interference on average. It would be up to the parties involved to decide whether this impact is significant compared with the benefit. As an example, a 10 km^2 suburban area might have 1000 people per km^2. All of them could be simultaneously operating cognitive radio devices and on average only 1 of the 10 000 people would suffer any TV interference (assuming one TV per person).

15F.3   Scenario Definition

15F.3.1   Study Question

What fraction F of TV receivers averaged over time would be subject to an interference event from lowpower cognitive radio devices operating in the TV bands? A second question is under what conditions F exceeds a threshold? A third question is what factors are under the cognitive radio designers control that can influence these questions?

15F.3.2   Benefits and Impacts of Proposal

Low-power devices with access to large bands of spectrum could provide opportunities for greater connectivity and opportunities for new applications. Consumers, equipment makers, and new service providers would all benefit from these opportunities. Consumers who still receive TV over the air could experience a greater level of interference than before causing them to not watch certain channels or programming. This in turn could impact TV broadcasters whose audience-based revenue is reduced.

15F.3.3 Scenario(s) and Usage Model

In this analysis, a TV receiver is the *recipient device*. The *recipient system* is the set of TV receivers in the coverage area of a TV broadcast station. The *source device* is the low-power cognitive radio. The *source system* is the set of low-power cognitive radios and supporting infrastructure in the coverage area defined for the recipient system.

15F.3.3.1 Frequency Relationships

The source devices are operating in the broadcast TV bands. The source devices only use the 29 UHF channels 21–51 (excluding channel 37, which is assigned to radio astronomy). This is for two reasons. First, this block forms a contiguous (except channel 37) region from 512 MHz to 698 MHz that simplifies the transceiver and antenna design on the source devices. Second, channels 20 and below have greater usage than channels above, and these lower channels also include secondary assignments for public safety and land mobile in some cities. Avoiding these lower channels will simplify the general problem of avoiding TV and other receivers. Although other recipient receivers are considered in defining our use of channels 21–51, further analysis in this section only considers interference with TV receivers. In particular, interference with wireless microphones is not considered.

Cochannel interference is considered in a later section in detail. Adjacent channel interference is based on ATSC (Advanced Television Systems Committee) A-74 guidelines,[38] which give the relative isolation between adjacent channels separated by i channels; see Table 15F.1.

From IEEE Recommended Practice for the Analysis of In-Band and Adjacent Band Interference and Coexistence Between Radio Systems, Mar 2018. IEEE.

Thus, if a signal of power T_s is sufficient to cause cochannel interference for a TV receiver on channel 24, then a power of $T_i = T_s + 74.2$ dB is sufficient to cause adjacent channel interference for a TV receiver on channel 26 (i.e., $i = 2$).

The source devices use a mechanism for detecting the usage of broadcast channels by the recipient devices in an area. A number of mechanisms have been proposed. For the analysis here, we simply assume that this mechanism can detect channel c with a probability of at least C_{det} independent of location in the area. If it does detect, then it will avoid using the same or adjacent channel. If it does not, it will use a random channel. Note that even if detection is perfect ($C_{det} = 1$), interference is still possible. At two or more channels away, a source transmitter sufficiently close to a recipient receiver can desensitize the receiver causing signal degradation.

We do not consider band edge and out-of-band issues.

Table 15F.1 Isolation between DTV (digital television) channels i channels apart.

i	T_i/T_s (dB)
0	0.0
±1	48.5
±2	74.2
±3	78.2
±4	84.2
±5	86.2
±6	80.2
±7	87.2
$|i| > 7$	90.2

---

38 Advanced Television Systems Committee [B1].

15F.3.3.2   Usage Model

The TV receivers can be connected to cable or operating over the air. We consider ordinary direct viewing or recording of signals by humans and not any machine monitoring of the audio, video, or hidden signals. Over-the-air TV antennas can be indoor or outdoor (roof or balcony) mounted. We consider in this model a dense suburban model. The recipient device density is not important since we focus on a single "typical" recipient and analyze the expected fraction of recipient devices impacted. We specifically avoid in this analysis three-dimensional urban environments. For simplicity of exposition so as to have a single example, we focus only on digital ATSC TV receivers. Analog (NTSC or PAL) and other digital (DVB-T or ISDB-T) receivers could be incorporated relatively easily. For the analysis, we will focus on a broadcast signal on a randomly chosen channel c in the range 21–51.

The source devices are assumed to be operating over relatively short ranges. They may be on continuously but have more or less channels active over time. The devices are portable and can be carried from place to place. A typical application is Web surfing. The traffic is sent via packet-based protocols similar to IEEE Std 802.11™ or IEEE Std 802.16™. The traffic may be point-to-point or point-to-multipoint. It may also be part of mesh network relaying. The spatial pattern of the traffic is not relevant here as we will focus on the impact of the source (low-power device) transmitters on the recipient (broadcast TV) receivers.

15F.3.3.3   Characteristics of Usage Model

For the TV receivers, we consider a mix of cable TV and over the air TVs. Although most households have cable TV, many households have additional TVs that are not connected to cable. Of those that are receiving over the air, only a fraction is receiving on channels 21–51. Finally, only a fraction of TVs are turned on and being viewed or recorded at a given time. We let F_R be the fraction of TVs (recipient devices) that are on, receiving via broadcast, and tuned to a channel in 21–51, $(0 \leq F_R \leq 1)$.

Source devices send data via packets. This implies noncontinuous transmission. Digital TV signals use error-correcting codes and can tolerate some errors. Occasional isolated packets from source devices are unlikely to cause a significant increase in decoded error rates. It is only when there is sustained activity that can potentially overwhelm the ability to correct errors and visible errors to the signal can be observed by the TV viewer. For typical Web browsing where activity downloading a page can be followed by many seconds or minutes of viewing, sustained activity can be a relatively small fraction of the time. We define F_S as the fraction of time that the source device is on and has significant activity that could cause observable interference to the TV signal, $(0 \leq F_S \leq 1)$.

Together the fraction $E = F_S F_R$ determines the number of potential interference events that are eligible. A case of interference is not counted if the source is off, the recipient is off, the source is not active, the recipient is not on channel 21–51, or the recipient is using cable instead of broadcast. Each of these events is considered independent of each other.

Spatial and Power Characteristics   The source and recipient devices are uniformly distributed in a given area A. The area A is defined by the nominal coverage of the randomly chosen channel c. The TV receivers are fixed, and the source devices can move and turn on at different places. The analysis does not use mobility and is based on a typical snapshot of locations at a particular time. The maximum transmit power of the source device is incorporated indirectly as is described below. Antennas can be directional on both the source and recipient devices. The source transmit antenna and the recipient receive antenna are assumed to be pointing at the horizon with random azimuth angles relative to each other. The source devices can use power control. The power control is used to minimize the power consumption of communication between source devices and is independent of the recipient devices.

Temporal Characteristics The usage of the source and recipient devices varies over time. In the model, this would be represented by a time varying fraction E of eligible interactions. In the analysis, we consider the maximum E averaged over the busiest (i.e., worst-case) hour in a typical week.

The recipient signals consist of a digital stream of image data. The source signal is packetized data sent from different sources on a shared channel over time. The cognitive radios may abandon channels and choose new channels over time. However, we consider the impact over a short interval that is assumed to be shorter than the typical channel hold time.

Frequency Characteristics Source and recipient channels are 6 MHz. The TV recipient system is the digital ATSC TV system using channels 2–51. The source system is restricted to channels 21–51 (excluding channel 37). The source system will use any channel that is not cochannel or adjacent channel with a detected recipient channel.

Other Orthogonal Variables The orthogonal parameters are not modeled directly but are captured in the parameter r_{min} below.

15F.3.3.4 System Relationships

Systems Considered This analysis only considers a TV recipient system and low-power cognitive radio source devices.

Protection Distance As a baseline we consider the minimum separation distance r_{min} required so that an active cognitive radio source will not cause visible or audible degradation to the TV signal in the worst case. The worst case is defined as follows:

a) The recipient device is at the edge of the defined broadcast coverage area.
b) The source device is transmitting at its maximum power.
c) The antennas of source and recipient device are pointed at each other (if directional).
d) The source device is transmitting on the same channel as the recipient device is receiving.

Lab measurements suggest that for a source device with noise-like signal, maximum EIRP power of 20 dBm and omnidirectional antenna communicating near the ground, $r_{min} = 100$ m.

An alternative way to look at r_{min} is that it is a specified limit on the cognitive radio device. The source radio designer must choose a max power level, modulation scheme, etc. so that in the worst case [in the sense of item a) through item d)] it will not cause signal degradation to a TV receiver tuned to channels 21–51 at a distance of r_{min} or greater. In this sense, r_{min} can be any specified value for our analysis.

Geographic Area for Analysis The area for analysis is the TV broadcast coverage area A of a hypothetical TV broadcast channel c. The area is assumed to be circular. The radius of the circle depends on the TV transmit power. However, in this analysis, what is important is the strength of the signal at the edge of the circle since this is where the TV is most susceptible. However, r_{min} already captures the susceptibility of the signal at the edge, and furthermore, our analysis is normalized to a dimensionless form so that the absolute size of A is not needed. The interaction distance between the source and recipient devices is less than r_{min}. Typical broadcast coverage areas are tens of kilometer and are much larger than r_{min}. As a result, we assume that all recipient devices capable of interfering with recipient devices are within A.

Impact of Interference A source device that is broadcasting continuously will cause the audio or video signal to degrade for recipient TV viewers if it is close enough. However, the source device traffic will be bursty and intermittent for typical Web browsing. The effect will be occasional lost or broken picture frames or gaps in the audio signal. Both of these would annoy TV viewers. If persistent, the viewer would switch channels.

Interference Mitigation TV's suffer interference from a variety of sources already. Devices such as power drills and computers can interfere if they are sufficiently close to the recipient antenna. Misconfigured wireless microphones can operate in the same band. Signals can also suffer from multipath fading (a form of self interference). In such cases, the source may be localized to the same household and thus under the recipient device user's control. Alternatively, the user can adjust the TV antenna.

These same remedies are available to resolve cognitive radio interference. Furthermore, the cognitive radio may over time detect the channel is in use and move off of the channel.

Baseline As noted in 15F.3.3.4.5, TV receivers are subject to interference from unlicensed devices. In addition, thermal noise, line noise, poorly positioned antennas, and the local environment (topology, vegetation, and man-made structures) can affect the signal. Ultimately the goal is to allow the TV user to watch over-the-air programming. Non-signal-related effects can interfere with this goal; examples include broadcast equipment malfunction, TV receiver malfunction, spousal screen obstructions, and power outages. To take the last of these, typical utilities average 50 min to 9 h of power outages per year (i.e., 99.9% to 99.99% availability).

15F.3.4   Case(s) for Analysis

This analysis only considers a TV recipient system and low-power cognitive radio source devices.

15F.4   Criteria for Interference

15F.4.1   Interference Characteristics

This analysis measures the fraction of time that the TV receiver experiences a desired-to-undesired ratio (D/U) below a threshold. For a given recipient device this is a function of the likelihood that a source device is nearby, the source transmit power level, whether the source is active, whether the source cognitive radio has detected the recipient TV channel, the relative orientation of recipient and source antennas, the strength of the intended TV signal at the recipient TV device, and the propagation between the source and the recipient device.

15F.4.1.1   Impacted Level

The D/U threshold is captured in the parameter r_{min} as described below. The specific conditions on picture and audio quality that would set the D/U threshold (and thus r_{min}) are outside the scope of this analysis. Recall that r_{min} can be viewed as a requirement on the cognitive radio device; i.e., it is the minimum distance in which the TV receiver can still achieve a desired output performance. Thus, the D/U threshold may depend on factors related to the cognitive radio source signal bandwidth, type of modulation, and communication protocols. By capturing these in r_{min}, this analysis leaves open a path for technology innovation that would allow a low-power device designer to trade more complex designs that have less impact on a TV reception in exchange for higher allowed transmit power.

15F.4.2   Measurement Event

The measurement event consists of a short time interval, 10 s. For a given recipient device, the relative location and activity of source devices are measured and the impact on the D/U ratio is measured at the recipient device between the antenna and the receiver input.

15F.4.3   Interference Event

A measurement event is an interference event if the signal strength at the recipient device from any one active source device causes the D/U to drop below a threshold for more than 100 ms (i.e., 1% of the measurement interval).

15F.4.4   Harmful Interference Criteria

We use the Probability of Interference Events model with $x_{\mathrm{pi}} = 0.0001$ (i.e., 1/10,000). This threshold is comparable with the reliability of the underlying utility power.

15F.5   Variables

The variables are listed here starting with the propagation parameters and ending with variables defined earlier:

ϵ_d pathloss exponent between the broadcast transmitter and the recipient device (desired).

ϵ_u pathloss exponent between the source and the recipient device (undesired).

σ_d log-normal shadow fading standard deviation between the broadcast transmitter and the recipient device (in decibels).

σ_u log-normal shadow fading standard deviations between the source and the recipient device (in decibels).

$g_\mathrm{s}(\theta)$ the source antenna gain as a function of azimuth angle θ (absolute gain and not in decibels).

$g_\mathrm{r}(\theta)$ the recipient antenna gain as a function of azimuth angle θ (absolute gain and not in decibels).

$p_\mathrm{s}(x)$ the density function for the distribution of transmit powers used by the source device, normalized by the maximum transmit power (absolute relative power and not in decibels).

D the density of source devices over the broadcast coverage area (devices/km^2).

F_S fraction of source devices that are active at a given time ($0 \le F_S \le 1$).

F_R fraction of recipient devices that are on and receiving over-the-air signals on channels 21–51 ($0 \le F_R \le 1$).

C_det probability that the cognitive radio detection successfully detects the recipient channel ($0 \le C_\mathrm{det} \le 1$).

r_min minimum source and recipient separation for no interference in the worst case (kilometers).

The likelihood of an interference event increases with the density of source devices D. The analysis will determine the maximum density that does not rise to harmful interference.

15F.6   Analysis—modeling, Simulation, Measurement, and Testing

15F.6.1   Selection of the Analysis Approach, Tools, and Techniques

15F.6.2   Matrix Reduction

Only one case is being analyzed.

15F.6.3 Performing the Analysis

This subclause provides the model details.

Model Derivation

1. Undesired received power at a recipient device from a source device is $P_u = K_u\, g_s\, g_r\, P_s S_u / r^{\epsilon u}$, where K_u is a constant related to antenna heights, cable losses, and other constants; g_s and g_r are the source and recipient device antenna gains along the path connecting them; P_s is the source transmit power; r is the separation between the source transmitter and the recipient receiver; ϵ_u is the pathloss exponent for signals between the source and the recipient device; and S_u is the shadow fading factor representing the variation in received power due to terrain, clutter, and other environmental factors.

2. Desired received power at a recipient device from a broadcast tower is $P_d = K_d S_d / R^{\text{ed}}$, where K_d is a constant related to broadcast power, antenna heights, cable losses, etc.; R is the separation between the transmitter and the receiver; ϵ_d is the pathloss exponent between the transmitter and the receiver; and S_d is a shadow fading factor representing the variation in received power due to terrain, clutter, and other environmental factors. The specific effects for the broadcast power and antenna gains are not broken out as they are fixed.

3. The recipient device is disrupted if $P_d/P_u < T$ for some defined threshold T. This threshold depends on the nature of the interference signal, and whether it is in the same channel as the recipient receiver or another nearby channel. Combining the previous assumptions, the signal-to-interference ratio is $P_d/P_u = K\, S\, r^{\epsilon u}/(g_s\, g_r\, P_s\, R^{\text{ed}})$, where $K = K_d/K_u$ and $S = S_d/S_u$.

4. The shadow fading S is well modeled by a log-normal distribution with a standard deviation of log S, σ. If S_d and S_u are both log normal with log standard deviations σ_d and σ_u, then their ratio is also log normal. In practice, S_d and S_u are correlated. A TV in the basement will receive weaker signals from both the broadcaster and the source device. Thus, σ^2 will be less than the sum of the desired and undesired variances.

The pathloss exponent is allowed to differ for the source and broadcast transmitters. It is expected that the desired signal from the broadcast transmitter will be close to a free-space pathloss model ($\epsilon_d = 2$). The undesired signal from the source device will be closer to the two-ray ground model ($\epsilon_u = 4$).

Shadow fading can have log-normal standard deviations as large as 10 dB for both S_d and S_u, suggesting a total of 14 dB for the log-normal standard deviation for their ratio. Because of correlations between them, we might expect a total variation equal to half of this value or 7 dB. For the analysis, we need to convert from decibels to absolute standard deviation, $\sigma = \sigma$ (dB) ln(10)/10.

There are three main random variables in this model. The distance of the recipient device to the broadcast transmitter R, the distance from the recipient device to the source transmitter r, and the shadow fading value S. Once these are accounted for, secondary random variables can be easily admitted.

We are interested in computing the probability a recipient device is interfered with by a source device. This analysis was carried out already[39] yielding

$$F = D r_{\min}^2 G_r G_s PCF_S F_R \frac{\pi \epsilon_u}{\epsilon_u + \epsilon_d} e^{\frac{2\sigma^2}{\epsilon_u^2}}$$

where in addition to variables already defined:

$$G_r = \frac{1}{2\pi} \int_0^{2\pi} \left(\frac{g_r(\theta)}{g_r^{\max}}\right)^{2/\epsilon_u} d\theta$$

$$G_s = \frac{1}{2\pi} \int_0^{2\pi} \left(\frac{g_s(\theta)}{g_s^{\max}}\right)^{2/\epsilon_u} d\theta$$

39 Brown [B6].

$$P = \int_0^{P_{\max}} \left(\frac{x}{P_{\max}}\right)^{2/\epsilon_u} p_s(x)dx$$

$$C = \sum_i p_i (T_i/T_S)^{2/\epsilon_u}$$

G captures the effect of the antenna pattern on the source or recipient, where $g^{\max}$ is the maximum gain of the antenna. Example values are as follows:

$G = 1$ if the antenna is omnidirectional

$G = w/360$ if the antenna is an ideal sectorized antenna of width w in degrees

P captures the effect of power control used by the source device, where $P^{\max}$ is the maximum transmit power. Example values are as follows:

$P = 1$ if the source device always transmits at maximum power

$P = \epsilon_u/(\epsilon_u + 2)$ if power is uniform between 0 and $P^{\max}$

C captures the effect that the detection algorithm induces a distribution on the channel used by the source device relative to the channel used by the recipient device. (T_i/T_s) is given by the table of isolation between different channels presented earlier (although the absolute values must be used and not the decibel values). The probability of using the ith adjacent channel is p_i. To compute p_i, we assume a channel near the center of channels 21–51. If the cognitive radio detection fails, then the source uses a random channel. If it is successful, then it uses a random channel that is two or more channels away. Under these assumptions (and noting that because of channel 37, there are only 30 channels in 21–51):

$$p_i = \begin{cases} \dfrac{1 - C_{\mathrm{det}}}{30} & i = 0 \\[2ex] \dfrac{1 - C_{\mathrm{det}}}{15} & i = 1 \\[2ex] \dfrac{1 - C_{\mathrm{det}}}{15} + \dfrac{2C_{\mathrm{det}}}{27} & i > 1 \end{cases}$$

As a worst case if $C_{\mathrm{det}} = 0$

$C = 0.034$

If $C_{\mathrm{det}} = 1$

$C = 5.9 \times 10^{-5}$

15F.6.4 Quantification of Benefits and Interference

We assume the following values for the model variables:

$\epsilon_d = 2$
$\epsilon_u = 4$
$\sigma = 1.6$ (i.e., 7 dB)
$g_s(\theta)$ is uniform
$g_r(\theta)$ is an ideal sectorized antenna with a beamwidth of 120 degrees
$p_s(x)$ is uniform
$F_S = 0.1$
$F_R = 0.1$
$C_{\mathrm{det}} = 0.95$
$r_{\min} = 0.1$ km

Combining these values, we get that the probability of interference is as follows:

$F = 1 \times 10^7 D$

The desired probability of interference is less than $x_{pi} = 0.0001$. Comparing, $F < x_{pi}$ if $D <$ 1000 source devices per square kilometer. The key tradeoff for the cognitive radio designer in this analysis is the relationship between r_{min} and C_{det}. Figure 15F.1 presents the missed detection probability that results when the factor is fixed at 1×10^{-7}. As Figure 15F.1 shows that as C_{det} approaches 1, the radio can be designed for an r_{min} in excess of 500 m and thus operate at higher powers. At the other extreme of $C_{det} = 0$ (i.e., no detection), r_{min} is 20 m and would require a very low-power operation.

15F.6.5 Analysis of Mitigation Options

If the fraction of recipient devices that are impacted is too high, several options are available. In specific instances of a known source causing interference to a recipient device, the source can be set to use lower power, relocated to reduce interference, or programmed by hand so that the recipient channel is avoided by the source device. The recipient device can also adjust its antenna or use an alternate channel. If the problem is more widespread, many cognitive radio concepts include a notion of policy information that must be periodically renewed. If there are excessive interference problems, policies can be tightened to use, for instance, lower maximum powers or more stringent criteria for deciding whether a channel is unoccupied.

15F.6.6 Analysis Uncertainty

The analysis here considers specific parameter instances. The pathloss exponent parameters are within a factor of 2 of any analysis that might be considered that, over this range, will vary the output by less than a factor of about 2. Generally higher pathloss will reduce the interference. The shadow fading standard deviation is within a few decibels of values that appear in the literature across a wide range of environments. Again, this range will vary the output by a factor of about 2. Smaller shadow fading standard deviation will reduce the interference. The

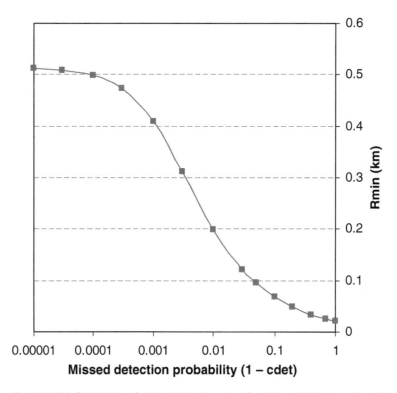

Figure 15F.1 Probability of missed detection as a function of distance with a factor of 1×10^{-7}. From IEEE Recommended Practice for the Analysis of In-Band and Adjacent Band Interference and Coexistence Between Radio Systems, Mar 2018. IEEE.

antenna pattern of the source and recipient devices might vary from what is assumed. Over typical ranges, this will vary the effect by a factor of about 2. More directional antennas will have lower G factors. However, these will be offset by increasing r_{min}. The power factor computes to be 2/3. More aggressive power control could reduce this significantly. In the worst case, this factor is 1. The fraction of time that the source and recipient devices are on and active over the air are only estimates, and further analysis would be needed to set them more accurately. The detection probability depends on the exact method of channel detection. The value here (95%) was set assuming a single-device TV signal RF detector approach. Other approaches can approach 99% or more, which would yield a factor of 4 lower output. Finally, r_{min} was set at 100 m. As noted already, this can be considered either an output result of a specific cognitive radio design or an input target specification of the design.

At a higher level, the model makes assumptions about the environment, e.g., uniformly distributed TV receivers that may not be accurate. TV receivers may be, because of geography and demographics, more concentrated at the limits of signal coverage. Such assumptions are made explicitly, and their validity would need to be assessed in specific situations.

15F.7   Conclusion and Summary

15F.7.1   Benefits and Impacts

The analysis indicates that a large number of cognitive radio devices can operate with an impact that is similar to existing interference effects. The relative benefit of the new cognitive radios versus the existing TVs that suffer harmful interference is not analyzed here specifically. However, the analysis establishes the tradeoff ratio. For the parameters considered here, each additional cognitive radio per square kilometer increases the fraction of TV receivers impacted by 10^{-7}.

15F.7.2   Summation

The analysis models the operation of a cognitive radio in broadcast TV band over channels 21–51. It provides a closed-form expression for the probability of interference to a TV as a function of propagation and cognitive radio parameters and the density of cognitve radio devices. Under the assumptions in this analysis, cognitive radio device densities as high as 1000 devices/km^2 can be supported at a probability of interference of 1 in 10,000 TVs. This number of devices is large and unlikely to be reached in the foreseeable future. The fraction of TVs impacted at this level is small and even smaller for lower cognitive radio densities.

The key elements to the analysis under the control of the cognitive radio designer are r_{min} and C_{det}. The first is the minimum distance away from a TV receiver that the cognitive radio must be able to operate in the worst case without causing signal degradation to the TV. As r_{min} decreases, the cognitive radio must use lower power or other means to avoid causing signal degradation in closer proximity. The detection accuracy C_{det} is a function of the mechanism used to detect the presence of TV signals. These elements trade off with each other with better detection accuracy allowing the radio to be designed to operate at a further distance from the TV without causing signal degradation.

Annex 15G

(informative)

Sample Analysis—RF Test Levels for ANSI C63.9 [B3]

[NOTE—This example shows a very different application for an interference and coexistence analysis. The task is to design a standard that will give a high level of assurance that compliant equipment is immune to interference from common mobile transmitters such as mobile phones. This analysis seeks to determine what immunity levels should be used in such a standard.]

15G.1 Executive Summary

This analysis is being prepared to explain and support the RF exposure levels and to test modulations required in ANSI C63.9 [B3]. Questions have been raised about the higher test levels required in this standard and about the modulations required by the standard. This analysis captures the thinking of the committee that prepared ANSI C63.9 and explains the reasons for their decisions on these matters.

15G.2 Findings

This analysis concludes that immunity levels on the order of 30 V/m to 35 V/m are necessary to assure immunity of office equipment from mobile phones. This conclusion assumes use cases in which the user can conveniently move a mobile transmitter 0.3 m (30 cm) from other equipment but may not be able to move it more than 0.3 m. Hence, the analysis is assuming a desired protection distance of 0.3 m.

15G.3 Scenario Definition

15G.3.1 Study Question

The committee responsible for preparing ANSI C63.9 was established after a series of situations arose in which interference from wireless devices to office equipment created significant problems. The committee was charged with developing a standard that would provide high levels of protection against such interference.

The standard was prepared with the anticipation that it would be used as a purchase specification by organizations that wanted high levels of protection for specific aspects of their operation. Examples might be the public address systems in important meeting rooms or conference call equipment used for critical business conferences.

The committee assumed that its intended users are capable of analyzing the costs and benefits of requiring that their vendors meet these requirements. Hence, the focus of the committee was on developing a test and limits that would provide high levels of assurance that interference would not occur, except in relatively extreme cases.

15G.3.2 Benefits and Impacts of Proposal

The benefit of this analysis is that users who desire or require a high level of immunity would have a standard to use in seeking that protection. The impact is that producers of equipment would be challenged to provide significantly more demanding levels of immunity.

15G.3.3 Scenario(s) and Usage Model

During the development of this standard, a number of usage scenarios were discussed by the committee. Information about various field experiences was presented by the participants. These presentations and deliberations took place over several years. In the end, a few common scenarios were identified as defining the focus for this effort.

The first scenario is of an investment trader at a major brokerage firm. Investment traders typically work in a physically small cubical but have a variety of computers, telephones, and wireless devices for their work. In this usage scenario, multiple devices operate in close proximity for most of the business day. It is almost impossible to separate the devices by more than 0.6 m and inconvenient to separate them by more than 0.3 m.

A second defining scenario is a public meeting room with a dais, equipped with microphones, such as are commonly used for government hearings of various types. In this scenario, each speaker commonly has a microphone in front of him or her and a small desk area to hold papers and wireless devices. It is not uncommon for participants to use wireless PDAs or mobile devices to confirm information or to query their staff during a proceeding. In this scenario, it is difficult to separate the wireless device from the microphone and associated preamplifier by more than 0.3 m.

A third scenario is a typical business conference room equipped with conference call equipment, including microphones distributed around the room. During conference calls, participants commonly use their wireless devices to check messages, obtain current information, or send queries seeking information for the meeting. On a crowded conference table, it may be difficult for participants to move their wireless devices more than 0.3 m from the conference call equipment, particularly from microphones that are distributed around the room in some situations.

15G.3.3.1 Frequency Relationships

The interference mechanism is assumed to be demodulation of an RF transmission across a nonlinear junction in the recipient equipment. The junction is assumed to be capable of demodulating all frequencies equally. Stated in a different way, given the variety of potential recipient circuits, nothing is assumed about their frequency response other than, if RF energy is present, it might be demodulated and cause interference.

ANSI C63.9 is intended to provide a measure of the immunity to common wireless devices. With a very popular cell phone band in the 800 MHz region and the increasingly used 5.8 GHz band, a range of 800 MHz to 6 GHz was a natural choice. This range captures the most common types of wireless transmitters. Although there are RF emitters above and below this range, they are far less common than those within this range.

15G.3.3.2 Usage Model

The RF attributes of a mobile transmitter in use close to office equipment are common to the usage models considered. The recipient equipment is assumed to be typical office equipment. The recipient circuit within this equipment is a nonlinear junction in some circuit. This scenario will typically respond to a wide range of frequencies with variations based on the particular characteristics of the recipient equipment. Hence, the analysis is restricted to the response of a nonlinear junction with unknown frequency response in the wiring or traces receiving the RF energy.

The RF transmitting systems are a wide range of wireless devices. The focus of the analysis is determined by the combination of transmit power, number of devices being used, and modulation characteristics. Cell phones became the primary focus for analysis because at 1 W to 2 W of transmit power, they produce more RF than many other common transmitters. Furthermore, they are very common and widely used. So the combination of use and RF power make cell phones a primary focus as potential interferes. Among cell phones, the 2 W devices in the 800 MHz band are of greater concern than the 1 W devices at 1900 MHz. GSM modulation drew attention in contrast to other modulations because its waveform demodulates into the audio band.

The probability of interference from other types of transmitters was considered. As an example, wireless networking devices, typically operating in the unlicensed bands, e.g., IEEE 802[R][40] devices, were discussed. Also higher power transmitters, like those commonly used by maintenance workers, were considered. In the case of the wireless networking device, their lower power made them of less interest than cell phones. The higher power devices drew less attention both because they are less common than cell phones and the typical separation distance is greater. Furthermore, such devices are not typically used routinely in the use scenarios being focused on of small cubicles and conference rooms.

The mobile transmitters with a transmission power between 0.1 W and 8.0 W were considered, with a special focus on mobile phones operating at 1 W or 2 W.

Only the condition when the transmitter is transmitting is considered. How frequently this occurs in close proximity to the recipient equipment is not analyzed. However, it is assumed to be relatively common.

> NOTE—In both the preceding subsection and this one, matrix reduction steps are incorporated. Furthermore, here several subclauses from the standard outlined are combined for brevity. Specifically the following subclauses are covered in this section:

15G.3.3.3 Characteristics of usage model
15G.3.3.3.1 Spatial and power characteristics
15G.3.3.3.2 Temporal characteristics
15G.3.3.3.3 Frequency relationships
15G.3.3.3.4 Other orthogonal variables

When incorporating subsections into a superior subsection of an analysis, care should be taken that the information required by the subclause is clearly provided.

15G.3.3.3   System Relationships

Systems Considered   The only system relationship considered is the potential for a mobile transmitter to impact office equipment.

Protection Distance   From the usage scenarios identified, 0.3 m is selected as the target protection distance required.

Geographic Area for Analysis   This analysis considers all geographic areas, but both field experience and analysis quickly brought the focus to the very close use scenario where equipment is within 1.0 m and, in particular, the situation at the target protection distance of 0.3 m.

Impact of Interference   The impact of interference can range from annoyance to significant business impact, such as being unable to complete important phone calls, conference calls, meetings, or hearings.

---

40 IEEE and 802 are registered trademarks in the U.S. Patent & Trademark Office, owned by The Institute of Electrical and Electronics Engineers, Incorporated.

Interference Mitigation Many techniques and protection mechanisms can be applied in the design of recipient equipment. As the standard measures the immunity of the equipment, consideration of what techniques might provide adequate protection is beyond the scope of this analysis. It is sufficient to know that it is possible to design equipment with sufficient immunity to withstand even the demanding environment created by a transmitter being used in close proximity.

For the user, few remedies exist for the scenarios being considered if the recipient equipment does not provide sufficient immunity. The offending device could be turned off or moved. However, these scenarios assume there is a requirement for simultaneous use of the devices in close proximity.

Baseline Interference from other sources is assumed to be typical. International standards recommend 3 V/m for home environments and 10 V/m for industrial environments to give adequate protection. Therefore, those levels are assumed to be reasonably common in those environments.

15G.3.4 Case(s) for Analysis

Transmitters operating at power levels of 0.1 W to 8.0 W will be considered. Separation distances down to the 0.3 m protection distance will be considered. Transmission modulations currently in common use and those that seem to have the potential for common use in the near future will be considered.

Line-of-sight is assumed between the transmitter and the recipient equipment.

15G.4 Criteria for Interference

15G.4.1 Interference Characteristics

The interference is assumed to be RF energy delivered through some reception path to a non-linear junction in the recipient equipment.

15G.4.1.1 Impacted Level

Of the various levels of operating impact, only audio interference, data loss, and operational upset were considered to be of interest for this issue. The analysis focuses on user perceptible interference. The interference energy is assumed to be received by circuitry wiring, demodulated at a nonlinear junction, and introduced into the following circuits. It is the user perceptible interference from this energy that will be measured.

15G.4.2 Measurement Event

For audio interference, the measurement event is a sample delivered from an RMS detector with a 120 mS time constant. The instrumentation used is assumed to have a very broadband front end, delivering the entire transmission signal to a nonlinear junction. The nonlinear junction serves as a detector. Post detection a band-pass filter with A-Weighting limits and weights the signal to the audio band. The energy into the audio band is measured after the detector using an RMS meter with a 120 mS time constant.

15G.4.3 Interference Event

Interference that lowers the S/N below 25 dB is counted as an interference event.

15G.4.4 Harmful Interference Criteria

Interference events that continue during the transmission are considered harmful interference. Interference events that last less than 2 s in a 30 s period are considered a click or transient and are allowed.

15G.5 Variables

The variables relevant to this problem are those that affect the RF environment or the response in an unintended, recipient circuit. The following variables are identified as being significant:

a) Frequency range
b) Field strength at recipient circuit
c) Emission bandwidth
d) Modulation waveform characteristics
e) Power control
f) Interference mitigation or tolerance
g) Demodulated waveform characteristics

In reviewing the existing RF immunity standards, all of these variables can have a significant impact on the problem and therefore were considered when developing ANSI C63.9 [B3].

15G.6 Analysis—modeling, Simulation, Measurement, and Testing

The committee responsible for ANSI C63.9 [B3] began its work by reviewing existing RF immunity standards and test methods. The committee agreed that they would not create a new document or introduce new test methods if existing RF immunity standards could be identified to serve the intended purpose.

Initially, the committee believed that its work might be accomplished by simply increasing the RF exposure requirements using existing test methods. In particular, several committee members voiced the opinion that RF sensitivity in an EUT could be adequately revealed using the classic 1 kHz 80% AM. During the initial discussions, it was felt that perhaps the bulk of the committee's work would focus on defining monitors of EUT performance and failure thresholds for such variables as near-end and far-end noise.

During the course of the committee's work, information from several sources was gathered. Information was provided about the types of wireless devices commonly used in close proximity to office equipment and the frequency of such occurrence. This body of information lead the committee to the conclusion that testing with more complex, realistic waveforms was required to accurately assess the immunity of office equipment.

Several participants reported that the traditional RF immunity tests simply did not adequately correlate to the field experience for their products. Some manufacturers were supplementing their standard RF immunity tests with ad hoc testing, using general market devices because this was the only method they had found satisfactory for predicting field performance. After much discussion and exchange of information, the committee decided that it was necessary to use modulations that more closely replicated the transmissions of real-world devices.

An example of information contributed to the committee is the following quote from a contribution provided by W. Schaefer of CISCO:

> *We still feel that the near-field testing should be performed with digitally modulated signals, not just pulse modulated signals, as currently stated. Attached is a summary of preliminary tests that were performed some years ago. These tests clearly indicate that the type of test signal used to perform the testing has a very significant impact on the observed functionality of an EUT. For*

example, an EUT was tested at a distance of 1 foot with field strengths between 50 V/m to 60 V/m, and no malfunction was observed. When using a cell phone with a field strength of 5 V/m, a malfunction was clearly detectable. This again shows empirically that standard test signals (i.e., AM, pulse modulation) will not properly exercise an EUT and thus will not adequately represent the threat an EUT is exposed to in a regular office environment. It is therefore suggested that the use of actual communication system signals be introduced in the document to ensure proper detection of malfunctions of EUTs.[41]

15G.6.1   Selection of the Analysis Approach, Tools, and Techniques

15G.6.2   Matrix Reduction

Of the various scenarios for evaluation, the critical scenario is as follows: Will this standard give adequate assurance that conforming equipment would have adequate immunity to withstand emissions from a GSM handset of 0.3 m. The equipment is assumed to be delivering human voice, e.g., conference call equipment or microphone. The symptom of primary interest is whether interference will be heard at a level that would degrade the ability to understand typical speech

Of less, but not insignificant, interest is whether the same protection from audio interference is provided for other common transmitters.

Although audio interference is the primary consideration, other kinds of interference, such as disturbance of video equipment and operational disruption, are of importance.

15G.6.3   Performing the Analysis

15G.6.3.1   Field Strength at Recipient Circuit

From the use scenarios, a protection distance of 0.3 m was selected. Table 15G.1 provides calculated field strength from various power levels, assuming a dipole radiator. For a 1 W cell phone, the estimated field strength at this distance is 24 V/m. For a 2 W cell phone, the estimated field strength is 34 V/m. That these calculations provide a realistic estimation was confirmed from several sources, including measurements of commercial devices. The committee selected 30 V/m as the test voltage in the handset transmit bands and 10 V/m in other frequency bands.

Table 15G.1 Calculated field strengths for common transmitters and protection distances.

RF Power (Watts)	W	0.1	0.5	1	2	4	8
RF Power (dBm)	dBm	20.0	27.0	30.0	33.0	36.0	39.0
		Dipole Radiator (assumed 2.4 dBi Gain)					
Distance / Field Strength	m	V/m	V/m	V/m	V/m	V/m	V/m
	0.10	22.8	51.1	72.2	102.1	144.4	204.2
	0.13	18.3	40.8	57.8	81.7	115.5	163.4
	0.15	15.2	34.0	48.1	68.1	96.3	136.1
	0.25	9.1	20.4	28.9	40.8	57.8	81.7
	0.30	7.61	17.02	24.07	34.04	48.14	68.07
	0.50	4.6	10.2	14.4	20.4	28.9	40.8
	0.75	3.0	6.8	9.6	13.6	19.3	27.2
	1.00	2.3	5.1	7.2	10.2	14.4	20.4

---

41 Quoted from communication to the working group for ANSI C63.9 [B3].

From IEEE Recommended Practice for the Analysis of In-Band and Adjacent Band Interference and Coexistence Between Radio Systems, Mar 2018. IEEE.

15G.6.3.2 Modulation Characteristics

The following variables will be dealt with together in this subsection:

a) Emission bandwidth
b) Modulation waveform characteristics
c) Power control
d) Demodulated waveform characteristics peak-to-average

Information presented to the committee indicated that at this time a generalized treatment of the impact of these variables is not available. Accordingly, it has been decided that testing for immunity should be performed using realistic modulation waveforms, such as those obtained by recording in-phase and quadrature (IQ) files using vector signal analyzers for recording and vector signal generators for recreation.

15G.6.4 Quantification of Benefits and Interference

This subsection is not directly applicable to this analysis. Each user of the standard will assess for themselves whether the cost is justified by the benefit provided.

15G.6.5 Analysis of Mitigation Options

It is assumed that the user of this standard wants sufficient immunity to avoid employing additional mitigations to achieve the desired immunity. Accordingly no additional mitigation options are assumed.

15G.6.6 Analysis Uncertainty

This analysis is intended to establish a single limit threshold. A high level of confidence that equipment meeting the threshold established will provide adequate immunity. At this time there is not sufficient quantitative data to provide a more precise conclusion other than the committee's consensus opinion that the threshold established achieves this objective.

Many variables exist, and equipment meeting this threshold will often provide immunity in excess of that intended. As one example, the most sensitive parts of the equipment may be on the far side from a transmitter in a specific situation and therefore be receiving less of its energy. As another example, the sensitive circuit and transmitter may be cross polarized. The goal of this analysis is not to predict interference but to establish a threshold that provides a high level of confidence that adequate immunity will be present.

15G.7 Conclusions and Summary

15G.7.1 Benefits and Impacts

The value of the standard that this analysis is supporting is to provide a tool for organizations seeking to assure that equipment they purchase provides the level of immunity they desire. To provide increased immunity can be expected to increase the cost and design complexity of the equipment. This analysis assumes that users of the standard find that the benefits of the increased immunity are sufficient to justify the increased cost.

15G.7.2 Summation

This analysis concludes that an immunity level of 30 V/m to 35 V/m is necessary to assure adequate resistance to mobile phones and other common transmitters. Furthermore, real-world waveforms, typical of the services used in various frequency bands, should be used when testing.

Annex 15H

(infnormative)

Glossary

For the purposes of this document, the following terms and definitions apply. These and other terms within IEEE standards are found in *The Authoritative Dictionary of IEEE Standards Terms* [B12].

adjacent channel: A channel whose frequency band is adjacent to that of another channel, known as the reference channel.

adjacent-channel interference: (data transmission) Interference, in a reference channel, caused by the operation of an adjacent channel.

channel: A repeated time and spectrum combination used for communications. In 47CFR15.323(c) the FCC uses the description a "combined time and spectrum window".

co-channel interference: Interference caused in one communication channel by a transmitter operating in the same channel. *See also:* **radio transmission**.

Erlang: (1) (telephone switching systems). Unit of traffic intensity, measured in number of arrivals per mean service time. For carried traffic measurements, the number of erlangs is the average number of simultaneous connections observed during a measurement period. (2) (data transmission). A term used in message loading of telephone leased facilities. An erlang is equal to the number of call-seconds divided by 3600 and is equal to a fully loaded circuit over a one-hour period.

fading: (1) (**A**) (**data transmission**) (Flat). That type of fading in which all frequency components of the received radio signal fluctuate in the same proportions simultaneously. *See also:* **selective fading**. (**B**) (**data transmission**) (Radio). The variation of radio field intensity caused by changes in the transmission medium, and transmission path, with time. *See also:* **selective fading**. (2) The temporal variation of received signal power caused by changes in the transmission medium or path(s).

far-field region: (1) (**land-mobile communications transmitters**) The region of the field of an antenna where the angular field distribution is essentially independent of the distance from the antenna.
Notes: 1. If the antenna has a maximum overall dimension (D) that is large compared to the wavelength (λ), the far-field region is commonly taken to exist at distances greater than $2D^2/\lambda$ from the antenna. 2. For an antenna focused at infinity, the far-field region is sometimes referred to as the Fraunhofer region on the basis of analogy to optical terminology.
(2) That region of the field of an antenna where the angular field distribution is essentially independent of the distance from a specified point in the antenna region.
Notes: 1. In free space, if the antenna has a maximum overall dimension, D, that is large compared to the wavelength, the far-field region is commonly taken to exist at distances greater than $2D^2/\lambda$ from the antenna, λ being the wavelength. The far-field patterns of certain antennas, such as multi-beam reflector antennas, are sensitive to variations in phase over their apertures. For these antennas, $2D^2/\lambda$ may be inadequate. 2. In physical media, if the antenna has a maximum overall dimension, D, that is large compared to π/γ,

the far-field region can be taken to begin approximately at a distance equal to $\gamma \, D^2/\pi$ from the antenna, γ being the propagation constant in the medium.

(3) That region of the field of an antenna array where the angular field distribution is essentially independent of the distance from the center of the array. A general far-field approximation is $2D^2/\lambda$, where d is the largest separation between elements in the array.

(4) That region of the field of an antenna where the angular field distribution is essentially independent of the distance from the antenna. In this region (also called the free space region), the field has a predominantly plane-wave character, i.e., locally uniform distributions of electric field strength and magnetic field strength in planes transverse to the direction of propagation.

(5) *See also:* **Fraunhofer region**.

far-field region in physical media: *See:* **far-field region**.

far-field region, radiating: *See:* **radiating far-field region**.

Fraunhofer region: (1) (**data transmission**) That region of the field in which the energy flow from an antenna proceeds essentially as though coming from a point source located in the vicinity of the antenna. Note: If the antenna has a well-defined aperture D in a given aspect, the Fraunhofer region in that aspect is commonly taken to exist at distances greater than $2D^2/\lambda$ from the aperture, λ being the wavelength.

(2) The region in which the field of an antenna is focused. Note: In the Fraunhofer region of an antenna focused at infinity, the values of the fields, when calculated from knowledge of the source distribution of an antenna, are sufficiently accurate when the quadratic phase terms (and higher order terms) are neglected. *See also:* **far-field region**.

(3) That region around an electromagnetic radiator or scatterer (maximum dimension D) where the fields can be described in terms of a radial distance and azimuthal and polar angles. Note: In this region, the distances of all points to the source's center are larger than $2D^2 \, / \, /\lambda$. *Syn.*: **far-field region**.

harmful interference: (1) Any emission, radiation, or induction that endangers the functioning, or seriously degrades, obstructs, or repeatedly interrupts a radiocommunication service or any other equipment or system operating in accordance with regulations. See also: electromagnetic compatibility.

(2) Interference which endangers the functioning of a radionavigation service or other safety services or seriously degrades, obstructs, or repeatedly interrupts a radiocommunication service operating in accordance with [international] Radio Regulations. ITU-R Radio Regulations (2004) 1.169 & FCC Part 2, 47 C.F.R. § 2.1 (2002)

interference: The effect of unwanted energy due to one or a combination of emissions, radiations, or inductions upon reception in a radio-communications system, manifested by any performance degradation, misinterpretation, or loss of information which could be extracted in the absence of such unwanted energy. ITU-R Radio Regulations (2004) 1.166 & FCC Part 2, 47 C.F.R. § 2.1 (2002)

near-field region: (1) That part of space between the antenna and far-field region. Note: In lossless media, the near-field may be further subdivided into reactive and radiating nearfield regions.

(2) That part of space between the antenna array and the far-field region. Refers to the field of a source at distances small compared to the wavelength. Note: The near-field includes the quasi-static and induction fields varying as r^{-3} and r^{-2}, respectively, but does not include the radiation field varying as r^{-1}.

(3) A region generally in proximity to an antenna or other radiating structure, in which the electric and magnetic fields do not have a substantially plane-wave character, but vary considerably from point to point. The near-field region is further subdivided into the reactive near-field region, which is closest to the radiating structure and that contains most or nearly all of the stored energy, and the radiating near-field region where the radiation field predominates over the reactive field, but lacks substantial plane-wave character and is complicated in structure. Note: For most antennas, the outer boundary of the reactive

near-field region is commonly taken to exist at a distance of one-half wavelength from the antenna surface.

near-field region, radiating: *See:* **radiating near-field region**.

near-field region, reactive: That portion of the near-field region immediately surrounding the antenna, wherein the reactive field predominates. Note: For a very short dipole, or equivalent radiator, the outer boundary is commonly taken to exist at a distance $\lambda/2\pi$ from the antenna surface, where λ is the wavelength.

pseudo code: (1) (**software**) A combination of programming language constructs and natural language used to express a computer program design. For example: IF the data arrives faster than expected, THEN reject every third input. ELSE process all data received.

(2) (**test, measurement, and diagnostic equipment**) An arbitrary code, independent of the hardware of a computer, which has the same general form as actual computer code but which must be translated into actual computer code if it is to direct the computer.

(3) A combination of programming language constructs and natural language used to express a computer program design.

For example:

    IF the data arrives faster than expected
    THEN reject every third input
    ELSE process all data received
    ENDIF

radiating far-field region: (land-mobile communications transmitters) Measurement is performed at or beyond a distance of 3λ, but not less than 1 meter (m). *See also:* **far-field region**.

radiating near-field region: (1) (land-mobile communications transmitters) Measurement is limited to the region external to the induction field and extending to the outer boundary of the reactive field that is commonly taken to exist at a distance of $\lambda/2\pi$. Either the electric or magnetic component of the radiated energy may be used to determine the magnitude of power present. *See also:* **near-field region**.

(2) That portion of the near-field region of an antenna between the farfield and the reactive portion of the near-field region, wherein the angular field distribution is dependent upon distance from the antenna. Notes: 1. If the antenna has a maximum overall dimension that is not large compared to the wavelength, this field region may not exist. 2. For an antenna focused at infinity, the radiating near-field region is sometimes referred to as the Fresnel region on the basis of analogy to optical terminology.

radio interference: (**overhead-power-line corona and radio noise**) Degradation of the reception of a wanted signal caused by RF disturbance.

Notes: 1. RF disturbance is an electromagnetic disturbance having components in the RF range. 2. The English words "interference" and "disturbance" are often used indiscriminately. The expression "radio frequency interference" is also commonly applied to an RF disturbance or an unwanted signal. Synonym: radio frequency interference.

radio noise: (1) (**radio noise from overhead power lines and substations**) Any unwanted disturbance within the radio frequency band, such as undesired electromagnetic waves in any transmission channel or device.

(2) (**radio noise from overhead power lines and substations**) An electromagnetic noise that may be superimposed upon a wanted signal and is within the radio-frequency range.

(3) (**overhead-power-line corona and radio noise**) Electromagnetic noise having components in the radio frequency range.

radio noise field strength: (**overhead-power-line corona and radio noise**) A measure of the field strength of the radiated radio noise at a given location.

Notes: 1. In practice, the quantity measured is not the electromagnetic field strength of the interfering waves but some quantity that is proportional to, or bears a known relation to, the electromagnetic field strength. 2. The radio noise field strength is measured in average, rms, quasi-peak, or peak values, according to which detector function of the radio noise meter is used. 3. The radio noise field strength is expressed either in mV/m, or in dB above 1 mV/m, per unit bandwidth, or in a specified bandwidth.

radio spectrum: (**radio-wave propagation**) The radio frequency portion of the electromagnetic spectrum. The frequency ranges are: ultra low frequency (ULF), lower than 3 Hz; extremely low frequency (ELF), 3 Hz to 3 kHz; very low frequency (VLF), 3 to 30 kHz; low frequency (LF), 30 to 300 kHz; medium frequency (MF), 300 kHz to 3 MHz; high frequency (HF), 3 to 30 MHz; very high frequency (VHF), 30 to 300 MHz; ultra high frequency (UHF), 300 MHz to 3 GHz; super high frequency (SHF), 3 to 30 GHz; extremely high frequency (EHF), 30 to 300 GHz; Submillimeter, 300 GHz to 1 THz.

Annex 15I

(infnormative)

Bibliography

B1 Advanced Television Systems Committee, DTV Receiver Performance Guidelines, ATSC A-74, June 18, 2004.

B2 Åkerberg, B., "On channel definitions and rules for continuous dynamic channel selection in coexistence etiquette radio systems," *1994 IEEE 44th Vehicular Technology Conference*, Stockholm, pp. 809–813, 1994.

B3 ANSI C63.9, RF Immunity of Office Equipment to General Use Transmission Devices with Transmission Power up to 8 Watts.[42]

B4 Barrios, A. E., A terrain parabolic equation model for propagation in the troposphere. *IEEE Transanctions on Antennas and Propagation*, vol. 42, no. 1, pp. 90–98, 1994.

B5 Brown, T. X., "A harmful interference model for unlicensed device operation in licensed service bands," *Journal of Communications*, vol. 1, no. 1, pp. 13–25, Apr. 2006.

B6 Brown, T. X., "An analysis of unlicensed device operation in licensed broadcast service bands," *Conference on Dynamic Spectrum Access Networks (DySPAN)*, 2005.

B7 CEPT ERC Report 100, Compatibility between certain radio communications technologies operating in adjacent bands—Evaluation of DECT/GSM1800 compatibility, Naples, Feb. 2000.

B8 Code of Federal Regulations, Title 47, Part 15 (47CFR15), Radio Frequency Devices.

B9 ETSI TR 101 310 v1.2.1 (2004), Digital Enhanced Cordless Telecommunications (DECT); Traffic capacity and spectrum requirements for multi-system and multi-service DECT applications coexisting in a common frequency band.

B10 Federal Communications Commission, "Report of the Spectrum Policy Task Force," ET Docket no. 02-135, Nov. 2002.[43]

B11 Hardin, R. H., and Tappert, F. D., "Application of the split-step Fourier method to the numerical solution of nonlinear and variable coeffiecient wave equations," *SIAM Review*, vol. 15, p. 423, 1973.

B12 IEEE 100, *The Authoritative Dictionary of IEEE Standards Terms*, Seventh Edition. New York: Institute of Electrical and Electronics Engineers, Inc.[44]

B13 IEEE Std 802.11™-1997, IEEE Standard for Information technology—Telecommunications and information exchange between systems—Local and metropolitan area networks—Specific requirements—Part 11: Wireless LAN Medium Access Control (MAC) and Physical Layer (PHY) Specifications.

B14 IEEE Std 802.16™-2004, IEEE Standard for Local and Metropolitan Area Networks Part 16: Air Interface for Fixed Broadband Wireless Access Systems.

B15 IEEE Std 1900.1™-2008, IEEE Standard Definitions and Concepts for Dynamic Spectrum Access: Terminology Relating to Emerging Wireless Networks, System Functionality, and Spectrum Management.

42 ANSI publications are available from the Sales Department, American National Standards Institute, 25 West 43rd Street, 4th Floor, New York, NY 10036, USA (http://www.ansi.org/).
43 Available at the FCC website, URL: http://fcc.gov/sptf/reports.html.
44 IEEE publications are available from the Institute of Electrical and Electronics Engineers, 445 Hoes Lane, Piscataway, NJ 08854, USA (http://standards/ieee.org/).

B16 Ishimaru, A., *Wave Propagation and Scattering in Random Media*, Vol. 2. New York: Academic Press, 1978.

B17 ISO/IEC 7498-1:1994, 2nd edition.[45]

B18 ITU-R Radio Regulations, 2004 edition, Appendix 5.[46]

B19 ITU-R Recommendation SM-1046-2, Definition of Spectrum Use and Efficiency of a Radio System.

B20 ITU-R Recommendation SM-1599, Determination of the Geographical and Frequency Distribution of the Spectrum Utilization Factor for Frequency Planning Purposes.

B21 Levy, M., Parabolic equation methods for electromagnetic wave propagation. IEE Electromagnetic Waves Series 45, Padstow, England, 2000.

B22 Margie, R. P., "Can You Hear Me Now? Getting Better Reception from the FCC's Spectrum Policy." *Stanford Technology Law Review*, 2003 STAN. TECH. L. REV. 5.[47]

B23 Mark, J. W., and Zhuang, W., *Wireless Communications and Networking*. Upper Saddle River, NJ: Pearson Education, 2003.

B24 Rappaport, T. S., *Wireless Communications, Principles and Practice*. Upper Saddle River, NJ: Pearson Education, 2002.

B25 Stuber, G., *Principles of Mobile Communications*. New York: Kluwer Academic, 2001.

B26 TIA TSB-86, Version 10.3, Criteria and Methodology to Assess Interference between Systems in the Fixed Service and the Mobile-Satellite Service in the Band 2165-2200 MHz, Telecommunications Industry Association, July 6, 1999.

B27 TIA TSB-88-B-1, Wireless Communications Systems—Performance in Noise and Interference—Limited Situations—Recommended Methods for Technology—Independent Modeling, Simulation, and Verifications, Telecommunications Industry Association, May 1, 2005.

B28 *User's Manual for Advanced Refractive Effects Prediction System*, Version 3.0, um-AREPS-30, San Diego, CA, Aug. 2003.[48]

45 ISO/IEC publications are available from the ISO Central Secretariat, Case Postale 56, 1 rue de Varembé, CH-1211, Genève 20, Switzerland/Suisse (http://www.iso.ch/). ISO/IEC publications are also available in the United States from Global Engineering Documents, 15 Inverness Way East, Englewood, Colorado 80112, USA (http://global.ihs.com/). Electronic copies are available in the United States from the American National Standards Institute, 25 West 43rd Street, 4th Floor, New York, NY 10036, USA (http://www.ansi.org/).

46 ITU-R publications are available from the International Telecommunications Union, Place des Nations, 1211 Geneva 20, Switzerland (http://www.itu.int/).

47 Available at the Stanford University website at URL: http://stlr.stanford.edu/STLR/Articles/03_STLR_5).

48 Available at the Space and Naval Warfare Systems Center, Atmospheric Propagation Branch website, URL: http://sunspot.spawar.navy.mil.

16

IEEE Standard for Architectural Building Blocks Enabling Network-Device Distributed Decision Making for Optimized Radio Resource Usage in Heterogeneous Wireless Access Networks

IMPORTANT NOTICE: This standard is not intended to ensure safety, security, health, or environmental protection in all circumstances. Implementers of the standard are responsible for determining appropriate safety, security, environmental, and health practices or regulatory requirements.

This IEEE document is made available for use subject to important notices and legal disclaimers. These notices and disclaimers appear in all publications containing this document and may be found under the heading "Important Notice" or "Important Notices and Disclaimers Concerning IEEE Documents." They can also be obtained on request from IEEE or viewed at http://standards.ieee.org/IPR/disclaimers.html.

16.1 Overview

16.1.1 Scope

The standard defines the building blocks comprising (i) network resource managers, (ii) device resource managers, and (iii) the information to be exchanged between the building blocks, for enabling coordinated network-device distributed decision making that will aid in the optimization of radio resource usage, including spectrum access control, in heterogeneous wireless access networks. The standard is limited to the architectural and functional definitions at a first stage. The corresponding protocols definition related to the information exchange will be addressed at a later stage.

16.1.2 Purpose

The purpose is to improve overall composite capacity and quality of service of wireless systems in a multiple Radio Access Technologies (RATs) environment, by defining an appropriate system architecture and protocols that will facilitate the optimization of radio resource usage, in particular, by exploiting information exchanged between network and mobile Terminals, whether or not they support multiple simultaneous links and dynamic spectrum access.

16.1.3 Document Overview

The structure of this document is as follows:

— Clause 16.2 lists normative references
— Clause 16.3 lists definitions and abbreviations used in this document
— Clause 16.4 presents overall system description of IEEE Std 1900.4, comprising system overview, summary of use cases, and assumptions
— Clause 16.5 specifies IEEE Std 1900.4 requirements, comprising system requirements, functional requirements, and information model requirements

Dynamic Spectrum Access Decisions: Local, Distributed, Centralized, and Hybrid Designs, First Edition. George F. Elmasry.
© 2021 John Wiley & Sons Ltd. Published 2021 by John Wiley & Sons Ltd.
Companion website: www.wiley.com/go/elmasry/dsad

— Clause 16.6 defines IEEE Std 1900.4 architecture. It is comprised of system description and functional description of an IEEE 1900.4 system
— Clause 16.7 defines the IEEE Std 1900.4 information model
— Clause 16.8 describes IEEE Std 1900.4 procedures
— Informative Annex 16A describes IEEE Std 1900.4 use cases
— Normative Annex 16B gives class definitions for the IEEE Std 1900.4 information model
— Normative Annex 16C provides data type definitions for the IEEE Std 1900.4 information model
— Informative Annex 16D describes the IEEE Std 1900.4 information model extensions and usage example
— Informative Annex 16E describes IEEE Std 1900.4 deployment examples
— Informative Annex 16F lists informative references related to IEEE Std 1900.4

16.2 Normative References

The following referenced documents are indispensable for the application of this document (i.e., they must be understood and used, so each referenced document is cited in text and its relationship to this document is explained). For dated references, only the edition cited applies. For undated references, the latest edition of the referenced document (including any amendments or corrigenda) applies.

IEEE Std 1900.1™-2008, IEEE Standard Definitions and Concepts for Dynamic Spectrum Access: Terminology Relating to Emerging Wireless Networks, System Functionality, and Spectrum Management.[1, 2]

ISO/IEC 8824, Information Processing Systems—Open Systems Interconnection—Specification of Abstract Syntax Notation One (ASN.1).[3]

ITU-T Recommendation X.701, System Management Overview.[4] Unified Modeling Language (UML), Version 2.1.2.[5]

16.3 Definitions, Acronyms, and Abbreviations

16.3.1 Definitions

For the purposes of this document, the following terms and definitions apply.

16.3.1.1 Base Station

This term is used to refer to any radio node on the network side from radio interface, independently of its commonly used name in a particular standard. Examples of common name are Base Station in IEEE Std 802.16, Base Transceiver System in cdma2000, Node B in UMTS, Access Point in IEEE Std 802.11, broadcasting transmitter, etc.

---

1 The IEEE standards or products referred to in this clause are trademarks of the Institute of Electrical and Electronics Engineers, Inc.
2 IEEE publications are available from the Institute of Electrical and Electronics Engineers, Inc., 445 Hoes Lane, Piscataway, NJ 08854, USA (http://standards.ieee.org/).
3 ISO/IEC publications are available from the ISO Central Secretariat, 1, ch. De la Voie-Creuse, Case Postale 56, CH-1211, Genève 20, Switzerland/Suisse (http://www.iso.ch/). ISO/IEC publications are also available in the United States from Global Engineering Documents, 15 Inverness Way East, Englewood, CO 80112, USA (http://global.ihs.com/). Electronic copies are available in the United States from the American National Standards Institute, 25 West 43rd Street, 4th Floor, New York, NY 10036, USA (http://www.ansi.org/).
4 ITU-T publications are available from the International Telecommunications Union, Place des Nations, CH-1211, Geneva 20, Switzerland/Suisse (http://www.itu.int/).
5 The UML is available from the Object Management Group Web site http://www.omg.org/spec/UML/2.1.2/June 2008.

16.3.1.2 Composite Wireless Network (CWN)

This term is used to refer to a network composed of several radio access networks with corresponding base stations, a packet-based core network connecting these radio access networks, and IEEE 1900.4 entities deployed in this network.[6]

NOTE—This definition does not exclude the case where some broadcasting system or future technology system is part of the composite wireless network.[7]

16.3.1.3 Context Information

This term is used to refer to any information that together with policies is needed for decision making on radio resource usage optimization in this standard. Radio access network (RAN) context information is distinguished from terminal context information. Also, context information is distinguished from policies.

16.3.1.4 Distributed Radio Resource Usage Optimization

The distributed optimization of radio resource usage by a composite wireless network to satisfy global network objectives and by terminals to satisfy local device and user objectives (see IEEE Std 1900.1).

16.3.1.5 Dynamic Spectrum Assignment

The dynamic assignment of frequency bands to radio access networks within a composite wireless network operating in a given region and time to optimize spectrum usage (see IEEE Std 1900.1).

16.3.1.6 Dynamic Spectrum Sharing

The process and mechanisms for a type of spectrum access that occurs when different radio access networks and Terminals dynamically access spectrum bands which are overlapping, in whole or in part, causing less than an admissible level of mutual interference, according to regulatory rules, and may be done with or without negotiation.

NOTE—The IEEE 1900.4 definition of dynamic spectrum sharing is intentionally more specific than the IEEE 1900.1 definition of dynamic frequency sharing (see 16A.2).

16.3.1.7 IEEE 1900.4 compliant terminal

This term is used to refer to any IEEE 1900.4 compliant radio node on the user side, that is, a reconfigurable terminal containing the Terminal Reconfiguration Manager, Terminal Measurement Collector, and Terminal Reconfiguration Controller IEEE 1900.4 entities.

16.3.1.8 Multi-homing Capability

This term is used to refer to a capability of a reconfigurable terminal to have more than one simultaneous active connections with radio access networks.

16.3.1.9 Network Reconfiguration Manager

The entity that manages the Composite Wireless Network and Terminals in terms of network-terminal distributed optimization of spectrum usage. This management is done within the framework of spectrum assignment policies conveyed by the Operator Spectrum Manager and in a manner consistent with available context information.

---

6 The IEEE Std 1900.4 definition of composite wireless network is intentionally more specific than the IEEE Std1900.1 definition of composite network. The IEEE 1900.4 definition describes the components of composite wireless networks in the context of the IEEE 1900.4 system architecture.

7 Notes in text, tables, and figures are given for information only and do not contain requirements needed to implement the standard.

16.3.1.10 Operator Spectrum Manager

The entity that enables the operator to control the dynamic spectrum assignment decisions of the Network Reconfiguration Manager.

16.3.1.11 Radio Access Network (RAN)

The network that connects base stations to the packet-based core network or external networks. If not specified, radio access network includes base stations.

16.3.1.12 Radio Access Network (RAN) Context Information

This term is used to refer to any information that together with policies is needed for decision making on radio resource usage optimization in this standard and has RANs as its source.

16.3.1.13 Radio Access Network (RAN) Measurement Collector

The entity that collects RAN context information and provides it to the Network Reconfiguration Manager.

16.3.1.14 Radio Access Network (RAN) Reconfiguration Controller

The entity that enables the Network Reconfiguration Manager to control reconfiguration of radio access networks.

16.3.1.15 Radio Enabler

A logical communication channel between NRM and TRM. Radio enabler may be mapped onto one or several radio access networks used for data transmission (in-band channel) and/or onto one or several dedicated radio access networks (out-of-band channel).

16.3.1.16 Radio Interface

This term is used to refer to an air interface specifications that shall be fulfilled to setup and maintain connection between terminal and base station. Radio interface may be characterized by multiple access method, modulation, etc. Examples are GSM, WCDMA, WiFi®,[8] WiMAX™,[9] etc radio interfaces.[10]

16.3.1.17 Radio Resource Selection Policy

A policy generated by the Network Reconfiguration Manager which guides the Terminal Reconfiguration Managers in terms of their radio resource usage optimization decisions.

16.3.1.18 Reconfigurable Terminal

This term is used to refer to any radio node on the user side that can reconfigure its hardware and/or software in order to change its operating parameters in the physical and link layers, such as carrier frequency, signal bandwidth, radio interface, etc. A reconfigurable terminal may have multi-homing capability.

16.3.1.19 Spectrum Assignment Policy

A policy generated by the Operator Spectrum Manager that guides the Network Reconfiguration Manager in terms of its radio resource usage optimization decisions.

---

8 WiFi® is a word mark of the WiFi Alliance.
9 WiMAX™ is a trademark of the WiMAX Forum.
10 This information is given for the convenience of users of this standard and does not constitute an endorsement by the IEEE of these products. Equivalent products may be used if they can be shown to lead to the same results.

16.3.1.20   Terminal

This is used as a short version of the term "**IEEE 1900.4 compliant terminal**."

16.3.1.21   Terminal Context Information

This term is used to refer to any information that has the user and/or Terminal as its source, which together with policies is needed in order for decision making on radio resource usage optimization within this standard to be possible.

16.3.1.22   Terminal Measurement Collector (TMC)

The entity that collects terminal context information and provides it to Terminal Reconfiguration Manager.

16.3.1.23   Terminal Reconfiguration Controller (TRC)

The entity that enables the Terminal Reconfiguration Manager to control reconfiguration of the Terminal.

16.3.1.24   Terminal Reconfiguration Manager (TRM)

The entity that manages the Terminal in terms of network-terminal distributed optimization of spectrum usage. This management is done within the framework of radio resource selection policies conveyed by the Network Reconfiguration Manager and in a manner consistent with the user's preferences and the available context information.

16.3.1.25   User Preferences

This term is used to refer to input parameters to decision making process on radio resource usage optimization originated from user and expressing his or her preferences. These parameters may describe, for example, preferred operator and radio interface, perceived audio/image/video quality, maximum cost, minimum data rate, etc.

16.3.2   Acronyms and Abbreviations

ASN	Abstract Syntax Notation
BS	base station
CWN	Composite Wireless Network
NRM	Network Reconfiguration Manager
OSI	Open Systems Interconnection
OSM	Operator Spectrum Manager
QoS	quality-of-service
RAN	radio access network
RMC	RAN Measurement Collector
RRC	RAN Reconfiguration Controller
SAP	service access point
SMAE	system management application entity
TMC	Terminal Measurement Collector
TRC	Terminal Reconfiguration Controller
TRM	Terminal Reconfiguration Manager
UML	Unified Modeling Language

16.4 Overall System Description

16.4.1 System Overview

The field of application of this standard is a heterogeneous wireless environment that might include the following (see Figure 16.1):

— Multiple operators
— Multiple radio access networks (RANs)
— Multiple radio interfaces
— Multiple Terminals

Within Figure 16.1, the Operator Spectrum Manager (OSM) may help the operator to coordinate the assignment of spectrum to the different RANs it owns in order to optimize radio resource usage within its Composite Wireless Network (CWN).

Within the stated field of application, the standard provides common means to

— Improve overall network capacity and quality of service
— Facilitate optimization of radio resource usage
— Support reconfiguration capabilities of RANs and Terminals
— Collect RAN and terminal context information
— Support exchange of information between the network and Terminals for radio resource usage optimization related distributed decision making
— Request and control reconfiguration of RANs and Terminals

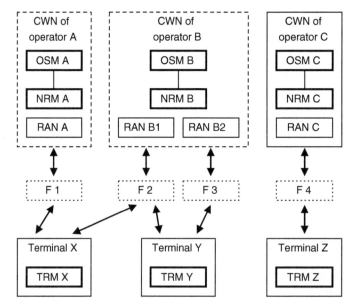

IEEE Std 1900.4-2009
IEEE standard for architectural building blocks enabling
network-device distributed decision making for optimized radio resource usage
in heterogeneous wireless access networks

CWN – Composite wireless network
OSM – Operator spectrum manager
RAN – Radio access network
NRM – Network reconfiguration manager
TRM – Terminal reconfiguration manager
F 1, 2, 3, and 4 – Frequency bands 1, 2, 3, and 4

Figure 16.1 Heterogeneous wireless environment considered in IEEE Std 1900.4. From IEEE Standard for Architectural Building Blocks Enabling Network-Device Distributed Decision Making for Optimized Radio Resource Usage in Heterogeneous Wireless Access Networks, Jan 2019. IEEE.

For this purpose, the standard defines the following:

— Network resource managers
— Device resource managers
— Interfaces between these building blocks

Within Figure 16.1, the Network Reconfiguration Manager (NRM) is the IEEE 1900.4 entity (representing a network resource manager) that manages the CWN and Terminals in terms of network-terminal distributed optimization of spectrum usage. This management is done within the framework of spectrum assignment policies conveyed by the OSM and in a manner consistent with available context information. Terminal Reconfiguration Manager (TRM) is an IEEE 1900.4 entity (representing a device resource manager) that manages the Terminal in terms of network-terminal distributed optimization of spectrum usage. This management is done within the framework of radio resource selection policies conveyed by the NRM and in a manner consistent with the user's preferences and the available context information.

16.4.2 Summary of use Cases

The following three use cases are defined within this standard:

— Dynamic spectrum assignment
— Dynamic spectrum sharing
— Distributed radio resource usage optimization

In the dynamic spectrum assignment use case, frequency bands are dynamically assigned to the RANs among the participating networks in order to optimize spectrum usage. In other words, the assigned frequency bands are not fixed, and can be dynamically changed.

OSMs generate spectrum assignment policies expressing the regulatory framework and operators objectives for spectrum usage optimization. The OSMs provide these spectrum assignment policies to the corresponding NRMs.

The NRMs analyze spectrum assignment policies and available context information and dynamically make spectrum assignment decisions to improve spectrum usage and quality of service.

After the new spectrum assignment decisions have been made, the NRMs request corresponding reconfiguration of their RANs. Following the RANs' reconfiguration, Terminals need to reconfigure correspondingly.

Example manifestations for this dynamic spectrum assignment use case are spectrum sharing, and spectrum renting between RANs. In the spectrum renting example, frequency bands of one RAN are assigned to another RAN on a temporary basis. In the spectrum sharing example, one frequency band is shared by several RANs.

Single operator and multiple operator scenarios are described within the dynamic spectrum assignment use case. Within the multiple operator scenario, there is either one NRM outside of operators CWNs or one NRM inside each operator CWN.

In the dynamic spectrum sharing use case, frequency bands assigned to RANs are fixed. However, a particular frequency band can be shared by several RANs. In other words, the dynamic spectrum sharing use case describes how fixed frequency bands are shared and/or used dynamically by RANs and Terminals.

NRMs analyze available context information and dynamically make spectrum access decisions to improve spectrum usage and quality of service. NRMs make these spectrum access decisions within the framework defined by spectrum assignment policies. Following these decisions, NRMs request corresponding reconfiguration of their RANs.

NRMs dynamically generate radio resource selection policies and send them to their TRMs. These radio resource selection policies will guide these TRMs in their spectrum access decisions.

TRMs analyze these radio resource selection policies and the available context information and dynamically make spectrum access decisions to improve spectrum usage and quality of service. These spectrum access decisions are made within the framework of the radio resource selection policies. Following these decisions, each TRM requests corresponding reconfiguration of its Terminal.

Dynamic spectrum sharing use case includes primary/secondary spectrum usage as a special case.

In the distributed radio resource usage optimization use case, frequency bands assigned to RANs are fixed. Reconfiguration of RANs is not involved in this use case.

Distributed radio resource usage optimization use case considers Terminals with or without multi-homing capability. Decision on Terminal reconfiguration is made by its TRM and is supported by NRM.

NRMs analyze available context information, dynamically generate radio resource selection policies, and send them to their TRMs. These radio resource selection policies will guide these TRMs in their reconfiguration decisions.

TRMs analyze radio resource selection policies and available context information and dynamically make decisions on reconfiguration of their Terminals to improve spectrum usage and quality of service. TRMs make these reconfiguration decisions within the framework defined by radio resource selection policies. Following these decisions, TRMs request corresponding reconfiguration of their Terminals.

Detailed description of these three use cases and their scenarios is given in Annex 16A.

16.4.3 Assumptions

16.4.3.1 General

This standard does not specify MAC and PHY layers of RANs and Terminals.

This standard does not specify the way measurements are done on network and terminal side.

A heterogeneous wireless environment exists, including one or several CWNs.

Terminals are present within this heterogeneous wireless environment.

These Terminals are reconfigurable with or without multi-homing capability.

16.4.3.2 Dynamic Spectrum Assignment

Assignment of spectrum to RANs can be dynamically changed, where spectrum assignment may be characterized by carrier frequency, signal bandwidth, and radio interface to be used in the assigned spectrum.

Concurrently with the new spectrum assignments, RANs can be reconfigured.

Concurrently with RAN reconfigurations, Terminals can be reconfigured.

16.4.3.3 Dynamic Spectrum Sharing

Assignment of spectrum to RANs is fixed.

Some RANs are allowed to concurrently operate in more than one spectrum assignment for dynamic spectrum sharing.

Some spectrum assignments are allowed to be shared by several RANs during dynamic spectrum sharing.

Some RANs can be reconfigured during dynamic spectrum sharing (maintaining their allocated spectrum assignments).

Some RANs cannot be reconfigured (for example, RANs of primary systems).

Concurrently with RAN reconfiguration, Terminals can be reconfigured during dynamic spectrum sharing.

16.4.3.4 Distributed Radio Resource Usage Optimization

Assignment of spectrum to RANs is fixed.

Reconfiguration of RANs is not involved in this use case.

Terminals with or without multi-homing capability are reconfigured during distributed radio resource usage optimization.

16.5 Requirements

16.5.1 System Requirements

16.5.1.1 Decision Making

There shall be an entity on network side, called Network Reconfiguration Manager (NRM), responsible for managing the CWN and Terminals for network-terminal-distributed optimization of spectrum usage.

There shall be an entity on terminal side, called Terminal Reconfiguration Manager (TRM), responsible for managing the Terminal for network-terminal-distributed optimization of spectrum usage.

The TRM shall manage the Terminal within the framework defined by the NRM and in a manner consistent with user's preferences and available context information.

Decision making in this standard is based on policy-based management framework.

There shall be an entity on network side responsible for generating spectrum assignment policies. The NRM shall be able to obtain spectrum assignment policies.

Spectrum assignment policies shall adhere to regulations.

Spectrum assignment policies shall express operator needs and/or radio resource usage objectives related to dynamic spectrum assignment.

The NRM shall make dynamic spectrum assignment decisions compliant with received spectrum assignment policies.

The NRM shall provide information on its dynamic spectrum assignment decisions to the entity on network side within the operator's control, which is responsible for generating spectrum assignment policies.

The NRM shall make dynamic spectrum sharing decisions compliant with its spectrum assignment decisions and with received spectrum assignment policies.

The NRM shall generate radio resource selection policies.

The NRM shall provide radio resource selection policies to TRM.

The NRM may have means to specify geo-location-based radio resource selection policies to TRM.

Radio resource selection policies and context information should be sent to an optimized set of TRMs in order not to overload selected RANs with broadcast. Target set of TRMs to be addressed may be based on combination of the following:

— RAN topology information: TRMs under coverage of list of base station IDs, such as call IDs, base transceiver station IDs, access points IDs, radio network controller IDs (if any), etc.
— TRMs inside given geo-localized area
— TRMs operating in given frequency bands
— Historical data related to Terminal radio resource usage patterns

The TRM shall make decisions on Terminal reconfiguration compliant with received radio resource selection policies.

If specified by radio resource selection policies, TRM of geo-localization-capable Terminal shall make decisions based on Terminal geo-location.

Several NRMs may collaborate in the process of decision making related to radio resource usage optimization.

If there are several NRMs, there may be interface between these NRMs.

This interface may be used to transmit the following:

— RAN context information
— Terminal context information
— Spectrum assignment policies
— RAN reconfiguration decisions
— Radio resource selection policies

16.5.1.2 Context Awareness

There shall be entities on network side and terminal side responsible for context information collection.

Context information collection entity on network side shall collect RAN context information.

RAN context information may include the following:

— RAN radio resource optimization objectives
— RAN radio capabilities
— RAN measurements
— RAN transport capabilities

The NRM shall be able to obtain RAN context information from context information collection entity on network side.

The NRM may receive this context information periodically and/or in response to request from the NRM and/or on event.

Context information collection entity on network side may be implemented in a distributed manner.

Context information collection entity on terminal side shall collect terminal context information.

Terminal context information may include the following:

— User preferences
— Required QoS levels
— Terminal capabilities
— Terminal measurements
— Terminal geo-location information
— Geo-location based terminal measurements

The TRM shall be able to obtain terminal context information from context information collection entity on terminal side.

The NRM and the TRM shall exchange context information.

The NRM shall send RAN context information to the TRM.

The NRM may send to the TRM's terminal context information related to other Terminals.

The NRM may send this context information to the TRM periodically and/or in response to request from the NRM and/or on event.

The TRM shall send terminal context information related to its Terminal to the NRM.

The TRM may send this context information to the NRM periodically and/or in response to request from the NRM and/or on event.

16.5.1.3 Reconfiguration

There shall be entities on network side and terminal side responsible for reconfiguration.

The NRM shall send reconfiguration requests to reconfiguration entity on network side.

Following received reconfiguration requests, reconfiguration entity on network side shall request and control reconfiguration of RANs.

Reconfiguration entity on network side may be implemented in a distributed manner.

The TRM shall send reconfiguration requests to reconfiguration entity on terminal side.

Following received reconfiguration requests, reconfiguration entity on terminal side shall request and control reconfiguration of Terminal.

If a maximum time interval for reconfiguration is specified by radio resource selection policies, reconfiguration of Terminal shall be performed within this time interval starting from the time when these radio resource selection policies are received.

16.5.2 Functional Requirements

16.5.2.1 NRM Functionality

The NRM shall have capability to make dynamic spectrum assignment, dynamic spectrum sharing, and distributed radio resource usage optimization decisions.

The NRM shall have capability to request reconfiguration of RANs corresponding to these dynamic spectrum assignment and dynamic spectrum sharing decisions. Actions should be taken by the NRM in the case where reconfiguration is not possible.

The NRM shall have the capability to evaluate the efficiency of spectrum usage under the current spectrum assignment.

The evaluation results shall be made available inside the NRM to assist in improving the efficiency of future dynamic spectrum assignment, dynamic spectrum sharing, and distributed radio resource usage optimization decisions.

The NRM shall have capability to generate radio resource selection policies.

There shall be no conflict in the radio resource selection policies generated by the NRM.

Radio resource selection policies shall be defined in a way that they correspond to a targeted group of Terminals (could be composed of any number of Terminals).

The targeted group of Terminals should be defined based on the needs of CWN radio resource usage optimization objectives, Terminal location, and radio resource usage patterns of the Terminals.

Radio resource selection policies shall guide TRMs in Terminals' reconfiguration decisions.

The NRM may specify radio resource selection policies referring to specific geo-location-based terminal measurements.

The NRM shall have capability to specify and control the time interval within which Terminal reconfiguration shall be performed.

The TRM shall perform Terminal reconfiguration within this time interval.

The NRM shall have capability to evaluate the efficiency of current radio resource selection policies.

These evaluation results shall be made available to the NRM to assist in improving the efficiency of future radio resource selection policies.

The NRM shall have capability to receive, process, and store the following context information:

— RAN context information (see 16.5.1.2)
— Terminal context information (see 16.5.1.2)

The NRM shall have the capability to use this context information for radio resource usage optimization purposes.

The NRM should have the capability to select RANs for exchanging radio resource selection policies and context information between the NRM and the TRM, that is, to map radio enabler onto specific RANs.

The NRM functions should have capability to perform the following in cooperation with each other:

— Make dynamic spectrum assignment, dynamic spectrum sharing, and distributed radio resource usage optimization decisions

— Request reconfiguration of RANs
— Evaluate the efficiency of spectrum usage
— Generate radio resource selection policies
— Evaluate the efficiency of current radio resource selection policies
— Select RANs for exchanging radio resource selection policies and context information between the NRM and the TRM

If there are several NRMs, there may be an interface between these NRMs.
This interface shall be used to exchange the following:

— RAN context information
— Terminal context information
— Spectrum assignment policies
— RAN reconfiguration decisions
— Radio resource selection policies

16.5.2.2 TRM Functionality

The TRM shall have the capability to make dynamic spectrum sharing and distributed radio resource usage optimization decisions, as well as, to support dynamic spectrum assignment decisions received from the NRM.

The TRM shall have the capability to request reconfiguration of its Terminal corresponding to these decisions. Actions should be taken by the TRM in cases where reconfiguration is not possible.

The TRM shall have the capability to receive, process, and store the following context information:

— Terminal context information
— RAN context information

The TRM shall have the capability to use this context information for radio resource usage optimization purposes.

The TRM shall have the capability to select RANs for exchanging radio resource selection policies and context information between the NRM and the TRM.

TRM functions should have the capability to perform the following in cooperation with each other:

— Make dynamic spectrum sharing and distributed radio resource usage optimization decisions
— Request reconfiguration of its Terminal
— Select RANs for exchanging radio resource selection policies and context information between the NRM and the TRM

16.5.3 Information Model Requirements

Information model shall provide a specified representation of information within the scope of this standard.

Information model shall consider two sets of managed objects, that is, CWN and Terminals.

CWN-related classes shall abstract operator, RAN, BS, and cell concepts within the scope of this standard.

Terminal-related classes shall abstract user, application, Terminal, frequency channel, and active connection concepts within the scope of this standard.

Information model shall abstract policy concept within the scope of this standard, including spectrum assignment policy and radio resource selection policy.

Information model may include time/duration reference related to the validity of the provided information. For instance the time at which measurements were made or the valid period in which they are to be taken.

Information model should provide geo-location related information items.

16.6 Architecture

16.6.1 System Description

According to system requirements, the following system architecture is defined in this standard (see Figure 16.2).

16.6.1.1 Entities

The following four entities are defined to represent network resource managers (see Figure 16.2):

— Operator Spectrum Manager (OSM)
— RAN Measurement Collector (RMC)
— Network Reconfiguration Manager (NRM)
— RAN Reconfiguration Controller (RRC)

The OSM is the entity that enables operator to control NRM dynamic spectrum assignment decisions.

The RMC is the entity that collects RAN context information and provides it to NRM. RMC may be implemented in a distributed manner.

The NRM is the entity that manages CWN and Terminals for network-terminal distributed optimization of spectrum usage. NRM may be implemented in a distributed manner.

The RRC is the entity that controls reconfiguration of RANs based on requests from NRM. RRC may be implemented in a distributed manner.

Three following entities are defined to represent device resource managers (see Figure 16.2):

— Terminal Measurement Collector (TMC)
— Terminal Reconfiguration Manager (TRM)
— Terminal Reconfiguration Controller (TRC)

The TMC is the entity that collects terminal context information and provides it to the TRM.

The TRM is the entity that manages the Terminal for network-terminal distributed optimization of spectrum usage within the framework defined by the NRM and in a manner consistent with user preferences and available context information.

The TRC is the entity that controls reconfiguration of Terminal based on requests from the TRM.

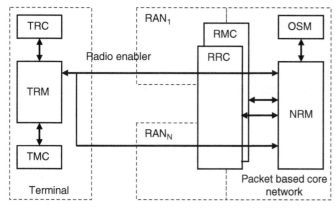

TRM – Terminal reconfiguration manager   NRM – Network reconfiguration manager
TRC – Terminal reconfiguration controller   RRC – RAN reconfiguration controller
TMC – Terminal measurement collector   RMC – RAN measurement collector
OSM – Operator spectrum manager   RAN – Radio access network

Figure 16.2 —System architecture. From IEEE Standard for Architectural Building Blocks Enabling Network-Device Distributed Decision Making for Optimized Radio Resource Usage in Heterogeneous Wireless Access Networks, Jan 2019. IEEE.

Radio enabler is the logical communication channel between the NRM and the TRM. Radio enabler may be mapped onto one or several RANs used for data transmission (in-band channel) and/or onto one or several dedicated RANs (out-of-band channel).

16.6.1.2 Interfaces Between Entities

The following key interfaces are defined (see Figure 16.2):

— Interface between the NRM and the TRM
— Interface between the TRM and the TRC
— Interface between the TRM and the TMC
— Interface between the NRM and the RRC
— Interface between the NRM and the RMC
— Interface between the NRM and the OSM

Interface Between the NRM and the TRM Interface between the NRM and the TRM is used to transmit the following:

— From NRM to TRM:
  — Radio resource selection policies
  — RAN context information
  — Terminal context information
— From TRM to NRM:
  — Terminal context information related to Terminal of this TRM

Interface Between the TRM and the TRC Interface between the TRM and the TRC is used to transmit the following:

— From TRM to TRC:
  — Terminal reconfiguration requests
— From TRC to TRM:
  — Terminal reconfiguration responses.

Interface Between the TRM and the TMC Interface between the TRM and the TMC is used to transmit the following:

— From TRM to TMC:
  — Terminal context information requests
— From TMC to TRM:
  — Terminal context information

Interface Between the NRM and the RRC Interface between the NRM and the RRC is used to transmit the following:

— From NRM to RRC:
  — RAN reconfiguration requests
— From RRC to NRM:
  — RAN reconfiguration responses

Interface Between the NRM and the RMC Interface between the NRM and the RMC is used to transmit the following:

— From NRM to RMC:
  — RAN context information requests
— From RMC to NRM:
  — RAN context information

Interface Between the NRM and the OSM Interface between the NRM and the OSM is used to transmit the following:

— From OSM to NRM:
  — Spectrum assignment policies
— From NRM to OSM:
  — Information on spectrum assignment decisions

Interface Between Several NRMs If there are several NRMs, a corresponding interface may be defined between these NRMs (not shown in Figure 16.2).

This interface is used to transmit the following:

— RAN context information
— Terminal context information
— Spectrum assignment policies
— RAN reconfiguration decisions
— Radio resource selection policies

16.6.1.3 Reference Model

In general, each IEEE 1900.4 entity (OSM, RMC, NRM, RRC, TMC, TRM, and TRC) has the reference model shown in Figure 16.3. IEEE 1900.4 entities are modeled as a system management application entity (SMAE) (see ITU-T X.701 for SMAE specification).[11] The IEEE 1900.4 entity, as SMAE, is located on the application layer and has access to any layer of the OSI model.

Each IEEE 1900.4 entity implements one or more of the following service access points (SAP):

— rCFG_TR_SAP – transport SAP
— rCFG_MEDIA_SAP – reconfiguration and measurement SAP
— rCFG_MNG_SAP – management SAP

Transport SAP provides transport service for message exchange between IEEE 1900.4 entities. It abstracts transport mechanisms from IEEE 1900.4 entities by providing a set of generic primitives and mapping these primitives on transport protocols.

For example, this SAP is used to exchange radio resource selection policies and context information between the NRM and the TRM over radio enabler.

If there are several NRMs and there is interface between them, this SAP is used to exchange context information, spectrum assignment policies, reconfiguration decisions, and radio resource selection policies between these NRMs.

Reconfiguration and Measurement SAP provides reconfiguration and measurement services for managing RANs and Terminals. It provides a set of generic primitives for IEEE 1900.4 entities to collect RAN and terminal context information, as well as, to control

Figure 16.3 IEEE 1900.4 reference model. From IEEE Standard for Architectural Building Blocks Enabling Network-Device Distributed Decision Making for Optimized Radio Resource Usage in Heterogeneous Wireless Access Networks, Jan 2019. IEEE.

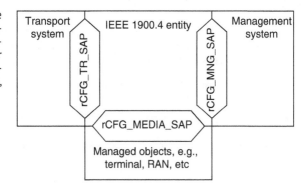

---

11 Information on references can be found in Clause 16.2.

reconfiguration of RANs and Terminals. These generic primitives are mapped onto specific protocols depending on the managed RANs and Terminals.

Management SAP provides management service for managing IEEE 1900.4 entities by legacy management systems. This SAP provides a set of generic primitives for IEEE 1900.4 entities to exchange information with these legacy management systems.

16.6.2 Functional Description

According to functional requirements, the following functional architecture is defined in this standard (see Figure 16.4).

16.6.2.1 NRM Functions

The following functions are defined inside NRM (see Figure 16.4):

— Policy Derivation
— Policy Efficiency Evaluation
— Network Reconfiguration Decision and Control
— Spectrum Assignment Evaluation
— Information Extraction, Collection, and Storage
— RAN Selection

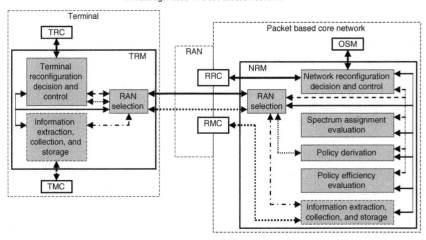

IEEE Std 1900.4-2009
IEEE Standard for architectural building blocks enabling
network-device distributed decision making for optimized radio resource usage
in heterogeneous wireless access networks

- IEEE 1900.4 entities

- Functions related to context awareness

- Functions related to decision making and reconfiguration

◀·····▶ - External interfaces of NRM and TRM functions related to context awareness

◀——▶ - External interfaces of NRM and TRM functions related to decision making and reconfiguration

◀——▶ - Internal interfaces of NRM and TRM functions related to context awareness, given as implementation example, and not defined in this standard

◀– –▶ - Internal interfaces of NRM and TRM functions related to decision making, given as implementation example, and not defined in this standard

◀·–▶ - Internal interfaces of NRM and TRM functions related to exchange of context information, given as implementation example, and not defined in this standard

◀·······▶ - Internal interfaces of NRM and TRM functions related to sending radio resource selection policies, given as implementation example, and not defined in this standard

Figure 16.4 Functional architecture. From IEEE Standard for Architectural Building Blocks Enabling Network-Device Distributed Decision Making for Optimized Radio Resource Usage in Heterogeneous Wireless Access Networks, Jan 2019. IEEE.

Policy Derivation function generates radio resource selection policies that guide TRMs in Terminals reconfiguration decisions.

The radio resource selection policies are derived according to context information from the Information Extraction, Collection, and Storage function.

The Policy Efficiency Evaluation function evaluates the efficiency of current radio resource selection policies.

Evaluation results may be used by the Policy Derivation function during generating radio resource selection policies.

The Network Reconfiguration Decision and Control function makes decisions on the RAN's reconfiguration compliant with spectrum assignment policies received from OSM. After making these decisions, Network Reconfiguration Decision and Control function sends corresponding reconfiguration requests to the RRC. Also, the Network Reconfiguration Decision and Control function sends information on the decisions that have been made to the OSM.

The Spectrum Assignment Evaluation function evaluates the efficiency of spectrum usage under the current spectrum assignment.

Evaluation results may be used by the Network Reconfiguration Decision and Control function while making decisions on the RAN's reconfiguration.

The Information Extraction, Collection, and Storage function receives, processes, and stores the following context information:

— RAN context information
— Terminal context information

RAN context information is received from the RMC periodically and/or by request and/or on event-basis.

Terminal context information is received from the TRM periodically and/or by request and/or on event-basis.

The Information Extraction, Collection, and Storage function provides information to functions inside the NRM.

The Information Extraction, Collection, and Storage function forwards RAN context information to the TRM.

The Information Extraction, Collection, and Storage function may forward terminal context information, related to other Terminals, to the TRM.

The RAN Selection function selects RANs for exchanging radio resource selection policies and context information between the NRM and one or several TRMs. Radio resource selection policies are sent from NRM to TRM. From NRM to TRM, RAN context information is sent and terminal context information may be sent. From TRM, NRM receives terminal context information related to Terminal managed by this TRM.

Policy Derivation, Policy Efficiency Evaluation, RAN Selection, Network Reconfiguration Decision and Control, and Spectrum Assignment Evaluation functions cooperate during their operation. They represent different aspects of decision making and reconfiguration. During their operation, these functions use information from the Information Extraction, Collection, and Storage function.

16.6.2.2 TRM Functions

The following functions are defined inside TRM (see Figure 16.4):

— Terminal Reconfiguration Decision and Control
— Information Extraction, Collection, and Storage
— RAN Selection

The Terminal Reconfiguration Decision and Control function makes decisions on Terminal reconfiguration. After making these decisions, the Terminal Reconfiguration Decision and Control function sends corresponding reconfiguration requests to the TRC.

The Information Extraction, Collection, and Storage function receives, processes, and stores the following context information:

— Terminal context information
— RAN context information

Terminal context information is received from the TMC periodically and/or by request and/or on event-basis.

Terminal context information regarding other Terminals is received from the NRM periodically and/or by request and/or on event-basis.

RAN context information is received from the NRM periodically and/or by request and/or on event-basis.

The Information Extraction, Collection, and Storage function provides information to functions inside the TRM.

The Information Extraction, Collection, and Storage function forwards terminal context information to the NRM.

The RAN Selection function selects RANs for exchanging radio resource selection policies and context information between the NRM and the TRM through the radio enabler. Radio resource selection policies are sent from NRM to TRM. From NRM to TRM, RAN context information and terminal context information are sent. From TRM to NRM, terminal context information is sent.

The Terminal Reconfiguration Decision and Control and the RAN Selection functions cooperate during their operation. They represent different aspects of decision making and reconfiguration. During their operation, these functions use information from the Information Extraction, Collection, and Storage function.

The Terminal Reconfiguration Decision and Control function makes reconfiguration decisions within the framework determined by the received radio resource selection policies.

16.6.2.3 Interfaces of NRM and TRM Functions

NRM Interfaces The following interfaces between the NRM and other IEEE 1900.4 entities on network side are defined (see Figure 16.4):

— Interface between Network Reconfiguration Decision and Control function and OSM
— Interface between Network Reconfiguration Decision and Control function and RRC
— Interface between Information Extraction, Collection, and Storage function and RMC

Interface between Network Reconfiguration Decision and Control function and OSM is used to transmit the following:

— From OSM to Network Reconfiguration Decision and Control function:
  — Spectrum assignment policies
— From Network Reconfiguration Decision and Control function to OSM:
  — NRM spectrum assignment decisions

Interface between Information Extraction, Collection, and Storage function and RMC is used to transmit the following:

— From Information Extraction, Collection, and Storage function to RMC:
  — RAN context information requests
— From RMC to Information Extraction, Collection, and Storage function:
  — RAN context information

Interface between Network Reconfiguration Decision and Control function and RRC is used to transmit the following:

— From Network Reconfiguration Decision and Control function to RRC:
  — RAN reconfiguration requests

— From RRC to Network Reconfiguration Decision and Control function:
  — RAN reconfiguration responses

If NRM is implemented in a distributed manner, that is, there are several NRMs, and there is interface between these NRMs, the following interfaces are additionally defined (not shown in Figure 4):

— Interface between Information Extraction, Collection, and Storage functions of several NRMs
— Interface between Network Reconfiguration Decision and Control functions of several NRMs
— Interface between Policy Derivation functions of several NRMs

Interface between Information Extraction, Collection, and Storage functions of several NRMs is used to exchange the following context information:

— RAN context information
— Terminal context information

Interface between Network Reconfiguration Decision and Control functions of several NRMs is used to exchange the following:

— Spectrum assignment policies
— RAN reconfiguration decisions

Interface between Policy Derivation functions of several NRMs is used to exchange the following:

— Radio resource selection policies

An interface between the RAN Selection function in the NRM and the RAN Selection function in the TRM is defined (see Figure 16.4). The RAN Selection functions distribute on this interface the radio resource selection policies to a specific defined group of Terminals, based on location and/or radio resource usage volume.

This interface is used by the NRM to transmit the following to the TRM:

— Radio resource selection policies
— RAN context information
— Terminal context information

This interface is used by the NRM to receive the following by the TRM:

— Terminal context information

TRM Interfaces The following interfaces between TRM and other IEEE 1900.4 entities on terminal side are defined (see Figure 16.4):

— Interface between Information Extraction, Collection, and Storage function and TMC
— Interface between Terminal Reconfiguration Decision and Control function and TRC

Interface between Information Extraction, Collection, and Storage function and TMC is used to transmit the following:

— From Information Extraction, Collection, and Storage function to TMC:
  — Terminal context information requests
— From TMC to Information Extraction, Collection, and Storage function:
  — Terminal context information

Interface between Terminal Reconfiguration Decision and Control function and TRC is used to transmit the following:

— From Terminal Reconfiguration Decision and Control function to TRC:
  — Terminal reconfiguration requests
— From TRC to Terminal Reconfiguration Decision and Control function:
  — Terminal reconfiguration responses

Interface between RAN Selection function in TRM and RAN Selection function in NRM is defined (see Figure 16.4). The RAN Selection functions distribute on this interface the radio resource selection policies to specific defined group of Terminals, based on location and/or radio resource usage volume.

This interface is used by the TRM to transmit the following to the NRM:

— Terminal context information

This interface is used by the TRM to receive the following from the NRM:

— Radio resource selection policies
— RAN context information
— Terminal context information

The system architecture defined in this standard can be deployed in different ways. Concrete deployment examples are given in Annex 16E, showing IEEE 1900.4 entities, interfaces, and SAPs.

16.7 Information Model

16.7.1 Introduction

IEEE 1900.4 uses an information model based on an object-oriented approach, whereby given that CWN and Terminals are controlled by an IEEE 1900.4 system, they are viewed as the two sets of managed objects.

To this end, the terminal-related classes abstract the user, application, device, and radio resource selection policy concepts, for instance structuring different profiles, capabilities, and measurements related to the Terminal.

The CWN-related classes present an abstract view of the CWN, capturing the operator and RANs, where operator concept includes assigned channels, regulatory rules, and spectrum assignment policies and RAN concept includes BSs and cells.

It must be noted that the presented conceptual abstraction is fully aligned to the scope of this standard. For example, **Application** class in the hierarchy of terminal-related classes does not incorporate generic application attributes; rather, it only incorporates those that have been identified within the standard scope.

Policy information as represented/abstracted by policy classes is a fundamental part of this standard ensuring the communication of policies to NRMs and TRMs to define framework of their operation. This standard therefore presents a concrete representation of events that trigger policies activation and execution, the conditions within which policies must act, and the precise actions that must be undertaken should, for example, a Terminal is found to be violating a policy.

In summary, the information model classes are grouped into the following categories:

— Common base class
— Policy classes
— Terminal-related classes
— CWN-related classes

For each of these categories, the simplified UML (see UML Version 2.1.2) diagrams are described in this clause. Detailed UML diagrams, as well as, attributes and methods of each class are given in precise detail in Annex 16B and Annex 16C. Annex 16D defines utility classes.

The rest of this clause is organized as follows. Subclause 16.7.2 specifies information modeling approach used in this standard, while 16.7.3 describes information model classes.

16.7.2 Information Modeling Approach

In order to fulfill the requirements, to keep the necessary level of information abstraction, to be prepared for future extensions, and to be easily used by different tools, the following properties have been taken into account for the information modeling:

a) The information model is developed in an extensible form in order to accommodate future radio access technologies and allow for custom extensions to existing data models.

b) The information model uses an object-oriented approach.

c) The information model supports sufficiently simple relationships between different classes.

d) The information model allows for inclusion of both uniform and non-uniform data structures (e.g., lists).

e) The information model allows for definition of new abstract data types to describe the information model items.

f) The information model allows providing information items allowing for specification of precision and accuracy.

g) The information model includes exclusivity or consistency relationships between objects to determine conflicts (for instance whether two different channels or radio technologies can be monitored at the same time).

h) The information model provides means for unique identification of managed objects.

i) The information model utilizes platform-independent unambiguous information/data type definitions.

j) The information model allows for inclusion of information about information objects distribution (e.g., to identify the targeted nodes in a multicasting case).

k) The information model is open to incorporate
   — Corresponding information elements towards developing a shared knowledge framework about the information objects themselves. Such framework may include information about the updates, status etc.
   — A notifications list, such as configuration changes, threshold crossings etc to align the shared knowledge framework.
   — Additional information elements to ensure alternative information retrieval for supporting an efficient retrieval mechanism to obtain performance, quality-of-service and related information and measurements data.
   — Information elements that can provide value (instantiate) through mechanisms such as statistical operations to reduce data transfers.
   — Managed objects in order to coordinate the measurements scheduling.

16.7.3 Information Model Classes

The IEEE 1900.4 information model classes are depicted together with the relations among them in order to present the breaking down of the conceptualization in the adopted several levels of abstraction/details. The corresponding cardinality information has been also included.

16.7.3.1 Common Base Class

Common base class is shown in Figure 16.5.

The following common base class is defined:

— **19004BaseClass**

This class comprises basic properties to be supported by all objects of the IEEE 1900.4 information model. It is considered as an abstract class that is only used for inheritance. The properties supported are, e.g., an attribute representing the class name of an instance, and three generic events that an instance of this class can report to a managing system. These three events are: creation of a new instance of this class, deletion of an instance of this class, change of an attribute value in an instance of this class.

Figure 16.5 Common base class. From IEEE Standard for Architectural Building Blocks Enabling Network-Device Distributed Decision Making for Optimized Radio Resource Usage in Heterogeneous Wireless Access Networks, Jan 2019. IEEE.

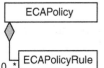

Figure 16.6 Policy classes. From IEEE Standard for Architectural Building Blocks Enabling Network-Device Distributed Decision Making for Optimized Radio Resource Usage in Heterogeneous Wireless Access Networks, Jan 2019. IEEE.

16.7.3.2 Policy Classes

Policy classes are shown in Figure 16.6.

The following policy classes are defined:

— **ECA Policy**

This class is used to describe policies of type Event-Condition-Action (ECA). An instance of this class comprises a set of Event-Condition-Action rules that have to be obeyed when applying the policy.

— **ECA Policy Rule**

An instance of this class describes (in terms of three attributes)

a) an event that triggers the evaluation of the policy condition

b) a condition that shall be fulfilled before applying the policy action

c) the action that has to be performed if the event has occurred and the condition is fulfilled

All these attributes refer to information entities that are available in the system that applies the policy.

16.7.3.3 Terminal-related Classes

A simplified UML diagram of terminal-related classes is shown in Figure 16.7. A detailed UML diagram of terminal-related classes is available in Annex 16B.

The following terminal-related classes are defined:

— **Terminal**

The instance of this class contains instances of all terminal-related classes using composition.

— **User**

This class describes information related to a user of the Terminal. Each instance of **Terminal** class can have one or several instances of **User** class as members.

— **User Profile**

This class contains general information about one user of the Terminal, for example, user ID. Each instance of **User** class can have only one instance of **User Profile** class as a member.

— **User Subscription**

This class contains information about one subscription of the user. It describes which RANs/services the user has been subscribed in and what is the associated cost. Each instance of **User Profile** class can have one or several instances of **User Subscription** class as members.

— **User Preference**

This class describes in a formalized form one preference of the user, for example, preferred operator and radio interface, perceived audio/image/video quality, maximum cost, minimum data rate, etc. Each instance of **User** class can have zero or several instances of **User Preference** class as members.

— **Application**

This class describes one currently active application. Each instance of **Terminal** class can have zero or several instances of **Application** class as members.

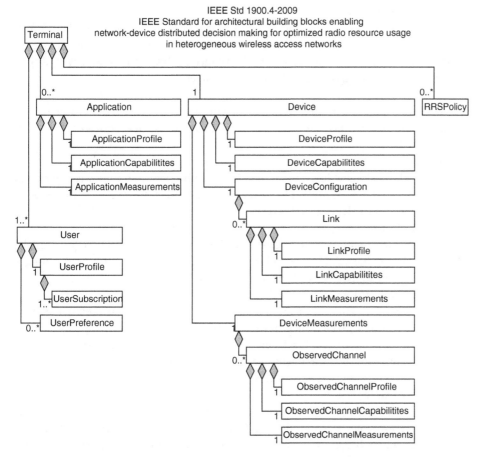

IEEE Std 1900.4-2009
IEEE Standard for architectural building blocks enabling
network-device distributed decision making for optimized radio resource usage
in heterogeneous wireless access networks

Figure 16.7 Terminal-related classes. From IEEE Standard for Architectural Building Blocks Enabling Network-Device Distributed Decision Making for Optimized Radio Resource Usage in Heterogeneous Wireless Access Networks, Jan 2019. IEEE.

— **Application Profile**

This class contains general information about the application, for example, application ID, traffic class, direction (downlink or uplink), links used to deliver this application, QoS requirements etc. Each instance of **Application** class can have only one instance of **Application Profile** class as a member.

— **Application Capabilities**

This class contains information about measurements (instantaneous measurement data and performance statistics derived from this data) supported by this application, for example, delay, loss, and bandwidth measurements. Each instance of **Application** class can have only one instance of **Application Capabilities** class as a member.

— **Application Measurements**

This class contains measurements (instantaneous measurement data and performance statistics derived from this data) performed by this application, such as delay, loss, and bandwidth measurements. Each instance of **Application** class can have only one instance of **Application Measurements** class as a member.

— **Device**

This class describes all radio interface related hardware and software of a Terminal, as well as, measurement information related to radio resources within the Terminal. Each instance of **Terminal** class can have only one instance of **Device** class as a member.

— **Device Profile**

This class contains general information about the Terminal, for example, Terminal ID. Each instance of **Device** class can have only one instance of **Device Profile** class as a member.

Device Capabilities

This class contains information about Terminal capabilities including both transmission and measurement capabilities, for example, supported radio interfaces, maximum transmission power, etc. Each instance of **Device** class can have only one instance of **Device Capabilities** class as a member.

— **Device Configuration**

This class contains information about the current configuration of Terminal. Each instance of **Device** class can have only one instance of **Device Configuration** class as a member.

— **Link**

This class contains information about one active connection between Terminal and RANs. Each instance of **Device Configuration** class can have zero or several instances of **Link** class as members.

— **Link Profile**

This class contains general information about this active connection, for example, link ID, serving cell ID, channel used, etc. Each instance of **Link** class can have only one instance of **Link Profile** class as a member.

— **Link Capabilities**

This class contains information about measurements (instantaneous measurement data and performance statistics derived from this data) supported on this active connection, such as block error rate, power, and signal-to-interference-plus-noise-ratio measurements. Each instance of **Link** class can have only one instance of **Link Capabilities** class as a member.

— **Link Measurements**

This class contains current measurements (instantaneous measurement data and performance statistics derived from this data) related to this active connection, such as block error rate, power, and signal-to-interference-plus-noise-ratio measurements. Each instance of **Link** class can have only one instance of **Link Measurements** class as a member.

— **Device Measurements**

This class contains current measurements (instantaneous measurement data and performance statistics derived from this data) related to Terminal, for example, battery capacity and Terminal location measurements, as well as, measurements related to observed channels not having active connections with the Terminal. Each instance of **Device** class can have only one instance of **Device Measurements** class as a member.

— **Observed Channel**

This class describes one frequency channel that does not have active connection with the Terminal, but is observed by this Terminal. Each instance of **Device Measurements** class can have zero or several instances of **Observed Channel** class as members.

— **Observed Channel Profile**

This class contains general information about this frequency channel, for example, channel ID, frequency range, etc. Each instance of **Observed Channel** class can have only one instance of **Observed Channel Profile** class as a member.

— **Observed Channel Capabilities**

This class contains information about measurements (instantaneous measurement data and performance statistics derived from this data) supported on this frequency channel, such as interference and load measurements. Each instance of **Observed Channel** class can have only one instance of **Observed Channel Capabilities** class as a member.

— **Observed Channel Measurements**

This class contains current measurements (instantaneous measurement data and performance statistics derived from this data) related to this frequency channel, such as interference and load measurements. Each instance of **Observed Channel** class can have only one instance of **Observed Channel Measurements** class as a member.

— **RRS Policy**

This class describes one radio resource selection (RRS) policy related to this Terminal. Each instance of **Terminal** class can have zero or several instances of **RRS Policy** class as members.

16.7.3.4 CWN-related Classes

A simplified UML diagram of CWN-related classes is shown in Figure 16.8. A detailed UML diagram of CWN-related classes is available in Annex 16B.

The following CWN-related classes are defined:

— **CWN**

The instance of this class contains instances of all CWN-related classes using composition.

— **Operator**

This class describes the operator of this CWN. Each instance of **CWN** class can have only one instance of **Operator** class as a member.

— **Operator Profile**

This class contains general information about the operator, for example, operator ID. Each instance of **Operator** class can have only one instance of **Operator Profile** class as a member.

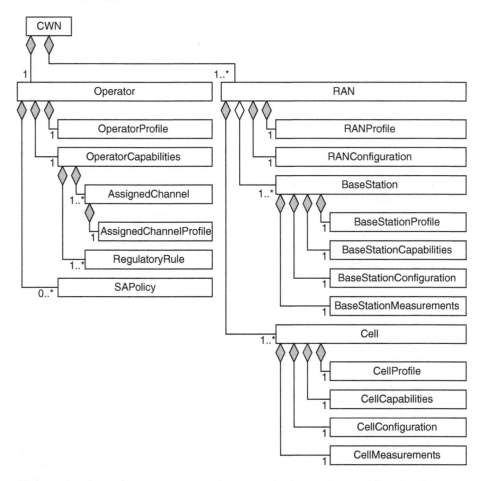

Figure 16.8 CWN-related classes. From IEEE Standard for Architectural Building Blocks Enabling Network-Device Distributed Decision Making for Optimized Radio Resource Usage in Heterogeneous Wireless Access Networks, Jan 2019. IEEE.

— **Operator Capabilities**

This class describes operator capabilities. Each instance of **Operator** class can have only one instance of **Operator Capabilities** class as a member.

— **Assigned Channel**

This class describes one frequency channel assigned to this operator. Each instance of **Operator** class can have one or several instances of **Assigned Channel** class as members.

— **Assigned Channel Profile**

This class contains general information about this frequency channel, for example, frequency channel ID, frequency range, and allowed radio interfaces. Each instance of **Assigned Channel** class can have only one instance of **Assigned Channel Profile** class as a member.

— **Regulatory Rule**

This class describes in a formalized form one regulatory rule to be applied to one or several assigned channels. Each instance of **Operator Capabilities** class can have one or several instances of **Regulatory Rule** class as members.

— **SA Policy**

This class describes one spectrum assignment (SA) policy specified by this operator. Each instance of **Operator** class can have zero or several instances of **SA Policy** class as members.

— **RAN**

This class describes one RAN of this CWN. Each instance of **CWN** class can have one or several instances of **RAN** class as members.

— **RAN Profile**

This class contains general information about this RAN, for example, RAN ID. Each instance of **RAN** class can have only one instance of **RAN Profile** class as a member.

— **RAN Configuration**

This class describes current configuration of this RAN, for example, RAN users. Each instance of **RAN** class can have only one instance of **RAN Configuration** class as a member.

— **Base Station**

This class describes one base station of the RAN. Each instance of **RAN** class can relate to one or several instances of **Base Station** class.

— **Base Station Profile**

This class contains general information about this base station, for example, base station ID, vendor, and location. Each instance of **Base Station** class can have only one instance of **Base Station Profile** class as a member.

— **Base Station Capabilities**

This class contains information about base station capabilities including both transmission and measurement capabilities, for example, supported radio interfaces, supported channels, transport capability, etc. Each instance of **Base Station** class can have only one instance of **Base Station Capabilities** class as a member.

— **Base Station Configuration**

This class contains information about the current configuration of the base station, for example, frequency channels and radio interfaces used. Each instance of **Base Station** class can have only one instance of **Base Station Configuration** class as a member.

— **Base Station Measurements**

This class contains current measurements (instantaneous measurement data and performance statistics derived from this data) performed by this base station, for example, transmission power and load measurements. Each instance of **Base Station** class can have only one instance of **Base Station Measurements** class as a member.

— Cell

This class describes one cell of the base station. Each instance of RAN class can have one or several instances of Cell class as members.

— Cell Profile

This class contains general information about this cell, for example, cell ID, location, coverage area, etc. Each instance of **Cell** class can have only one instance of **Cell Profile** class as a member.

— Cell Capabilities

This class contains information about cell capabilities, for example, supported radio interfaces, supported channels, supported measurements, etc. Each instance of **Cell** class can have only one instance of **Cell Capabilities** class as a member.

— Cell Configuration

This class contains information about the current configuration of the cell, for example, Terminals served and transport service used. Each instance of **Cell** class can have only one instance of **Cell Configuration** class as a member.

— Cell Measurements

This class contains current measurements (instantaneous measurement data and performance statistics derived from this data) related to this cell, for example, transmission power, cell and traffic loads, throughput, and interference measurements. Each instance of **Cell** class can have only one instance of **Cell Measurements** class as a member.

16.8   Procedures

16.8.1   Introduction

This clause describes the IEEE 1900.4 generic procedures.

As described in the reference use cases, three scenarios for heterogeneous wireless environment are considered in this standard:

— Single operator
— Multiple operator 1 (NRM is inside operator)
— Multiple operator 2 (NRM is outside operators)

Subclause 16.8.2 defines the following IEEE 1900.4 generic procedures:

— Collecting context information (see 16.8.2.1)
— Generating spectrum assignment policies (see 16.8.2.2)
— Making spectrum assignment decision (see 16.8.2.3)
— Performing spectrum access on network side (see 16.8.2.4)
— Generating radio resource selection policies (see 16.8.2.5)
— Performing reconfiguration on terminal side (see 16.8.2.6)

Each of these IEEE 1900.4 generic procedures has three descriptions, corresponding to three reference network architectures.

Based on the IEEE 1900.4 generic procedures defined in 16.8.2, subclause 16.8.3 gives examples of realization of the IEEE 1900.4 reference use cases.

Figure 16.9, Figure 16.10, and Figure 16.11 describe how IEEE 1900.4 system architecture (see Clause 16.6) can be applied to single operator, multiple operator 1 (NRM is inside operator), and multiple operator 2 (NRM is outside operators) scenarios. These figures show differences between these scenarios and are required for understanding the IEEE 1900.4 generic procedures described in this clause. More deployment examples can be found in Annex 16E.

Figure 16.9 illustrates the single operator scenario.

Single operator scenario considers one CWN and one Terminal (multiple Terminals are not required to define generic procedures).

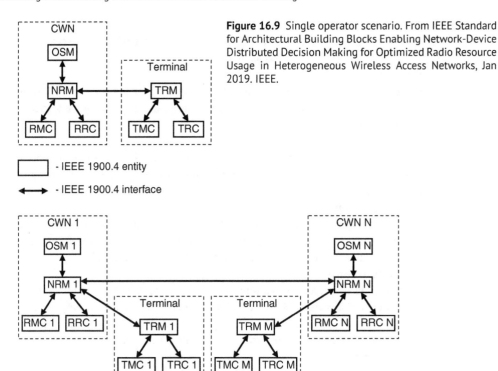

Figure 16.9 Single operator scenario. From IEEE Standard for Architectural Building Blocks Enabling Network-Device Distributed Decision Making for Optimized Radio Resource Usage in Heterogeneous Wireless Access Networks, Jan 2019. IEEE.

Figure 16.10 Multiple operator scenario 1 (NRM is inside operator). From IEEE Standard for Architectural Building Blocks Enabling Network-Device Distributed Decision Making for Optimized Radio Resource Usage in Heterogeneous Wireless Access Networks, Jan 2019. IEEE.

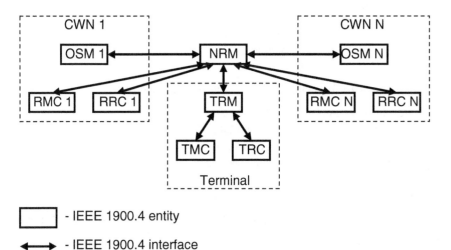

Figure 16.11 Multiple operator scenario 2 (NRM is outside operators). From IEEE Standard for Architectural Building Blocks Enabling Network-Device Distributed Decision Making for Optimized Radio Resource Usage in Heterogeneous Wireless Access Networks, Jan 2019. IEEE.

The following IEEE 1900.4 entities are used:

— OSM
— NRM
— RMC
— RRC

— TRM
— TMC
— TRC

The following IEEE 1900.4 interfaces are used:

— Interface between OSM and NRM
— Interface between NRM and RMC
— Interface between NRM and RRC
— Interface between NRM and TRM
— Interface between TRM and TMC
— Interface between TRM and TRC

Figure 16.10 illustrates the multiple operator scenario 1 (NRM is inside operator).

Multiple operator scenario 1 (NRM is inside operator) considers multiple CWNs and multiple Terminals. The number of CWNs is denoted as N and the number of Terminals is denoted as M.

The following IEEE 1900.4 entities are used:

— OSM 1, ..., OSM N
— NRM 1, ..., NRM N
— RMC 1, ..., RMC N
— RRC 1, ..., RRC N
— TRM 1, ..., TRM M
— TMC 1, ..., TMC M
— TRC 1, ..., TRC M

The following IEEE 1900.4 interfaces are used:

— Interface between OSM i and NRM i, i = 1, ..., N
— Interface between NRM i and RMC i
— Interface between NRM i and RRC i
— Interface between NRM i and NRM j, j = 1, ..., N, where i ≠ j
— Interface between NRM i and TRM k, k = 1, ..., M
— Interface between TRM k and TMC k
— Interface between TRM k and TRC k

Figure 16.11 illustrates the multiple operator scenario 2 (NRM is outside operators).

Multiple operator scenario 2 (NRM is outside operators) considers multiple CWNs and one Terminal (multiple Terminals are not required to define generic procedures). The number of CWNs is denoted as N.

The following IEEE 1900.4 entities are used in this reference network architecture:

— OSM 1, ..., OSM N
— NRM
— RMC 1, ..., RMC N
— RRC 1, ..., RRC N
— TRM
— TMC
— TRC

The following IEEE 1900.4 interfaces are used in this reference network architecture:

— Between OSM i and NRM, i = 1, ..., N
— Between NRM and RMC i
— Between NRM and RRC i
— Between NRM and TRM
— Between TRM and TMC
— Between TRM and TRC

16.8.2 Generic Procedures

16.8.2.1 Collecting Context Information

Single Operator Collecting context information procedure for single operator scenario is shown in Figure 16.12.

Collecting context information procedure for single operator scenario is as follows:

— TMC forwards terminal context information about its Terminal to TRM
— RMC forwards RAN context information about its RANs to NRM
— TRM sends terminal context information about its Terminal to NRM
— NRM sends RAN context information about its RANs to TRM
— NRM may send terminal context information about other Terminals to TRM

Multiple Operator 1 (NRM is Inside Operator) Collecting context information procedure for multiple operator scenario 1 (NRM is inside operator) is shown in Figure 16.13.

Collecting context information procedure for multiple operator scenario 1 (NRM is inside operator) is as follows:

— TMC k forwards terminal context information about its Terminal to TRM k
— RMC i forwards RAN context information about RANs of operator i to NRM i
— TRM k sends terminal context information about its Terminal to NRM i
— NRM i sends RAN context information about RANs of operator i to TRM k
— NRM i may send terminal context information about other Terminals of operator i to TRM k
— TMC l forwards terminal context information about its Terminal to TRM l
— RMC j forwards RAN context information about RANs of operator j to NRM j
— TRM l sends terminal context information about its Terminal to NRM j
— NRM j sends RAN context information about RANs of operator j to TRM l
— NRM j may send terminal context information about other Terminals of operator j to TRM l
— NRM i sends terminal and RAN context information about Terminals and RANs of operator i to NRM j
— NRM j sends terminal and RAN context information about Terminals and RANs of operator j to NRM i
— NRM i sends RAN context information about RANs of operator j to TRM k
— NRM i may send terminal context information about Terminals of operator j to TRM k
— NRM j sends RAN context information about RANs of operator i to TRM l
— NRM j may send terminal context information about Terminals of operator i to TRM l

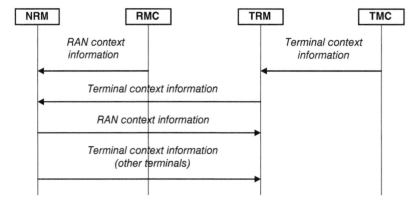

Figure 16.12 Collecting context information procedure for single operator scenario. From IEEE Standard for Architectural Building Blocks Enabling Network-Device Distributed Decision Making for Optimized Radio Resource Usage in Heterogeneous Wireless Access Networks, Jan 2019. IEEE.

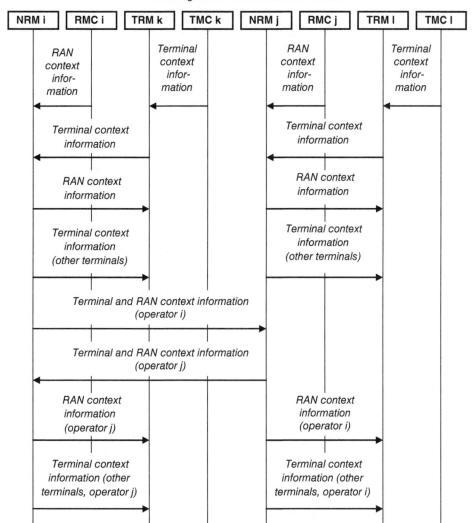

Figure 16.13 Collecting context information procedure for multiple operator scenario 1 (NRM is inside operator). From IEEE Standard for Architectural Building Blocks Enabling Network-Device Distributed Decision Making for Optimized Radio Resource Usage in Heterogeneous Wireless Access Networks, Jan 2019. IEEE.

Multiple Operator 2 (NRM is Outside Operators) Collecting context information procedure for multiple operator scenario 2 (NRM is outside operators) is shown in Figure 16.14.

Collecting context information procedure for multiple operator scenario 2 (NRM is outside operators) is as follows:

— TMC forwards terminal context information about its Terminal to TRM
— RMC i forwards RAN context information about RANs of operator i to NRM
— TRM sends terminal context information about its Terminal to NRM
— NRM sends RAN context information about RANs of operators 1, ..., N to TRM
— NRM may send terminal context information about other Terminals to TRM

16.8.2.2 Generating Spectrum Assignment Policies

Single Operator Generating spectrum assignment policies procedure for single operator scenario is shown in Figure 16.15.

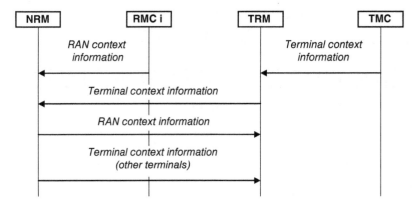

Figure 16.14 —Collecting context information procedure for multiple operator scenario 2 (NRM is outside operators). From IEEE Standard for Architectural Building Blocks Enabling Network-Device Distributed Decision Making for Optimized Radio Resource Usage in Heterogeneous Wireless Access Networks, Jan 2019. IEEE.

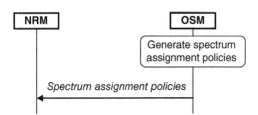

Figure 16.15 —Generating spectrum assignment policies procedure for single operator scenario. From IEEE Standard for Architectural Building Blocks Enabling Network-Device Distributed Decision Making for Optimized Radio Resource Usage in Heterogeneous Wireless Access Networks, Jan 2019. IEEE.

Generating spectrum assignment policies procedure for single operator scenario is as follows:

— OSM generates spectrum assignment policies
— OSM sends spectrum assignment policies to its NRM

Multiple Operator 1 (NRM is Inside Operator) Generating spectrum assignment policies procedure for multiple operator scenario 1 (NRM is inside operator) is shown in Figure 16.16.

Generating spectrum assignment policies procedure for multiple operator scenario 1 (NRM is inside operator) is as follows:

— OSM i generates spectrum assignment policies on behalf of operator i
— OSM i sends spectrum assignment policies to NRM i

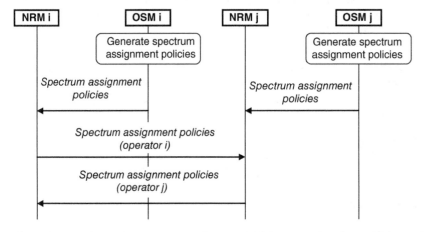

Figure 16.16 —Generating spectrum assignment policies procedure for multiple operator scenario 1 (NRM is inside operator). From IEEE Standard for Architectural Building Blocks Enabling Network-Device Distributed Decision Making for Optimized Radio Resource Usage in Heterogeneous Wireless Access Networks, Jan 2019. IEEE.

— OSM j generates spectrum assignment policies on behalf of operator j
— OSM j sends spectrum assignment policies to NRM j
— NRM i sends spectrum assignment policies of operator i to NRM j
— NRM j sends spectrum assignment policies of operator j to NRM i

Multiple Operator 2 (NRM is Outside Operators) The generating spectrum assignment policies procedure for multiple operator scenario 2 (NRM is outside operators) is shown in Figure 16.17.

The generating spectrum assignment policies procedure for multiple operator scenario 2 (NRM is outside operators) is as follows:

— OSMs 1, …, N generate spectrum assignment policies on behalf of operators 1, …, N
— OSMs 1, …, N send spectrum assignment policies to its NRM

16.8.2.3 Making Spectrum Assignment Decision

Single Operator The making spectrum assignment decision procedure for single operator scenario is shown in Figure 16.18.

The making spectrum assignment decision procedure for single operator scenario is as follows:

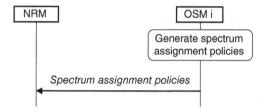

Figure 16.17 —Generating spectrum assignment policies procedure for multiple operator scenario 2 (NRM is outside operators). From IEEE Standard for Architectural Building Blocks Enabling Network-Device Distributed Decision Making for Optimized Radio Resource Usage in Heterogeneous Wireless Access Networks, Jan 2019. IEEE.

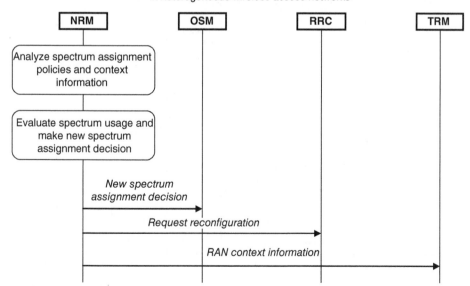

Figure 16.18 —Making spectrum assignment decision procedure for single operator scenario. From IEEE Standard for Architectural Building Blocks Enabling Network-Device Distributed Decision Making for Optimized Radio Resource Usage in Heterogeneous Wireless Access Networks, Jan 2019. IEEE.

— NRM analyzes spectrum assignment policies and context information
— NRM evaluates spectrum usage and makes new spectrum assignment decision
— NRM reports new spectrum assignment decision to its OSM
— NRM requests corresponding reconfiguration of its RANs to its RRC
— NRM sends new RAN context information to its TRM

Multiple Operator 1 (NRM is Inside Operator) The making spectrum assignment decision procedure for multiple operator scenario 1 (NRM is inside operator) is shown in Figure 16.19.

The making spectrum assignment decision procedure for multiple operator scenario 1 (NRM is inside operator) is as follows:

— NRM i, and NRM j negotiate regarding dynamic spectrum assignment
— NRM i analyzes spectrum assignment policies and context information
— NRM i evaluates spectrum usage and makes new spectrum assignment decision
— NRM i reports new spectrum assignment decision to OSM i
— NRM i requests corresponding reconfiguration of RANs of operator i to RRC i

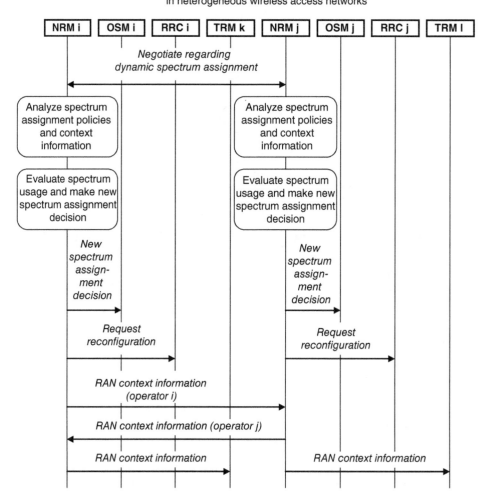

Figure 16.19 —Making spectrum assignment decision procedure for multiple operator scenario 1 (NRM is inside operator). From IEEE Standard for Architectural Building Blocks Enabling Network-Device Distributed Decision Making for Optimized Radio Resource Usage in Heterogeneous Wireless Access Networks, Jan 2019. IEEE.

— NRM j analyzes spectrum assignment policies and context information
— NRM j evaluates spectrum usage and makes new spectrum assignment decision
— NRM j reports new spectrum assignment decision to OSM j
— NRM j requests corresponding reconfiguration of RANs of operator j to RRC j
— NRM i sends new RAN context information about RANs of operator i to NRM j
— NRM j sends new RAN context information about RANs of operator j to NRM i
— NRM i sends new RAN context information to TRM k
— NRM j sends new RAN context information to TRM l

Multiple Operator 2 (NRM is Outside Operators) The making spectrum assignment decision procedure for multiple operator scenario 2 (NRM is outside operators) is shown in Figure 16.20.

Making spectrum assignment decision procedure for multiple operator scenario 2 (NRM is outside operators) is as follows:

— NRM analyzes spectrum assignment policies and context information
— NRM evaluates spectrum usage and makes new spectrum assignment decision
— NRM reports new spectrum assignment decision to OSM 1, …, N
— NRM requests corresponding reconfiguration of RANs to RRC i
— NRM sends new RAN context information to its TRM

16.8.2.4 Performing Spectrum Access on Network Side

Single Operator The performing spectrum access on network side procedure for single operator scenario is shown in Figure 16.21.

The performing spectrum access on network side procedure for single operator scenario is as follows:

— NRM analyzes spectrum assignment policies and context information
— NRM evaluates spectrum usage and makes new spectrum access decision
— NRM requests corresponding reconfiguration of its RANs to its RRC
— NRM sends new RAN context information to its TRM

Multiple Operator 1 (NRM is Inside Operator) The performing spectrum access on network side procedure for multiple operator scenario 1 (NRM is inside operator) is shown in Figure 16.22.

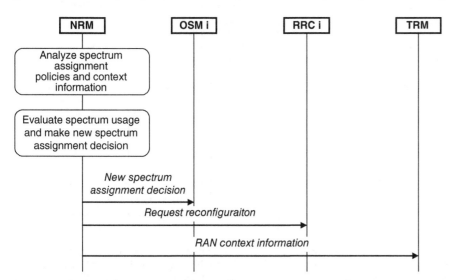

Figure 16.20 — Making spectrum assignment decision procedure for multiple operator scenario 2 (NRM is outside operators). From IEEE Standard for Architectural Building Blocks Enabling Network-Device Distributed Decision Making for Optimized Radio Resource Usage in Heterogeneous Wireless Access Networks, Jan 2019. IEEE.

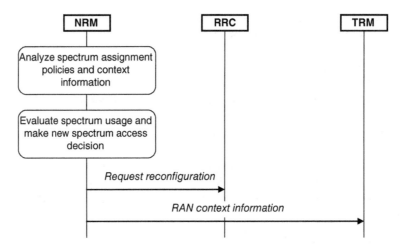

Figure 16.21 —Performing spectrum access on network side procedure for single operator scenario. From IEEE Standard for Architectural Building Blocks Enabling Network-Device Distributed Decision Making for Optimized Radio Resource Usage in Heterogeneous Wireless Access Networks, Jan 2019. IEEE.

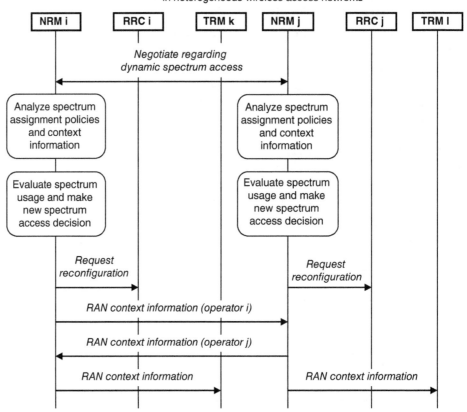

Figure 16.22 —Performing spectrum access on network side procedure for multiple operator scenario 1 (NRM is inside operator). From IEEE Standard for Architectural Building Blocks Enabling Network-Device Distributed Decision Making for Optimized Radio Resource Usage in Heterogeneous Wireless Access Networks, Jan 2019. IEEE.

The performing spectrum access on network side procedure for multiple operator scenario 1 (NRM is inside operator) is as follows:

— NRM i and NRM j negotiate regarding dynamic spectrum sharing
— NRM i analyzes spectrum assignment policies and context information
— NRM i evaluates spectrum usage and makes new spectrum access decision
— NRM i requests corresponding reconfiguration of RANs of operator i to RRC i
— NRM j analyzes spectrum assignment policies and context information
— NRM j evaluates spectrum usage and makes new spectrum access decision
— NRM j requests corresponding reconfiguration of RANs of operator j to RRC j
— NRM i sends new RAN context information about RANs of operator i to NRM j
— NRM j sends new RAN context information about RANs of operator j to NRM i
— NRM i sends new RAN context information to TRM k
— NRM j sends new RAN context information to TRM l

Multiple Operator 2 (NRM is Outside Operators) The performing spectrum access on network side procedure for multiple operator scenario 2 (NRM is outside operators) is shown on Figure 16.23.

The performing spectrum access on network side procedure for multiple operator scenario 2 (NRM is outside operators) is as follows:

— NRM analyzes spectrum assignment policies and context information
— NRM evaluates spectrum usage and makes new spectrum access decision
— NRM requests corresponding reconfiguration of RANs of corresponding operators i to corresponding RRCs i
— NRM sends new RAN context information to its TRM

16.8.2.5 Generating Radio Resource Selection Policies

Single Operator The generating radio resource selection policies procedure for single operator scenario is shown in Figure 16.24.

Generating radio resource selection policies procedure for single operator scenario is as follows:

— NRM analyzes spectrum assignment policies and context information
— NRM evaluates spectrum usage and generates new radio resource selection policies
— NRM sends radio resource selection policies to its TRM

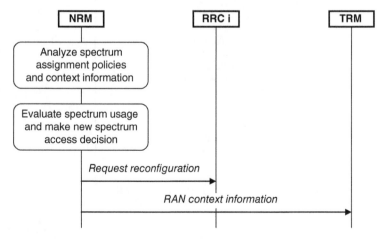

Figure 16.23 —Performing spectrum access on network side procedure for multiple operator scenario 2 (NRM is outside operators). From IEEE Standard for Architectural Building Blocks Enabling Network-Device Distributed Decision Making for Optimized Radio Resource Usage in Heterogeneous Wireless Access Networks, Jan 2019. IEEE.

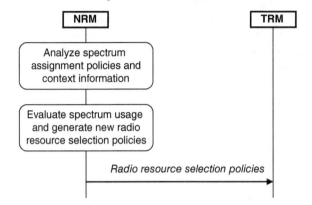

IEEE Std 1900.4-2009
IEEE Standard for Architectural Building Blocks Enabling
Network-Device Distributed Decision Making for Optimized Radio Resource Usage
in Heterogeneous Wireless Access Networks

Figure 16.24 —Generating radio resource selection policies procedure for single operator scenario. From IEEE Standard for Architectural Building Blocks Enabling Network-Device Distributed Decision Making for Optimized Radio Resource Usage in Heterogeneous Wireless Access Networks, Jan 2019. IEEE.

Multiple Operator 1 (NRM is Inside Operator) The generating radio resource selection policies procedure for multiple operator scenario 1 (NRM is inside operator) is shown in Figure 16.25.

The generating radio resource selection policies procedure for multiple operator scenario 1 (NRM is inside operator) is as follows:

— NRM i and NRM j negotiate regarding generating new radio resource selection policies
— NRM i analyzes spectrum assignment policies and context information
— NRM i evaluates spectrum usage and generates new radio resource selection policies on behalf of operator i
— NRM i sends radio resource selection policies to its TRM
— NRM j analyzes spectrum assignment policies and context information

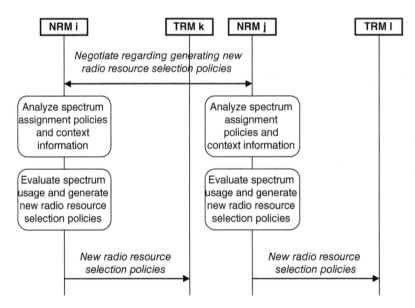

Figure 16.25 —Generating radio resource selection policies procedure for multiple operator scenario 1 (NRM is inside operator). From IEEE Standard for Architectural Building Blocks Enabling Network-Device Distributed Decision Making for Optimized Radio Resource Usage in Heterogeneous Wireless Access Networks, Jan 2019. IEEE.

— NRM j evaluates spectrum usage and generates new radio resource selection policies on behalf of operator j
— NRM j sends radio resource selection policies to its TRM

Multiple Operator 2 (NRM is Outside Operators) See the procedure for single operator scenario in 16.8.2.5.1.

16.8.2.6 Performing Reconfiguration on Terminal Side

Single Operator The performing reconfiguration on terminal side procedure for single operator scenario is shown in Figure 16.26.

The performing reconfiguration on terminal side procedure for single operator scenario is as follows:

— Upon reception of new radio resource selection policies from NRM, TRM performs the following:
  — If maximum time interval for reconfiguration is specified, then within this time interval TRM performs the following:
    — TRM analyzes radio resource selection policies and context information
    — TRM evaluates spectrum usage and makes new decision on its Terminal reconfiguration
    — TRM requests corresponding reconfiguration of its Terminal to its TRC
  — If maximum time interval for reconfiguration is not specified, then TRM performs the same sequence of actions but without time constraint

Multiple Operator 1 (NRM is Inside Operator) See the procedure for single operator scenario in 16.8.2.6.1.

Multiple Operator 2 (NRM is Outside Operators) See the procedure for single operator scenario in 16.8.2.6.1.

16.8.3 Examples of use Case Realization

16.8.3.1 Dynamic Spectrum Assignment

An example of dynamic spectrum assignment use case realization using the defined IEEE 1900.4 generic procedures is shown in Figure 16.27.

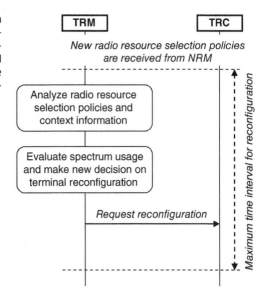

Figure 16.26 −Performing reconfiguration on terminal side procedure for single operator scenario. From IEEE Standard for Architectural Building Blocks Enabling Network-Device Distributed Decision Making for Optimized Radio Resource Usage in Heterogeneous Wireless Access Networks, Jan 2019. IEEE.

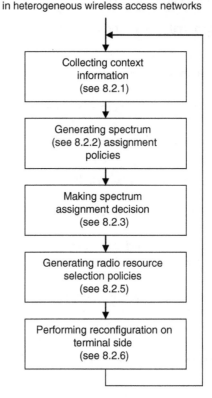

IEEE Std 1900.4-2009
IEEE Standard for architectural building blocks enabling
network-device distributed decision making for optimized radio resource usage
in heterogeneous wireless access networks

Figure 16.27 —Example of dynamic spectrum assignment use case realization. From IEEE Standard for Architectural Building Blocks Enabling Network-Device Distributed Decision Making for Optimized Radio Resource Usage in Heterogeneous Wireless Access Networks, Jan 2019. IEEE.

Dynamic spectrum assignment use case can be realized as follows using the IEEE 1900.4 procedures:

— Collecting context information procedure is performed (16.8.2.1)
— Generating spectrum assignment policies procedure is preformed (16.8.2.2)
— Making spectrum assignment decision procedure is preformed (16.8.2.3)
— Generating radio resource selection policies procedure is preformed (16.8.2.5)
— Performing reconfiguration on terminal side procedure is performed (16.8.2.6)

16.8.3.2 Dynamic Spectrum Sharing

An example of dynamic spectrum sharing use case realization using the defined IEEE 1900.4 generic procedures is shown in Figure 16.28.

Dynamic spectrum sharing use case can be realized as follows using the IEEE 1900.4 procedures:

— Collecting context information procedure is performed (16.8.2.1)
— Performing spectrum access on network side procedure is preformed (16.8.2.4)
— Generating radio resource selection policies procedure is preformed (16.8.2.5)
— Performing reconfiguration on terminal side procedure is performed (16.8.2.6)

16.8.3.3 Distributed Radio Resource Usage Optimization

An example of distributed radio resource usage optimization use case realization using the defined IEEE 1900.4 generic procedures is shown in Figure 16.29.

IEEE Std 1900.4-2009
IEEE Standard for architectural building blocks enabling
network-device distributed decision Making for optimized radio resource usage
in heterogeneous wireless access networks

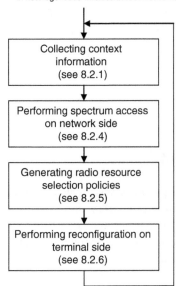

Figure 16.28 —Example of dynamic spectrum sharing use case realization. From IEEE Standard for Architectural Building Blocks Enabling Network-Device Distributed Decision Making for Optimized Radio Resource Usage in Heterogeneous Wireless Access Networks, Jan 2019. IEEE.

IEEE Std 1900.4-2009
IEEE Standard for architectural building blocks enabling
network-device distributed decision making for optimized radio resource usage
in heterogeneous wireless access networks

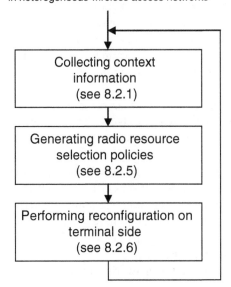

Figure 16.29 —Example of distributed radio resource usage optimization use case realization. From IEEE Standard for Architectural Building Blocks Enabling Network-Device Distributed Decision Making for Optimized Radio Resource Usage in Heterogeneous Wireless Access Networks, Jan 2019. IEEE.

Distributed radio resource usage optimization use case can be realized as follows using the IEEE 1900.4 procedures:

— Collecting context information procedure is performed (16.8.2.1)
— Generating radio resource selection policies procedure is preformed (16.8.2.5)
— Performing reconfiguration on terminal side procedure is performed (16.8.2.6)

Annex 16A

(informative)

Use Cases

16A.1 Dynamic Spectrum Assignment

In the dynamic spectrum assignment use case, frequency bands are dynamically assigned to the RANs among the participating networks in order to optimize spectrum usage. In other words, the assigned frequency bands are not fixed, and can be dynamically changed.

Following the dynamic spectrum assignment decisions, corresponding RANs are reconfigured. Following the RANs' reconfiguration, Terminals need to reconfigure correspondingly.

The following three scenarios are defined to illustrate dynamic spectrum assignment use case:

— Single operator scenario
— Multiple operator scenario 1 (NRM is inside operator)
— Multiple operator scenario 2 (NRM is outside operators)

16A.1.1 Single Operator Scenario

The single operator scenario assumes that several frequency bands are assigned to an operator that has several RANs. This operator has flexibility for distributing these frequency bands between its RANs. OSM will enable the operator to evaluate efficiency of current spectrum assignment; however, it cannot facilitate dynamic spectrum reconfiguration for its RAN. NRM introduced in this standard enables dynamic reconfiguration of RANs inside the operator to improve usage of its frequency bands.

A single operator dynamic spectrum assignment example is shown in Figure 16A.1. The operator operates RAN 1 and RAN 2 in two frequency bands. The frequency band assigned to RAN 1 is overused. The frequency band assigned to RAN 2 is underused. Spectrum usage is unbalanced.

OSM of this operator sends spectrum assignment policy to NRM requesting reconfiguration of its RAN 1 and RAN 2 networks to allow RAN 1 network to use part of RAN 2 frequency band.

NRM requests and controls the corresponding reconfiguration of RAN 1 and RAN 2 networks. Following the RANs' reconfiguration, Terminals need to reconfigure correspondingly.

After reconfiguration, part of RAN 1 starts to use part of RAN 2 frequency band. The usage of frequency bands of the operator is now balanced.

For the operator, this allows reconfiguring its networks to balance usage of its frequency bands.

The single operator dynamic spectrum assignment scenario is enabled by NRM. NRM allows reconfiguration of RANs of operator. This improves spectrum usage and increases quality of service for the operator.

The single operator dynamic spectrum assignment scenario is realized as follows:

a) OSM of operator generates spectrum assignment policies and sends them to NRM
b) NRM of operator receives these spectrum assignment policies
c) NRM obtains RAN context information
d) TRMs obtain terminal context information

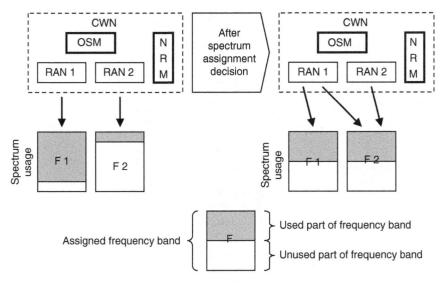

IEEE Std 1900.4-2009
IEEE Standard for architectural building blocks enabling
network-device distributed decision making for optimized radio resource usage
in heterogeneous wireless access networks

Figure 16A.1 —Dynamic spectrum assignment: single operator scenario. From IEEE Standard for Architectural Building Blocks Enabling Network-Device Distributed Decision Making for Optimized Radio Resource Usage in Heterogeneous Wireless Access Networks, Jan 2019. IEEE.

e) NRM and TRMs exchange context information
f) NRM analyzes spectrum assignment policies and context information
g) NRM evaluates current spectrum assignment inside operator and makes new spectrum assignment decision
h) NRM informs OSM about its spectrum assignment decision
i) NRM requests and controls corresponding reconfiguration of RANs of operator
j) Following the RANs' reconfiguration, Terminals reconfigure correspondingly

16A.1.2   Multiple Operator Scenario 1 (NRM is Inside Operator)

Multiple operator scenario 1 assumes that several frequency bands are allocated to several operators and operators have some level of flexibility for renting or sharing these frequency bands. OSM evaluates efficiency of current spectrum assignment for each operator. NRM decides dynamic spectrum reconfiguration for RANs of this operator. Cross-operator collaboration is performed via NRMs and/or OSMs of different operators. NRM introduced in this standard enables cross-operator optimization of spectrum usage by performing dynamic spectrum assignment.

A spectrum sharing example is shown in Figure 16A.2. Operator A and B operate RAN 1 and RAN 2 in their frequency bands. The frequency band assigned to operator A is overused. The frequency band assigned to operator B is underused. Spectrum usage is unbalanced.

OSMs and/or NRMs of operators A and B negotiate to allow operator A to use frequency band of operator B. After negotiation, NRM of operator A requests and controls reconfiguration of RAN 1 and NRM of operator B requests and controls reconfiguration of RAN 2. Following RANs' reconfiguration, the Terminals need to reconfigure correspondingly.

After reconfiguration, part of BSs of RAN 1 starts operation in the frequency band of RAN 2. The usage of the frequency bands of operators A and B is now balanced.

For operator A, this allows the use of the frequency band of operator B to improve radio resource usage and quality of service. Operator B can get some revenue for sharing its frequency band with operator A.

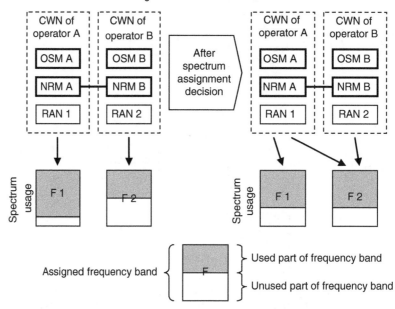

IEEE Std 1900.4-2009
IEEE Standard for architectural building blocks enabling
network-device distributed decision making for optimized radio resource usage
in heterogeneous wireless access networks

Figure 16A.2 —Dynamic spectrum assignment: multiple operator scenario 1 (NRM is inside operator): spectrum sharing example. From IEEE Standard for Architectural Building Blocks Enabling Network-Device Distributed Decision Making for Optimized Radio Resource Usage in Heterogeneous Wireless Access Networks, Jan 2019. IEEE.

A spectrum renting example is shown in Figure 16A.3. Operator A operates RAN 1 network in its frequency band. This frequency band is currently overused. Operator B has frequency band for future RAN 2. Currently, only a small part of this frequency band is occasionally used for trial. Spectrum usage is unbalanced.

OSMs and/or NRMs of operators A and B negotiate to rent part of RAN 2 frequency band of operator B to operator A. After negotiation, NRM of operator A requests and controls reconfiguration of RAN 1 of operator A. Following RANs' reconfiguration, the Terminals need to reconfigure correspondingly.

After reconfiguration, part of the BSs of RAN 1 of operator A starts operation in RAN 2 frequency band of operator B. The usage of frequency bands of operators A and B is now balanced.

For operator A, this allows the use of the frequency band of operator B to improve radio resource usage and quality of service. Operator B can get some revenue for renting its frequency band to operator A.

The multiple operator dynamic spectrum assignment scenario is enabled by NRM. NRM allows crossoperator balancing of spectrum usage. This increases quality of service of all involved operators and provides new mechanisms to receive additional revenue.

Multiple operator scenario 1 (NRM is inside operator) is realized as follows:

a) OSMs and/or NRMs of operators perform interactions between each other regarding dynamic spectrum assignment
b) OSM of each operator generates spectrum assignment policies and sends them to its NRM
c) NRM of each operator receives these spectrum assignment policies
d) NRMs obtain RAN context information
e) TRMs obtain terminal context information
f) NRMs and their TRMs exchange context information
g) Each NRM analyzes spectrum assignment policies and context information
h) Each NRM evaluates current spectrum assignment inside operator and makes new spectrum assignment decision

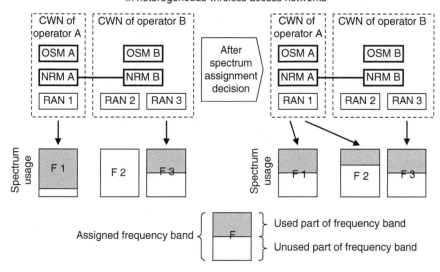

IEEE Std 1900.4-2009
IEEE Standard for architectural building blocks enabling
network-device distributed decision making for optimized radio resource usage
in heterogeneous wireless access networks

Figure 16A.3 —Dynamic spectrum assignment: multiple operator scenario 1 (NRM is inside operator): spectrum renting example. From IEEE Standard for Architectural Building Blocks Enabling Network-Device Distributed Decision Making for Optimized Radio Resource Usage in Heterogeneous Wireless Access Networks, Jan 2019. IEEE.

i) Each NRM informs OSM about its spectrum assignment decision
j) NRM of each operator requests and controls corresponding reconfiguration of RANs of this operator
k) Following the RANs' reconfiguration, Terminals reconfigure correspondingly

16A.1.3   Multiple Operator Scenario 2 (NRM is Outside Operators)

Multiple operator scenario 2 assumes that several frequency bands are allocated to several operators and operators have some level of flexibility for renting or sharing these frequency bands. OSM evaluates efficiency of current spectrum assignment for each operator. However, OSMs of different operators cannot negotiate with each other and cross-operator dynamic spectrum sharing or renting is not possible. NRM introduced in this standard enables cross-operator optimization of spectrum usage by performing dynamic spectrum assignment.

A spectrum sharing example is shown in Figure 16A.4. Operator A and B operate RAN 1 and RAN 2 in their frequency bands. The frequency band assigned to operator A is overused. The frequency band assigned to operator B is underused. Spectrum usage is unbalanced.

OSM of operator A sends spectrum assignment policy to NRM expressing need for additional spectrum resources. OSM of operator B sends spectrum assignment policy to NRM expressing possibility to share its frequency band.

Based on the analysis of spectrum assignment policies received from OSMs of operators A and B, as well as, on analysis of terminal and network context information NRM makes new spectrum assignment decision. NRM allows operator A to use the frequency band of operator B.

NRM informs OSMs of operators A and B about this spectrum assignment decision. Also, NRM requests and controls reconfiguration of RAN 1 of operator A. Following the RANs' reconfiguration, Terminals need to reconfigure correspondingly.

After reconfiguration, part of the BSs of RAN 1 of operator A starts operation in frequency band of operator B. The usage of frequency bands of operators A and B is now balanced.

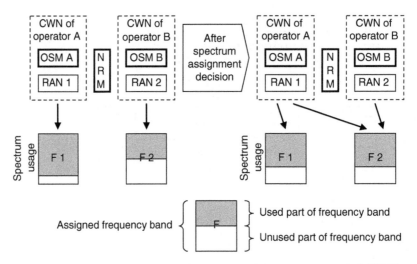

Figure 16A.4 —Dynamic spectrum assignment: multiple operator scenario 2 (NRM is outside operators): spectrum sharing example. From IEEE Standard for Architectural Building Blocks Enabling Network-Device Distributed Decision Making for Optimized Radio Resource Usage in Heterogeneous Wireless Access Networks, Jan 2019. IEEE.

For operator A, this allows the use of the frequency band of operator B to improve radio resource usage and quality of service. Operator B can get some revenue for sharing its frequency band with operator A.

A spectrum renting example is shown in Figure 16A.5. Operator A operates RAN 1 in its frequency band. This frequency band is currently overused. Operator B has a frequency band for future RAN 2. Currently, only a small part of this frequency band is occasionally used for trial. Spectrum usage is unbalanced.

The OSM of operator A sends spectrum assignment policy to NRM expressing the need for additional spectrum resources. The OSM of operator B sends spectrum assignment policy to NRM expressing the possibility to rent its RAN 2 frequency band.

Based on the analysis of the spectrum assignment policies received from the OSMs of operators A and B, as well as, on analysis of terminal and network context information, the NRM makes a new spectrum assignment decision. NRM decides to rent part of the RAN 2 frequency band of operator B to operator A.

The NRM informs the OSMs of operators A and B about this spectrum assignment decision. Also, the NRM requests and controls reconfiguration of RAN 1 of operator A. Following the RANs' reconfiguration, Terminals need to reconfigure correspondingly.

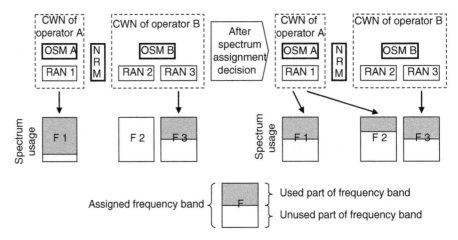

Figure 16A.5 —Dynamic spectrum assignment: multiple operator scenario 2 (NRM is outside operators): spectrum renting example. From IEEE Standard for Architectural Building Blocks Enabling Network-Device Distributed Decision Making for Optimized Radio Resource Usage in Heterogeneous Wireless Access Networks, Jan 2019. IEEE.

After reconfiguration, part of the BSs of RAN 1 of operator A starts operation in the RAN 2 frequency band of operator B. The usage of frequency bands of operators A and B is now balanced.

For operator A, this allows the use of the frequency band of operator B to improve radio resource usage and quality of service. Operator B can get some revenue for renting its frequency band to operator A.

The multiple operator dynamic spectrum assignment scenario is enabled by NRM. NRM allows crossoperator balancing of spectrum usage. This increases quality of service of all involved operators and provides new mechanisms to receive additional revenue.

Multiple operator scenario 2 (NRM is outside operators) is realized as follows:

a) OSMs of operators generate spectrum assignment policies and send them to NRM
b) NRM receives these spectrum assignment policies from the OSMs
c) NRM obtains RAN context information
d) TRMs obtain terminal context information
e) NRM and TRMs exchange context information
f) NRM analyzes spectrum assignment policies and context information
g) NRM evaluates current spectrum assignment inside multiple operators and makes new spectrum assignment decision
h) NRM informs OSMs about its spectrum assignment decision
i) NRM requests and controls corresponding reconfiguration of RANs of multiple operators
j) Following the RANs' reconfiguration, Terminals reconfigure correspondingly

16A.2 Dynamic Spectrum Sharing

In the dynamic spectrum sharing use case, frequency bands assigned to the RANs are fixed. However, a particular frequency band can be shared by several RANs. In other words, the dynamic spectrum sharing use case describes how fixed frequency bands are shared and/or used dynamically.

Following the dynamic spectrum sharing decisions, corresponding RANs and Terminals are reconfigured.

One or several frequency bands are available for joint use by several RANs. These RANs and their Terminals dynamically access these frequency bands for improving spectrum usage and quality of service. Decisions on this dynamic spectrum sharing are jointly made by the NRM and the TRMs in a distributed manner. After the decisions have been made, NRM facilitates corresponding reconfiguration of RANs and TRMs facilitate corresponding reconfiguration of their Terminals.

An example of dynamic spectrum sharing is shown in Figure 16A.6. Operators A and B operate RAN 1 and RAN 2 in two frequency bands. These frequency bands are available for joint use by both operators according to regulatory rules. Currently, the frequency band of operator B is underused. Spectrum usage is unbalanced.

NRMs obtain RAN context information from the RANs of operators A and B. TRMs obtain terminal context information from their Terminals. NRMs and TRMs exchange context information with each other.

Based on the analysis of context information, NRMs detect that frequency band of operator A is overused while the frequency band of operator B is underused. As a result, NRMs make new dynamic spectrum sharing decision. It decides that part of the BSs of RAN 1 of operator A shall access the frequency band of operator B.

NRMs request and control corresponding reconfiguration of RAN 1 and RAN 2. Also, NRMs generate radio resource selection policies that guide Terminals to dynamically access the frequency band of operator B. NRMs send these radio resource selection policies to corresponding TRMs.

TRMs analyze received radio resource selection policies and available context information. Based on the analysis, some TRMs make the decision to access the frequency band of operator B. These TRMs request and control the corresponding reconfiguration of their Terminals.

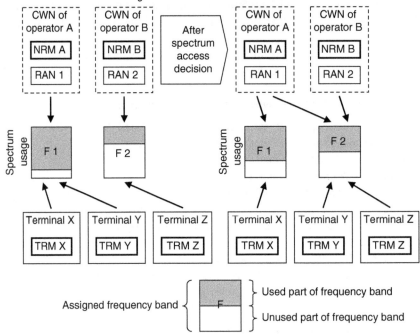

IEEE Std 1900.4-2009
IEEE Standard for architectural building blocks enabling
network-device distributed decision making for optimized radio resource usage
in heterogeneous wireless access networks

Figure 16A.6 —Dynamic spectrum sharing. From IEEE Standard for Architectural Building Blocks Enabling Network-Device Distributed Decision Making for Optimized Radio Resource Usage in Heterogeneous Wireless Access Networks, Jan 2019. IEEE.

After part of RAN 1 and part of its Terminals of operator A start operation in the frequency band of operator B, usage of the frequency bands of operators A and B is balanced.

Dynamic spectrum sharing is enabled by NRMs, TRMs, and collaboration between NRMs and TRMs. NRMs make spectrum access decisions for RANs, as well as, requests and controls corresponding reconfiguration of RANs. Also, NRMs generate radio resource selection policies and send them to TRMs. TRMs make final decision on spectrum access for Terminals, as well as, request and control corresponding reconfiguration of their Terminals. Finally, NRMs and TRMs obtain and exchange context information used for decision making. Dynamic spectrum sharing improves spectrum usage and increases quality of service.

Dynamic spectrum sharing use case is realized as follows:

a) NRMs obtain RAN context information
b) TRMs obtain terminal context information
c) NRMs and TRMs exchange context information
d) NRMs exchange context information
e) NRMs analyzes context information
f) NRMs make decisions on spectrum access to improve spectrum usage and quality of service
g) NRMs request and control corresponding reconfiguration of RANs
h) NRMs generate radio resource selection policies to guide TRMs in their spectrum access decisions and send them to their TRMs
i) TRMs analyze received radio resource selection policies and available context information
j) TRMs make decisions on spectrum access to improve spectrum usage and quality of service
k) TRMs request and control corresponding reconfiguration of their Terminals

16A.3   Distributed Radio Resource Usage Optimization

Distributed radio resource usage optimization use case demonstrates how the IEEE 1900.4 system can be applied to legacy RANs in order to optimize radio resource usage and improve quality of service.

In the distributed radio resource usage optimization use case, frequency bands assigned to RANs are fixed. Also, reconfiguration of RANs is not involved in this use case. Decision making on reconfiguration of Terminals is performed in a distributed manner.

Distributed radio resource usage optimization use case considers Terminals with or without multi-homing capability.

An example of distributed radio resource usage optimization use case is shown in Figure 16A.7. Operators A and B operate RAN 1, RAN 2, and RAN 3 in three frequency bands. Terminal 1 with multi-homing capability can have simultaneous connections with these RANs. Currently, Terminal 1 is connected to RAN 1 of operator A and RAN 2 of operator B. Terminal 2 without multi-homing capability can have one active connection with any of these RANs. Currently, Terminal 2 is connected to RAN 2 of operator B. Currently, the RAN 2 frequency band of operator B is overused while the RAN 3 frequency band of operator B is underused. Spectrum usage is unbalanced.

NRMs of operators A and B analyze available context information and detect imbalance in spectrum usage. NRMs generate radio resource selection policies that recommend changing some connections from RAN 2 to RAN 3. These radio resource selection policies are sent from NRMs of operators A and B to TRMs of Terminals 1 and 2. Together with these radio resource selection policies, NRMs specify and send time intervals for reconfiguration of Terminals. TRMs of Terminals 1 and 2 analyze received radio resource selection policies and available context information. They detect imbalance in spectrum usage. The TRM of Terminal 1 makes the decision to change one of its two connections from RAN 2 to RAN 3. The TRM of Terminal 2 makes the decision to change its connection from RAN 2 to RAN 3.

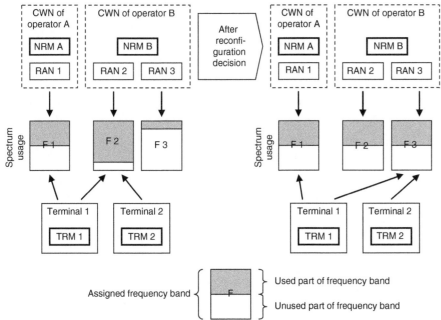

Figure 16A.7 —Distributed radio resource usage optimization. From IEEE Standard for Architectural Building Blocks Enabling Network-Device Distributed Decision Making for Optimized Radio Resource Usage in Heterogeneous Wireless Access Networks, Jan 2019. IEEE.

The TRMs request and control the corresponding reconfiguration of their Terminals. After reconfiguration, Terminal 1 is connected to RAN 1 of operator A and RAN 3 of operator B, while Terminal 2 is connected to RAN 3 of operator B. The usage of RAN 2 and RAN 3 frequency bands is now balanced.

Distributed radio resource usage optimization is enabled by NRMs, TRMs, and the collaboration between NRMs and TRMs. NRMs and TRMs obtain and exchange context information used for decision making. NRMs generate radio resource selection policies that guide TRMs in their decisions. TRMs make final decisions on reconfiguration of their Terminals, as well as, request and control corresponding reconfiguration of their Terminals. Distributed radio resource usage optimization improves spectrum usage and increases quality of service.

Distributed radio resource usage optimization use case is realized as follows:

a) NRMs obtain RAN context information
b) TRMs obtain terminal context information
c) NRMs and TRMs exchange context information
d) NRMs exchange context information
e) NRMs analyze context information
f) NRMs generate radio resource selection policies and time intervals for reconfiguration of Terminals and send them to their TRMs. These radio resource selection policies should correspond to specific groups of Terminals
g) TRMs analyze received radio resource selection policies and available context information
h) TRMs make decisions on their Terminals reconfiguration, within the specified reconfiguration time intervals to improve radio resource usage and quality of service
i) TRMs request and control corresponding reconfiguration of their Terminals within the provided time intervals

Annex 16B

(normative)

Class Definitions for Information Model

16B.1 Notational Tools

The tables defining the classes use the following template (see Table 16B.1).
A description of the template is provided within the following list:

— <Class name> is the name of the Class as it is appeared in the corresponding model. Additional information is also included in case the class in question has been specified as an abstract one.
— DERIVED FROM field identifies the super class of the class in case of sub-classing.
— ATTRIBUTES field describes the attributes that have been defined in the class. More specifically:
  — <Attribute name> identifies the name of an attribute, as it is included in the class definition.
  — <Attribute value type> holds the type of the attribute specified in ASN.1. Readers shall refer to the ASN.1 module for details (see Annex 16C).
  — <Attribute access qualifier> provides information about the level of accessibility of the attribute. This may include: 'Read,' 'Write,' 'Read-Write,' 'Add-Remove' (for list-type attributes), 'Read-Add-Remove,' and 'None' (for internal access only).

Table 16B.1 — Template of table of class definition.

Class \<Class name> *[(abstract class)]*			
<Description of the class>			
DERIVED FROM	**<List of super-classes>**		
ATTRIBUTES			
<Attribute name> *[<optional>]*	*Value type:* <Attribute value type>	*Possible access:* <Attribute access qualifier>	*Default value:* <Default value>
<Description of the attribute>			
CONTAINED IN	**<List of classes, whose instances may contain an instance of this class. If this class is an abstract class, that is, it is used for further refinement only and will never be instantiated, then this list is empty.>**		
CONTAINS	**<List of classes, whose instances may be contained in an instance of this class. Constraints used are: [*] – zero or more instances, [+] – one or more instances, [\<n>] – exactly n instances, [\<m> – \<n>] – not less than m and not more than n instances.>**		
SUPPORTED EVENTS	<List of event names that are detected by this class and lead potentially to a corresponding event report>		

From IEEE Standard for Architectural Building Blocks Enabling Network-Device Distributed Decision Making for Optimized Radio Resource Usage in Heterogeneous Wireless Access Networks, Jan 2019. IEEE.

— CONTAINED IN field includes a list of classes whose instances may contain an instance of this class; containment is a strong aggregation relationship, that is, a contained instance is for its lifetime bound to its container object and it is contained only in this one container.

— CONTAINS field provides a list of classes whose instances may be contained in an instance of the class in question.

— SUPPORTED EVENTS field includes a list of event names that are detected by this class and lead potentially to a corresponding event report. Possible usage of these properties is explained in Annex 16D.

16B.2   Common Base Class

UML class diagram for common base class without inheritance relations is shown in Figure 16B.1.

Table 16B.2 describes the 19004BaseClass class.

19004BaseClass
className: PrintableString reportingDelay_ObjectCreation: INTEGER reportingDelay_AttributeValueChanged: INTEGER

Figure 16B.1 —UML class diagram for common base class. From IEEE Standard for Architectural Building Blocks Enabling Network-Device Distributed Decision Making for Optimized Radio Resource Usage in Heterogeneous Wireless Access Networks, Jan 2019. IEEE.

Table 16B.2 —19004BaseClass class definition.

Class **19004BaseClass** *(abstract class)*			
This class provides base interface for IEEE 1900.4 class definitions. Each class defined for IEEE 1900.4 shall be derived from this class if it can be instantiated more than once in the scope of the same immediate superior object.			
DERIVED FROM			
ATTRIBUTES			
className	*Value type:* PrintableString	*Possible access:* Read	*Default value:* - *not specified* -
This attribute allows to retrieve the name of the class an object belongs to.			
reportingDelay_Object Creation	*Value type:* INTEGER	*Possible access:* Read-Write	*Default value:* 0
Requested delay (in ms) between internal event detection and its (potential) event reporting. This is useful to report sustainable configuration changes only. By default, no delay is specified.			
reportingDelay_Attribute ValueChanged	*Value type:* INTEGER	*Possible access:* Read-Write	*Default value:* 0
Requested delay (in ms) between internal event detection and its (potential) event reporting. This is useful to report sustainable configuration changes only. By default, no delay is specified.			
CONTAINED IN			
CONTAINS			
SUPPORTED EVENTS	objectCreation, objectDeletion, attributeValueChanged		

From IEEE Standard for Architectural Building Blocks Enabling Network-Device Distributed Decision Making for Optimized Radio Resource Usage in Heterogeneous Wireless Access Networks, Jan 2019. IEEE.

16B.3   Policy Classes

UML class diagram for policy classes without inheritance relations is shown in Figure 16B.2. Table 16B.3 describes the ECAPolicy class.

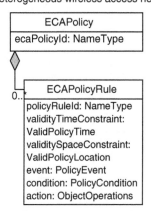

IEEE Std 1900.4-2009
IEEE Standard for architectural building blocks enabling
network-device distributed decision making for optimized radio resource usage
in heterogeneous wireless access networks

Figure 16B.2 —UML class diagram for policy classes. From IEEE Standard for Architectural Building Blocks Enabling Network-Device Distributed Decision Making for Optimized Radio Resource Usage in Heterogeneous Wireless Access Networks, Jan 2019. IEEE.

Table 16B.3 —ECAPolicy class definition.

Class **ECAPolicy**			
The instance of this class contains a set of policy rules governing the behavior of a decision making function. Within this model only ECA policy rules are defined.			
DERIVED FROM	**19004BaseClass**		
ATTRIBUTES			
ecaPolicyId	*Value type:* NameType	*Possible access:* Read	*Default value:* *- not specified -*
This attribute contains string or number assigned to uniquely identify the ECAPolicy object.			
CONTAINED IN			
CONTAINS	**ECAPolicyRule[*]**		
SUPPORTED EVENTS			

From IEEE Standard for Architectural Building Blocks Enabling Network-Device Distributed Decision Making for Optimized Radio Resource Usage in Heterogeneous Wireless Access Networks, Jan 2019. IEEE.

Table 16B.4 —ECAPolicyRule class definition.

Class **ECAPolicyRule**			
This class describes an event condition action based policy rule.			
DERIVED FROM	**19004BaseClass**		
ATTRIBUTES			
policyRuleId	*Value type:* NameType	*Possible access:* Read	*Default value:* *- not specified -*
This attribute contains string or number assigned to uniquely identify the policy rule object.			
validityTimeConstraint	*Value type:* ValidPolicyTime	*Possible access:* Read-Write	*Default value:* NULL
If this time is reached the policy rule object has to be deleted at the Terminal. The type definition for this attribute is optional, that is, possibility not to define a time constraint is supported. Default: no time constraint			
validitySpaceConstraint	*Value type:* ValidPolicyLocation	*Possible access:* Read-Write	*Default value:* NULL
If the Terminal is outside this location area the policy rule has to be deleted at the Terminal. The type definition for this attribute is optional, that is, possibility not to define a space constraint is supported. Default: no space constraint			
event	*Value type:* PolicyEvent	*Possible access:* Read-Write	*Default value:* *- not specified -*
This attribute describes the event that triggers the evaluation of this policy rule for decision making.			
condition	*Value type:* PolicyCondition	*Possible access:* Read-Write	*Default value:* *- not specified -*
This attribute describes the condition that shall be true to apply the action of this policy rule.			
action	*Value type:* ObjectOperations	*Possible access:* Read-Write	*Default value:* *- not specified -*
This attribute describes the action to be performed in case the event has occurred and the condition holds.			
CONTAINED IN	**ECAPolicy**		
CONTAINS			
SUPPORTED EVENTS			

From IEEE Standard for Architectural Building Blocks Enabling Network-Device Distributed Decision Making for Optimized Radio Resource Usage in Heterogeneous Wireless Access Networks, Jan 2019. IEEE.

Table 16B.4 describes the ECAPolicyRule class.

16B.4 Terminal Classes

UML class diagram for terminal classes without inheritance relations is shown in Figure 16B.3.

In addition to relations between terminal-related classes described in Clause 16.7, the following relations are defined:

— Each instance of **Application** class can be associated to zero or one instances of **Application** class.
— Each instance of **Application** class can be associated to one or several instances of **Link** class.

Table 16B.5 through Table 16B.27 describe each Terminal class in detail.

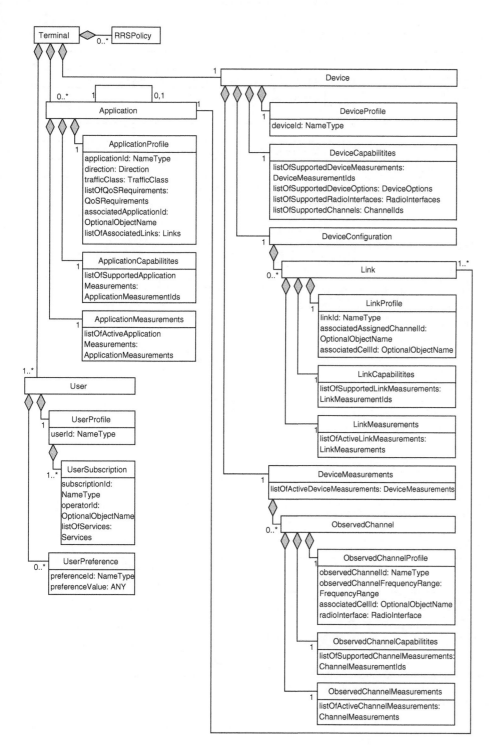

Figure 16B.3 —UML class diagram for terminal classes. From IEEE Standard for Architectural Building Blocks Enabling Network-Device Distributed Decision Making for Optimized Radio Resource Usage in Heterogeneous Wireless Access Networks, Jan 2019. IEEE.

Table 16B.5 — Terminal class definition.

Class **Terminal**	
The instance of this class contains instances of all terminal-related classes using composition.	
DERIVED FROM	**19004BaseClass**
ATTRIBUTES	
CONTAINED IN	
CONTAINS	**User[+], Application[*], Device[1], RRSPolicy[*]**
SUPPORTED EVENTS	

From IEEE Standard for Architectural Building Blocks Enabling Network-Device Distributed Decision Making for Optimized Radio Resource Usage in Heterogeneous Wireless Access Networks, Jan 2019. IEEE.

Table 16B.6 — User class definition.

Class **User**	
This class describes information related to a user of the Terminal.	
DERIVED FROM	**19004BaseClass**
ATTRIBUTES	
CONTAINED IN	**Terminal**
CONTAINS	**UserProfile[1], UserPreference[*]**
SUPPORTED EVENTS	

From IEEE Standard for Architectural Building Blocks Enabling Network-Device Distributed Decision Making for Optimized Radio Resource Usage in Heterogeneous Wireless Access Networks, Jan 2019. IEEE.

Table 16B.7 — UserProfile class definition.

Class **UserProfile**			
This class contains general information about one user of the Terminal.			
DERIVED FROM			
ATTRIBUTES			
userId	*Value type:* NameType	*Possible access:* Read	*Default value:* - not specified -
This attribute contains string or number assigned to uniquely identify the user.			
CONTAINED IN	**User**		
CONTAINS	**UserSubscription[+]**		
SUPPORTED EVENTS			

From IEEE Standard for Architectural Building Blocks Enabling Network-Device Distributed Decision Making for Optimized Radio Resource Usage in Heterogeneous Wireless Access Networks, Jan 2019. IEEE.

Table 16B.8 —UserSubscription class definition.

Class **UserSubscription**			
This class contains information about one subscription of the user.			
DERIVED FROM	**19004BaseClass**		
ATTRIBUTES			
subscriptionId	*Value type:* NameType	*Possible access:* Read	*Default value:* - not specified -
This attribute contains string or number assigned to uniquely identify one subscription of the user.			
operatorId	*Value type:* OptionalObjectName	*Possible access:* Read	*Default value:* - not specified -
This optional attribute contains string or number assigned to uniquely identify the operator providing service within this subscription.			
listOfServices	*Value type:* Services	*Possible access:* Read	*Default value:* - not specified -
This attributes describes services provided within this subscription.			
CONTAINED IN	**UserProfile**		
CONTAINS			
SUPPORTED EVENTS			

From IEEE Standard for Architectural Building Blocks Enabling Network-Device Distributed Decision Making for Optimized Radio Resource Usage in Heterogeneous Wireless Access Networks, Jan 2019. IEEE.

Table 16B.9 —UserPreference class definition.

Class **UserPreference**			
This class describes in a formalized form one preference of the user, for example, preferred operator and radio interface, perceived audio/image/video quality, maximum cost, minimum data rate, etc.			
DERIVED FROM	**19004BaseClass**		
ATTRIBUTES			
preferenceId	*Value type:* NameType	*Possible access:* Read	*Default value:* - not specified -
This attribute contains string or number assigned to uniquely identify one preference of the user.			
preferenceValue	*Value type:* ANY	*Possible access:* Read	*Default value:* - not specified -
This attribute described in a formalized form one preference of the user.			
CONTAINED IN	**User**		
CONTAINS			
SUPPORTED EVENTS			

From IEEE Standard for Architectural Building Blocks Enabling Network-Device Distributed Decision Making for Optimized Radio Resource Usage in Heterogeneous Wireless Access Networks, Jan 2019. IEEE.

Table 16B.10 — Application class definition.

Class **Application**	
This class describes one currently active application.	
DERIVED FROM	**19004BaseClass**
ATTRIBUTES	
CONTAINED IN	**Terminal**
CONTAINS	**ApplicationProfile[1], ApplicationCapabilities[1], ApplicationMeasurements[1]**
SUPPORTED EVENTS	

From IEEE Standard for Architectural Building Blocks Enabling Network-Device Distributed Decision Making for Optimized Radio Resource Usage in Heterogeneous Wireless Access Networks, Jan 2019. IEEE.

Table 16B.11 — ApplicationProfile class definition.

Class **ApplicationProfile**			
This class contains general information about the application.			
DERIVED FROM			
ATTRIBUTES			
applicationId	*Value type:* NameType	*Possible access:* Read	*Default value:* - not specified -
This attribute contains string or number assigned to uniquely identify one application.			
direction	*Value type:* Direction	*Possible access:* Read	*Default value:* - not specified -
This attribute describes whether this application is downlink or uplink application.			
trafficClass	*Value type:* TrafficClass	*Possible access:* Read	*Default value:* - not specified -
This attributes describes traffic class of the application.			
listOfQoSRequirements	*Value type:* QoSRequirements	*Possible access:* Read	*Default value:* - not specified -
This attributes describes QoS requirements of the application.			
associatedApplicationId	*Value type:* OptionalObjectName	*Possible access:* Read	*Default value:* - not specified -
This attribute contains ID of associated application having other direction if any.			
listOfAssociatedLinks	*Value type:* Links	*Possible access:* Read	*Default value:* - not specified -
This attributes contains list of IDs of links used to transmit this application.			
CONTAINED IN	**Application**		
CONTAINS			
SUPPORTED EVENTS			

From IEEE Standard for Architectural Building Blocks Enabling Network-Device Distributed Decision Making for Optimized Radio Resource Usage in Heterogeneous Wireless Access Networks, Jan 2019. IEEE.

Table 16B.12 —ApplicationCapabilities class definition.

Class **ApplicationCapabilities**			
This class contains information about measurements (instantaneous measurement data and performance statistics derived from this data) supported by this application.			
DERIVED FROM			
ATTRIBUTES			
listOfSupportedApplication Measurements	*Value type:* ApplicationMeasurementIds	*Possible access:* Read	*Default value:* - *not specified* -
This attribute describes measurements supported by this application.			
CONTAINED IN	**Application**		
CONTAINS			
SUPPORTED EVENTS			

From IEEE Standard for Architectural Building Blocks Enabling Network-Device Distributed Decision Making for Optimized Radio Resource Usage in Heterogeneous Wireless Access Networks, Jan 2019. IEEE.

Table 16B.13 —ApplicationMeasurements class definition.

Class **ApplicationMeasurements**			
This class contains measurements (instantaneous measurement data and performance statistics derived from this data) performed by this application.			
DERIVED FROM			
ATTRIBUTES			
listOfActiveApplication Measurements	*Value type:* ApplicationMeasurements	*Possible access:* Read-Add-Remove	*Default value:* - *not specified* -
This attribute describes measurements that are currently performed by the application.			
CONTAINED IN	**Application**		
CONTAINS			
SUPPORTED EVENTS			

From IEEE Standard for Architectural Building Blocks Enabling Network-Device Distributed Decision Making for Optimized Radio Resource Usage in Heterogeneous Wireless Access Networks, Jan 2019. IEEE.

Table 16B.14 —Device class definition.

Class **Device**	
This class describes all radio interface related hardware and software of a Terminal, as well as, measurement information related to radio resources within the Terminal.	
DERIVED FROM	**19004BaseClass**
ATTRIBUTES	
CONTAINED IN	**Terminal**
CONTAINS	**DeviceProfile[1], DeviceCapabilities[1], DeviceConfiguration[1], DeviceMeasurements[1]**
SUPPORTED EVENTS	

From IEEE Standard for Architectural Building Blocks Enabling Network-Device Distributed Decision Making for Optimized Radio Resource Usage in Heterogeneous Wireless Access Networks, Jan 2019. IEEE.

Table 16B.15 — DeviceProfile class definition.

Class **DeviceProfile**			
This class contains general information about the Terminal.			
DERIVED FROM			
ATTRIBUTES			
deviceId	*Value type:* NameType	*Possible access:* Read	*Default value:* - *not specified* -
This attribute contains string or number assigned to uniquely identify the Terminal.			
CONTAINED IN	**Device**		
CONTAINS			
SUPPORTED EVENTS			

From IEEE Standard for Architectural Building Blocks Enabling Network-Device Distributed Decision Making for Optimized Radio Resource Usage in Heterogeneous Wireless Access Networks, Jan 2019. IEEE.

Table 16B.16 — DeviceCapabilities class definition.

Class **DeviceCapabilities**			
This class contains information about the Terminal capabilities including both transmission and measurement capabilities.			
DERIVED FROM			
ATTRIBUTES			
listOfSupportedDevice Measurements	*Value type:* DeviceMeasurementIds	*Possible access:* Read	*Default value:* - *not specified* -
This attribute describes measurements supported by the Terminal that are not related to any link or observed channel.			
listOfSupportedDeviceOptions	*Value type:* DeviceOptions	*Possible access:* Read	*Default value:* - *not specified* -
This attribute describes options supported by the Terminal.			
listOfSupportedRadioInterfaces	*Value type:* RadioInterfaces	*Possible access:* Read	*Default value:* - *not specified* -
This attributes describes radio interfaces supported by this Terminal.			
listOfSupportedChannels	*Value type:* ChannelIds	*Possible access:* Read	*Default value:* - *not specified* -
This attributes describes frequency channels supported by the Terminal.			
CONTAINED IN	**Device**		
CONTAINS			
SUPPORTED EVENTS			

From IEEE Standard for Architectural Building Blocks Enabling Network-Device Distributed Decision Making for Optimized Radio Resource Usage in Heterogeneous Wireless Access Networks, Jan 2019. IEEE.

Table 16B.17 —DeviceConfiguration class definition.

Class **DeviceConfiguration**	
This class contains information about the current configuration of the Terminal.	
DERIVED FROM	
ATTRIBUTES	
CONTAINED IN	**Device**
CONTAINS	**Link[*]**
SUPPORTED EVENTS	

From IEEE Standard for Architectural Building Blocks Enabling Network-Device Distributed Decision Making for Optimized Radio Resource Usage in Heterogeneous Wireless Access Networks, Jan 2019. IEEE.

Table 16B.18 —Link class definition.

Class **Link**	
This class contains information about one active connection between the Terminal and RANs.	
DERIVED FROM	**19004BaseClass**
ATTRIBUTES	
CONTAINED IN	**DeviceConfiguration**
CONTAINS	**LinkProfile[1], LinkCapabilities[1], LinkMeasurements[1]**
SUPPORTED EVENTS	

From IEEE Standard for Architectural Building Blocks Enabling Network-Device Distributed Decision Making for Optimized Radio Resource Usage in Heterogeneous Wireless Access Networks, Jan 2019. IEEE.

Table 16B.19 —LinkProfile class definition.

Class **LinkProfile**			
This class contains general information about this active connection.			
DERIVED FROM			
ATTRIBUTES			
linkId	*Value type:* NameType	*Possible access:* Read	*Default value:* - not specified -
This attribute contains string or number assigned to uniquely identify this link.			
associatedAssignedChannelId	*Value type:* OptionalObjectName	*Possible access:* Read	*Default value:* - not specified -
This attribute contains ID of frequency channel used by this link.			
associatedCellId	*Value type:* OptionalObjectName	*Possible access:* Read	*Default value:* - not specified -
This attribute contains ID of cell used by this link.			
CONTAINED IN	**Link**		
CONTAINS			
SUPPORTED EVENTS			

From IEEE Standard for Architectural Building Blocks Enabling Network-Device Distributed Decision Making for Optimized Radio Resource Usage in Heterogeneous Wireless Access Networks, Jan 2019. IEEE.

Table 16B.20 —LinkCapabilities class definition.

Class **LinkCapabilities**			
This class contains information about measurements (instantaneous measurement data and performance statistics derived from this data) supported on this active connection.			
DERIVED FROM			
ATTRIBUTES			
listOfSupportedLink Measurements	*Value type:* LinkMeasurementIds	*Possible access:* Read	*Default value:* - *not specified -*
This attribute describes measurements supported on this links.			
CONTAINED IN	**Link**		
CONTAINS			
SUPPORTED EVENTS			

From IEEE Standard for Architectural Building Blocks Enabling Network-Device Distributed Decision Making for Optimized Radio Resource Usage in Heterogeneous Wireless Access Networks, Jan 2019. IEEE.

Table 16B.21 —LinkMeasurements class definition.

Class **LinkMeasurements**			
This class contains current measurements (instantaneous measurement data and performance statistics derived from this data) related to this active connection.			
DERIVED FROM			
ATTRIBUTES			
listOfActiveLink Measurements	*Value type:* LinkMeasurements	*Possible access:* Read-Add-Remove	*Default value:* - *not specified -*
This attribute describes measurements that are currently performed on this link.			
CONTAINED IN	**Link**		
CONTAINS			
SUPPORTED EVENTS			

From IEEE Standard for Architectural Building Blocks Enabling Network-Device Distributed Decision Making for Optimized Radio Resource Usage in Heterogeneous Wireless Access Networks, Jan 2019. IEEE.

Table 16B.22 —DeviceMeasurements class definition.

Class **DeviceMeasurements**			
This class contains current measurements (instantaneous measurement data and performance statistics derived from this data) related to the Terminal.			
DERIVED FROM			
ATTRIBUTES			
listOfActiveDevice Measurements	*Value type:* DeviceMeasurements	*Possible access:* Read-Add-Remove	*Default value:* - *not specified -*
This attribute describes measurements that are currently performed by the Terminal and are not related to any link or observed channel.			
CONTAINED IN	**Device**		
CONTAINS	**ObservedChannel[*]**		
SUPPORTED EVENTS			

From IEEE Standard for Architectural Building Blocks Enabling Network-Device Distributed Decision Making for Optimized Radio Resource Usage in Heterogeneous Wireless Access Networks, Jan 2019. IEEE.

Table 16B.23 —ObservedChannel class definition.

Class **ObservedChannel**	
This class describes one frequency channel that does not have active connection with the Terminal, but is observed by this Terminal.	
DERIVED FROM	**19004BaseClass**
ATTRIBUTES	
CONTAINED IN	**DeviceMeasurements**
CONTAINS	**ObservedChannelProfile[1], ObservedChannelCapabilities[1], ObservedChannelMeasurements[1]**
SUPPORTED EVENTS	

From IEEE Standard for Architectural Building Blocks Enabling Network-Device Distributed Decision Making for Optimized Radio Resource Usage in Heterogeneous Wireless Access Networks, Jan 2019. IEEE.

Table 16B.24 —ObservedChannelProfile class definition.

Class **ObservedChannelProfile**			
This class contains general information about this frequency channel.			
DERIVED FROM			
ATTRIBUTES			
observedChannelId	*Value type:* NameType	*Possible access:* Read	*Default value:* - not specified -
This attribute contains string or number assigned to uniquely identify this frequency channel.			
observedChannelFrequency Range	*Value type:* FrequencyRange	*Possible access:* Read	*Default value:* - not specified -
This attribute describes frequency range used by this channel.			
associatedCellId	*Value type:* OptionalObjectName	*Possible access:* Read	*Default value:* - not specified -
This attributes contains ID of cell using this frequency channel if any.			
radioInterface	*Value type:* RadioInterface	*Possible access:* Read	*Default value:* - not specified -
This attributes describes radio interface used in this frequency channel if any.			
CONTAINED IN	**ObservedChannel**		
CONTAINS			
SUPPORTED EVENTS			

From IEEE Standard for Architectural Building Blocks Enabling Network-Device Distributed Decision Making for Optimized Radio Resource Usage in Heterogeneous Wireless Access Networks, Jan 2019. IEEE.

Table 16B.25 —ObservedChannelCapabilities class definition.

Class **ObservedChannelCapabilities**			
This class contains information about measurements (instantaneous measurement data and performance statistics derived from this data) supported on this frequency channel.			
DERIVED FROM			
ATTRIBUTES			
listOfSupportedChannelMeasurements	*Value type:* ChannelMeasurementIds	*Possible access:* Read	*Default value:* - *not specified* -
This attribute describes measurements supported on this frequency channel.			
CONTAINED IN	**ObservedChannel**		
CONTAINS			
SUPPORTED EVENTS			

From IEEE Standard for Architectural Building Blocks Enabling Network-Device Distributed Decision Making for Optimized Radio Resource Usage in Heterogeneous Wireless Access Networks, Jan 2019. IEEE.

Table 16B.26 —ObservedChannelMeasurements class definition.

Class **ObservedChannelMeasurements**			
This class contains current measurements (instantaneous measurement data and performance statistics derived from this data) related to this frequency channel.			
DERIVED FROM			
ATTRIBUTES			
listOfActiveChannelMeasurements	*Value type:* ChannelMeasurements	*Possible access:* Read-Add-Remove	*Default value:* - *not specified* -
This attribute describes measurements that are currently performed on this frequency channel.			
CONTAINED IN	**ObservedChannel**		
CONTAINS			
SUPPORTED EVENTS			

From IEEE Standard for Architectural Building Blocks Enabling Network-Device Distributed Decision Making for Optimized Radio Resource Usage in Heterogeneous Wireless Access Networks, Jan 2019. IEEE.

Table 16B.27 —RRSPolicy class definition.

Class **RRSPolicy**	
This class describes one radio resource selection policy related to this Terminal.	
DERIVED FROM	**ECAPolicy**
ATTRIBUTES	
CONTAINED IN	**Terminal**
CONTAINS	
SUPPORTED EVENTS	

16B.5   CWN Classes

UML class diagram for CWN classes without inheritance relations is shown in Figure 16B.4.

In addition to relations between the CWN-related classes described in Clause 16.7, the following relations are defined:

— Each instance of **Base Station** class can be associated to one or several instances of **Cell** class.

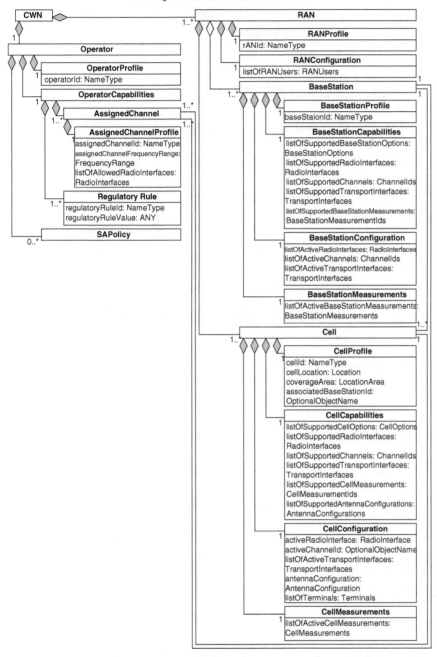

IEEE Std 1900.4-2009
IEEE Standard for architectural building blocks enabling
network-device distributed decision making for optimized radio resource usage
in heterogeneous Wireless Access Networks

Figure 16B.4 —UML class diagram for CWN classes. From IEEE Standard for Architectural Building Blocks Enabling Network-Device Distributed Decision Making for Optimized Radio Resource Usage in Heterogeneous Wireless Access Networks, Jan 2019. IEEE.

— Each instance of **Base Station** class can be associated to one or several instances of **Assigned Channel** class.

— Each instance of **Cell** class can be associated to one or several instances of **Assigned Channel** class.

Table 16B.28 through Table 16B.48 describe each CWN class in detail.

Table 16B.28 — CWN class definition.

Class **CWN**	
The instance of this class contains instances of all CWN-related classes using composition.	
DERIVED FROM	**19004BaseClass**
ATTRIBUTES	
CONTAINED IN	
CONTAINS	**Operator[1], RAN[+]**
SUPPORTED EVENTS	

From IEEE Standard for Architectural Building Blocks Enabling Network-Device Distributed Decision Making for Optimized Radio Resource Usage in Heterogeneous Wireless Access Networks, Jan 2019. IEEE.

Table 16B.29 — Operator class definition.

Class **Operator**	
This class describes operator of this CWN.	
DERIVED FROM	**19004BaseClass**
ATTRIBUTES	
CONTAINED IN	**CWN**
CONTAINS	**OperatorProfile[1], OperatorCapabilities[1], SAPolicy[*]**
SUPPORTED EVENTS	

From IEEE Standard for Architectural Building Blocks Enabling Network-Device Distributed Decision Making for Optimized Radio Resource Usage in Heterogeneous Wireless Access Networks, Jan 2019. IEEE.

Table 16B.30 — OperatorProfile class definition.

Class **OperatorProfile**			
This class contains general information about the operator.			
DERIVED FROM			
ATTRIBUTES			
operatorId	*Value type:* NameType	*Possible access:* Read	*Default value:* - not specified -
This attribute contains string or number assigned to uniquely identify this operator.			
CONTAINED IN	**Operator**		
CONTAINS			
SUPPORTED EVENTS			

From IEEE Standard for Architectural Building Blocks Enabling Network-Device Distributed Decision Making for Optimized Radio Resource Usage in Heterogeneous Wireless Access Networks, Jan 2019. IEEE.

Table 16B.31 — OperatorCapabilities class definition.

Class **OperatorCapabilities**	
This class describes operator capabilities.	
DERIVED FROM	
ATTRIBUTES	
CONTAINED IN	**Operator**
CONTAINS	**AssignedChannel[+], RegulatoryRule[+]**
SUPPORTED EVENTS	

From IEEE Standard for Architectural Building Blocks Enabling
Network-Device Distributed Decision Making for Optimized Radio Resource
Usage in Heterogeneous Wireless Access Networks, Jan 2019. IEEE.

Table 16B.32 — AssignedChannel class definition.

Class **AssignedChannel**	
This class describes one frequency channel assigned to this operator.	
DERIVED FROM	**19004BaseClass**
ATTRIBUTES	
CONTAINED IN	**OperatorCapabilities**
CONTAINS	**AssignedChannelProfile[1]**
SUPPORTED EVENTS	

From IEEE Standard for Architectural Building Blocks Enabling Network-
Device Distributed Decision Making for Optimized Radio Resource Usage
in Heterogeneous Wireless Access Networks, Jan 2019. IEEE.

Table 16B.33 — AssignedChannelProfile class definition.

Class **AssignedChannelProfile**			
This class contains general information about this frequency channel.			
DERIVED FROM			
ATTRIBUTES			
assignedChannelId	*Value type:* NameType	*Possible access:* Read	*Default value:* - not specified -
This attribute contains string or number assigned to uniquely identify this frequency channel.			
assignedChannelFrequency Range	*Value type:* FrequencyRange	*Possible access:* Read	*Default value:* - not specified -
This attribute describes frequency range used by this channel.			
listOfAllowedRadioInterfaces	*Value type:* RadioInterfaces	*Possible access:* Read	*Default value:* - not specified -
This attributes contains list of IDs of radio interfaces allowed to be used on this frequency channel.			
CONTAINED IN	**AssignedChannel**		
CONTAINS			
SUPPORTED EVENTS			

From IEEE Standard for Architectural Building Blocks Enabling Network-Device Distributed Decision
Making for Optimized Radio Resource Usage in Heterogeneous Wireless Access Networks, Jan 2019. IEEE.

Table 16B.34 — RegulatoryRule class definition.

Class **RegulatoryRule**			
This class describes in a formalized form one regulatory rule to be applied to one or several assigned channels.			
DERIVED FROM	**19004BaseClass**		
ATTRIBUTES			
regulatoryRuleId	*Value type:* NameType	*Possible access:* Read	*Default value:* - not specified -
This attribute contains string or number assigned to uniquely identify this regulatory rule			
regulatoryRuleValue	*Value type:* ANY	*Possible access:* Read	*Default value:* - not specified -
This attribute describes this regulatory rule.			
CONTAINED IN	**OperatorCapabilities**		
CONTAINS			
SUPPORTED EVENTS			

From IEEE Standard for Architectural Building Blocks Enabling Network-Device Distributed Decision Making for Optimized Radio Resource Usage in Heterogeneous Wireless Access Networks, Jan 2019. IEEE.

Table 16B.35 — SAPolicy class definition.

Class **SAPolicy**	
This class describes one spectrum assignment policy specified by the operator.	
DERIVED FROM	**ECAPolicy**
ATTRIBUTES	
CONTAINED IN	**Operator**
CONTAINS	
SUPPORTED EVENTS	

From IEEE Standard for Architectural Building Blocks Enabling Network-Device Distributed Decision Making for Optimized Radio Resource Usage in Heterogeneous Wireless Access Networks, Jan 2019. IEEE.

Table 16B.36 — RAN class definition.

Class **RAN**	
This class describes one RAN of this CWN.	
DERIVED FROM	**19004BaseClass**
ATTRIBUTES	
CONTAINED IN	**CWN**
CONTAINS	**RANProfile[1], RANConfiguration[1], BaseStation[+], Cell[+]**
SUPPORTED EVENTS	

From IEEE Standard for Architectural Building Blocks Enabling Network-Device Distributed Decision Making for Optimized Radio Resource Usage in Heterogeneous Wireless Access Networks, Jan 2019. IEEE.

Table 16B.37 —RANProfile class definition.

Class **RANProfile**			
This class contains general information about this RAN.			
DERIVED FROM			
ATTRIBUTES			
rANId	*Value type:* NameType	*Possible access:* Read	*Default value:* - not specified -
This attribute contains string or number assigned to uniquely identify this RAN.			
CONTAINED IN	**RAN**		
CONTAINS			
SUPPORTED EVENTS			

From IEEE Standard for Architectural Building Blocks Enabling Network-Device Distributed Decision Making for Optimized Radio Resource Usage in Heterogeneous Wireless Access Networks, Jan 2019. IEEE.

Table 16B.38 —RANConfiguration class definition.

Class **RANConfiguration**			
This class describes current configuration of this RAN.			
DERIVED FROM			
ATTRIBUTES			
listOfRANUsers	*Value type:* RANUsers	*Possible access:* Read	*Default value:* - not specified -
This attribute contains list of users of this RAN.			
CONTAINED IN	**RAN**		
CONTAINS			
SUPPORTED EVENTS			

From IEEE Standard for Architectural Building Blocks Enabling Network-Device Distributed Decision Making for Optimized Radio Resource Usage in Heterogeneous Wireless Access Networks, Jan 2019. IEEE.

Table 16B.39 —BaseStation class definition.

Class **BaseStation**	
This class describes one base station of the RAN.	
DERIVED FROM	**19004BaseClass**
ATTRIBUTES	
CONTAINED IN	**RAN**
CONTAINS	**BaseStationProfile[1], BaseStationCapabilities[1], BaseStationConfiguration[1], BaseStationMeasurements[1]**
SUPPORTED EVENTS	

From IEEE Standard for Architectural Building Blocks Enabling Network-Device Distributed Decision Making for Optimized Radio Resource Usage in Heterogeneous Wireless Access Networks, Jan 2019. IEEE.

Table 16B.40 — BaseStationProfile class definition.

Class **BaseStationProfile**			
This class contains general information about this base station.			
DERIVED FROM			
ATTRIBUTES			
baseStationId	*Value type:* NameType	*Possible access:* Read	*Default value:* - *not specified* -
This attribute contains string or number assigned to uniquely identify this base station.			
CONTAINED IN	**BaseStation**		
CONTAINS			
SUPPORTED EVENTS			

From IEEE Standard for Architectural Building Blocks Enabling Network-Device Distributed Decision Making for Optimized Radio Resource Usage in Heterogeneous Wireless Access Networks, Jan 2019. IEEE.

Table 16B.41 — BaseStationCapabilities class definition.

Class **BaseStationCapabilities**			
This class contains information about base station capabilities including both transmission and measurement capabilities.			
DERIVED FROM			
ATTRIBUTES			
listOfSupportedBaseStation Options	*Value type:* BaseStationOptions	*Possible access:* Read	*Default value:* - *not specified* -
This attribute describes options supported by this base station.			
listOfSupportedRadio Interfaces	*Value type:* RadioInterfaces	*Possible access:* Read	*Default value:* - *not specified* -
This attribute contains list of radio interfaces supported by this base station.			
listOfSupportedChannels	*Value type:* ChannelIds	*Possible access:* Read	*Default value:* - *not specified* -
This attributes contains list of IDs of frequency channels supported by this base station			
listOfSupportedTransport Interfaces	*Value type:* TransportInterfaces	*Possible access:* Read	*Default value:* - *not specified* -
This attributes describes transport interfaces supported by this base station.			
listOfSupportedBaseStation Measurements	*Value type:* BaseStationMeasurementIds	*Possible access:* Read	*Default value:* - *not specified* -
This attributes describes measurements supported by this base station and not related to any cell.			
CONTAINED IN	**BaseStation**		
CONTAINS			
SUPPORTED EVENTS			

From IEEE Standard for Architectural Building Blocks Enabling Network-Device Distributed Decision Making for Optimized Radio Resource Usage in Heterogeneous Wireless Access Networks, Jan 2019. IEEE.

Table 16B.42 —BaseStationConfiguration class definition.

Class **BaseStationConfiguration**			
This class contains information about the current configuration of the base station.			
DERIVED FROM			
ATTRIBUTES			
listOfActiveRadioInterfaces	*Value type:* RadioInterfaces	*Possible access:* Read	*Default value:* - not specified -
This attribute contains list of radio interfaces that are currently used by this base station.			
listOfActiveChannels	*Value type:* ChannelIds	*Possible access:* Read	*Default value:* - not specified -
This attributes contains list of IDs of frequency channels that are currently used by this base station.			
listOfActiveTransportInterface s	*Value type:* TransportInterfaces	*Possible access:* Read	*Default value:* - not specified -
This attributes describes transport interfaces that are currently used by this base station.			
CONTAINED IN	**BaseStation**		
CONTAINS			
SUPPORTED EVENTS			

From IEEE Standard for Architectural Building Blocks Enabling Network-Device Distributed Decision Making for Optimized Radio Resource Usage in Heterogeneous Wireless Access Networks, Jan 2019. IEEE.

Table 16B.43 —BaseStationMeasurements class definition.

Class **BaseStationMeasurements**			
This class contains current measurements (instantaneous measurement data and performance statistics derived from this data) performed by this base station.			
DERIVED FROM			
ATTRIBUTES			
listOfActiveBaseStation Measurements	*Value type:* BaseStationMeasurements	*Possible access:* Read-Add-Remove	*Default value:* - not specified -
This attributes describes measurements that are currently performed by this base station and not related to any cell.			
CONTAINED IN	**BaseStation**		
CONTAINS			
SUPPORTED EVENTS			

From IEEE Standard for Architectural Building Blocks Enabling Network-Device Distributed Decision Making for Optimized Radio Resource Usage in Heterogeneous Wireless Access Networks, Jan 2019. IEEE.

Table 16B.44 —Cell class definition.

Class **Cell**	
This class describes one cell of the base station.	
DERIVED FROM	**19004BaseClass**
ATTRIBUTES	
CONTAINED IN	**RAN**
CONTAINS	**CellProfile[1], CellCapabilities[1], CellConfiguration[1], CellMeasurements[1]**
SUPPORTED EVENTS	

From IEEE Standard for Architectural Building Blocks Enabling Network-Device Distributed Decision Making for Optimized Radio Resource Usage in Heterogeneous Wireless Access Networks, Jan 2019. IEEE.

Table 16B.45 —CellProfile class definition.

Class **CellProfile**			
This class contains general information about this cell.			
DERIVED FROM			
ATTRIBUTES			
cellId	*Value type:* NameType	*Possible access:* Read	*Default value:* - *not specified* -
This attribute contains string or number assigned to uniquely identify this cell.			
cellLocation	*Value type:* Location	*Possible access:* Read	*Default value:* - *not specified* -
This attribute describes location of the cell.			
coverageArea	*Value type:* LocationArea	*Possible access:*	*Default value:* - *not specified* -
This attributes describes coverage area of the cell			
associatedBaseStationId	*Value type:* OptionalObjectName	*Possible access:* Read	*Default value:* - *not specified* -
This attributes contains ID of the base station to which this cell belongs.			
CONTAINED IN	**Cell**		
CONTAINS			
SUPPORTED EVENTS			

From IEEE Standard for Architectural Building Blocks Enabling Network-Device Distributed Decision Making for Optimized Radio Resource Usage in Heterogeneous Wireless Access Networks, Jan 2019. IEEE.

Table 16B.46 —CellCapabilities class definition.

Class **CellCapabilities**			
This class contains information about cell capabilities.			
DERIVED FROM			
ATTRIBUTES			
listOfSupportedCellOptions	*Value type:* CellOptions	*Possible access:* Read	*Default value:* - *not specified* -
This attribute describes options supported by this cell.			
listOfSupportedRadio Interfaces	*Value type:* RadioInterfaces	*Possible access:* Read	*Default value:* - *not specified* -
This attribute contains list of radio interfaces supported by this cell.			
listOfSupportedChannels	*Value type:* ChannelIds	*Possible access:* Read	*Default value:* - *not specified* -
This attributes contains list of IDs of frequency channels supported by this cell.			
listOfSupportedTransport Interfaces	*Value type:* TransportInterfaces	*Possible access:* Read	*Default value:* - *not specified* -
This attributes describes transport interfaces supported by this cell.			
listOfSupportedCell Measurements	*Value type:* CellMeasurementIds	*Possible access:* Read	*Default value:* - *not specified* -

(continued)

Table 16B.46 (Continued)

Class **CellCapabilities**			
This attributes describes measurements supported by this cell.			
listOfSupportedAntenna Configurations	*Value type:* AntennaConfigurations	*Possible access:* Read	*Default value:* - *not specified* -
This attributes describes antenna configurations supported by this cell			
CONTAINED IN	**Cell**		
CONTAINS			
SUPPORTED EVENTS			

From IEEE Standard for Architectural Building Blocks Enabling Network-Device Distributed Decision Making for Optimized Radio Resource Usage in Heterogeneous Wireless Access Networks, Jan 2019. IEEE.

Table 16B.47 —CellConfiguration class definition.

Class **CellConfiguration**			
This class contains information about the current configuration of the cell.			
DERIVED FROM			
ATTRIBUTES			
activeRadioInterface	*Value type:* RadioInterface	*Possible access:* Read	*Default value:* - *not specified* -
This attribute describes radio interface that is currently used by this cell.			
activeChannelId	*Value type:* OptionalObjectName	*Possible access:* Read	*Default value:* - *not specified* -
This attributes contains ID of frequency channel that is currently used by this cell.			
listOfActiveTransportInterfaces	*Value type:* TransportIntefaces	*Possible access:* Read	*Default value:* - *not specified* -
This attributes describes transport interfaces that are currently used by this cell.			
antennaConfiguration	*Value type:* AntennaConfiguration	*Possible access:* Read	*Default value:* - *not specified* -
This attributes describes current antenna configuration of this cell.			
listOfTerminals	*Value type:* Terminals	*Possible access:* Read	*Default value:* - *not specified* -
This attributes contains list of Terminals connected to this cell.			
CONTAINED IN	**Cell**		
CONTAINS			
SUPPORTED EVENTS			

From IEEE Standard for Architectural Building Blocks Enabling Network-Device Distributed Decision Making for Optimized Radio Resource Usage in Heterogeneous Wireless Access Networks, Jan 2019. IEEE.

Table 16B.48 —CellMeasurements class definition.

Class **CellMeasurements**			
This class contains current measurements (instantaneous measurement data and performance statistics derived from this data) related to this cell.			
DERIVED FROM			
ATTRIBUTES			
listOfCellMeasurements	*Value type:* CellMeasurements	*Possible access:* Read-Add-Remove	*Default value:* - *not specified* -
This attributes describes measurements that are currently performed by this cell.			
CONTAINED IN	**Cell**		
CONTAINS			
SUPPORTED EVENTS			

From IEEE Standard for Architectural Building Blocks Enabling Network-Device Distributed Decision Making for Optimized Radio Resource Usage in Heterogeneous Wireless Access Networks, Jan 2019. IEEE.

16B.6   Relations Between Terminal and CWN Classes

UML class diagram defining relations between terminal and CWN classes is shown in Figure 16B.5. This figure shows only a part of terminal related and CWN-related classes.
    The following relations are defined:

— Each instance of **User** class can be associated to one or several instances of **Operator** class.
— Each instance of **RAN** class can be associated to zero or several instances of **User** class.
— Each instance of **Cell** class can be associated to zero or several instances of **Terminal** class.
— Each instance of **Observed Channel** class can be associated to zero or several instances of **Cell** class.
— Each instance of **Cell** class can be associated to zero or several instances of **Link** class.
— Each instance of **Assigned Channel** class can be associated to zero or several instances of **Link** class.

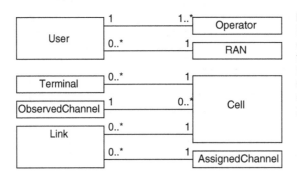

Figure 16B.5 —UML class diagram defining relations between terminal and CWN classes. From IEEE Standard for Architectural Building Blocks Enabling Network-Device Distributed Decision Making for Optimized Radio Resource Usage in Heterogeneous Wireless Access Networks, Jan 2019. IEEE.

Annex 16C

(normative)

Data Type Definitions for Information Model

16C.1   Function Definitions

The managed system is modeled as a tree of objects that are instances of the classes defined within the information model. A number of functions are assumed to be available at the managed system side that work on this object tree and fulfill the required (by the managing system) actions or provide the requested information (to the managing system). These functions are described below in order to define the behavior of a managed system with respect to policy data (and their types) it has received from a managing system.

The following functions are assumed to be available locally (not necessarily exposed to an outside interface) at the managed system.

Select a set of objects:

```
ObjectSet (baseObject, level, filter)
```

Returns: List of object names
Parameters:

baseObject:	'root' or name of an object
level:	'self' | 'directContained' | 'contained' | 'allSubtree'
filter:	logical expression combined from simple attribute value expressions

Comments:

baseObject and level together define the scope of objects (pre-selection step): 'self' means just the baseObject, 'directContained' means all objects that are directly contained in the baseObject, 'contained' also those objects that are indirectly contained (i.e., also contained in contained etc.) objects and, finally 'allSubtree' includes on top of 'contained' also the baseObject itself.

Simple attribute value expressions are defined as

```
<attributeName>  '<' | '=' | '>' | '!=' | '<=' | '>='
                                    <attributeName> | <constantValue>
```

If a simple attribute value expression is evaluated for an object found in the pre-selection step and the name of the attribute(s) is not contained in the definition of the object's class, the expression is treated as false. All objects of the pre-selection step that pass the filter form the resulting object set.

Count objects of an object set:

```
ObjectCount (baseObject, level, filter)
```

Returns: Number of objects contained in ObjectSet (baseObject, level, filter).
Parameters: See function ObjectSet
Comments: None

Count events occurring in an object set:

```
EventCount (eventType, timeInterval, baseObject, level, filter)
```

Returns:

Number of events of type "eventType" occurring in the set of objects returned by "Object-Set(baseObject, level, filter)." Only those events are counted that occurred during the last "timeInterval" seconds.

Parameters:

eventType: Type of events to be counted
timeInterval: Event count is the number of events occurring during the last timeInterval milliseconds
See function ObjectSet for the other parameters

Comments: None
The following function is needed for describing policy actions.
Ensure that attribute values are in a given value set or range:

```
Ensure (baseObject, level, filter, valueAssertion, timeslot)
```

Returns: Boolean value indicating success of the operation
Parameters:

valueAssertion: Type (`set` or `avoid`), attributeName and set(range) of values
timeslot: all necessary attribute settings are done in <timeslot> ms after this function has been called
See function ObjectSet for the parameters baseObject, level, and filter

Comments:

The function considers each object from ObjectSet (baseObject, level, filter) and applies to it the value assertion. The managed system has to ensure that the value of this attribute (providing this attribute is defined for the object) is set (`set`) or not set (`avoid`) to one of those values.

16C.2   ASN.1 Type Definitions

The following ASN.1 (see ISO/IEC 8824 for ASN.1 specification) module contains all necessary abstract data definitions used in the attribute definitions in Annex 16B.

```
1900-4-Type-Definitions DEFINITIONS ::= BEGIN
            ------------------------------------------------------------
            ------------------------------------------------------------
            -- START Common Data Types
            ------------------------------------------------------------
            -- START Name Related Data Types
            NameType    ::=    CHOICE    {
                number    INTEGER,
                string    PrintableString
            }
            ObjectName    ::=    SEQUENCE OF NameType
            OptionalObjectName    ::=    CHOICE    {
                id    ObjectName,
                void    NULL
            }
            -- END Name Related Data Types
            ------------------------------------------------------------
            ------------------------------------------------------------
            -- START Radio Interface Related Data Types
            RadioInterfaceId    ::=    ENUMERATED    {
                umts, hsdpa, wimax, lte, wifi, gsm, ...
            }
            RadioInterface    ::= CHOICE    {
                id        RadioInterfaceId
```

```
    void    NULL
}
RadioInterfaces   ::=    SEQUENCE OF    RadioInterfaceId
-- END Radio Interface Related Data Types
------------------------------------------------------------
------------------------------------------------------------
-- START Channel Related Data Types
ChannelIds   ::=    SEQUENCE OF    OptionalObjectName
FrequencyRange    ::=     SEQUENCE   {
    centralFrequency    REAL,
    frequencyBand       REAL
}
-- END Channel Related Data Types
------------------------------------------------------------
------------------------------------------------------------
-- START Location Related Data Types
Location    ::= SEQUENCE    {
    latitude     REAL,
    longitude    REAL,
    height       REAL OPTIONAL
}
LocationArea     ::=    SEQUENCE (SIZE(3 .. MAX)) OF SEQUENCE   {
    latitude     REAL,
    longitude    REAL
}
-- END Common Data Types
------------------------------------------------------------
------------------------------------------------------------
------------------------------------------------------------
------------------------------------------------------------
-- START Policy Related Data Types
------------------------------------------------------------
-- START Policy Event Related Data Types
PolicyEvent    ::=     SEQUENCE    {
    eventType    ENUMERATED    {
        objectCreation, objectDeletion, stateChanged,
        attributeChanged, immediatelyOnce, scheduledTimer, ...
    }
    -- Additional data qualifying the event.
    -- Type is different and depends on eventType.
    eventQualifier    ANY
}
-- END Policy Event Related Data Types
------------------------------------------------------------
------------------------------------------------------------
-- START Policy Condition Related Data Types
ValidPolicyTime ::= CHOICE {
    validUntil    GeneralizedTime,
    unspecified    NULL
}
ValidPolicyLocation ::= CHOICE {
    validLocationArea    LocationArea,
    unspecified          NULL
}
LogicalOperator    ::=     ENUMERATED {
    equal, non-equal, less, less-or-equal, greater,
    greater-or-equal
}
AttributeValueAssertion    ::=      SEQUENCE    {
    object           OptionalObjectName OPTIONAL,
    attributeName    PrintableString,
    fieldName        PrintableString OPTIONAL,
    attributeValue   ANY
}
ValueRange     ::=      CHOICE   {
    intRange    SEQUENCE{
```

```
                               low       INTEGER,
                               high      INTEGER
                           }
                           floatRange     SEQUENCE{
                               low     REAL,
                               high    REAL
                           }
                           stringRange    SEQUENCE{
                               low     PrintableString,
                               high    PrintableString
                           }
                   }
                   ElementaryLogicalExpression    ::=
         SEQUENCE  {
                       attributeName      PrintableString,
                       fieldname          PrintableString OPTIONAL,
                       operation    CHOICE    {
                           valueComparison     SEQUENCE    {
                               operator     LogicalOperator,
                               comparedWith    CHOICE    {
                                   definedValue           ANY,
                                   otherAttributeValue    AttributeValueAssertion
                               },
                               isContainedIn   SET OF CHOICE    {
                                   definedValue          ANY,
                                   definedRange          ValueRange,
                                   otherAttributeValue   AttributeValueAssertion
                               }
                           }
                       }
                   }
                   Filter    ::=   CHOICE    {
                       elementary       ElementaryLogicalExpression,
                       andExpression  SEQUENCE    {
                           operand1   Filter,
                           operand2   Filter
                       },
                       orExpression   SEQUENCE   {
                           operand1   Filter,
                           operand2   Filter
                       },
                       negExpression   Filter
                   }
                   ObjectSetParameters   ::=   {
                       -- This type fits to provide a call of the ObjectSet
                       -- function (see C.1) with parameter values
                       object    ObjectName,
                       level     ENUMERATED    {
                           self, directContained, contained, allSubtree
                       },
                       filter    Filter
                   }
                   EventCountParameters    ::=    SEQUENCE    {
                       -- This type fits to provide a call of the EventCount
                       -- function (see C.1) with parameter values
                       objectSet    ObjectSetParameters,
                       eventType    ENUMERATED    {
                           objectCreation, objectDeletion, stateChnaged,
                           attributeChanged, ...
                       },
                       timeslot    INTEGER
                   }
                   CountItem    ::=    CHOICE    {
                       objects    ObjectSetParameters,
                       events    EventCountParameters
                   }
```

```
PolicyCondition      ::=      SEQUENCE OF SEQUENCE      {
    -- Number of objects/events
    countItem        CountItem,
    operator         LogicalOperator,
    comparedWith     CHOICE     {
        definedValue      INTEGER,
        otherCountItem     CountItem
    }
}
-- END Policy Condition Related Data Types
-------------------------------------------------------------
-------------------------------------------------------------
-- START Policy Action Related Data Types
ObjectOperation    ::=     SEQUENCE    {
    baseObject      OptionalObjectName,
    level    ENUMERATED      {
        self, directContained, contained, allSubtree
    },
    filter       Filter,
    operation    CHOICE     {
        valueAssertion      SEQUENCE    {
            operationType      ENUMERATED    {
                guaranteeValues, avoidValues
            },
            attributeName    PrintableString OPTIONAL,
            valueList    SET OF CHOICE    {
                definedValue     ANY,
                definedRange     ValueRange
            }
        },
        objectPresence    SEQUENCE     {
            operationType     ENUMERATED     {
                mustExist, mustNotExist
            },
            className    PrintableString,
            attributeValueConstraints    SET OF SEQUENCE    {
                attributeName      PrintableString,
                fieldName          PrintableString OPTIONAL,
                attributeValue     ANY
            }
        }
    },
    timeSlot    INTEGER
}
ObjectOperations    ::=    SEQUENCE OF   ObjectOperation
-- END Policy Action Related Data Types
-------------------------------------------------------------
-- END Policy Related Data Types
-------------------------------------------------------------
-------------------------------------------------------------
-------------------------------------------------------------
-------------------------------------------------------------
-- START Terminal Related Data Types
-------------------------------------------------------------
-- START Services Related Data Types
Services    ::=     SEQUENCE OF SEQUENCE     {
    serviceName    PrintableString
    serviceCost    ANY
}
-- END Services Related Data Types
-------------------------------------------------------------
-------------------------------------------------------------
-- START Appliction Related Data Types
Direction    ::=     ENUMERATED     {
    downlink, uplink
}
```

```
TrafficClassId    ::=    ENUMERATED         {
    conversational, streaming, interactive, background, ...
}
TrafficClass    ::= CHOICE    {
    id    TrafficClassId
    void    NULL
}
QoSRequirementId    ::=    ENUMERATED    {
    maximumDelay, maximumDelayVariation, maximumPacketLoss,
    minimumBandwidth, preferredBandwidth, ...
}
QoSRequirements    ::=    SEQUENCE OF SEQUENCE    {
    qoSRequirementName    QoSRequirementId
    qoSRequirementValue    ANY
}
Links    ::=    SEQUENCE OF OptionalObjectName
ApplicationMeasurementId    ::=    ENUMERATED    {
    observedDelay, observedDelayVariation, observedPacketLoss,
    observedBandwidth, ...
}
ApplicationMeasurementIds    ::=    SEQUENCE OF {
    ApplicationMeasurementId
}
ApplicationMeasurements    ::=    SEQUENCE OF SEQUENCE    {
    applicationMeasurementName    ApplicationMeasurementId
    applicationMeasurementValue    ANY
}
-- END Application Related Data Types
------------------------------------------------------------
------------------------------------------------------------
-- START Device Related Data Types
DeviceOptionId    ::=    ENUMERATED    {
    maximumTxPower, maximumNumberOfRadioInterfaces, ...
}
DeviceOptions    ::=    SEQUENCE OF SEQUENCE    {
    deviceOptionName    DeviceOptionId
    deviceOptionValue    ANY
}
DeviceMeasurementId    ::=    ENUMERATED    {
    deviceLocation, batteryPower, ...
}
DeviceMeasurementIds    ::=    SEQUENCE OF {
    DeviceMeasurementId
}
DeviceMeasurements    ::=    SEQUENCE OF SEQUENCE    {
    deviceMeasurementName    DeviceMeasurementId
    deviceMeasurementValue    ANY
}
LinkMeasurementId    ::=    ENUMERATED    {
    receivedPower, receivedSINR, ...
}
LinkMeasurementIds    ::=    SEQUENCE OF {
    LinkMeasurementId
}
LinkMeasurements    ::=    SEQUENCE OF SEQUENCE    {
    linkMeasurementName    LinkMeasurementId
    linkMeasurementValue    ANY
}
ChannelMeasurementId    ::=    ENUMERATED    {
    channelInterference, channelLoad, ...
}
ChannelMeasurementIds    ::=    SEQUENCE OF {
    ChannelMeasurementId
}
ChannelMeasurements    ::=    SEQUENCE OF SEQUENCE    {
    channelMeasurementName    ChannelMeasurementId
```

```
        channelMeasurementValue    ANY
}
-- END Device Related Data Types
-----------------------------------------------------------
-- END Terminal Related Data Types
-----------------------------------------------------------
-----------------------------------------------------------
-----------------------------------------------------------
-----------------------------------------------------------
-- START CWN Related Data Types
-----------------------------------------------------------
-- START RAN Related Data Types
RANUsers    ::=     SEQUENCE OF SEQUENCE      {
    userId      OptionalObjectName
    userData    ANY
}
-- END RAN Related Data Types
-----------------------------------------------------------
-----------------------------------------------------------
-- START Transport Interface Related Data Types
TransportInterface    ::=     SEQUENCE   {
    transportTechnology     PrintableString,
    bandwidth               REAL,
    userPlaneBandwidth      REAL OPTIONAL,
    controlPlaneBadwidth    REAL OPTIONAL,
    omBandwidth             REAL OPTIONAL
}
TransportInterfaces    ::=    SEQUENCE OF TransportInterface
-- END Transport Interface Related Data Types
-----------------------------------------------------------
-----------------------------------------------------------
-- START Base Station Related Data Types
BaseStationOptionId    ::=     ENUMERATED    {
    maximumTxPower, maximumNumberOfRadioInterfaces, ...
}
BaseStationOptions    ::=     SEQUENCE OF SEQUENCE    {
    baseStationOptionName     BaseStationOptionId
    baseStationOptionValue    ANY
}
BaseStationMeasurementId    ::=    ENUMERATED    {
    transmitPower, transportLoad, processingLoad, ...
}
BaseStationMeasurementIds    ::=    SEQUENCE OF {
    BaseStationMeasurementId
}
BaseStationMeasurements    ::=     SEQUENCE OF SEQUENCE    {
    baseStationMeasurementName     BaseStationMeasurementId
    baseStationMeasurementValue    ANY
}
-- END Base Station Related Data Types
-----------------------------------------------------------
-----------------------------------------------------------
-- START Cell Related Data Types
CellOptionId    ::=    ENUMERATED      {
    maximumTxPower, ...
}
CellOptions    ::=     SEQUENCE OF SEQUENCE    {
    cellOptionName     CellOptionId
    cellOptionValue    ANY
}
CellMeasurementId    ::=     ENUMERATED     {
    transmitPower, cellLoad, trafficLoad, cellThroughput,
    cellInterference, ...
}
CellMeasurementIds    ::=     SEQUENCE OF {
    CellMeasurementId
```

```
}
CellMeasurements     ::=     SEQUENCE OF SEQUENCE     {
    cellMeasurementName     CellMeasurementId
    cellMeasurementValue    ANY
}
AntennaConfiguration     ::=    CHOICE     {
    omnidirectional     NULL,
    beamforming         REAL,
    ...
}
AntennaConfigurations   ::=     SEQUENCE OF     {
    AntennaConfiguration
}
Terminals    ::=     SEQUENCE OF    OptionalObjectName
-- END Cell Related Data Types
------------------------------------------------------------
-- END CWN Related Data Types
------------------------------------------------------------
------------------------------------------------------------
END
```

Annex 16D

(informative)

Information Model Extensions and Usage Example

16D.1   Functions for External Management Interface

For the purpose of describing the usage example, the NRM and TRM entities are considered to be managing and managed systems, respectively. Both systems share the knowledge on terminal and policyrelated classes of the information model. It is assumed that the managed system provides at its external interface a set of primitives representing typical management functions. These functions are used by the managing system assuming that an underlying communication protocol between managing and managed systems is available. It is important to note that the functions are designed in a way where in one invocation they perform operations on a set of managed objects.

The description of the function is as follows.

Set attribute values in a set of objects:

```
SetAttributeValue (baseObject, level, filter [, attributeName, value]+)
```

Sets the attributes from the list to their respective values in all objects selected by ObjectSet (baseObject, level, filter).

Get attribute values in a set of objects:

```
GetAttributeValue (baseObject, level, filter [, attributeName]+)
```

Returns a list of triples (objectName, attributeName, value) with the values for all requested attributes in all objects selected by ObjectSet (baseObject, level, filter).

Create an object:

```
CreateObject (baseObject, level, filter, className [, attribute-
Name, value]+)
```

Creates an object of class given by className contained in all objects selected by ObjectSet (baseObject, level, filter). Initializes attribute values as given in the parameter list. Returns the name of the new object (relative to the baseObject's name).

Delete an object:

```
DeleteObject (baseObject, level, filter)
```

Deletes all objects selected by ObjectSet (baseObject, level, filter). In case one of these objects contains further objects, those are also deleted by the operation.

Finally, the managing system provides at its external interface a primitive representing the function of the managing system to receive a spontaneous event report from the managed system. The content of this message is defined in a corresponding class provided for each type of event report.

Send a report:

```
EventReport(reportType, reportingObject, reportData)
```

16D.2 Additional Utility Classes

To facilitate efficient exchange of information, it is useful to incorporate common utility classes to apply the required statistical operations, filters (that is, selection criteria), trigger thresholds, and other mechanisms that can optimize the efficiency of information exchanges.

The following utility classes are used within this annex in order to control on behalf of their instances common functions in a managed system:

— **Threshold**

An object of this class, once instantiated in the scope of another (its superior) object, provides the possibility to define threshold values for an attribute of the superior object. Attribute thresholds can be use to generate corresponding threshold crossed events.

— **Value characteristic**

An object of this class, once instantiated in the scope of another (its superior) object, provides the possibility to provide for the values of an attribute of this superior a characteristic over a given interval of time (window) taking into account those values collected each time a smaller sampling interval time is elapsed. Characteristics are defined as one of min, max, mean value, standard deviation, or just the values sequence (for collecting the value history). The window may be fixed (meaning the next window starts after the previous window time completely elapsed) or sliding (meaning that after each sampling interval the window is shifted by the sampling interval time). Further, accuracy and precision may be defined for the measured values considered in the calculation of the characteristic. The latest available characteristic is provided as the 'current value' attribute of this class.

— **Scheduler**

An object of this class causes the generation of timeout events in the managed system in a periodic fashion. By specifying a stop time, only a certain number (even one) of timeout events may be specified. Other objects may refer to a scheduler object with the purpose to get triggered by the timeout event for specific actions.

A start time for a scheduler object may be specified (to allow starting timeout generation later than at creation time), an operational state is added in order to enable/disable the timeout generation from the managing system if needed.

— **Measurement reporter**

An object of this class, once instantiated in the scope of another (its superior) object, provides the possibility to end a collection of current values of a specified set of attributes of the superior or contained therein objects to the managing system in form of a measurement report. Therefore, it refers to a scheduler object in order to get triggers for sending the reports. The possibility to specify the managing system is foreseen; while currently only NRM is assumed as such managing system. Further, an attribute 'timeEllapsedSinceLastReport' is defined. This may especially be used to trigger an immediate measurement report (by setting its value to 0).

— **Report Generator**

An object of this class is used to dynamically control the event reports sent to a managing system. Generic event types are defined for the objects in the managed system, but it might be worth spontaneously reporting them to a managing system in some circumstances or at special time intervals. Therefore, a **Report Generator** must be created that specifies the event type, the target system, and (optionally) filter criteria for the data members of the report. Otherwise, no event reports are sent (except measurement reports that are not an accepted event type for **Report Generator** objects).

— **Event Report** (and derived specific reports)

Objects of these classes are just data objects that define what data are expected to be delivered to the managing system in case an event of a given type occurred and the managed system is requested to send an event report on that. Derived reports include the following:

— **Object Creation Report**
— **Object Deletion Report**
— **Attribute Value Change Report**
— **Measurement Report**
— **High Threshold Crossed Report**
— **Low Threshold Crossed Report**

Table 16D.1 through Table 16D.12 describe each utility class in detail.
The UML class diagram for utility classes is shown in Figure 16D.1.

Table 16D.1 —EventReport class definition.

Class **EventReport** *(abstract class)*			
This is the mandatory base class for all IEEE 1900.4 event reports. An event report object is created if a) an event of the specified event type has occurred, and b) an EventReporter object or an ECAPolicy object are requiring the creation of the event report.			
DERIVED FROM			
ATTRIBUTES			
eventType	*Value type:* PrintableString	*Possible access:* Read	*Default value:* - *not specified* -
This attribute defines the type of the reported event.			
eventTime	*Value type:* GeneralizedTime	*Possible access:* Read	*Default value:* - *not specified* -
This attribute defines the time at that the event occurred.			
eventId	*Value type:* NameType	*Possible access:* Read	*Default value:* - *not specified* -
This attribute contains string or number assigned to uniquely identify the event report object.			
reportingObjectName	*Value type:* ObjectName	*Possible access:* Read	*Default value:* - *not specified* -
This attribute defines the object that reported the occurring event.			
CONTAINED IN			
CONTAINS			
SUPPORTED EVENTS			

From IEEE Standard for Architectural Building Blocks Enabling Network-Device Distributed Decision Making for Optimized Radio Resource Usage in Heterogeneous Wireless Access Networks, Jan 2019. IEEE.

Table 16D.2 —ObjectCreationReport class definition.

Class **ObjectCreationReport**			
The ObjectCreationReport object is generated if a new object has been created (objectCreation event type). This may be of interest in case an object occurs during normal operation and a managing system needs to be aware of it.			
DERIVED FROM	**EventReport**		
ATTRIBUTES			
objectClass	*Value type:* PrintableString	*Possible access:* Read	*Default value:* - *not specified* -
This attribute reports the class of the created object.			
CONTAINED IN			
CONTAINS			
SUPPORTED EVENTS			

From IEEE Standard for Architectural Building Blocks Enabling Network-Device Distributed Decision Making for Optimized Radio Resource Usage in Heterogeneous Wireless Access Networks, Jan 2019. IEEE.

Table 16D.3 —ObjectDeletionReport class definition.

Class **ObjectDeletionReport**			
The ObjectDeletionReport is generated if a object has been deleted (objectDeletion event type). This may be of interest in case an object is deleted during normal operation and a managing system needs to be aware of it.			
DERIVED FROM	**EventReport**		
ATTRIBUTES			
objectClass	*Value type:* PrintableString	*Possible access:* Read	*Default value:* *- not specified -*
This attribute reports the class of the deleted object.			
CONTAINED IN			
CONTAINS			
SUPPORTED EVENTS			

From IEEE Standard for Architectural Building Blocks Enabling Network-Device Distributed Decision Making for Optimized Radio Resource Usage in Heterogeneous Wireless Access Networks, Jan 2019. IEEE.

Table 16D.4 —AttributeValueChangedReport class definition.

Class **AttributeValueChangedReport**			
The AttributeValueChangedReport is generated after the value of an attribute has been changed (attributeValueChanged event type). This may be of interest in case the attribute value is changed during normal operation and a managing system needs to be aware of this value change.			
DERIVED FROM	**EventReport**		
ATTRIBUTES			
attributeName	*Value type:* PrintableString	*Possible access:* Read	*Default value:* *- not specified -*
This attribute reports the name of the attribute that changed its value.			
formerValue	*Value type:* ReportedValue	*Possible access:* Read	*Default value:* *- not specified -*
This attribute reports the former attribute value.			
currentValue	*Value type:* ReportedValue	*Possible access:* Read	*Default value:* *- not specified -*
This attribute reports the new attribute value.			
CONTAINED IN			
CONTAINS			
SUPPORTED EVENTS			

From IEEE Standard for Architectural Building Blocks Enabling Network-Device Distributed Decision Making for Optimized Radio Resource Usage in Heterogeneous Wireless Access Networks, Jan 2019. IEEE.

Table 16D.5 —MeasurementReport class definition.

Class **MeasurementReport**			
The MeasurementReport is generated by a MeasurementReporter object after a measurement period is over (measurementReport event type). It contains all values defined in the MeasurementReporter object.			
DERIVED FROM	**EventReport**		
ATTRIBUTES			
valueCollection	*Value type:* ValueCollection	*Possible access:* Read	*Default value:* - *not specified* -
This attribute reports the list of measured values.			
CONTAINED IN			
CONTAINS			
SUPPORTED EVENTS			

From IEEE Standard for Architectural Building Blocks Enabling Network-Device Distributed Decision Making for Optimized Radio Resource Usage in Heterogeneous Wireless Access Networks, Jan 2019. IEEE.

Table 16D.6 —HighThresholdCrossedReport class definition.

Class **HighThresholdCrossedReport**	
The HighThresholdCrossedReport is generated if the attribute value of the monitored attribute in a Threshold object has exceeded the highThresholdValue (highThresholdCrossed event type).	
DERIVED FROM	**EventReport**
ATTRIBUTES	
CONTAINED IN	
CONTAINS	
SUPPORTED EVENTS	

From IEEE Standard for Architectural Building Blocks Enabling Network-Device Distributed Decision Making for Optimized Radio Resource Usage in Heterogeneous Wireless Access Networks, Jan 2019. IEEE.

Table 16D.7 —LowThresholdCrossedReport class definition.

Class **LowThresholdCrossedReport**	
The LowThresholdCrossedReport is generated if the attribute value of the monitored attribute in a Threshold object has gone below the lowThresholdValue (lowThresholdCrossed event type).	
DERIVED FROM	**EventReport**
ATTRIBUTES	
CONTAINED IN	
CONTAINS	
SUPPORTED EVENTS	

From IEEE Standard for Architectural Building Blocks Enabling Network-Device Distributed Decision Making for Optimized Radio Resource Usage in Heterogeneous Wireless Access Networks, Jan 2019. IEEE.

Table 16D.8 —Threshold class definition.

Class **Threshold**			
This class provides the possibility to define various threshold crossed event reports for an object.			
DERIVED FROM	**19004BaseClass**		
ATTRIBUTES			
name	*Value type:* NameType	*Possible access:* Read	*Default value:* *- not specified -*
This attribute contains string or number assigned to uniquely identify the Threshold object.			
observedAttributeAndField	*Value type:* PrintableString	*Possible access:* Read-Write	*Default value:* *- not specified -*
This attribute defines what attribute (and optionally, field) of the superior object has to be monitored. Its value must allow for lower/higher comparison.			
highThresholdValue	*Value type:* ThresholdValue	*Possible access:* Read-Write	*Default value:* *- not specified -*
This attribute defines the high threshold value. If NULL – no high threshold passed event is generated.			
lowThresholdValue	*Value type:* ThresholdValue	*Possible access:* Read-Write	*Default value:* *- not specified -*
This attribute defines the low threshold value. If NULL – no low threshold passed event is generated.			
reportingDelay_High ThresholdCrossed	*Value type:* INTEGER	*Possible access:* Read-Write	*Default value:* 0
This attribute defines a delay between internal event detection and event reporting (if the monitored value goes again below the threshold during this time – no report will be generated)			
reportingDelay_Low Threshold Crossed	*Value type:* INTEGER	*Possible access:* Read-Write	*Default value:* 0
This attribute defines the delay between internal event detection and event reporting (if the monitored value goes again above the threshold during this time – no report will be generated)			
CONTAINED IN	**ApplicationMeasurements, DeviceMeasurements, LinkMeasurements, ChannelMeasurements, BaseStationMeasurements, CellMeasurements**		
CONTAINS			
SUPPORTED EVENTS	highThresholdCrossed, lowThresholdCrossed		

From IEEE Standard for Architectural Building Blocks Enabling Network-Device Distributed Decision Making for Optimized Radio Resource Usage in Heterogeneous Wireless Access Networks, Jan 2019. IEEE.

Table 16D.9 —ValueCharacteristic class definition.

Class **ValueCharacteristic**			
This class allows for defining different statistics to be calculated for attribute values of a managed object. It may be contained (created) within an arbitrary object, depending on the operational needs.			
DERIVED FROM	**19004BaseClass**		
ATTRIBUTES			
name	*Value type:* NameType	*Possible access:*	*Default value:* *- not specified -*
This attribute contains string or number assigned to uniquely identify the ValueCharacteristic object.			
monitoredValue	*Value type:* MonitoringTarget	*Possible access:* Read-Write	*Default value:* *- not specified -*

(continued)

Table 16D.9 (Continued)

Class **ValueCharacteristic**			
This attribute defined what has to be monitored in the monitored object. This may be simply the value of an attribute, the number of occurrences of a given event type or the number of objects of a given class contained in the monitored object.			
characteristicsKind	*Value type:* StatisticsType	*Possible access:* Read-Write	*Default value:* *- not specified -*
This attribute defines what calculation shall be applied to the monitored values.			
accuracy	*Value type:* REAL	*Possible access:* Read-Write	*Default value:* 0
This optional attribute defines the accuracy applied for measuring. Default: not specified			
precision	*Value type:* REAL	*Possible access:* Read-Write	*Default value:* 0
This optional attribute defines the precision applied for for measuring. Default: not specified			
window	*Value type:* CalculationTiming	*Possible access:* Read-Write	*Default value:* *- not specified -*
This attribute defines the time interval (in ms) during that the calculation shall be applied. Further, the sampling period can be defined in this attribute.			
windowType	*Value type:* Window	*Possible access:* Read-Write	*Default value:* *- not specified -*
This attribute defines if the currentValue calculation is updated after each sampling time interval or after each window time interval.			
currentValue	*Value type:* NumericValue	*Possible access:* Read	*Default value:* *- not specified -*
This attribute provides the latest calculated value			
CONTAINED IN	**ApplicationMeasurements, DeviceMeasurements, LinkMeasurements, ChannelMeasurements, BaseStationMeasurements, CellMeasurements**		
CONTAINS			
SUPPORTED EVENTS			

From IEEE Standard for Architectural Building Blocks Enabling Network-Device Distributed Decision Making for Optimized Radio Resource Usage in Heterogeneous Wireless Access Networks, Jan 2019. IEEE.

Table 16D.10 −MeasurementReporter class definition.

Class **MeasurementReporter**			
This class providing a flexible way to specify various measurement reports containing a list of different attribute values observed at a certain time (usually these values represent measurement results). Further, this class includes the ability to send the reports to defined targets in accordance with a specified scheduler.			
DERIVED FROM	**19004BaseClass**		
ATTRIBUTES			
name	*Value type:* NameType	*Possible access:* Read	*Default value:* *- not specified -*
This attribute contains string or number assigned to uniquely identify the MeasurementReporter object.			
reportTarget	*Value type:* SABlock	*Possible access:* Read-Write	*Default value:* *- not specified -*
This attribute defines the target entity where the measurement reports send to.			
relatedScheduler	*Value type:* RelatedScheduler	*Possible access:* Read-Write	*Default value:* *- not specified -*

(continued)

Table 16D.10 (Continued)

Class **MeasurementReporter**			
This attribute defines the scheduler that controls the sending of measurement reports. If NULL – no measurement reporting times are defined.			
requestedValues	*Value type:* AttributeList	*Possible access:* Read, Add-Remove	*Default value:* *- not specified -*
This attribute defines all values that must be included into in the measured report. For example, currentValue attributes of ValueCharacteristic objects can be included into this list.			
timeEllapsedSinceReport	*Value type:* INTEGER	*Possible access:* Read-Write	*Default value:* *- not specified -*
This attribute provides the time ellapsed since the last report. Measured unit is ms. If no report has been done yet - will be reported as −1. Setting to 0 causes immediate measurement report.			
CONTAINED IN	**BaseStationMeasurements, CellMeasurements, ApplicationMeasurements, DeviceConfiguration, DeviceMeasurements**		
CONTAINS			
SUPPORTED EVENTS	MeasurementReport		

From IEEE Standard for Architectural Building Blocks Enabling Network-Device Distributed Decision Making for Optimized Radio Resource Usage in Heterogeneous Wireless Access Networks, Jan 2019. IEEE.

Table 16D.11 −Scheduler class definition.

Class **Scheduler**			
This class provides the possibility to define various schedulers to schedule different operations in a common way.			
DERIVED FROM	**19004BaseClass**		
ATTRIBUTES			
name	*Value type:* NameType	*Possible access:* Read	*Default value:* - not specified -
This attribute contains string or number assigned to uniquely identify the Scheduler object.			
startTime	*Value type:* SchedulerStartTime	*Possible access:* Read-Write	*Default value:* - not specified -
This attribute defines the time when the scheduled object starts the periodic timer for the scheduler.			
stopTime	*Value type:* SchedulerStopTime	*Possible access:* Read-Write	*Default value:* - not specified -
This attribute defines the time when the scheduled object stops the periodic timer for the scheduler. On behalf of this a scheduler may be used as a single timer. If NULL - no stop time defined.			
periodicInterval	*Value type:* INTEGER	*Possible access:* Read-Write	*Default value:* - not specified -
This attribute defines the time (in ms) for one interval			
operationalState	*Value type:* OperationalState	*Possible access:* Read-Write	*Default value:* enabled
This attribute allows to stop or re-start the scheduler.			
CONTAINED IN	**BaseStationMeasurements, CellMeasurements, ApplicationMeasurements, DeviceConfiguration, DeviceMeasurements, …**		
CONTAINS			
SUPPORTED EVENTS			

From IEEE Standard for Architectural Building Blocks Enabling Network-Device Distributed Decision Making for Optimized Radio Resource Usage in Heterogeneous Wireless Access Networks, Jan 2019. IEEE.

Table 16D.12 — ReportGenerator class definition.

Class **ReportGenerator**			
This class provides the possibility to define various reports on reportable events of an object. Reports will be sent on event occurrence.			
DERIVED FROM	**19004BaseClass**		
ATTRIBUTES			
name	*Value type:* NameType	*Possible access:* Read	*Default value:* - not specified -
This attribute contains string or number assigned to uniquely identify the ReportGenerator object.			
reportTarget	*Value type:* SABlock	*Possible access:* Read-Write	*Default value:* - not specified -
This attribute defines the target where the reports send to.			
eventFilter	*Value type:* Filter	*Possible access:* Read-Write	*Default value:* - not specified -
This attribute defines the filter to be applied. It prevents too many event or undesired reports.			
eventType	*Value type:* PrintableString	*Possible access:* Read-Write	*Default value:* - not specified -
This attribute defines the event type to be reported. Values of all report parameters defined for this event type must be included into the report.			
CONTAINED IN	**Terminal, BaseStation, Cell**		
CONTAINS			
SUPPORTED EVENTS			

16D.3 Additional ASN.1 Type Definitions for Utility Classes

```
1900-4-Utilities-Type-Definitions DEFINITIONS ::= BEGIN
        IMPORTS ObjectName, OptionalObjectName, Filter FROM 1900-4-Type-
Definitions;
        ReportedValue ::= ANY
        ThresholdValue ::= CHOICE {
            Undefined           NULL,
            wholeNumber         INTEGER,
            fractionalNumber    REAL
        }
        MonitoringTarget ::= CHOICE {
            attributeOrFieldValue SEQUENCE {
                -- target of monitoring is an attribute or field value
                attributeName    PrintableString,
                fieldName        PrintableString OPTIONAL
            },
            eventName   PrintableString,
            -- target of monitoring is the number of occurring events
            className   PrintableString
            -- target of monitoring is the number of objects
            -- of this class contained
            -- in the monitored object
        }
        StatisticsType ::= ENUMERATED {min, max, mean,
            standardDeviation, history}
        -- history: store the list of all measured values
        CalculationTiming ::= SEQUENCE {
            Total            INTEGER,     -- in ms
            sampleInterval   INTEGER OPTIONAL      -- in ms
        }
```

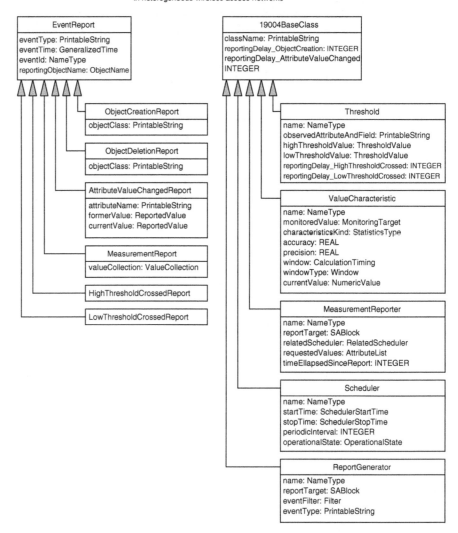

IEEE Std 1900.4-2009
IEEE Standard for architectural building blocks enabling
network-device distributed decision making for optimized radio resource usage
in heterogeneous wireless access networks

Figure 16D.1 —UML class diagram for utility classes. From IEEE Standard for Architectural Building Blocks Enabling Network-Device Distributed Decision Making for Optimized Radio Resource Usage in Heterogeneous Wireless Access Networks, Jan 2019. IEEE.

```
Window ::= ENUMERATED {fix, sliding}
NumericValue ::= CHOICE {
    wholeNumber         INTEGER,
    fractionalNumber    REAL
}
SABlock ::= ENUMERATED {nRM, tRM, oSM}
RelatedScheduler ::= OptionalObjectName
AttributeList ::= SEQUENCE OF SEQUENCE {
    level   ENUMERATED    { -- level + filter define a set of
        objects self, directContained, contained, allSubtree
        },
    filter    Filter,
```

```
          attributeName      PrintableString,
          fieldName      PrintableString OPTIONAL
                                        -- pick up non-constructed
                                        -- values from attributes
      }
      ValueCollection ::= SEQUENCE OF SEQUENCE {
          Object      ObjectName,
          attributeName      PrintableString,
          fieldName      PrintableString OPTIONAL,
                                        -- pick up non-constructed
                                        -- values from attributes
          value      ANY
      }
      SchedulerStartTime ::= CHOICE {
          absoluteTime     GeneralizedTime,
          relativeTime     INTEGER    -- relative to scheduler cre-
ation time
      }
      SchedulerStopTime ::= CHOICE {
          Undefined      NULL,
          absoluteTime     GeneralizedTime,
          relativeTime     INTEGER     -- relative to start time
      }
      OperationalState ::= ENUMERATED {enabled, disabled}
END
```

16D.4 Example for Distributed Radio Resource Usage Optimization Use Case

To implement distributed radio resource usage optimization use case, it is necessary to obtain context information and to generate radio resource selection policies. This example gives illustration of how to apply the IEEE 1900.4 information model for these purposes.

In particular, the following is described:

— Radio resource selection policy, generated by NRM, is received by TRM
— According to this radio resource selection policy TRM performs to following two actions:
  — Action 1: TRM is allowed to select periodically new RAN to which to connect.
  — Action 2: TRM periodically obtains link measurements from TMC and sends measurement report to NRM.

In this example the following is assumed:

— NRM and TRM offer the functions described in 16D.1
— NRM and TRM use the information model extensions given in 16D.1, 16D.2, and 16D.3

Figure 16D.2 describes this example. In Figure 16D.2, objects are described using legend <object name>:<class name>, where object names are selected arbitrarily, while class names are defined in the IEEE 1900.4 information model.

Steps (A)–(E) in Figure 16D.2 can be performed on behalf of the function CreateObject, while for the reporting of measurement results the function EventReport can be used. Both functions are defined in 16D.1. Steps (A)–(E) are described in details in the following tables.

Some objects used for the description of steps (A)–(E) in Table 16D.13 through Table 16D.17 are not shown in Figure 16D.2 for simplicity (each of these cases is commented).

To keep the example short, some attribute values used in Table 16D.17 are described in a simplified manner compared to their ASN.1 description given in Annex 16C.

IEEE Std 1900.4-2009
IEEE Standard for architectural building blocks enabling
network-device distributed decision making for optimized radio resource usage
in heterogeneous wireless access networks

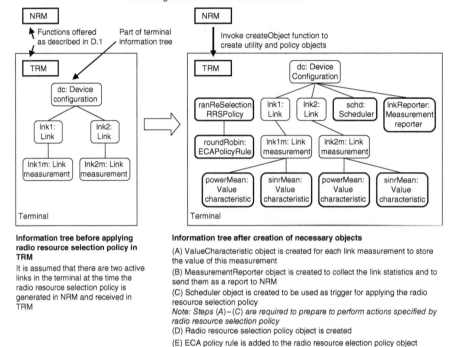

Information tree before applying radio resource selection policy in TRM

It is assumed that there are two active links in the terminal at the time the radio resource selection policy is generated in NRM and received in TRM

Information tree after creation of necessary objects

(A) ValueCharacteristic object is created for each link measurement to store the value of this measurement

(B) MeasurementReporter object is created to collect the link statistics and to send them as a report to NRM

(C) Scheduler object is created to be used as trigger for applying the radio resource selection policy
Note: Steps (A)–(C) are required to prepare to perform actions specified by radio resource selection policy

(D) Radio resource selection policy object is created

(E) ECA policy rule is added to the radio resource election policy object
Note: During policy creation TRM arranges (on behalf of the specified attribute values) the associations between the radio resource selection policy, the scheduler, and the measurement reporter objects

Figure 16D.2 —Description of example. From IEEE Standard for Architectural Building Blocks Enabling Network-Device Distributed Decision Making for Optimized Radio Resource Usage in Heterogeneous Wireless Access Networks, Jan 2019. IEEE.

Table 16D.13 —Description of step (A).

Function	Formal parameter	Actual parameter	Comment
CreateObject	baseObject	trm/dc	Device configuration object
	level	contained	Consider all objects contained in the base object
	filter	className == "LinkMeasurement"	This value representation is not an ASN1 value, it is used here for simplicity
	className	ValueCharacteristic	Include object of this class
	attributeName	name	
	value	sinrMean	
	attributeName	monitoredValue	
	value	*attributeOrFieldValue:* *attributeName:* listOfActiveLinkMeasurements *fieldName:* linkMeasurementValue [linkMeasurementId = receivedSINR]	From monitoring target the attributeOrFieldValue option is chosen. The field name uses the following syntax: From the list of active link measurements select that with linkMeasurementId is equal to receivedSINR and take the value from its linkMeasurementValue field.

(continued)

Table 16D.13 (Continued)

Function	Formal parameter	Actual parameter	Comment
	attributeName	characteristicsKind	
	value	mean	
	attributeName	window	
	value	*total:* 10000	
	attributeName	windowType	
	value	fix	
CreateObject	baseObject	trm/dc	The device configuration object
	level	contained	Consider all objects contained in the base object
	filter	className == "LinkMeasurement"	This value representation is not an ASN1 value, it is used here for simplicity
	className	ValueCharacteristic	Include object of this class
	attributeName	name	
	value	powerMean	
	attributeName	monitoredValue	
	value	*attributeOrFieldValue:* *attributeName:* listOfActiveLinkMeasurements *fieldName:* linkMeasurementValue [linkMeasurementId = receivedPower]	From monitoring target the attributeOrFieldValue option is chosen. The field name uses the following syntax: From the list of active link measurements select that with linkMeasurementId is equal to receivedPower and take the value from its linkMeasurementValue field.
	attributeName	characteristicsKind	
	value	mean	
	attributeName	window	
	value	*total:* 10000	
	attributeName	windowType	
	value	fix	

From IEEE Standard for Architectural Building Blocks Enabling Network-Device Distributed Decision Making for Optimized Radio Resource Usage in Heterogeneous Wireless Access Networks, Jan 2019. IEEE.

Table 16D.14 — Description of step (B).

Function	Formal parameter	Actual parameter	Comment
CreateObject	baseObject	trm/dc	
	level	contained	Consider only the device configuration object
	filter		Empty filter
	className	MeasurementReporter	Create object of this class
	attributeName	name	
	value	linkReporter	
	attributeName	reportTarget	

(continued)

Table 16D.14 (Continued)

Function	Formal parameter	Actual parameter	Comment
	value	nRM	
	attributeName	relatedSceduler	
	value	NULL	
	attributeName	valueCollection	
	value	*level:* contained *filter:* className == "ValueCharacteristic" *attributeName:* curentValue	

From IEEE Standard for Architectural Building Blocks Enabling Network-Device Distributed Decision Making for Optimized Radio Resource Usage in Heterogeneous Wireless Access Networks, Jan 2019. IEEE.

Table 16D.15 —Description of step (C).

Function	Formal parameter	Actual parameter	Comment
CreateObject	baseObject	trm/dc	
	level	contained	Consider only the device configuration object
	filter		Empty filter
	className	Scheduler	Create object of this class
	attributeName	name	
	value	schd	
	attributeName	startTime	
	value	*absoluteTime* 8:00:00 + random(10000)	Initializing scheduler
	attributeName	stopTime	
	value	undefined	
	attributeName	periodicInterval	
	value	10000	

From IEEE Standard for Architectural Building Blocks Enabling Network-Device Distributed Decision Making for Optimized Radio Resource Usage in Heterogeneous Wireless Access Networks, Jan 2019. IEEE.

Table 16D.16 —Description of step (D).

Function	Formal parameter	Actual parameter	Comment
CreateObject	baseObject	trm/dc	
	Level	contained	Consider only the device configuration object
	Filter		Empty filter
	className	RRSPolicy	Create object of this class
	attributeName	policyRuleSetId	
	value	ranReSelection	

From IEEE Standard for Architectural Building Blocks Enabling Network-Device Distributed Decision Making for Optimized Radio Resource Usage in Heterogeneous Wireless Access Networks, Jan 2019. IEEE.

Table 16D.17 — Description of step (E).

Function	Formal parameter	Actual parameter	Comment
CreateObject	baseObject	trm/ranReSelection	
	level	contained	Consider only the RRSPolicy object
	filter		Empty filter
	className	ECAPolicyRule	Create object of this class
	attributeName	policyRuleId	
	value	roundRobin	
	attributeName	validityTimeConstraint	
	value		
	attributeName	validitySpaceConstraint	
	value		
	attributeName	event	
	value	*eventType:* scheduledTimer *eventQualifier:* trm/dc/schd	
	attributeName	condition	
	value	true	
	attributeName	action	
	value	[*baseObject:* trm *level:* contained *filter:* className == "LinkProfile" *operationType:* guarateeValues *attributeName:* Cell(associatedCellId). CellConfiguration. activeRadioInterface *valueSet:* [lte, wifi] *timeslot:* 3], [*baseObject:* trm/dc/lnkReporter *level:* self *filter:* true *operationType:* guarateeValues *attributeName:* timeEllapsedSinceReport *valueSet:* [0] *timeslot:* 1]	Action 1: Selecting new RAN to connect to Note: LinkProfile object inside TRM is not shown in Figure D.2 for simplicity. Cell and CellConfiguraiton objects inside NRM are not shown in Figure D.2 for simplicity. Action 2: Sending measurement report

From IEEE Standard for Architectural Building Blocks Enabling Network-Device Distributed Decision Making for Optimized Radio Resource Usage in Heterogeneous Wireless Access Networks, Jan 2019. IEEE.

Annex 16E

(informative)

Deployment Examples

16E.1 Introduction

Referring to the system architecture in Clause 16.6, this standard specifies the following interfaces:

On the terminal side, interfaces between TRM and TMC for terminal context information collection and between TRM and TRC for reconfiguration management are specified.

Interface between NRM and TRM for policy-based management is specified.

On the network side, interfaces between OSM and NRM for policy-based management, between NRM and RMC for RAN context information collection, and between NRM and RRC for reconfiguration management are specified.

If there are several NRMs, interface between these NRMs for context information exchange and coordination of decision making is specified.

This annex presents examples of deployment of IEEE 1900.4 entities in heterogeneous wireless environment.

The following three scenarios, corresponding to three scenarios of dynamic spectrum assignment use case, are considered:

— Single operator scenario
— Multiple operator scenario 1 (NRM is inside operator)
— Multiple operator scenario 2 (NRM is outside operator)

For each of these scenarios, seven different variants of deployment of RMC and RRC entities are possible:

— In packet-based core network
— In RAN
— In BS
— In a combination of the above

NRM, RMC, and RRC may be implemented in one or several separate network nodes.

All these variants are shown for the single operator scenario. To avoid repetition, only a part of these variants is shown for two multiple operator scenarios. Deployment examples for the multiple operator scenario mainly show differences with the single operator scenario.

For each of the presented deployment examples, this annex clearly shows IEEE 1900.4 entities. It also clearly highlights interfaces specified in this standard and interfaces that are not specified.

16E.2 Deployment Examples for Single Operator Scenario

Figure 16E.1 and Figure 16E.2 show deployment examples 1 and 2 for single operator scenario.

In both figures, RMC and RRC are deployed in packet-based core network only.

In example 1 (Figure 16E.1) NRM, RMC, and RRC are deployed in packet-based core network. They are implemented in one network node.

Figure 16E.1 —Single operator scenario, deployment example 1. From IEEE Standard for Architectural Building Blocks Enabling Network-Device Distributed Decision Making for Optimized Radio Resource Usage in Heterogeneous Wireless Access Networks, Jan 2019. IEEE.

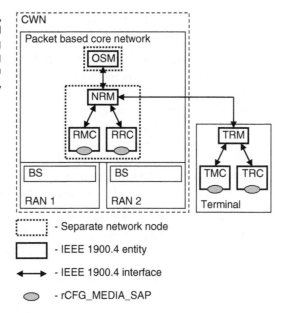

IEEE Std 1900.4-2009
IEEE Standard for architectural building blocks enabling
Network-device distributed decision making for optimized radio resource usage
in heterogeneous wireless access networks

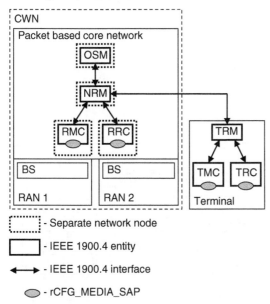

Figure 16E.2 —Single operator scenario, deployment example 2. From IEEE Standard for Architectural Building Blocks Enabling Network-Device Distributed Decision Making for Optimized Radio Resource Usage in Heterogeneous Wireless Access Networks, Jan 2019. IEEE.

In example 2 (Figure 16E.2) NRM, RMC, and RRC are also deployed in packet-based core network. But they are implemented in three different network nodes.

In these two deployment examples, the following interfaces specified in this standard are used:

— Interface between OSM and NRM
— Interface between NRM and RMC
— Interface between NRM and RRC
— Interface between NRM and TRM
— Interface between TRM and TMC

— Interface between TRM and TRC

RMC deployed in packet-based core network obtains RAN context information via Reconfiguration and Measurement SAP (rCFG_MEDIA_SAP), provided by packet-based core network. TMC obtains terminal context information via Reconfiguration and Measurement SAP (rCFG_MEDIA_SAP), provided by Terminal.

RRC deployed in packet-based core network controls reconfiguration of RAN via Reconfiguration and Measurement SAP (rCFG_MEDIA_SAP), provided by packet-based core network. TRC controls reconfiguration of Terminal via Reconfiguration and Measurement SAP (rCFG_MEDIA_SAP), provided by Terminal.

Figure 16E.3 shows deployment example 3 for single operator scenario.

In this example, RMC and RRC are deployed as follows:

— RMC and RRC are deployed in RAN only for RAN 1, while BSs of RAN 1 are legacy BSs
— RMC and RRC are deployed in BSs only for RAN 2
— RMC and RRC are deployed in both RAN and BSs for RAN 3

In RAN 1, RMC and RRC are implemented in one network node.
In RAN 3, RAN part of RMC and RRC are implemented in two different network nodes.
In this deployment example, the following interfaces specified in this standard are used:

— Interface between OSM and NRM
— Interface between NRM and RMC
— Interface between NRM and RRC
— Interface between NRM and TRM
— Interface between TRM and TMC

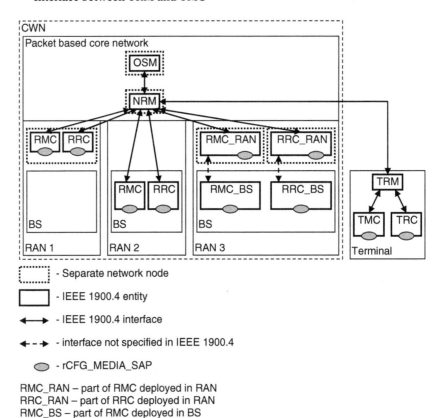

RMC_RAN – part of RMC deployed in RAN
RRC_RAN – part of RRC deployed in RAN
RMC_BS – part of RMC deployed in BS
RRC_BS – part of RRC deployed in BS

Figure 16E.3 —Single operator scenario, deployment example 3. From IEEE Standard for Architectural Building Blocks Enabling Network-Device Distributed Decision Making for Optimized Radio Resource Usage in Heterogeneous Wireless Access Networks, Jan 2019. IEEE.

— Interface between TRM and TRC

RMCs deployed in RAN and BSs obtain RAN context information via Reconfiguration and Measurement SAP (rCFG_MEDIA_SAP), provided by RAN and BSs. TMC obtains terminal context information via Reconfiguration and Measurement SAP (rCFG_MEDIA_SAP), provided by Terminal.

RRCs deployed in RAN and BSs control reconfiguration of RAN and BSs via Reconfiguration and Measurement SAP (rCFG_MEDIA_SAP), provided by RAN and BSs. TRC controls reconfiguration of Terminal via Reconfiguration and Measurement SAP (rCFG_MEDIA_SAP), provided by Terminal.

Figure 16E.4 shows deployment example 4 for single operator scenario.

In this example RMC and RRC are deployed as follows:

— Part of RRC and RMC are deployed in packet-based core network
— Part of RRC and RMC are deployed in RAN for RAN 1
— Part of RRC and RMC are deployed in BSs for RAN 2
— Part of RRC and RMC are deployed in RAN and BSs for RAN 3

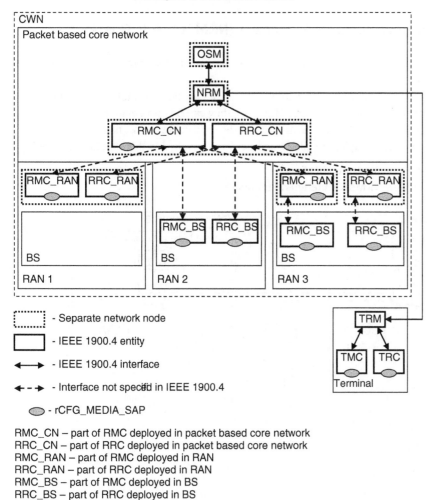

RMC_CN – part of RMC deployed in packet based core network
RRC_CN – part of RRC deployed in packet based core network
RMC_RAN – part of RMC deployed in RAN
RRC_RAN – part of RRC deployed in RAN
RMC_BS – part of RMC deployed in BS
RRC_BS – part of RRC deployed in BS

Figure 16E.4 —Single operator scenario, deployment example 4. From IEEE Standard for Architectural Building Blocks Enabling Network-Device Distributed Decision Making for Optimized Radio Resource Usage in Heterogeneous Wireless Access Networks, Jan 2019. IEEE.

Packet-based core network part of RMC and RRC is implemented in one network node.
In RAN 1, RAN part of RMC and RRC is implemented in one network node.
In RAN 3, RAN part of RMC and RRC is implemented in two different network nodes.
In this deployment example, the following interfaces defined in this standard are used:

— Interface between OSM and NRM
— Interface between NRM and RMC
— Interface between NRM and RRC
— Interface between NRM and TRM
— Interface between TRM and TMC
— Interface between TRM and TRC

Parts of RMC deployed in packet-based core network, RAN, and BSs obtain RAN context information via Reconfiguration and Measurement SAP (rCFG_MEDIA_SAP), provided by packet-based core network, RAN, and BSs. TMC obtains terminal context information via Reconfiguration and Measurement SAP (rCFG_MEDIA_SAP), provided by Terminal.

Parts of RRCs deployed in packet-based core network, RAN, and BSs control reconfiguration of RAN and BSs via Reconfiguration and Measurement SAP (rCFG_MEDIA_SAP), provided by packet-based core network, RAN, and BSs. TRC controls reconfiguration of Terminal via Reconfiguration and Measurement SAP (rCFG_MEDIA_SAP), provided by Terminal.

16E.3   Multiple Operator Scenario 1 (NRM is Inside Operator)

Figure 16E.5 shows deployment example for multiple operator scenario 1, where NRMs are inside operators.

To avoid repetition, in multiple operator scenario 1 deployment example RMC and RRC are deployed in packet-based core network. However, they can be also deployed in a distributed manner as described in deployment examples for the single operator scenario.

In this deployment example, the following interfaces specified in this standard are used:

— Interface between OSM and NRM
— Interface between NRMs of different operators

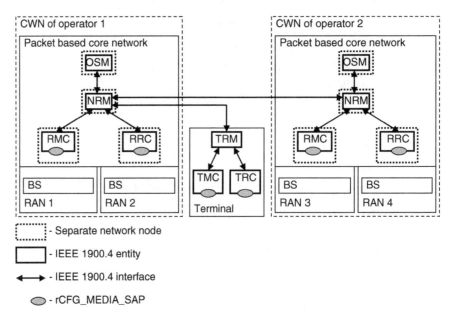

Figure 16E.5 —Deployment example for multiple operator scenario 1. From IEEE Standard for Architectural Building Blocks Enabling Network-Device Distributed Decision Making for Optimized Radio Resource Usage in Heterogeneous Wireless Access Networks, Jan 2019. IEEE.

— Interface between NRM and RMC
— Interface between NRM and RRC
— Interface between NRM and TRM
— Interface between TRM and TMC
— Interface between TRM and TRC

RMC deployed in packet-based core network obtains RAN context information via Reconfiguration and Measurement SAP (rCFG_MEDIA_SAP), provided by packet-based core network. TMC obtains terminal context information via Reconfiguration and Measurement SAP (rCFG_MEDIA_SAP), provided by the Terminal.

RRC deployed in packet-based core network controls reconfiguration of RAN via Reconfiguration and Measurement SAP (rCFG_MEDIA_SAP), provided by packet-based core network. TRC controls reconfiguration of Terminal via Reconfiguration and Measurement SAP (rCFG_MEDIA_SAP), provided by the Terminal.

16E.4 Multiple Operator Scenario 2 (NRM is Outside Operator)

Figure 16E.6 shows deployment example for multiple operator scenario 2, where NRM is outside operators. To avoid repetition, in multiple operator scenario 2 deployment example RMC and RRC are deployed in packet-based core network. However, they can be also deployed in a distributed manner as described in deployment examples for single operator scenario.

In this deployment example, the following interfaces specified in this standard are used:

— Interface between OSM and NRM
— Interface between NRM and RMC
— Interface between NRM and RRC
— Interface between NRM and TRM
— Interface between TRM and TMC
— Interface between TRM and TRC

RMC deployed in packet-based core network obtains RAN context information via Reconfiguration and Measurement SAP (rCFG_MEDIA_SAP), provided by packet-based core

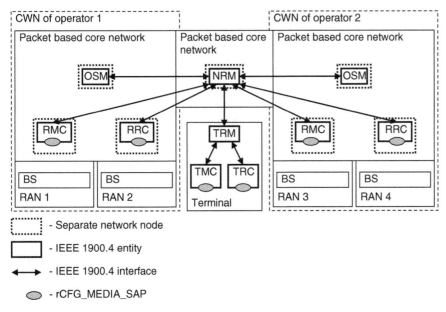

Figure 16E.6 —Deployment example for multiple operator scenario 2. From IEEE Standard for Architectural Building Blocks Enabling Network-Device Distributed Decision Making for Optimized Radio Resource Usage in Heterogeneous Wireless Access Networks, Jan 2019. IEEE.

network. TMC obtains terminal context information via Reconfiguration and Measurement SAP (rCFG_MEDIA_SAP), provided by the Terminal.

RRC deployed in packet-based core network controls reconfiguration of RAN via Reconfiguration and Measurement SAP (rCFG_MEDIA_SAP), provided by packet-based core network. TRC controls reconfiguration of Terminal via Reconfiguration and Measurement SAP (rCFG_MEDIA_SAP), provided by the Terminal.

Annex 16F

(informative)

Bibliography

B1 3GPP TS 23.251 V6.6.0, Network Sharing; Architecture and functional description (Release 6), Mar. 2006.

B2 BBN Technologies, "The XG Vision," Version 2.0, 2004.[12]

B3 Bourse, D., *et al* "FP7 E3 Project: Introducing Cognitive Wireless Systems in the B3G World," *ICT Mobile Summit 2008*, June 2008.

B4 Buljore, S., *et al* "IEEE P1900.4 Standard: Reconfiguration of Multi-Radio Systems," *IEEE SIBIRCON 2008*, July. 2008.

B5 Buljore, S., *et al*, "IEEE P1900.4 System Overview on Architecture and Enablers for Optimised Radio and Spectrum resource usage," *IEEE DySPAN 2008*, Oct. 2008.

B6 Buljore, S., and Martigne, P., "Proposed System Concept," P1900.4 Working Group, doc: P1900.4-07-04-2007, Apr. 2007.

B7 Buljore, S., *et al* "Introduction to IEEE P1900.4 Activities," *IEICE Transactions on Communications*, vol. E91-B, no. 1, pp. 2–9, Jan. 2008.

B8 Buljore, S., and Martigne, P., "SCC41 Plenary Meeting Working Group 4 Overview and Report," IEEE SCC41 Plenary, Apr. 2007.

B9 Cordeiro, C., Challapali, K., and Birru, D., "IEEE 802.22: An Introduction to the First Wireless Standard based on Cognitive Radios," *Journal of Communications*, vol. 1, no. 1, pp. 38–47, Apr. 2006.

B10 European Parliament Resolution, "Towards a European policy on radio spectrum," ITRE/6/37236, Feb. 2007.

B11 FCC ET Docket No. 04-186, "Unlicensed Operation in the TV Broadcast Bands," 2004.

B12 Filin, S., *et al* "Dynamic Spectrum Assignment and Access Scenarios, System Architecture, and Procedures for IEEE P1900.4 Management System," *Third International Conference on Cognitive Radio Oriented Wireless Networks and Communications (CrownCom 2008)*, May 2008.

B13 Filin, S., Harada, H., Hasegawa, M., and Kato, S., "QoS-Guaranteed Load-Balancing Dynamic Spectrum Access Algorithm," *IEEE PIMRC 2008*, Sept. 2008.

B14 Guenin, J., "IEEE Standards Coordinating Committee 41 on Dynamic Spectrum Access," *ITU-R WP5A SDR/CR Seminar*, Feb. 2008.

B15 Hanaoka, S., Yamamoto, J., and Yano, M., "Platform for Load Balancing and Throughput Enhancement with Cognitive Radio," *IEICE Transactions on Communications*, vol. E91-B, no. 8, pp. 2501–2508, Aug. 2008.

B16 Hanaoka, S., Yano, M., and Hirata,T., "Testbed System of Inter-Radio System Switching for Cognitive Radio," *IEICE Transactions on Communications*, vol. E91-B, no. 1 pp. 14–21, Jan. 2008.

B17 Harada, H., "Software defined radio prototype toward Cognitive Radio Communication Systems," *IEEE DySPAN 2005*, vol. 1, pp. 539–547, Nov. 2005.

B18 Harada, H., *et al* "A Software Defined Cognitive Radio System," *IEEE Globecom 2007*, pp. 294–299, Nov. 2007.

12 This publication is accessible from http://www.ir.bbn.com/projects/xmac/rfc/rfc-vision.pdf.

B19 Hase, Y., Okada, K., and Wu, G., "A Novel Mobile Basic Access System Using Mobile Access Signaling Card On Telecommunication Systems (MASCOT)," *IPSJ SIG Notes. MBL*, vol. 97, no. 72, pp. 37–42, July 1997.

B20 Holland, O., Attar, A., Olaziregi, N., Sattari, N., and Aghvami, A. H., "A Universal Resource Awareness Channel for Cognitive Radio," *IEEE PIMRC 2006*, Sept. 2006.

B21 Holland, O. *et al* "Development of a Radio Enabler for Reconfiguration Management within the IEEE P1900.4 Working Group," *IEEE DySPAN 2007*, pp. 232–239, Apr. 2007.

B22 IEEE Std 802.21™-2008, IEEE Standard for Local and Metropolitan Area Networks: Media Independent Handover Services.

B23 IEEE P802.11y™/D11.0, June 2008, Draft Standard for Information Technology—Telecommunications and Information Exchange between systems—Local and Metropolitan Area Networks—Specific Requirements—Part 11:Wireless LAN Medium Access Control (MAC) and Physical Layer (PHY) Specifications—Amendment 3: 3650–3700 MHz Operation in USA.[13]

B24 IEEE P802.22™/D6.0, April 2006, Draft Standard for Wireless Regional Area Networks Part 22: Cognitive Wireless RAN Medium Access Control (MAC) and Physical Layer (PHY) specifications: Policies and procedures for operation in the TV Bands.

B25 Inoue, M., Mahmud, K., Murakami, H., Hasegawa, M., and Morikawa, "Seamless Handover Using Out-Of-Band Signaling in Wireless Overlay Networks," *WPMC 2003*, vol. 1, pp. 186–190, Oct. 2003.

B26 Inoue, M., Mahmud, K., Murakami, H., Hasegawa, M., and Morikawa, "Novel Out-Of-Band Signaling for Seamless Interworking between Heterogeneous Networks," *IEEE Wireless Communication Magazine*, vol. 11, no. 2, pp. 56–63, Apr. 2004.

B27 Inoue, M., Mahmud, K., Murakami, H., Hasegawa, M., and Morikawa, H., "Design and Implementation of Out-Of-Band Signaling for Seamless Handover in Wireless Overlay Networks," *IEEE ICC 2004*, pp. 3932–3936, June 2004.

B28 Ishizu, K., *et al* "Design and Implementation of Cognitive Wireless Network based on IEEE P1900.4," *Third IEEE Workshop on Networking Technologies for Software Defined Radio (SDR) Networks 2008*, June 2008.

B29 Muck, M., *et al* "IEEE P1900.B: Coexistence Support for Reconfigurable, Heterogeneous Air Interfaces," *IEEE DySPAN 2007*, pp. 381–389, Apr. 2007.

B30 Muck, M., *et al* "End-to-End Reconfigurability in Heterogeneous Wireless Systems—Software and Cognitive Radio Solutions," *IST Mobile and Wireless Summit 2007*, July 2007.

B31 Murakami, H., Mahmud, K., Hasegawa, M., and Inoue, M., "On the Methods of Provisioning Basic Access Signaling for MIRAI," *IEICE Society Conference*, B-5-125, p. 422, Sept. 2002.

B32 Ofcom, "Spectrum Usage Rights: Technology and Usage Neutral Access to the Spectrum" Consultation, 2006.[14]

B33 Prasad, R. V., Pawelczak, P., Hoffmeyer, J. A., and Berger, H. S., "Cognitive functionality in next generation wireless networks: standardization efforts," *IEEE Communications Magazine*, vol. 46, no. 4, pp. 72–78, Apr. 2008.

B34 Seelig, F., "A Description of the August 2006 XG Demonstrations at Fort A.P. Hill," *IEEE DySPAN 2007*, pp. 1–12, Apr. 2007.

B35 Sherman, M., Mody, A., Martinez, R., Rodriguez, C., and Reddy, R., "IEEE Standards Supporting Cognitive Radio and Networks, Dynamic Spectrum Access, and Coexistence," *IEEE Communications Magazine*, vol. 46, no. 7, pp. 72–79, July 2008.

---

13 Numbers preceded by P are IEEE authorized standards projects that were not approved by the IEEE-SA Standards Board at the time this publication went to press. For information about obtaining drafts, contact the IEEE.

14 This publication can be accessed at http://www.ofcom.org.uk/consult/condocs/sur/.

B36 Sherman, M., Mody, A., Martinez, R., Reddy, R., and Kiernan, T., "A Survey Of IEEE Standards Supporting Cognitive Radio And Dynamic Spectrum Access," *IEEE MILCOM 2008*, Nov. 2008.

B37 Strassner, J., "Policy-based network management: solutions for the next generation," Morgan Kaufmann (series in networking), 2005.

B38 Wu, G., Havinga, P. J. M., and Mizuno, M., "Architecture of Multimedia Integrated network by Radio Access Innovation (MIRAI)," *Technical report of IEICE RCS*, vol. 100, no. 664, pp. 111–119, Mar. 2001.

B39 Wu, G., Mizuno, M., and Havinga, P. J. M., "MIRAI Architecture for Heterogeneous Network," *IEEE Communications Magazine*, vol. 40, no. 2, pp. 126–134, Feb. 2002.

17

IEEE Standard for Policy Language Requirements and System Architectures for Dynamic Spectrum Access Systems

IMPORTANT NOTICE: This standard is not intended to ensure safety, security, health, or environmental protection. Implementers of the standard are responsible for determining appropriate safety, security, environmental, and health practices or regulatory requirements.

This IEEE document is made available for use subject to important notices and legal disclaimers. These notices and disclaimers appear in all publications containing this document and may be found under the heading "Important Notice" or "Important Notices and Disclaimers Concerning IEEE Documents." They can also be obtained on request from IEEE or viewed at http://standards.ieee.org/IPR/disclaimers.html.

17.1 Overview

17.1.1 Scope

This standard defines a vendor-independent set of policy-based control architectures and corresponding policy language requirements for managing the functionality and behavior of Dynamic Spectrum Access networks.

17.1.2 Purpose

The purpose of this standard is to define policy language requirements and associated architecture requirements for interoperable, vendor-independent control of Dynamic Spectrum Access functionality and behavior in radio systems and wireless networks. This standard will also define the relationship of policy language and architecture to the needs of at least the following constituencies: the regulator, the operator, the user, and the network equipment manufacturer.

17.1.3 Document Overview

This standard specifies policy language (PL) requirements and policy architecture(s) for policy-based Dynamic Spectrum Access (DSA) radio systems.

In this standard, a distinction is made between the policy reasoning that is accomplished within the Policy-Based Radio (PBR) node and policy generation and validation that is accomplished through a policy generation system prior to provision of the policy to the PBR node. Policy reasoning may be distributed, i.e., it may take place either within a PBR node or in other elements of a policy-based radio communications network.

17.2 Normative References

The following referenced documents are indispensable for the application of this document (i.e., they must be understood and used, so each referenced document is cited in text and its

Dynamic Spectrum Access Decisions: Local, Distributed, Centralized, and Hybrid Designs, First Edition. George F. Elmasry.
© 2021 John Wiley & Sons Ltd. Published 2021 by John Wiley & Sons Ltd.
Companion website: www.wiley.com/go/elmasry/dsad

relationship to this document is explained). For dated references, only the edition cited applies. For undated references, the latest edition of the referenced document (including any amendments or corrigenda) applies.

IEEE Std 1900.1™, IEEE Standard Definitions and Concepts for Dynamic Spectrum Access Terminology Relating to Emerging Wireless Networks, System Functionality, and Spectrum Management.[1,2]

17.3 Definitions, Acronyms, and Abbreviations

For the purposes of this document, the following terms and definitions apply. The *IEEE Standards Dictionary: Glossary of Terms & Definitions*[3] should be referenced for terms not defined in this clause.

17.3.1 Definitions

awareness: The ability of a radio to gain (or acquire) the knowledge of the state of itself and of its environment for the purpose of assessing the implications of such state on its current and future operating behavior.

binding: Associating a value with a variable. A variable may be bound or unbound. The value of variable may be unbound (unknown) when the policy is written, for example, values that must be provided by a policy-based DSA radio system (PBDRS) when the policy is processed.

coexistence policy: Policy specifying coexistence constraints and parameters.

coexistence: The ability of two or more spectrum-dependent devices or networks to operate without harmful interference.[4]

> *Note 1.* Coexistence policy may be specified by the regulator as a subset of the regulatory policy, or may be specified by the spectrum manager/planner or the system administrator.[5]

> *Note 2.* As an example, a coexistence policy might specify a listen-before-talk coexistence mechanism and might specify the sensor detection threshold (e.g., -90 dBm in a 10 kHz bandwidth).

cognitive radio: A) A type of radio in which communication systems are aware of their environment and internal state and can make decisions about their radio operating behavior based on that information and predefined objectives. B) Cognitive radio [as defined in item a)] that uses software-defined radio, adaptive radio, and other technologies to adjust automatically its behavior or operations to achieve desired objectives.[6,7,8]

> *Note.* The environmental information may or may not include location information related to communication systems.

---

1 The IEEE standards or products referred to in this clause are trademarks of the Institute of Electrical and Electronics Engineers, Inc.

2 IEEE publications are available from the Institute of Electrical and Electronics Engineers, Inc., 445 Hoes Lane, Piscataway, NJ 08854, USA (http://standards.ieee.org/).

3 *IEEE Standards Dictionary: Glossary of Terms and Definitions* is available at http://shop.ieee.org.

4 From IEEE Std 1900.1™.

5 Notes in text, tables, and figures are given for information only and do not contain requirements needed to implement the standard.

6 See Footnote 4.

7 IEEE recognizes that the terminology commonly used is "cognitive radio." However, generally the cognitive functionality may be outside the boundary normally associated with a radio (e.g., environment sensing is a cognitive function that is not normally part of a radio).

8 IEEE notes that the terms "dumb radio," "aware radio," and "smart radio" are used in the technical literature, but IEEE does not define these terms at this time. They are additional descriptive terms that are sometimes applied to radios.

domain: An area of knowledge or activity characterized by a set of concepts and terminology understood by practitioners in that area.

dynamic spectrum access (DSA) policy language: A formal system for representing information that includes both grammar (syntax) and meaning (semantics) that has been created for the purpose of communicating DSA policy primarily between machines (computers). There are two types of languages, declarative and imperative. Declarative language may be implemented with imperative extensions to address language pragmatics.

dynamic spectrum access (DSA): The real-time adjustment of spectrum utilization in response to changing circumstances and objectives.[9]

> **Note.** Changing circumstances and objectives include (and are not limited to) energy-conservation, changes of the radio's state (operational mode, battery life, location, etc.), interference-avoidance (either suffered or inflicted), changes in environmental/external constraints (spectrum, propagation, operational policies, etc.), spectrum-usage efficiency targets, quality of service (QoS), graceful degradation guidelines, and maximization of radio lifetime.

fact: A formula that is either asserted or proved to be true under a given interpretation.

formal policy: A set of formulas in the logic associated with the policy language that specify how a resource (e.g., radio spectrum) may be used.

formal semantics: The interpretation or mapping from the vocabulary of the language to a mathematical domain.

machine learning: The capability to use experience and reasoning to adapt the decision-making process to improve subsequent performance relative to predefined objectives.[10]

> **Note.** This notion of learning corresponds most closely to the subdiscipline of machine learning known as reinforcement learning. Learning in the intelligent radio context is meant to exclude learning "by being told"—e.g., acquiring information, messages received from other systems, configuration files, and initialization parameters. Rather learning implies the adaptation of decision making based on direct experience resulting from previous actions.

machine-understandable policies: Policies expressed in a form that allows for a policy-based radio to read and "interpret" them automatically (without requiring human intervention). That is, an automated procedure exists by which the implications of the constraints expressed by the policies are reflected in the actions of the radio.[11]

meta-policy: One or more assertions in the policy language that state relationships between policies.

model-theoretic computational semantics: Defines logical consequents by relating statements in the language to entities in a given structure, the so-called model.

> **Note.** The importance of the model-theoretic semantics lies in its simplicity that allows us to understand the meaning of a policy without the need to understand the reasoning process.

ontology: Definitions that associate the names of entities and concepts in a problem domain (e.g., objects, classes, relations, functions, objects) with text describing what the names mean, and axioms (expressed in a formal language) that constrain the interpretation of these entities and concepts.

> **Note.** Modified from IEEE Std 1900.1™.

policy: A) A set of rules governing the behavior of a system.[12]

> **Note 1.** Policies may originate from regulators, manufacturers, developers, network and system operators, and system users. A policy may define, for example, allowed frequency bands, waveforms, power levels, and secondary user protocols.

---

9 See Footnote 4.
10 See Footnote 4.
11 See Footnote 4.
12 See Footnote 4.

B) A machine-understandable instantiation of policy such as defined in Definition A).

Note 2. Policies are normally applied post manufacturing of the radio as a configuration to a specific service application.

Note 3. Definition B) recognizes that in some contexts the term "policy" is assumed to refer to machineunderstandable policies.

Note 4. *See also:* **formal policy**, **meta-policy**, **machine-understandable policies**, and **coexistence policy**.

policy authority: An entity that has jurisdiction over spectrum usage and is authorized to create policy for that jurisdiction.

Note. An authority may be, for example, a regulatory agency or a primary user who is authorized to lease their spectrum to other users.

policy-based radio (PBR): A type of radio in which the behavior of communications systems is governed by a policy-based control mechanism.[13] *See also:* **policy-based control mechanism**.

Note 1. Policies may restrict behaviors (e.g., policies constraining time, power, or frequency use) associated with a specific set of radio functions, but they do not necessarily change the functional capability of a radio. Because policies often do not change basic radio functionality, a policy-based radio need not also be a reconfigurable radio.

Note2 . Because the definition for the term policy-based control mechanism considers radio policy to be a type of radio control software, the policy-based radio is considered a subset of software-controlled radio.

policy-based control mechanism: A mechanism that governs radio behavior by sets of rules, expressed in a machine-understandable format, that is independent of the radio implementation regardless of whether the radio implementation is in hardware or software.[14]

Note 1. The definition of rules and associated modification of radio functionality can occur:

— during manufacture or reconfiguration
— during configuration of a device by the user or service provider
— during over-the-air provisioning
— by over-the-air or other real-time control

Note 2. As implied by the scope of this standard, the control of radio dynamic spectrum access behavior is expected to be a typical application of a policy-based control mechanism. However, the concepts of policy-based control could be applied to network management policies as well. Policy sources include spectrum regulators, manufacturers, and network operators.

policy conformance reasoner (PCR): The IEEE 1900.5 system component that evaluates the policy compliance of transmission requests.

Note. The PCR is software capable of making logical inferences from a set of asserted facts and rules (i.e., policies). It is able to formally prove or disprove a hypothesis (e.g., that a transmission request is policy compliant), and is capable of inferring additional knowledge (e.g., identifying transmission opportunities for unbound transmission requests).

policy enforcer (PE): The Policy Enforcer is the realization of the Policy Enforcement Point in an IEEE 1900.5 system. It examines transmission control commands received from the System Strategy Reasoning Capability (SSRC) to ensure the Policy Conformance Reasoner (PCR) has found them to be policy compliant and only outputs policy-compliant commands to the transmitter for execution.

---

13 See Footnote 4.
14 See Footnote 4.

Note. The Policy Enforcer may maintain a record of previous SSRC transmission decision requests, PCR transmission decision replies, and validity periods to support its policy enforcement function.

policy rule: A policy rule is a formula in the policy language that has the form of an implication (e.g., A logically implies B).

policy traceability: The ability to provide evidence of the source of a policy that cannot be repudiated.[15]

Note. Policy includes machine-understandable policy.

procedural attachments: An executable procedure (code) written in an imperative language and invoked from within a reasoner.

proof-theoretic semantics (PTS): In proof-theoretic semantics (PTS), proofs are entities in terms of which meaning and logical consequences can be explained. PTS are the axioms and inference rules.

Note. Proof-theoretic semantics is an alternative to model-theoretic semantics. In PTS, meanings are assigned to expressions of the languages as proofs rather than as truth. In this sense, PTS is inferential rather than denotational in spirit.

reasoner: The decision making entity which uses a logical system to infer formal conclusions from logical assertions. It is able to formally prove or disprove a hypothesis and is capable of inferring additional knowledge.

regulatory policy: A policy that is specified by a regulatory authority [such as the Federal Communications Commission (FCC) or the National Telecommunications and Information Administration (NTIA) in the U.S.A.]. Typically, these describe what constitutes valid use of spectrum without having specific knowledge of the environment.

semantics: Defines the meaning of expressions (formulas) in the language.

Note. There are various types of semantics including model-theoretic computational semantics and proof-theoretic semantics.

system policy: A policy, specified by a system administrator, which typically specifies inputs beyond those available in regulatory policy. It can provide specific strategies or instructions to the radio. System policy depends on the knowledge of the operational environment.

system strategy reasoning capability (SSRC): The functional capability in an IEEE 1900.5 system that generates transmission requests and proposes the parameters of each transmission.

Note 1. The system strategy reasoning capability gathers data to optimize radio operation, formulates communications strategies, and coordinates these strategies with the policy conformance reasoner to evaluate compliance with the active policy set.

Note 2. The system strategy reasoning capability may maintain a record of previous SSRC transmission decision requests, PCR transmission decision replies, and validity periods.

transmission request: A query expressed in the policy language about the validity of a specific transmission with respect to the policies currently in use. A transmission request may include parameters of a transmission that are necessary for the PCR to establish whether the proposed transmission satisfies the conditions defined by the current policies.

Note. A transmission request may regard a specific transmission, or a strategy for transmissions (unbounded requests).

transmission opportunity reply: A set of transmission parameters (e.g., tuned frequency, power, bandwidth, coexistence protocol, etc.), evidence (e.g., sensor data and other awareness

---

15 See Footnote 4.

knowledge), and the validity period associated with a Transmission Opportunity Request or Transmission Decision Refresh Request that the PCR has found to be policy compliant.

transmission opportunity request: A set of transmission parameters (e.g., tuned frequency, power, bandwidth, coexistence protocol, etc.) and evidence (e.g., sensor data and other awareness knowledge) that is sent as a request from the SSRC to the PCR.

variable: A variable is a symbol in the policy language that may be associated with (bound to) a value. The value may change over time.

17.3.2 Acronyms and Abbreviations

AGC	Automatic Gain Control
API	Application Programming Interface
ARQ	Automatic Repeat reQuest
BER	Bit Error Rate
BW	Bandwidth
CPE	Common Platform Enumeration
CRS	Cognitive Radio System
CSMA	Carrier Sense Multiple Access
DSA	Dynamic Spectrum Access
DTED	Digital Terrain Elevation Data
DMTF	Distributed Management Task Force
EIRP	Equivalent Isotropically Radiated Power
FCC	Federal Communications Commission
ID	Identifier
IETF	Internet Engineering Task Force
INR	Interference to Noise Ratio
ISO	International Organization for Standardization
ITU	International Telecommunications Union
MAC	Media Access Control
NAF	Negation as Failure
NIST	National Institute of Standards and Technology
NTIA	National Telecommunications and Information Administration
OWL	Web Ontology Language
P2P	Peer to Peer
PBR	Policy-based Radio
PBDRS	Policy-based Dynamic Spectrum Access Radio System
PCIM	Policy Core Information Model
PCR	Policy Conformance Reasoner
PDP	Policy Decision Point
PE	Policy Enforcer
PEP	Policy Enforcement Point

PL	Policy Language
PMP	Policy Management Point
PTS	Proof-theoretic Semantics
QoS	Quality of Service
REM	RF Environment Map
RF	Radio Frequency
RSSI	Received Signal Strength Indicator
RX	Receiver
SINAD	Signal-plus-Interference-plus-Noise-plus-Distortion to Interference-plus-Noise-plus-Distortion Ratio
SINR	Signal to Interference plus Noise Ratio
SIR	Signal to Interference Ratio
SNR	Signal to Noise Ratio
SSRC	System Strategy Reasoning Capability
SQL	Structured Query Language
SWRL	Semantic Web Rule Language
TDMA	Time Division Multiple Access
TVWS	Television White Space
TX	Transmitter
UML	Unified Modeling Language

17.4 Architecture Requirements for Policy-based Control of DSA Radio Systems

The term "architecture" in this document refers to the policy-based DSA radio system (PBDRS) functional architecture. This architecture is defined in terms of a combination of functions and components. Functions refer to required capabilities that may be distributed across multiple components within the architecture. Components refer to specific elements of the architecture which have defined interfaces and functionality.

17.4.1 General Architecture Requirements

This subclause includes generic requirements placed against the architecture as a whole.

17.4.1.1 Accreditation

To simplify the PBDRS device authorization process and accreditation of its policy conformance mechanism, the policy shall be separable from the detailed system behavior. Regulators should be able to accredit radios based on the ability to interpret policies correctly to obtain desired behavior rather than verifying the conformance of a fixed set of behaviors programmed into the radio at the time of manufacture. The accredited portion of the radio shall enforce the regulatory policy, while the system policy, which is enforced outside the accreditation boundary, will provide opportunities for innovation, including proprietary optimization techniques or other added value, by a radio manufacturer or service provider. The IEEE 1900.5 system components that check policy conformance are separate from those that are radio-specific and optimize performance.

17.4.1.2 Policy Conformance Reasoner (PCR)

The architecture shall define a component that allows for policy conformance reasoning. The PCR evaluates the radio transmission commands to determine if they are in compliance with the active policy set. If the policy compliance evaluation fails, the radio shall be precluded from executing the associated transmission command.

17.4.1.3 System Strategy Reasoning Capability (SSRC)

The architecture shall define a functional capability that allows for system strategy reasoning. Refer to Figure 1and associated notes for more details about functionality.

17.4.2 Policy Management Requirements

17.4.2.1 Policy Management

The architecture shall define a policy management component that allows for the management and distribution of policies.

17.4.2.2 Revocability

It shall be possible for a policy authority to revoke or amend any type of policy or invoke new policy.

17.4.2.3 Effectivity

It shall be possible to associate the identity of the policy authority and information about its jurisdiction, such as a time period and geographic area with a policy to identify the time period and/or geographic policy domain in which the policy is in force.

17.4.2.4 Non-repudiation of Policy Traceability

The policy shall provide evidence of the source of a policy that cannot be repudiated.

17.4.2.5 Policies and Policy Messaging

The architecture shall include a language to express policies and an interface to specify messages between the PE, SSRC, and PCR.

17.4.2.6 Policy Matching and Comparison

The architecture shall be structured such that policies from different sources can be compared and shall permit the resolution of differences in the case of incompatibility or inconsistency of policies that originate from different sources.

17.4.2.7 Logging

The logging mechanism shall provide information to assist in verifying that all policies have been followed correctly.

17.4.2.8 Security Requirements

Security requirements for policy languages and architectures include the security of the policy authority, the security of policy downloads from the policy authority, the security of the local information on the Cognitive Radio, and the security of the policy enforcer. The network based sharing of data supporting dynamic spectrum operations shall provide authentication and non-repudiation of users, limited permissions for the modification of spectrum use data on the dynamic spectrum access (DSA) database, and audit logging of actions of all DSA participants so that information integrity and other security problems may be quickly and accurately diagnosed and resolved. The tools employed to create the database(s), related network(s), network service(s), and/or web site(s), and the related DSA devices should employ anti-malware measures.

17.4.2.9 Notification Feedback Channels

The policy architecture shall support the requirement to provide acknowledgement messages to the Policy Management Point (PMP).

17.5 Architecture Components and Interfaces for Policy-based Control of DSA Radio Systems

The policy architecture specified in this standard provides a high-level framework of the functional capabilities required to control the behavior of a policy-based DSA radio system. This standard makes no assumptions as to the physical implementation of those functional components.

Note. The objectives of the IEEE 1900.5 DSA policy architecture are to:

a) Minimize vendor dependence; minimize design detail.
b) Clearly identify components, interfaces, and functionality to support accreditation.
c) Clearly separate functionalities of the major components and not have functionality distributed across major components.
d) Clearly describe and specify the functions, interfaces, and information that each normative component needs to receive and which it provides to other components.
e) Specify the architecture only to the level of detail needed for implementation of language requirements.

The components of the IEEE 1900.5 DSA policy architecture as shown in Figure 17.1 are:

— Policy Management Point (PMP)
— System Strategy Reasoning Capability (SSRC)
— Policy Conformance Reasoner (PCR)
— Policy Enforcer (PE)
— Policy Repository (optional)

The above components and their relationships are depicted in Figure 17.1 and specified in this subclause.

Note 1. The designs of the components and interfaces in Figure 1 shown in thin black lines are left for the system designer to decide. The focus of this architecture is on the Policy Conformance Reasoner, the Policy Enforcer, the System Strategy Reasoning Capability, and the communications interfaces shown in heavy black lines.

Note 2. The depiction of the SSRC and the PCR as individual boxes does not imply that either is a single entity within the radio node. The functionality of the SSRC and the PCR can be distributed within the DSA communications network. It is more likely that the PCR may be a single component for the purposes of certification. The functionality can also be distributed within multiple components of a radio. The distribution of this functionality is left to the PBDRS designer.

Note 3. The PBDRS may include a Radio Frequency Environment Map (REM). If implemented, the REM may contain information about the radio device's RF environment that the SSRC uses to identify spectrum that may form a suitable basis for a transmission opportunity. Information available from the REM may include the results of spectrum sensing (local or remote) and databases that describe current usage and spectrum opportunities. The SSRC may incorporate some of this information into transmission requests sent to the PCR to provide evidence to the PCR that the request is policy compliant.

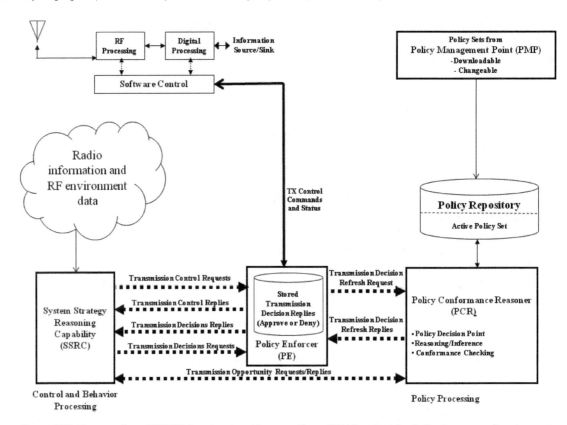

Figure 17.1 Presentation of PBDRS functional architecture. From IEEE Standard for Policy Language Requirements and System Architectures for Dynamic Spectrum Access Systems, Dec 2011. IEEE.

Note 4. The PBDRS may incorporate a Radio Information Base. If implemented, the Radio Information Base shall include information about the capabilities and status of the radio, the radio identifier (ID), and radio location and time of update of the radio information.

Note 5. The role of the policy enforcer only applies to transmission control commands. It is important to note that in any radio system there can be complex waveform behaviors and protocols that govern both the transmit and receive functions. While Figure 1 illustrates only the transmit command path, it is assumed that the full waveform functionality, to include receiver control and other aspects of waveform behaviors, is present in the radio system illustrated.

17.5.1 Policy Management Point

The IEEE 1900.5 system shall include a policy management point.

17.5.1.1 Functions Performed

The PMP shall take high-level policy information, such as spectrum access policy, and convert it to policies in a standard policy language that can be applied to various devices in the network. The PMP may perform additional functions not required by this standard such as managing policy sets, validating and error checking of policy sets, and policy distribution and download.

17.5.1.2 Inputs

Policies expressed in human language from administrators or spectrum owners.

17.5.1.3 Outputs

The output is a policy or set of policies represented in a standard policy language.

17.5.1.4 Interfaces

The PMP shall interface with the PCR. This interface may be either a direct interface, or it may be an indirect interface via the policy repository.

17.5.2 Policy Conformance Reasoner

The policy conformance decision making in the IEEE 1900.5 system is done by the PCR, which uses a logical system to infer formal conclusions from logical assumptions. It is able to formally prove or disprove a hypothesis and is capable of inferring additional knowledge. The PCR functionality may be located at the policy-based radio node, or it may be distributed to other points in the network.

17.5.2.1 Functions Performed

The architecture shall include a component that allows for conformance reasoning. The following functions are performed by the PCR:

— The PCR may accept policy sets either directly from the Policy Management Point or via the Policy Repository.
— The PCR shall make all of the policy conformance decisions.
— The PCR shall accept transmission opportunity requests from the SSRC and determine if the parameter values that define the transmission requests comply with the policies comprising the active policy set.
— Transmission decision replies shall be provided to both the SSRC and PE in response to SSRC transmission opportunity requests. The outcome is that both SSRC and PE know which transmission control commands should be accepted or denied.
— The PCR may execute on a processor located on the radio platform being managed or may be accessed remotely over the radio network.
— The PCR shall accept transmission decision refresh requests from the PE, determine if the parameter values that define the transmission requests comply with the policies comprising the active policy set, and provide responses back to the PE. See 17.5.3 for more detail on policy enforcer.

> *Note.* An active policy is a policy which is required to be included in policy conformance reasoning when loaded in the PBDRS.

17.5.2.2 Inputs

Inputs to the PCR include:

— Transmission opportunity requests from the SSRC
— Transmission decision requests originated by the SSRC and passed via the PE
— Transmission decision refresh requests from the PE
— Policy management sets from the PMP

17.5.2.3 Outputs

Outputs from the PCR include:

— Transmission opportunity replies to the SSRC
— Transmission opportunity decisions to the SSRC and PE
— Transmission decision refresh replies to the PE

17.5.2.4 Interfaces

The PCR interfaces with the following components:

— PMP
— SSRC
— PE

17.5.3 Policy Enforcer (PE)

The PE is the only component that can cause transmission control commands from the SSRC to be executed. These are commands that configure the transmitter's properties, including transmission frequencies, hopping scheme, bandwidth (BW), power, Automatic Gain Control (AGC), modulation, etc., and that turn on/off the transmitter. Consequently, the PE shall be instantiated such that the transmitter (TX) of the policy-based radio node can be configured to the policy-compliant parameters before the expiration of the corresponding Transmission Opportunity Reply validity period.

> **Note.** Since this imposes severe requirements on the PE response latency, the PE function cannot be easily implemented in a distributed manner and will usually be located at the policy-based radio node.

17.5.3.1 Functions Performed

The Policy Enforcer (PE) shall be the point where the policy decisions are actually enforced.

> **Note.** The IEEE 1900.5 system follows the definition introduced by IETF RFC 2753.

The Policy Enforcer (PE) shall perform the following functions:

— The Policy Enforcer shall examine each transmission command issued by the SSRC to ensure that the command is policy compliant, as determined by PCR, and causes policy compliant commands to be delivered to the "Software Control" component and shall provide a reply (e.g. Approve/Deny response) to the SSRC, along with any feedback it receives from the "Software Control" component.
— If the PE cannot accept the transmission control command, the PE shall reply to SSRC with an error code.
— If the validity period of an entry in the PE's cache of SSRC decisions has expired, the PE may ask the PCR to refresh its decision for that transmission request. Basically this implies resending the transmission decision request that the SSRC sent previously, which caused the PE to receive a decision from PCR in the first place.
— The Policy Enforcer may maintain a record of previous SSRC transmission decision requests, PCR transmission decision replies, and the validity periods associated with the replies. When the PCR has determined that a particular transmit configuration is policy compliant, the actual transmissions corresponding to that configuration shall be initiated within a time window referred to as the validity period. If the PE receives a transmission command from the SSRC for which the validity period has expired, the PE cannot allow the transmission to occur until it determines that the transmission parameters continue to be policy compliant. The PE may accomplish this by resending the associated transmission opportunity request to the PCR in an attempt to extend the validity period. The expired requests that are resent by the PE to the PCR are referred to as transmission decision "refresh" requests.

17.5.3.2 Inputs

The inputs to the PE include allowed transmission parameters provided in response to either SSRC transmission opportunity requests to the PCR or PE transmission decision refresh requests to the PCR.

17.5.3.3 Outputs

The outputs from the PE include:

— Radio control commands to the "Software Control" component
— Transmission decision refresh requests to the PCR
— Transmission control replies to the SSRC
— Transmission decision replies to the SSRC

17.5.3.4 Interfaces

The PE interfaces with the following components:

— PCR
— Radio "Software Control" component
— SSRC

17.5.4 Policy Repository

The IEEE 1900.5 system may include a policy repository. If included, the policy repository is a logical container for policies.

17.5.4.1 Functions Performed

The policy repository shall contain policies needed by the PCR to ensure policy compliance.

17.5.4.2 Inputs

The policy repository requires policies from the PMP.

17.5.4.3 Outputs

The outputs of the policy repository are policies compatible with the standard policy language whose requirements are specified in this standard.

17.5.4.4 Interfaces

The policy repository interfaces with the following components:

— PMP
— PCR

17.5.5 System Strategy Reasoning Capability (SSRC)

The architecture shall define a component that allows for system strategy reasoning. SSRC is responsible for configuring and re-configuring the system. All SSRC transmitter control commands flow through the PE to enforce policy compliance. The SSRC also coordinates with the policy conformance reasoner to evaluate the policy compliance of spectrum opportunities. The SSRC functionality may be located at the policy-based radio node or it may be distributed to other points in the network.

17.5.5.1 Functions Performed

The following functions shall be performed by the SSRC:

— The SSRC is aware of the state of the radio and the RF environment and uses that information to formulate transmission strategies. The SSRC shall send transmission opportunity requests to the PCR to identify transmission opportunities that are policy compliant.
— The SSRC submits transmission decision requests to the PE. If the PE accepts its Transmission Control Commands, the SSRC should pre-verify the policy compliance of the associated transmission decision requests with the PCR.
— The SSRC may revise its transmission strategies in response to replies from the PCR.
— The SSRC sends transmitter control requests to the "Software Control" component via the PE.
— The SSRC receives responses from the PE that indicate the success or failure status of submitted transmitter control commands.

— The SSRC may include a machine learning capability by which it learns from the responses from the PCR to the SSRC's transmission opportunity request (e.g., if a previous request for a specific transmission frequency was denied by the PCR, the SSRC may learn not to ask for this frequency again unless the radio environment, geo-location, or time of day change).

17.5.5.2 Inputs

Inputs into the SSRC include:

a) Radio information
   1) Radio capabilities
   2) User identification
   3) Network information may be required (e.g., if the policies are distributed within the network and not centralized at each radio node)
b) RF environment
   1) Local sensor information (i.e., from an RF sensor at the policy-based radio node)
   2) Sensor information from other nodes in the cooperative sensor network
   3) Databases (e.g., TV station information for the Television White Space (TVWS) application of DSA)
c) Responses from the PCR to transmission requests that the SSRC has sent to the PCR
   1) Approve
   2) Deny with reason for denial
d) Transmission opportunities from the PCR
   1) The PCR can provide the SSRC with policy derived information to assist it in constructing policy compliant transmission requests. The SSRC uses this information along with information about the radio and the RF environment to develop new transmission requests and to revise previously denied requests to make them policy compliant.
e) Transmission Information

17.5.5.3 Outputs

Outputs from the SSRC include:

— Partially specified (i.e., unbound) Transmission opportunity requests to the PCR
— Fully specified (i.e. bound) Transmission opportunity requests to the PCR
— Transmission Control Requests

17.5.5.4 Interfaces

The SSRC interfaces with the following components:

— PCR
— PE
— Radio information
— RF environment

17.6 Policy Language and Reasoning Requirements

NOTE 1—The motivation for developing a Policy Language may be explained as follows. One of the fundamental characteristics of policy-based dynamic spectrum access systems including cognitive radio systems (CRS) is the ability to adapt their behavior during operation to new circumstances. Because not all possible future circumstances can be foreseen, approaches that code alternative behaviors as functions or procedures into radio firmware and allow selection of appropriate behaviors during runtime will fall short when situations arise, for which no appropriate behaviors have been encoded since the situation is new and unforeseen. A policy language approach allows uploading of new policies to the radio during runtime and take effect without the need to recompile radio software components or the need to reboot the radio.

Note 2. A policy-based approach to radio control satisfies the requirement to adapt dynamically to new situations. Policies expressed in a high-level, abstract policy language can be loaded onto the radio during operation and take immediate effect. In the following, we state requirements of such policy language.

A standardized policy language adds value in scenarios where flexibility is needed, policies may need to be changed frequently, and it is difficult or impossible to anticipate all decision outcomes in advance.

Note 3. Policies in the policy language contain information about various concepts in the radio and DSA domains. Ontologies are a means to provide vocabulary of terms and relationships between terms in the radio and DSA domain. Thus, statements in the policy language use ontological terms. A policy conformance reasoner processes policy statements to decide spectrum access. In summary, there is a relationship among ontologies, policy languages, and policy conformance reasoner. Ontologies and policies are expressed in formal languages. However, ontologies do not have to be completed before the language is specified. A policy conformance reasoner is an interpreter of the formal language. In developing a specific policy language and implementation, the computational complexity and logical grounding requirements of reasoning about policies need to be considered by the system designer but are not part of this standard.

Note 4. Using a standard language representation is preferred as it enables reuse of libraries, applications, and tools previously developed. These tools reduce the cost of developing and using the language and increase interoperability between different tools. It is also expected that using a standard language representation will facilitate the international acceptance of the policy language.

17.6.1 Language Expressiveness

17.6.1.1 General Expressiveness Requirements

Policies should implement policy requirements generated by regulators and may implement policies generated by system operators and users. The policies should contain a pointer to, or the full text of, the source document.

Note. DSA radio system policies have a complex structure with many dimensions and layers of exceptions that are difficult even for human interpretation. The policy language shall be capable of capturing and potentially simplifying a number of aspects of this complex structure.

Declarative Language A declarative language shall be used.
  A policy language shall have clear and unambiguous syntax and semantics.

Note 1. The policy language serves as an interface between at least two different viewpoints, namely that of the regulators and that of the radio engineers.

Note 2. The main interest of regulators has been the specification of permissible transmission behavior. Regulators have historically not been interested in how policy conformance is checked as long as the check is correctly implemented and the radio enforces the policy. This is referred to as the soundness of the check. They are not interested in the strategy used to discover opportunities, assuming that policy conformance is ultimately enforced. Furthermore, they are not interested in verifying if a radio's strategy engine can exploit all transmission opportunities. Various trade-offs (e.g., cost of sensing versus need for spectrum), radio capabilities (e.g., ability to sense the spectrum), and the quality (degree of completeness) of the strategy engine itself will all affect which opportunities are exploited.

Note 3. The main interest of radio engineers, in contrast, is to exploit as many policy conforming transmission opportunities as possible. Thus, they have an incentive to enhance capabilities of both the strategy engine and the policy conformance engine.

Note 4. The foremost objective is to be able to specify—as opposed to implement—policy conforming behavior. Thus, a declarative language is a considerably better fit than an imperative language such as C, since a declarative language expresses "what" needs to be done, but not "how" it should be implemented. Policy does not tell the radio what to do; it only defines what constitutes authorized behavior. However, enforcing a policy may require the results of a function that may be implemented in the policy-based DSA radio system. Using a declarative language, the policy can describe the function and specify rules based on its inputs and outputs without specifying the particular implementation of the function (see 17.6.1.1.6).

Natural Language Annotations to Formal Statements of the Policy Language The policy language shall have the capability for annotations.

Note. Many policies are promulgated by policy makers via laws, treaties, orders, and business policies. Typically these are expressed in natural language. Since the policy language statements are intended to implement policies, it is likely that the formal language statements need to be augmented with annotations regarding intent. The annotation ability provides a formal mechanism for the association of natural language with formal language, akin to comments capabilities in most computer languages.

Machine-understandable Syntax The policy language shall have a machine-understandable syntax.

The policy language may have an easily human understandable syntax.

Permissive and Restrictive Policies The policy language shall support permissive policies and restrictive policies.

Note. Permissive policies describe conditions under which spectrum usage is allowed, and restrictive policies describe conditions under which spectrum usage is not allowed.

Inheritance The policy language shall be capable of expressing inheritance and extension of policies.

Note 1. Spectrum policy is very large and complex. The property of inheritance helps manage this complexity by enabling policy rules and properties to extend others and reduce the need for duplication.

Note 2. For instance, rules for the 2.4 GHz unlicensed band can inherit and extend rules for general unlicensed use.

Capable of Specifying the Dynamics The language shall be capable of specifying the dynamics (behaviors), including temporal aspects, of PBDRS components.

Capable of Defining New Functions The policy language shall include the ability to introduce definitions of new functions in terms of other known functions and allow inferring relationships between two functions, such as whether two functions are equivalent or not.

Note. For instance, the regulator might want to specify how to verify a specific condition, e.g., that the transmit power would be within a given power mask. The policy conformance reasoner should be able to verify that a given transmission request satisfies the condition. Moreover, in case the policy gives a prescription on how to check this condition, the policy conformance reasoner shall be able to infer whether the procedure it is going to use is equivalent with the policy specification or not.

Language Features The expressivity of the policy language shall include the following:

— Classes
— Individuals
— Binary relations
— Composition of relations
— Functions

— Temporal aspects of the system
— Behavioral descriptions
— State of the system
— Rules

Two Types of Negation The policy language shall be explicit about which negation [logical negation or negation as failure (NAF)] it supports (see also 17.6.2.2).

Types of Policies The language shall support a variety of policy types. As a minimum, the following is a list of policy types that shall be supported:

a) Geospatial – Policies that are based on distance or position as their primary trigger factors.
b) Time based – Policies for which time or relative time are the primary trigger factors.
c) Identity based – Policies that are triggered by the particular identifier (ID) of the device, the user, or the network.
d) Frequency based – Policies that regulate the selection of frequency.
e) Radio parameter enforcement – Policies that regulate or are triggered by the operating characteristics of the device.
f) Directive control – Policies that are triggered by entities external to the device and which are often temporary in nature.
g) Group behavior – Policies that govern how DSA devices interact and coordinate with each other.
h) Monitoring behavior – Policies that dictate how the sensing capabilities of a DSA device are utilized.
i) Network specific – Policies that would most often be implemented by DSA devices connected to a network with a Network Management System.
j) Policy source specific – Policies that originate from different sources, including, but not limited to regulatory agency, manufacturer, operational doctrine, and user.
k) Transmission types specific – Policies that govern transmission parameters based on the nature of the data being transmitted.

Meta-policies The policy language shall be capable of expressing meta-policies.

Meta-policies may state relationships between policies. The following types of meta-policy facts should be provided: grouping, precedence, and disjunction. The grouping requirement should include the following:

— The policy language should support a mechanism for combining policies according to different composition operators.
— The policy language shall be accompanied by a clear definition of the possible composition operators and their semantics.

The policy language shall define at least one meta-policy that expresses how to combine restrictive and permissive policies.

> ***Note 1.*** Meta-policies are expressions in the policy language that relate two or more policies to each other and change how they are processed. For example, we can make a policy governing when or where a set of policies will apply.

> ***Note 2.*** Grouping simply creates a named set of policies so they can be referenced as a group. The group may be described either by explicitly listing the member policy rules or by creating an expression that describes the policies that are members of the group. The policy language should enable aggregation of policies when multiple policies from different sources refer to a specific piece of data. Aggregation, in this context, refers to a combination of policies that does not result in a conflict. In practice, aggregation is represented by a union of a set of policies. Enforcing this set of policies should not result in incoherent behavior. Combination, in this context, refers to handling conflicting rules. In simple cases, combination can be automated

as the intersection of rights and union of obligations. Depending on the scenario, combination may result in more restrictive or less restrictive policies. For instance, depending on the manner in which conflict is resolved (e.g., prioritization, weighting), some of the conflicting rules may be deleted.

Note 3. Two policy groups may be disjunctive. If policy groups are disjunctive, then the policies in one or more of those otherwise applicable groups may be selected for application.

Note 4. Because the policy language shall support restrictive and permissive policies, it is possible to have contradicting policies in the policy repository. For example, a restrictive policy may disallow transmission between 800-940 MHz and a permissive policy may allow transmission between 750 MHz and 850 MHz. The two policies contradict each other in the overlapping frequency range 800-850 MHz. A meta-policy should express what constitutes a transmission opportunity. For example, a meta-policy may express that transmission opportunities are those for which one can find at least one permissive policy and for which there exists no restrictive policy that disallows the transmission opportunity.

Note 5. Precedence provides information on how to interpret conflicting policies. In the event that two policies represent a transmission opportunity but have different constraints on the variables, the policy with the higher precedence is applied. Precedence may be defined between two policies or two policy groups. If one policy group has higher precedence than another, then all its member policies have a higher precedence than the member polices of the other group.

Policy Composition   The language shall be capable of expressing relations among policies (e.g., policy P1 is a specialization of policy P2) and composition of policies. The policy language shall define the semantics of composition operators so that it is possible to infer the effects of composition on the resulting (composed) policy.

Policy Templates   The language shall support the definition of policy templates.

Note. A "template" or "schema" is a formula in the language of an axiomatic system in which one or more schematic variables appear. These variables may be required to satisfy certain conditions. Since the number of possible subformulas or terms that can be inserted in place of a schematic variable is countably infinite, a schema stands for a countably infinite set of instantiations. A "policy template" is a formula in the policy language that is a template, as described above.

Nested Policies   The policy language shall support nested policies.

Note. Nested policies are policies within policies. This allows complex policies to be constructed from simpler policies.

17.6.1.2   Ontology Language Requirements

The ontology shall be sufficient to express concepts required to encode the regulatory, coexistence, and system policies as indicated in 17.6.1.1.10.

Note. Ontologies can be a powerful tool to adapt the language to particular DSA applications and contexts while maintaining interoperability. For example, a common ontology on the structure of spectrum management regulations can be used to allow different policy authorities to author DSA radio policies that can be used in any DSA-capable device or network that adheres to the ontology. This allows the policy authority to create policies without the need of being aware of the specific characteristics and capabilities of each device or network in which the policy may be used. Another example is the use of a standard ontology for wireless communications concepts to facilitate the vendor-independent negotiation of air interfaces between cognitive, DSA-capable devices. The Wireless Innovation Forum [B3] has proposed an ontology for this purpose.

Domain Ontologies   At a minimum, information from the following three domains must be available to authorize a request for transmission:

— The capabilities of the radio;
— The current environment of the radio;
— And the characteristics of the requested transmission.

The policy language shall provide the ontological concepts to describe information from these domains through domain specific concepts further detailed in 17.6.1.2.2 and Table 17.1 by choosing an appropriate information representation further detailed by 17.6.1.4.

The policy language shall be extensible in a way to provide concepts to describe information not explicitly covered by Table 17.1 but required to authorize a transmission request in a specific application.

Ontologies of domain concepts shall be common between the SSRC and the PCR.

> **Note.** In general, before a request for a transmission can be authorized, three types of information need to be available: the capabilities of the radio, the current environment of the radio, and the characteristics of the requested transmission. Such information is heavily dependent on domain concepts.

Domain Concepts The policy language shall provide syntax and semantics for the representation of at least the following domain specific concepts as detailed further by Table 17.1:

— Radio and spectrum regulatory concepts
— Evidence
— Parametric constraints
— Security concepts
— Reliability concepts
— Network concepts
— Policy authority and authority delegation
— User defined concepts that can be added as needed and used in policy specification

The ontology shall be sufficient to express domain specific concepts as given in Table 17.1 and shall prescribe appropriate information representations to encode the regulatory, coexistence, and system policies.

> **Note 1.** Radio and spectrum regulatory concepts include channel, co-channel, adjacent channel, modulation, waveform, data rate, type of filtering, block size, device characteristics and attributes (e.g., FCC ID, serial number, transmitter, receiver, detector), Equivalent Isotropically Radiated Power (EIRP), mean EIRP, peak receiver (RX) power, Time-To-Live, frequency, center frequency, powermasks, bandwidth, duty cycle, signal detector [including detection threshold, frequency range, sample rate, precision, Signal to Noise Ratio (SNR), and Received Signal Strength Indicator (RSSI)], and signal type.

> **Note 2.** Evidence includes signal evidence (detected signal, sensed frequency interval, peak sensed power, detected time, scan time/duration, and count), location evidence, and time evidence.

> **Note 3.** Numerical constraints include restrictions on frequencies, time, and date.

> **Note 4.** Security concepts include type of encryption, keys, key exchange, credential, security mechanism (e.g., integrity, confidentiality, authentication, authorization), and security/classification level.

> **Note 5.** Reliability concepts include Forward Error Control.

> **Note 6.** Network concepts include node identity, network membership, and type of network [e.g., Peer to Peer (P2P)].

> **Note 7.** Policy authority and authority delegation include primary and secondary spectrum markets and primary and non-primary users.

The table below lists a minimal set of domain specific concepts that the policy language shall be able to represent. The policy language may use forms of representation as provided in Table 17.1.

Table 17.1 Domain specific concepts and representations.

Concept	Potential forms of representation for a single entity (e.g. single radio or single frequency record)
Date and time concepts	
Time of year, seasonal period	Enumeration
Time of operation	Data/Time range
Elapsed time	Single value
Location/Position concepts	
Fixed at one or multiple points	Single point or collection of points (Lat/Lon or conversion from other formats)
Route	Polyline
Relative position	Distance and bearing
Within a defined area	Circle (point and radius), rectangle, ellipse, polygon
Altitude	Single value or range
Mobility (Maneuver) concepts	
Speed	Single value or range
Velocity	Speed and direction or range of directions
Acceleration	Value and direction or range
Allocation/Allotment/Assignment concepts	
Service	Enumeration
Class of station	Enumeration
Allocation	Associative array
Allotments	Associative array
Frequency plans	Associative array
Assigned frequency	Scalar or array (hopset)
Response frequency	Scalar or array
User	Enumeration
License detail as appropriate	Enumeration
Technical criteria	
Emission masks	Vector
Antenna masks	Vector
Geographical concepts	
Political boundaries (e.g. countries)	Polygons
Administrative or legal boundaries (e.g. states, counties, cities)	Polygons
Roads	Lines
Bodies of water	Polygons
Emission	
Modulation	Enumeration
Occupied bandwidth	Scalar
Power (average or peak)	Scalar
Power spectral density	Associative array (value, frequency offset)
Spurious emission	Associative array (value, frequency offset)
Data rate	Scalar
Time access	Continuous vs. pulsed or slotted

Table 17.1 (Continued)

Concept	Potential forms of representation for a single entity (e.g. single radio or single frequency record)
Error correction	Enumeration
Interference protection criteria	
Interference to Noise Ratio (INR)	Scalar
Permissible outage time	Scalar
Receiver sensitivity	Associative arrays (value, modulation type)
Sensitivity criteria	Enumeration
Bit Error Rate (BER)	Multidimensional associative array (value, modulation type, SNR, SIR, etc.) (This could be a function of the SNR and SINR or SNR and SIR depending on the data available.)
Signal-plus-Interference-plus-Noise-plus-Distortion to Interference-plus-Noise-plus-Distortion Ratio (SINAD)	Scalar
Signal to Interference Ratio (SIR)	Scalar
Signal to Noise Ratio (SNR)	Scalar
Processing gain	Scalar
Noise figure	Scalar
Receiver selectivity (frequency-dependent vector quantity)	Associative array (value, frequency offset)
Spurious rejection	Scalar
Image frequency	Scalar
Image rejection	Scalar
Antenna concepts	
Maximum gain	Scalar Should be frequency-dependent vector for wideband antennas
Antenna pattern	Multidimensional associative array (value, frequency offset, polarization)
Elevation angle	Scalar
Azimuth angle	Scalar
Equivalent Isotropically Radiated Power (EIRP)	Scalar
Height of antenna above ground	Scalar
Polarization	Enumeration
Beam width	Scalar
Number of sectors	Scalar
Bandwidth	Scalar
Adaptive/Smart antenna concepts	
Maximum sweep angle	Scalar
Number of elements	Scalar
Type of system	Enumeration
Sensing concepts	
Bandwidth	Scalar
Duration	Scalar
Time stamp	Scalar
Type of detector	Enumeration
Type of signal	Enumeration

Table 17.1 (Continued)

Concept	Potential forms of representation for a single entity (e.g. single radio or single frequency record)
Sensed values	Associative array (value, frequency, time, position)
Noise estimate	Scalar
Sample rate	Scalar
Precision	Scalar
Position stamp	Point
Policy concepts	
Priority	Enumeration
Validity constraints	
Time	
Authority	Enumeration
Transmission data type	Enumeration
Spectrum access concepts	
Threshold	Associative array (value, frequency)
Frequency separation	Multidimensional associative array (value, frequency, sensed power, type of signal)
Cosite frequency separation	Scalar
Network concepts	
Device ID	Scalar
Network ID	Scalar
User role	Enumeration
Context parameters	Enumeration
Propagation conditions	
Weather	Enumeration
Type of terrain	Enumeration
Terrain irregularity	Scalar
Terrain profile	Array
Type of built environment	Enumeration
Electrical ground constants	Scalars
Radio climatic conditions	Enumeration
Surface refractivity	Scalar

From IEEE Standard for Policy Language Requirements and System Architectures for Dynamic Spectrum Access Systems, Dec 2011. IEEE.

17.6.1.3 Expressions

The policy language shall provide operators to form expressions.

The following is a list of operators that the policy language shall provide:

— Logical operators: These operators shall include: and, or, not, and exists.
— Relational operators: These operators shall include the data type comparison operators: less than, greater than, greater than or equal to, less than or equal to, equal to, not equal to, and the interval comparison operators: before, after, within, contains, overlaps-start, overlaps-end, just-before (before and overlaps-start), just-after (after and overlaps-end), at-start-of (within and overlaps-start), at-end-of (within and overlaps-end), starts-with

(contains and overlaps-start), ends-with (contains and overlaps-end), and overlaps (within and contains and overlaps-start and overlaps-end).

— Mathematical operators: These operators shall include: addition, subtraction, unary negation, multiplication, division, modulus, exponent, assignment, and logarithm.

— Spatial operators: These operators shall include the data type spatial operators: inside, distance (the distance between points or the shortest possible distance between a point and a geometry such as a polygon), and overlap. Spatial operators that take into account the curvature of the Earth's surface when performing calculations on geodetic data shall be provided.

— Set operators: Union, intersection, and null.

— Invocation of procedural attachments: An operator for procedural attachments shall be provided.

> **Note.** Any additional domain-specific functions may be invoked as procedural attachments. Procedural attachments may be defined for arbitrary functions such as mathematical operators, signal processing primitives, and other system operations such as the sensing of signals, and database retrieval. Such procedural attachments may be implemented in the radio platform to be successfully invoked.

17.6.1.4 Policy Language Support of Data Types

The policy language shall support the following data types:

a) Scalars of various types including
   1) Numbers (integers and floats)
   2) Dates/times
   3) Boolean
   4) Characters
   5) Text
b) Enumerations
c) Arrays
d) Associative arrays
e) Terrain Elevation Data (e.g., Digital Terrain Elevation Data – DTED)
f) Spatial Objects, such as:
   1) Points
   2) Lines
   3) Shapes (e.g, Polygons, circles, etc.)
   4) Polylines

17.6.2 Reasoning About Policies

A reasoning capability shall be provided that is capable of returning constraints for underspecified requests.

The reasoning capability shall be able to infer facts that are not explicitly stated but follow from one or more policies or facts.

> **Note 1.** The ability to return constraints (or something more than just yes or no) is essential to make full use of a policy-based DSA radio system: Policies may depend on a large variety of information. It is not realistic to assume that the policy reasoner is provided with all the information to come to a conclusive decision about spectrum access because there is often a cost associated with gathering information. For example, some policies may have geolocation constraints, or constraints on sensed power in order to avoid primary spectrum users. A radio may have a GPS and power sensors, but not use them all the time (e.g., to preserve battery life). With a policy reasoner that returns constraints, the PBDRS can choose to activate its sensors only when there is a constraint that requires it. Furthermore, the constraint may include information on which band to sense so the radio can perform a focused sensing operation on a small band.

Note 2. Reasoning example: For instance, given that TV channels are 6 MHz wide, and that a policy exists to permit PBDRS devices to transmit within a locally unallocated TV channel, one may infer that the maximum bandwidth for a PBDRS device using an unallocated TV channel is 6 MHz.

17.6.2.1 Formal Language System Requirements

A formal language (i.e., a formal syntax) and formal semantics shall be defined for the policy language.

The formal language and formal semantics shall satisfy the following conditions:

— A formal syntax shall be defined. This implies that the language definition includes rules for deciding whether a given expression is in the language.
— The policy language representation will be in a machine-understandable form and support machine-based reasoning.
— Formal semantics shall be defined for this language.
— Inference rules shall be defined for this language.

17.6.2.2 Negation

The policy language semantics shall include the semantics of negation.

Note 1. Logical Negation: Facts that have not explicitly been asserted to be true are not presumed to be false, they are simply unknown.

Note 2. Negation as Failure (NAF): Statements that have not been asserted to be true and cannot be proven to be true are regarded as false.

17.6.2.3 Policy Language Semantics

A formal model theoretic computational semantics shall be defined for the language. A proof theoretic semantics shall be defined and it shall be sound. Proof theoretic semantics shall be compatible with model theoretic semantics.

The semantics shall be capable of reasoning about numerical constraints.

Note 1. Completeness is not a requirement but is desirable. Completeness is the ability to infer all possible true sentences using the rules of inference.

Note 2. Soundness: Only true sentences can be derived (inferred) from true sentences.

Note 3. Reasoning about numerical constraints. Policies often define conditions about transmission opportunities in terms of numerical constraints. For example, a policy may restrict transmissions to carrier frequencies within a certain range or to situations in which the peak transmitted power is lower than a given threshold. These conditions are defined as numerical constraints that the policy language can represent and about which the PE has to adequately reason. For example, if a policy defines the allowable range of carrier frequencies to be between 3100 MHz and 3300 MHz, then the PE can recognize whether a given carrier frequency is within this range. This problem is made more complex by combinations of interacting constraints from multiple conditions and policies.

17.6.2.4 Types of Reasoning

The reasoning capability shall reason about all the required language features in 17.6.1.

The following reasoning functionality should be supported:

— Monitor for non-monotonic changes. The reasoner should be able to remove assertions that are contradictory, i.e., assertions which have been overridden by new knowledge.
— The reasoner should explicitly represent and reason about incomplete, inexact, and contradictory information, including the sources of conflict, confidence level, estimated accuracy of measurements, and parties responsible for conflicts.
— Closed world reasoning

Note 1. A semantic using negation as failure can support closed world reasoning.

— Express and reason about time dependent and location and other context dependent information (as in policies using Temporal Logic)
— Reason about abstract specifications
— Prove conjectures on functional equivalence of components

Note 2. Example: On-the-fly generation of waveform conforms to required waveform.

17.6.2.5 Reflection and Awareness

The reflection and awareness requirements are that:

— The language shall support the PBDRS's ability to express queries about its own state and its operational environment.
— The language shall provide access to the appropriate variables holding the state information.
— The reasoning capability associated with the language shall support returning the query results expressed in the language.
— The language and reasoning shall support the requirement for the PBDRS to be able to draw conclusions from its knowledge base facts.

Annex 17A

(informative)

Use Cases

17A.1   Example DSA Policy Management Architecture

This annex describes an example policy management approach for DSA policies as shown in Figure 17A.1. DSA policy management includes policy generation, policy storage, policy validation, and policy distribution/updating. DSA policy generation is a process of converting high-level policy information, usually expressed in a natural, human-understandable language, and converting it to a more detailed, low-level form suitable for machine reasoning by the PCR. Policy sources include spectrum regulators, network or system operators, and users. Using a policy authoring and management tool, the Policy Administrator extracts information from human-readable policy sources to create machine-understandable policy. The administrator selects the policy sources and individual policies that are appropriate to the anticipated use of the DSA radio and network and the environment they will operate in. This includes consideration of the spectrum usage and regulations applicable to the frequency bands and geographic region of operation and additional policy constraints that the operating organization may wish to impose. An interface with an external spectrum management tool may be useful in optimizing the DSA policies to the anticipated electromagnetic environment and to coordinate planned dynamic sharing of spectrum with external spectrum management systems. The policy authoring tool may provide templates and wizards to assist the administrator with the task of converting human-readable policies into the formal language that a policy reasoner within the DSA radio can use. This will allow individuals who are not experts in the syntax of the formal language to author policies. The output of the policy generation process is a set of policies represented in the formal policy language that can be applied to the control of various devices in the network.

The PMP may interface with a policy repository to store policies encoded in the formal policy language. The DSA radio nodes that participate together in the same DSA network each need to operate from a common set of policies. However, the management entity may need to manage several DSA networks or manage multiple policy configurations for a single network. Consequently, the PMP may need to maintain several policy sets. This might be accomplished by maintaining a global repository of all policies and specific subsets configured for specific networks or for multiple DSA scenarios/modes in a single network.

Because policies may originate from multiple sources and seemingly independent policies may include unanticipated dependencies, it is possible for policies to conflict with each other. To address this issue, the PMP and/or reasoning capability may include policy analysis and validation capabilities to identify errors and conflicts. The capability to flag potential policy problems allows the administrator to address them before the policy sets are downloaded to the radios in a DSA network.

An interface exists between the PMP and DSA radios/network to allow managed policy sets to be loaded into each radio in the network. Various interfaces are possible such as an RF link or a wired connection from a fill device. Depending on the type of network [e.g., peer-to-peer, base station-handset/Common Platform Enumeration (CPE)], connectivity with the policy download source may exist only at certain limited times. The PMP may be physically connected to

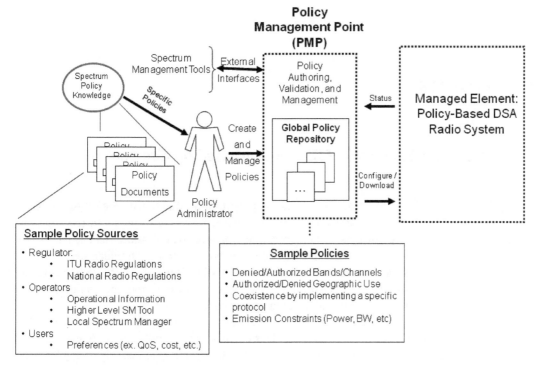

Figure 17A.1 DSA policy management example. From IEEE Standard for Policy Language Requirements and System Architectures for Dynamic Spectrum Access Systems, Dec 2011. IEEE.

or integrated with one of the DSA radio nodes to provide RF connectivity with the DSA network and allow the PMP to query the status of network nodes and remotely manage radio policies. Due to the variability of radio propagation, such connectivity may be intermittent or nonexistent at times.

Annex 17B

(informative)

Illustrative Examples of DSA Policy-based Architecture

Figure 17B.1 shows the cognitive engine mapped into a "traditional" Media Access Control (MAC) layer on the Data/Control plane. Alternatively, the elements in the cognitive engine may be represented as part of a separate "cognitive plane" rather than as part of the data/control plane. It could even be outside the radio all together, in which case it would operate across the MAC Layer Management Entity (MLME) to control the elements of the MAC. It is also expected that there may be cross layer connections from the cognitive engine (e.g. to the physical layer to control modulation and coding parameters).

An example set of typical MAC mechanisms is shown in Figure 17B.1. They are shown here to be within a single device, but could be distributed across multiple devices (e.g. a base station and subscriber). The "Tx Control Commands" shown in Figure 1 as originating from the Policy Enforcer would be interpreted as "rule sets" or parameters for the MAC. For instance, a particular strategy might require the use of Carrier Sense Multiple Access (CSMA) in the Access State Machine, or alternatively Time Domain Multiple Access (TDMA). The framing rules would likely have to change as well for different access strategies. Adaptation and classification rules might also be impacted, such as whether header compression is permitted, and how particular classes of data are mapped to scheduling service disciplines. In extreme cases within a software defined radio, actual "code" might be downloaded and executed to express the rules determined by the System Strategy Reasoner. However, it is expected that parameters rather than rule sets will normally be used to control the configuration of a MAC. For instance, the MAC might have the ability to implement Automatic Repeat reQuest (ARQ). A parameter could be set by the cognitive engine that would turn this capability on or off for different classes of traffic.

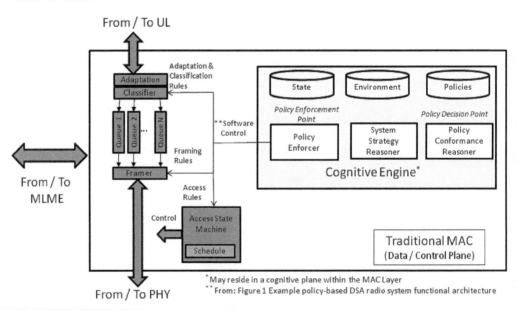

Figure 17B.1 Example cognitive engine mapped to MAC layer. From IEEE Standard for Policy Language Requirements and System Architectures for Dynamic Spectrum Access Systems, Dec 2011. IEEE

Annex 17C

(informative)

Relation of IEEE 1900.5 Policy Architecture to Other Policy Architectures

Network policy implementation is described in a variety of models; the most common defined by the Internet Engineering Task Force (IETF) and the Distributed Management Task Force (DMTF) policy framework described in the Policy Core Information Model (PCIM) specification (RFC3060/RFC3460) [B4], [B5]. This framework consists of four basic elements:

1. A policy management point (PMP)
2. A policy repository
3. A policy decision point (PDP)
4. A policy enforcement point (PEP)

PCIM specifications depict a policy architecture with the PMP, Policy Repository, PDP, and PEP components. Figure 17C.1 shows how the reasoner fits into that architecture. The additional reasoning functionality updates the PCIM architecture for future generation radio applications. A policy management tool is used to create policies for control of DSA networks. This tool takes high-level policy information and converts it to a form that can be evaluated by the policy reasoner. The resulting detailed policy description is stored in the policy repository. Policy enforcement points (PEP) enforce and execute the different policies. When an event in the system indicates that a policy must be applied, the PDP takes the event trigger and retrieves the policy information from the repository. This information may be converted to a format the PEP

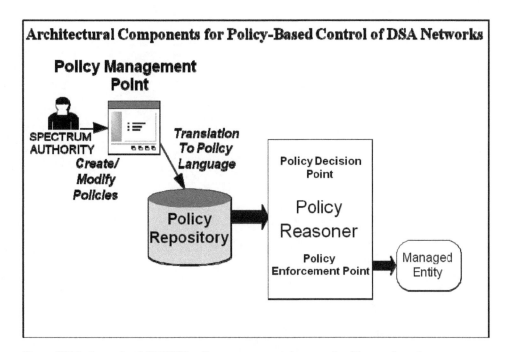

Figure 17C.1 Example of RFC3060 policy management framework with mapping of reasoner component. From IEEE Standard for Policy Language Requirements and System Architectures for Dynamic Spectrum Access Systems, Dec 2011. IEEE

understands. In most cases, the conversion is from policy language to device configuration. The PEP then executes the policy. Some policy architectures combine the PDP and PEP functions. Common examples of policy enforcement points are routers, firewalls, or gateways. When the radio needs to change or update its transmission strategy, the SSRC generates a transmission request for evaluation by the PCR. In formulating the request, the SSRC may need to convert the parameters of a transmission strategy from an internal device-unique format to a standardized format conforming to the language and ontology requirements of this standard. After the SSRC has negotiated a policy compliant transmission strategy with the PCR, the PEP function within the SSRC issues the policy compliant transmission commands to the transmitter.

All aspects of network configuration should be taken into account when deciding how these components are deployed. If the radio network is disconnected from the management point, how then will the policy changes propagate through the network? Is this for opportunistic spectrum usage? Optimized use of pre-allocated spectrum? These considerations, in addition to the needs of the stakeholders, such as the user, regulatory agency, manufacturer, and operator all have to be considered when defining the IEEE 1900.5 architecture.

Note. PEP is equivalent to the Policy Enforcer (PE) in the IEEE 1900.5 architecture.

Annex 17D

(informative)

Characteristics of Imperative (procedural) and Declarative Languages for Satisfying Language Requirements for Cognitive Radio Systems

17D.1 Types of Languages

Kokar [B2] provides an overview of the differences between two types of computer languages, namely imperative (or procedural) and declarative. The paper makes the argument that the functional flexibility inherent in cognitive radio systems requires the inference power of a formal declarative language with formal, computer-processable semantics. The arguments made in supporting this conclusion are not repeated here but may be found in Kokar [B2].

17D.1.1 Imperative (or Procedural)

Examples of imperative languages are C, C++, Java, and Fortran. A program in an imperative or procedural language consists of a list of statements that manipulate the state of the program and are executed in a sequence dictated by a control structure. The program includes the information on what needs to be done and how it needs to be accomplished (in what sequence).

The information in Carlson [B1] is relevant to consideration of the use of an imperative language for cognitive radio systems as well as the arguments put forth in Kokar [B2] as to why cognitive radio systems, because of their functional flexibility, should utilize a declarative type of language.

17D.1.2 Declarative

Examples of declarative languages are Prolog, Structured Query Language (SQL), Web Ontology Language (OWL), and Semantic Web Rule Language (SWRL). Declarative languages express what is to be accomplished rather than how it is to be accomplished. Thus a declarative program is a collection of facts known as clauses and a goal provided by the user. The inference engine finds the solution to the goal.

Annex 17E

(informative)

Example Sequence Diagrams of IEEE 1900.5 System

17E.1 Overview

Figure 17E.1 illustrates the interactions between the PCR, PE, SSRC, and Software Control. The ordering of interactions between components is not completely specified in Clause 17.5. The sequence diagrams in Figure 17E.1 and Figure 17E.2 show a possible implementation that is consistent with the descriptions in that clause. Furthermore, while there is some level of asynchronous operation between the components noted above, there are portions which do imply a specific ordering. Figure 17E.1 sequence diagram shows policy installation portion of policy management, and Figure 17E.2 shows policy component interaction during policy activation.

The sequence diagrams are intended to:

— reflect the interactions identified in Figure 17E.1 based on their general description in Clause 17.5
— show the interaction dependencies between the PCR, SSEC, and PE in terms of the actions of one component and the response of another
— visually describe the interactions in a temporal order to illustrate pairwise interactions between components
— identify any implicit synchronous or asynchronous behavior of the interaction between the components

17E.2 Assumptions

The sequence diagrams reflect the following assumptions which are implied by Figure 17E.1 and the interface descriptions in Clause 17.5.

a) The SSRC and PCR may be external to the radio. Based on this assertion, the sequence diagram shows the SSRC and its associated Radio Information Base and the PCR as outside the bounding box of the radio system. However, the Radio Environment Map may have portions that may be external to the radio (e.g., sensors network and database). However, the Local Sensor may be embedded within the radio or an external sensor geographically co-located with the radio. For this reason, the Radio Environment Map is shown as existing within the radio bounding box.
b) The PE is the only component that has a binding to the Software Control interface set in the radio. This assumes that the PE is the only element that is resident in the radio and has the interface to the Software Control component. Thus, it shall have a binding that describes the form and content of the configuration request.
c) The Software Control Application Programming Interface (API) set of the radio is the only component that can assure that each API call from the PE in response to a TxControlCommands is valid. The rationale for this assumption is that the Software Control interface is the core set of control APIs for the radio. Reconfiguration of the radio may also mandate

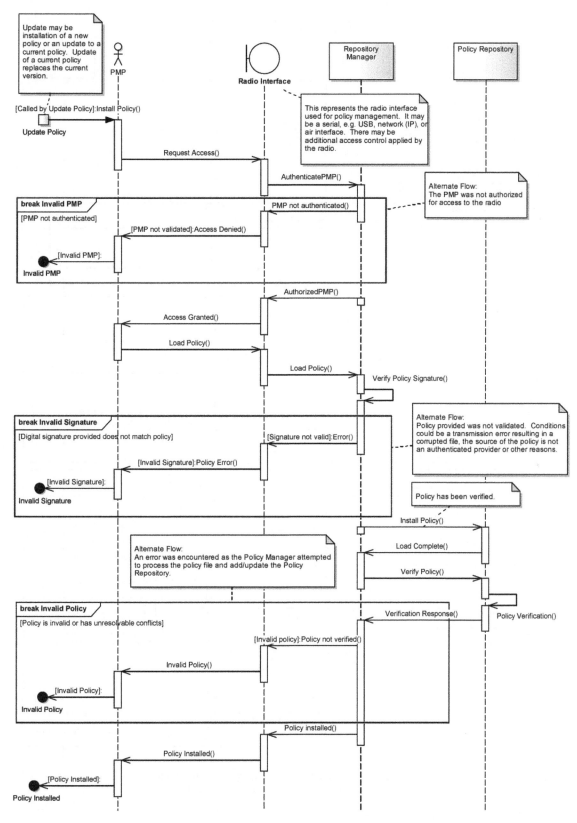

Figure 17E.1 Example policy management sequence diagram. From IEEE Standard for Policy Language Requirements and System Architectures for Dynamic Spectrum Access Systems, Dec 2011. IEEE

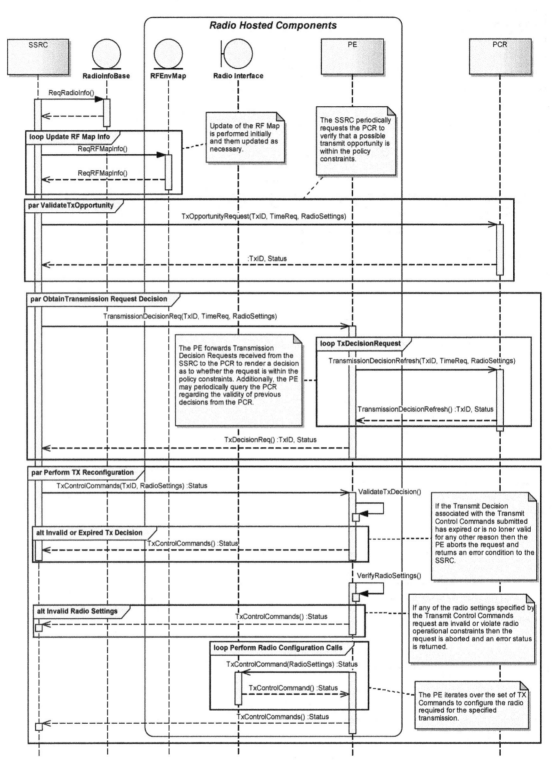

Figure 17E.2 Example radio configuration sequence diagram. From IEEE Standard for Policy Language Requirements and System Architectures for Dynamic Spectrum Access Systems, Dec 2011. IEEE

that the current operational state of the radio be considered prior to initiating any requested configuration. Although the PE is not prohibited from maintaining state information, there is no indication that the PE shall maintain local state of the radio.

d) The PE has the responsibility for assuring that the radio settings requested in a Tx Configuration are valid for the radio. This may be accomplished in a number of ways. The PE may maintain a set of configurable settings, ranges, and limits or it may query the Software Control interface within the radio to validate the parameter settings prior to issuing the commands. The former provides a quick check that the settings provided are valid but requires the PE to maintain an internal table of settings and allowable parameter values. The latter ensures that the radio itself is the final arbiter as to whether a setting is allowable or not but may incur an operational impact on the radio due to the interrogation of the radio by the PE.

e) TxControlCommands request may only be described at a higher level of abstraction than the Software Control radio API. The PE is the only component that will have an actual binding to the Software Control API set of the radio. Thus, the SSRC will only be able to provide a set of control parameters required to the PE for reconfiguration of the radio.

f) A unique transaction identifier is needed to assure traceability from TxOpportunityRequest to TxDecisionRequest to TxControlCommands. The standard requires no consistent method for maintaining traceability from the request to validate an opportunity, to a decision request, and finally to a set of control commands. However, the need for a unique ID is implicit in the description of how the process flows. Since the PCR, SSRC, and PE may be geographically separated and the interaction is largely asynchronous, some method of maintaining traceability is necessary to ensure that the delayed response from the PCR to the SSRC, for example, can be correctly mapped to the corresponding opportunity within the SSRC. Associated with each unique ID is also a time identifying the requested or planned time to perform the configuration.

17E.3 Sequence Diagram Organization

Within the sequence diagram, there are four high level blocks representing sub-flows that signify a logical set of interactions, each of which may be performed independently of each other and function in an asynchronous manner. These are:

a) Update Radio RF Map – In this simple interaction, the SSRC queries the RfEnvMap for the current set of relevant environment for a given radio. This is performed once on start up for each radio managed by the SSRC and then periodically as determined by the SSRC to update the RF data within the SSRC.[16] At the moment, there is no specification of the request call or the data provided back to the SSRC. This requires some additional thought. Presumably the response back from the RfEnvMap would be a sequence of frequencies and occupancies data, e.g., Receive Signal Strength Indicator (RSSI) from the RF section of the radio or other local sensor data.

b) ValidateTxOpportunity – This flow is a simple interaction between the SSRC and the PCR in which the SSRC has identified a candidate list of transmission opportunities and has forwarded the sequence to the PCR for policy determination. The data provided is TxID, TimeReq, and RadioSettings. These represent:

1) TxID – A unique identifier for each TXOpportunity sent to the PCR. Since there is no guarantee of immediate or synchronous response, the reply back from the PCR may be delayed. Thus, a unique ID would be required in order for the SSRC to update its internal information on return.

---

16 There is nothing the specification that indicates the SSRC is limited to the operational control of a single radio and, since it may reside separately from the radio, a single SSRC could be applied to some logical group of radios to coordinate the reconfiguration across the group.

  2) TimeReq – The time requested for the TXOpportunity. This may be a start time and duration. All Tx requests have some shelf life, or time to live, in order to be useful. Thus, the requested opportunity may only be operationally useful for the given time period. In addition, although it is not described, the PCR may have some temporal policy constraints defining valid times of usage for specific frequencies.

  3) RadioSettings – This is a sequence of the essential information required by the PCR to adjudicate the request. This may be as simple as a frequency or it may contain a frequency range, power settings, and other information, such as evidence (e.g., RF environment, time of day, or geographic location sensor data) that may be required to show that the specific conditions required by a policy will be satisfied by the request.

 c) ObtainTransmissionRequestDecisions – This sub-flow depicts the interaction between the SSRC and the PE to obtain decisions regarding intended radio transmission reconfigurations. The SSRC forwards a TXDecisionRequest to the PE containing essentially the same information it provided to the PCR in requesting approval for a Tx opportunity, TxID, TimeReq, and RadioSettings. The PE receives the request and forwards the request to the PCR for confirmation receiving a TxDecisionReply back from the PCR.[17],[18] Also, the PE may periodically request the PCR to refresh the current TxDecisionRequest information within the PE. This periodic refresh may be initiated asynchronously by the PE and is represented in the sub-flow.

 d) PerformTxConfiguration – Finally, this sub-flow describes the process of actually reconfiguring the radio. It has the most interactions and embedded sub-flows. The SSRC selects an approved TxDecisionRequest and initiates a TxControlCommands request to the PE to reconfigure the radio. The parameters provided consist of the TxID associated with the approved TxDecisionRequest, and the sequence of RadioSettings associated with the reconfiguration. The PE validates the request against its internal set of TxDecisionRequest data. If the request cannot be validated, e.g. the TxID is invalid, the time of the request has expired, etc., then a failure status is returned to the SSRC. If the request is valid, then the PE verifies the radio settings. For TxControlCommands that are valid, the PE issues corresponding commands to reconfigure the transmitter via the Software Control Interface.

Note. Some of the material covered in Annex 17E is outside of the scope of IEEE Std 1900.5-2011.

---

17 Transmission decision requests originated by the SSRC and passed (to the PCR) via the PE.
18 If each TxDecisionRequest received from the SSRC must be forwarded to the PCR, this could cause significant overhead.

Annex 17F

(informative)

Bibliography

B1 Carlson, M., "Policy Presentation–Work of the DMTF Policy Working Group," Presentation to the IEEE 1900.5 Working Group on 8 December 2008.

B2 Kokar, M. M., and Lechowicz, L., "Language Issues for Cognitive Radio," accepted for publication in IEEE Proceedings, April 2009.

B3 Modeling Language for Mobility Work Group, "Description of the Cognitive Radio Ontology," Document WINNF-10-S-0007, Version V1.0.0, The Software Defined Radio Forum Inc., 30 September 2010.

B4 Moore, Ellesson, *Strassner, Westerinen, RFC3060: Policy Core Information Model–Version 1 Specification*, The Internet Society, 2001.

B5 Moore, RFC3460: Policy Core Information Model (PCIM) extensions, The Internet Society, January 2003.

18

IEEE Standard for Spectrum Sensing Interfaces and Data Structures for Dynamic Spectrum Access and Other Advanced Radio Communication Systems

IMPORTANT NOTICE: This standard is not intended to ensure safety, security, health, or environmental protection. Implementers of the standard are responsible for determining appropriate safety, security, environmental, and health practices or regulatory requirements.

This IEEE document is made available for use subject to important notices and legal disclaimers. These notices and disclaimers appear in all publications containing this document and may be found under the heading "Important Notice" or "Important Notices and Disclaimers Concerning IEEE Documents." They can also be obtained on request from IEEE or viewed at http://standards.ieee.org/IPR/disclaimers.html.

18.1 Overview

Background Given the increasing proliferation of devices that use radio spectrum and the resulting shortage of capacity in allocated spectrum bands, new technologies are being introduced that allow devices to access unused spectrum bands dynamically to serve traffic demands. These new technologies require reliable, dependable, and trusted spectrum sensing capabilities in order to make accurate assessments of spectrum availability in the surrounding operational area. Such capabilities will assist devices and associated radio equipment in identifying locally/temporally available spectrum that can be accessed without causing harmful interference to the incumbent users of that spectrum.

Problem Recently proposed advanced radio systems based on sensing technology (e.g., those being worked on within IEEE P802.22™ [B2][1]) combine sensing and the protocols and cognitive engines (CEs) that use the sensing results into proprietary architectures. This model of development reduces innovation and limits the opportunities for integrating new component technologies for better system performance. Furthermore, the results of sensing extend beyond the activities of a single system and are ideally integrated into the larger spectrum management process including the development of spectrum use monitoring and enforcement activities.

Many different sensing techniques have been defined and implemented, yet there has been no effort to provide interoperability between sensors and clients developed by different manufacturers.

Solution This standard defines the interfaces and data structures required to exchange sensing-related information for increasing interoperability between sensors and their clients developed by different manufacturers. The clients can be cognitive engines as in the focus of this standard or can be any other type of algorithms or devices (e.g., adaptive radio) that use sensing-related information. Being aware of evolving technologies, interfaces are developed to accommodate future extensions, new service primitives, and parameters.

---

1 The numbers in brackets correspond to those of the bibliography in Annex F.

Dynamic Spectrum Access Decisions: Local, Distributed, Centralized, and Hybrid Designs, First Edition. George F. Elmasry.
© 2021 John Wiley & Sons Ltd. Published 2021 by John Wiley & Sons Ltd.
Companion website: www.wiley.com/go/elmasry/dsad

How this Standard applies This standard provides a formal definition of data structures and interfaces for exchange of spectrum sensing-related information.

18.1.1 Scope

This standard defines the information exchange between spectrum sensors and their clients in radio communication systems. The logical interface and supporting data structures used for information exchange are defined abstractly without constraining the sensing technology, client design, or data link between the sensor and client.

18.1.2 Purpose

The purpose of this standard is to define spectrum sensing interfaces and data structures for dynamic spectrum access (DSA) and other advanced radio communications systems that will facilitate interoperability between independently developed devices and thus allow for separate evolution of spectrum sensors and other system functions.

18.1.3 Interfaces and Sample Application Areas

18.1.3.1 IEEE 1900.6 Interfaces

Figure 18.1 illustrates IEEE 1900.6 interfaces between spectrum sensors and their clients (cf. 3.1).

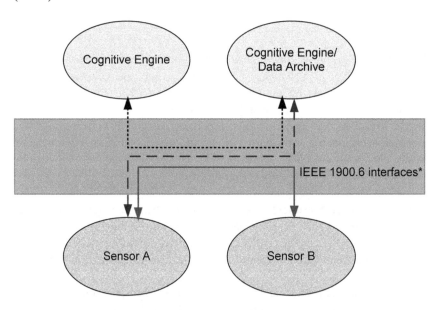

*The client role can be taken by Cognitive Engine, Sensor and Data Archive

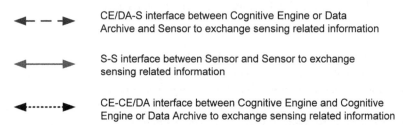

CE/DA-S interface between Cognitive Engine or Data Archive and Sensor to exchange sensing related information

S-S interface between Sensor and Sensor to exchange sensing related information

CE-CE/DA interface between Cognitive Engine and Cognitive Engine or Data Archive to exchange sensing related information

Figure 18.1 —IEEE 1900.6 interfaces between spectrum sensors and their clients. From IEEE Standard for Spectrum Sensing Interfaces and Data Structures for Dynamic Spectrum Access and Other Advanced Radio Communication Systems, Feb 2011. IEEE.

As shown in Figure 18.1, there are three possible instances of the logical interface within the scope of this standard depending on the logical entities using the IEEE 1900.6 interface:

— CE/DA–S^2 interface between CE or data archive (DA) and Sensor
— S–S interface between Sensor and Sensor
— CE–CE/DA interface between CE and CE or DA

The CE/DA–S interface is used for exchanging sensing-related information between a CE or DA and a Sensor. As an example, the CE/DA–S interface is used in scenarios where a given CE or DA obtains sensing-related information from one or several Sensors or a given Sensor provides sensing-related information to one or several CEs or DAs.

The S–S interface is used for exchanging sensing-related information between Sensors. As an example, the S–S interface is used in scenarios when multiple Sensors exchange sensing-related information for distributed sensing (see the definition of distributed sensing in 18.3.1).

The CE–CE/DA interface is used for exchanging sensing-related information between CEs or between CE and DA. The CE–CE/DA interface is used in scenarios where CEs exchange sensing-related information for distributed sensing. The CE–CE/DA interface is also used in scenarios where a CE obtains sensing-related information and or policy/regulatory information from a DA. For CE–CE/DA communication, a CE/DA shall be able to take the role of Sensors in terms of providing the CE/DA–S interface for the duration and purpose of exchanging sensing-related information.

The following subclauses detail sample application areas of this standard.

18.1.3.2 Reporting Data Format of the Sensing-related Information

The standard defines the set of sensing parameters that will allow for spectral opportunities exploitation. For example, binary values "0" for idle and "1" for occupied band or soft parameters (e.g., detected received power, cyclo-stationary features level, and direction of arrival of the licensed user signal) shall be reported by the Sensors to their clients. The standard specifies the data formats of these parameters in a protocol-independent manner.

18.1.3.3 Single-device Sensing and Multiple-device Sensing

Single device sensing denotes spectrum sensing capabilities or performance measurements based on the information obtained from a single sensor. This can be considered as the reference case for more sophisticated sensing techniques based on multiple sensing devices. Multiple device sensing includes distributed sensing between different radio nodes or between sensors. Sensing information obtained from distributed sensors shall be qualified either implicitly or explicitly by time and location of the acquisition of measurement data as well as by the attributes or characteristics of the acquisition method utilized. Some concise examples are given in 18A.2. The standard has been developed considering different use cases which rely on the exchange of sensing-related information between multiple sensors and a client by referring to existing sensing methods, techniques, and their approaches (see Noguet et al. [B7]). The detailed study of protocols related to multiple device sensing is outside the scope of this standard.

18.1.3.4 Usage of Spectrum Sensing Cognitive Radio (CR) for the Investigation of Policy Violations

The standard defines a logical interface to support exchange of a complete set of sensing-related information between Sensors and their clients. The logical interface defined in this standard may be applied to use cases such as the radio scene analysis and investigation of policy violations. Verification algorithms are outside the scope of this standard.

---

2 "CE/DA" denotes CE or DA.

18.1.4 Conformance Keywords

In this document, the word *shall* is used to indicate a mandatory requirement. The word *should* is used to indicate a recommendation. The word *may* is used to indicate a permissible action. The word *can* is used for statements of possibility and capability.

18.2 Normative References

The following referenced documents and URLs are indispensable for the application of this document (i.e., they must be understood and used, so each referenced document is cited in text and its relationship to this document is explained). For dated references, only the edition cited applies. For undated references, the latest edition of the referenced document (including any amendments or corrigenda) applies.

IEEE Std 1900.1™-2008, IEEE Standard Definitions and Concepts for Dynamic Spectrum Access: Terminology Relating to Emerging Wireless Networks, System Functionality, and Spectrum Management.[3],[4]

IEEE Std 1900.4™-2009, IEEE Standard for Architectural Building Blocks Enabling Network-Device Distributed Decision Making for Optimized Radio Resource Usage in Heterogeneous Wireless Access Networks.

18.3 Definitions, Acronyms, and Abbreviations

18.3.1 Definitions

For the purposes of this document, the following terms and definitions apply. *The IEEE Standards Dictionary: Glossary of Terms & Definitions* should be consulted for terms not defined in this clause.[5] For any definition not given in this subclause, IEEE Std 1900.1™-2008 shall apply.

cognitive engine (CE): A logical entity that contains cognitive control mechanisms, which may include policy-based control mechanisms, and uses sensing information to assess spectrum availability.

data archive (DA): A logical entity storing systematically sensing-related information.

distributed sensing: The process of sensing where sensors are distributed in acquisition time, space, frequency, and function.

IEEE 1900.6 application service access point (A-SAP): A service access point used by applications accessing services of an IEEE 1900.6 compliant server.

IEEE 1900.6 client[6]**:** A logical entity, application, or device (compliant to this standard) that receives sensing information and spectrum usage-related information from an IEEE 1900.6 compliant server. In general, the information exchange between IEEE 1900.6 compliant clients and servers applies to sensing-related information.

> **Note.** The term "client" is used in the context of describing the IEEE 1900.6 service and denotes a conceptual functional role of an IEEE 1900.6 logical entity. It is the counterpart to an IEEE 1900.6 server.[7]

---

3 The IEEE standards or products referred to in this clause are trademarks owned by the Institute of Electrical and Electronics Engineers, Incorporated.

4 IEEE publications are available from the Institute of Electrical and Electronics Engineers, 445 Hoes Lane, Piscataway, NJ 08854, USA (http://standards.ieee.org/).

5 *The IEEE Standards Dictionary: Glossary of Terms & Definitions* is available at http://shop.ieee.org/.

6 The terms "client" and "servers" refer to roles of IEEE 1900.6 logical entities that are defined by the direction of the flow of sensing-related information. These terms are used when addressing the exchange of chunks of information between logical entities and are not related to the use of service primitives and subsequent exchange of protocol messages required to realize the exchange of sensing-related information.

7 Notes in text, tables, and figures of a standard are given for information only and do not contain requirements needed to implement this standard.

IEEE 1900.6 communication service access point (C-SAP): Used for the exchange of sensing-related information between remote IEEE 1900.6 servers and clients by means of a communication subsystem.

IEEE 1900.6 logical entity: Part of the underlying logical entity–relationship model. It is defined by its functional role(s) and interfaces with other IEEE 1900.6 logical entities. The three types of logical entities in the IEEE 1900.6 logical model are Sensor, CE, and DA. Logical entities are realized through an IEEE 1900.6 service.

IEEE 1900.6 logical interface: A conceptual boundary between two or more IEEE 1900.6 logical entities. Logical interfaces exist between Sensors (S–S interface), between CE and/or DA and Sensors (CE/DA–S interface) and between CE and DA (CE–CE/DA interface). These interfaces are realized by the services or by a subset of the services available at the IEEE 1900.6 SAPs.

IEEE 1900.6 measurement service access point (M-SAP): Used for the exchange of sensing-related information between the spectrum measurement module and its M-SAP user.

IEEE 1900.6 server[6]: An IEEE 1900.6 logical entity, application, or device that provides sensing information and spectrum usage-related information to IEEE 1900.6 clients. In general, the information exchange between IEEE 1900.6 clients and IEEE 1900.6 compliant servers applies to sensing-related information.

> *Note.* The term "IEEE 1900.6 server" is used in the context of describing the IEEE 1900.6 service and denotes a conceptual functional role of an IEEE 1900.6 logical entity. It is the counterpart to an IEEE 1900.6 client.

IEEE 1900.6 service: An abstraction of the totality of those functional blocks inside IEEE 1900.6 servers and clients realizing the IEEE 1900.6 logical interface.

IEEE 1900.6 service access point (SAP): A conceptual location at which an IEEE 1900.6 service user can interact with the IEEE 1900.6 service provider by means of utilizing IEEE 1900.6 service primitives. A service access point makes an IEEE 1900.6 service accessible to an IEEE 1900.6 client. The three types of service access points are IEEE 1900.6 communication service access point (C-SAP), IEEE 1900.6 application service access point (A-SAP), and IEEE 1900.6 measurement service access point (M-SAP).

IEEE 1900.6 service primitive: Describes the interaction with an IEEE 1900.6 service provider through an IEEE 1900.6 service access point (SAP) in an abstract implementation-independent way.

> *Note.* The term "IEEE 1900.6 service primitive" is adopted from ITU-T Recommendation X.210 [B5].

IEEE 1900.6 service provider: An abstraction of the totality of those entities that provide an IEEE 1900.6 service to the IEEE 1900.6 service user. The IEEE 1900.6 service provider includes the IEEE 1900.6 service and the associated IEEE 1900.6 service access points (SAPs) that instantiate the logical interface for the exchange of sensing-related information between IEEE 1900.6 servers and clients.

> *Note.* The term "service provider" is used in the context of protocol and procedure specifications and denotes a conceptual functional role of an IEEE 1900.6 logical entity. It is the counterpart to a service user.

IEEE 1900.6 service user: The IEEE 1900.6 logical entity that shall use the IEEE 1900.6 service and the associated IEEE 1900.6 SAPs that instantiate the logical interface with the purpose to exchange and use sensing-related information.

> *Note.* The term "service user" is used in the context of protocol and procedure specifications and denotes a conceptual functional role of an IEEE 1900.6 logical entity. It is the counterpart to a service provider.

message transport service: A communication service commonly associated with the functionality of an open systems interconnection (OSI) transport layer protocol that supports transaction based exchange of messages between communication peers. In this standard, the message transport service implements the exchange of sensing-related information between remote IEEE 1900.6 servers and clients across a platform-provided communication subsystem.

regulatory requirements: A category of sensing-related information derived from regulations, including definitions of the sensing range, accuracy requirements, granularity, required measurement bandwidth, periodicity, and detection sensitivity, as well as information about sensing permission and synchronization. Regulatory requirements are expressed in the form of sensing and sensing control parameters.

> *Note.* The term "regulatory information" is used synonymously in conjunction with the exchange of sensing-related information.

sensing: In the context of radio frequency spectrum, it refers to the act of measuring information indicative of spectrum occupancy (information may include frequency ranges, signal power levels, bandwidth, location information, etc.). Sensing may include determining how the sensed spectrum is used (cf. Clause 18.6).

sensing control information: Information describing the status and configuration of IEEE 1900.6 logical entities as well as information controlling and configuring the acquisition and processing of sensing information.

sensing information: Any information acquired by and obtained from sensors, including related spatiotemporal state information such as position, time, and confidence of acquisition.

sensor: A logical entity that performs sensing (see the definition of sensing in 18.3.1) within a radio system. Sensors may also act as clients of other Sensors.

> *Note.* For the remainder of the document, any entity described as "IEEE 1900.6 logical entity" denotes a logical entity compliant to this standard.

18.3.2 Acronyms and Abbreviations

A-SAP	application service access point
A/D	analog-to-digital
ADC	analog-to-digital conversion
AP	access point
BS	base station
C-SAP	communication service access point
CE	cognitive engine
CR	cognitive radio
D/A	digital-to-analog
DA	data archive
DAC	digital-to-analog conversion
DLC	data link control
DRRUO	distributed radio resource usage optimization
DSA	dynamic spectrum access
DSM	distributed sensing models
DSS	dynamic spectrum sharing
ISM	industrial, scientific, and medical
LTE	long -term evolution
M-SAP	measurement service access point
MAC	media access control
MCM	mobile node spectrum coordination model
NCM	network node spectrum coordination model
NRM	network reconfiguration manager
OSI	open systems interconnection

PHY	physical layer
PDA	personal digital assistant
PDU	protocol data unit
RAN	radio access network
RAT	radio access technology
RF	radio frequency
SAP	service access point
SeEM	sensing enhanced model
SEM	service enhanced model
SME	station management entity
SDU	service data unit
SPOLD	substitute for primary operation of longer duration
SPOSD	substitute for primary operation of shorter duration
TRM	terminal reconfiguration manager
ULME	upper layer management entity
UML	Unified Modeling Language
UTC	Coordinated Universal Time
WLAN	wireless local area network

18.4   System Model

The IEEE 1900.6 system model describes the relationship of logical entities and interfaces detailed by subsequent clauses of this standard. The description of the interaction of these entities across their interfaces shall be the purpose of the IEEE 1900.6 reference model (Clause 18.5).

The system model has been derived from the requirements implied by the use cases given in Annex 18A. It respects various spectrum usage models, distributed sensing models, and sensing topologies as elaborated further by Annex 18A. The scenarios given by this clause complement each other and shall be seen as applications of the IEEE 1900.6 system model, each of them satisfying a given purpose and a certain subset of requirements as detailed by Annex 18A.

The two main categories of spectrum usage models are a long-term spectrum usage model and a short-term spectrum usage model. The terms "long term" and "short term" are understood within the context of use cases as related to the observed period of activity of the licensed/primary spectrum user.

— Long-term spectrum usage model: Spectrum is used over a relative long period (e.g., emergency services; please refer to 18A.1.1.1).
— Short-term spectrum usage model: Spectrum is used over short periods (e.g., ad hoc licensee service; please refer to 18A.1.2.2).

The abstract, high-level system model presented in this subclause is not dependent on topology or the spectrum usage model. Based on this system model, the use cases detailed in Annex 18A have been classified within each usage as shown in 18.4.1.

18.4.1   Scenario 1: Single CE/DA and Single Sensor

In this scenario as shown in Figure 18.2, a single Sensor provides sensing-related information to one CE/DA. It is denoted as "1:1 scenario." The following instance of the IEEE 1900.6 logical interface is involved:

— CE/DA–S interface

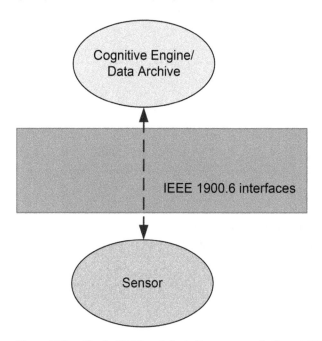

Figure 18.2 —Single CE/DA and single Sensor scenario. From IEEE Standard for Spectrum Sensing Interfaces and Data Structures for Dynamic Spectrum Access and Other Advanced Radio Communication Systems, Feb 2011. IEEE.

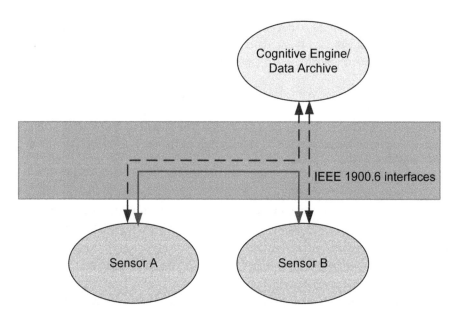

Figure 18.3 —Single CE/DA and multiple Sensors scenario. From IEEE Standard for Spectrum Sensing Interfaces and Data Structures for Dynamic Spectrum Access and Other Advanced Radio Communication Systems, Feb 2011. IEEE.

18.4.2   Scenario 2: Single CE/DA and Multiple Sensors

In this scenario as shown in Figure 18.3, multiple Sensors provide sensing-related information to a CE/DA. One CE/DA can access sensing-related information from multiple Sensors. It is denoted as "1:N scenario" where N is the number of Sensors.

The following instances of the IEEE 1900.6 logical interface are involved:

— CE/DA–S interface
— S–S interface

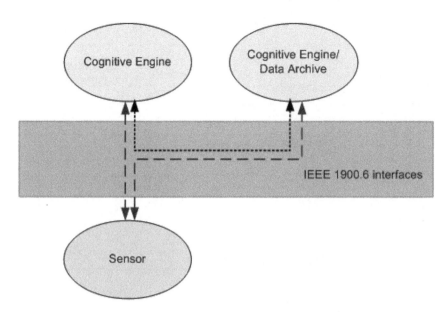

Figure 18.4 —Multiple CE/DA and single Sensor scenario. From IEEE Standard for Spectrum Sensing Interfaces and Data Structures for Dynamic Spectrum Access and Other Advanced Radio Communication Systems, Feb 2011. IEEE.

In distributed sensing, the CE/DA can access multiple Sensors and make dynamic spectrum access decisions based on sensing-related information from distributed Sensors. Also, multiple Sensors can exchange information and provide the CE/DA with an improved sensing result.

18.4.3 Scenario 3: Multiple CE/DA and Single Sensor

In this scenario as shown in Figure 18.4, multiple CE/DAs access sensing-related information from one Sensor. One Sensor provides sensing-related information to multiple CE/DAs. If some of the multiple CE/DAs are capable of accessing a Sensor, the CE/DA can exchange sensing-related information with other CE/DAs. This scenario is denoted as "M:1 scenario" where M is the number of CE/DAs.

The following instances of the IEEE 1900.6 logical interface are involved:

— CE/DA–S interface
— CE–CE/DA interface

Note that for the case of multiple CE/DAs and multiple Sensors, (i.e., the M:N scenario referred to in A.2.3), the system can be decomposed into a combination of the above three scenarios. This is understood within the context of model description. This does not imply decomposition at the implementation level.

18.5 The IEEE 1900.6 Reference Model

18.5.1 General Description

Figure 18.5 shows the reference model for the IEEE 1900.6 interface and defined entities (i.e., Sensor, CE, and DA). The three SAPs shown in Figure 18.5 are a realization of the IEEE 1900.6 interface. Sensors and their clients may have all of the three SAPs or a subset of the SAPs depending on the implementation. The IEEE 1900.6 logical entities may also follow reference model views as depicted in Figure 18.6 through Figure 18.8, which show the instantiation of the IEEE 1900.6 interface as CE/DA–S, CE–CE/DA, S–S given in Figure 18.1. They are utilizing distinct SAPs to realize the IEEE 1900.6 logical interface. A compliant realization of the interface shall be sufficient to comply with the standard. No assumption is made on the

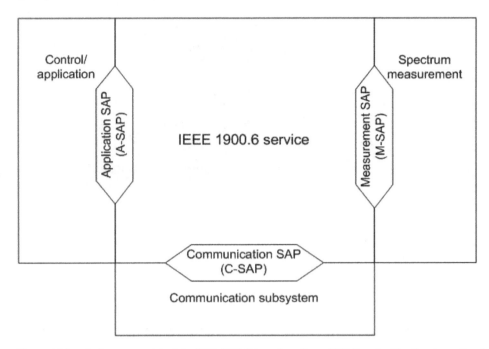

Figure 18.5 —Reference model of the IEEE 1900.6 interface. From IEEE Standard for Spectrum Sensing Interfaces and Data Structures for Dynamic Spectrum Access and Other Advanced Radio Communication Systems, Feb 2011. IEEE.

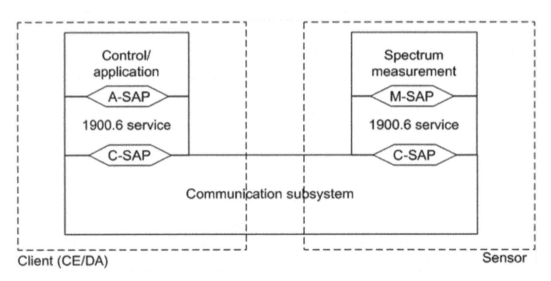

Figure 18.6 —View of reference model when the client of Sensor is a CE or DA. From IEEE Standard for Spectrum Sensing Interfaces and Data Structures for Dynamic Spectrum Access and Other Advanced Radio Communication Systems, Feb 2011. IEEE.

realization of other functions of the device or devices involved. The figures show how the previously mentioned SAPs are involved in the sensing-related information exchange between Sensors and their clients. Subclause 18.5.1 provides formal definitions for each of the entities identified in Figure 18.5 through Figure 18.8.

The IEEE 1900.6 service is responsible for realizing IEEE 1900.6 service primitives and for generating data structures defined by this standard for the exchange of sensing-related information between IEEE 1900.6 servers and clients.

An **IEEE 1900.6 service access point (SAP)** is a conceptual location where an IEEE 1900.6 logical entity or an IEEE 1900.6 service user, which is not an IEEE 1900.6 logical entity by itself, can request the services provided by another IEEE 1900.6 logical entity or by an IEEE 1900.6

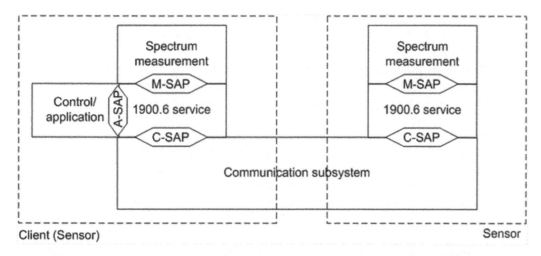

Figure 18.7 —View of reference model when the client of Sensor is a Sensor. From IEEE Standard for Spectrum Sensing Interfaces and Data Structures for Dynamic Spectrum Access and Other Advanced Radio Communication Systems, Feb 2011. IEEE.

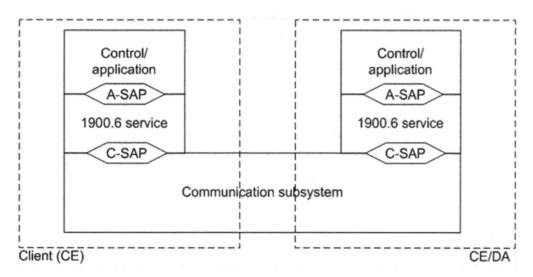

Figure 18.8 —View of reference model for sensing-related information exchange between CE and CE/DA. From IEEE Standard for Spectrum Sensing Interfaces and Data Structures for Dynamic Spectrum Access and Other Advanced Radio Communication Systems, Feb 2011. IEEE.

compliant station service (see 18.3.1). For example, an IEEE 1900.6 logical entity "Sensor" can utilize physical radio frequency (RF) sensor capabilities provided by the station (e.g., by the terminal radio hardware and firmware) and can provide RF measurement data to an IEEE 1900.6 logical entity CE upon request. The term "IEEE 1900.6 compliant service" here refers to the fact that primitives, parameters, and data structures defined throughout this standard are understood by this service and that measurement data returned are formatted accordingly.

The application service access point (A-SAP) is used by applications to access services of an IEEE 1900.6 compliant server. It defines a set of generic primitives and data structures to control the IEEE 1900.6 logical entity and/or to obtain the sensing results for application purposes. The IEEE 1900.6 client utilizing the A-SAP takes the role of an information consumer and of a control application.

The A-SAP is used by an application (i.e., the user of the A-SAP) to utilize sensing-related information for its purpose (e.g., for policy investigation and analysis of spectrum usage). The A-SAP may provide functions to set up a configuration of IEEE 1900.6 logical entities (e.g., Sensors and CE), to configure these for collaborative sensing, to start the data acquisition and processing (e.g., policy processing), and to obtain the results of the processing in order to configure the RF interface accordingly. The A-SAP shall be instantiated by the IEEE 1900.6 logical entity.

The measurement service access point (M-SAP) is used for the exchange of sensing-related information between the spectrum measurement module and its M-SAP user. The user of the measurement module is realized by dedicated functions of the IEEE 1900.6 service. It is used by IEEE 1900.6 logical entities to access IEEE 1900.6 compliant services provided by the station's hardware and/or firmware to control the spectrum measurement module (such as a collocated physical spectrum measurement module, i.e., ADC/DAC, filtering, signal conditioning, etc.) and to acquire spectrum measurement data. For example, a station (terminal) utilizes its RF interface during idle times for spectrum measurement and provides RF spectrum data to collocated IEEE 1900.6 "sensor" entities that are registered at the local M-SAP. The M-SAP shall be instantiated by the station's IEEE 1900.6 compliant measurement function.

The communication service access point (C-SAP) exchanges sensing-related information (sensing information, sensor information, control information, and requirements derived from regulation) between remote IEEE 1900.6 servers and their clients. It abstracts communication mechanisms for use by IEEE 1900.6 service providers and service users through a dedicated set of generic service primitives. The client role can be taken by a Sensor, CE, or DA. It abstracts communication mechanisms for use by IEEE 1900.6 services through defining a set of generic primitives and mapping these primitives to transport protocols.

An IEEE 1900.6 compliant message transport service shall be provided by the station in order to locate a remote IEEE 1900.6 peer entity and to establish a communication link with this entity. Message exchange then takes place as defined by this standard and it is the responsibility of the transport service to map these message transfers to a suitable transport, network, or link layer communication. For example, an IEEE 1900.6 logical entity CE can take the role of a client to a remote IEEE 1900.6 Sensor by sending a request to configure the sensor for delivering measurement data to the requestor utilizing its C-SAP functionality. The C-SAP shall be instantiated by the IEEE 1900.6 compliant message transport service.

Control/application in the reference model refers to a hardware/software module that utilizes functions of the A-SAP to obtain sensing information or other information or to control and configure the behavior and functions of other IEEE 1900.6 logical entities and functions.

Communication subsystem in the reference model refers to a hardware/software module that provides the levels of the communications protocol stack and other communication services required by the IEEE 1900.6 services. The communication subsystem can be implemented with the IEEE 1900.6 compliant message transport service.

Spectrum measurement in the reference model is realized by the spectrum measurement module, a hardware/software module that provides radio spectrum measurement functions (e.g., analog-to-digital conversion [ADC]/digital-to-analog conversion [DAC], filtering, and signal conditioning).

18.5.2 An Implementation Example of the IEEE 1900.6 Reference Model

This subclause presents an informative example of realizing the IEEE 1900.6 reference model presented in 18.5.1 based on the reference model for an IEEE 802.11 station (see IEEE Std 802.11™-2007 [B3]). Figure 18.9 shows a case where the IEEE 1900.6 service and corresponding SAPs are implemented by the station management entity (SME). The IEEE 1900.6 application (control/application) is implemented on the application layer and the sensor (spectrum measurement) is implemented on the physical layer (PHY). The IEEE 1900.6 SAPs may be provided through different sublayer management entities and their related SAPs. This figure applies to instantiation of both (IEEE 1900.6) client and server.

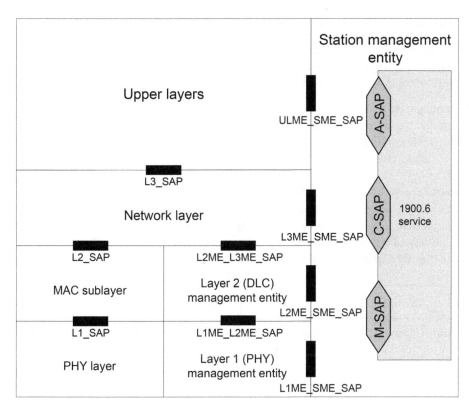

Figure 18.9 —An implementation example of the IEEE 1900.6 service. From IEEE Standard for Spectrum Sensing Interfaces and Data Structures for Dynamic Spectrum Access and Other Advanced Radio Communication Systems, Feb 2011. IEEE.

18.5.3 Service Access Points

This standard defines the following SAPs that are involved in the exchange of sensing-related information between Sensors and their clients. Abstract service primitives describe the interaction with the IEEE 1900.6 services through SAPs in an implementation independent way.

> **Note.** This standard does not distinguish between the SAPs at the local and remote entities. Hence, interaction at the SAP is only described using request and response primitives. If for implementation purposes, distinguishing of local and remote entities is necessary, the additional primitives, indication, and confirmation have the same semantics of the request and response primitives, correspondingly.

A number of parameters to primitives used throughout this clause are specified but not qualified nor detailed further in their effect on the behavior of a potential implementation. This has been done intentionally to remain neutral with respect to specific technologies and the demand for specific functions and algorithms of entities and services considered. Detailed descriptions of parameters and data representation are given in 18.6.3 and 18.6.4, respectively.

This standard specifies the information flow between sensors and their clients. The specification of a related message exchange protocol is left to future versions. Subclause 18.6.1 addresses the information exchange by giving the structure of the commands and parameters in that flow. The information structure given in this subclause defines how the primitives can be implemented for the information exchange. The current standard does not cover addressing of logical entities. It is assumed that a platform service exists that performs the mapping between the IDs of logical entities and proper device or network addresses.

18.5.3.1 Measurement Service Access Point

The M-SAP provides a set of generic primitives or method for IEEE 1900.6 SAP users to control spectrum sensing and to obtain sensing-related information from spectrum sensing.

The IEEE 1900.6 logical entities obtain the following services from the M-SAP:

— Measurement capabilities discovery services
— Measurement configuration discovery services
— Measurement configuration services
— Information services

Measurement Capabilities Discovery Services The measurement capabilities discovery services provide a set of primitives or method through which the IEEE 1900.6 SAP user obtains the information related to the measurement capabilities of the associated measurement module.

The primitives of the measurement capabilities discovery services defined as part of the M-SAP are described in the following table.

Primitives	Description
Get_Supported_Spectrum_Measurement_Description	This primitive is used by the IEEE 1900.6 SAP user to obtain the information related to the capability of the spectrum measurement module.

Get_Supported_Spectrum_Measurement_Description.request

Function This primitive is used by the IEEE 1900.6 SAP user to obtain the supported measurement of the measurement module.

Semantics of the Service Primitive

Get_Supported_Spectrum_Measurement_Description.request

(

MeasurementCapabilityTransID

)

Parameters

Name	Type	Description
MeasurementCapabilityTransID	Unsigned integer	MeasurementCapabilityTransID uniquely identifies one transaction of requesting a measurement feature (type and range) within the capability of the measurement module.

When Used This primitive is used by the IEEE 1900.6 SAP user when it needs to discover the spectrum measurement capabilities of the entity.

Effect of Receipt The spectrum measurement module subsequently uses a Get_Supported_ Spectrum_Measurement_Description.reponse to reflect the results of the request.

Get_Supported_Spectrum_Measurement_Description.response

Function This primitive returns the results of the request to obtain the description of spectrum measurement capabilities.

Semantics of the Service Primitive

Get_Supported_Spectrum_Measurement_Description.response

    (

    MeasurementCapabilityTransID,

    Status,

    MeasuRange,

    SensingMode,

    DataSheet.ADDAResolution,

    DataSheet.AngleResolution,

    DataSheet.FrequencyResolution,

    DataSheet.LocationTimeCapability,

    DataSheet.LoggingFunctions,

    DataSheet.RecordingCapability,

    DataSheet.SweepTime,

    LockStatus

)

Parameters

Name	Type	Description
MeasurementCapabilityTransID	Unsigned integer	MeasurementCapabilityTransID uniquely identifies one transaction of requesting a measurement feature (type and range) within the capability of the measurement module.
Status	Enumeration	Status of operation. 0: success 1: unspecified failure 2: rejection 3: authorization failure
MeasuRange	Bandwidth	The MeasuRange parameter indicates the range of frequency that can be measured according to the manufacturer specification (cf. 18.6.3.27).
SensingMode	Enumeration	The SensingMode parameter indicates the numbers and types of standard sensing methods supported by the sensor (cf. 18.6.3.28).
DataSheet.ADDAResolution	Unsigned integer	The resolution of AD/DA convertor of the spectrum measurement module (cf. 18.6.3.32).
DataSheet.AngleResolution	Angle	The angle resolution of the spectrum measurement module (cf. 18.6.3.32).
DataSheet.FrequencyResolution	Frequency	Frequency resolution of the spectrum measurement module (cf. 18.6.3.32).
DataSheet.LocationTimeCapability	String	The capability of spectrum module in determining the location and time of the measurement (cf. 18.6.3.32).
DataSheet.LoggingFunctions	Unsigned integer	The ability of access log files (cf. 18.6.3.32).
DataSheet.RecordingCapability	String	The capability to recording data at the spectrum measurement module (cf. 18.6.3.32).
DataSheet.SweepTime	Microsecond	The time to sweep between two frequencies (cf. 18.6.3.32).

Name	Type	Description
LockStatus	Enumeration	The LockStatus indicates whether the entity (i.e., the sender of this primitive) has been locked by a client for exclusive service.
		1: The status of the entity is locked. The entity is locked by a client for exclusive service.
		0: The status of the entity is unlocked. The entity has been unlocked by its client, or has been unlocked by a self-generated timeout event.

When Used   The primitive is used in response to the Get_Supported_Spectrum_Measurement_ Description .request primitive.

Effect of Receipt   Upon receipt, the IEEE 1900.6 SAP user obtains the information related to the spectrum measurement capabilities of the spectrum measurement module.

Measurement Configuration Discovery Services   The measurement configuration discovery services provide a set of primitives or method through which the IEEE 1900.6 SAP user obtains the information related to the measurement configuration of the spectrum measurement module. The following information is discovered through these services:

— Sensor PHY profile
— Sensor antenna profile
— Sensor location

The primitives of the measurement configuration services defined as part of the M-SAP are described in the following table.

Primitives	Description
	Sensor PHY profile
Get_Sensor_PHY _Description	This primitive is used by the IEEE 1900.6 SAP user to obtain the information related to physical configuration of the spectrum measurement module.
	Sensor antenna profile
Get_Sensor_Antenna _Description	This primitive is used by the IEEE 1900.6 SAP user to obtain the information related to the antenna configuration of the spectrum measurement module.
	Sensor location
Get_Sensor_Location _Description	This primitive is used by the IEEE 1900.6 SAP user to obtain the information related to the location of the spectrum measurement module. Note that the spectrum measurement module represents the location of the measurement.

Get_Sensor_PHY_Description.request

Function   This primitive is used by the IEEE 1900.6 SAP user to obtain the PHY description of the spectrum measurement module.

Semantics of the Service Primitive

Get_Sensor_PHY_Description.request

$$($$
SensorPHYProfileID
$$)$$

Parameters

Name	Type	Description
SensorPHYProfileID	Unsigned integer	SensorPHYProfileID uniquely defines the PHY profile of the spectrum measurement module.

When Used   This primitive is used by the IEEE 1900.6 SAP user when it needs to discover the PHY configuration of the spectrum measurement module.

Effect of Receipt   The spectrum measurement module subsequently uses a Get_Sensor_PHY_ Description. response to reflect the results of the request.

Get_Sensor_PHY_Description.response

Function   This primitive returns the results of the request to obtain the description of the PHY profile of the spectrum measurement module.

Semantics of the Service Primitive

Get_Sensor_PHY_Description.response

$$($$
SensorPHYProfileID, Status,
Datasheet.CalibrationData,
Datasheet.CalibrationMethod,
Datasheet.ChannelFiltering,
Datasheet.DynamicRange,
Datasheet.NoiseFactor,
Datasheet.PhaseNoise,
LockStatus
$$)$$

Parameters

Name	Type	Description
SensorPHYProfileID	Unsigned integer	SensorPHYProfileID uniquely defines the PHY profile of the spectrum measurement module.
Status	Enumeration	Status of operation. 0: success 1: unspecified failure 2: rejection 3: authorization failure
Datasheet.CalibrationData	String	Data used to calibrate the spectrum measurement module (cf. 18.6.3.32).
Datasheet.CalibrationMethod	String	Method to calibrate the spectrum measurement module. (cf. 18.6.3.32).

Name	Type	Description
Datasheet.ChannelFiltering	String	The filtering function of the spectrum measurement module (cf. 18.6.3.32).
Datasheet.DynamicRange	Float	Dynamic range in dB of signal detection of the spectrum measurement module (cf. 18.6.3.32).
Datasheet.NoiseFactor	Float	The ratio in linear scale of the noise produced by the spectrum measurement module to the thermal noise (cf. 18.6.3.32).
Datasheet.PhaseNoise	Float	Phase noise in dBc/Hz produced by the spectrum measurement module (cf. 18.6.3.32).
LockStatus	Enumeration	The LockStatus indicates whether the entity (i.e., the sender of this primitive) has been locked by a client for exclusive service.
		1: The status of the entity is locked. The entity is locked by a client for exclusive service.
		0: The status of the entity is unlocked. The entity has been unlocked by its client, or has been unlocked by a self-generated timeout event.

When Used The primitive is used in response to the Get_Sensor_PHY_Description.request primitive.

Effect of Receipt Upon receipt, the IEEE 1900.6 SAP user obtains the information related to the PHY profile of the spectrum measurement module.

Get_Sensor_Antenna_Description.request

Function This primitive is used by the IEEE 1900.6 SAP user to obtain the antenna description of the spectrum measurement module.

Semantics of the Service Primitive

Get_Sensor_Antenna_Description.request

(

SensorAntennaProfileID

)

Parameters

Name	Type	Description
SensorAntennaProfileID	Unsigned integer	SensorAntennaProfileID uniquely defines the antenna profile of the spectrum measurement module.

When Used This primitive is used by the IEEE 1900.6 SAP user when it needs to discover the antenna configuration of the spectrum measurement module.

Effect of Receipt The spectrum measurement module subsequently uses a Get_Sensor_Antenna_Description. response to reflect the results of the request.

Get_Sensor_Antenna_Description.response

Function This primitive returns the results of the request to obtain the description of the antenna profile of the spectrum measurement module.

Semantics of the Service Primitive

Get_Sensor_Antenna_Description.response

(

SensorAntennaProfileID, Status,

Datasheet.AntennaBandwidth,

Datasheet.AntennaBeamPointing,

Datasheet.AntennaBeamwidth,

Datasheet.AntennaDirectivityGain,

Datasheet.AntennaGain,

Datasheet.AntennaHeight,

Datasheet.AntennaPolarization,

LockStatus

)

Parameters

Name	Type	Description
SensorAntennaProfileID	Unsigned integer	SensorAntennaProfileID uniquely defines the antenna profile of the spectrum measurement module.
Status	Enumeration	Status of operation. 0: success 1: unspecified failure 2: rejection 3: authorization failure
Datasheet.AntennaBandwidth	Array(Frequency)	Bandwidth of the antenna used at the spectrum measurement module (cf. 18.6.3.32).
Datasheet.AntennaBeamPointing	Array(Angle)	The DataSeet.AntennaBeamPointing parameter specifies the beam pointing direction of the antenna used at the spectrum measurement module by giving the azimuthal angle with respect to North and elevation angle with respect to the horizon (cf. 18.6.3.32).
Datasheet.AntennaBeamwidth	Array(Angle)	Beamwidth of the antenna used at the spectrum measurement module, normally specified as half-power horizontal and vertical beamwidth (cf. 18.6.3.32).
Datasheet.AntennaDirectivityGain	Float	Directivity gain in dBi of the antenna radiation pattern at the spectrum measurement module (cf. 18.6.3.32).
Datasheet.AntennaGain	Float	Power gain in dB of the antenna used at the spectrum measurement module (cf. 18.6.3.32).
Datasheet.AntennaHeight	Float	Height of the antenna in meters with respect to sea level (cf. 18.6.3.32).

Name	Type	Description
Datasheet.AntennaPolarization	Enumeration	Polarization of the antenna used at the spectrum measurement module (cf. 18.6.3.32). 0: Linear polarization 1: Circular polarization 2: Elliptical polarization
LockStatus	Enumeration	The LockStatus indicates whether the entity (i.e., the sender of this primitive) has been locked by a client for exclusive service. 1: The status of the entity is locked. The entity is locked by a client for exclusive service. 0: The status of the entity is unlocked. The entity has been unlocked by its client, or has been unlocked by a self-generated timeout event.

When Used   The primitive is used in response to the Get_Sensor_Antenna_Description.request primitive.

Effect of Receipt   Upon receipt, the IEEE 1900.6 SAP user obtains the information related to the antenna profile of the spectrum measurement module.

Get_Sensor_Location_Description.request

Function   This primitive is used by the IEEE 1900.6 SAP user to obtain the location information of the spectrum measurement module.

Semantics of the Service Primitive

Get_Sensor_Location_Description.request

$$(
$$

SensorLocationTransID

$$)
$$

Parameters

Name	Type	Description
SensorLocationTransID	Unsigned integer	SensorLocationTransID uniquely defines one transaction of a requesting the location of the spectrum measurement module.

When Used   This primitive is used by the IEEE 1900.6 SAP user when it needs to discover the location of the spectrum measurement module.

Effect of Receipt   The spectrum measurement module subsequently uses a Get_Sensor_Location.response to reflect the results of the request.

Get_Sensor_Location_Description.response

Function   This primitive returns the results of the request to obtain the location information of the spectrum measurement module.

Semantics of the Service Primitive

Get_Sensor_Location_Description.response

       (

       SensorLocationTransID,

       Status,

       AbsSensorLocation,

       RelSensorLocation,

       LockStatus

)

Parameters

Name	Type	Description
SensorLocationTransID	Unsigned integer	SensorLocationTransID uniquely defines one transaction of a requesting the location of the spectrum measurement module.
Status	Enumeration	Status of operation. 0: success 1: unspecified failure 2: rejection 3: authorization failure
AbsSensorLocation	RGeolocation	The absolute location of the spectrum measurement module (cf. 18.6.3.25).
RelSensorLocation	RGeolocation	The RelSensorLocation parameter is used when the position of the sensor is measured by triangulation with respect to known reference positions (cf. 18.6.3.26).
LockStatus	Enumeration	The LockStatus indicates whether the entity (i.e., the sender of this primitive) has been locked by a client for exclusive service. 1: The status of the entity is locked. The entity is locked by a client for exclusive service. 0: The status of the entity is unlocked. The entity has been unlocked by its client, or has been unlocked by a self-generated timeout event.

When Used   The primitive is used in response to the Get_Sensor_Location_Description.request primitive.

Effect of Receipt   Upon receipt, the IEEE 1900.6 SAP user obtains the location information of the spectrum measurement module. The sensor location can be expressed by either AbsSensorLocation or RelSensorLocation. If RelLocation is used, the reference positions are assumed to be known.

Measurement Configuration Services   The measurement configuration services provide a set of primitives or method through which the IEEE 1900.6 SAP user configures the spectrum measurement module. The following configuration is set through these services:

— Measurement objective
— Measurement profile
— Measurement performance

The primitives of the measurement configuration services are defined as part of the M-SAP and are described in the following table.

Primitives	Description
	Measurement objective
Set_Sensor_Measurement_Obj	This primitive is used by the IEEE 1900.6 SAP user to set the measurement objective of the spectrum measurement module.
	Measurement profile
Set_Sensor_Measurement_Profile	This primitive is used by the IEEE 1900.6 SAP user to set the measurement profile of the spectrum measurement module.
	Measurement performance
Set_Sensor_Measurement_Performance	This primitive is used by the IEEE 1900.6 SAP user to set the target performance of the spectrum measurement module.

Set_Sensor_Measurement_Obj.request

Function This primitive is used by the IEEE 1900.6 SAP user to set the measurement objective of the spectrum measurement module.

Semantics of the Service Primitive

Set_Sensor_Measurement_Obj.request

$$($$
$$SensorMeasurementObjID,$$
$$StartTime,$$
$$EndTime,$$
$$Bandwidth$$
$$)$$

Parameters

Name	Type	Description
SensorMeasurementObjID	Unsigned integer	SensorMeasurementObjID uniquely defines the measurement objective of the spectrum measurement module.
StartTime	TimeStamp	The StartTime parameter specifies the time stamp when spectrum measurement starts (cf. 18.6.3.8).
EndTime	TimeStamp	The EndTime parameter specifies the time stamp when spectrum measurement finishes (cf. 18.6.3.8).
Bandwidth	Array(Frequency)	Bandwidth parameter specifies the bandwidth, by giving the start frequency and stop frequency, to perform spectrum measurement (cf. 18.6.3.11).

When Used This primitive is used by the IEEE 1900.6 SAP user to set the measurement objective of the spectrum measurement module.

Effect of Receipt Parameters of the spectrum measurement module are subsequently set to the values given by the primitive. The spectrum measurement module subsequently uses a Set_Sensor_ Measurement_Obj.response primitive to reflect the results of the request.

Set_Sensor_Measurement_Obj.response

Function This primitive returns the results of the request to set the measurement objective of the spectrum measurement module.

Semantics of the Service Primitive

Set_Sensor_Measurement_Obj.response

(

SensorMeasurementObjID,

Status,

LockStatus

)

Parameters

Name	Type	Description
SensorMeasurementObjID	Unsigned integer	SensorMeasurementObjID uniquely defines the measurement objective of the spectrum measurement module.
Status	Enumeration	Status of operation. 0: success 1: unspecified failure 2: rejection 3: authorization failure
LockStatus	Enumeration	The LockStatus indicates whether the entity (i.e., the sender of this primitive) has been locked by a client for exclusive service. 1: The status of the entity is locked. The entity is locked by a client for exclusive service. 0: The status of the entity is locked. The entity has been unlocked by its client, or has been unlocked by a self-generated timeout event.

When Used The primitive is used in response to the Set_Sensor_Measurement_Obj.request primitive.

Effect of Receipt Upon receipt, the spectrum measurement module sets the measurement objective.

Set_Sensor_Measurement_Profile.request

Function This primitive is used by the IEEE 1900.6 SAP user to set the measurement profile of the spectrum measurement module.

Semantics of the Service Primitive

Set_Sensor_Measurement_Profile.request

(

SensorMeasurementProfileID,

SensingMode,

ChOrder,

ChList,

ReportRate,

Scan.LowerThreshold, ReportingMode

)

Parameters

Name	Type	Description
SensorMeasurementProfileID	Unsigned integer	SensorMeasurementProfileID uniquely defines the measurement profile of the spectrum measurement module.
SensingMode	Enumeration	The SensingMode parameter indicates the numbers and types of standard sensing methods supported by the sensor (cf. 18.6.3.28).
ChOrder	Vector(Unsigned integer)	The ChOrder specifies the scanning order of the channels. The channel order is represented by a predefined vector of step indices (sequence vector should use a bounded type rather than the generic Vector type). The default for this parameter is "sequential scanning in ascending order" (cf. 18.6.3.13).
ChList	Structured	The ChList specifies a list of frequency bands by providing a vector of channel numbers along with a corresponding vector of bandwidth parameters (cf. 18.6.3.32).
ReportRate	Structured	The ReportRate parameter is issued by the IEEE 1900.6 client to specify how often the measurement result should be reported or updated (e.g.,every 2 s or every 100 ms) (cf. 18.6.3.14).
Scan.LowerThreshold	NoisePower	The Scan.LowerThreshold parameter specifies the noise floor (cf. 18.6.3.24).
ReportingMode	Enumeration	The ReportMode parameter is issued by the IEEE 1900.6 client to indicate in what form the sensing-related information should be reported (cf. 18.6.3.15).

When Used   This primitive is used by the IEEE 1900.6 SAP user to set the measurement profile of the spectrum measurement module.

Effect of Receipt   Parameters of the spectrum measurement module are subsequently set to the values given by the primitive. The spectrum measurement module subsequently uses a Set_Sensor_Measurement_Profile.response primitive to reflect the results of the request.

Set_Sensor_Measurement_Profile.response

Function   This primitive returns the results of the request to set the measurement profile of the spectrum measurement module.

Semantics of the Service Primitive

Set_Sensor_Measurement_Profile.response

                    (
                    SensorMeasurementProfileID,
                    Status,
                    LockStatus
)

Parameters

Name	Type	Description
SensorMeasurementProfileID	Unsigned integer	SensorMeasurementProfileID uniquely defines the measurement profile of the spectrum measurement module.
Status	Enumeration	Status of operation.
		0: success
		1: unspecified failure
		2: rejection
		3: authorization failure
LockStatus	Enumeration	The LockStatus indicates whether the entity (i.e., the sender of this primitive) has been locked by a client for exclusive service.
		1: The status of the entity is locked. The entity is locked by a client for exclusive service.
		0: The status of the entity is unlocked. The entity has been unlocked by its client, or has been unlocked by a self-generated timeout event.

When Used   The primitive is used in response to the Set_Sensor_Measurement_Profile.request primitive.

Effect of Receipt   Upon receipt, the IEEE 1900.6 SAP user sets the measurement profile of the spectrum measurement module.

Set_Sensor_Measurement_Performance.request

Function   This primitive is used by the IEEE 1900.6 SAP user to set the measurement performance of the spectrum measurement module.

Semantics of the Service Primitive

Set_Sensor_Measurement_Performance.request

(
SensorMeasurementPerformanceID,
PerfMetric.Pd,
PerfMetric.Pf
)

Parameters

Name	Type	Description
SensorMeasurementPerformanceID	Unsigned integer	SensorMeasurementPerformanceID uniquely defines the measurement performance of the spectrum measurement module.
PerfMetric.Pd	Unsigned integer	The PerfMetric.Pd specifies the rate of detection that the spectrum measurement should achieve (cf. 18.6.3.16).
PerfMetric.Pf	Unsigned integer	The PerfMetric.Pf specifies the rate of false alarm that the spectrum measurement should achieve (cf. 18.6.3.16).

When Used This primitive is used by the IEEE 1900.6 SAP user to set the measurement performance of the spectrum measurement module.

Effect of Receipt Parameters of the spectrum measurement module are subsequently set to the values given by the primitive. The spectrum measurement module subsequently uses a Set_Sensor_ Measurement_Performance.response to reflect the results of the request.

Set_Sensor_Measurement_Performance.response

Function This primitive returns the results of the request to set the measurement performance of the spectrum measurement module.

Semantics of the Service Primitive

Set_Sensor_Measurement_Performance.response

(

SensorMeasurementPerformanceID,

Status,

LockStatus

)

Parameters

Name	Type	Description
SensorMeasurementPerformanceID	Unsigned integer	SensorMeasurementPerformanceID uniquely defines the measurement performance of the spectrum measurement module.
Status	Enumeration	Status of operation. 0: success 1: unspecified failure 2: rejection 3: authorization failure 4: function not supported
LockStatus	Enumeration	The LockStatus indicates whether the entity (i.e., the sender of this primitive) has been locked by a client for exclusive service. 1: The status of the entity is locked. The entity is locked by a client for exclusive service. 0: The status of the entity is unlocked. The entity has been unlocked by its client, or has been unlocked by a self-generated timeout event.

When Used The primitive is used in response to the Set_Sensor_Measurement_Performance .request primitive.

Effect of Receipt Upon receipt, the spectrum measurement module sets the measurement performance.

Information Services The information services provide a set of primitives or methods through which the IEEE 1900.6 SAP user collects the information related to the measurement configuration of the spectrum measurement module. The following information is collected through these services:

— Sensor profile
— Measurement profile
— Signal measurement
— Channel measurement
— Radio access technology (RAT) measurement
— Management

The primitives of information services are defined as part of the M-SAP in the following table.

Primitives	Description
Sensor profile collection	
Get_Sensor_Manufacturer_Profile	This primitive is used by the IEEE 1900.6 SAP user to collect the information related to the manufacturer of the spectrum measurement module.
Get_Sensor_Power_Profile	This primitive is used by the IEEE 1900.6 SAP user to collect the information related to the power consumption of the spectrum measurement module.
Measurement profile collection	
Get_Measurement_Profile	This primitive is used by the IEEE 1900.6 SAP user to obtain the information related to the spectrum measurement that has been carried out.
Get_Measurement_Location_Information	This primitive is used by the IEEE 1900.6 SAP user to collect the information related to the location where the spectrum measurement has been carried out.
Signal measurement collection	
Get_Signal_Measurement_Value	This primitive is used by the IEEE 1900.6 SAP user to collect the measured value related to the measured signals.
Channel measurement collection	
Get_Channel_Measurement_Value	This primitive is used by the IEEE 1900.6 SAP user to collect the measured value related to the measured channels.
RAT measurement collection	
Get_RAT_ID_Value	This primitive is used by the IEEE 1900.6 SAP user to collect the measured value related to the measured RAT.
Management	
Notify	This primitive is used by the measurement module to report its status to IEEE 1900.6 SAP user.

Get_Sensor_ Manufacturer_Profile.request

Function This primitive is used by the IEEE 1900.6 SAP user to obtain the manufacturer information of the spectrum measurement module.

Semantics of the Service Primitive

Get_Sensor_Manufacturer_Profile.request

(

SensorManufacturerProfileID,

)

Parameters

Name	Type	Description
SensorManufacturerProfileID	Unsigned integer	SensorManufacturerProfileID uniquely defines the manufacturer profile of the spectrum measurement module.

When Used This primitive is used by the IEEE 1900.6 SAP user when it needs to collect the manufacturer profile of the spectrum measurement module.

Effect of Receipt The spectrum measurement module subsequently uses a Get_Sensor_Manufacturer_Profile.response primitive to reflect the results of the request.

Get_Sensor_Manufacturer_Profile.response

Function This primitive returns the results of the request to collect the manufacturer profile of the spectrum measurement module.

Semantics of the Service Primitive

Get_Sensor_Manufacturer_Profile.response

    (

    SensorManufacturerProfileID,

    Status,

    SensorID.VendorID,

    SensorID.ProductID,

    LockStatus

)

Parameters

Name	Type	Description
SensorManufacturerProfileID	Unsigned integer	SensorManufacturerProfileID uniquely defines the manufacturer profile of the spectrum measurement module.
Status	Enumeration	Status of operation. 0: success 1: unspecified failure 2: rejection 3: authorization failure
SensorID.VendorID	String	The SensorID.VendorID uniquely identifies the manufacturer of the sensor (cf. 18.6.3.29).
SensorID.ProductID	String	The SensorID.ProductID uniquely identifies the product of a manufacturer (cf. 18.6.3.29).
LockStatus	Enumeration	The LockStatus indicates whether the entity (i.e., the sender of this primitive) has been locked by a client for exclusive service. 1: The status of the entity is locked. The entity is locked by a client for exclusive service. 0: The status of the entity is unlocked. The entity has been unlocked by its client, or has been unlocked by a self-generated timeout event.

When Used   The primitive is used in response to the Get_Sensor_Manufacturer_Profile.request primitive.

Effect of Receipt   Upon receipt, the IEEE 1900.6 SAP user collects the manufacturer profile of the spectrum measurement module.

Get_Sensor_ Power_Profile.request

Function   This primitive is used by the IEEE 1900.6 SAP user to obtain the power profile of the spectrum measurement module.

Semantics of the Service Primitive

Get_Sensor_Power_Profile.request

> (
> SensorPowerProfileID,
>)

Parameters

Name	Type	Description
SensorPowerProfileID	Unsigned integer	SensorPowerProfileID uniquely defines the power profile of the spectrum measurement module.

When Used   This primitive is used by the IEEE 1900.6 SAP user when it needs to collect the power profile of the spectrum measurement module.

Effect of Receipt   The spectrum measurement module subsequently uses a Get_Sensor_Power_Profile.response primitive to reflect the results of the request.

Get_Sensor_Power_Profile.response

Function   This primitive returns the results of the request to collect the power profile of the spectrum measurement module.

Semantics of the Service Primitive

Get_Sensor_Power_Profile.response

> (
> SensorPowerProfileID,
> Status,
> BatteryStatus,
> Datasheet.PowerConsumption,
> LockStatus
>)

Parameters

Name	Type	Description
SensorPowerProfileID	Unsigned integer	SensorPowerProfileID uniquely defines the power profile of the spectrum measurement module.
Status	Enumeration	Status of operation. 0: success 1: unspecified failure 2: rejection 3: authorization failure

Name	Type	Description
BatteryStatus	Unsigned integer	The BatteryStatus parameter indicates the remaining battery power in percentage (cf. 18.6.3.31).
Datasheet.PowerConsumption	String	The Datasheet.PowerConsumption parameter indicates the power consumption profile of the spectrum measurement module (cf. 18.6.3.32).
LockStatus	Enumeration	The LockStatus indicates whether the entity (i.e., the sender of this primitive) has been locked by a client for exclusive service.
		1: The status of the entity is locked. The entity is locked by a client for exclusive service. 0: The status of the entity is unlocked. The entity is has been unlocked by its client, or has been unlocked by a self-generated timeout event.

When Used   The primitive is used in response to the Get_Sensor_Power_Profile.request primitive.

Effect of Receipt   Upon receipt, the IEEE 1900.6 SAP user collects the power profile of the spectrum measurement module.

Get_Sensor_Measurement_Profile.request

Function   This primitive is used by the IEEE 1900.6 SAP user to obtain the measurement profile of the spectrum measurement module.

Semantics of the Service Primitive

Get_Sensor_Measurement_Profile.request

$$($$

SensorMeasurementProfileID,

$$)$$

Parameters

Name	Type	Description
SensorMeasurementProfileID	Unsigned integer	SensorMeasurementProfileID uniquely defines the profile of spectrum measurement that has been carried out by the spectrum measurement module.

When Used   This primitive is used by the IEEE 1900.6 SAP user when it needs to collect the measurement profile of the spectrum measurement module.

Effect of Receipt   The spectrum measurement module subsequently uses a Get_Sensor_ Measurement _Profile.response to reflect the results of the request.

Get_Sensor_Measurement_Profile.response

Function   This primitive returns the results of the request to collect the measurement profile of the spectrum measurement module.

Semantics of the Service Primitive

Get_Sensor_Measurement_Profile.response

> (
> SensorMeasurementProfileID,
> Status,
> ConfidenceLevel,
> ReportMode,
> SensingMode,
> ReportRate,
> TimeStamp,
> Scan.LowerThreshold,
> MeasuBandwidth,
> LockStatus
>)

Parameters

Name	Type	Description
SensorMeasurementProfileID	Unsigned integer	SensorMeasurementProfileID uniquely defines the profile of spectrum measurement that has been carried out by the spectrum measurement module.
Status	Enumeration	Status of operation. 0: success 1: unspecified failure 2: rejection 3: authorization failure
ConfidenceLevel	Structured	ConfidenceLevel indicates the degree of certainty of estimated value (cf. 18.6.3.33).
ReportMode	Enumeration	The ReportMode indicates either hard information or soft information (cf. 18.6.3.15).
SensingMode	Enumeration	The SensingMode parameter indicates what sensing method is used to carry out the spectrum measurement (cf. 18.6.3.28).
ReportRate	Structured	The ReportRate parameter is issued by the IEEE 1900.6 client to specify how often the measurement result should be reported or updated (e.g., every 2 s or every 100 ms) (cf. 18.6.3.14).
TimeStamp	Structured	This parameter indicates the time that the spectrum measurement has been carried out (cf. 18.6.3.8).
Scan.LowerThreshold	NoisePower	The Scan.LowerThreshold parameter indicates the noise threshold used in the spectrum experiment. This value is set to satisfy certain detection requirement (cf. 18.6.3.24).
MeasuBandwidth	Array(Frequency)	The MeasuBandwidth parameter is the same as the bandwidth used by the control parameter, but when the value is returned by the sensor, it indicates the measurement bandwidth that is actually used by the sensor (cf. 18.6.3.34).
LockStatus	Enumeration	The LockStatus indicates whether the entity (i.e., the sender of this primitive) has been locked by a client for exclusive service. 1: The status of the entity is locked. The entity is locked by a client for exclusive service. 0: The status of the entity is unlocked. The entity has been unlocked by its client, or has been unlocked by a self-generated timeout event.

When Used   The primitive is used in response to the Get_Sensor_Measurement_Profile.request primitive.

Effect of Receipt   Upon receipt, the IEEE 1900.6 SAP user collects the measurement profile of the spectrum measurement module.

Get_Measurement_Location_Information.request

Function   This primitive is used by the IEEE 1900.6 SAP user to obtain the measurement location information of the spectrum measurement module.

Semantics of the Service Primitive

Get_Measurement_Location_Information.request
(
MeasurementLocationTransID,
)

Parameters

Name	Type	Description
MeasurementLocationTransID	Unsigned integer	MeasurementLocationTransID uniquely defines one transaction of requesting the location where the spectrum measurement has been carried out by the spectrum measurement module.

When Used   This primitive is used by the IEEE 1900.6 SAP user when it needs to collect the information related to the location where the spectrum measurement has been carried out by the spectrum measurement module.

Effect of Receipt   The spectrum measurement module subsequently uses a Get_Measurement _Location_Information.response primitive to reflect the results of the request.

Get_Measurement_Location_Information.response

Function   This primitive returns the results of the request to collect the measurement location information of the spectrum measurement module.

Semantics of the Service Primitive

Get_Measurement_Location_Information.response
(
MeasurementLocationTransID,
Status,
AbsSensorLocation,
RelSensorLocation,
LockStatus
)

Parameters

Name	Type	Description
MeasurementLocationTransID	Unsigned integer	MeasurementLocationTransID uniquely defines one transaction of requesting the location where the spectrum measurement has been carried out by the spectrum measurement module.
Status	Enumeration	Status of operation. 0: success 1: unspecified failure 2: rejection 3: authorization failure
AbsSensorLocation	RGeolocation	The AbsSensorLocation parameter is the absolute location that the experiment has carried out (cf. 18.6.3.25).
RelSensorLocation	RGeolocation	The RelSensorLocation parameter is the relative location with respect to a known reference location that the experiment has been carried out (cf. 18.6.3.26).
LockStatus	Enumeration	The LockStatus indicates whether the entity (i.e., the sender of this primitive) has been locked by a client for exclusive service. 1: The status of the entity is locked. The entity is locked by a client for exclusive service. 0: The status of the entity is unlocked. The entity has been unlocked by its client, or has been unlocked by a self-generated timeout event.

When Used The primitive is used in response to the Get_Measurement_Location_Information. request primitive.

Effect of Receipt Upon receipt, the IEEE 1900.6 SAP user collects the measurement location information of the spectrum measurement module.

Get_Signal_Measurement_Value.request

Function This primitive is used by the IEEE 1900.6 SAP user to collect signal measurement values of the spectrum measurement module.

Semantics of the Service Primitive

Get_Signal_Measurement_Value.request

$$($$

SignalMeasurementID

$$)$$

Parameters

Name	Type	Description
SignalMeasurementID	Unsigned integer	SignalMeasurementID uniquely defines the set of measurement values of a signal obtained by the spectrum measurement module.

When Used This primitive is used by the IEEE 1900.6 SAP user when it needs to collect the signal measurement values of the spectrum measurement module.

Effect of Receipt The spectrum measurement module subsequently uses a Get_Signal_Measurement_Value.response primitive to reflect the results of the request.

Get_Signal_Measurement_Value.response

Function This primitive returns the results of the request to collect the signal measurement values of the spectrum measurement module.

Semantics of the Service Primitive

Get_Signal_Measurement_Value.response

(

SignalMeasurementID,

Status,

SignalDesc,

LockStatus

)

Parameters

Name	Type	Description
SignalMeasurementID	Unsigned integer	SignalMeasurementID uniquely defines the set of measurement values of the signal obtained by the spectrum measurement module.
Status	Enumeration	Status of operation. 0: success 1: unspecified failure 2: rejection 3: authorization failure
SignalDesc	Structured	The signal description parameter represents a set of information elements that describe the behavior of the signal (e.g., traffic pattern type and modulation scheme (cf. 18.6.3.41).
LockStatus	Enumeration	The LockStatus indicates whether the entity (i.e., the sender of this primitive) has been locked by a client for exclusive service. 1: The status of the entity is locked. The entity is locked by a client for exclusive service. 0: The status of the entity is unlocked. The entity has been unlocked by its client or has been unlocked by a self-generated timeout event.

When Used The primitive is used in response to the Get_Signal_Measurement_Value.request primitive.

Effect of Receipt Upon receipt, the IEEE 1900.6 SAP user collects the measurement values of signals by the spectrum measurement module.

Get_Channel_Measurement_Value.request

Function This primitive is used by the IEEE 1900.6 SAP user to collect channel measurement values of the spectrum measurement module.

Semantics of the Service Primitive

Get_Channel_Measurement_Value.request

(

ChannelMeasurementID

)

Parameters

Name	Type	Description
ChannelMeasurementID	Unsigned integer	ChannelMeasurementID uniquely defines the set of measurement values of the channel obtained by the spectrum measurement.

When Used   This primitive is used by the IEEE 1900.6 SAP user when it needs to collect the channel measurement values of the spectrum measurement module.

Effect of Receipt   The spectrum measurement module subsequently uses a Get_Channel _Measurement_Value.response primitive to reflect the results of the request.

Get_Channel_Measurement_Value.response

Function   This primitive returns the results of the request to collect the channel measurement values of the spectrum measurement module.

Semantics of the Service Primitive

Get_Channel_Measurement_Value.response

(

ChannelMeasurementID,

Status,

Bandwidth,

NoisePower,

SignalLevel,

LockStatus

)

Parameters

Name	Type	Description
ChannelMeasurementID	Unsigned integer	ChannelMeasurementID uniquely defines the set of measurement values of the channel obtained by the spectrum measurement module.
Status	Enumeration	Status of operation. 0: success 1: unspecified failure 2: rejection 3: authorization failure
Bandwidth	Array(Frequency)	The bandwidth parameter describes the estimated bandwidth based on spectrum measurement (cf. 18.6.3.11).
NoisePower	NoisePower	The NoisePower parameter describes the noise power over the measured channel bandwidth (cf. 18.6.3.35).

Name	Type	Description
SignalLevel	Structured	The SignalLevel parameter designates the amplitude of measured signal power within the measured bandwidth (cf. 18.6.3.36).
LockStatus	Enumeration	The LockStatus indicates whether the entity (i.e., the sender of this primitive) has been locked by a client for exclusive service. 1: The status of the entity is locked. The entity is locked by a client for exclusive service. 0: The status of the entity is unlocked. The entity has been unlocked by its client, or has been unlocked by a self-generated timeout event.

When Used The primitive is used in response to the Get_Channel_Measurement_Value .request primitive.

Effect of Receipt Upon receipt, the IEEE 1900.6 SAP user collects the measurement values of the spectrum measurement module.

Get_RAT_ID_Value.request

Function This primitive is used by the IEEE 1900.6 SAP user to collect RAT ID values of the spectrum measurement module.

Semantics of the Service Primitive

Get_RAT_ID_Value.request

(

RATIDValueRequestTransID

)

Parameters

Name	Type	Description
RATIDValueRequestTransID	Unsigned integer	RATIDValueRequestTransID uniquely defines one transaction of requesting the set of measurement values of the RAT obtained by the spectrum measurement module.

When Used This primitive is used by the IEEE 1900.6 SAP user when it needs to collect the RAT ID values from the spectrum measurement module.

Effect of Receipt The spectrum measurement module subsequently uses a Get_RAT_ID_Value .response primitive to reflect the results of the request.

Get_RAT_ID_Value.response

Function This primitive returns the results of the request to collect the RAT ID values from the spectrum measurement module.

Semantics of the Service Primitive

Get_RAT_ID_Value.response

> (
> RATIDValueRequestTransID,
> Status,
> RATID,
> LockStatus
>)

Parameters

Name	Type	Description
RATIDValueRequestTransID	Unsigned integer	RATIDValueRequestTransID uniquely defines one transaction of requesting the set of measurement values of the RAT obtained by the spectrum measurement module.
Status	Enumeration	Status of operation. 0: success 1: unspecified failure 2: rejection 3: authorization failure
RATID	Enumeration	The RATID parameter uniquely identifies the existing standard RATs known by both spectrum sensors and their clients, for example, Global System for Mobile (GSM), etc. (cf. 18.6.3.42).
LockStatus	Enumeration	The LockStatus indicates whether the entity (i.e., the sender of this primitive) has been locked by a client for exclusive service. 1: The status of the entity is locked. The entity is locked by a client for exclusive service. 0: The status of the entity is unlocked. The entity has been unlocked by its client, or has been unlocked by a self-generated timeout event.

When Used   The primitive is used in response to the Get_RAT_ID_Value.request primitive.

Effect of Receipt   Upon receipt, the IEEE 1900.6 SAP user collects the RAT ID values from the spectrum measurement module.

Notify

Function   This primitive is used by the measurement module to notify a status change to the IEEE 1900.6 SAP user through the M-SAP.

Semantics of the Service Primitive

Notify

> (
> Type,
> Status,
> Reason
>)

Parameters

Name	Type	Description
Type	String	The Type parameter specifies the type of the notification.
Status	Enumeration	Status of operation. 0: success 1: unspecified failure 2: rejection 3: authorization failure
Reason	String	The Reason parameter expresses the reason of the notification.

When Used This primitive is used by the spectrum measurement module to report a status change to the IEEE 1900.6 SAP user.

Effect of Receipt The IEEE 1900.6 SAP user is notified the status change of the spectrum measurement module.

18.5.3.2 Communication Service Access Point

The C-SAP is used for the exchange of sensing-related information (cf. 6.1) between Sensors and their clients. The client role can be taken by a Sensor, CE, or DA. It abstracts services of the communication by providing a set of generic primitives or method and mapping these primitives to transport protocols.

The IEEE 1900.6 SAP users obtain the following services from the C-SAP:

— Sensing-related information send service
— Sensing-related information receive service
— Information services

Sensing-related Information Send Service The sensing-related information send service provides a set of primitives or method through which the IEEE 1900.6 SAP users send sensing-related information to another IEEE 1900.6 SAP user utilizing their local C-SAP. Because both SAP users are collocated with their respective IEEE 1900.6 service (see the definition of IEEE 1900.6 service in 18.3.1), this is the general method to carry out the exchange of sensing-related information between IEEE 1900.6 client and server (see the definitions of IEEE 1900.6 client and IEEE 1900.6 server in 18.3.1), or between remote IEEE 1900.6 logical entities, respectively (see the definition of IEEE 1900.6 logical entity in 18.3.1).

The sensing-related information send service primitives are defined as part of the C-SAP and are described in the following table.

Primitives	Description
Sensing_Related_Information_Send	This primitive is used by the IEEE 1900.6 SAP user to send sensing-related information to another IEEE 1900.6 SAP user.

Sensing_Related_Information_Send.request

Function This primitive is used by the IEEE 1900.6 SAP user to send sensing-related information through its C-SAP to another IEEE 1900.6 SAP user that has C-SAP.

Semantics of the Service Primitive

Sensing_Related_Information_Send.request

```
(
InfoSource,
InfoDestination,
Route,
ReportMode,
SecLevel,
ReportRate,
SensingRelatedInformation
)
```

Parameters

Name	Type	Description
InfoSource	String	InfoSource gives an ID that defines one IEEE 1900.6 logical entity as the source of sensing-related information (cf. 18.6.3.22 and 18.6.3.30).
InfoDestination	String	InfoDestination gives an ID that uniquely defines one IEEE 1900.6 logical entity as the destination of sensing-related information (cf. 18.6.3.22 and 18.6.3.30).
Route	Array(String)	Route gives a list of IDs of sensors and clients through which the sensing-related information flows from remote sensors to the client (cf. 18.6.3.17, 18.6.3.22, and 18.6.3.30).
ReportMode	Enumeration	The ReportMode parameter indicates either hard information or soft information (cf. 18.6.3.15).
SecLevel	Unsigned integer	The Seclevel parameter indicates the level of security depending on the available levels of security for authentication and data verification (cf. 18.6.3.20).
ReportRate	Structured	The ReportRate parameter specifies how often the measurement result should be reported (e.g., every 2 s or every 100 ms) (cf. 18.6.3.14).
SensingRelatedInformation	Structured	SensingRelatedInformation describes the sensing-related information that is sent through the C-SAP. Primitives/commands can also be sent as a payload via this C-SAP (cf. Clause 18.6).

When Used This primitive is used by the IEEE 1900.6 SAP user when it needs to send sensing-related information to another IEEE 1900.6 SAP user through the C-SAP.

Effect of Receipt The communication subsystem subsequently uses a Sensing_Related_ Information _Send.response primitive to reflect the results of the request.

Sensing_Related_Information_Send.response

Function This primitive returns the results of the request to send sensing-related information through the C-SAP.

Semantics of the Service Primitive

Sensing_Related_Information_Send.response

    (
    InfoSource,
    InfoDestination,
    Route,
    ReportMode,
    SecLevel,
    ReportRate,
    Status,
    Timeout,
    LockStatus
)

Parameters

Name	Type	Description
InfoSource	String	InfoSource gives an ID that uniquely defines one IEEE 1900.6 logical entity as the source of sensing-related information (cf. 18.6.3.22 and 18.6.3.30).
InfoDestination	String	InfoDestination gives an ID that uniquely defines one IEEE 1900.6 logical entity as the destination of sensing-related information (cf. 18.6.3.22 and 18.6.3.30).
Route	Array(String)	Route gives a list of IDs of sensors and clients through which the sensing-related information flows from remote sensors to the client (cf. 18.6.3.17, 18.6.3.22, and 18.6.3.30).
ReportMode	Enumeration	The ReportMode parameter indicates either hard information or soft information (cf. 18.6.3.15).
SecLevel	Unsigned integer	The Seclevel parameter indicates the level of security depending on the available levels of security for authentication and data verification (cf. 18.6.3.20).
ReportRate	Structured	The ReportRate parameter specifies how often the measurement result should be reported (e.g., every 2 s or every 100 ms) (cf. 18.6.3.14).
Status	Enumeration	Status of operation. 0: success 1: unspecified failure 2: rejection 3: authorization failure
Timeout	Unsigned fixed-point	The Timeout value in second indicates the timeout of the link.
LockStatus	Enumeration	The LockStatus indicates whether the entity (i.e., the sender of this primitive) has been locked by a client for exclusive service. 1: The status of the entity is locked. The entity is locked by a client for exclusive service. 0: The status of the entity is unlocked. The entity has been unlocked by its client, or has been unlocked by a self-generated timeout event.

When Used The primitive is used in response to the Sensing_Related_Information_Send.request primitive.

Effect of Receipt Upon receipt, the IEEE 1900.6 SAP user obtains the results of sending sensing-related information by the communication subsystem.

Sensing-related Information Receive Service The sensing-related information receive service provides a set of primitives or method through which the IEEE 1900.6 SAP user receives sensing-related information from another IEEE 1900.6 SAP user through the C-SAP.

The primitives for receiving sensing-related information are defined as a part of the C-SAP and are described in the following table.

Primitives	Description
Sensing_Related_Information_Receive	This primitive is used by the IEEE 1900.6 SAP user to receive sensing-related information from another IEEE 1900.6 SAP user.

Sensing_Related_Information_Receive.request

Function This primitive is used by the IEEE 1900.6 SAP user to receive sensing-related information through its C-SAP.

Semantics of the Service Primitive

Sensing_Related_Information_Receive.request

(
InfoSource,
InfoDestination,
Route,
ReportMode,
SecLevel,
ReportRate
)

Parameters

Name	Type	Description
InfoSource	String	InfoSource gives an ID that uniquely defines one IEEE 1900.6 logical entity as the source of sensing-related information (cf. 18.6.3.22 and 18.6.3.30).
InfoDestination	String	InfoDestination gives an ID that uniquely defines one IEEE 1900.6 logical entity as the destination of sensing-related information (cf. 18.6.3.22 and 18.6.3.30).
Route	Array(String)	Route gives a list of IDs of sensors and clients through which the sensing-related information flows from remote sensors to the client (cf. 18.6.3.17, 18.6.3.22, and 18.6.3.30).
ReportMode	Enumeration	The ReportMode parameter indicates either hard information or soft information (cf. 18.6.3.15).
SecLevel	Unsigned integer	The Seclevel parameter indicates the level of security depending on the available levels of security for authentication and data verification (cf. 18.6.3.20).
ReportRate	Structured	The ReportRate parameter specifies how often the measurement result should be reported (e.g., every 2 s or every 100 ms) (cf. 18.6.3.14).

When Used This primitive is used by the IEEE 1900.6 SAP user when it needs to receive sensing-related information from another IEEE 1900.6 SAP user through a C-SAP.

Effect of Receipt The communication subsystem subsequently uses a Sensing_Related_Information_Receive.response primitive to reflect the results of the request.

Sensing_Related_Information_Receive.response

Function This primitive returns the results of the request to send sensing-related information through the C-SAP.

Semantics of the Service Primitive

Sensing_Related_Information_Receive.response

(
InfoSource,
InfoDestination,
Route,
ReportMode,
SecLevel,
ReportRate,
Status,
Timeout,
LockStatus
)

Parameters

Name	Type	Description
InfoSource	String	InfoSource gives an ID that uniquely defines one IEEE 1900.6 logical entity as the source of sensing-related information (cf. 18.6.3.22 and 18.6.3.30).
InfoDestination	String	InfoDestination gives an ID that uniquely defines one IEEE 1900.6 logical entity as the destination of sensing-related information (cf. 18.6.3.22 and 18.6.3.30).
Route	Array(String)	Route gives a list of IDs of sensors and clients through which the sensing-related information flows from remote sensors to the client (cf. 18.6.3.17, 18.6.3.22, and 18.6.3.30).
ReportMode	Enumeration	The ReportMode parameter indicates either hard information or soft information (cf. 18.6.3.15).
SecLevel	Unsigned integer	The Seclevel parameter indicates the level of security depending on the available levels of security for authentication and data verification (cf. 18.6.3.20).
ReportRate	Structured	The ReportRate parameter specifies how often the measurement result should be reported (e.g., every 2 s or every 100 ms) (cf. 18.6.3.14).
Status	Enumeration	Status of operation. 0: success 1: unspecified failure 2: rejection 3: authorization failure
Timeout	Unsigned fixed-point	The Timeout value in second indicates the timeout of the link.
LockStatus	Enumeration	The LockStatus indicates whether the entity (i.e., the sender of this primitive) has been locked by a client for exclusive service. 1: The status of the entity is locked. The entity is locked by a client for exclusive service. 0: The status of the entity is unlocked. The entity has been unlocked by its client, or has been unlocked by a self-generated timeout event.

When Used The primitive is used in response to the Sensing_Related_Information_Receive .request primitive.

Effect of Receipt Upon receipt, the IEEE 1900.6 SAP user who sends the request obtains the results of requesting to receive sensing-related information through the communication subsystem.

Information Services The sensing information services provide a set of primitives or method through which the IEEE 1900.6 SAP users obtain information such as IDs and capabilities of the communication subsystem.

The primitives of the information services are defined as a part of the C-SAP and are described in the following table.

Primitives	Description
Get_CommSubsys_Profile	This primitive is used by the IEEE 1900.6 SAP user to obtain the information related to the communication subsystem.
Notify	This primitive is used by the communication subsystem to notify a status change to the IEEE 1900.6 SAP user.

Get_CommSubsys_Profile.request

Function This primitive is used by the IEEE 1900.6 SAP user to obtain the information related to the capabilities of a communication subsystem.

Semantics of the Service Primitive Get_CommSubsys_Profile.request ()
   No parameters are used in this primitive.

When Used This primitive is used by the IEEE 1900.6 SAP user when it needs to obtain the information related to the capabilities of the communication subsystem.

Effect of Receipt The communication subsystem subsequently uses a Get_CommSubsys_ Profile.response primitive to reflect the results of the request.

Get_CommSubsys_Profile.response

Function This primitive returns the results of the request to get the communication subsystem profile.

Semantics of the Service Primitive

Get_CommSubsys_Profile.response

(
Status,
Comm_Subsys_ID,
Comm_Subsys_Capability,
LockStatus
)

Parameters

Name	Type	Description
Status	Enumeration	Status of operation. 0: success 1: unspecified failure 2: rejection 3: authorization failure
Comm_Subsys_ID	Unsigned integer	Comm_Subsys_ID uniquely defines one compliant communication subsystem.
Comm_Subsys_Capability	String	Comm_Subsys_Capability describes the capability of the communication subsystem.
LockStatus	Enumeration	The LockStatus indicates whether the entity (i.e., the sender of this primitive) has been locked by a client for exclusive service. 1: The status of the entity is locked. The entity is locked by a client for exclusive service. 0: The status of the entity is unlocked. The entity has been unlocked by its client, or has been unlocked by a self-generated timeout event.

When Used The primitive is used in response to the Get_CommSubsys_Profile.request primitive.

Effect of Receipt Upon receipt, the IEEE 1900.6 SAP user sending the request obtains the profile of the communication subsystem.

Notify

Function This primitive is used by the communication subsystem to notify a status change to the IEEE 1900.6 SAP user through the C-SAP.

Semantics of the Service Primitive

Notify

       (
       Type,
       Status,
       Reason
)

Parameters

Name	Type	Description
Type	String	The Type parameter specifies the type of the notification.
Status	Enumeration	Status of operation. 0: success 1: unspecified failure 2: rejection 3: authorization failure
Reason	String	The Reason parameter expresses the reason of the notification.

When Used   This primitive is used by the communication subsystem to report a status change to an IEEE SAP user.

Effect of Receipt   The IEEE 1900.6 SAP user obtains the status of the communication subsystem.

18.5.3.3   Application Service Access Point

The A-SAP is used by the control/application (cf. 18.5.1) to interact with IEEE 1900.6 service. It provides a set of primitives or method for IEEE 1900.6 SAP users to control spectrum sensing and obtain sensingrelated information from spectrum sensing.

The IEEE 1900.6 SAP users obtain the following services from the A-SAP:

— Sensor discovery service
— Sensing-related information access service
— Management and configuration service
— Information services

Sensor discovery service   The sensor discovery service provides a set of primitives or method through which the control/application discovers available spectrum measurement modules. The service also provides a set of primitives or method through which the control/application discovers available communication subsystems to provide communication services for sensing-related information exchange with another IEEE 1900.6 logical entity that uses the communication subsystem. The communication subsystem provides services to IEEE 1900.6 logical entities through the C-SAP (refer to 18.5.1). The following information is discovered in this service:

— Sensor logical ID
— Communication subsystem ID

The primitives of the sensor discovery service are defined as a part of the A-SAP and are defined in the following table. They provide a means to obtain available sensors.

Primitives	Description
	Sensor logical ID
Get_ Sensor_Logical_ID	This primitive is used by the control/application to obtain the list of sensors identified by the sensor logical IDs (cf. 18.6.3.30).
	Communication subsystem ID
Get_CommSubsys_ID	This primitive is used by the control/application to obtain the list of communication subsystems identified by the communication subsystem IDs.

Get_ Sensor_Logical_ID.request

Function   This primitive is used by the control/application to obtain a list of available sensors.

Semantics of the Service Primitive   Get_Sensor_Logical_ID.request ()
   No parameters are used in this primitive.

When Used   This primitive is used by the control/application when it needs to discover available sensors.

Effect of Receipt   Upon receipt, the IEEE 1900.6 service provider uses a Get_Sensor_Logical_ID .reponse primitive to reflect the results of the request sent by the control/application.

Get_Sensor_Logical_ID.response

Function   This primitive returns the results of the request to obtain the list of available sensors.

Semantics of the Service Primitive

Get_ Sensor_Logical_ID.response

(

Status,

ListOfSensors,

LockStatus

)

Parameters

Name	Type	Description
Status	Enumeration	Status of operation. 0: success 1: unspecified failure 2: rejection 3: authorization failure
ListOfSensors	Vector(String)	List of sensors identified by the sensor logical IDs (cf. 18.6.3.30).
LockStatus	Enumeration	The LockStatus indicates whether the entity (i.e., the sender of this primitive) has been locked by a client for exclusive service. 1: The status of the entity is locked. The entity is locked by a client for exclusive service. 0: The status of the entity is unlocked. The entity has been unlocked by its client or has been unlocked by a self-generated timeout event.

When Used   The primitive is used in response to the Get_Sensor_Logical_ID.request primitive.

Effect of Receipt   Upon receipt, the control/application obtains the logical IDs of available sensors.

Get_CommSubsys_ID.request

Function   This primitive is used by the control/application to obtain a list of available communication subsystems.

Semantics of the Service Primitive   Get_CommSubsys_ID.request ()
   No parameters are used in this primitive.

When Used   This primitive is used by the control/application when it needs to discover the available communication subsystems.

Effect of Receipt   Upon receipt, the IEEE 1900.6 service provider uses a Get_CommSubsys_ID .response primitive to reflect the results of the request.

Get_CommSubsys_ID.response

Function   This primitive returns the results of the request to obtain the list of available communication subsystems.

Semantics of the Service Primitive

Get_CommSubsys_ID.response

       (

       Status,

       ListOfCommSubsys,

       LockStatus

)

Parameters

Name	Type	Description
Status	Enumeration	Status of operation. 0: success 1: unspecified failure 2: rejection 3: authorization failure
ListOfCommSubsys	Vector(Unsigned integer)	List of available communication subsystem IDs.
LockStatus	Enumeration	The LockStatus indicates whether the entity (i.e., the sender of this primitive) has been locked by a client for exclusive service. 1: The status of the entity is locked. The entity is locked by a client for exclusive service. 0: The status of the entity is unlocked. The entity has been unlocked by its client or has been unlocked by a self-generated timeout event.

When Used The primitive is used in response to the Get_CommSubsys_ID.request primitive.

Effect of Receipt Upon receipt, the control/application obtains the IDs of available communication subsystems.

Sensing-related Information Access Services The sensing-related information access services provide a set of primitives or method through which the control/application accesses or issues sensing-related information. By these services, the control/application performs the following:

— Read sensing-related information
— Write sensing-related information

The primitives of sensing-related information services are defined as part of the A-SAP and are defined in the following table.

Primitives	Description
Read sensing-related information	
Read_Sensing_Related_Info	This primitive is used by control/application to read sensing-related information.
Write sensing-related information	
Write_Sensing_Related_Info	This primitive is used by control/application to write sensing-related information.

Read_Sensing_Related_Info.request

Function This primitive is used by the control/application to read sensing-related information.

Semantics of the Service Primitive

Read_Sensing_Related_Info.request

(
SensingInfoList,
ClientLogID,
ClientPriorityFlag,
SensorLogID,
SensorPriority
)

Parameters

Name	Type	Description
SensingInfoList	Vector(Unsigned integer)	SensingInfoList gives a list of parameter IDs of sensing-related information that the control/application wants to read (cf. Table 18.2).
ClientLogID	Unsigned integer	ClientLogID uniquely specifies an IEEE 1900.6 client (cf. 18.6.3.22).
ClientPriorityFlag	Unsigned integer	The ClientPriorityFlag parameter indicates a priority level flag. Such priority level flag is assigned based on the application for which the client exploits the sensing-related information (cf. 18.6.3.18).
SensorLogID	Unsigned integer	Unique logical identification of a sensor within a certain area (cf. 18.6.3.30).
SensorPriority	Vector(String)	Sensor priority indicates the list of selected sensors according to priority ranking (cf. 18.6.3.19).

When Used This primitive is used by the control/application when it needs to read sensing-related information from one or more IEEE 1900.6 logical entities.

Effect of Receipt Upon receipt, the IEEE 1900.6 service provider uses a Read_Sensing_Related_Info.response primitive to reflect the results of the request.

Read_Sensing_Related_Info.response

Function This primitive returns the results of the request to obtain the sensing-related information issued by the control/application.

Semantics of the Service Primitive

Read_Sensing_Related_Info.response

(
Status,
Information,
LockStatus
)

Parameters

Name	Type	Description
Status	Enumeration	Status of operation. 0: success 1: unspecified failure 2: rejection 3: authorization failure
Information	Structured	Information describes the sensing-related information that is read by the control/application (cf. Table 18.2).
LockStatus	Enumeration	The LockStatus indicates whether the entity (i.e., the sender of this primitive) has been locked by a client for exclusive service. 1: The status of the entity is locked. The entity is locked by a client for exclusive service. 0: The status of the entity is unlocked. The entity has been unlocked by its client, or has been unlocked by a self-generated timeout event.

When Used   The primitive is used in response to the Read_Sensing_Related_Info.request primitive.

Effect of Receipt   Upon receipt, the control/application obtains the desired sensing-related information.

Write_Sensing_Related_Info.request

Function   This primitive is used by the control/application to write sensing-related information.

Semantics of the Service Primitive

Write_Sensing_Related_Info.request

    (
    SensingInfoList,
    Information,
    ClientLogID,
    ClientPriorityFlag,
    SensorLogID,
    SensorPriority
)

Parameters

Name	Type	Description
SensingInfoList	Vector(Unsigned integer)	SensingInfoList gives a list of parameter IDs of sensing-related information that the control/application wants to read (cf. Table 18.2).
Information	Structured	Information gives a list of sensing-related information that the control/application wants to write (cf. Table 18.2).

Name	Type	Description
ClientLogID	Unsigned integer	ClientLogID uniquely specifies an IEEE 1900.6 client (cf. 18.6.3.22).
ClientPriorityFlag	Unsigned integer	The ClientPriorityFlag parameter indicates a priority level flag. Such priority level flag is assigned based on the application for which the client exploits the sensing-related information (cf. 18.6.3.18).
SensorLogID	Unsigned integer	Unique logical identification of a sensor within a certain area (cf. 18.6.3.30).
SensorPriority	Vector(String)	Sensor priority indicates the list of selected sensors according to priority ranking (cf. 18.6.3.19).

When Used This primitive is used by the control/application when it needs to write sensing-related information to one or more IEEE 1900.6 logical entities.

Effect of Receipt Upon receipt, the IEEE 1900.6 service provider uses a Write_Sensing_Related_Info.response primitive to reflect the results of the request.

Write_Sensing_Related_Info.response

Function This primitive returns the results of the request to write the sensing-related information issued by the control/application.

Semantics of the Service Primitive

Write_Sensing_Related_Info.response

(

Status,

LockStatus

)

Parameters

Name	Type	Description
Status	Enumeration	Status of operation. 0: success 1: unspecified failure 2: rejection 3: authorization failure
LockStatus	Enumeration	The LockStatus indicates whether the entity (i.e., the sender of this primitive) has been locked by a client for exclusive service. 1: The status of the entity is locked. The entity is locked by a client for exclusive service. 0: The status of the entity is unlocked. The entity has been unlocked by its client, or has been unlocked by a self-generated timeout event.

When Used The primitive is used in response to the Write_Sensing_Related_Info.request primitive.

Effect of Receipt Upon receipt, the control/application obtains the results of the request to write sensing-related information to one or more IEEE 1900.6 logical entities.

Management and Configuration Services The management and configuration services provide a set of primitives or method through which the control/application manages IEEE 1900.6 logical entities and configures communication among IEEE 1900.6 logical entities. Through these services, the control/application performs the following:

— IEEE 1900.6 logical entity management
— Communication configuration

The primitives of management and configuration services are defined as a part of the A-SAP and are listed in the following table.

Primitives	Description
IEEE 1900.6 logical entity management	
Lock	This primitive is used by the control/application to lock IEEE 1900.6 logical entities and communication subsystems to prevent other controls/applications from accessing those resources.
Unlock	This primitive is used by the control/application to unlock IEEE 1900.6 logical entities and communication subsystems so that other controls/applications can access those resources.
BreakLock	This primitive is used by the control/application to break the lock so that other controls/applications can access those resources.
Trigger	This primitive is used by the control/application to trigger a specific action.
Communication configuration	
Comm_Manage	This primitive is used by the control/application to manage communications of IEEE 1900.6 logical entities.

Lock.request

Function This primitive is used by the control/application to lock IEEE 1900.6 logical entities or communication subsystems for exclusive use and to prevent other controls/applications from accessing those resources.

Semantics of the Service Primitive

Lock.request

   (

   EntityID,

   CommSubsysID

)

Parameters

Name	Type	Description
EntityID	String	EntityID uniquely identifies an IEEE 1900.6 logical entity. This EntityID can be client logical ID (cf. 18.6.3.22) or sensor logical ID (cf. 18.6.3.30).
CommSubsysID	Unsigned integer	CommSubsysID uniquely identifies a communication subsystem.

When Used   This primitive is used by the control/application when it needs to lock IEEE 1900.6 logical entities or communication subsystems for exclusive use.

Effect of Receipt   Upon receipt, the IEEE 1900.6 service uses a Lock.response to reflect the results of the request.

Lock.response

Function   This primitive returns the results of the request to lock IEEE 1900.6 logical entities or communication subsystems.

Semantics of the Service Primitive

Lock.response

> (
> Status,
> EntityID,
> CommSubsysID
>)

Parameters

Name	Type	Description
Status	Enumeration	Status of operation. 0: success 1: unspecified failure 2: rejection 3: authorization failure
EntityID	String	EntityID uniquely identifies an IEEE 1900.6 logical entity. This EntityID can be client logical ID (cf. 18.6.3.22) or sensor logical ID (cf. 18.6.3.30).
CommSubsysID	Unsigned integer	CommSubsysID uniquely identifies a communication subsystem.

When Used   The primitive is used in response to the Lock.request primitive.

Effect of Receipt   Upon receipt, the control/application obtains the results of the request to lock IEEE 1900.6 logical entities or communication subsystems.

Unlock

Unlock.request

Function   This primitive is used by the control/application to unlock IEEE 1900.6 logical entities or communication subsystems from exclusive use.

Semantics of the Service Primitive

Unlock.request

> (
> EntityID,
> CommSubsysID
>)

Parameters

Name	Type	Description
EntityID	String	EntityID uniquely identifies an IEEE 1900.6 logical entity. This EntityID can be client logical ID (cf. 18.6.3.22) or sensor logical ID (cf. 18.6.3.30).
CommSubsysID	Unsigned integer	CommSubsysID uniquely identifies a communication subsystem.

When Used   This primitive is used by the control/application when it needs to unlock the IEEE 1900.6 logical entities or communication subsystems from exclusive use.

Effect of Receipt   Upon receipt, the IEEE 1900.6 service uses an Unlock.response primitive to reflect the results of the request.

Unlock.response

Function   This primitive returns the results of the request to unlock IEEE 1900.6 logical entities or communication subsystems.

Semantics of the Service Primitive

Unlock.response

> (
> Status,
> EntityID,
> CommSubsysID
>)

Parameters

Name	Type	Description
Status	Enumeration	Status of operation. 0: success 1: unspecified failure 2: rejection 3: authorization failure
EntityID	String	EntityID uniquely identifies an IEEE 1900.6 logical entity. This EntityID can be client logical ID (cf. 18.6.3.22) or sensor logical ID (cf. 18.6.3.30).
CommSubsysID	Unsigned integer	CommSubsysID uniquely identify a communication subsystem at an IEEE 1900.6 logical entity.

When Used   The primitive is used in response to the Unlock.request primitive.

Effect of Receipt   Upon receipt, the control/application can obtain the results of the request to unlock IEEE 1900.6 logical entities or communication subsystems.

BreakLock.request

Function   This primitive is used by the control/application to break the lock of IEEE 1900.6 logical entities or communication subsystems so that it can access those resources.

Semantics of the Service Primitive

BreakLock.request

(

EntityID,

CommSubsysID

)

Parameters

Name	Type	Description
EntityID	String	EntityID uniquely identifies an IEEE 1900.6 logical entity. This EntityID can be client logical ID (cf. 18.6.3.22 or sensor logical ID (cf. 18.6.3.30).
CommSubsysID	Unsigned integer	CommSubsysID uniquely identifies a communication subsystem.

When Used   This primitive is used by the control/application when it needs to override a lock set on IEEE 1900.6 logical entities or communication subsystems so that it can use those resources.

Effect of Receipt   Upon receipt, the IEEE 1900.6 service uses a BreakLock.response primitive to reflect the results of the request.

BreakLock.response

Function   This primitive returns the results of the request to override the lock of IEEE 1900.6 logical entities or communication subsystems.

Semantics of the Service Primitive

Unlock.response

(

Status,

EntityID,

CommSubsysID

)

Parameters

Name	Type	Description
Status	Enumeration	Status of operation. 0: success 1: unspecified failure 2: rejection 3: authorization failure
EntityID	String	EntityID uniquely identifies an IEEE 1900.6 logical entity. This EntityID can be client logical ID (cf. 18.6.3.22) or sensor logical ID (cf. 18.6.3.30).
CommSubsysID	Unsigned integer	CommSubsysID uniquely identifies a communication subsystem.

When Used   The primitive is used in response to the BreakLock.request.primitive.

Effect of Receipt   Upon receipt, the control/application obtains the results of the request to override the lock set on IEEE 1900.6 logical entities or communication subsystems.

Trigger.request

Function   This primitive is used by the control/application to trigger an event.

Semantics of the Service Primitive

Trigger.request

    (
    EventID,
    TriggerTime,
    Timeout
)

Parameters

Name	Type	Description
EventID	Unsigned integer	EventID uniquely identifies a particular triggered event.
TriggerTime	Unsigned integer	TriggerTime specifies when to begin operation.
Timeout	Unsigned fixed-point	Timeout in second is the maximum time to wait for a time-out error.

When Used   This primitive is used by the control/application when it needs to trigger a particular event.

Effect of Receipt   Upon receipt, the IEEE 1900.6 service uses a Trigger.response primitive to reflect the results of the request.

Trigger.response

Function   This primitive is used by the control/application to return the result of trigger.

Semantics of the Service Primitive

Trigger.response

    (
    EventID,
    Status,
    LockStatus
)

Parameters

Name	Type	Description
EventID	Unsigned integer	EventID uniquely identifies a particular triggered event.
Status	Enumeration	Status of operation. 0: success 1: unspecified failure 2: rejection 3: authorization failure

Name	Type	Description
LockStatus	Enumeration	The LockStatus indicates whether the entity (i.e., the sender of this primitive) has been locked by a client for exclusive service. 1: The status of the entity is locked. The entity is locked by a client for exclusive service. 0: The status of the entity is unlocked. The entity has been unlocked by its client, or has been unlocked by a self-generated timeout event.

When Used This primitive is used by IEEE 1900.6 service in response to the Trigger.request primitive.

Effect of Receipt The control/application obtains the result of the trigger request.

Comm_Management.request

Function This primitive is used by the control/application to manage communications among IEEE 1900.6 logical entities.

Semantics of the Service Primitive

Comm_Management.request

(

Status,

CommManagementTransID

)

Parameters

Name	Type	Description
Status	Enumeration	Status of operation. 0: success 1: unspecified failure 2: rejection 3: authorization failure
CommManagementTransID	Unsigned integer	CommManagementTransID uniquely identifies one transaction of a communication management.

When Used This primitive is used by the control/application when it needs to obtain the communication type IEEE 1900.6 logical entities.

Effect of Receipt Upon receipt, the IEEE 1900.6 service uses a Comm_Management.response primitive to reflect the results of the request.

Comm_Management.response

Function This primitive returns the results of the communication management request issued by the control/application.

Semantics of the Service Primitive

Comm_Management.response

> (
> Status,
> CommManagementTransID,
> NetworkTopology,
> LockStatus
>)

Parameters

Name	Type	Description
Status	Enumeration	Status of operation. 0: success 1: unspecified failure 2: rejection 3: authorization failure
CommManagementTransID	Unsigned integer	CommManagementTransID uniquely identifies one transaction of a communication management.
NetworkTopology	String	NetworkTopology uniquely describes the communication type of IEEE 1900.6 logical entities, such as P2P, star, etc.
LockStatus	Enumeration	The LockStatus indicates whether the entity (i.e., the sender of this primitive) has been locked by a client for exclusive service. 1: The status of the entity is locked. The entity is locked by a client for exclusive service. 0: The status of the entity is unlocked. The entity has been unlocked by its client, or has been unlocked by a self-generated timeout event.

When Used　The primitive is used in response to the Comm_Management.request primitive.

Effect of Receipt　Upon receipt, the control/application obtains the results of the request.

Information services　The sensing information service provides a set of primitives or method through which the control/application obtains information such as ID and capability of the IEEE 1900.6 clients through the A-SAP.

The primitives of the information services are defined as a part of the A-SAP and are described in the following table.

Primitives	Description
Get_Client_Profile	This primitive is used by the control/application to obtain the information related to IEEE 1900.6 clients.
Notify	This primitive is used by the control/application to report its status.

Get_ Client_Profile.request

Function　This primitive is used by the control/application to obtain the profile such as ID and capability of IEEE 1900.6 clients.

Semantics of the Service Primitive Get_Client_Profile.request ()
No parameters are used in this primitive.

When Used This primitive is used by the control/application when it needs to obtain the profile of the IEEE 1900.6 clients.

Effect of Receipt The IEEE 1900.6 service uses a Get_AppContr_Profile.response primitive to reflect the results of the request.

Get_Client_Profile.response

Function This primitive returns the results of the request to get IEEE 1900.6 client profile.

Semantics of the Service Primitive

Get_ Client_ Profile.response

(
Status,
ClientLogID,
Client_Capability,
LockStatus
)

Parameters

Name	Type	Description
Status	Enumeration	Status of operation. 0: success 1: unspecified failure 2: rejection 3: authorization failure
ClientLogID	String	ClientLogID uniquely defines one IEEE 1900.6 client (cf. 18.6.3.22).
Client_Capability	String	Client_Capability describes the capability of IEEE 1900.6 clients.
LockStatus	Enumeration	The LockStatus indicates whether the entity (i.e., the sender of this primitive) has been locked by a client for exclusive service. 1: The status of the entity is locked. The entity is locked by a client for exclusive service. 0: The status of the entity is unlocked. The entity has been unlocked by its client, or has been unlocked by a self-generated timeout event.

When Used The primitive is used in response to the Get_AppContr_Profile.request primitive.

Effect of Receipt Upon receipt, the control/application obtains the profile of IEEE 1900.6 logical entities.

Notify

Function This primitive is used by the control/application to notify a status change to IEEE 1900.6 service.

Semantics of the Service Primitive

Notify

   (

   Type,

   Status,

   Reason

)

Parameters

Name	Type	Description
Type	String	The Type parameter specifies the type of the notification.
Status	Enumeration	Status of operation. 0: success 1: unspecified failure 2: rejection 3: authorization failure
Reason	String	The Reason parameter expresses the reason of the notification.

When Used This primitive is used by the control/application to report a status change to IEEE service.

Effect of Receipt The IEEE 1900.6 service is notified the status change of the control/application.

18.6 Information Description

This clause presents the structure of information exchanged between 1900.6 clients and sensors or, more generic, between providers and users of sensing-related information. First, 18.6.2 elaborates on the main categories of sensing-related information (i.e., sensing information, sensing control information, sensor information, and requirements derived from regulation). Additionally, the structure of control commands is addressed. Subclause 18.6.2 elaborates on data types and structures required in formally describing the parameters exchanged. Subclause 18.6.3 then formally describes these parameters exchanged based on the data type descriptions given by 18.6.2. Finally, 18.6.4 provides an object model for all sensing-related parameters and commands exchange between client and sensor.

18.6.1 Information Categories

This subclause addresses the main categories of sensing-related information exchanged between sensors and their clients. Sensing-related information comprises sensing information, sensing control information, control commands, sensor information, and requirements derived from regulation.

18.6.1.1 Sensing Information

Sensing information may include information about frequency band, energy, channel condition, time stamp of sensing, and local detection results, etc. In addition, related meta-information (i.e., the information required for describing, understanding, and evaluating information) is also considered sensing information. Meta-information may be both a priori knowledge or may be obtained from sensors. For example, the update rate of a sensing parameter is considered meta-information because it provides information on the temporal resolution of another sensing parameter. Consequently, relative positioning of sensors is providing the necessary information on the spatial resolution of the sensing information obtained without communicating sensor positions along with every sensing information report.

18.6.1.2 Sensing Control Information

Sensing control information may include information about the data to sense, the nature of the data (whether it should be raw data, not yet processed data obtained from a measurement, preprocessed data, relayed data, or fused data), the expected target performance, sensing duration, start/stop time of sensing activity, synchronization control information, power control, priority control, etc.

18.6.1.3 Control Commands

Control commands are exchanged between IEEE 1900.6 logical entities to obtain information about a logical entity or about an ongoing sensing or processing activity. Additionally, control commands are used to set parameters for a sensing activity and to configure the logical entity. In general, two types of control commands are available: (1) standard control commands and (2) proprietary (manufacturer-specific) control commands.

— Standard control commands
  Standard control commands are defined by this document. They may include commands to read the status of a sensor, to request information about sensing capabilities, to request sensing information, to identify the sensor information source, etc.
— Proprietary control commands
  Proprietary control commands can be included for manufacturer or device specific purposes such as sensor calibration or sensor diagnostics. Proprietary control commands are not specified by this standard.

Structure of Control Commands A control command should follow a predefined structure in such a way that it is understood by any entities implementing IEEE 1900.6 logical interfaces.

As shown in Figure 18.10, a control command consists of the command class, command function, and command version fields.

The command class is an identifier of a group of command functions. Command groups are either associated with a certain 1900.6 logical entity, are generic for all entities, or are manufacturer specific. The latter group shall use a specific command structure so that commands of this group are only recognized within the scope set by the device manufacturer.

The command function unambiguously defines the purpose and scope of validity of a command as well as the subsequent command structure (i.e., if it requires additional command parameters).

The command version indicates the version of the command and the command structure following the command class, function, and version fields. This field is used to provide backward compatibility for further releases of this standard.

Command class	Command function	Command version

Command class	Command function	Command version	Command parameter

Figure 18.10 — Structure of control commands without (top) and with (bottom) control parameters. From IEEE Standard for Spectrum Sensing Interfaces and Data Structures for Dynamic Spectrum Access and Other Advanced Radio Communication Systems, Feb 2011. IEEE.

Some control commands might be accompanied by parameters to narrow the required action. For example, starting a sensing activity might require setting start time, start frequency, scanning bandwidth, frequency increment, etc. to specify completely the action requested. Requesting the activity status of a sensor (i.e., if it is busy or idle) may not need any command parameter. Vice versa commands, such as TimeSync (cf. 18.6.3.23) and Scan (cf. 18.6.3.24), do require a certain number of parameters.

The structure of the command parameter field is specific for each command and depends on the command class, function, and version. It may consist of zero, one, or more sensing control parameters.

Chaining of control commands (Figure 18.11) shall be supported for communication efficiency and for implementing atomic control commands with and without command parameters. The command structure given by Figure 18.10 shall be preserved in the process of chaining commands.

Control Command Parameters In general, information exchange between IEEE 1900.6 logical entities is following a request/response scheme:

— A client requests a sensing parameter and a sensor responds by providing the parameter value.
— A client requests to set a parameter to a new value and a sensor (or DA) responds by providing the set value.

More complex schemes can be derived from this basic behavior as addressed by the description of service primitives in 18.5.3.

Control command parameters thus consist of (a vector of) zero, one, or more sensing control parameter IDs and, depending on the control command issued, the corresponding sensing control parameter value in a sequence as shown in Figure 18.12.

Herein, the sensing control parameter ID is considered a tag value that unambiguously identifies the parameter requested (or provided) and the parameter value. Thus, any sensing parameter and any sensing control parameter need to have their own unique identifiers as defined in the parameter description in 18.6.3.

This scheme is also valid for the exchange of sensing information. The distinction between sensing parameters and sensing control parameters is not necessary in the scope of information exchange by control commands. For example, setting a sensing parameter can be understood as a control parameter. When the entity is responding by confirming the action and by providing

Figure 18.11 —Control command chaining. From IEEE Standard for Spectrum Sensing Interfaces and Data Structures for Dynamic Spectrum Access and Other Advanced Radio Communication Systems, Feb 2011. IEEE.

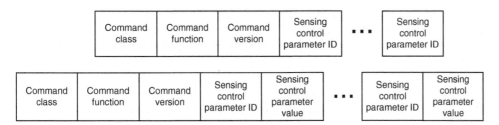

Figure 18.12 —Control command parameter structure without including parameter values (top) and with parameter values (bottom). From IEEE Standard for Spectrum Sensing Interfaces and Data Structures for Dynamic Spectrum Access and Other Advanced Radio Communication Systems, Feb 2011. IEEE.

back the updated sensing parameter, the same parameter can be considered a sensing parameter.

18.6.1.4 Sensor Information

Sensor information in this document is related to information about sensors and properties of sensors (i.e., sensor specification, sensor capabilities, or sensor identity). Sensor information, thus, is closely related to meta-information in the context of sensing information. Knowledge about a sensor's properties enables the correct understanding of sensing information obtained from this sensor.

Sensor properties may include information about the types of sensing techniques that can be performed by the sensor, measurement range, data accuracy, calibration information, analog-to-digital (A/D) and digitalto-analog (D/A) resolution, communication channel utilized, sensor's address, sensor's ID (logical or manufacturer ID), battery wear level, sensing cost, etc.

The term "sensor information" has been chosen here to emphasize the focus set to describe the sensor's attributes. The basic principle applies to other IEEE 1900.6 logical entities as well.

18.6.1.5 Requirements Derived from Regulation

National and international regulations specify spectrum allocation and spectrum usage in each country. Regulations may specify the operating requirements of devices, and this category of sensing-related information includes definitions of the sensing range, accuracy requirements, granularity, required measurement bandwidth, repeat frequency, and detection sensitivity. Furthermore, it may include information about sensing permission and synchronization. Although types of requirements are mostly identical, the range of the parameters may vary depending on the system, region, or country of deployment. The detailed fields, data types, and parameter ranges to meet the requirements from national regulatory bodies are defined in 18.6.3.

18.6.2 Data Types

This subclause defines the primitive data types, simple data types, and derived data types used in the formal definition of IEEE 1900.6 parameters defined in 18.6.3. The physical units used in this definition are based on the International System of Units (SI) and are summarized in Table 18.1.

From IEEE Standard for Spectrum Sensing Interfaces and Data Structures for Dynamic Spectrum Access and Other Advanced Radio Communication Systems, Feb 2011. IEEE.

18.6.2.1 Primitive and Simple Data Types

This subclause summarizes primitive and simple data types (i.e., scalars and arrays made up of a single base type).

Table 18.1 Units used in the description of types and parameters.

Unit	Unit symbol	Value	Note
second	s		SI unit
meter	m		SI unit
Hertz	Hz	$1 \, Hz = 1 / s$	SI derived unit
radian	rad	$1 \, rad = 180 / \pi$	SI derived unit, dimensionless
Watt	W	$kg \cdot m^2 / s^3$	SI derived unit
degree of arc	°	$1° = \pi / 180 \, rad$	dimensionless
power ratio	dBm	$[dBm] = 10 \times \log_{10} ([W] / 1 \, mW)$	dimensionless

Boolean A primitive logical data type having one of two values of "true" (1, nonzero) or "false" (0, zero).

Example: Indication that a signal level is above a certain threshold (cf. 18.6.3.36).

Integer A primitive integral data type representing natural numbers and their negatives. Note that common binary representations limit the number range due to machine word length restrictions. In this standard, the integer length is defined as 32 bits.

Example: Indication of priority level of an IEEE 1900.6 client (cf. 18.6.3.18).

Unsigned Integer A primitive data type representing non-negative integrals.

Example: Priority level of a sensor in relation to a given default priority (cf. 18.6.3.19).

Float A primitive data type storing real numbers, usually as floating-point numbers. Floating-point number representations as defined in IEEE Std 754™-2008 [B3] can be taken as an example.

Example: Indication of the amplitude of a signal (cf. 18.6.3.41).

String A simple data type storing a sequence of data values, usually bytes or characters. A string is a special use of a one-dimensional array or vector.

Example: Logical identification of the 1900.6 client (cf. 18.6.3.22).

Vector A simple data type storing a sequence of data values of a specified type. The notation vector(type) is used to specify the type of vector elements. A vector is a one-dimensional array.

Example: A list of channel numbers (cf. 18.6.3.13).

Array A simple data type storing a collection of data values of a specified type. The notation array(type) is used to specify the type of array elements.

Example: Indication of different paths for the flow of sensing-related information (cf. 18.6.3.17).

18.6.2.2 Complex and Derived Data Types

This subclause summarizes complex and derived data types such as structured types or types that rely on specific interpretation or restriction of the underlying primitive or simple type.

Enumeration An enumeration is a listing of elements of a set in a way that maps to an index set consisting of natural numbers. That is, each element of the set is unambiguously represented by an ordinal.

Example: One out of a well-defined selection of possible configurations of a sensor (cf. 18.6.3.15).

Fixed-point Fixed-point numbers are rational numbers with a fixed length mantissa and a fixed exponent. In contrast to a floating-point representation utilizing a fixed length but variable exponent, the value range is limited by the mantissa length, but the resolution is constant over the value range. They can be realized by using an integer value in conjunction with an implicit multiplier.

Example: A single relative time value in μs (cf. 18.6.3.4)

Unsigned Fixed-point Unsigned fixed-point numbers represent non-negative fixed-point numbers.

Example: A single frequency value in Hz (cf. 18.6.3.1).

Structured A complex data type that aggregates a fixed set of labeled elements, possibly of different primitive or simple types, into a single element.

Example: The geographical location of a sensor (cf. 18.6.3.6).

18.6.3 Description of Sensing-related Parameters

Parameter descriptions are given in a tabular form throughout 18.6.3.1 through 18.6.3.42. They consist of the parameter name and ID, a short textual description, and a type and size specification, if needed.

— Parameter name and ID
The parameter's name provides a unique identification of the parameter in human readable form, whereas the numerical ID is given to unambiguously identify the parameter in the process of information exchange between IEEE 1900.6 logical entities.
— Parameter type and size
Parameters are of one of the types defined by 18.6.2 (primitive, simple, complex, and derived types). Some parameters may be further restricted in their value range or magnitude. The size field of the parameter description is supplementary information that is either a fixed value, determined by the number of elements contained (subparameters), or variable if at least one of the elements contained is optional, or is of variable size itself. To avoid implementation-dependent specifications for parameters, the size of a parameter is always given in terms of the underlying type.
For parameters based on array types, the size is given as the number of elements stored in the array or as "variable" if the size of the array is unspecified. Note that variable size arrays demand for an implicit array length value in information exchange.
For structured types, the aggregated size depends on the implementation of the elements enclosed and thus is omitted in the parameter description. The implementation then will decide on the binary representation, encoding, and size in terms of bits or bytes and any tag or length values needed.

Table 18.2 provides a summary of sensing-related parameters and categorizes these into parameters for sensing, sensing control, sensor information, and requirements derived from regulation (see the definition of regulatory requirements in 18.3.1).

— Sensing information (see 18.6.1.1)
Sensing parameters indicate the measurement output at the spectrum sensor and other associated parameters that augment the measurement data. Some parameters may appear as a control parameter when issued by the client to configure the sensor, and they may also be represented as a sensing parameter when they are issued as measurement output by sensor. Bandwidth and time stamp are examples of such parameters.
— Sensing control information (see 18.6.1.2 and 18.6.1.3)
Sensing control information is used to optimize the spectrum sensing and the procedure to obtain sensing information. Sensing control parameters are generated by the IEEE 1900.6 client. These parameters shall be used to realize the following two major functions:
— Sensor configuration according to the demands of the application and to the measurement process.
— Topological configuration of multiple sensors according to the demands of information exchange between sensors and between sensors and client.
One or more of the sensing control parameters given by Table 2 can be sent to the sensor for the preceding two purposes at the same time.
— Sensor information (see 18.6.1.4)
To optimize the measurement request or to configure the sensor, an IEEE 1900.6 client may require information describing a sensor's capabilities. This information may be known to the client if the sensor's capabilities conform to a standard or if the information is available in a local database on the client. In any other case, the client may also request this information directly from the sensor. Depending on the application, selected sensor profile information can be requested.
— Requirements derived from regulation (see 18.6.1.5)
Regulation may require specification of sensing range, accuracy, granularity, measurement bandwidth, repeat frequency, detection sensitivity, sensing permission, and synchronization information. The values of these parameters may vary according to regulation depending on the region or country of deployment.

Table 18.2 —Summary and taxonomy of sensing-related parameters.

ID	Parameter name	Subclause	Sensing	Sensing control	Sensor	Regulatory
001	Frequency	18.6.3.1	Yes	Yes	Yes	Yes
002	Second	18.6.3.2	Yes	Yes	Yes	Yes
003	Time reference (RSecond)	18.6.3.3	Yes	Yes	Yes	Yes
004	Microsecond	18.6.3.4	Yes	Yes	Yes	Yes
005	Angle	18.6.3.5	Yes	Yes	Yes	Yes
006	Reference geolocation (RGeolocation)	18.6.3.6	Yes	Yes	Yes	Yes
007	Power	18.6.3.7	Yes	Yes	Yes	Yes
008	Time stamp (TimeStamp)	18.6.3.8	Yes	Yes	Yes	Yes
009	Time duration (TimeDuration)	18.6.3.9	Yes	Yes	Yes	Yes
010	Channel list (ChList)	18.6.3.10	Yes	Yes	Yes	Yes
…						
101	Bandwidth	18.6.3.11	Yes	Yes		Yes
102	Total measurement duration (TotMeasuDur)	18.6.3.12	Yes	Yes		
103	Channel order (ChOrder)	18.6.3.13	Yes	Yes		
104	Reporting rate (ReportRate)	18.6.3.14	Yes	Yes		
105	Reporting mode (ReportMode)	18.6.3.15	Yes	Yes		
106	Performance metric (PerfMetric)	18.6.3.16	Yes	Yes		
107	Route	18.6.3.17		Yes		
108	ClientPriorityFlag	18.6.3.18		Yes		
109	SensorPriority	18.6.3.19		Yes	Yes	
110	Security level (SecLevel)	18.6.3.20			Yes	
111	DataKey	18.6.3.21	Yes	Yes		
112	ClientLogID	18.6.3.22		Yes		
…						
201	Time synchronization (TimeSync)	18.6.3.23		Yes		
202	Scan	18.6.3.24		Yes		
…						
301	Absolute sensor location (AbsSensorLocation)	18.6.3.25	Yes		Yes	
302	Relative sensor location (RelSensorLocation)	18.6.3.26	Yes		Yes	
303	Measurement range (MeasuRange)	18.6.3.27	Yes	Yes	Yes	
304	SensingMode	18.6.3.28	Yes	Yes	Yes	
305	SensorID	18.6.3.29		Yes	Yes	
306	Sensor logical ID (SensorLogID)	18.6.3.30		Yes	Yes	
307	BatteryStatus	18.6.3.31	Yes		Yes	
308	DataSheet	18.6.3.32			Yes	
309	ConfidenceLevel	18.6.3.33	Yes			
…						
401	Measurement bandwidth (MeasuBandwidth)	18.6.3.34	Yes			
402	NoisePower	18.6.3.35	Yes			
403	SignalLevel	18.6.3.36	Yes			
404	ModulationType (ModuType)	18.6.3.37	Yes			
405	TrafficPattern	18.6.3.38	Yes			
406	TrafficInformation	18.6.3.39	Yes			
407	SignalType	18.6.3.40	Yes			
408	SignalDesc	18.6.3.41	Yes			
409	RATID	18.6.3.42	Yes			

Table 18.2 categorizes the sensing parameters into each of the four sensing-related information categories described earlier. It indicates most of the parameters as allocated to more than one category. This is especially true for simple parameter descriptions because these are sufficiently generic to be used, for example, when exchanging sensing parameters as well as for sensing control.

From IEEE Standard for Spectrum Sensing Interfaces and Data Structures for Dynamic Spectrum Access and Other Advanced Radio Communication Systems, Feb 2011. IEEE.

18.6.3.1 Frequency

Frequency denotes a single generic frequency parameter given in Hz. The parameter is realized as an unsigned fixed-point value assuming a resolution of 1 Hz and a maximum range of 0 to $2^{32}-1$ Hz (i.e., the equivalent of a 32-bit mantissa). Derived parameters, relaxing or setting constraints to range and resolution, for example, may extend this definition and may modify the range, type, or physical unit fields accordingly.

Name:	Frequency	Phys. unit	Hz	Extends:	—
ID:	001	Size:	1	Type:	Unsigned fixed-point
Desc:	Basic unbounded frequency parameter.				
	Range (min/resolution/max):		0	1 Hz	$(2^{32}-1)$Hz

18.6.3.2 Second

Second denotes a single generic time parameter given in s. The parameter is realized as a signed integer value assuming a resolution of 1 s and a maximum range of -2^{31} to $+2^{31}-1$ s (i.e., the equivalent of a 32-bit two's complement integer). This parameter can be used to realize time difference values. Derived parameters, relaxing or setting constraints to range and resolution, for example, may extend this definition and may modify the range, type, or physical unit fields accordingly.

Name:	Second	Phys. unit	s	Extends:	—
ID:	002	Size:	1	Type:	Signed integer
Desc:	Basic time value in seconds.				
	Range (min/resolution/max):		-2^{31}	1	$2^{31}-1$

18.6.3.3 Time Reference (RSecond)

RSecond denotes a single time parameter given in s. It extends Second (cf. 18.6.3.2) to realize time difference values with respect to midnight (UTC) of January 1, 1970 not counting leap seconds. The parameter is realized as an unsigned integer assuming a maximum range of 0 to $2^{32}-1$ s (i.e., the equivalent of a 32-bit integer).

Name:	RSecond		Phys. unit	s	Extends:	Second
ID:	003		Size:	1	Type:	Unsigned integer
Desc:	Basic reference time value. Seconds since midnight (UTC) of January 1, 1970 absolute time.					
	Range (min/resolution/max):		0		1	$2^{32}-1$

18.6.3.4 Microsecond

Microsecond denotes a single generic time parameter given in μs. It can be used in conjunction with Second (cf. 18.6.3.2) or derived parameters to form a high-resolution time parameter. The parameter is realized as a fixed-point value assuming a resolution of 1 ns and a maximum range of $-10^6 + 10^{-3}$ to $10^6 - 10^{-3}$ μs. Derived parameters, relaxing or setting constraints to range and resolution, for example, may extend this definition and may modify the range, type, or physical unit fields accordingly.

Name:	Microsecond	Phys. unit	µs	Extends:	–
ID:	004	**Size:**	1	**Type:**	Fixed-point
Desc:	Basic time value in microseconds. Resolution is nanoseconds, maximum value is 1 s.				
	Range (min/resolution/max):		$-10^6 + 10^{-3}$	10^{-3}	$10^6 - 10^{-3}$

18.6.3.5 Angle

The Angle parameter is used to specify angular values such as longitude or latitude of a specific geolocation. It can also be used to specify antenna related parameters such as the elevation angle.

Name:	Angle	Phys. Unit	° (degree)	Extends:	–
ID:	005	**Size:**	1	**Type:**	Fixed-point
Desc:	Generic parameter describing an angular value or measurement.				
	Range (min/resolution/max):		-180	$1\,\mu°$	180

18.6.3.6 Reference Geolocation (RGeolocation)

The RGeolocation parameter indicates an absolute geolocation based on the WGS 84 reference coordinate system or its successors (see National Imagery and Mapping Agency [B6]). It may serve as a reference location for relative positions to specify, for example, geographical areas.

Name:	RGeolocation	Phys. Unit	–	Extends:	–
ID:	006	**Size:**	3	**Type:**	Structured
Desc:	Basic reference geolocation parameter indicating the absolute geolocation of an IEEE 1900.6 logical entity based on the WGS 84 reference coordinate system or its successors.				
.0	**RGeolocation.elev**		**Type**	Signed integer	See NOTE
.1	**RGeolocation.lat**		**Type**	Angle	
.2	**RGeolocation.long**		**Type**	Angle	
NOTE—Elevation is measured as altitude in m with respect to sea level.					

18.6.3.7 Power

Power denotes a single generic parameter given in dBm. It indicates a basic RF power. Derived parameters related to power, relaxing or setting constraints to range and resolution, for example, may extend this definition and may modify the range, type, or physical unit fields accordingly.

Name:	Power	Phys. unit	dBm	Extends:	–
ID:	007	**Size:**	1	**Type:**	Fixed-point
Desc:	Basic RF power value.				
	Range (min/resolution/max):		-160	10^{-1}	70

18.6.3.8 Time Stamp (TimeStamp)

Name:	TimeStamp	Phys. unit	s, µs	Extends:	–
ID:	008	**Size:**	2	**Type:**	Structured
Desc:	Timestamp value. Seconds and microseconds since midnight (UTC) of January 1, 1970.				
.0	**Timestamp.seconds**		**Type:**	RSeconds	
.1	**Timestamp.usec**		**Type:**	Microseconds	

18.6.3.9 Time Duration (TimeDuration)

Name:	TimeDuration	Phys. unit	s, µs	Extends:	–
ID:	009	Size:	2	Type:	Structured
Desc:	Time distance value in seconds and microseconds.				
.0	**TimeDuration.seconds**		Type:	Seconds	
.1	**TimeDuration.usec**		Type:	Microseconds	

18.6.3.10 Channel List (ChList)

The ChList specifies a list of frequency bands by providing a vector of channel numbers along with a corresponding vector of bandwidth parameters.

Name:	ChList	Phys. unit	–	Extends:	–
ID:	010	Size:	2	Type:	Structured
Desc:	Parameter that describes a set of frequency bands.				
.0	**ChList.channel**		Type:	Vector(Unsigned integer)	
.1	**ChList.band**		Type:	Vector(Bandwidth)	

18.6.3.11 Bandwidth

The Bandwidth parameter defines lower and upper frequencies (bandstop and bandstart). The starting frequency can be, for example, 30 MHz for military, 54 MHz for fixed TV White Space, 470 MHz for portable TV White Space, as required for cellular and WiMAX™,[8] and 2400 MHz for Unlicensed bands.[9] To specify fully the bandwidth, the bandstop frequency shall also be specified.

Name:	Bandwidth	Phys. unit	Hz	Extends:	–
ID:	101	Size:	2	Type:	Array(Frequency)
Desc:	Basic unbounded bandwidth parameter.				
.0	**Bandwidth.bandstart**		Type:	Frequency	
.1	**Bandwidth.bandstop**		Type:	Frequency	

18.6.3.12 Total Measurement Duration (TotMeasuDur)

The TotMeasuDur parameter is issued by the IEEE 1900.6 client to the sensor to specify the total duration of time required to measure the spectrum. The actual unit measurement period will be specified by the sensor based on the quality of sensing. Multiple measurements can be reported during the total measurement duration.

Name:	TotMeasuDur	Phys. Unit	s, µs	Extends:	–
ID:	102	Size:	2	Type:	Structured
Desc:	Total duration in seconds and microseconds.				
.0	**TotMeasuDur.seconds**		Type:	Seconds	
.1	**TotMeasuDur.usec**		Type:	Microseconds	

---

8 WiMAX is a trademark in the U.S. Patent & Trademark Office, owned by the WiMAX Forum.
9 This information is given for the convenience of users of this standard and does not constitute an endorsement by the IEEE of these products. Equivalent products may be used if they can be shown to lead to the same results.

18.6.3.13 Channel Order (ChOrder)

The ChOrder parameter specifies an index set. In order to utilize the information stored in ChOrder, the start and stop frequencies of each channel shall be known to both client and sensor. This can be achieved, for example, by providing a ChList parameter along with this parameter.

Name:	ChOrder	Phys. unit	–	Extends:	–
ID:	103	**Size:**	Variable	**Type:**	Vector(Unsigned integer)
Desc:	Parameter that indicates the sequence of channels to be scanned.				

18.6.3.14 Reporting Rate (ReportRate)

The ReportRate parameter is issued by the IEEE 1900.6 client to specify how often the measurement result should be reported or updated (e.g., every 2 s or every 100 ms).

Name:	ReportRate	Phys. unit	s, µs	Extends:	–
ID:	104	**Size:**	2	**Type:**	Structured
Desc:	Time distance value in seconds and microseconds between two consecutive measurement reports.				
.0	**ReportRate.seconds**	**Type:**	Seconds		
.1	**ReportRate.usec**	**Type:**	Microseconds		

18.6.3.15 Reporting Mode (ReportMode)

The ReportMode parameter is issued by the IEEE 1900.6 client to indicate in what form the sensing information should be reported. Two reporting modes are anticipated, namely "hard information" and "soft information." The reporting mode can take a form of one bit hard decision information or soft information in a form of quantized value of signal strength.

Name:	ReportMode	Phys. unit	–	Extends:	–
ID:	105	**Size:**	Variable	**Type:**	Enumeration
Desc:	Reporting mode to indicate either hard information or soft information.				
Enumerator	**ReportMode.hard**	**Value:**	0	See NOTE 1	
Enumerator	**ReportMode.soft**	**Value:**	1	See NOTE 2	
NOTE 1—When hard information reporting mode is selected, sensors provide information that only indicates the spectral occupancy as vacant or occupied. NOTE 2—When soft information reporting mode is selected, sensors provide information that indicates the RF signal level (signal strength) in the band of interest.					

18.6.3.16 Performance Metric (PerfMetric)

The PerfMetric parameter is issued by the IEEE 1900.6 client to the sensors to indicate the quality of sensing. In particular, this value is needed at the sensor side if local decision is carried. Sensors perform measurement configuration to obtain the desired performance metric as specified by the IEEE 1900.6 client.

Name:	PerfMetric	Phys. unit	%	Extends:	–
ID:	106	**Size:**	Variable[a]	**Type:**	Array(Unsigned integer)
Desc:	Parameter that indicates the quality of sensing.				
.0	**PerfMetric.Pd**	**Type:**	Unsigned integer	See NOTE 1	
.1	**PerfMetric.Pfa**	**Type:**	Unsigned integer	See NOTE 2	
NOTE 1—When the PerformanceMetric.pd is specified, sensors perform sensing by setting the rate of detection according to this value. Rate of detection is expressed as a percentage bounded between 0% and 100%. NOTE 2—When the PerformanceMetric.pfa is specified, sensors perform sensing by setting the rate of false alarm according this value. Rate of false alarm is expressed as a percentage bounded between 0% and 100%.					

a) In some cases, only one of the information elements can be sufficient to describe the performance.

18.6.3.17   Route

The Route parameter carries information of the path in which the measurement data flows from remote sensors to the client. Routing information may indicate only one sensor to be used as an intermediate node between the remote sensors and the client where this sensor can act as a gateway between the client and multiple sensors. Data fusion at intermediate nodes or gateways can reduce the traffic load, can offload computational demands, and may reduce sensor transmit power toward the IEEE 1900.6 client, potentially reducing communication overhead and extending sensor battery life.

Name:	Route	Phys. Unit	–	Extends:	–
ID:	107	Size:	Variable	Type:	Array(String)
Desc:	Route command indicates the information flow path.				
.0	**Route.path1**		Type:	String	See NOTE 1
.1	**Route.path2**		Type:	String	See NOTE 2
NOTE 1—Route.path1 is the client address. NOTE 2—Route.path2 is the IEEE 1900.6 logical entity that may act as a gateway. Depending on the path length, additional addresses may be specified.					

18.6.3.18   ClientPriorityFlag

The ClientPriorityFlag parameter indicates a priority-level flag. Such a priority-level flag is assigned based on the application for which the client exploits the sensing information.

Name:	ClientPriorityFlag	Phys. Unit	–	Extends:	–
ID:	108	Size:	1	Type:	Unsigned integer	See NOTE
Desc:	Client priority indicator where integer values represent different priority levels.				
NOTE—It is assumed that priority levels are defined and understood by both the client and the sensor.					

18.6.3.19   SensorPriority

The SensorPriority parameter indicates which sensors are given priority to report their measurement result. After the client obtains the necessary information about the potential sensing information sources, it will assign priority to each sensor. Only information sources indicated by the sensor priority will be valid to provide sensing information. This parameter is useful when implementing distributed sensing where sensors with an acceptable link quality are allowed to take part in the cooperation. Priority ranking may also be used to implement weighted combining.

Name:	SensorPriority	Phys. Unit	--	Extends:	--
ID:	109	Size:	Variable	Type:	Vector(String)
Desc:	Sensor priority indicates the list of selected sensors according to priority ranking				
NOTE—When the sensor priority flag is on, it contains a list of sensor's IDs according to a descending priority rank. This information is broadcasted back to the sensors to indicate which sensors are selected for the measurement report.					

18.6.3.20   Security Level (SecLevel)

The Seclevel parameter indicates the level of security depending on the available levels of security for authentication and data verification. Once the security level preference is indicated by the application, the corresponding authentication procedure is carried out by the communication subsystem.

Name:	SecLevel	Phys. Unit	–	Extends:	–
ID:	110	**Size:**	1	**Type:**	Unsigned integer
Desc:	Security level indicator.				
NOTE—Depending on the application, a client may specify a high or low level of security from the available authentication mechanisms supported by both the client and the server.					

18.6.3.21 DataKey

The DataKey parameter is an information ID that identifies the type of database information stored in the DA. This ID is used by the client to filter the type of information needed from the DA. Database information identifiers (a set of IDs and corresponding database information type) should be specified and known by clients.

Name:	DataKey	Phys. Unit	–	Extends:	–
ID:	111	**Size:**	1	**Type:**	Unsigned integer
Desc:	Indicates the type of database information stored in the DA.				

18.6.3.22 ClientLogID

ClientLogID represents the unique logical identification of the IEEE 1900.6 client. It can be used to identify the control entity during a request for sensing information.

Name:	ClientLogID	Phys. unit	–	Extends:	–
ID:	112	**Size:**	1	**Type:**	String
Desc:	Unique logical identification of an IEEE 1900.6 client.				

18.6.3.23 Time Synchronization (TimeSync)

The TimeSync command provides the system to be synchronized. The command described here is with a minimum parameter space. Unnecessary synchronization wastes resources, and some sensors may not have the resource to implement advanced protocols. Insufficient synchronization also leads to poor application performance. This command can be extended by incorporating more information elements depending on the choice of synchronization scheme.

Name:	TimeSync	Phys. Unit		–	Extends:	–
ID:	201	**Size:**		3	**Type:**	Structured
Desc:	Sync command for synchronization.					
.0	**TimeSync. ClockAddress (ClientLogID)**			**Type:**	---	
.1	**TimeSync.MaxError**			**Type:**	integer	See NOTE
.2	**TimeSync.RSecond**			**Type:**	RSecond	
NOTE—Max error can be defined as tolerable error where it is given by the ratio of time offset divided by reference time.						

18.6.3.24 Scan

The Scan command instructs the sensor to scan between a lower and a upper frequency bound with a given measurement bandwidth and temporal resolution. This command can be extended by strategies that are more complex.

Name:	Scan	Phys. Unit	—	Extends:	—	
ID:	202	**Size:**	6	**Type:**	Structured	
Desc:	Basic scan command.					
.0	**Scan.Range**		**Type:**	Bandwidth		
.1	**Scan.Stepbandwidth**		**Type:**	Bandwidth		
.2	**Scan.Resolution**		**Type:**	Frequency		
.3	**Scan.TimePerStep**		**Type:**	TimeDuration		
.4	**Scan.LowerThreshold**		**Type:**	NoisePower		
.5	**Scan.ChannelOrder**		**Type**	ChOrder	See NOTE	
NOTE—ChannelOrder specifies the scan order according to a predefined vector of step indices (sequence vector should use a bounded type rather than the generic IntVector type). When this parameter is omitted or set to 0, the scan order will be assumed linear or sequential as the default scan order.						

18.6.3.25   Absolute Sensor Location (AbsSensorLocation)

The AbsSensorLocation parameter indicates the absolute location of the sensor.

Name:	AbsSensorLocation	Phys. unit	—	Extends:	—
ID:	301	**Size:**	1	**Type:**	RGeolocation
Desc:	The absolute sensor location parameter indicates the absolute location of the sensor.				

18.6.3.26   Relative Sensor Location (RelSensorLocation)

The RelSensorLocation parameter is used when the position of the sensor is measured by triangulation with respect to known reference positions.

Name:	RelSensorLocation	Phys. unit	—	Extends:	—
ID:	302	**Size:**	1	**Type:**	RGeolocation
Desc:	Relative sensor location parameter describes the location with respect to a known reference point.				

18.6.3.27   Measurement Range (MeasuRange)

The MeasuRange parameter indicates the range of frequencies that can be measured according to the manufacturer specification.

Name:	MeasuRange	Phys. unit	—	Extends:	—
ID:	303	**Size:**	1	**Type:**	Bandwidth
Desc:	Range of frequency supported by the sensor according to the manufacturer specification.				

18.6.3.28   SensingMode

The SensingMode parameter indicates the numbers and types of standard sensing methods supported by the sensor. Note that the size of SensingMode is variable, which allows the user to choose between multiple sensing modes. These sensing techniques are indentified in Noguet et al. [B7] and Pucker [B8].

Name:	SensingMode	Phys. unit		--	Extends:	--
ID:	304	Size:		Variable	Type:	Enumeration
Desc:	List of sensing modes supported by the sensor.					
Enumerator		SensingMode.EnergyDetection			Value	0
Enumerator		SensingMode.MatchedFilter			Value	1
Enumerator		SensingMode.Cyclostationary			Value	2
Reserved						

18.6.3.29 SensorID

The SensorID parameter describes information about the sensor identity as assigned by the manufacturer. It contains both the vendor identification and product identification.

Name:	SensorID	Phys. unit		–	Extends:	–
ID:	305	Size:		2	Type:	Array(String)
Desc:	Unique identification of the sensor.					
.0	**SensorID.VendorID**		Type:		String	
.1	**SensorID.ProductID**		Type:		String	

18.6.3.30 Sensor Logical ID (SensorLogID)

SensorLogID is a parameter indicating the logical ID of a sensor within a certain area.

Name:	SensorLogID	Phys. unit		–	Extends:	–
ID:	306	Size:		1	Type:	String
Desc:	Unique logical identification of a sensor within a certain area.					

18.6.3.31 BatteryStatus

When the sensor is battery operated, the BatteryStatus parameter indicates the remaining power.

Name:	BatteryStatus	Phys. unit	%	Extends:	–	
ID:	307	Size:	1	Type:	Unsigned integer	
Desc:	The remaining power in percentage.					
	Range (min/resolution/max):		0	1	100	

18.6.3.32 DataSheet

The DataSheet parameter indicates standard manufacturer specifications, which are also known by the client device.

> **Note.** The manufacturer ultimately defines the content provided by the datasheet parameter. Some of the subparameters thus might be void; that is, it is assumed that most/all of the parameters are optional. The primitive retrieves all subparameters given that the corresponding capability is provided by the sensor. Otherwise, the subparameter is omitted. The list of datasheet parameters given in the parameter description thus is the best case.

Name:	DataSheet	Phys. unit	–	Extends:	–
ID:	308	Size:	Variable	Type:	Structured
Desc:	List of datasheet items according to manufacturer specification.				
.0	**DataSheet.ADDAResolution**		Type:		Unsigned integer
.1	**DataSheet.AmplitudeSensitivity**		Type:		Float
.2	**DataSheet.AngleResolution**		Type:		Angle
.3	**DataSheet.AntennaBandwidth**		Type:		Frequency
.4	**DataSheet.AntennaBeamPointing**		Type:		Array(Angle)
.5	**DataSheet.AntennaBeamwidth**		Type:		Array(Angle)
.6	**DataSheet.AntennaDirectivityGain**		Type:		Float
.7	**DataSheet.AntennaGain**		Type:		Float
.8	**DataSheet.AntennaHeight**		Type:		Float
.9	**DataSheet.AntennaPolarization**		Type:		Enumeration
.10	**DataSheet.CalibrationData**		Type:		String
.11	**DataSheet.CalibrationMethod**		Type:		Unsigned integer
.12	**DataSheet.ChannelFiltering**		Type:		String
.13	**DataSheet.DynamicRange**		Type:		Float
.14	**DataSheet.FrequencyResolution**		Type:		Frequency
.15	**DataSheet.LocationTimeCapability**		Type:		String
.16	**DataSheet.LoggingFunctions**		Type:		String
.17	**DataSheet.NoiseFactor**		Type:		Float
.18	**DataSheet.PhaseNoise**		Type:		Float
.19	**DataSheet.PowerConsumption**		Type:		String
.20	**DataSheet.RecordingCapability**		Type:		String
.21	**DataSheet.SweepTime**		Type:		Time
.22	**Reserve**				

18.6.3.33 ConfidenceLevel

The ConfidenceLevel parameter indicates the degree of certainty that a statistical estimation of the spectrum measurement (e.g., frequency, amplitude, or phase) is accurate. A new confidence level can be computed whenever the current confidence level needs to be updated, for instance, when the sample size changes.

Name:	ConfidenceLevel	Phys. unit	–	Extends:	–
ID:	309	Size:	2	Type:	Structured
Desc:	The ConfidenceLevel parameter should be accompanied by the confidence interval in which the true value may be located with respect to the estimated value.				
.0	**ConfidenceLevel.Value**	Type:	Float		
.1	**ConfidenceLevel.Interval**	Type:	Array(Float)		

18.6.3.34 Measurement Bandwidth (MeasuBandwidth)

The MeasuBandwidth parameter is the same as the bandwidth used by the control parameter, but when the value is returned by the sensor, it indicates the measurement bandwidth that has actually been used by the sensor.

Name:	MeasuBandwidth	Phys. unit	Hz	Extends:	--
ID:	401	Size:	2	Type:	Array(Frequency)
Desc:	Describes the bandwidth being used by the sensor.				
.0	**MeasuBandwidth.BandStart**	Type:	Frequency		
.1	**MeasuBandwidth.BandStop**	Type:	Frequency		

18.6.3.35 NoisePower

Noise power is thermal noise in accordance with the measurement bandwidth.

Name:	NoisePower	Phys. unit	dBm	Extends:	Power
ID:	402	Size:	1	Type:	Fixed point
Desc:	Basic RF noise power value.				
	Range (min/resolution/max):		−173	10^{-2}	0

18.6.3.36 SignalLevel

The SignalLevel parameter designates the amplitude of the measured signal power within a specific bandwidth.

Name:	SignalLevel	Phys. unit	—	Extends:	Power
ID:	403	Size:	Variable	Type:	Structured
Desc:	SignalLevel indicating the RF power of measured signal.				
.0	**SignalLevel.Soft**		Type	Power	
.1	**SignalLevel.Hard**		Type	Boolean	

18.6.3.37 Modulation Type (ModuType)

The ModuType parameter specifies the type of modulation that the detected signal exhibits.

Name:	ModuType	Phys. unit	—	Extends:	—
ID:	404	Size:	1	Type:	Enumeration
Desc:	List of known modulation types for RF signal characterization.				
Enumerator	**ModuType.Unknown**		Value:	0	
Enumerator	**ModuType.Analog**		Value:	1	
Enumerator	**ModuType.Digital**		Value:	2	

18.6.3.38 TrafficPattern

The TrafficPattern parameter classifies the detected signal in temporal domain as periodic burst, random burst, continuous, or unknown. This parameter can be helpful in identifying a signal by comparing its content with known traffic patterns of existing standard air interfaces.

Name:	TrafficPattern	Phys. unit	—	Extends:	—
ID:	405	Size:	1	Type:	Enumeration
Desc:	List of traffic patterns.				
Enumerator	**TrafficPattern.Unknown**		Value:	0	
Enumerator	**TrafficPattern.Continuous**		Value:	1	
Enumerator	**TrafficPattern.PeriodicBurst**		Value:	2	
Enumerator	**TrafficPattern.Random**		Value:	3	

18.6.3.39 TrafficInformation

The TrafficInformation parameter indicates channel load (e.g., code allocation ratio, time slot allocation ratio, and subcarrier allocation ratio). This parameter is helpful for balancing traffic among channels (e.g., off loading peak traffic into other channels).

Name:	TrafficInformation	Phys. unit	—	Extends:	—
ID:	406	Size:	1	Type:	Float
Desc:	The traffic information parameter indicates channel load.				

18.6.3.40 SignalType

The SignalType parameter specifies the detected signal in terms of its RF characteristics when possible. Such characterization can be very useful to perform a coarse sensing first followed by a second level of fine sensing once the signal type is known.

Name:	SignalType	Phys. unit	--	Extends:	--		
ID:	407	Size:	1	Type:	Enumeration		
Desc:	List of known signal types for RF signal characterization.						
Enumerator	SignalType.Unknown		Value:	0	NOTE	1	
Enumerator	SignalType.Compound		Value:	1	NOTE	2	
Enumerator	SignalType.Noise		Value:	2	NOTE	3	
Enumerator	SignalType.Multicarrier		Value:	3	NOTE	—	
Enumerator	SignalType.Singlecarrier		Value:	4	NOTE	—	

NOTE 1—*Unknown* is selected if the sensor cannot determine a valid signal type.
NOTE 2—*Compound* is selected if the sensor detects a signal too complex to describe by a single parameter (e.g.,OFDM with time-variant subcarrier modulation).
NOTE 3—*Noise* is selected if the sensor detects structured noise (e.g., correlated) but cannot determine the underlying signal type (e.g., due to low power level).

18.6.3.41 SignalDesc

The signal description parameter represents a set of information elements that describe the behavior of the signal such as traffic pattern, type of signal, and modulation scheme.

Name:	SignalDesc	Phys. unit	--	Extends:	--
ID:	408	Size:	variable	Type:	Structured
Desc:	Detail description of a detected RF signal.				
.0	SignalDesc.Amplitude		Type:	Float	
.1	SignalDesc.AngleOfArrival		Type:	Angle	
.2	SignalDesc.Bandwidth		Type:	Bandwidth	
.3	SignalDesc.Dutycycle		Type:	Unsigned integer	
.4	SignalDesc.Frequency		Type:	Frequency	
.5	SignalDesc.Modulation		Type:	ModulationType	
.6	SignalDesc.Phase		Type:	Angle	
.7	SignalDesc.Power		Type:	Power	
.8	SignalDesc.TrafficInformation		Type:	TrafficInformation	
.9	SignalDesc.TrafficPattern		Type:	TrafficPattern	
.10	SignalDesc.Type		Type:	SignalType	
	Reserve				

18.6.3.42 RATID

The RATID parameter gives an ID of an existing standard RAT, for example, Global System for Mobile (GSM), etc.

Name:	RATID	Phys. unit	—	Extends:	—
ID:	409	Size:	1	Type:	Enumerated
Desc:	Unique identification of existing standard RATs known by both spectrum sensors and their clients.				
Enumerator	RATID.GSM		Value:	0	
Enumerator	RATID.WiMAX		Value:	1	
Enumerator	RATID.CDMA2000		Value:	2	
Enumerator	RATID.WiFi		Value:	3	

18.6.4 Data Representation

This subclause gives the data representation for this standard using Unified Modeling Language (UML). The sensing-related information described in 18.6.3 is grouped into the following four classes:

— ControlInformation
— SensorInformation
— SensingInformation
— RegulatoryRequirement

18.6.4.1 ControlInformation Class

As shown in Figure 18.13, this class describes control parameters for realizing sensing information exchange between sensors and their clients.

— Transport
This class contains control parameters for transport of sensing information. Each instance of ControlInformation class can have 0 or 1 instance of Transport class.
  — TransportNetwork
  This class contains parameters to control network for the transportation of sensing information. The instance of Transport class can have only one instance of TransportNetwork class.
  — TransportMAC
  This class contains parameters to control MAC for the transportation of sensing information. The instance of Transport class can have only one instance of TransportMAC class.
  — TransportMode
  This class contains parameters to control the transportation mode for sensing information exchange, for example, information type such as soft information and hard information. An instance of Transport class can have only one instance of TransportMode class.
— Measurement
This class contains control parameters for spectrum measurement. Each instance of ControlInformation class can have zero or several instances of Measurement class.
  — MeasurementObj
  This class contains parameters to control the measurement objective, for example, the start frequency and end frequency of a radio frequency to be measured. Any instance of Measurement class can have one or several instances of MeasurementObj class.
  MeasurementProfile
  This class contains parameters to control the behavior of measurement, for example, determining the technique that shall be used for spectrum sensing. Any instance of Measurement class can have one or several instances of MeasurementProfile class.
  — MeasurementPerformance
  This class contains parameters to control the performance of radio frequency measurement, (e.g., the desired probability of false alarm as the output of spectrum sensing). Any instance of Measurement class can have one or several instances of MeasurementPerformance class.
— Application
This class contains control parameters for sensing information request and sensing information provision. Each instance of ControlInformation class can have zero or several instances of Application class.
  — SensingInformationRequest
  This class contains parameters to set priority levels for the clients to request sensing information. Each instance of Application class has one or multiple instances of SensingInformationRequest class.

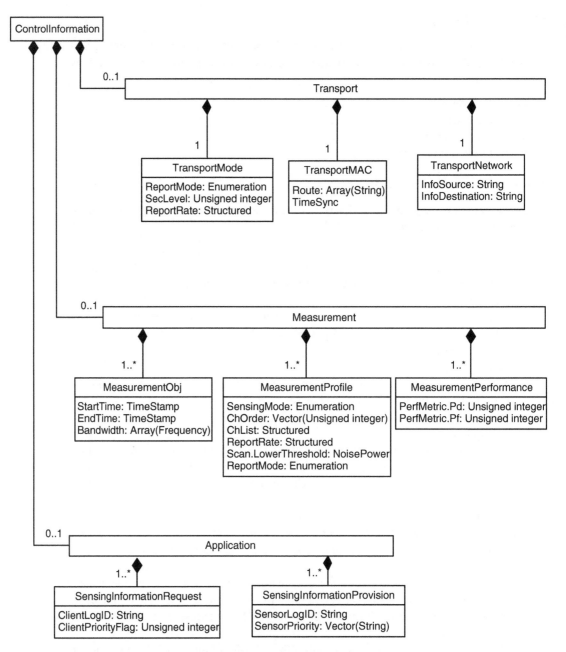

Figure 18.13 —UML class diagram of control information classes. From IEEE Standard for Spectrum Sensing Interfaces and Data Structures for Dynamic Spectrum Access and Other Advanced Radio Communication Systems, Feb 2011. IEEE.

— SensingInformationProvision

This class contains parameters to set priority levels for the sensors when providing sensing information to their clients. Each instance of Application class has one or multiple instances of SensingInformationRequest class.

18.6.4.2 SensorInformation Class

As shown in Figure 18.14, this class describes parameters related to sensors.

— SensorPHYProfile

This class contains information about PHY of sensors, for example, phase noise. Each instance of SensorInformation can have only one instance of SensorPHYProfile class.

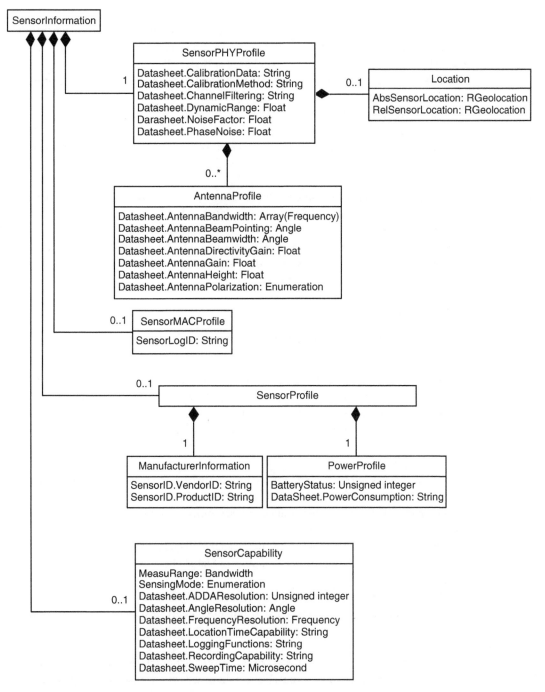

Figure 18.14 —UML class diagram of sensor information classes. From IEEE Standard for Spectrum Sensing Interfaces and Data Structures for Dynamic Spectrum Access and Other Advanced Radio Communication Systems, Feb 2011. IEEE.

— Location

This class contains information about sensors location, for example, absolute location and relative location. Each instance of SensorPHYProfile can have zero or one instance of LocationProfile class.

— AntennaProfile

This class contains information about sensors' antennas, for example, antenna hight, beam pattern, etc. Each instance of SensorPHYProfile can have zero or multiple instances of AntennaProfile class.

— SensorMACProfile

This class contains information about MAC of sensors, for example, MAC ID. Each instance of SensorInformation can have zero or one instance of SensorMACProfile class.

— SensorProfile

This class contains general information about sensors. Each instance of SensorInformation can have zero or one instance of SensorProfile class.

— ManufacturerInformation

This class contains manufacturer information of sensors such as manufacturer's ID and product ID. Each instance of SensorProfile class has one instance of ManufacturerInformation class.

— PowerProfile

This class contains power information of sensors such as battery level and power consumption. Each instance of SensorProfile class has one instance of PowerProfile class.

— SensorCapability

This class contains information about sensors' capability of radio spectrum measurement, such as measurement range, sensitivity, etc. Each instance of SensorInformation can have zero or one instance of SensorCapability class.

18.6.4.3 SensingInformation Class

As shown in Figure 18.15, this class describes parameters of sensing results.

— SensingInformationProfile

This class contains general information about the measurement that is carried out to produce sensing information. For example, the information can be confidence level of the measurement, measurement feature, threshold used in the measurement, etc. One instance of SensingInformation class has only one instance of SensingInformationProfile class.

— LocationInformation

This class contains information about the location where the measurement has been carried out, for example, absolute location and relative location. Each instance of SensingInformation class has zero or one instance of LocationInformation class.

— Signal

This class contains information about the measured signals. Each instance of SensingInformation class has zero or multiple instances of Signal class.

— SignalProfile

This class contains general information about the measured signals, such as signal level, estimated amplitude, etc. Each instance of Signal class can have only one instance of SignalProfile class.

— SignalBehavior

This class contains information describing the behavior of the measured signals such as duty cycle. One instance of Signal class can have zero or one instance of SignalBehavior class.

— Channel

This class contains information about the measured channel. Each instance of SensingInformation class has zero or multiple instances of Channel class.

— ChannelProfile

This class contains general information about the measured channel such as start frequency and stop frequency. Each instance of Channel class has only one instance of ChannelProfile class.

— ChannelMeasurement

This class contains information about the measurement results of the channel such as measured bandwidth and noise power. Each instance of Channel class has zero or one instance ChannelMeasurement class.

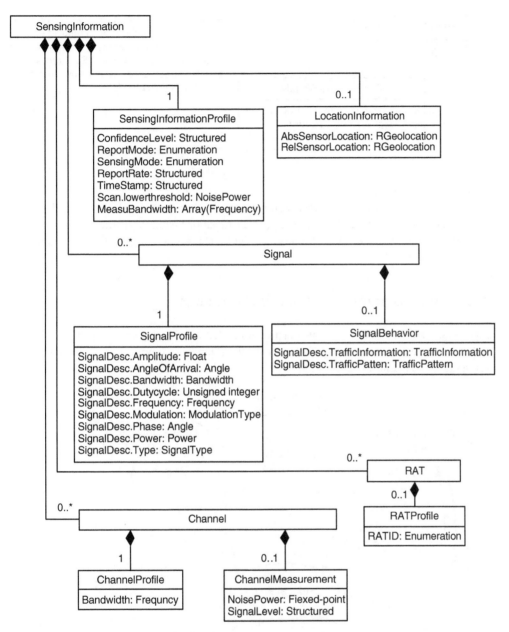

Figure 18.15 —UML class diagram of sensing information classes. From IEEE Standard for Spectrum Sensing Interfaces and Data Structures for Dynamic Spectrum Access and Other Advanced Radio Communication Systems, Feb 2011. IEEE.

— RAT

This class contains information about RAT IDs. Each instance of SensingInformation class has zero or multiple instances of RAT class.

— RATProfile

This class contains information about the IDs of RAT. Each instance of RAT class has zero or one instance of RATProfile class.

18.6.4.4 RegulatoryRequirement Class

As shown in Figure 18.16, this class contains requirements derived from regulation according to different primary user signals at different locations. The types of signals are different in terms of regulations.

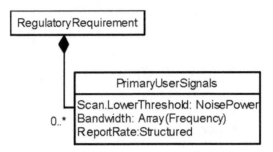

Figure 18.16 —UML class diagram of regulatory requirement classes. From IEEE Standard for Spectrum Sensing Interfaces and Data Structures for Dynamic Spectrum Access and Other Advanced Radio Communication Systems, Feb 2011. IEEE.

— PrimaryUserSignals
This class contains the regulatory requirements of primary user signals such as detection threshold, detection bandwidth, etc. Each instance of RegulatoryRequirement class has zero or multiple instances of PrimaryUserSignals class depending on different regulatory requirements.

18.7 State Diagram and Generic Procedures

Figure 18.17 shows a high-level state diagram for an IEEE 1900.6 logical entity. It includes initialization state, idle state, data gathering state, simultaneous communication and data gathering state, and communication state.

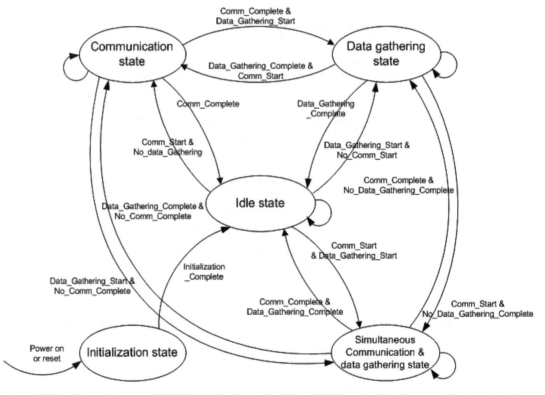

Figure 18.17 —State diagram of IEEE 1900.6 logical entity. From IEEE Standard for Spectrum Sensing Interfaces and Data Structures for Dynamic Spectrum Access and Other Advanced Radio Communication Systems, Feb 2011. IEEE.

18.7.1 State Description

18.7.1.1 Initialization State

The initialization state includes functions that shall be executed before sensing information exchange, for example, obtaining the control/application ID.

18.7.1.2 Idle State

This is the state when there is no usage of the logical interface.

18.7.1.3 Data Gathering State

The data gathering state includes functions that shall be executed for obtaining measurement results. It also includes functions that shall be executed for obtaining sensing information for the control/application.

18.7.1.4 Communication State

The communication state includes functions that shall be executed for transportation of sensing-related information from an IEEE 1900.6 logical entity to another IEEE 1900.6 logical entity using the communication subsystem.

18.7.1.5 Simultaneous Communication and Data Gathering State

In the simultaneous communication and data gathering state, communication for sensing information exchange and data gathering take place at the same time. For example, the measurement module performs spectrum measurement at frequency f1 and the communication subsystem is transporting sensing information at another frequency f2. The sensor may also receive sensing control information during execution of the measurement.

Note that this state may not be present in simple systems performing only one action at a time (communication or data gathering).

18.7.2 State Transition Description

18.7.2.1 Initialization State→Idle State

After power on or reset, IEEE 1900.6 logical entities go to idle state waiting for other commands.

18.7.2.2 Idle State→Data Gathering State

The IEEE 1900.6 logical entity is triggered to collect sensing information without communication request. This excludes the transition from the idle state to the communication state. For example, the control/application initiates spectrum sensing by an internal spectrum measurement module and starts to collect sensing information.

18.7.2.3 Idle State←Data Gathering State

The IEEE 1900.6 logical entity acknowledges that sensing information has been obtained, which indicates the completion of data gathering.

18.7.2.4 Idle State→Communication State

The IEEE 1900.6 logical entity requests to send available sensing information and commands/primitives remote entity without data gathering request. This excludes the transition from the idle state to the data gathering state.

18.7.2.5 Idle State←Communication State

The IEEE 1900.6 logical entity acknowledges that the sensing information or commands/primitives payload has been received.

18.7.2.6 Data Gathering State→Communication State

The IEEE 1900.6 logical entity requests to send available sensing information and commands/primitives remote entity upon completion of data gathering.

18.7.2.7 Data Gathering State←Communication State

The IEEE 1900.6 logical entity starts data gathering after confirming that the sensing information or commands/primitives payload have been received.

18.7.2.8 Idle→Simultaneous Communication and Data Gathering State

The IEEE 1900.6 logical entity starts both data gathering and communication at the same time.

18.7.2.9 Idle←Simultaneous Communication and Data Gathering State

The IEEE 1900.6 logical entity completes a data gathering task and communication task at the same time.

18.7.2.10 Simultaneous Communication and Data Gathering State→Communication State

The IEEE 1900.6 logical entity ends the data gathering while maintaining the communication status; i.e., the communication task is ongoing.

18.7.2.11 Simultaneous Communication and Data Gathering State←Communication State

The IEEE 1900.6 logical entity is requested to initiate a data gathering task while the previously initiated communications task continues.

18.7.2.12 Simultaneous Communication and Data Gathering State→Data Gathering State

The IEEE 1900.6 logical entity completes its communications task while continuing its data gathering task.

18.7.2.13 Simultaneous Communication and Data Gathering State←Data Gathering State

The IEEE 1900.6 logical entity is requested to start the communication for sending information to another entity as payload while maintaining the data gathering status. The IEEE 1900.6 logical entity may also receive configuration commands while executing measurement.

18.7.3 Generic Procedures

18.7.3.1 Purpose

This subclause provides generic procedures and usage examples for the exchange of sensing-related information between an IEEE 1900.6 logical entity and its client.

An IEEE 1900.6 logical entity shall be a CE, DA, or Sensor as defined in 18.3.1. Any IEEE 1900.6 logical entity can be a client to one or more of these. Additionally, any other entity not explicitly defined by the standard can act as a client provided that the interface procedures are used in conformance with the standard.

Without loss of generality, for the following discussion, a Sensor is assumed as the primary source of information. CE and DA are assumed to act primarily as clients but may also provide information within specific scenarios.

To express the procedures, the service user and service provider model concept as specified by ITU-T X.210 [B5] is used as shown in Figure 18.18 (cf. 18.3.1). The interface between the IEEE 1900.6 logical entity and its client is assumed transparent in this information flow with respect to any kind of transport service that may be needed to mediate between the service user side and service provider side of the interface.

18.7.3.2 Generic Procedure and Notations

In the description of the generic procedure, the following terms are used along with the terms known in the document.

IEEE 1900.6 Service Provider: An abstraction of the totality of those entities that provide an IEEE 1900.6 service to the service user. The IEEE 1900.6 service provider includes the IEEE 1900.6 service (as indicated in the reference model in Clause 18.5) and the associated 1900.6 SAPs that instantiate the logical interface for exchange of sensing-related information between the IEEE 1900.6 servers and clients. (This is a logical mechanism that should be seen different from the IEEE 1900.6 logical entities.)

IEEE 1900.6 Service User: The IEEE 1900.6 logical entities that shall implement or use the IEEE 1900.6 logical interface to exchange sensing information. For example, the IEEE 1900.6 logical entity that shall use the logical interface to obtain sensing-related information plays the client role. While the IEEE 1900.6 logical entity that shall use the logical interface to provide sensing information plays a server role. This means, the client controls the initiation and termination of the sensing information exchange.

The basic communication flow for a synchronous, blocking primitive is described as follows:

— When a client IEEE 1900.6 logical entity needs to communicate with another IEEE 1900.6 logical entity (server) (e.g., to acquire sensing information), it will create a request (request message) to an IEEE 1900.6 service provider.

— In the following message exchange procedure, the IEEE 1900.6 service forwards the request along with its parameters toward the remote IEEE 1900.6 logical entity. The parameters from the IEEE 1900.6 logical interface forms the service data unit (SDU), which shall be communicated between client and server as the payload of a protocol data unit (PDU) used by the protocol layers beneath. Upon reception of an SDU, the service provider generates an indication toward the remote IEEE 1900.6 logical entity (server). The actual procedures and protocols used in the communication subsystem are out of the scope of this standard. Upon receiving the indication, the server performs further processing to consider or reject the indication.

— In response to the indication, the IEEE 1900.6 logical entity (server) generates a response to the IEEE 1900.6 service provider to communicate the results (e.g., the sensing parameters requested), toward the client (response message) via the logical interface and supporting C-SAP. It is also possible for either an error message or an acknowledgement message to be conveyed in the response message.

— Upon successful reception of the response message on the client side, the IEEE 1900.6 service provider generates a confirmation toward the initiating client providing the SDU as originated by the remote IEEE 1900.6 logical entity (server). For this type of synchronous communication flow, the confirmation indicates that the IEEE 1900.6 request has been completed and the client may now check if it obtained valid results.

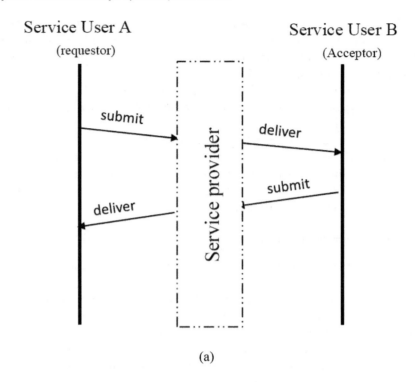

(a)

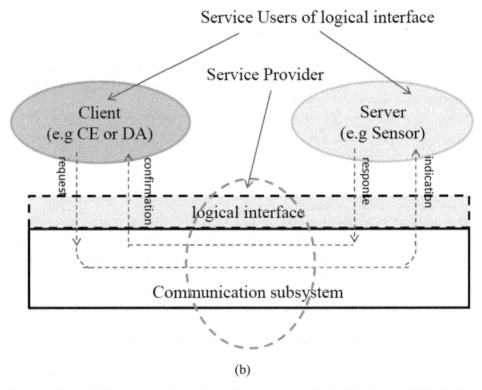

(b)

Figure 18.18 —(a) Information exchange between service users of logical interface and (b) logical interface on top of communication subsystem acting as service provider for the exchange of information between client and server. From IEEE Standard for Spectrum Sensing Interfaces and Data Structures for Dynamic Spectrum Access and Other Advanced Radio Communication Systems, Feb 2011. IEEE.

Note that, in the preceding communication message flow (also as depicted in Figure 18.18(a) and Figure 18.18(b)), only the service request from service user A or client is depicted. In this case, the service user B or the server is stated as "acceptor," whereas the service user is stated as "requestor." The requestor and acceptor role changes depending on who is initiating the service request.

— Requestor: A service user from whom a service request originates
— Acceptor: A service user to whom a service request is targeted

Following these roles of the service users, the four primitives [depicted in Figure 18.18(a) and Figure 18.18(b)] can be written only by using two basic service-primitives known as submit and deliver, as follows:

— Service-primitive:
  Submit: a basic service primitive issued by service user
  Deliver: a basic service primitive issued by service provider
— Primitives:
  Request, or requestor.submit
  Indication, or acceptor.deliver
  Response, or acceptor.submit
  Confirm, or requestor.deliver

For the purpose of this standard, indication primitives are equivalent to their request counterparts, and confirmation primitives are equivalent to their response counterparts.

18.7.4 Example Procedures for Use Cases

This subclause presents multistep procedures by using the generic procedure shown in 18.7.3. These examples are based on different instances of the IEEE 1900.6 logical entities paired for sensing information exchange involved in different scenarios in Clause 18.4.

18.7.4.1 CE–Sensor Procedure

In this example, as depicted in Figure 18.19, a sequence of the generic procedure is used to portray a typical information exchange between CE and Sensor. The procedure is initialized by the spectrum sensing demand by the CE. The complete cycle to obtain sensing information can follow the following steps:

a) CE detects a spectrum sensing demand (internal)
b) CE performs spectrum sensing if it has embedded sensor (internal)
c) CE detects control channel (to obtain additional information from externally located sensor)
d) CE identifies spectrum sensors (sending control signal and receiving response)
   1) Request: CE submits a request on the presence of spectrum sensor/s
   2) Indication: Deliver primitive is issued to the sensor so that it responds on its presence
   3) Response: Sensor submits its presence status information
   4) Confirm: Deliver primitive is issued to CE to confirm successful reception of message from sensor
e) CE identifies capability of the sensor
   1) Request: CE submits a request to provide sensor's profile
   2) Indication: Deliver primitive is issued to the sensor so that it responds to the request
   3) Respond: Sensor submits its profile information
   4) Confirm: Deliver primitive is issued to the CE to confirm successful reception of the sensor's message

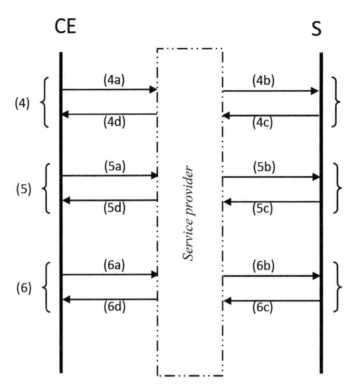

Figure 18.19 —Sensing information exchange procedure between CE and Sensor (S). From IEEE Standard for Spectrum Sensing Interfaces and Data Structures for Dynamic Spectrum Access and Other Advanced Radio Communication Systems, Feb 2011. IEEE.

f) CE requests sensing information
  1) Request: CE submits a request to provide sensing information (by specifying control parameters)
  2) Indication: Deliver primitive is issued to the sensor so that it responds to the request
  3) Respond: Sensor submits sensing information
  4) Confirm: Deliver primitive is issued to CE to confirm successful reception of sensor's message

18.7.4.2 DA–Sensor Procedure

In this example, as depicted in Figure 18.20, a sequence of the generic procedure is used to portray a typical information exchange between DE and Sensor. The procedure is initialized by the spectrum sensing demand in the DA. The complete cycle to obtain sensing information might follow the following steps:

a) DA regularly updates sensing and related information (internal)
b) DA detects control channel (to obtain sensing information from externally located sensor/s)
c) DA identifies spectrum sensors (sending control signal and receiving response)
  1) Request: DA submits a request on the presence of spectrum sensor(s)
  2) Indication: Deliver primitive is issued to the sensor so that it responds on its presence
  3) Response: Sensor submits its presence status information
  4) Confirm: Deliver primitive is issued to DA to confirm successful reception of message from sensor
d) DA identifies capability of the sensor
  1) Request: DA submits a request to provide sensor's profile
  2) Indication: Deliver primitive is issued to the sensor so that it responds to the request
  3) Respond: Sensor submits its profile information
  4) Confirm: Deliver primitive is issued to DA to confirm successful reception of the sensor's message

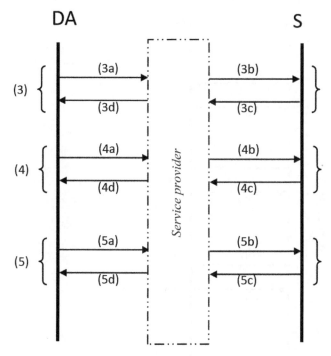

Figure 18.20 —Sensing information exchange procedure between DA and Sensor (S). From IEEE Standard for Spectrum Sensing Interfaces and Data Structures for Dynamic Spectrum Access and Other Advanced Radio Communication Systems, Feb 2011. IEEE.

e) DA requests sensing information
1) Request: DA submits a request to provide sensing information (by specifying control parameters)
2) Indication: Deliver primitive is issued to the sensor so that it responds to the request
3) Respond: Sensor submits sensing information
4) Confirm: Deliver primitive is issued to DA to confirm successful reception of sensor's message

18.7.4.3 CE–DA Procedure

In this example, as depicted in Figure 18.21, a sequence of the generic procedure is used to portray a typical information exchange between CE and DA. The procedure is initialized by the spectrum sensing demand in the CE. The complete cycle to obtain sensing information might follow the following steps:

a) CE detects a spectrum sensing demand (internal)
b) CE performs spectrum sensing if it has embedded sensor (internal)
c) CE detects control channel (to obtain additional information from externally located data source or DA)
d) CE identifies DA (sending control signal and receiving response)
1) Request: CE submits a request on the presence of DA
2) Indication: Deliver primitive is issued to the DA so that it responds on its presence
3) Response: DA submits its presence status information
4) Confirm: Deliver primitive is issued to CE to confirm successful reception of message from CE
e) CE identifies type (list) of available information on the DA
1) Request: CE submits a request to provide type of information available on DA
2) Indication: Deliver primitive is issued to the DA so that it responds to the request
3) Respond: DA submits type of available information
4) Confirm: Deliver primitive is issued to CE to confirm successful reception of the sensor's message

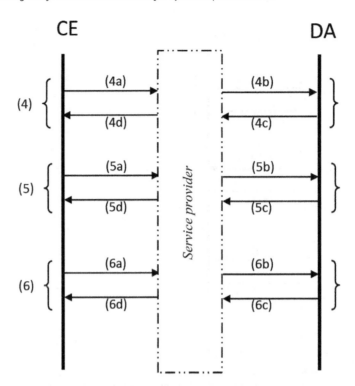

Figure 18.21 —Sensing information exchange procedure between CE and DA. From IEEE Standard for Spectrum Sensing Interfaces and Data Structures for Dynamic Spectrum Access and Other Advanced Radio Communication Systems, Feb 2011. IEEE.

f) CE requests sensing information
  1) Request: CE submits a request to provide sensing information by specifying control parameters, (e.g., for filtering).
  2) Indication: Deliver primitive is issued to the DA so that it responds for the request
  3) Respond: DA submits sensing information
  4) Confirm: Deliver primitive is issued to CE to confirm successful reception of the DA's message

18.7.4.4 CE–CE Procedure

In this example, as depicted in Figure 18.22, a sequence of the generic procedure is used to portray a typical information exchange between two cooperating CEs. The procedure is initialized by the spectrum sensing demand in the client CE. The complete cycle to obtain sensing information might follow the following steps:

a) CE_A detects a spectrum sensing demand (internal)
b) CE_A performs spectrum sensing if it has embedded sensor (internal)
c) CE_A detects control channel (to obtain additional information from CE_B)
d) CE_A identifies CE_B (sending control signal and receiving response)
  1) Request: CEA submits a request on the presence of CEB
  2) Indication: Deliver primitive is issued to the CEB so that it responds on its presence
  3) Response: CEB submits its presence status information
  4) Confirm: Deliver primitive is issued to CEA to confirm successful reception of message from CEB
e) CE_A identifies sensing capability of CE_B and its appropriateness for distributed sensing
  1) Request: CEA submits a request to provide capability of CEB for distributed sensing
  2) Indication: Deliver primitive is issued to the CEB so that it responds for the request
  3) Respond: CEB submits its capability (availability) for distributed sensing
  4) Confirm: Deliver primitive is issued to CEA to confirm successful reception of sensor's message

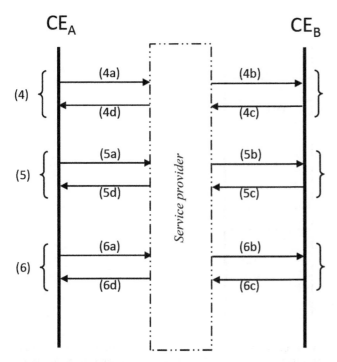

Figure 18.22 —Sensing information exchange procedure between CE$_A$ and CE$_B$. From IEEE Standard for Spectrum Sensing Interfaces and Data Structures for Dynamic Spectrum Access and Other Advanced Radio Communication Systems, Feb 2011. IEEE.

f) CE$_A$ requests sensing information

  1) Request: CEA submits a request to provide sensing information (by specifying control parameters...)
  2) Indication: Deliver primitive is issued to the CEB so that it responds for the request
  3) Respond: CEB submits sensing information
  4) Confirm: Deliver primitive is issued to CEA to confirm successful reception of CEB's message

18.7.4.5 Sensor–Sensor Procedure

In this example, as depicted in Figure 18.23, a sequence of the generic procedure is used to portray a typical information exchange between two cooperating Sensors. The procedure is initialized by the client Sensor request for sensing information. The complete cycle to obtain sensing information might follow the following steps:

a) S$_A$ detects a spectrum sensing request (multiple-device spectrum sensing mode)
b) S$_A$ performs spectrum sensing (internal)
c) S$_A$ detects control channel (to obtain sensing information from S$_B$ either for distributed sensing or for simple relaying)
d) S$_A$ identifies S$_B$ (sending control signal and receiving response)

  1) Request: SA submits a request on the presence of SB
  2) Indication: Deliver primitive is issued to the SB so that it responds on its presence
  3) Response: SB submits its presence status information
  4) Confirm: Deliver primitive is issued to SA to confirm successful reception of message from SB

e) S$_A$ requests sensing information

  1) Request: S$_A$ submits a request to provide sensing information (e.g., to relay the sensing information to another client)
  2) Indication: Deliver primitive is issued to the S$_B$ so that it responds to the request
  3) Respond: S$_B$ submits sensing information
  4) Confirm: Deliver primitive is issued to S$_A$ to confirm successful reception of S$_B$'s message

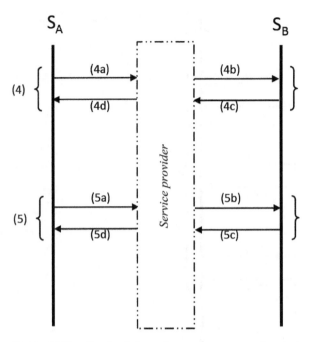

Figure 18.23 —Sensing information exchange procedure between S_A and S_B. From IEEE Standard for Spectrum Sensing Interfaces and Data Structures for Dynamic Spectrum Access and Other Advanced Radio Communication Systems, Feb 2011. IEEE.

Annex 18A

(informative)

Use cases

CRs should be able to detect independently the presence of primary users through continuous spectrum sensing. The technical characteristics of the primary users will require varying sensing requirements depending on the use case. For example, sensitivity requirements or sensing rates may vary depending on the use cases. Detecting GPS signals, for example, is more difficult than detecting TV signals because the GPS receiver sensitivity is higher than that of TV receivers. Thus, for each use case, a set of sensing requirements consistent with the scope of this standard is identified.

Subclauses 18A.1 and 18A.2 define spectrum sensing use cases from different perspectives. A detailed analysis of these use cases with respect to the underlying spectrum usage model and the sensing requirements that can be derived from these use cases is given after each use case description.

18A.1 Use Cases from the Perspective of Spectrum Usage

Subclauses 18A.1.1 to 18A.1.4 examine use cases from the perspective of spectrum usage. The usage models that use spectrum sensing for purposes of opportunistic access to unutilized spectrum are considered. Service type alignment is one of many criteria for categorization of use cases. From the industry point of view, how the spectrum is used (such as who owns the spectrum) in each use cases is a major factor.

CR usage models can be broadly categorized into two models: substitute for primary operation of longer duration (SPOLD) and substitute for primary operation of shorter duration (SPOSD). These describe the relative long-term or short-term usage of spectrum as in the case of the primary system's mode of operation.

18A.1.1 SPOLD

In these usage models, CR usage of spectrum is as in the case of the primary system's mode of operation. The spectrum usage is of relatively longer duration. The following subclauses provide examples of each substitute primary operation identified.

18A.1.1.1 Emergency Services (SPOLD1)

Use Case After a major disaster, much of the existing communications infrastructure is damaged, and emergency personnel from around the country bring their own communications equipment to assist with the response. CR allows all of the emergency personnel to find common usage channels throughout the disaster area in which the CR can operate without causing interference to the surviving legacy communications systems.

Use Case Analysis
— Spectrum usage model and scenario classification
  — Long-term spectrum usage::1:1
— Sensing requirements
  — Reconfigurable sensing performance.
  — Sensing of multiple channels simultaneously.
  — Only information on signal occupancy can be sufficient (band is free or occupied).
  — Sensing is performed only at initial stage (during or after the catastrophe).
  — Sensing is performed by embedded sensors with in radios (if infrastructure and fixed sensors are destroyed) or temporary deployment of distributed sensors.
— Usage of interface
  — CE–S
— Key sensing information
  — Target detection performance (Pd, Pf, etc.)
  — Sensing method, techniques
  — Sensing duration (channel latency issue)

The key feature of this use case is that the priority level of the emergency system might increase and the emergency system might become primary users. In this situation, the emergency system, even as a secondary system, may have more aggressive usage of spectrum and less protection to primary users. The sensing system should have a reconfiguration capability such that their probability of detection and probability of false alarm can be changed or their sensing techniques can be changed.

To detect spectrum opportunities as quickly as possible, multiple channels should be sensed simultaneously.

Because the target of sensing is to identify occupancy not usage analysis, information on signal occupancy might be enough.

Sensing might be performed at the initial stage when the disaster happened. Once the emergency system has obtained the spectrum, it will take control of the spectrum over a long period.

18A.1.1.2 Load Sharing to reduce blocking at Peak Traffic Times (SPOLD2)

Use Case Two or more services (e.g., IEEE 802.11 and Cellular) exhibit peak traffic loads at different times and locations. The IEEE 802.11 service may peak during working hours or during the evenings, whereas the Cellular system may exhibit peaks during commute time. Off loading peak traffic into the spectrum of the other service would be possible through the use of CR.

Use Case Analysis
— Spectrum usage model and scenario classification
  — Long-term spectrum usage::1:1
— Sensing requirements
  — Traffic sensing capability
  — Sensing to identify service channels
  — Sense the most suitable channel first
— Usage of interface
  — CE–S, CE–DA
— Key sensing information
  — Traffic information (resource allocation ratio)
  — Service channels (RATs) ID

To perform load sharing, the network needs to know the traffic of different services.

The CR needs to identify which service channels are available in their environment.

The history of the traffic information might be stored at the DA. Based on this information, the terminals can sense the service channel that is more likely to have low traffic.

18A.1.1.3 Sharing of Spectrum to reduce blocking at Peak Traffic Times (SPOLD3)

Use Case Two or more services (e.g., WiMAX™8 and long-term evolution [LTE]) may exhibit peak loads at different times and locations. Offloading peak traffic into the spectrum of the other service would be possible through the use of CR.[9]

Use Case Analysis
— Spectrum usage model and scenario classification
  — Long-term spectrum usage::1:N
  — Sensing requirements
  — Traffic sensing capability (mostly at base stations [BSs])
  — Sensing of white space (mostly at BSs)
  — Sensing to identify service channels
  — Check with DA
  — Sense and vacate immediately if a primary signal is detected
— Usage of interface
  — CE–S, DA–S, S–S
— Key sensing information
  — Traffic information (resource allocation ratio)
  — Service channel ID

Part of the spectrum of a service channel might be shared with another service channel to reduce blocking at peak traffic times. In this case, the traffic information should be known. The white space (unused spectrum) of each service should be identified so that the spectrum can be shared among different services. Note that "white space" is not limited to TV white space but refers to all spectrum opportunities.

18A.1.1.4 Self-management of Uncoordinated Spectrum (SPOLD4)

Use Case In the unlicensed bands, the high density of privately owned networks, together with freedom of movement requirements for certain transitory operations, makes the central coordination of frequency assignments highly impractical. CR provides an effective means of self-coordination to avoid interference with other networks while providing useful throughput.

This is an example of interference mitigation technique on frequency assignment. This can be in the case of public safety, conferences, etc., as the type of user is immaterial. The central factor is the temporary application of the bands.

Use Case Analysis
— Spectrum usage model and scenario classification
  — Long-term spectrum usage::M:1
— Sensing requirements
  — Sensing of multiple channels (used and unused white space)
  — Sensing to prevent interference
  — Provide sensing information to multiple clients
— Usage of interface
  — CE–S, CE–CE, DA–S
— Key sensing information
  — Client priority (priority flag is set upon coordination of participating engines)
  — Duty cycle (to predict potential interference)

The key feature of self-coordination is sensing to avoid interference. Multiple channels should be sensed to identify unused spectrum and the duty cycle of spectrum usage. Multiple CRs might use the same sensor (for example, a standalone sensor to provide sensing services over an application area).

18A.1.1.5 Introduction of New Users or Services (SPOLD5)

Use Case The near 100% assignment of frequencies in high-density areas makes the introduction of new users or services almost impossible because very few existing licensees are willing to give up what has become a valuable asset. As a result, it may take many months or even years for a new user or service to obtain a frequency assignment in high-density locations. Using CR to find and utilize spectrum that is underutilized at a specific time and location would allow the introduction of new users or services without significant delay. This case is similar to selection of TV whitespaces.

Use Case Analysis
— Spectrum usage model and scenario classification
  — Long-term spectrum usage::1:N
— Sensing requirements
  — Traffic sensing capability
  — Sensing of white space
  — Sensing to identify service channels
  — Check with DA
  — Sense and vacate immediately if a primary signal is detected
— Usage of interface
  — CE–S, DA–S, S–S
— Key sensing information
  — Traffic information (resource allocation ratio)
  — Service channel ID
The requirements are similar to SPOLD3.

18A.1.1.6 Worldwide Mobility (SPOLD6)

Use Case Because radio frequency assignments are not harmonized around the world, a traveler may need to change the frequency of operation when moving from country to country. A CR radio can do this automatically, thus avoiding missed calls in the event that the user forgets to select appropriate frequencies upon arrival.

Use Case Analysis
— Spectrum usage model and scenario classification
  — Long-term spectrum usage::1:1
— Sensing requirements
  — Traffic sensing capability
  — Sense the most suitable channel first
  — Sensing to identify service channels
— Usage of interface
  — CE–S, CE–DA
— Key sensing information
  — Traffic information (resource allocation ratio)
  — Service channel (RATs) ID
The requirements are the same as for SPOLD2.

18A.1.2 SPOSD

In these usage models, the CR is operating as a secondary user in an ad hoc manner, for a relatively short duration of time.

18A.1.2.1 Tamper-resistant Services (SPOSD1)

Use Case In an upscale apartment block, a common server is widely used to provide an Internet connection, as well as to store and manage video and Internet content that is being displayed on multiple monitors throughout each apartment. Each apartment operates its own IEEE 802.11n wireless network at near maximum capacity. All IEEE 802.11n channels are assigned to minimize interference between apartments and between buildings. In addition, building security uses event-triggered video cameras to monitor suspicious activities throughout the building. A CR-based IEEE 802.11n type connection reduces the risk of interruption of the security link due to the other IEEE 802.11n usage or by deliberate jamming. In this use case, the main intent is not for the priority access but to prevent the malicious access.

Use Case Analysis
— Spectrum usage model and scenario classification
  — Short-term spectrum usage::M:1
— Sensing requirements
  — Sensing of multiple channels simultaneously
  — Traffic sensing capability
  — Provide sensing information to multiple clients (negotiation to maintain normal operation of security devices).
— Usage of interface
  — CE–S, S–S
— Key sensing information
  — Client priority (Priority flag is set upon coordination of participating engines)
  — Duty cycle (to predict potential interference)
  — Traffic information (resource allocation ratio)
  This is limited to WLAN. It is similar to both SPOLD2 (load sharing) and SPOLD4 (self-management).

18A.1.2.2 Ad hoc licensee service—virtual spectrum coordination model (SPOSD2)

Use Case A licensed ad hoc secondary user is operating inside the primary service area. Other secondary users may be inside or outside of the primary service area. The licensed ad hoc secondary node takes the role of a virtual access point (AP)/BS when other secondary users want to connect to the IP network and acts as an interference controller for the primary service. The secondary system may not always follow the primary system's configuration (i.e., using different RATs). The ad hoc licensee node can be defined by either a virtual AP/BS controlled by the primary AP/BS or a self-controlled virtual AP/BS.

Use Case Analysis
— Spectrum usage model and scenario classification
  — Short-term spectrum usage::1:N
— Sensing requirements
  — Sense and vacate immediately when a primary user signal is detected
  — Sensing to identify service channels
  — Sensing information sharing with primary resource controller for validation of measurement results
  — Sensing to prevent interference
  — Check with DA
  — Sensing the most suitable channel list
  — Synchronization among sensors or radios

— Sensing of multiple channels (used and unused white space)
— Sensing of RAT
— Usage of interface
    — CE–S, S–S, DA–CE
— Key sensing information
    — Service channel (RATs) ID
    — Power level of primary signals
    — Policy information to enable underlay operation
    — Sensing method, techniques
    — Sensing duration (channel latency issue)
    — Detection threshold
    — Information to be sent to DA should be as complete as possible
    — Client ID

Two spectrum coordination methods are defined in this case. One possibility is to define it as a centric secondary spectrum coordination method via a mobile node, and a target spectrum for secondary spectrum utilization is selected by itself. The other is to define it as a spectrum sharing case controlled by a wired-network-node.

18A.1.2.3   Ad hoc licensee service–underlay spectrum coordination model (SPOSD3)

Use Case   This ad hoc networking model is composed of an ad hoc connection in the primary service area. Two kinds of secondary spectrum usage models may be assumed: first, a model where the ad hoc connection uses an underlay spectrum and where the received power of the primary signal at the ad hoc networking area is larger than the transmitted power of the secondary signal, and second, a model where the ad hoc connection uses a specific underlay spectrum and where the received power of the primary signal at the ad hoc networking area is smaller than the transmitted power of the secondary signal. These cases should assume perfect interference control for primary users, and the secondary spectrum coordination can be a centralized or a decentralized operation.

Use Case Analysis
— Spectrum usage model and scenario classification
    — Short-term spectrum usage::1:N
— Sensing requirements
    — Sense and vacate immediately if a primary signal is detected
    — Sensing to identify service channels
    — Sensing to identify primary system feature
    — Sensing to prevent interference
    — Sensing of RAT
— Usage of interface
    — CE–S, S–S
— Key sensing information
    — Service channel (RATs) ID
    — Sensing method, techniques
    — Sensing duration (channel latency issue)
    — Detection threshold
    — Client ID
    — Policy information to enable underlay operation

The requirements are similar to those of spectrum sharing (SPOLD3) but in short-term fashion. It is an underlay system. The power level of primary signals can be used to indicate the presence or absence of primary signals so that the secondary user can vacate immediately when a primary user signal is detected.

18A.1.2.4 · Self-management of Uncoordinated Spectrum (SPOSD4)

Use Case In the unlicensed bands, and to a lesser extent in some licensed bands, the high density of privately owned networks together with freedom-of-movement requirements for certain transitory operations makes central coordination of frequency assignments highly impractical. CR provides an effective means of self-coordination so as to avoid interference with other networks while providing useful throughput.

Use Case Analysis
— Spectrum usage model and scenario classification
  — Short-term spectrum usage::M:1
— Sensing requirements
  — Sensing of multiple channels (used and unused white space)
  — Sensing to prevent interference
  — Provide sensing information to multiple clients
— Usage of interface
  — CE–S, CE–CE, DA–S
— Key sensing information
  — Client priority (Priority flag is set upon coordination of participating engines)
  — Duty cycle (to predict potential interference)
This use case is similar to SPOLD4.

18A.1.3 Service Enhanced Models (SEM)

These CR usage models look into assisting CR technology in terms of service enhancements. The first four models look into applying CR technology to enhance the communications capabilities of public safety responders. These use cases are in line with the identifications of SDR Forum relating to the Scenario of 7/7 bombing of the London underground (see SDR Forum [B9]). The fifth model is related to assisting CR technology for the investigation of policy violation.

18A.1.3.1 Network Extension (SEM1)

Use Case For coverage extension, CR capabilities can be used to reconfigure radios automatically to include a repeater capability to extend network coverage to areas where radios are otherwise cut off from their infrastructure, particularly during initial response to an incident prior to additional communications resources being deployed.

Use Case Analysis
— Spectrum usage model and scenario classification
  — Short-term spectrum usage::1:N
— Sensing requirements
  — Sense and vacate immediately if a primary signal is detected
  — Sensing to identify service channels
— Usage of interface
  — CE–S, S–S, DA–CE
— Key sensing information
  — Service channel (RATs) ID
  — Power level of primary signals
  — Policy information to enable underlay operation
The requirements are the same as for SPOSD2.

18A.1.3.2 Dynamic Access Additional Spectrum (SEM2)

Use Case At several points in the London bombing scenario, there were communications difficulties because of the sheer volume of calls on the voice communications networks. Dynamic spectrum access, or the ability for CRs to identify unused or underutilized spectrum, can be a solution in this scenario and provide a means for expanding capacity when needed. In this case, the objective is to create an additional capacity for emergency personnel to communicate.

Use Case Analysis
— A combination of load sharing (SPOLD2) and spectrum sharing (SPOLD3) use cases

18A.1.3.3 Temporary Reconfiguration (SEM3)

Use Case In Temporarily Reconfigure First Responder Communication Device Priorities use case, CRs (referring to cell phones) might be able to be temporarily reconfigured with higher priorities based on the circumstances of the emergency responder.

Use Case Analysis
— Sensing is not required

18A.1.3.4 Interface to Non-first Responders (SEM4)

Use Case CRs can allow non-first responders communication access to first responders in specific situations in which the non-first responders are actively participating in the response while avoiding impact on mission-critical public safety networks.

Use Case Analysis
— Spectrum usage model and scenario classification
  — Short-term spectrum usage::M:1
— Sensing requirements
  — Reconfigurable protection to public safety network
  — Sensing of multiple channels simultaneously
  — Provide sensing information to multiple clients
— Usage of interface
  — CE–S, CE–CE
— Key sensing information
  — Client ID
  — Detection threshold
  — Sensing model
  — Sensing duration
  No distinct difference between the first responder and the non-first responder. No unique sensing requirements exist.

18A.1.3.5 Policy Investigation (SEM5)

Use Case This use case focuses on the usage of spectrum sensing to assist investigation of policy violations in CR usage. Spectrum sensing functionalities can be deployed for policy verifications as in the case of rouge radio identification.

Use Case Analysis
— Spectrum usage model and scenario classification
  — Short-term spectrum usage::1:M
— Sensing requirements
  — As complete as possible
  — On regular basis

— Usage of interface
  — DA–S, S–S
— Key sensing information
  — Information to be sent to DA should be as complete as possible (RAT or modulation, geolocation, time stamp, operator ID, signal level, center frequency, bandwidth, Sensing methods, etc.).

The DA plays an important role in this use case. Because the goal of this use case is to analyze spectrum usage for policy investigation, information on spectrum usage should be as complete as possible for usage analysis. Furthermore, information should be collected regularly.

18A.1.3.6 An ad hoc network model composed of multimode virtual AP/BS in several primary service areas (SEM6)

Use Case This model is an extension model of SPOSD2. There is a virtual AP/BS in the overlapping area of at least two kinds of different primary services. The licensed ad hoc node may select a link that has a better performance or lower traffic from the connectable primary services when other secondary users want to connect to an IP network. Furthermore, the node should be able to control interference to primary services. Other secondary users may be inside or outside of these primary service areas. The other assumptions are the same as the SPOSD2.

Use Case Analysis
— Spectrum usage model and scenario classification
  — Short-term spectrum usage::1:N
  — Sensing requirements
  — Traffic sensing capability
  — Sensing identify service channels
  — Sensing of multiple RATs
  — Sensing information sharing with primary resource controller for validation of measurement results
  — Sensing to prevent interference
  — Check with DA
  — Sensing the most suitable channel list
  — Synchronization among sensors or radios
  — Sense and vacate immediately if a primary signal is detected
  — Sensing of multiple channels (used and unused white space)
— Usage of interface
  — CE–S, S–S
— Key sensing information
  — Traffic information (resource allocation ratio)
  — Service channel (RATs) ID
  — Sensing method, techniques
  — Sensing duration (channel latency issue)
  — Detection threshold
  — Information to be sent to DA should be as complete as possible
  — Client ID

The end effect is load sharing, but the use case looks into a different implementation scenario.

18A.1.4 IEEE 1900.4-enabled Dynamic Spectrum Access use Cases

IEEE Std 1900.4™-2009 considers a heterogeneous wireless environment shown in Figure 18A.1. Such a heterogeneous wireless environment may include multiple operators, multiple radio access networks (RANs), multiple RATs, or multiple terminals.

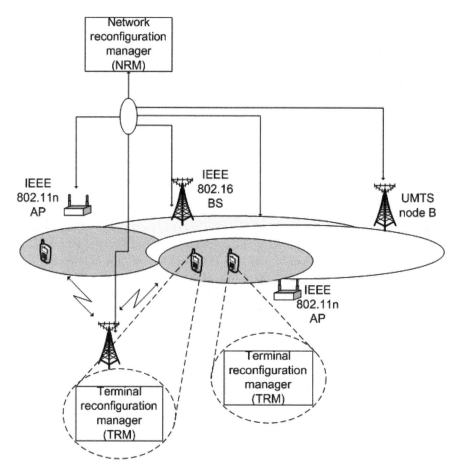

Figure 18A.1 —Heterogeneous wireless environment considered in IEEE Std 1900.4-2009. From IEEE Standard for Spectrum Sensing Interfaces and Data Structures for Dynamic Spectrum Access and Other Advanced Radio Communication Systems, Feb 2011. IEEE.

Within Figure 18A.1:

Network Reconfiguration Manager (NRM) is the entity that manages the composite wireless network and terminals in terms of network-terminal distributed optimization of spectrum usage. This management is done within the regulatory framework and in a manner consistent with available context information.

Terminal Reconfiguration Manager (TRM) is the entity that manages the terminal in terms of network-terminal distributed optimization of spectrum usage. This management is done within the framework of radio resource selection policies conveyed by the NRM and in a manner consistent with the user's preferences and the available context information.

For this heterogeneous wireless environment, IEEE Std 1900.4-2009 defines three use cases representing different types of dynamic spectrum access (see Figure 18A.2):

— Dynamic spectrum assignment
— Dynamic spectrum sharing
— Distributed radio resource usage optimization

Note. NRM and TRM may require sensing-related information for resource management.

18A.1.4.1   Dynamic Spectrum Assignment (DS Assignment)

Use Case   In the dynamic spectrum assignment (DS Assignment) use case, frequency bands are dynamically assigned to the RANs in order to optimize radio resource usage and improve quality of service (see Figure 18A.2).

Spectrum

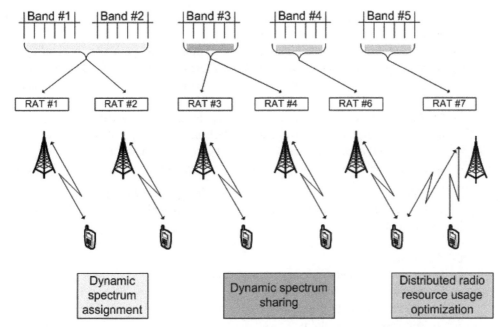

Figure 18A.2 —IEEE 1900.4-2009 use cases. From IEEE Standard for Spectrum Sensing Interfaces and Data Structures for Dynamic Spectrum Access and Other Advanced Radio Communication Systems, Feb 2011. IEEE.

A short summary of the DS Assignment use case is as follows:

— The NRM analyzes available context information and dynamically makes spectrum assignment decisions.
— After the new spectrum assignment decisions have been made, the NRM requests corresponding reconfiguration of its RANs.
— After the RANs reconfiguration, terminals may need to reconfigure correspondingly.

Decision making for DS Assignment requires context information, which could include spectrum sensing information.

Use Case Analysis
— Spectrum usage model and scenario classification
  — Long-term spectrum usage::1:N
— Sensing requirements
  — Traffic sensing capability
  — Sensing of service channel
  — Sensing of multiple channels
— Usage of interface
  — CE–S, CE–DA, S–S
— Key sensing information
— Resource allocation ratio
— Service channel
  This is a general use case of CR.

18A.1.4.2 Dynamic Spectrum Sharing (DSS)

Use Case In the DSS use case, frequency bands assigned to RANs are fixed. However, a particular frequency band can be shared by several RANs to optimize radio resource usage and improve quality of service (see Figure 18A.2).

A short summary of the dynamic spectrum assignment use case is as follows:

— NRM analyzes available context information and dynamically makes spectrum access decisions.
— After these decisions, NRM requests corresponding reconfiguration of its RANs.
— NRM dynamically generates radio resource selection policies and sends them to its TRMs. These radio resource selection policies will guide these TRMs in their spectrum access decisions.
— TRMs analyze these radio resource selection policies and the available context information and dynamically make spectrum access decisions. These spectrum access decisions are made within the framework of the radio resource selection policies.
— After these decisions, each TRM requests corresponding reconfiguration of its terminal.

The DSS use case can also include the case of primary/secondary spectrum access.

Decision making for DSS requires context information, which could include spectrum sensing information.

Use Case Analysis
— Spectrum usage model and scenario classification
    — Long-term spectrum usage::M:1
— Sensing requirements
    — Traffic sensing capability
    — Sensing of white space
    — Sensing of service channel
    — Check with DA
— Usage of interface
    — CE–S, CE–DA, S–S
— Key sensing information
    — Traffic information (resource allocation ratio)
    — Service channel ID
This use case is similar to SPOLD3 in terms of sensing requirements.

18A.1.4.3 Distributed Radio Resource Usage Optimization (DRRUO)

Use Case In the distributed radio resource usage optimization (DRRUO) use case, frequency bands assigned to RANs are fixed. Also, reconfiguration of RANs is not considered in this use case. Instead, reconfigurable terminals with or without multihoming capability are considered (see Figure 18A.2).

A short summary of the DRRUO use case is as follows:

— NRM analyzes available context information, dynamically generates radio resource selection policies, and sends them to its TRMs. These radio resource selection policies will guide these TRMs in their reconfiguration decisions.
— TRMs analyze radio resource selection policies and available context information and dynamically make decisions on reconfiguration of their terminals to improve radio resource usage and quality of service.
— Following these decisions, TRMs request corresponding reconfiguration of their terminals.

Decision making for DRRUO requires context information, which may include spectrum sensing information.

Use Case Analysis
— Spectrum usage model and scenario classification
    — Long-term spectrum usage::1:N

— Sensing requirements
  — Traffic sensing capability
  — Sensing to identify service channels
  — Sensing of multiple channels
— Usage of interface
  — CE–S, CE–DA, S–S
— Key sensing information
  — Resource allocation ratio
  — Service channel IDs

The key feature of this use case is that the terminals can perform spectrum usage optimization. Therefore, multiple sensors can be used to sense as many channels as possible.

18A.2 Perspective of Spectrum Sensing

In 18A.1, the use cases are given from the perspective of spectrum usage. This annex introduces use cases from the perspective of how the sensing information is used. Each use case is analyzed, and the requirements are extracted.

18A.2.1 Distributed Sensing Models (DSMs)

Multiple sensing devices (sensors) that are located at different physical positions in a service area can fulfill the sensing functionality of a radio system either independently or in a coordinated manner. Note that this is different from distributed processing. These usage models look into the CR technology in which the sensing information is obtained from spatially distributed sensors.

18A.2.1.1 CR System with Distributed Sensors (DSM1)

Use Case The CR system analyzes spectrum usage and makes spectrum access decision based on the sensing-related information provided by sensors that are located at different physical positions (i.e., distributed sensing [cf. 18.3.1]). This type of sensing usage can be found in applications such as using CR to increase the capacity of communication for commercial purpose in hot-spot areas such as a university campus or a central business district area in a city. This type of sensing usage can also be found in SEM2, but the purpose and priority levels are different.

A CR terminal (such as a cognitive mobile phone) can initiate the sensing function of the sensing network once it enters the service area of the sensor network. If the CR terminal has an embedded sensing functional block, it can use the sensing information from the sensor network to enhance its sensing capability. This would ultimately help DSA decision making.

In another example, multiple CR nodes form a radio network, such as a CR-based ad hoc/multihop network for building up an efficient communication. Some of the sensors in individual nodes can form a virtual sensor network. Furthermore, these nodes can use the sensing information from this virtual sensor networks to analyze spectrum usage and make the decision of spectrum access.

Distributed sensing can be further classified into peer-to-peer sensing, cooperative sensing, collaborative sensing, or selective sensing based on the way the sensing-related information is conveyed from sensors to the CE. Here, sensing channel denotes the frequency channel on which the sensors perform spectrum measurement, and sensor channel denotes the frequency channel through which distributed sensors exchange sensing-related information. Note that the sensing channel and sensor channel may be of different frequencies, types, and media.

Peer-to-peer Sensing This is a special case where closely located CRs can negotiate and exchange sensing-related information to improve the sensing quality at each CR terminal. This case does not require multiple access schemes, whereas in collaborative and cooperative schemes, multiple access schemes should be provided for exchanging sensing-related information among multiple sensors in the sensor channel. Initial negotiation between the peer radios to exchange the sensing-related information can be carried out through handshaking.

Cooperative Sensing In this case, multiple sensors report sensing-related information independently to a CE. Such local processing can be performed in hybrid fashion. When multiple local processed data are received, the CE still needs to perform data fusion operations. This scenario is a combination of the examples shown in Figure 18A.3. The CE collects and combines the information from the sensor network to obtain a resultant output. The different sensors may provide information on different frequency bands, or the same frequency but collect data from a different location. The resultant output is an aggregate of information based on the information from individual sensors, as well as the underlying combining algorithm or data fusion.

Collaborative Sensing In this case, multiple sensors exchange sensing-related information to perform local processing before providing the sensing-related information as well as sensing results to the CE. To achieve this, the sensors should have a common goal and should provide similar sensing information.

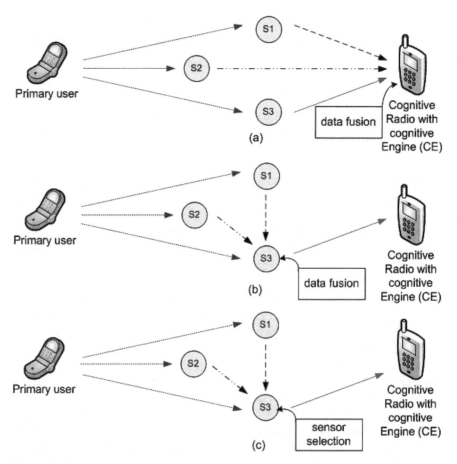

Figure 18A.3 —Examples of distributed sensing with three spectrum sensors: (a) cooperative sensing, (b) collaborative sensing, and (c) selective sensing. From IEEE Standard for Spectrum Sensing Interfaces and Data Structures for Dynamic Spectrum Access and Other Advanced Radio Communication Systems, Feb 2011. IEEE.

Selective Sensing In this case, one sensor is selected to perform sensing. The sensing information is sent to a CR. Selection of sensors can be performed on the basis of a priori information regarding to the quality of the sensor channels, sensors' previous performance, energy level, etc. One of the benefits of this scheme is that only one sensor is active during sensing period and the others are in standby mode. This reduces the energy consumption of these sensors.

Use Case Analysis
— Spectrum usage model and scenario classification
  — Short-term spectrum usage::1:N
— Sensing requirements
  — Sensing of multiple channels simultaneously
  — Synchronization among sensors or radios
— Usage of interface
  — CE–S, S–S
— Key sensing information
  — Sensor information (ID, geolocation, battery, specification, etc.)

This is a unique case for distributed sensing (cooperative sensing, collaborative sensing, and selective sensing). Because multiple sensors exchange sensing information, their IDs should be known. Synchronization among sensors or radios is needed for sensing information exchange. Note that this use case is applicable at both the network side and the terminal.

18A.2.1.2 Peer-to-peer CRs (DSM2)

Use Case In this scenario, at least one terminal has a built-in spectrum sensor. When such terminals are located in close range and wish to establish a communication link, each terminal can use its own sensor (if it has a sensor) or the sensor in the other terminal can perform sensing and decide spectrum usage. Note that in the latter case, the sensor is distributed to a location that is different from that of the terminal. In the situation when both terminals have built-in sensors, the sensors can exchange sensing-related information for better exploitation of spectrum opportunity. Furthermore, the peer radios can also exploit any relevant sensing information from other nonpeer radios in their vicinity to enhance their sensing information.

This kind of use of sensing scenario can be utilized in the following types of applications.

Power Efficient Communication In most of the existing communication systems, mobile terminals communicate with other mobile terminals through a BS with other mobile terminals. There are many situations where two mobile terminals are located very close to each other and communication via BSs becomes power inefficient. In this situation, the power of the mobile terminal can be efficiently used by setting up peer-to-peer communication. Otherwise, a large amount of power is wasted for transmitting signals to BSs.

Short-range Data Transfer This application includes short-range data transfer (e.g., a cordless projector using the ISM band). Spectrum sensing can be used to identify a frequency band appropriate for the data transfer.

Cognitive Wireless Network In a cognitive wireless network, multiple CR terminals form a wireless ad hoc network or mesh network with multiple peer-to-peer links. Each terminal in the peer-to-peer link detects the behavior of the spectrum occupancy in multiple channels and provides efficient communication links and paths. Peers also can exploit multiple channels at a time through different interfaces to form multichannel mesh networks. The sensing-related information from a sensor at one terminal can be shared by multiple terminals in a wireless network.

As an example, consider a simple cognitive relay network consisting of nodes A, B, and C. A relay can be set up on the basis of dynamic spectrum access between any two CR terminals, such as the service model SEM1. Data are then relayed from node A to node B and then to node C. The link between A and B can be considered as a peer-to-peer communication, which can be set up relying on the sensing results at node A and node B. The same goes to nodes B and C. Note that the example given here is to show where distributed sensing can be used for peer-to-peer communication. It is not related to the routing protocol.

Use Case Analysis
— Spectrum usage model and scenario classification
    — Short-term spectrum usage::M:1
— Sensing requirements
    — Sensing of multiple channels simultaneously
    — Sense to avoid interference
    — Synchronization among radios
— Usage of interface
    — CE–S, CE–CE
— Key sensing information
    — Client ID
    — Duty cycle

This is a use case for the cooperation among radios where the CE–CE interface is needed. Furthermore, the CE is equally distributed between two radios. CRs should sense to avoid interference with each other. Multiple CEs can access the same sensors. Synchronization among sensors or radios is needed for sensing information exchange.

18A.2.2 Sensing Enhanced Models (SeEM)

These usage models look into applications where enhanced sensing function can be provided to the CR terminals.

18A.2.2.1 Faulty Sensing Prevention (SeEM1)

Use Case Some sensors located in a blocked area might fail to detect the presence of the primary users. The CR terminals relying on these sensors might make the wrong decision of spectrum access and thus interfere with the primary users. Using enhanced sensing, the cognitive terminal can be provided with the most accurate sensing information to make a spectrum access decision. In this sense, the application prevents faulty sensing information by switching off the faulty sensor.

The reasons for a sensor producing faulty sensing information are many. For example, it might be due to the hidden terminal problem, such as the application of CR technologies in a highly scattering environment, including an office or urban area. Buildings in an urban area or office partitions, desk, closets, etc., prevent the line-of-sight propagation channel between primary users and sensors.

Use Case Analysis
— Spectrum usage model and scenario classification
    — Short-term spectrum usage::1:N
— Sensing requirements
    — Provide sensing information with performance measure
— Usage of interface
    — CE–S, S–S
— Key sensing information
    — Sensor ID
    — Sensing methods
    — Sensing performance information (e.g., confidence level)

This is a use case where sensors should provide their sensing information with Sensor ID, information regarding sensing methods, as well as sensing confidence level so that those sensors that providing wrong sensing information can be identified.

18A.2.2.2 Power Constrained Sensing (SeEM2)

Use Case There are many applications where a sensor(s) is battery powered. It is desirable to increase the service life of these sensors to account for longer duration spectrum sensing. The most efficient way to reduce power consumption is only to turn on a sensor when necessary. This way of using sensing could be applied to applications described in use case such SPOSD1.

Use Case Analysis
— Spectrum usage model and scenario classification
   — Short-term spectrum usage::1:1
— Sensing requirements
   — Provide sensing information with power consumption
   — Information exchange method should be chosen for low power consumption
— Usage of interface
   — CE–S
— Key sensing information
   — Sensor ID
   — Sensor power consumption
   — Reporting mode (soft or hard information)

Sensors should provide sensing information with limited power consumption. Furthermore, different ways of exchanging sensing information have different power requirements. For example, sending soft information takes a relatively longer time and more power than sending hard information. In power constrained sensing, power-efficient ways of exchanging sensing information should be chosen.

18A.2.2.3 Priority based Sensing Information Provision (SeEM3)

Use Case Multiple clients may try to access the sensing-related information from a resource (e.g., Sensor and DA) at the same time. Therefore, competition might occur among these clients, and a priority scheme may be introduced. The client that has a higher priority level can access the sensing-related information first. This way of using sensing information reduces system overhead for transporting sensing-related information to multiple clients.

Use Case Analysis
— Spectrum usage model and scenario classification
   — Short-term spectrum usage::M:1
— Sensing requirements
   — Provide sensing information to multiple clients
— Usage of interface
   — CE–S, CE–CE
— Key sensing information
   — Client ID
   — Client Priority

The priority levels of sensors' clients should be identified when there are multiple clients. This priority characteristic can also be applied at sensor side. The use case is applicable to all M:1 and 1:M scenarios.

18A.2.3 Sensing Models for Spectrum Coordination

This subclause identifies sensing models for spectrum coordination systems. First, two spectrum coordination models are defined. One is the network node spectrum coordination model (NCM), where the final decision-making function of the secondary spectrum usage is in a single or multiple network nodes. The other is the mobile node spectrum coordination model (MCM), where the final decision-making function of the secondary spectrum usage

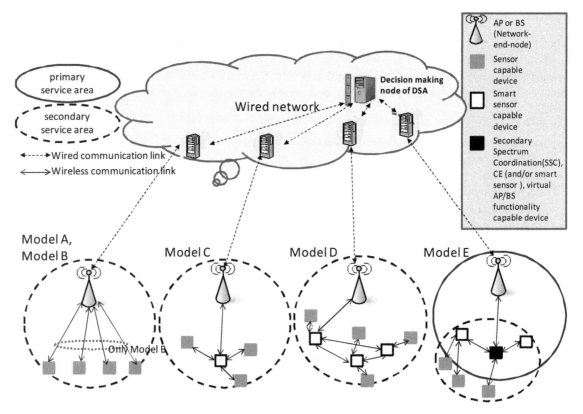

Figure 18A.4 —Sensing models of NCM. From IEEE Standard for Spectrum Sensing Interfaces and Data Structures for Dynamic Spectrum Access and Other Advanced Radio Communication Systems, Feb 2011. IEEE.

is in a single or multiple terminal nodes. Second, on the basis of the spectrum coordination models, several sensing models are defined as shown in Figure 18A.4 through Figure 18A.6. Figure 18A.4 shows the sensing models of NCM, and Figure 18A.5 shows the sensing models for MCM. Furthermore, Figure 18A.6 shows extension models of NCM and MCM.

18A.2.3.1   Network Node Spectrum Coordination Model

Sensing models based on NCM are introduced in Figure 18A.4. The virtual AP/BS acts as a repeater for the secondary users when they want to connect to the IP network. The smart sensor can combine sensingrelated information from multiple sensors and can relay the information to others. The Secondary Spectrum Coordination can make the decision for dynamic spectrum usage, or it can make the decision for spectrum usage through the permission of a spectrum coordinator node in a network.

Sensing Model A: A Sensing Model for an Interface Definition in a Node (NCM-A)   NCM-A defines an internal interface between the internal functional blocks in each node, which may come from different manufacturers.

The following points need to be considered:

— Sensing/Control information exchange between the smart sensor and the sensors
— Sensing/Control information exchange between the CE and the smart sensor
— Sensing/Control information exchange between the CE and the sensors

Sensing Model B: A Sensing Model between the CE in a Network-node and the Sensors (NCM-B)
Use Case   This is a distributed sensing model in a network centralized coordination system such as IEEE P802.22 [B2]. Each sensor informs its measurement results to the coordinator.

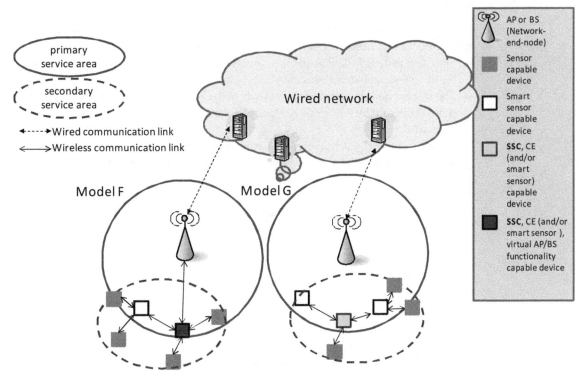

Figure 18A.5 —Sensing models of MCM. From IEEE Standard for Spectrum Sensing Interfaces and Data Structures for Dynamic Spectrum Access and Other Advanced Radio Communication Systems, Feb 2011. IEEE.

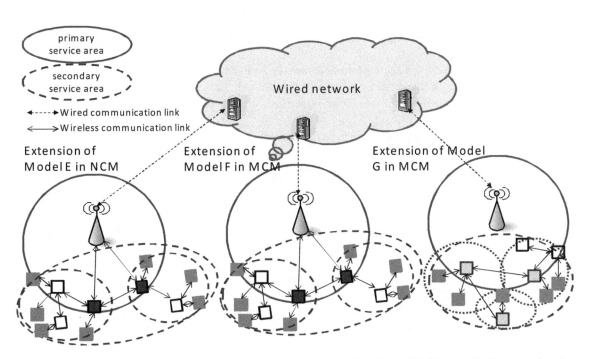

Figure 18A.6 —Extension of model E in NCM and model F and G in MCM. From IEEE Standard for Spectrum Sensing Interfaces and Data Structures for Dynamic Spectrum Access and Other Advanced Radio Communication Systems, Feb 2011. IEEE.

The following points need to be considered:

— Sensing/Control information exchange between the network-node and the sensors
— Specific points that need to be considered for this model may be summarized as follows:
  — The client can easily download the information about the spectrum usage environment based on geographical information via a wired link.
  — The volume and the number of elements of the sensing information that needs to be shared between the network-end-node and the sensors may be the smallest among all the models.
  — If the client needs just the received power of the primary signal, the information may be hard information (one bit) produced based on a regulation determined level.

Use Case Analysis
— Spectrum usage model and scenario classification
  — Long-term spectrum usage::1:N
— Sensing requirements
  — Sensing of multiple channels
  — Sensing of white space
  — Check with DA
— Usage of interface
  — CE–S, S–S, DA–CE
— Key sensing information
  — Sensor information (to distinguish sensor *vs.* smart sensor)
  — Sensing method, techniques
  — Sensing duration (channel latency issue)
  — Detection threshold
  — Reporting mode
  — Client ID

These use cases employ distributed sensing. The CE is located at the network side.

Sensing Model C: A Sensing Model between the Smart Sensor and the Sensors (NCM-C)

Use Case In this model, each sensor transmits the measurement results to a smart sensor. The smart sensor then combines the results and sends the merged sensing information to the coordinator.

The following points need to be considered:

— Sensing/Control information exchange between the smart sensor and the sensors
— Sensing/Control information exchange between the smart sensor and the CE in wired network/node
— Specific points that need to be considered for this model may be summarized as follows:
  — The volume and the number of elements of the sensing information that needs to be shared between the smart sensor and the sensors may be the smallest among all the models.
  — The control information between the smart sensor and the sensors should be well considered.
  — The control information between the smart sensor and the CE in wired-network/fixed-node should be well considered.
  — The other features may be the same as those of sensing model B.

Use Case Analysis
— Spectrum usage model and scenario classification
  — Long-term spectrum usage::1:N
— Sensing requirements
  — Sensing of multiple channels
  — Sensing of white space
  — Check with DA
— Usage of interface
  — CE–S, S–S, DA–CE
— Key sensing information
  — Sensor information (to distinguish sensor *vs.* smart sensor)
  — Sensing method, techniques
  — Sensing duration (channel latency issue)
  — Detection threshold
  — Reporting mode
  — Client ID
These use cases employ distributed sensing. The CE is located at the network side.

Sensing Model D: A Sensing Model between Smart Sensors (NCM-D)

Use Case   This model has smart sensors that collect the sensing information and combine it into a merged result. This merged result is sent to the coordinator.
  The following points need to be considered:

— Sensing/Control information exchange between the smart sensors
— Specific points that need to be considered for this model may be summarized as follows:
— The volume and the number of elements of the sensing information that needs to be shared between the smart sensors may be the smallest compared with the other models described.
— The control information between the smart sensors should be well considered.
— The other features may be the same as those of the sensing model B.

Use Case Analysis
— Spectrum usage model and scenario classification
  — Long-term spectrum usage::1:N
— Sensing requirements
  — Sensing of multiple channels
  — Sensing of white space
  — Check with DA
  Usage of interface
  — CE–S, S–S, DA–CE
— Key sensing information
  — Sensor information (to distinguish sensor vs. smart sensor)
  — Sensing method, techniques
  — Sensing duration (channel latency issue)
  — Detection threshold
  — Reporting mode
  — Client ID
These use cases employ distributed sensing. The CE is located at the network side.

Sensing Model E: A Sensing Model of a Secondary Spectrum Coordination with a Primary AP/BS Connection (NCM-E)

Use Case   This model has a CE capable device in the primary service area, where the device is a secondary spectrum coordinator authorized by a spectrum coordinator node in a wired network. The device collects sensing information from several other devices and combines it into a merged result. This merged result is sent to the spectrum coordinator node. The device also acts as a virtual AP/BS when a secondary user wants to connect to the IP network.

The following points need to be considered:

— Sensing/Control information exchange between the CE and the smart sensors.
— Sensing/Control information exchange between the CE and the sensors.
— Specific points that need to be considered for this model may be summarized as follows:
— The client can easily download the information about the spectrum usage environment based on geographical information via a primary wireless link.
— The volume and the number of elements of the sensing information that needs to be shared between the smart sensor and the CE of the secondary spectrum coordinator may be relatively smaller than other models.
— The volume and the number of elements of the sensing information that needs to be shared between the CE of the secondary spectrum coordinator and the sensors may be relatively smaller than other models.
— The control information between the CE of the secondary spectrum coordinator and the sensors should be well considered.
— The control information between the CE of the secondary spectrum coordinator and the smart sensor should be also well considered.
— The other features may be the same as those of the sensing model D.

Use Case Analysis

— Spectrum usage model and scenario classification
  — Short-term spectrum usage::1:N
— Sensing requirements
  — Sensing of multiple channels
  — Sensing information integration
  — Sensing information sharing control
  — Sensing information sharing with primary resource controller for the validation of measurement results
  — Distributed device control
  — Check with DA
  — Sensing the most suitable channel list
  — Provide sensing information to multiple clients
— Usage of interface
  — CE–S, S–S, DA–CE
— Key sensing information
  — Sensor information (to distinguish sensor vs. smart sensor)
  — Target detection performance
  — Sensing method, techniques
  — Sensing duration
  — Client priority
  — Duty cycle
  — Detection threshold
  — Information to be sent to DA should be as complete as possible
  — Reporting mode
  — Client ID

These use cases employ distributed sensing. Furthermore, CEs are located at a wired-network-node and a mobile, respectively.

18A.2.3.2 Mobile Node Spectrum Coordination Model

Figure 18A.5 introduces the sensing models based on MCM.

Sensing Model F: A Sensing Model of a Secondary Spectrum Coordination with a Primary AP/BS Connection (MCM-F)

Use Case This model has the CE-capable device acting as a permanent secondary spectrum coordinator operating in the primary service area. The difference between this model and model E is that in this model, the spectrum coordinator has a permanent spectrum usage right. The device collects and combines sensing information from several sensors. This merged result is transmitted to the coordinator. The device also acts as a virtual AP/BS when a secondary user wants to connect to the IP network.

The following points need to be considered:

— Sensing/Control information exchange between the smart sensor and the sensors with a primary AP/BS connection
— Sensing/Control information exchange between the CE and the sensors with a primary AP/BS connection
— Sensing/Control information exchange between the CE and the smart sensor with a primary AP/BS connection
— Specific points that need to be considered for this model may be summarized as follows:
  — The secondary spectrum coordinator can download the information about the spectrum usage environment based on geographical information via a primary wireless link, but the download may be difficult depending on the condition of the primary link.
  — The volume and number of elements of sensing information that needs to be shared between the smart sensor and the CE of the secondary spectrum coordinator may be relatively larger than other models.
  — The volume and the number of elements of sensing information to be shared between the CE of the secondary spectrum coordinator and the sensors may be relatively larger than other models.
  — The volume and the number of elements of sensing information to be shared between the smart sensor and the sensors may be relatively larger than other models.
  — The control information between the CE of the secondary spectrum coordinator and the sensors should be well considered.
  — The control information between the CE of the secondary spectrum coordinator and the smart sensor should be well considered.
  — The control information between the sensors and the smart sensor should be well considered.

Use Case Analysis

— Spectrum usage model and scenario classification
  — Long-term spectrum usage::1:N
— Sensing requirements
  — Sensing of multiple channels
  — Sensing information integration
  — Sensing information sharing control
  — Sensing information sharing with primary resource controller for validation of measurement results
  — Distributed device control
  — Sensing to prevent interference
  — Check with DA
  — Sensing the most suitable channel list
  — Synchronization among sensors or radios
  — Sensing to identify service channels

- — Sense and vacate immediately if a primary signal is detected
- — Provide sensing information to multiple clients
- — Sensing of multiple channels
- — Sensing of RAT
- Usage of interface
  - — CE–S, S–S, CE–DA
- Key sensing information
  - — Sensor information (to distinguish sensor *vs.* smart sensor)
  - — Target detection performance
  - — Sensing method, techniques
  - — Sensing duration
  - — Client priority
  - — Duty cycle
  - — Detection threshold
  - — Information to be sent to DA should be as complete as possible
  - — Reporting mode
  - — Client ID

The spectrum coordinator has permanent spectrum usage. Distributed sensing is applied with CE located at the terminal side.

Sensing Model G: A Sensing Model of a Secondary Spectrum Coordination without a Primary AP/BS Connection (MCM-G)

Use Case In this model, the CE capable device, acting as a permanent secondary spectrum coordinator, is operating in the primary service area. The difference between this model and model F is that in this scenario, the secondary spectrum coordinator does not have the capability to act as a virtual AP/BS. The device collects and combines sensing information from several sensors into combined sensing information. This merged result is sent to the secondary spectrum coordinator.

The following points need to be considered:

- Sensing/Control information exchange between the smart sensor and the sensors
- Sensing/Control information exchange between the CE and the sensors
- Sensing/Control information exchange between the CE and the smart sensor
- Specific points that need to be considered for this model may be summarized as follows:
  - The secondary spectrum coordinator cannot download the information about the spectrum usage environment based on geographical information.
  - The volume and the number of elements of sensing information that needs to be shared between the smart sensor and the CE of the secondary spectrum coordinator may be the largest among all models.
  - The volume and number of elements of sensing information that needs to be shared between the CE of the secondary spectrum coordinator and the sensors may be the largest among all models.
  - The volume and number of elements of sensing information that needs to be shared between the smart sensor and the sensors may be the largest amongst all models.
  - The control information between the CE of the secondary spectrum coordinator and the sensors should be well considered.
  - The control information between the CE of the secondary spectrum coordinator and the smart sensor should be well considered.
  - The control information between the sensors and the smart sensor should be well considered.

Use Case Analysis
— Spectrum usage model and scenario classification
    — Long-term spectrum usage::1:N
— Sensing requirements
    — Sensing of multiple channels
    — Sensing to identify primary system feature
    — Sensing information integration
    — Sensing information sharing control
    — Distributed device control
    — Sensing to prevent interference
    — Synchronization among sensors or radios
    — Sensing to identify service channels
    — Sense and vacate immediately if a primary signal is detected
    — Provide sensing information to multiple clients
    — Sensing of multiple channels
    — Sensing of RAT
— Usage of interface
    — CE–S, S–S
— Key sensing information
    — Sensor information (to distinguish sensor *vs.* smart sensor)
    — Target detection performance
    — Sensing method, techniques
    — Sensing duration
    — Client priority
    — Duty cycle
    — Detection threshold
    — Information to be sent to DA should be as complete as possible
    — Reporting mode
    — Client ID
    — Policy information to enable underlay operation

Spectrum coordinator has permanent spectrum usage. Distributed sensing is applied with CE being located at the terminal side.

18A.2.3.3 Extension Models

The sensing models shown in Figure 18A.6 are described in this subclause and are based on extensions of NCM model E and extensions of MCM models F and G.

Extension Model of Sensing Model E (NCM-E-extension)

Use Case   In this model, the CE capable devices are located in or near the primary service area. The devices function as secondary spectrum coordinators authorized by a spectrum coordinator node in a wired network. The device also acts as a virtual AP/BS when a secondary user wants to connect to the IP network. This model may be essential for an ad hoc secondary spectrum usage case in NCM because the secondary service can extend the communication range more than the single network cooperated by secondary spectrum coordinators.

The following points need to be considered:

— Sensing/Control information exchange between CEs
— Sensing/Control information exchange between smart sensors
    — Specific points that need to be considered for this model may be summarized as follows:

— The clients can easily download the information about the spectrum usage environment based on geographical information via a primary wireless link.
— The volume and the number of elements of the sensing information that needs to be shared between the smart sensors may be relatively smaller than other models.
— The volume and the number of elements of the sensing information that needs to be shared between the CEs of the secondary spectrum coordinators may be relatively smaller than other models.
— The control information between the CEs of the secondary spectrum coordinator should be well considered.
— The control information between the smart sensors should be well considered.
— The other features may be the same as those of the sensing model E.

Use Case Analysis
— Spectrum usage model and scenario classification
    — Short-term spectrum usage::M:1
— Sensing requirements
    — Providing sensing information to multiple clients
    — Sensing information integration
    — Sensing information sharing control
    — Sensing information sharing with primary resource controller for validation of measurement results
    — Distributed device control
    — Check with DA
    — Sensing the most suitable channel list
    — Provide sensing information to multiple clients
— Usage of interface
    — CE–S, CE–CE, CE–DA
— Key sensing information
    — Client ID
    — Client priority
    — Target detection performance
    — Sensing method, techniques
    — Sensing duration
    — Client priority
    — Duty cycle
    — Detection threshold
    — Information to be sent to DA should be as complete as possible
    — Sensor information
    — Reporting mode
This use case extends NCM-E by using multiple CEs. Thus, CE–CE interface is required.

Extension Model of Sensing Model F (MCM-F-extension)

Use Case In this model, the CE capable devices are located in or near the primary service area. The devices function as permanent secondary spectrum coordinators. The difference between this case and the extension model of sensing model E is that in this case the secondary spectrum coordinator has a permanent spectrum usage right. The device also acts as a virtual AP/BS when a secondary user wants to connect to the IP network. This model may be essential for an ad hoc secondary spectrum usage case in MCM because the secondary service coverage area is extended by allowing multiple individual networks that are operated by different secondary spectrum coordinators.

The following points need to be considered:

— Sensing/Control information exchange between CEs
— Sensing/Control information exchange between smart sensors

— Specific points that need to be considered for this model may be summarized as follows:

- — The secondary spectrum coordinators can download the information about the spectrum usage environment based on geographical information via a primary wireless link, but the download may be difficult depending on the condition of the primary link.
- — The volume and number of elements of sensing information that needs to be shared between the smart sensors may be relatively larger than other models.
- — The volume and the number of elements of sensing information to be shared between the CEs of the secondary spectrum coordinators may be relatively larger than other models.
- — The volume and the number of elements of sensing information to be shared between the smart sensor and the sensors may be relatively larger than other models.
- — The control information between the CEs of the secondary spectrum coordinator should be well considered.
- — The control information between the smart sensors should be well considered.
- — The other features may be the same as those for the sensing model F.

Use Case Analysis
- — Spectrum usage model and scenario classification
  - — Long-term spectrum usage::M:1
- — Sensing requirements
- — Providing sensing information to multiple clients
- — Sensing to identify primary system feature
- — Sensing information integration
- — Sensing information sharing control
- — Sensing information sharing with primary resource controller for validation of measurement results
- — Distributed device control
- — Sensing to prevent interference
- — Check with DA
- — Sensing the most suitable channel list
- — Synchronization among sensors or radios
- — Sensing to identify service channels
- — Sense and vacate immediately if a primary signal is detected
- — Provide sensing information to multiple clients
- — Sensing of multiple channels
- — Sensing of RAT
- — Usage of interface
  - — CE–S, CE–CE, CE–DA
- — Key sensing information
  - — Client ID
  - — Client priority
  - — Target detection performance
  - — Sensing method, techniques
  - — Sensing duration
  - — Client priority
  - — Duty cycle
  - — Detection threshold
  - — Information to be sent to DA should be as complete as possible
  - — Sensor information
  - — Reporting mode

MCM-F is extended by using multiple CEs.

Extension Model of Sensing Model G (MCM-G-extension)

Use Case This model has the CE capable devices located in or near the primary service area. The devices function as permanent secondary spectrum coordinators. The difference from the extension model of sensing model F is that in this case, the secondary spectrum coordinator does not have the capability to act as a virtual AP/BS. This model may be essential for an ad hoc secondary spectrum usage case in MCM because the secondary service coverage area is extended by allowing multiple individual networks that are operated by different secondary spectrum coordinators.

The following points need to be considered:

— Sensing/Control information exchange between CEs
— Sensing/Control information exchange between smart sensors
— Specific points that need to be considered for this model may be summarized as follows:
  — The client cannot download the information about the spectrum usage environment based on geographical information.
  — Therefore, the volume and the number of elements of sensing information that needs to be shared between the smart sensor and the CE of the secondary spectrum coordinator may be the largest among all models.
  — Also, the volume and the number of elements of sensing information that needs to be shared between the CE of the secondary spectrum coordinator and the sensors may be the largest among all models.
  — Also, the volume and the number of elements of sensing information to be shared between the smart sensor and the sensors may be the largest among all models.
  — The control information between the CEs of the secondary spectrum coordinator should be well considered.
  — The control information between the smart sensors should be well considered.
  — The other features may be the same as those of the sensing model G.

Use Case Analysis
— Spectrum usage model and scenario classification
  — Long-term spectrum usage::M:1
— Sensing requirements
  — Providing sensing information to multiple clients
  — Sensing to identify primary system feature
  — Sensing information integration
  — Sensing information sharing control
  — Distributed device control
  — Sensing to prevent interference
  — Synchronization among sensors or radios
  — Sensing to identify service channels
  — Sense and vacate immediately if a primary signal is detected
  — Provide sensing information to multiple clients
  — Sensing of multiple channels
  — Sensing of RAT
— Usage of interface
  — CE–S, CE–CE
— Key sensing information
  — Client ID
  — Client priority
  — Target detection performance
  — Sensing method, techniques
  — Sensing duration
  — Client priority
  — Duty cycle

— Detection threshold
— Information to be sent to DA should be as complete as possible
— Sensor information
— Reporting mode
— Policy information to provide underlay operation

MCM-G is extended by using multiple CEs.

Annex 18B

(informative)

Use case classification

Table 18B.1 — Use case classification.

Spectrum usage models	Scenarios of system model	Use cases and application examples
Long-term spectrum usage	1:1	Emergency services (SPOLD1) Load Sharing to reduce blocking at peak traffic times (SPOLD2)
	1:N	Sharing of spectrum to reduce blocking at peak traffic times (SPOLD3) NCM-B,NCM-C , NCM-D, NCM-E, MCM-F, MCM-G IEEE 1900.4 Dynamic spectrum assignment (IEEE 1900.4-DSS)
	M:1	Self management of uncoordinated spectrum (SPOLD4) MCM-F-extension MCM-G-extension IEEE 1900.4 Dynamic spectrum assignment (IEEE 1900.4-DSS)
Short-term spectrum usage	1:1	Power constrained sensing (SeEM2)
	1:N	Policy Investigation (SEM5) Distributed sensing (DSM1) Faulty sensing prevention (SeEM1) NCM-E IEEE 1900.4 Distributed radio resource usage optimization (1900.4-DRRUO) IEEE 1900.4 Dynamic spectrum sharing (IEEE 1900.4-DSS)
	M:1	Peer-to-peer Communication (DSM2) Priority based sensing (SeEM3) NCM-E-Extension IEEE 1900.4 Distributed radio resource usage optimization (1990.4-DRRUO) IEEE 1900.4 Dynamic spectrum sharing (IEEE 1900.4-DSS)

From IEEE Standard for Spectrum Sensing Interfaces and Data Structures for Dynamic Spectrum Access and Other Advanced Radio Communication Systems, Feb 2011. IEEE.

Table 18B.2 — Mapping between use cases and sensing requirements.

Reconfigurable sensing performance	SPOLD1, SEM4
Sensing of multiple channels simultaneously	SPOLD1, SPOLD4, SPOSD1, SPOSD2,SPOSD4, SEM4, SEM6, IEEE 1900.4-DS Assignment, IEEE 1900.4-DRRUO, DSM1, DSM2, NCM-B, NCM-C, NCM-D, NCM-E, MCM-F(-extension), MCM-G(-extension)
Only information on signal occupancy can be sufficient (band is free or occupied)	SPOLD1
Sensing is performed only at initial stage (during or after the catastrophe)	SPOLD1
Sensing is performed by embedded sensors with in radios (if infrastructure and fixed sensors are destroyed) or temporary deployment of distributed sensors	SPOLD1
Sensing of RAT(s)	SPOSD2,SPOSD3, SEM6
Traffic sensing capability	SPOLD2, SPOLD3, SPOLD5, SPOLD6, SPOSD1, SEM6, IEEE 1900.4-DS Assignment, IEEE 1900.4-DSS, IEEE 1900.4-DRRUO
Sensing to identify service channels	SPOLD2, SPOLD3, SPOLD5, SPOLD6, SPOSD2, SPOSD3, SEM1, SEM6, IEEE 1900.4-DS Assignment, IEEE 1900.4-DSS, IEEE 1900.4-DRRUO, MCM-F(-extension)
Sense the most suitable channel first	SPOLD2, SPOLD6, SPOSD2, SEM6, NCM-E(-extension), MCM-F(-extension)
Sensing of white space	SPOLD3, SPOLD5, IEEE 1900.4-DSS, NCM-B, NCM-C, NCM-D
Check with DA	SPOLD3, SPOLD5, SPOSD2, SEM6, IEEE 1900.4-DSS, NCM-B, NCM-C, NCM-D, NCM-E(-extension), MCM-F(-extension),
Sense and vacate immediately if a primary signal is detected	SPOLD3, SPOLD5, SPOSD2, SPOSD3, SEM1, SEM6, MCM-F(-extension) , MCM-F(-extension), MCM-F(-extension)
Sensing to prevent interference	SPOLD4, SPOLD4, SPOSD2, SPOSD3, SEM6, DSM2, MCM-F(-extension), MCM-F(-extension)
Provide sensing information to multiple clients	SPOLD4, SPOSD1, SPOSD4, SEM4, SeEM3, NCM-E-extension, MCM-F-extension, MCM-G-extension
As complete as possible	SEM5
On regular basis	SEM5
Synchronization among sensors or radios	SPOSD2, SEM6, DSM1, DSM2, MCM-F(-extension), MCM-F(-extension)
Provide sensing information with performance measure	SeEM1,
Provide sensing information with power consumption	SeEM2
Information exchange method should be chosen for low power consumption	SeEM2
Sensing to identify primary system feature	SPOSD3, MCM-F(-extension), MCM-G(-extension)
Sensing information integration	NCM-E-extension, MCM-F-extension, MCM-G-extension, DSM1, DSM2
Sensing information sharing control	NCM-E-extension, MCM-F-extension, MCM-G-extension, DSM1, DSM2, SPOSD3
Sensing information sharing with primary resource controller for validation of measurement results	SPOSD2, SEM6, NCM-E(-extension), MCM-F(-extension)
Distributed device control	NCM C, D, E(-extension), MCM-F(-extension), MCM-G(+extension), DSM1, DSM2

From IEEE Standard for Spectrum Sensing Interfaces and Data Structures for Dynamic Spectrum Access and Other Advanced Radio Communication Systems, Feb 2011. IEEE.

Table 18B.3 — Mapping between use cases and key sensing information.

Target detection performance (Pd, Pf, etc.)	SPOLD1
Sensing method, techniques	SPOLD1, SPOSD2, SPOSD3, SEM6, NCM-B, NCM-C, NCM-D, NCM-E(-extension) , MCM-F(-extension),
Sensing duration (channel latency issue)	SPOLD1, SPOSD2, SPOSD3, SEM6, NCM-B, NCM-C, NCM-D, NCM-E(-extension) , MCM-F(-extension), MCM-F(-extension)
Traffic information (resource allocation ratio)	SPOLD2, SPOLD3, SPOLD5, SPOLD6, SPOSD1, SEM6, IEEE 1900.4-DS Assignment, IEEE 1900.4-DSS, IEEE 1900.4-DRRUO
Service channels (RATs) ID	SPOLD2, SPOLD3, SPOLD5, SPOLD6, SPOSD2, SPOSD3, SEM1, SEM6, IEEE 1900.4-DS Assignment, IEEE 1900.4-DSS, IEEE 1900.4-DRRUO
Client priority, ID	SPOLD4, SPOSD1, SPOSD4, SEM4, DSM2, SeEM3, NCM-E-extension, MCM-F-extension, MCM-G-extension
Duty cycle (to predict potential interference)	SPOLD4, SPOSD1, SPOSD4, DSM2, NCM-E(-extension) , MCM-F(-extension) , MCM-F(-extension)
Power level of primary signals	SPOSD2, SEM1
Policy information	SPOSD2, SEM1
Sensing mode, duration	SEM4
Information to be sent to DA should be as complete as possible (RAT or modulation, geolocation, timestamp, operator ID, signal level, center frequency, bandwidth, Sensing methods, etc.)	SEM5, SPOSD2, SEM6, NCM-E(-extension) , MCM-F(-extension),
Sensor information (ID, geolocation, battery, specification, etc.)	DSM1, DSM2, SeEM2, NCM-B, NCM-C, NCM-D, NCM-E, MCM-F, MCM-G
Sensing performance (confidence level)	SeEM1
Sensor power consumption	SeEM2
Reporting mode (soft or hard information)	SeEM2, NCM-B, NCM-C, NCM-D, NCM-E(-extension) , MCM-F(-extension) , MCM-G(-extension)
primary system feature	SPOSD3, MCM-F(-extension), MCM-G(-extension)
Sensing information integration	NCM-E-extension, MCM-F-extension, MCM-G-extension, DSM1, DSM2
Sensing information sharing control	NCM-E-extension, MCM-F-extension, MCM-G-extension, DSM1, DSM2, SPOSD3
Sensing information sharing with primary resource controller for validation of measurement results	SPOSD2, SEM6, NCM-E(-extension), MCM-F(-extension)
Distributed device control	NCM C, D, E(-extension), MCM-F(-extension), MCM-G(+extension), DSM1, DSM2

From IEEE Standard for Spectrum Sensing Interfaces and Data Structures for Dynamic Spectrum Access and Other Advanced Radio Communication Systems, Feb 2011. IEEE.

Table 18B.4 —Spectrum sensing system model and its relation to IEEE 1900.6 interfaces.

System models	Interfaces involved	Control information	Sensing-related information
1:1 scenario Single device spectrum sensing model (reference model)	CE/DA–S interface	Command + sensing parameters	Sensing + sensor information
1:N scenario Distributed sensors spectrum sensing model	CE/DA–S interface	Command + sensing parameters	Sensing + sensor information
	S–S interface	Command + sensing parameters	Sensing + sensor information
M:1 scenario Multiple CEs	CE–CE/DA interface	Command + sensing parameters	Sensing + sensor information
	CE/DA–S interface	Command + sensing parameters	Sensing + sensor information

From IEEE Standard for Spectrum Sensing Interfaces and Data Structures for Dynamic Spectrum Access and Other Advanced Radio Communication Systems, Feb 2011. IEEE.

Table 18B.5 —IEEE 1900.6 interfaces and sensing-related information.

Sensing-related information	CE–CE/DA network	CE–CE/DA terminal	CE/DA–S single sensor	CE/DA–S distributed sensors	S–S distributed sensors
Sensing control information	Sensing information request	Sensing information request	—Sensing information request —Measurement data request —Sensing duration —Termination .desired Pd or Pf —Types of sensing information —Priority control —Sensing information security level —Database information request —Client identification	—Sensing information request —Measurement data request —Sensing duration —Termination .desired Pd or Pf —Types of sensing information —Routing —Sensor gateway selection —Sensing information security level —Client identification —Location information	—Sensing information request —Measurement data request —Sensing duration —Termination —types of sensing information —Routing —Sensor gateway selection —Sensing information security level —Client identification
Sensor information	Not required	Not required	—Sensor logical ID —Battery status —Noise power —Sensor capability —Manufacturer ID —Types of sensing technique —Measurement range —A/D, D/A resolution —Calibration data	—Sensor logical ID —Battery status Noise power —Sensor capability —Manufacturer ID —Types of sensing technique —Measurement range —A/D, D/A resolution —Calibration data	—Sensor logical ID —Battery status —Noise power —Manufacturer ID —Types of sensing technique —Measurement range —A/D, D/A resolution —Calibration data
Sensing information	—Band occupancy —Frequency band —Confidence level	—Band occupancy —Frequency band —Confidence level	—Band occupancy —Signal energy —Confidence level —Frequency band —Bandwidth —Center frequency —Cyclic frequency —Location information —Local threshold	—Band occupancy —Signal energy —Confidence level —Frequency band —Bandwidth —Center frequency —Cyclic frequency —Location information —Local threshold	—Band occupancy —Signal energy —Confidence level —Frequency band —Bandwidth —Cyclic frequency —Location information —Local threshold

From IEEE Standard for Spectrum Sensing Interfaces and Data Structures for Dynamic Spectrum Access and Other Advanced Radio Communication Systems, Feb 2011. IEEE.

Table 18B.6 —IEEE 1900.6 interfaces and locations of sensors.

Sensing-related information	CE–CE/DA network	CE–CE/DA terminal	CE/DA–S single sensor	CE/DA–S distributed sensors	S–S distributed sensors
Sensor location	—Distributed sensors —Smart - sensors —Sensing capable device	—Distributed sensors —Smart - sensors —Sensing capable device	—Base station —AP —CR terminal	—Base station —AP —CR Terminal —Non-CR devices (personal digital assistant [PDA], smart phone, laptop, etc.) —Distributed standalone fixed sensors —Smart sensors	—Distributed standalone fixed sensors —Smart sensors

From IEEE Standard for Spectrum Sensing Interfaces and Data Structures for Dynamic Spectrum Access and Other Advanced Radio Communication Systems, Feb 2011. IEEE.

Annex 18C

(informative)

Implementation of distributed sensing

18C.1.1   General Description

Sensors and their clients can exchange sensing-related information through approved radio channels. The clients can be sensors and CE. The approved radio for sensing-related information exchange can be WiFi®,[10] ZigBee™,[11] UWB, etc.[12] This sensing-related information exchange is involved in distributed sensing models (DSMs) as described in Annex 18A.

18C.1.2   Assumptions/Requirement

For a given terminal or network node to communicate with multiple spectrum sensors, multiple access support is assumed on the client side. Furthermore, sensors are assumed to be equipped with onboard transceivers. In the case of distributed sensors spectrum sensing, sensors may need to execute directly sensing algorithms locally; hence, they need to be equipped with digital signal processing chip to perform sensing. Sensing algorithms may require estimating some parameters adaptively at certain interval such as noise power. Therefore, a memory chip is needed to store the information temporarily. Figure 18C.1 shows and example of the OSI model for sensing-related information exchange between a sensor and a client.

Figure 18C.2 shows the architecture of a sensor that can exchange sensing-related information. The figure shows where the IEEE 1900.6 interface is applied in the sensing-related information exchange.

18C.1.3   Examples

18C.1.3.1   Embedded/Plugged Sensor Type I (Spectrum Sensor Collocated with CE)

Embedded or plugged sensor type I referrers to a sensor that is physically wired to the CE capable device with a communication application programming interface. This is a typical example of where the IEEE 1900.6 logical interface is applicable. Interoperability between devices coming from different vendors can be provided by implementing a functional entity at both the client and the sensors that provide the IEEE 1900.6 logical interface as depicted in Figure 18C.3 and Figure 18C.4.

18C.1.3.2   Embedded/Plugged Sensor Type II (Spectrum Sensor not Collocated with CE)

Embedded or plugged sensor type II referrers a sensor that is physically connected to a non-CE capable device. A non-CE capable device is any electronic device that has memory, computational resource for spectrum sensing application, and transceiver to communicate with other

---

10 WiFi is a registered trademark in the U.S. Patent & Trademark Office, owned by the WiFi Alliance.
11 ZigBee is a trademark in the U.S. Patent & Trademark Office, owned by the ZigBee Alliance.
12 This information is given for the convenience of users of this standard and does not constitute an endorsement by the IEEE of these products. Equivalent products may be used if they can be shown to lead to the same results.

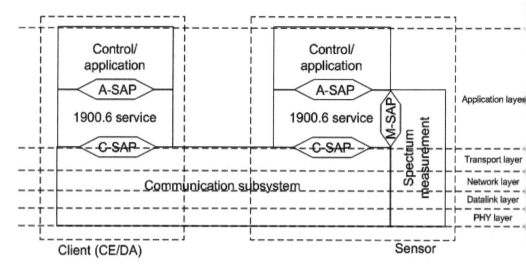

Figure 18C.1 —Example OSI model for sensing-related information exchange between sensor and client. From IEEE Standard for Spectrum Sensing Interfaces and Data Structures for Dynamic Spectrum Access and Other Advanced Radio Communication Systems, Feb 2011. IEEE.

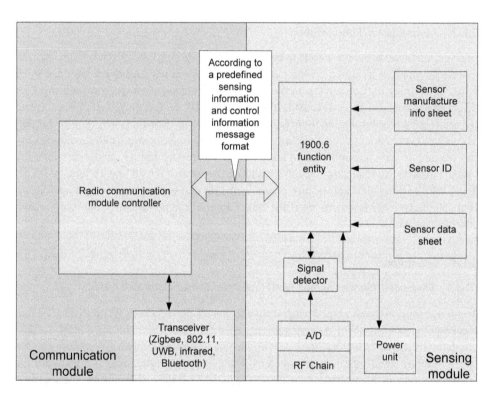

Figure 18C.2 —Architecture of a sensor that can exchange sensing-related information through approved radio channels. From IEEE Standard for Spectrum Sensing Interfaces and Data Structures for Dynamic Spectrum Access and Other Advanced Radio Communication Systems, Feb 2011. IEEE.

devices. Such devices can include but are not limited to legacy terminal such as cell phone, PDA, notebook, personal computer, AP, etc. The entire combined set of the spectrum sensing module and the client can be viewed as a smart sensing capable device. For example, smart sensing algorithms can be implemented on the client side for distributed sensing. One interesting aspect of the smart sensor capable device is that an existing office infrastructure such as laptops and APs can be exploited to deploy distributed spectrum sensing. Interoperability between devices coming from different vendors can be provided by implementing a functional

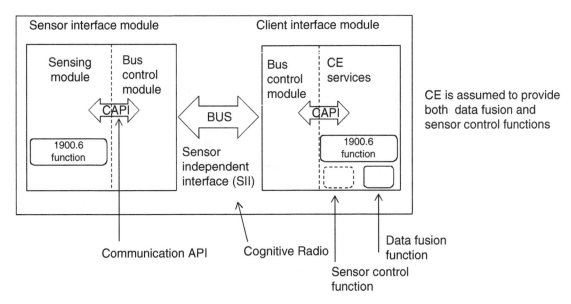

Figure 18C.3 —Block diagram showing a wired spectrum sensor to a CE capable device. From IEEE Standard for Spectrum Sensing Interfaces and Data Structures for Dynamic Spectrum Access and Other Advanced Radio Communication Systems, Feb 2011. IEEE.

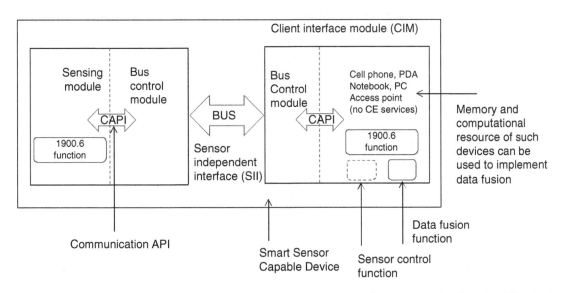

Figure 18C.4 —Block diagram showing a spectrum sensor wired to a non-CE capable device. From IEEE Standard for Spectrum Sensing Interfaces and Data Structures for Dynamic Spectrum Access and Other Advanced Radio Communication Systems, Feb 2011. IEEE.

entity at both the client and sensors that provide the IEEE 1900.6 logical interface. Figure 18C.4 depicts the block diagram of the smart sensor capable device.

18C.1.3.3 Distributed Standalone Sensor Type I (Information Exchange between CE or DA and a Sensor)

In distributed standalone sensor type I, sensors provide their sensing information to CE/DA through a wireless medium. Sensors in such an implementation example do not require any smart functionality, such as data fusion or control over other sensors. Simple spectrum sensors with wireless transceiver capability can be used to realize distributed sensing such as cooperative sensing, collaborative sensing, and selective sensing. Data fusion or sensor control is performed at the client side. Interoperability between devices coming from different vendors

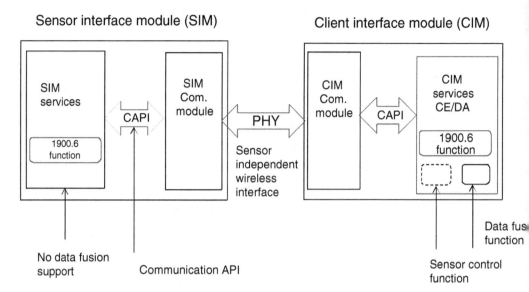

Figure 18C.5 —Block diagram showing a spectrum sensor connected in a wireless medium to a client that is ei'
a CE or DA. From IEEE Standard for Spectrum Sensing Interfaces and Data Structures for Dynamic Spectrum Ac·
and Other Advanced Radio Communication Systems, Feb 2011. IEEE.

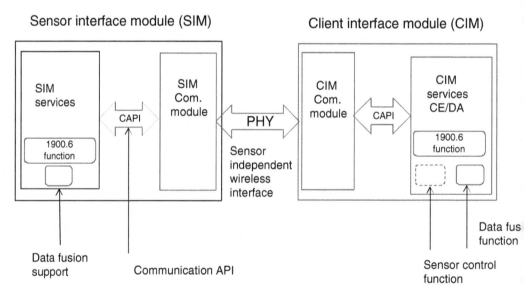

Figure 18C.6 —Block diagram showing a smart spectrum sensor connected in a wireless medium to a client th·
either a CE or DA. From IEEE Standard for Spectrum Sensing Interfaces and Data Structures for Dynamic Spect·
Access and Other Advanced Radio Communication Systems, Feb 2011. IEEE.

can be provided by implementing a functional entity at both the client and the sensors that
provide the IEEE 1900.6 logical interface. Figure 18C.5 depicts a block diagram of a client.

18C.1.3.4   Distributed Standalone Sensor Type II (Information Exchange between CE or DA and Smart Sensor)

Distributed standalone sensor type II, as depicted in Figure 18C.6, is an example of using the
IEEE 1900.6 logical interface to exchange sensing control and sensing information between a
CE/DA and a smart sensor.. Interoperability between devices coming from different vendors
can be provided by implementing a functional entity at both the client and the sensors that provides the IEEE 1900.6 logical interface. The smart sensor can provide an aggregate of sensing
information obtained from other sensors to the client to enhance sensing quality.

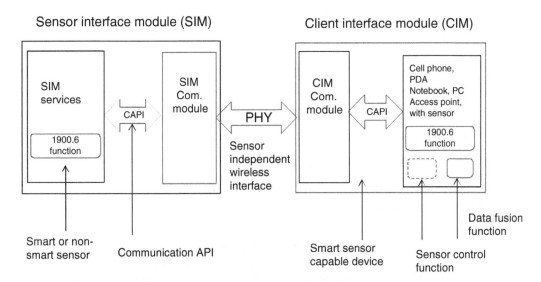

Figure 18C.7 —Block diagram showing a spectrum sensor connected in a wireless medium to a client spectrum sensor. From IEEE Standard for Spectrum Sensing Interfaces and Data Structures for Dynamic Spectrum Access and Other Advanced Radio Communication Systems, Feb 2011. IEEE.

18C.1.3.5   Distributed Standalone Sensor Type III (Information Exchange between Sensor and Smart Sensor or between Smart Sensors)

Distributed standalone sensor type III, as depicted in Figure 18C.7, is an example of using the IEEE 1900.6 logical interface to exchange sensing control and sensing information between a client spectrum sensor and another spectrum sensor. The client sensor is a sensor with application for either data fusion or relaying. It obtains sensing information from other sensors, merges it with its own sensing information, and forwards it to a CE. The smart sensor can be an integrated sensor manufactured to provide the preceding functionality or a smart sensor capable device introduced in Figure 18C.4. This particular implementation of the IEEE 1900.6 logical interface assists distributed spectrum sensing where sensors share their sensing information to make optimum local decisions before forwarding the final result to the CE. Furthermore, it can assist the relaying function of sensing-related information.

Annex 18D

(informative)

IEEE 1900.6 DA: Scope and usage

18D.1 Introduction

The DA is a logical entity in which sensing-related information obtained from spectrum sensors or other sources, as well as regulatory and policy information, are processed and stored systematically. The DA processing capability is limited to storing, retrieving, data format conversion, and querying (fundamental data processing). Analyzing sensing-related information for decision making is done by the CE.

Annex 18E clarifies the scope and possible usage of the DA by describing the information to be managed and to be provided by the DA, minimum functionality of the DA, as well as possible locations of the DA. The scope and usage of DA are shown in Figure 18D.1.

Note that the DA may contain more information and may have interfaces to other entities that are not discussed in this standard. Also, the details of data processing algorithms inside the DA as well as administrative issues for the DA are not included.

18D.2 Scope and Usage of the DA

The DA can be a part of a CR system; it stores sensing-related information obtained from multiple sources such as spectrum sensors, CEs, other DAs, or other external data repositories. The DA also stores regulatory and policy information obtained from regulatory repositories. An example of a regulatory repository is the TV bands database defined in FCC ET Docket No. 08-260 [B1]. Information stored in the DA can be provided to other DAs or CEs. This information can be provided in the original form (as it is) or after processing by some predefined algorithms. The IEEE 1900.6 logical interface can be used to convey sensing-related information and processed information. However, regulatory information exchange may need another interface.

18D.3 Categories of Information Stored at DA

Examples of information stored at a DA are shown in 18D.3.1 and 18D.3.2.

18D.3.1 Sensing-related Information

Sensing-related information includes sensing information, sensing control information and control commands, sensor information, and requirements derived from regulation (cf. Clause 18.7 for parameter description):

— Spectrum measurement information and/or local decision of spectrum usage
— Location and time stamp of the spectrum measurement information and spectrum usage information
— Sensing method, sensor specifications, and information accuracy such as confidence level

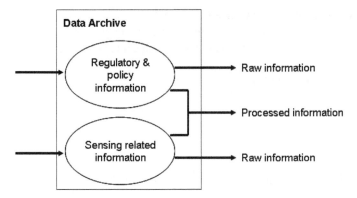

Figure 18D.1 —Scope and usage of DA. From IEEE Standard for Spectrum Sensing Interfaces and Data Structures for Dynamic Spectrum Access and Other Advanced Radio Communication Systems, Feb 2011. IEEE.

Note that sensing-related information is dynamically changed and that its validity is in short time and for a relative small geographical area.

18D.3.2   Regulatory and Policy Information

This information can be obtained from a database containing information on license holders, facility operation parameters (e.g., frequency, location, etc.), and any special conditions that apply. Based on this information, requirement as sensing-related information can be derived (cf. 18.7.1.5):

— Transmitter location or geographic area of operation
— Effective radiated power
— Transmitter height above average terrain
— Antenna height above ground level
— Call sign
— License holders spectrum masks,
— License holders receiver sensitivity, etc.

Note that regulatory and policy information is relative stable information. The information does not change frequently for a relative large area.

18D.4   DA Requirements

— Information acquisition capability
  The DA shall have the capability to obtain sensing information from spectrum sensors or other sources and shall have the capability to obtain regulatory and policy information from regulatory repositories or other sources.
— Information storage capability
  The DA shall have the capability to store sensing information and regulatory/policy information.
— Information provision capability
  The DA shall have the capability to provide storage information to other clients (e.g., CE or another DA).
— IEEE 1900.6 compliance
  The DA shall comply with the IEEE 1900.6 reference model (cf. Clause 18.5) and shall be able to exchange information according to the IEEE 1900.6 logical interface.
— Systematic organization of information
  Sensing information and regulatory/policy information shall be processed and stored systematically in the DA. To improve the performance of information exchange, stored data should be classified, sorted, and indexed.

— Preliminary data processing capability of raw sensing information
  The DA may provide storage information in the original form, or it may provide the processed information as a result of regulatory information and sensing information according to some algorithms.
  The input for these processes can be as follows:

— Enforcing regulatory body
— Protected users and their area
— Current location of the DA and CRs
— Sensing information

The output of these processes can be as follows:

— Reliable sensing information collected over a long period of time possibly obtained from multiple sensors.

18D.5 Necessary Interfaces

The following interfaces are required for a DA to exchange information:

— Interface to spectrum sensors
— Interface to CEs
— Interface to other DAs
— Interface to regulatory and policy repositories

18D.6 DA Services

The DA may provide the following services to other IEEE 1900.6 logical entities:

— The repository for sensing information from spectrum sensors.
— The register for CEs, sensors, and other DAs. Registered information may include the logical ID and geolocation of the entities. For example, the DA may keep the logical ID of sensors and provide this information to the CE, where this information can be used to compute the confidence level.
— The information source for CEs or other DAs. To support this service, the DA may have the capability to respond to queries from CE/DAs.

18D.7 Deployment Examples

One possibility is to implement the DA as shown in Figure 18D.2. In this case, the DA needs to have an interface to the regulatory repository to obtain regulatory and policy information, as shown in Figure 18D.2.

Another possible deployment is shown in Figure 18D.3. If the regulatory repository has the capability to store and exchange sensing information and has the IEEE 1900.6 interface, it is considered an IEEE 1900.6 DA.

Figure 18D.2 —DA has interface to regulatory repository. From IEEE Standard for Spectrum Sensing Interfaces and Data Structures for Dynamic Spectrum Access and Other Advanced Radio Communication Systems, Feb 2011. IEEE.

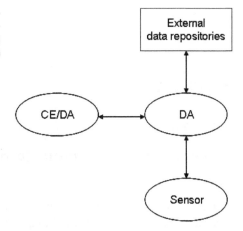

Figure 18D.3 —DA includes the regulatory repository. From IEEE Standard for Spectrum Sensing Interfaces and Data Structures for Dynamic Spectrum Access and Other Advanced Radio Communication Systems, Feb 2011. IEEE.

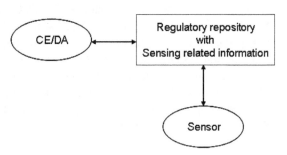

Annex 18E

(informative)

Analysis of available/future technologies

To have an overview of the spectrum sensing technologies and techniques, two surveys of the algorithms being researched and the equipment available on the market have been edited.[13]

The first of these documents, titled *Review of Contemporary Spectrum Sensing Technologies*, catalogues a representative sample of available spectrum sensing technologies and then through analysis ascertains the interface features between the "sensing system" and the "CE" that are common among them (see Pucker et al. [B8]).

The second document, titled *Sensing Techniques for Cognitive Radio—State of the Art and Trends,* identifies the spectrum sensing techniques being used and researched in the field of cognitive radio. It provides a comprehensive overview on the techniques that have been proposed and the ones that may emerge in the next few years. Links to an in-depth bibliography are provided systematically to enable the reader to get more technical details (see Noguet et al. [B7]).

---

13 These reports are available at http://www.scc41.org (IEEE 1900.6 folder).

Annex 18F

(informative)

Bibliography

B1 FCC ET Docket No. 08-260, "Second report and order and memorandum opinion and order," 2008.[14,15]

B2 IEEE P802.22™/D2.08, May 2009, Draft Standard for Wireless Regional Area Networks—Part 22: Cognitive Wireless RAN Medium Access Control (MAC) and Physical Layer (PHY) Specifications: Policies and Procedures for Operation in the TV Bands.[16,17,18]

B3 IEEE Std 754™-2008, IEEE Standard for Floating-Point Arithmetic.

B4 IEEE Std 802.11™-2007, IEEE Standard for Information Technology—Telecommunications and Information Exchange between Systems—Local and Metropolitan Area Networks—Specific Requirements Part 11: Wireless LAN Medium Access Control (MAC) and Physical Layer (PHY) Specifications.

B5 ITU-T Recommendation X.210, Information Technology—Open Systems Interconnection—Basic Reference Model: Conventions for the Definition of OSI Services.[19]

B6 National Imagery and Mapping Agency, *Department of Defense World Geodetic System 1984—Its Definition and Relationships with Local Geodetic Systems*, Technical Report TR8350.2, Bethesda, MD, 2004.[20]

B7 Noguet, D., Demessie, Y. A., Biard L., Bouzegzi A., Debbah M., Haghighi K., Jallon P., Laugeois M., Marques P., Murroni M., Palicot J., Sun C., Thilakawardana S., and Yamaguchi A., *Sensing Techniques for Cognitive Radio—State of the Art and Trends*, IEEE SCC41—IEEE 1900.6 Working Group, White Paper, Apr. 2009.[21]

B8 Pucker L., *Review of Contemporary Spectrum Sensing Technologies*, IEEE SCC41—IEEE 1900.6 Working Group, White Paper, Sept. 2008.[22]

B9 SDR Forum, "Use Cases for Cognitive Applications in Public Safety Communications Systems—Volume 1: Review of the 7 July Bombing of the London Underground," Approved 8 November 2007, SDRF-07-P-0019-V1.0.0.[23]

---

14 Available at: http://hraunfoss.fcc.gov/edocs_public/attachmatch/FCC-08-260A1.pdf.

15 FCC publications are available from the Federal Communications Commission, 445 12th Street SW, Washington, DC 20554 (http://www.fcc.gov).

16 IEEE publications are available from the Institute of Electrical and Electronics Engineers, 445 Hoes Lane, Piscataway, NJ 08854, USA (http://standards.ieee.org/).

17 The IEEE standards or products referred to in this clause are trademarks owned by the Institute of Electrical and Electronics Engineers, Incorporated.

18 This IEEE standards project was not approved by the IEEE-SA Standards Board at the time this publication went to press. For information about obtaining a draft, contact the IEEE.

19 ITU-T publications are available from the International Telecommunications Union, Place des Nations, CH-1211, Geneva 20, Switzerland/Suisse (http://www.itu.int/).

20 Available at: http://earth-info.nga.mil/GandG/publications/tr8350.2/tr8350_2.html.

21 Available at: http://grouper.ieee.org/groups/scc41/6/documents/white_papers/P1900.6_WhitePaper_Sensing_final.pdf.

22 Available at: http://grouper.ieee.org/groups/scc41/6/documents/white_papers/P1900.6_Sensor_Survey.pdf.

23 Available at: http://groups.sdrforum.org/download.php?sid=769.

19

IEEE Standard for Radio Interface for White Space Dynamic Spectrum Access Radio Systems Supporting Fixed and Mobile Operation

IMPORTANT NOTICE: IEEE Standards documents are not intended to ensure safety, security, health, or environmental protection, or ensure against interference with or from other devices or networks. Implementers of IEEE Standards documents are responsible for determining and complying with all appropriate safety, security, environmental, health, and interference protection practices and all applicable laws and regulations.

This IEEE document is made available for use subject to important notices and legal disclaimers. These notices and disclaimers appear in all publications containing this document and may be found under the heading "Important Notice" or "Important Notices and Disclaimers Concerning IEEE Documents." They can also be obtained on request from IEEE or viewed at http://standards.ieee.org/IPR/disclaimers.html.

19.1 Overview

19.1.1 Scope

This standard specifies a radio interface including medium access control (MAC) sublayer(s) and physical (PHY) layer(s) of white space dynamic spectrum access radio systems supporting fixed and mobile operation in white space frequency bands, while avoiding causing harmful interference to incumbent users in these frequency bands.

19.1.2 Purpose

This standard enables the development of cost-effective, multi-vendor white space dynamic spectrum access radio systems capable of interoperable operation in white space frequency bands on a non-interfering basis to incumbent users in these frequency bands. This standard facilitates a variety of applications, including the ones capable to support mobility, both low-power and high-power, short-, medium-, and long-range.

19.2 Definitions, Acronyms, and Abbreviations

19.2.1 Definitions

The *IEEE Standards Dictionary Online* should be consulted for terms not defined in this clause.[1]

19.2.2 Acronyms and Abbreviations

ACK   acknowledgement
BP   beacon period
BPSK   binary PSK

---

1 *IEEE Standards Dictionary Online* subscription is available at: http://ieeexplore.ieee.org/xpls/dictionary.jsp.

Dynamic Spectrum Access Decisions: Local, Distributed, Centralized, and Hybrid Designs, First Edition. George F. Elmasry.
© 2021 John Wiley & Sons Ltd. Published 2021 by John Wiley & Sons Ltd.
Companion website: www.wiley.com/go/elmasry/dsad

BPST	beacon period start time
BSN	beacon sequence number
CAP	contention access period
CCA	clear channel assessment
CRC	cyclic redundancy check
CS	convergence sublayer
CSMA-CA	carrier sense multiple access with collision avoidance
CTS	clear to send
CW	contention window
DIFS	distributed coordination function IFS
DSN	data sequence number
FBMC	filter bank multi-carrier
IFS	inter frame space
Imm-ACK	immediate ACK
MAC	medium access control
MAS	medium access slot
MCS	modulation and coding scheme
MIFS	minimum IFS
MLME	MAC sublayer management entity
MPDU	MAC sublayer protocol data unit
MSDU	MAC sublayer service data unit
NAV	network allocation vector
No-ACK	no ACK
PD	physical layer data service
PHY	physical layer
PLCP	physical layer convergence protocol
PLME	physical layer management entity
PPDU	physical layer protocol data unit
PRBS	pseudo random binary sequence
PSDU	physical layer service data unit
PSK	phase shift keying
QAM	quadrature amplitude modulation
QPSK	quadrature PSK
RTS	request to send
SAP	service access point
SIFS	short IFS
TXOP	transmit opportunity
WS	white space

19.3 Reference Model

Figure 19.1 illustrates the reference model of an IEEE 1900.7 device.

The IEEE 1900.7 radio interface comprises MAC sublayer and PHY layer.

MAC sublayer provides MAC data service and MAC management service further specified in Clause 19.4.

PHY layer provides PHY data service and PHY management service further specified in Clause 19.5.

Convergence sublayer (CS) is responsible for translation of external network data received via the CS service access point (CS_SAP) into MAC service data units (SDU) received by the MAC sublayer through the MAC_SAP. CS is not specified in this standard.

Figure 19.1 Reference model of an IEEE 1900.7 device. From IEEE Standard for Radio Interface for White Space Dynamic Spectrum Access Radio Systems Supporting Fixed and Mobile Operation, Dec 2015. IEEE.

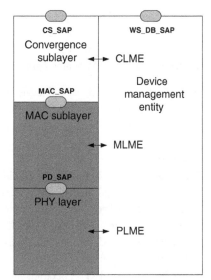

CS, MAC sublayer, and PHY layer are controlled by convergence sublayer management entity (CLME), MAC sublayer management entity (MLME), and PHY layer management entity (PLME) respectively. CLME, MLME, and PLME are part of a device management entity. The device management entity is not specified in this standard.

The white space database SAP (WS_DB_SAP) provides the interface to a white space (WS) database. This interface is not specified in this standard.

19.4 MAC Sublayer

19.4.1 Architecture of the MAC Sublayer

The reference model is provided in Figure 19.1.

The MAC sublayer provides the following two services:

— The MAC data service and the MAC management service interfacing to the MAC sublayer management entity (MLME) service access point (SAP) (known as MLME-SAP).
— The MAC data service (MAC) enables the transmission and reception of MAC protocol data units (MPDUs) across the PHY data service.

A network operates in a master-slave mode. In the master-slave mode, a device is designated as master (also referred to as network coordinator hereafter) and others are associated with the master as slaves. The master coordinates channel access in the master-slave mode. Communication is normally established between slave devices and the master device.

19.4.2 Type Definition

```
Boolean          1 if TRUE, 0 otherwise
DeviceAddress    16/48 bit integer (see 4.3.2.1.1)
Int32            32 bit integer
Array            Array of bytes
```

19.4.3 MAC Frame Formats

This subclause specifies the format of the MAC frame (MPDU).

19.4.3.1 Frame Format Convention

MAC frames are described as a sequence of fields in a specific order. Unless otherwise noted, fields longer than a single byte are delivered to the PHY SAP in order from the byte containing the least-significant bits to the byte containing the most-significant bits.

Values specified in integer are encoded in unsigned binary unless otherwise stated.

Reserved fields and subfields are set to ZERO on transmission and ignored on reception.

19.4.3.2 General MAC Frame Format

A MAC frame consists of a fixed-length MAC header and an optional variable-length MAC Frame Body. The general MAC frame shall be formatted as illustrated in Figure 19.2.

CRC refers to the cyclic redundancy check.

MAC Header

FrameControl   The FrameControl field is specified in Table 19.1.

The protocol version field specifies the version of the protocol. For this revision of the standard, the protocol version is set to zero. All other values are reserved.

The Address format field specifies if short (16 bits) or long (48 bits) addresses are considered. If 0, short addressed are used for DestAddr and SrcAddr fields.

The ACK policy filed is specified in Table 19.2.

The frame type filed is specified in Table 19.3.

The sequence number field is 1 byte in length and specifies the sequence identifier for the frame. For a beacon frame, the sequence number field shall specify a beacon sequence number (BSN). For a data acknowledgment, or a MAC command frame, the sequence number field shall specify a data sequence number (DSN) that is used to match an acknowledgment frame to the data or MAC command frame.

24 bits	16/48 bits	16/48 bits	Variable	16 bits
FrameControl	DestAddr	ScrAddr	FramePayload	CRC
MAC Header				

Figure 19.2 General MAC frame format. From IEEE Standard for Radio Interface for White Space Dynamic Spectrum Access Radio Systems Supporting Fixed and Mobile Operation, Dec 2015. IEEE.

Table 19.1 FrameControl Field Format.

Name	Size
Protocol version	3 bits
Address format	1 bit
ACK policy	2 bits
Frame type	3 bits
Sequence number	8 bits
Reserved	7 bits

From IEEE Standard for Radio Interface for White Space Dynamic Spectrum Access Radio Systems Supporting Fixed and Mobile Operation, Dec2015. IEEE.

Table 19.2 ACK policy field encoding.

Value	ACK policy type	Description
0	No-ACK	The recipient(s) do not acknowledge the transmission, and the sender treats the transmission as successful without retransmission.
1	Imm-ACK	The addressed recipient returns an Imm-ACK frame after correct reception.
2–3	Reserved	—

From IEEE Standard for Radio Interface for White Space Dynamic Spectrum Access Radio Systems Supporting Fixed and Mobile Operation, Dec2015. IEEE.

Table 19.3 Frame type field encoding.

Value	Frame type
0	Beacon frame
1	RTS frame
2	CTS frame
3	Data frame
4	ACK frame
5–7	Reserved

From IEEE Standard for Radio Interface for White Space
Dynamic Spectrum Access Radio Systems Supporting
Fixed and Mobile Operation, Dec2015. IEEE.

DestAddr The DestAddr field is the destination address. If frame is broadcast to the entire
network, the DestAddr field shall be set to 0xFFFF when Address format is set to 0, and shall
be set to 0xFFFFFFFFFFFF otherwise.

SrcAddr The SrcAddr field is the source address.

MAC Payload The MAC payload format depends on Frame type. Description is detailed in the
following subclause.

Cyclic Redundancy Check (CRC) The CRC field is 16 bits in length and contains a 16-bit ITU-T
CRC. The CRC is calculated over the MAC header and MAC payload parts of the frame.

The CRC shall be calculated using the standard generator polynomial of degree 16 defined
in Equation (19.1):

$$G_{16}(x) = x^{16} + x^{12} + x^5 + 1 \tag{19.1}$$

As an example, consider the following 3-byte frame:

```
0100  0000  0000  0000  0101  0110
b0                            b23
```

Leftmost bit (b0) transmitted first in time.
The CRC for this case would be the following:

```
0010    0111    1001    1110
r0                      r15
```

Leftmost bit (r0) transmitted first in time.
b0 is considered as MSB in the calculation of the CRC.

19.4.3.3 Beacon Frame

The beacon frame payload is specified in Table 19.4.

The SuperframeNumber increments once per superframe, following a modulo 256 counter.

The BeaconSlotNumber field is set to the number of the beacon slot where the beacon is sent
within the beacon period (BP), in the range of [0, $mBPslot$ −1] (see 19.4.5.1.1).

The OccupiedBeaconSlot is the number of beacon slots that are not sensed idle by the device.
For each occupied slot, the Beacon Slot Number and the device address is informed.

The ChannelMapMask field indicates the current/future subband mask. Entry index [0]
corresponds to the first 2 MHz subband, relative to the lower bound frequency defined by
the LbFreq parameter. Each entry corresponds to the maximal allowed power in the 2 MHz
subband in dBm with an offset of −50 dBm for 0x00. The ChannelMapMask is provided by
the Device Management Entity and computed based on information queried from a database
and/or sensors for instance.

Table 19.4 Beacon frame payload format.

Syntax	Size
SuperframeNumber	8 bits
OccupiedBeaconSlot	4 bits
for $i = 1$ to OccupiedBeaconSlot	
BeaconSlotNumber	4 bits
AddressFormat	1 bit
DeviceAddress	16/48 bits
end	
LbFreq	16 bits
ChannelMapMask	128 byte table
ChannelMapSwitch	1 bit
SuperFrameIndexChannelMapSwitch	8 bits

From IEEE Standard for Radio Interface for White Space Dynamic Spectrum
Access Radio Systems Supporting Fixed and Mobile Operation, Dec2015. IEEE.

ChannelMapSwitch indicates when the ChannelMapMask must be taken into account by
the master and slaves:

— When the field ChannelMapSwitch is set to 1, SuperFrameIndexChannelMapSwitch field
indicates the index of the super frame corresponding where ChannelMapMask needs to be
taken into account by the master and slaves.

— When the field ChannelMapSwitch is 0, the ChannelMapMask is applied immediately.

19.4.3.4   Request to Send (RTS) Frame

In RTS frames, the DestAddr field is set to the device address of the device to receive the
following frame from the transmitter. The RTS frame payload is specified in Table 19.5.

The TransportFormatIndicator field indicates the PHY layer configuration (modulation and
coding scheme, see 19.5.3.6) that will be used to transmit the data frame.

Duration field contains value in μs of time needed to transmit data + CTS + ACK [including
short inter frame space (SIFS) times].

19.4.3.5   Clear to Send (CTS) Frame

In CTS frames, the DestAddr field is set to the SrcAddr of the received RTS frame. The CTS
frame payload is specified in Table 19.6.

Duration field contains value in μs obtained by previous RTS minus time needed to transmit
CTS and it SIFS interval.

19.4.3.6   Data Frame

The data frame payload is specified in Table 19.7.

Table 19.5 RTS frame payload format.

Syntax	Size
TransportFormatIndicator	5 bits
Duration	16 bits
Reserved	11 bits

From IEEE Standard for Radio Interface for White Space
Dynamic Spectrum Access Radio Systems Supporting
Fixed and Mobile Operation, Dec2015. IEEE.

Table 19.6 CTS Frame Payload Format.

Syntax	Size
Duration	16 bits
Reserved	3 bits

From IEEE Standard for Radio Interface for White Space Dynamic Spectrum Access Radio Systems Supporting Fixed and Mobile Operation, Dec2015. IEEE.

Table 19.7 Data Frame Payload Format.

Syntax	Size
Duration	16 bits
PayloadSize	16 bits
Reserved	3 bits
Payload	variable

From IEEE Standard for Radio Interface for White Space Dynamic Spectrum Access Radio Systems Supporting Fixed and Mobile Operation, Dec2015. IEEE.

Duration field contains value in μs obtained by previous CTS minus time needed to transmit the payload frame and it SIFS interval.

PayloadSize indicates the size of the payload in bytes. The Payload field is the payload of size PayloadSize bytes.

19.4.3.7 ACK Frame

In ACK frames, the DestAddr field is set to the SrcAddr of the received Data frame. The ACK frame payload is specified in Table 19.8.

Duration field contains value in μs obtained by previous data frame minus time needed to transmit the ACK frame and it SIFS interval.

The ACKStatus field indicates if the data frame has been received. If the bit is set to 1, the data frame has been correctly received. If the bit is set to 0, the data frame has been received with errors.

19.4.4 MAC Sublayer Service Specification

19.4.4.1 MAC Data Service

Data Service Message Sequence Chart Figure 19.3 illustrates a sequence of messages necessary for a successful data transfer between two devices.

Table 19.8 ACK Frame Payload Format.

Syntax	Size
Duration	16 bits
ACKStatus	1 bit
Reserved	7 bits

From IEEE Standard for Radio Interface for White Space Dynamic Spectrum Access Radio Systems Supporting Fixed and Mobile Operation, Dec2015. IEEE.

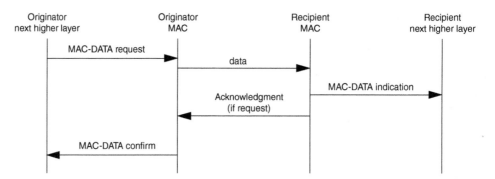

Figure 19.3 Message sequence chart describing MAC data service. From IEEE Standard for Radio Interface for White Space Dynamic Spectrum Access Radio Systems Supporting Fixed and Mobile Operation, Dec 2015. IEEE.

When an acknowledgement is requested by the originator MAC, MAC-DATA.confirm primitive is generated upon receipt of the acknowledge message with the appropriated status; otherwise, the MACDATA.confirm primitive is sent back with status SUCCESS. The recipient MAC generates an acknowledgement (if requested) upon receipt of a valid message from the originator MAC.

MAC-DATA.request The MAC-DATA.request primitive requests the transfer of a data. The MAC-DATA_request fields are shown in Table 19.9. The semantics of the MACDATA.request primitive are as follows:

```
MAC-DATA.request(
            AddressFormat
            SourceAddress
            DestinationAddress
            NbBytesPayload
            Payload
            TXOption
)
```

MAC-DATA.confirm The MAC-DATA_confirm filed is shown in Table 19.10. The semantics of the MAC-DATA.confirm primitive are as follows:

Table 19.9 MAC-DATA.request fields.

Name	Type	Description
AddressFormat	Boolean	If 0, 16-bit addresses are used; otherwise, 48-bit addresses are used.
SourceAddress	DeviceAddress	The individual device address of the entity from which the payload is being transferred.
DestinationAddress	DeviceAddress	The individual device address of the entity to which the payload is being transferred.
NbBytesPayload	Int32	The number of bytes contained in the payload to be transmitted by the MAC sublayer entity.
Payload	Array	The set of NbBytesPayload bytes forming the payload to be transmitted by the MAC sublayer entity.
TXOption	Boolean	If 1, acknowledged transmission is used, if 0, unacknowledged transmission is used.

From IEEE Standard for Radio Interface for White Space Dynamic Spectrum Access Radio Systems Supporting Fixed and Mobile Operation, Dec2015. IEEE.

Table 19.10 MAC-DATA.confirm Field.

Name	Type	Description
Status	Enumeration	Status of the request to transmit: SUCCESS, TRANSACTION_OVERFLOW, TRANSACTION_EXPIRED, CHANNEL_ACCESS_FAILURE, INVALID_PARAMETER

From IEEE Standard for Radio Interface for White Space Dynamic Spectrum Access Radio Systems Supporting Fixed and Mobile Operation, Dec2015. IEEE.

```
MAC-DATA.confirm(
            Status
)
```

The MAC-DATA.confirm primitive is generated by the MAC sublayer entity (originator MAC) in response to a MAC-DATA.request primitive. The MAC-DATA.confirm primitive returns a status of either SUCCESS (when the originator MAC receives an acknowledge from the recipient MAC, see Figure 19.3), indicating that the request to transmit was successful, or the appropriate error code.

MAC-DATA.indication The semantics of the MAC-DATA.indication primitive are as follows:

```
MAC-DATA.indication(
            AddressFormat
            SourceAddress
            DestinationAddress
            NbBytesPayload
            Payload
)
```

The MAC-DATA.indication primitive is generated by the MAC sublayer and issued to the CS on receipt of a data frame at the local MAC sublayer entity that passes the appropriate message filtering operations. The MAC-DATA_indication fields are shown in Table 19.11.

19.4.4.2   MAC Management Service

The MLME-SAP allows the transport of management commands between the next higher layer and the MLME.

Table 19.11 MAC-DATA.indication fields.

Name	Type	Description
AddressFormat	Boolean	If 0, 16-bit addresses are used; otherwise, 48-bit addresses are used.
SourceAddress	DeviceAddress	The individual device address of the entity from which the payload is being transferred.
DestinationAddress	DeviceAddress	The individual device address of the entity to which the payload is being transferred.
NbBytesPayload	Int32	The number of bytes contained in the payload to be transmitted by the MAC sublayer entity.
Payload	Array	The set of bytes forming the payload to be transmitted by the MAC sublayer entity.

From IEEE Standard for Radio Interface for White Space Dynamic Spectrum Access Radio Systems Supporting Fixed and Mobile Operation, Dec2015. IEEE.

Table 19.12 MLME-ASSOCIATE.request fields.

Name	Type	Description
AddressFormat	Boolean	If 0, 16-bit addresses are used; otherwise, 48-bit addresses are used.
MasterAddress	DeviceAddress	The master device address.

From IEEE Standard for Radio Interface for White Space Dynamic Spectrum Access Radio Systems Supporting Fixed and Mobile Operation, Dec2015. IEEE.

Association Message

MLME-ASSOCIATE.request The MLME-ASSOCIATE.request primitive allows a device to request an association with a coordinator. The MLME-ASSOCIATE_request fields are shown in Table 19.12. The semantics of the MLME-ASSOCIATE.request primitive as listed below Tables 19.11 and 19.12.

```
MLME-ASSOCIATE.request(
        AddressFormat
        MasterAddress
)
```

The MLME-ASSOCIATE.request primitive is generated by the next higher layer of an unassociated device and issued to its MLME to request an association.

On receipt of the association request command, the MLME of the coordinator issues the MLMEASSOCIATE.indication primitive.

Upon receipt of the MLME-ASSOCIATE.response primitive, the coordinator attempts to add the information contained in the primitive to its list of pending transactions. If there is no capacity to store the transaction, the MAC sublayer will discard the frame and issue the MLME-COMM-STATUS.indication primitive with a status of TRANSACTION_OVERFLOW. If there is capacity to store the transaction, the coordinator will add the information to the list. If the transaction is not handled within mMacTransactionPersistenceTime (see Table 19.29), the transaction information will be discarded and the MAC sublayer will issue the MLME-COMM-STATUS.indication primitive with a status of TRANSACTION_EXPIRED.

MLME-ASSOCIATE.confirm The MLME-ASSOCIATE.confirm primitive is used to inform the next higher layer of the initiating device whether its request to associate was successful or unsuccessful.

```
MLME-ASSOCIATE.confirm(
        Status
)
```

The MLME-ASSOCIATE.request fields are shown in Table 19.13. If the MLME-ASSOCIATE. request primitive cannot be sent to the coordinator due to the Carrier Sense Multiple Access with Collision Avoidance (CSMA-CA) algorithm indicating a busy channel, the MLME will issue the MLME-ASSOCIATE.confirm primitive with a status of CHANNEL_ACCESS_FAILURE.

Table 19.13 MLME-ASSOCIATE.confirm fields.

Name	Type	Description
Status	Enumeration	Status of the request to associate: SUCCESS, TRANSACTION_EXPIRED, CHANNEL_ACCESS_FAILURE, INVALID_PARAMETER, NO_ACK, NO_DATA

From IEEE Standard for Radio Interface for White Space Dynamic Spectrum Access Radio Systems Supporting Fixed and Mobile Operation, Dec2015. IEEE.

If the MLME successfully transmits an association request command, the MLME will expect an acknowledgment in return. If an acknowledgment is not received, the MLME will issue the MLMEASSOCIATE.confirm primitive with a status of NO_ACK.

If the MLME of an unassociated device successfully receives an acknowledgment to its association request command, the MLME will wait for a response to the request. If the MLME of the device does not receive a response, it will issue the MLME-ASSOCIATE.confirm primitive with a status of NO_DATA.

If the MLME of the device extracts an association response command frame from the coordinator, it will then issue the MLME-ASSOCIATE.confirm primitive with a status equal to the contents of the association Status field in the association response command.

If any parameter in the MLME-ASSOCIATE.request primitive is either not supported or out of range, the MLME will issue the MLME-ASSOCIATE.confirm primitive with a status of INVALID_PARAMETER.

The MLME-ASSOCIATE.confirm primitive is generated by the initiating MLME and issued to its next higher layer in response to an MLME-ASSOCIATE.request primitive. If the request was successful, the Status parameter will indicate a successful association, as contained in the Status field of the association response command. Otherwise, the Status parameter indicates either an error code from the received association response command.

MLME-ASSOCIATE.indication The MLME-ASSOCIATE.indication primitive is used to indicate the reception of an association request command.

```
MLME-ASSOCIATE.indication(
            AddressFormat
            Address
)
```

The MLME-ASSOCIATE.indication primitive is generated by the MLME of the coordinator and issued to its next higher layer to indicate the reception of an association request command. The MLME-ASSOCIATE_indication fields are shown in Table 19.14.

When the next higher layer of a coordinator receives the MLME-ASSOCIATE.indication primitive, the Coordinator determines whether to accept or reject the unassociated device using an algorithm outside the scope of this standard. The next higher layer of the coordinator then issues the MLMEASSOCIATE.response primitive to its MLME.

MLME-ASSOCIATE.response The MLME-ASSOCIATE.response primitive is used to initiate a response to an MLME-ASSOCIATE.indication primitive. The MLME-ASSOCIATE_response fields are shown in Table 19.15. The semantics of the MLME-ASSOCIATE.response primitive are as follows:

Table 19.14 MLME-ASSOCIATE.indication fields.

Name	Type	Description
AddressFormat	Boolean	If 0, 16-bit addresses are used; otherwise, 48-bit addresses are used.
Address	DeviceAddress	The address of the device requesting association.

From IEEE Standard for Radio Interface for White Space Dynamic Spectrum Access Radio Systems Supporting Fixed and Mobile Operation, Dec2015. IEEE.

Table 19.15 MLME-ASSOCIATE.response fields.

Name	Type	Description
Status	Enumeration	Status of the request to transmit: SUCCESS, ACCESS_DENIED

From IEEE Standard for Radio Interface for White Space Dynamic Spectrum Access Radio Systems Supporting Fixed and Mobile Operation, Dec2015. IEEE.

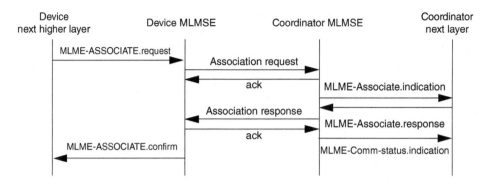

Figure 19.4 Message sequence chart describing association procedure. From IEEE Standard for Radio Interface for White Space Dynamic Spectrum Access Radio Systems Supporting Fixed and Mobile Operation, Dec 2015. IEEE.

```
MLME-ASSOCIATE.response(
            Status
)
```

Association Message Sequence Chart   The association message sequence chart is depicted in Figure 19.4.

Disassociation Primitives   The MLME disassociation primitives define how a device can disassociate from the network.

MLME-DISASSOCIATE.request   The MLME-DISASSOCIATE.request primitive is used by an associated device to notify the coordinator of its intent to leave the network. It is also used by the coordinator to instruct an associated device to leave the network. The MLME-DISASSOCIATE_request fields are shown in Table 19.16 and the semantics of MLME_DISASSOCIATE.request are:

```
MLME-DISASSOCIATE.request(
            AddressFormat
            SourceAddress
            DestinationAddress
            DisassociateReason
)
```

If there is no capacity to store the transaction, the MLME will discard the frame and issue the MLME-DISASSOCIATE.confirm primitive with a status of TRANSACTION_OVERFLOW. If there is capacity to store the transaction, the coordinator will add the information to the list.

Table 19.16 MLME- DISASSOCIATE.request fields.

Name	Type	Description
AddressFormat	Boolean	If 0, 16-bit addresses are used; otherwise 48-bit addresses are used.
SourceAddress	DeviceAddress	The address of the device requesting disassociation.
DestinationAddress	DeviceAddress	The address of the device receiving disassociation .command
DisassociateReason	0x0–0xf	Disassociate reason: 0x0: reserved, 0x1: the master wishes the device to leave the network, 0x2: the device wishes to leave the network, Others: reserved.

From IEEE Standard for Radio Interface for White Space Dynamic Spectrum Access Radio Systems Supporting Fixed and Mobile Operation, Dec2015. IEEE.

If the transaction is not handled within mMacTransactionPersistenceTime, the transaction information will be discarded and the MLME will issue the MLME-DISASSOCIATE.confirm with a status of TRANSACTION_EXPIRED.

If the disassociation notification command cannot be sent due to a CSMA-CA algorithm failure, and this primitive was either received by the MLME of a coordinator or it was received by the MLME of a device, the MLME will issue the MLME-DISASSOCIATE.confirm primitive with a status of CHANNEL_ACCESS_FAILURE.

If the MLME successfully transmits a disassociation notification command, the MLME will expect an acknowledgment in return. If an acknowledgment is not received, and this primitive was either received by the MLME of a coordinator or it was received by the MLME of a device, the MLME will issue the MLMEDISASSOCIATE.confirm primitive with a status of NO_ACK.

If the MLME successfully transmits a disassociation notification command and receives an acknowledgment in return, the MLME will issue the MLME-DISASSOCIATE.confirm primitive with a status of SUCCESS.

On receipt of the disassociation notification command, the MLME of the recipient issues the MLMEDISASSOCIATE.indication primitive.

If any parameter in the MLME-DISASSOCIATE.request primitive is not supported or is out of range, the MLME will issue the MLME-DISASSOCIATE.confirm primitive with a status of INVALID_PARAMETER.

MLME- DISASSOCIATE.indication The MLME-DISASSOCIATE.indication primitive is used to indicate the reception of a disassociation notification command. The MLME-DISASSOCIATE. indication fields are shown in Table 19.17.

```
MLME-DISASSOCIATE.indication(
         AddressFormat
         Address
         DissassociateReason
)
```

MLME-DISASSOCIATE.confirm The MLME-DISASSOCIATE.confirm primitive reports the results of an MLME-DISASSOCIATE.request primitive.

```
MLME-DISASSOCIATE.confirm(
         Status
)
```

The MLME-DISASSOCIATE.confirm primitive is generated by the initiating MLME and issued to its next higher layer in response to an MLME-DISASSOCIATE.request primitive. The MLME-DISASSOCIATE_confirm fields are shown in Table 19.18. This primitive returns a status of either SUCCESS, indicating that the disassociation request was successful, or the appropriate error code.

Table 19.17 MLME- DISASSOCIATE.indication fields.

Name	Type	Description
AddressFormat	Boolean	If 0, 16-bit addresses are used; otherwise 48-bit addresses are used.
Address	DeviceAddress	The address of the device sending indication.
DissassociateReason	0x0–0xf	Disassociate reason: 0x0: reserved, 0x1: the master wishes the device to leave the network, 0x2: the device wishes to leave the network, Others: reserved.

From IEEE Standard for Radio Interface for White Space Dynamic Spectrum Access Radio Systems Supporting Fixed and Mobile Operation, Dec2015. IEEE.

Table 19.18 MLME-DISASSOCIATE.confirm fields.

Name	Type	Description
Status	Enumeration	Status of the disassociate request: SUCCESS, TRANSACTION_OVERFLOW, TRANSACTION_EXPIRED, NO_ACK, CHANNEL_ACCESS_FAILURE, INVALID_PARAMETER.

From IEEE Standard for Radio Interface for White Space Dynamic Spectrum Access Radio Systems Supporting Fixed and Mobile Operation, Dec2015. IEEE.

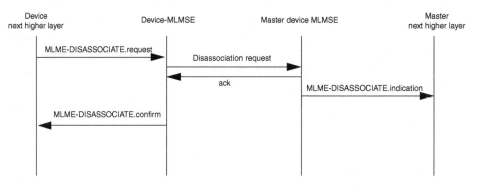

Figure 19.5 Message sequence chart describing disassociation procedure. From IEEE Standard for Radio Interface for White Space Dynamic Spectrum Access Radio Systems Supporting Fixed and Mobile Operation, Dec 2015. IEEE.

Disassociation Message Sequence Charts The request to disassociate may originate either from a device or the coordinator through which the device has associated. Figure 19.5 illustrates the sequence of messages necessary for a device to successfully disassociate itself from the network.

MLME-BEACON-NOTIFY.indication The MLME-SAP beacon notification primitive defines how a device may be notified when a beacon is received during normal operating conditions.

The MLME-BEACON-NOTIFY.indication indicates the reception of a beacon in the specified BeaconSlotNumber. The first set of field (index 0 of the loop) indicates parameters corresponding to the specified BeaconSlotNumber to scan (see 19.4.4.2.8.1). Fields ChannelMapMask, ChannelMapSwitch, and SuperFrameIndexChannelMapSwitch are those corresponding to the first set of the list (index 0).

The MLME-BEACON-NOTIFY.indication primitive is used to send parameters contained within a beacon frame received by the MAC sublayer to the next higher layer.

The semantics of the MLME-BEACON-NOTIFY.indication primitive are as follows:

```
MLME-BEACON-NOTIFY.indication(
          SuperFrameNumber
          OccupiedBeaconSlot
          for i=0 to 'OccupiedBeaconSlot-1'
                    BeaconSlotNumber
                    AddressFormat
                    Address
          End
          LbFreq
          ChannelMapMask
          ChannelMapSwitch
          SuperFrameIndexChannelMapSwitch
)
```

Table 19.19 MLME-BEACON-NOTIFY.indication fields.

Name	Type	Description
SuperFrameNumber	Int32	Index of super frame as defined in 19.4.5.
OccupiedBeaconSlot	Int32	Number of occupied beacon slot.
BeaconSlotNumber	Int32	Index of beacon slot.
AddressFormat	Boolean	If 0, 16-bit addresses are used; otherwise 48-bit addresses are used.
Address	DeviceAddress	The address of the device sending beacon.
LbFreq	16 bits	The lower bound frequency is defined by LbFreq × 0.1 MHz.
ChannelMapMask	128 byte table	It indicates the current/future subband mask. Entry index [0] corresponds to the first 2 MHz subband, relative to the channel that has been set. Each entry corresponds to the maximal allowed power in the 2 MHz subband in dBm with an offset of −50 dBm for 0x00. The subband characteristics are provided in Table 41.
ChannelMapSwitch	Boolean	Indicates when the ChannelMapMask must be taken into account by the master and slaves.
SuperFrameIndex ChannelMapSwitch	Int32	When the field ChannelMapSwitch is set to 1, SuperFrameIndexChannelMapSwitch field indicates the index of the super frame corresponding where ChannelMapMask needs to be taken into account by the master and slaves.

From IEEE Standard for Radio Interface for White Space Dynamic Spectrum Access Radio Systems Supporting Fixed and Mobile Operation, Dec2015. IEEE.

The MLME-BEACON-NOTIFY.indication fields are shown in Table 19.19. The MLME-BEACON-NOTIFY.indication primitive is generated by the MLME and issued to its next higher layer upon receipt of a beacon frame.

Primitives for Resetting the MAC Sublayer MLME reset primitives specify how to reset the MAC sublayer to its default values.

MLME-RESET.request The MLME-RESET.request primitive allows the next higher layer to request that the MLME performs a reset operation.

The MLME-RESET.request primitive has no parameter. The MLME-RESET.request primitive is issued prior to the use of the MLME-START.request or the MLME-ASSOCIATE.request primitives. On receipt of the MLME-RESET.request primitive, the MLME issues the PLME-SET-TRX-STATE.request primitive with a state of FORCE_TRX_OFF. On receipt of the PLME-SET-TRX-STATE.confirm primitive, the MAC sublayer is then set to its initial conditions, clearing all internal variables to their default values (see Table 19.29).

MLME-RESET.confirm The MLME-RESET.confirm primitive reports the results of the reset operation.

The MLME-RESET.confirm primitive is generated by the MLME and issued to its next higher layer in response to an MLME-RESET.request primitive and following the receipt of the PLME-SET-TRX-STATE confirm primitive. On receipt of the MLME-RESET.confirm primitive, the next higher layer is notified of its request to reset the MAC sublayer. This primitive returns a status of SUCCESS indicating that the request to reset the MAC sublayer was successful.

Primitives for Channel Scanning MLME scan primitives define how a device can determine the energy usage or the presence or absence of network in a communications channel.

Table 19.20 MLME-SCAN.request fields.

Name	Type	Description
ScanChannels	128 bits field	The 128 bits (b0, b1,... b127) indicate which 2MHz channels are to be scanned (1 = scan, 0 = do not scan), starting from the lower bound frequency as defined by the LbFreq parameter.
LbFreq	16 bits	The lower bound frequency is defined by LbFreq × 0.1 MHz.

From IEEE Standard for Radio Interface for White Space Dynamic Spectrum Access Radio Systems Supporting Fixed and Mobile Operation, Dec2015. IEEE.

MLME-SCAN.request   The MLME-SCAN.request primitive is used to initiate a channel scan over a given list of channels. The semantics of the MLME-SCAN.request primitive are as follows:

```
MLME-SCAN.request(
            ScanChannels
            LbFreq
)
```

The MLME-SCAN request fields are shown in Table 19.20. If the MLME receives the MLME-SCAN.request primitive while performing a previously initiated scan operation, it issues the MLME-SCAN.confirm primitive with a status of SCAN_IN_PROGRESS. Otherwise, the MLME initiates a scan in all channels specified in the ScanChannels parameter.

The scan is performed on each channel by the MLME repeatedly issuing the PLME-SCAN .request primitive to the PHY until [mSuperframeDuration × (2n + 1)] symbols, where n is the value of the mScanDuration parameter, have elapsed. The MLME notes the maximum energy measurement and moves on to the next channel in the channel list.

The results of a scan are reported to the next higher layer through the MLME-SCAN.confirm primitive. If any parameter in the MLME-SCAN.request primitive is not supported or is out of range, the MAC sublayer will issue the MLME-SCAN.confirm primitive with a status of INVALID_PARAMETER.

MLME-SCAN.confirm   The MLME-SCAN.confirm primitive reports the result of the channel scan request. The MLME-SCAN confirm fields are shown in Table 19.21. The MLMES-CAN.confirm primitive is generated by the MLME and issued to its next higher layer when the channel scan initiated with the MLME-SCAN.request primitive has completed.

The semantics of the MLME-SCAN.confirm primitive are as follows:

```
MLME-SCAN.confirm(
            Status
            ListSize
            NetworkDescriptorList
)
```

Communication Status Primitive   The MLME-SAP communication status primitive defines how the MLME communicates to the next higher layer about transmission status.

The semantics of the MLME-COMM-STATUS.indication primitive are as follows:

```
MLME-COMM-STATUS.indication(
            Status
)
```

The MLME-COMM-STATUS indication fields are shown in Table 19.23. The MLME-COMM-STATUS.indication primitive is generated by the MAC sublayer entity following either the

Table 19.21 MLME-SCAN.confirm fields.

Name	Type	Description
Status	Enumeration	Status of the channel scan request: SUCCESS, LIMIT_REACHED, NO_BEACON, SCAN_IN_PROGRESS, INVALID_PARAMETER.
ListSize	Int32	The number of elements of the NetworkDescriptorList.
NetworkDescriptorList	List of network descriptors	The list of Network descriptors, one for each beacon found. The network description list fields are in Table 19.22.

From IEEE Standard for Radio Interface for White Space Dynamic Spectrum Access Radio Systems Supporting Fixed and Mobile Operation, Dec2015. IEEE.

Table 19.22 NetworkDescriptorList fields.

Name	Type	Description
AddressFormat	Boolean	If 0, 16-bit addresses are used; otherwise 48-bit addresses are used.
DeviceAddress	DeviceAddress	Address of master in the beacon slot.
BeaconSlotNumber	Int32	Index of beacon slot.
LbFreq	16 bits	The lower bound frequency is defined by LbFreq $\times$ 0.1 MHz.
ChannelMapMask	128 byte table	It indicates the current/future subband mask. Entry index [0] corresponds to the first 2 MHz subband, relative to the channel that has been set. Each entry corresponds to the maximal allowed power in the 2 MHz subband in dBm with an offset of −50 dBm for 0x00. The subband characteristics are provided in Table 41.

From IEEE Standard for Radio Interface for White Space Dynamic Spectrum Access Radio Systems Supporting Fixed and Mobile Operation, Dec2015. IEEE.

Table 19.23 MLME-COMM-STATUS.indication fields.

Name	Type	Description
Status	Enumeration	Status of the request to transmit: SUCCESS, TRANSACTION_OVERFLOW, TRANSACTION_EXPIRED, CHANNEL_ACCESS_FAILURE, NO_ACK, INVALID_PARAMETER

From IEEE Standard for Radio Interface for White Space Dynamic Spectrum Access Radio Systems Supporting Fixed and Mobile Operation, Dec2015. IEEE.

MLME-ASSOCIATE.response primitive. This primitive returns a status of either SUCCESS, indicating that the request to transmit was successful, or an error code of TRANSACTION_OVERFLOW, TRANSACTION_EXPIRED, CHANNEL_ACCESS_FAILURE, NO_ACK or INVALID_PARAMETER.

Primitive for Updating the Superframe Configuration MLME START primitives define how a device can request to start using a new superframe (as defined in 19.4.5) configuration in order to initiate a network, begin transmitting beacons on an already existing network, thus facilitating device discovery, or to stop transmitting beacons.

Table 19.24 MLME-START.request fields.

Name	Type	Description
LbFreq	16 bits	The lower bound frequency is defined by LbFreq × 0.1 MHz.
ChannelMapMask	Array of 128 bytes	It indicates the current/future subband mask. Entry index [0] corresponds to the first 2 MHz subband, relative to the channel that has been set. Each entry corresponds to the maximal allowed power in the 2 MHz subband in dBm with an offset of −50 dBm for 0x00. The subband characteristics are provided in Table 19.41.
AddressFormat	Boolean	If 0, 16-bit addresses are used; otherwise 48-bit addresses are used.
DeviceAddress	DeviceAddress	Address of master in the beacon slot.
BeaconSlotNumber	Int32	Index of the beacon slot on which to start.
ListSize	Int32	Size of the NetworkDescriptorList.
NetworkDescriptorList	List of network description	The list of network descriptors, one for each beacon found during the previous discovery process.

From IEEE Standard for Radio Interface for White Space Dynamic Spectrum Access Radio Systems Supporting Fixed and Mobile Operation, Dec2015. IEEE.

MLME-START.request The MLME-START.request primitive allows the network coordinator to initiate a new network or to begin using a new superframe configuration. This primitive may also be used by a device already associated with an existing network to begin using a new superframe configuration.

The MLME-START request fields are shown in Table 19.24. The semantics of the MLME-START.request primitive are as follows:

```
MLME-START.request(
          LbFreq
          ChannelMapMask
          AddressFormat
          DeviceAddress
          BeaconSlotNumber
          ListSize
          NetworkDescriptorList
)
```

The MLME responds with the MLME-START.confirm primitive. If the attempt to start using a new superframe configuration was successful, the Status parameter will be set to SUCCESS. If any parameter is not supported or is out of range, the Status parameter will be set to INVALID_PARAMETER.

MLME-START.confirm The MLME-START.confirm primitive reports the results of the attempt to start using a new superframe configuration.

The semantics of the MLME-START.confirm primitive are as follows:

```
MLME-START.confirm (
          Status
)
```

The MLME-START confirm fields are shown in Table 19.25. On receipt of the MLME-START.confirm primitive, the next higher layer is notified of the result of its request to start using a new superframe configuration. If the MLME-START.request has been successful, the Status parameter will be set to SUCCESS. Otherwise, the Status parameter indicates the error, INVALID_PARAMETER.

Table 19.25 MLME-START.confirm fields.

Name	Type	Description
Status	Enumeration	State of the request: SUCCESS, INVALID_PARAMETER.

From IEEE Standard for Radio Interface for White Space Dynamic Spectrum Access Radio Systems Supporting Fixed and Mobile Operation, Dec2015. IEEE.

Primitives for Synchronizing with a Coordinator MLME synchronization primitives define how synchronization with a coordinator may be achieved and how a loss of synchronization is communicated to the next higher layer. The MLME-SYNC request fields are shown in Table 19.26.

MLME-SYNC.request The semantics of the MLME-SYNC.request primitive are as follows:

```
MLME-SYNC.request(
              LbFreq
              ChannelMapMask
              AddressFormat
              DeviceAddress
              BeaconSlotNumber
)
```

If this primitive is received by the MLME while it is currently tracking the beacon, the MLME will not discard the primitive, but rather treat it as a new MLME-SYNC.request.

If the beacon could not be located either on its initial search or during tracking, the MLME will issue the MLME-SYNC-LOSS.indication primitive with a loss reason of BEACON_LOST.

MLME-SYNC-LOSS.indication The MLME-SYNC-LOSS.indication primitive indicates the loss of synchronization with a coordinator.

Table 19.26 MLME-SYNC.request fields.

Name	Type	Description
LbFreq	16 bits	The lower bound frequency is defined by LbFreq × 0.1 MHz.
ChannelMapMask	Array of 128 bytes	It indicates the current/future subband mask. Entry index [0] corresponds to the first 2 MHz subband, relative to the channel that has been set. Each entry corresponds to the maximal allowed power in the 2 MHz subband in dBm with an offset of −50dBm for 0x00. The subband characteristics are provided in Table 19.41.
AddressFormat	Boolean	If 0, 16-bit addresses are used; otherwise, 48-bit addresses are used.
LbFreq	16 bits	The lower bound frequency is defined by LbFreq × 0.1 MHz.
DeviceAddress	DeviceAddress	Address of master in the beacon slot.
BeaconSlotNumber	Int32	Index of the beacon slot on which to start.

From IEEE Standard for Radio Interface for White Space Dynamic Spectrum Access Radio Systems Supporting Fixed and Mobile Operation, Dec2015. IEEE.

```
MLME-SYNC-LOSS.indication(
              LbFreq
              ChannelMapMask
              AddressFormat
              DeviceAddress
              LossReason
)
```

The MLME-SYNC-LOSS.indication primitive is generated by the MLME of a device and issued to its next higher layer in the event of a loss of synchronization with the coordinator.

If a device has not heard the beacon for mMaxLostBeacons consecutive superframes following an MLMESYNC.request primitive, either initially or during tracking, the MLME will issue this primitive with the LossReason parameter set to BEACON_LOST.

MAC Enumeration Description   This subclause explains the meaning of the enumerations used in the primitives defined in the MAC sublayer specification.

19.4.5   MAC Functional Description

This subclause provides a detailed description of the MAC functionality.

19.4.5.1   Channel Access

Timing Structure   The basic timing structure for frame exchange is a superframe. The superframe duration is specified as mSuperframeDuration. The superframe is composed of 256 medium access slots (MASs), where each MAS duration is mMASDuration. The timing structure is shown in Figure 19.6.

Each superframe starts with a beacon period (BP). The BP is composed of mBPslot beacon slot, where each beacon slot duration is mBeaconSlotDuration. The start of the first MAS in the BP, and the superframe, is called the beacon period start time (BPST).

A recurring superframe consists of a BP and a contention access period (CAP).

All devices that share the same channel shall follow the same superframe structure.

CSMA-CA Algorithm   This standard is based on the CSMA-CA with RTS/CTS mechanism. The MLME-SYNC-LOSS indication fields are shown in Table 19.27. The MAC enumeration description is shown in Table 19.28.

Table 19.27 MLME-SYNC-LOSS.indication fields.

Name	Type	Description
LbFreq	16 bits	The lower bound frequency is defined by LbFreq × 0.1 MHz.
ChannelMapMask	Array of 128 bytes	It indicates the current/future subband mask. Entry index [0] corresponds to the first 2 MHz subband, relative to the channel that has been set. Each entry corresponds to the maximal allowed power in the 2 MHz subband in dBm with an offset of −50 dBm for 0x00. The subband characteristics are provided in Table 19.41.
AddressFormat	Boolean	If 0, 16-bit addresses are used; otherwise 48-bit addresses are used.
DeviceAddress	DeviceAddress	Address of master in the beacon slot.
LossReason	Enumeration	BEACON_LOST

Table 19.28 MAC enumeration description.

Enumeration	Value	Description
SUCCESS	0x00	Confirms successful completion of the request (this applies to any of the confirm primitives described previously).
BEACON_LOST	0x01	The beacon was lost.
CHANNEL_ACCESS_FAILURE	0x02	A transmission could not take place due to activity on the channel.
INVALID_PARAMETER	0x03	A parameter is invalid or is out of the valid range.
LIMIT_REACHED	0x04	A scan operation terminated prematurely, because the number of network descriptors stored reached an implementation-specified maximum.
NO_ACK	0x05	Acknowledgment was not received.
NO_BEACON	0x06	A scan operation failed to find any network beacons.
SCAN_IN_PROGRESS	0x07	A request to perform scan failed because the MLME was in the process of performing a previously initiated scan operation.
TRANSACTION_EXPIRED	0x08	The transaction has expired.
TRANSACTION_OVERFLOW	0x09	There is no capacity to store the transaction.
ACCESS_DENIED	0x0A	Association is denied.

From IEEE Standard for Radio Interface for White Space Dynamic Spectrum Access Radio Systems Supporting Fixed and Mobile Operation, Dec2015. IEEE.

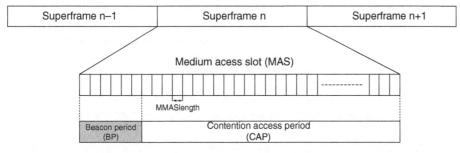

Figure 19.6 Timing structure of MAC frame. From IEEE Standard for Radio Interface for White Space Dynamic Spectrum Access Radio Systems Supporting Fixed and Mobile Operation, Dec 2015. IEEE.

Network Allocation Vector A device that transmits or receives frames shall maintain a network allocation vector (NAV) that contains the remaining time that a neighbor device has indicated it will access the medium. A device that receives a MAC header not addressed to it shall update its NAV with the received Duration field if the new NAV value is greater than the current NAV value. A device shall consider the updated NAV value to start at the end of the header on the medium.

A device that receives a MAC header with invalid header error indicator outside its unreleased reservation blocks shall update its NAV as if the frame were correctly received with Duration equal to mAccessDelay.

A device shall reduce its NAV as time elapses until it reaches zero. The NAV shall be maintained during at least mClockResolution.

Medium Status A device shall consider the medium to be busy for any of the following conditions:

— When its clear channel assessment (CCA) mechanism indicates that the medium is busy
— When the device's NAV is greater than zero

— When the device is transmitting or receiving a frame on the medium
— When the Duration announced in a previously transmitted frame has not yet expired

At all other times a device shall consider the medium to be idle.

Parameters A device shall use the set of parameters to obtain a transmit opportunity (TXOP) or perform backoff (see 19.4.5.1.2.6). These parameters are summarized in 19.4.5.1.2.3.1 through 19.4.5.1.2.3.3.

Distributed Coordination Function Inter Frame Space (DIFS) A device shall wait for the medium to become idle for DIFS before obtaining a TXOP or starting/resuming decrementing the backoff counter. DIFS is defined as follows:

```
DIFS = SIFS + 2*SlotTime
```

Here SIFS is the short inter frame space.

CWmin and CWmax A device shall set the contention window (CW) to an appropriate integer in the range [CWmin, CWmax] after invoking a backoff, and shall set the backoff counter to an integer sampled from a random variable uniformly distributed over the interval [0, CW].

TXOPLimit A device shall not initiate a frame transaction in a TXOP unless the frame transaction can be completed within TXOPLimit of the start of the TXOP and SIFS plus GuardTime before the medium becomes unavailable.

Obtaining a TXOP A device shall consider itself to have obtained a TXOP if it meets the following conditions:

— The device has one or more newly arrived data frames or newly generated command frames;
— The device had a backoff counter of zero value and had no frames prior to the arrival or generation of the new frames; and
— The device determines that the medium has been idle for DIFS or longer.

The device shall start transmitting a frame, which may be an RTS frame, as soon as the above conditions are satisfied. The device shall treat the start of the frame transmission on the wireless medium as the start of the TXOP.

A device shall also consider itself to have obtained a TXOP if it meets the following conditions:

— The device has one or more frames buffered for transmission, including retry; and
— The device set the backoff counter to zero in the last backoff and determines that the medium has been idle for DIFS since that backoff at the end of the current backoff slot, or the device decrements its backoff counter from one to zero in the current backoff slot.

The TXOP shall start at the end of the current backoff slot, i.e., the start of the next backoff slot.

A device shall check that the TXOP it has obtained is not longer than TXOPLimit and shall end SIFS plus GuardTime before the medium becomes unavailable.

Using a TXOP A device that has obtained a TXOP is referred to as a TXOP owner. A frame transmission, including a retry, is conducted as part of a frame transaction.

A TXOP owner shall initiate a frame transaction, and continue with one or more frame transactions without backoff, subject to the following criteria:

— Each transaction in the TXOP will be completed within the obtained TXOP
— The recipient device will be available to receive and respond during that frame transmission

A device may transmit a frame in a new TXOP that will result in the frame transaction that exceeds the TXOPLimit restriction under the following circumstances:

— The frame contains a fragment of a MAC Service Data Unit (MSDU)
— The frame is the sole frame transmitted by the device in the current TXOP
— The frame transaction will be completed SIFS and GuardTime before the medium becomes unavailable

A recipient device shall not transmit a CTS frame in response to a received RTS frame if its NAV is greater than zero. A recipient device shall not transmit a CTS or Imm-ACK response to a received frame requiring such a response if the response will not be completed SIFS before the medium becomes unavailable.

Under the rules stated above, the following timings apply to transmissions, including responses, in a TXOP (these timings are referenced with respect to transmission to or reception from the wireless medium):

— The TXOP owner shall transmit the first frame of the first or sole frame transaction in the TXOP at the start of the TXOP.
— After transmitting a frame with the ACK Policy set to No-ACK, the TXOP owner shall transmit a subsequent frame SIFS after the end of that transmitted frame.
— After receiving an RTS frame or a non-RTS frame with the ACK Policy set to Imm-ACK request, the recipient device shall transmit a CTS frame or an Imm-ACK frame SIFS after the end of the received frame.
— After receiving an expected CTS, Imm-ACK response to the preceding frame it transmitted, the TXOP owner shall transmit the next frame, SIFS after the end of the received frame.

Invoking a Backoff Procedure A device shall maintain a backoff counter to transmit frames.

A device shall set the backoff counter to an integer sampled from a random variable uniformly distributed over the range [0, CW], inclusive, when it invokes a backoff. The device shall initialize CW to CWmin before invoking any backoff, adjusting CW in the range [CWmin, CWmax].

The device shall set CW back to CWmin after receiving a CTS or Imm-ACK frame or upon transmitting a frame with ACK Policy set to No-ACK. A device shall also set CW back to CWmin, but shall not select a new backoff counter value, after discarding a buffered frame.

A device shall invoke a backoff procedure and draw a new backoff counter value as specified below.

A device shall invoke a backoff, with CW set to CWmin, when it has an MSDU arriving at its MAC SAP, or at the end of transmitting a frame with the ACK policy set to No-ACK, or at the end of receiving an expected Imm-ACK response to its last transmitted frame under the following conditions:

— The device had a backoff counter of zero value but is not in the middle of a frame transaction
— The device determines that the medium is busy
— The device has no other frames for transmission in the current TXOP

A device shall invoke a backoff, with CW set to the smaller of CWmax or $2 \times CW + 1$ (the latter CW being the last CW value), at the end of the current backoff slot under the following conditions:

— The device has one or more frames buffered for transmission, including retry
— The device set the backoff counter to zero in the last backoff and determines that the medium has been idle for DIFS since that backoff at the end of the current backoff slot, or the device decrements its backoff counter from one to zero in the current backoff slot
— The device does not receive an expected CTS or Imm-ACK frame

Decrementing a Backoff Counter Upon invoking a backoff, a device shall check that the medium is idle for DIFS before starting to decrement the backoff counter. To this end, a device shall define the first backoff slot to start at the time when the medium has been idle for DIFS after the backoff invocation, with subsequent backoff slots following successively until the medium becomes busy. All backoff slots have a length of SlotTime.

A device shall treat the Clear Chanel Assessment (CCA) result CCADetectTime after the start of a backoff slot to be the CCA result for this backoff slot. After the medium has been idle for DIFS the device shall decrement the backoff counter by one CCADetectTime after the start of the backoff slot if it finds the CCA result to be idle at this time and determines the medium to be idle for the backoff slot, unless the backoff counter is already at ZERO value.

The device shall freeze the backoff counter once the medium becomes busy. The device shall treat the residual backoff counter value as if the value were set due to the invocation of a backoff, following the above procedure to resume decrementing the backoff counter.

Frame Processing   This subclause provides rules on preparing MAC frames for transmission and processing them on reception.

Frame Reception   A MAC header is considered to be received by the device if it indicates a protocol version supported by the device, regardless of the CRC validation.

Frame Transaction   A frame transaction consists of a RTS/CTS frame exchange, a single frame, and the associated acknowledgement frame if requested by the ACK policy (see Figure 19.7).

Inter-Frame Space (IFS)   Three types of IFS are used in this Standard: the minimum inter-frame space (MIFS), the short inter-frame space (SIFS), and the DIFS. The actual values of the MIFS, SIFS, and DIFS are PHY-dependent.

A device shall not start transmission of a frame on the medium with non-zero length payload earlier than MIFS, or with zero length payload earlier than SIFS, after the end of a frame it transmitted previously on the medium. A device shall not start transmission of a frame on the medium earlier than SIFS duration after the end of a previously received frame on the medium.

MIFS   Burst frame transmissions are those frames transmitted from the same device where the timing of each frame in the burst after the first is related to the preceding frame through use of the PHY burst mode. In this case, a MIFS duration will occur between frames in the burst. All frames in a burst except the last frame shall be sent with the ACK Policy field. The last frame in a burst may be sent with any ACK Policy.

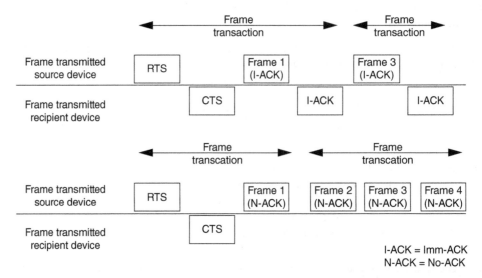

Figure 19.7 Frame transaction examples. From IEEE Standard for Radio Interface for White Space Dynamic Spectrum Access Radio Systems Supporting Fixed and Mobile Operation, Dec 2015. IEEE.

Within a burst, the Duration field shall cover only consecutive frames addressed to the same destination. If the burst continues after the Duration is exhausted, the next frame shall use a standard preamble.

SIFS   Within a frame transaction, all frames shall be separated by a SIFS interval.

DIFS   The DIFS is the minimum time that a device defers access to the medium after it determines the medium to have become idle.

Duplicate Detection   Because a device may not receive an Imm-ACK response for a frame it transmitted, it may send duplicate frames even though the intended recipient has already received and acknowledged the frame. A recipient device shall consider a received frame to be a duplicate if the sequence number field has the same value as the previous frame received with the same SrcAddr, DestAddr. A recipient device shall not release a duplicate frame to the MAC client.

RTS/CTS use   An RTS/CTS exchange precedes data to be transferred from a source device to a recipient device. Without a frame body, the RTS frame allows the source device to regain medium access relatively quickly in case of an unsuccessful transmission.

With an appropriately set Duration field, the RTS and CTS frames prevent the neighbors of the source and recipient devices from accessing the medium while the source and recipient are exchanging the following frames.

A source device may transmit an RTS frame as part of one or more frame transactions with another device in an obtained TXOP. In a TXOP, a device should transmit an RTS frame prior to transmitting a sequence of frames using the No-ACK acknowledgment policy if those frame transmissions would otherwise not be covered by the Duration field contained in a frame transmitted previously between the same source and recipient devices.

If a reservation target receives an RTS frame addressed to it in the reservation block, from the reservation owner, it shall transmit a CTS frame SIFS after the end of the received frame, regardless of its NAV setting. If a device receives an RTS frame addressed to it outside a reservation block, it shall transmit a CTS frame SIFS after the end of the received frame if and only if its NAV is zero and the CTS frame transmission will be completed SIFS before the start of the next CAP.

On receiving an expected CTS response, the source device shall transmit the frame, or the first of the frames, for which it transmitted the preceding RTS frame SIFS after the end of the received CTS frame. If the source device does not receive the expected CTS frame SIFS plus the CTS frame transmission time after the end of the RTS frame transmission, and it transmitted the RTS frame in a TXOP, it shall invoke a backoff. If it transmitted the RTS frame in one of its reservation blocks, it shall not retransmit the RTS frame or transmit another frame earlier than SIFS after the end of the expected CTS frame.

Duration Header Fields   A device shall set the Duration field in RTS, to the sum of the following:

— The transmission time of the frame body of the current frame
— The transmission time of the expected response frame for the current frame
— The transmission time of subsequent frames, if any, to be sent to the same recipient up to and including a) the next RTS frame or frame with ACK Policy set to Imm-ACK or b) the last frame, whichever is earlier; or, alternatively, the transmission time of the next frame TXOP to be sent to the same recipient, if any
— All the IFSs separating the frames included in the Duration calculation

A device shall round a fractional calculated value for Duration in microseconds up to the next integer.

A device shall set the Duration field in CTS and Imm-ACK frames to the larger of zero or a value equal to the duration value contained in the previous frame minus SIFS, minus the transmission time of the frame body of the received frame to which the CTS or Imm-ACK responding, minus the transmission time up to the end of the PHY Convergence Protocol (PLCP) header of this CTS or Imm-ACK frame.

19.4.5.2 Starting and Maintaining Network

This subclause specifies the procedures for scanning through channels, and starting network.

Scanning All devices shall be capable of performing scans across a specified list of channels. The higher layer should submit a scan request. A device is instructed to begin a channel scan through the MLME-SCAN.request primitive. For the duration of the scan, the device shall suspend beacon transmissions, if applicable, and shall only accept frames received over the PHY data services that are relevant to the scan being performed. Upon the conclusion of the scan, the device shall recommence beacon transmissions. The results of the scan shall be returned via the MLME-SCAN.confirm primitive.

Starting a Network A network should be started by a device only after having first performed a MAC sublayer reset, by issuing the MLME-RESET.request primitive, and a channel scan.

After completing this, the MAC sublayer shall issue the MLME-START.confirm primitive with a status of SUCCESS and begin operating as the Network coordinator.

Beacon The network coordinator device shall use the MLME-START.request primitive to begin transmitting beacons.

If the device receives no beacon frame during the scan, it shall create a new BP and send a beacon in the first beacon slot.

If the device receives one or more beacons during the scan, it shall not create a new BP. The device shall transmit a beacon in an unoccupied beacon slot. If all beacons slots are occupied, the MLME-START.confirm primitive returns the INVALID_PARAMETER value and the device shall not send beacon frame.

Device Discovery The network coordinator device indicates its presence on a network to other devices by transmitting beacon frames. This allows the other devices to perform device discovery. It should be mentioned that data carried by the beacon frame should be considered to speed up the device discovery process.

19.4.5.3 Synchronization

This subclause specifies the procedures for coordinator devices to generate beacon frames and for devices to synchronize with a coordinator. Synchronization is performed by receiving and decoding the beacon frames.

Synchronization with Beacons All devices shall be able to acquire synchronization beacons. A device is instructed to attempt to acquire the beacon through the MLME-SYNC.request primitive. When the MLME-SYNC.request primitive is invoked, the device shall attempt to acquire the beacon and keep track of it by regular and timely activation of its receiver.

To acquire beacon synchronization, a device shall enable its receiver and search for at most n MAC superframes symbols, where n is the value of mMacBeaconOrder. If a beacon frame is not received, the MLME shall repeat this search. Once the number of missed beacons reaches mMaxLostBeacons, the MLME shall notify the next higher layer by issuing the MLME-SYNC-LOSS.indication primitive with a loss reason of BEACON_LOST.

The MLME shall timestamp each received beacon. The timestamp value shall be that of the local clock of the device at the time of the symbol boundary. The timestamp is intended to be a relative time measurement that may or may not be made absolute, at the discretion of the implementer.

If a beacon frame is received, the MLME shall discard the beacon frame if the source address identifier fields of the MAC header of the beacon frame do not match the coordinator source address.

Orphaned Device Realignment If the next higher layer receives repeated communications failures following its requests to transmit data, it may conclude that it has been orphaned. A single communications failure occurs when a device transaction fails to reach the coordinator device, i.e., an acknowledgment is not received after mMacMaxFrameRetries attempts at sending the data. If the next higher layer concludes that it has been orphaned, it may instruct the MLME to reset the MAC sublayer and then perform the association procedure.

19.4.5.4 Association and Disassociation

This subclause specifies the procedures for association and disassociation.

Association A device shall attempt to associate only after having first performed a MAC sublayer reset, by issuing the MLME-RESET.request primitive, and then having completed the channel scan. The results of the channel scan are used to choose a suitable network. The algorithm for selecting a suitable network with which to associate is out of the scope of this standard.

Following the selection of a network with which to associate, the next higher layers shall request through the MLME-ASSOCIATE.request primitive that the MLME configures the following PHY and MAC attributes to the values necessary for association.

The MAC sublayer of an unassociated device shall initiate the association procedure by sending an association request command to the coordinator device of an existing network; if the association request command cannot be sent due to a channel access failure, the MAC sublayer shall notify the next higher layer. Since the association request command contains an acknowledgment request, the coordinator device shall confirm its receipt by sending an acknowledgment frame.

If the next higher layer of the coordinator finds that the device was previously associated on its network, all previously obtained device-specific information should be removed.

If sufficient resources are not available, the next higher layer of the coordinator should inform the MAC sublayer, and the MLME shall generate an association response command containing a status indicating a failure.

Disassociation The disassociation procedure is initiated by the next higher layer by issuing the MLME-DISASSOCIATE.request primitive to the MLME. When a coordinator device wants one of its associated devices to leave the network, the MLME of the coordinator device shall send the disassociation notification command.

If the command frame is not successfully extracted by the device, the coordinator device should consider the device disassociated. Otherwise, the MLME shall send the disassociation notification command to the device directly. In this case, if the disassociation notification command cannot be sent due to a channel access failure, the MAC sublayer shall notify the next higher layer.

Since the disassociation command contains an acknowledgment request, the receiving device shall confirm its receipt by sending an acknowledgment frame. If the transmission fails, the coordinator should consider the device disassociated.

19.4.5.5 Transmission, Reception, and Acknowledgment

This subclause describes the fundamental procedures for transmission, reception, and acknowledgment.

Transmission The source address field shall contain the address of the device sending the frame, which may be either a 16-bit address or a 48-bit address. The destination address field shall contain the address of the intended recipient of the frame, which may be either a 16-bit address or a 48-bit address.

If the frame is to be transmitted, the transmitting device shall attempt to find the beacon before transmitting. Transmissions in the CAP shall follow a successful application of the slotted version of the CSMA-CA algorithm.

If the CSMA-CA algorithm fails, the next higher layer shall be notified and the frame shall remain in the transaction queue until it is requested again and successfully transmitted or until the transaction expires.

To transmit the frame, the MAC sublayer shall first enable the transmitter by issuing the PLME-SET-TRX-STATE.request primitive with a state of TX_ON to the PHY. On receipt of the PLME-SET-TRX-STATE.confirm primitive with a status of either SUCCESS or TX_ON, the constructed frame shall then be transmitted by issuing the PD-DATA.request primitive. Finally, on receipt of the PD-DATA.confirm primitive, the MAC sublayer shall disable the transmitter by issuing the PLME-SET-TRX-STATE.request primitive with a state of RX_ON or TRX_OFF to the PHY, depending on whether the receiver is to be enabled following the transmission. In the case where the acknowledgment request subfield (ack policy) is set to zero, the MAC sublayer shall enable the receiver immediately following the transmission of the frame by issuing the PLME-SET-TRX-STATE.request primitive with a state of RX_ON to the PHY.

Each device initialize it DSN; the algorithm for choosing a random number is out of the scope of this standard. Each time a data or a MAC command frame is generated, the MAC sublayer shall copy the value of the current DSN into the sequence number field of the MAC header of the outgoing frame and then increment it by one.

A data or MAC command frame shall be sent with the acknowledgment request subfield of its frame control field set appropriately for the frame. A beacon or acknowledgment frame shall always be sent with the acknowledgment request subfield set to zero.

Reception and Rejection A transceiver task shall be defined as a transmission request with acknowledgment reception, if required, or a reception request. On completion of each transceiver task, the MAC sublayer shall request that the PHY enables or disables its receiver.

Due to the nature of radio communications, a device with its receiver enabled will be able to receive and decode transmissions from all devices complying with the current standard that are currently operating on the same channel, along with interference from other sources. The MAC sublayer shall, therefore, be able to filter incoming frames and present only the frames that are of interest to the upper layers.

The MAC sublayer shall discard all received frames that do not contain a correct value in their CRC field. The CRC field shall be verified on reception by recalculating the purported frame check sequence over the MAC header and MAC payload of the received frame and by subsequently comparing this value with the received CRC field. The CRC field of the received frame shall be considered to be correct if these values are the same and incorrect otherwise.

If the valid frame is a data frame, the MAC sublayer shall pass the frame to the next higher layer. This is achieved by issuing the MAC-DATA.indication primitive containing the frame information.

If the valid frame is a MAC command or beacon frame, it shall be processed by the MAC sublayer accordingly and a corresponding confirm or indication primitive may be sent to the next higher layer.

Acknowledgment A frame transmitted with the acknowledgment request subfield set to one shall be acknowledged by the recipient. If the intended recipient correctly receives the frame, it shall generate and send an acknowledgment frame containing the same DSN from the data or MAC command frame that is being acknowledged.

Retransmissions A device that sends a frame with the acknowledgment request subfield set to zero shall assume that the transmission was successfully received and shall hence not perform the retransmission procedure.

A device that sends a data or MAC command frame with its acknowledgment request subfield set to one shall wait for at most mMacAckWaitDuration for the corresponding acknowledgment frame to be received. If an acknowledgment frame is received within mMacAckWaitDuration and contains the same DSN as the original transmission, the transmission is considered successful, and no further action regarding retransmission shall be taken by the device. If an acknowledgment is not received or an acknowledgment is received containing a DSN that was not the same as the original transmission, the device shall conclude that the single transmission attempt has failed.

If a single transmission attempt has failed, the device shall repeat the process of transmitting the data or MAC command frame and waiting for the acknowledgment, up to a maximum of mMacMaxFrameRetries times. The retransmitted frame shall contain the same DSN as was used in the original transmission. Each retransmission shall only be attempted if it can be completed within the same portion of the superframe, i.e., the CAP in which the original transmission was attempted. If this timing is not possible, the retransmission shall be deferred until the same portion in the next superframe. If an acknowledgment is still not received after mMacMaxFrameRetries retransmissions, the MAC sublayer shall assume the transmission has failed and notify the next higher layer of the failure.

Transmission Scenarios Due to the imperfect nature of the radio medium, a transmitted frame does not always reach its intended destination. Figure 19.8 illustrates the following three different data transmission scenarios:

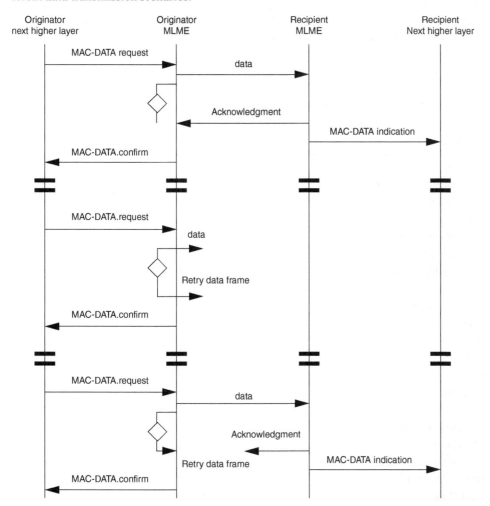

Figure 19.8 Transmission scenarios. From IEEE Standard for Radio Interface for White Space Dynamic Spectrum Access Radio Systems Supporting Fixed and Mobile Operation, Dec 2015. IEEE.

— Successful data transmission. The originator MAC sublayer transmits the data frame to the recipient via the PHY data service. In waiting for an acknowledgment, the originator MAC sublayer starts a timer that will expire after mMacAckWaitDuration symbols. The recipient MAC sublayer receives the data frame, sends an acknowledgment back to the originator, and passes the data frame to the next higher layer. The originator MAC sublayer receives the acknowledgment from the recipient before its timer expires and then disables and resets the timer. The data transfer is now complete, and the originator MAC sublayer issues a success confirmation to the next higher layer.

— Lost data frame. The originator MAC sublayer transmits the data frame to the recipient via the PHY data service. In waiting for an acknowledgment, the originator MAC sublayer starts a timer that will expire after mMacAckWaitDuration symbols. The recipient MAC sublayer does not receive the data frame and so does not respond with an acknowledgment. The timer of the originator MAC sublayer expires before an acknowledgment is received, and therefore the data transfer has failed. The originator retransmits the data, and this entire sequence may be repeated up to a maximum of macMaxFrameRetries times; if a data transfer attempt fails, the originator MAC sublayer will issue a failure confirmation to the next higher layer.

— Lost acknowledgment frame. The originator MAC sublayer transmits the data frame to the recipient via the PHY data service. In waiting for an acknowledgment, the originator MAC sublayer starts a timer that will expire after mMacAckWaitDuration symbols. The recipient MAC sublayer receives the data frame, sends an acknowledgment back to the originator, and passes the data frame to the next higher layer. The originator MAC sublayer does not receive the acknowledgment frame and its timer expires, and therefore the data transfer has failed. The originator retransmits the data, and this entire sequence may be repeated up to a maximum of mMacMaxFrameRetries times; if a data transfer attempt fails, the originator MAC sublayer will issue a failure confirmation to the next higher layer.

Table 19.29 MAC sublayer parameters.

Parameter	Value
mMacTransactionPersistenceTime	200 ms
mScanDuration	4
mMaxLostBeacons	4
mBPslot	16
mSuperframeDuration	153,6 ms
mMASDuration	600 μs
mBeaconSlotDuration	600 μs
mAccessDelay	5 ms
mClockResolution	1 μs
SlotTime	40 μs
CWmin	16
CWmax	1024
SIFS	10 μs
CCADetectTime	30 μs
MIFS	15 μs
mMacBeaconOrder	4
mMaxLostBeacons	3
mMacMaxFrameRetries	9
mMacAckWaitDuration	70 μs
TXOPLimit	10 ms

From IEEE Standard for Radio Interface for White Space Dynamic Spectrum Access Radio Systems Supporting Fixed and Mobile Operation, Dec2015. IEEE.

19.4.5.6   MAC Sublayer Parameters

The MAC sublayer parameters are specified Table 19.29.

19.5   PHY Layer

19.5.1   PHY Layer Service Specification

The PHY layer provides an interface between the MAC sublayer and the physical radio channel. The PHY layer conceptually includes a management entity called the PLME. This entity provides the layer management service interfaces through which layer management functions may be invoked.

The PHY layer provides the following two services:

— The PHY data service and the PHY management service interfacing to the PHY layer management entity (PLME) service access point (SAP) (known as the PLME-SAP).
— The PHY data service (PD) enables the transmission and reception of PHY protocol data units (PPDUs) across the physical radio channel.

19.5.1.1   PHY Data Service

The PD-SAP supports the transport of MPDUs between peer MAC sublayer entities.

PD-DATA.request   The PD-DATA.request primitive requests the transfer of an MPDU from the MAC sublayer to the local PHY entity (i.e., PHY service Data UnitPSDU).

The receipt of the PD-DATA.request primitive by the PHY entity will cause the transmission of the supplied PSDU to be attempted. Provided the transmitter is enabled (TX_ON state), the PHY will first construct a frame, and then transmit the PPDU. When the PHY entity has completed the transmission, it will issue the PD-DATA.confirm primitive with a status of SUCCESS.

If the PD-DATA.request primitive is received while the receiver is enabled (RX_ON state) the PHY entity will discard the PSDU and issue the PD-DATA.confirm primitive with a status of RX_ON. If the PD-DATA.request primitive is received while the transceiver is disabled (TRX_OFF state), the PHY entity will discard the PSDU and issue the PD-DATA.confirm primitive with a status of TRX_OFF. If the PD-DATA.request primitive is received while the transmitter is already busy transmitting (BUSY_TX state) the PHY entity will discard the PSDU and issue the PD-DATA.confirm primitive with a status of BUSY_TX. The PD-DATA request fields are shown in Table 19.30 while the PD-DATA confirm fields are shown in Table 19.31.

The semantics of the PD-DATA.request primitive is as follows:

```
PD-DATA.request(
          psduLength
          psdu
)
```

Table 19.30 PD-DATA.request fields.

Name	Type	Description
psduLength	Integer	The number of octets contained in the PSDU to be transmitted by the PHY entity.
psdu	Array of bytes	The array of bytes forming the PSDU to be transmitted by the PHY entity.

From IEEE Standard for Radio Interface for White Space Dynamic Spectrum Access Radio Systems Supporting Fixed and Mobile Operation, Dec2015. IEEE.

Table 19.31 PD-DATA.confirm fields.

Name	Type	Description
Status	Enumeration	SUCCESS, RX_ON, TRX_OFF, BUSY_TX

From IEEE Standard for Radio Interface for White Space
Dynamic Spectrum Access Radio Systems Supporting
Fixed and Mobile Operation, Dec2015. IEEE.

PD-DATA.confirm   The PD-DATA.confirm primitive is generated by the PHY entity and issued
to its MAC sublayer entity in response to a PD-DATA.request primitive. The PD-DATA.confirm
primitive will return a status of either SUCCESS, indicating that the request to transmit was
successful, or an error code of RX_ON, TRX_OFF, or BUSY_TX.

On receipt of the PD-DATA.confirm primitive, the MAC sublayer entity is notified of the
result of its request to transmit. If the transmission attempt was successful, the Status parameter is set to SUCCESS. Otherwise, the Status parameter will indicate the error.

The semantics of the PD-DATA.confirm primitive is as follows:

```
PD-DATA.confirm(
        Status
)
```

PD-DATA.indication   The PD-DATA.indication primitive is generated by the PHY entity and
issued to its MAC sublayer entity to transfer a received PSDU. This primitive will not be
generated if the received psduLength field is zero or greater than aMaxPHYPacketSize. The
PA-DATA indication fields are shown in Table 19.32.

The semantics of the PD-DATA.indication primitive is as follows:

```
PD-DATA.indication(
        psduLength
        psdu
)
```

19.5.1.2   PHY Management Service

The PLME-SAP allows the transport of management commands between the MLME and
the PLME.

PLME-CCA.request   The PLME-CCA.request primitive requests that the PLME perform a CCA.
The PLME-CCA.request primitive is generated by the MLME and issued to its PLME whenever the CSMA-CA algorithm requires an assessment of the channel. The PLME-CCA.request
primitive has no parameter.

Table 19.32 PD-DATA.indication fields.

Name	Type	Description
psduLength	Integer	The number of octets contained in the PSDU to be received by the PHY entity.
psdu	Array of bytes	The array of bytes forming the PSDU to be received by the PHY entity.

From IEEE Standard for Radio Interface for White Space Dynamic Spectrum Access Radio Systems Supporting
Fixed and Mobile Operation, Dec2015. IEEE.

Table 19.33 PLME-CCA.confirm fields.

Name	Type	Description
Status	Enumeration	Status of the request to transmit: BUSY, IDLE, TRX_OFF

From IEEE Standard for Radio Interface for White Space Dynamic Spectrum Access Radio Systems Supporting Fixed and Mobile Operation, Dec2015. IEEE.

If the receiver is enabled on receipt of the PLME-CCA.request primitive, the PLME will cause the PHY to perform a CCA. When the PHY has completed the CCA, the PLME will issue the PLME-CCA.confirm primitive with a status of either BUSY or IDLE, depending on the result of the CCA. The PLME-CCA confirm fields are shown in Table 19.33.

If the PLME-CCA.request primitive is received while the transceiver is disabled (TRX_OFF state) or if the transmitter is enabled (TX_ON state), the PLME will issue the PLME-CCA.confirm primitive with a status of TRX_OFF or BUSY, respectively.

PLME-CCA.confirm The PLME-CCA.confirm primitive reports the results of a CCA.

The semantics of the PLME-CCA.confirm primitive are as follows:

```
PLME-CCA.confirm(
            Status
)
```

The PLME-CCA.confirm primitive is generated by the PLME and issued to its MLME in response to a PLME-CCA.request primitive. The PLME-CCA.confirm primitive will return a status of either BUSY or IDLE, indicating a successful CCA, or an error code of TRX_OFF.

On receipt of the PLME-CCA.confirm primitive, the MLME is notified of the results of the CCA. If the CCA attempt was successful, the Status parameter is set to either BUSY or IDLE. Otherwise, the Status parameter will indicate the error.

PLME-SET-TRX-STATE.request The PLME-SET-TRX-STATE.request primitive requests that the PHY entity change the internal operating state of the transceiver. The transceiver will have the following three main states:

— Transceiver disabled (TRX_OFF)
— Transmitter enabled (TX_ON)
— Receiver enabled (RX_ON)

The PLME-SET-TRX-STATE.request primitive is generated by the MLME and issued to its PLME when the current operational state of the receiver needs to be changed. The PLME-SET-TRX-STATE request fields are shown in Table 19.34.

The semantics of the PLME-SET-TRX-STATE.request primitive are as follows:

```
PLME-SET-TRX-STATE.request (
            State
)
```

On receipt of the PLME-SET-TRX-STATE.request primitive, the PLME will cause the PHY to change to the requested state. If the state change is accepted, the PHY will issue the PLME-SET-TRX-STATE.confirm primitive with a status of SUCCESS.

If this primitive requests a state that the transceiver is already configured, the PHY will issue the PLME-SET-TRX-STATE.confirm primitive with a status indicating the current state, i.e., RX_ON, TRX_OFF, or TX_ON.

If this primitive is issued with RX_ON or TRX_OFF argument and the PHY is busy transmitting a PPDU, at the end of transmission the state change will occur and then the PHY will issue the PLME-SET-TRX-STATE.confirm primitive with a status of SUCCESS.

Table 19.34 PLME-SET-TRX-STATE.request fields.

Name	Type	Description
State	Enumeration	State of the request: TRX_OFF, TX_ON, RX_ON

From IEEE Standard for Radio Interface for White Space Dynamic Spectrum Access Radio Systems Supporting Fixed and Mobile Operation, Dec2015. IEEE.

If this primitive is issued with TX_ON, the PHY will cause the PHY to go to the TX_ON state irrespective of the state the PHY is in, the PHY will issue the PLME-SET-TRX-STATE.confirm primitive with a status of SUCCESS. If this primitive is issued with FORCE_TRX_OFF, the PHY will cause the PHY to go to the TRX_OFF state irrespective of the state the PHY is in, the PHY will issue the PLME-SET-TRX-STATE.confirm primitive with a status of SUCCESS. The PLME-SET-TRX-STATE confirm fields are shown in Table 19.35.

PLME-SET-TRX-STATE.confirm The PLME-SET-TRX-STATE.confirm primitive reports the result of a request to change the internal operating state of the transceiver.

The semantics of the PLME-SET-TRX-STATE.confirm primitive are as follows:

```
PLME-SET-TRX-STATE.confirm(
        Status
)
```

On receipt of the PLME-SET-TRX-STATE.confirm primitive, the MLME is notified of the result of its request to change the internal operating state of the transceiver. A Status value of SUCCESS indicates that the internal operating state of the transceiver was accepted. A Status value of RX_ON, TRX_OFF, or TX_ON indicates that the transceiver is already in the requested internal operating state.

19.5.1.3 PHY Enumerations Description

Table 19.36 shows the PHY enumeration description.

19.5.2 CRC Method

Error detection is provided on bitstream through a cyclic redundancy check (CRC). The size of the CRC is 16 bits. The bits before CRC attachment are denoted as $a_1, a_2, \ldots, a_{NiCRC}$, where *NiCRC* is the size of the bitstream. NiCRC shall be even and in the range 64 bits to 4072 bits.

The entire bitstream is used to calculate the CRC parity bits. The parity bits are generated by the cyclic generator polynomial defined in Equation (19.2).

$$g_{CRC16}(D) = D^{16} + D^{12} + D^5 + 1 \tag{19.2}$$

Table 19.35 PLME-SET-TRX-STATE.confirm fields.

Name	Type	Description
Status	Enumeration	Status of the confirmation: SUCCESS, TRX_OFF, TX_ON, RX_ON

From IEEE Standard for Radio Interface for White Space Dynamic Spectrum Access Radio Systems Supporting Fixed and Mobile Operation, Dec2015. IEEE.

Table 19.36 PHY enumerations description.

Enumeration	Value	Description
BUSY	0x00	The CCA attempt has detected a busy channel.
BUSY_RX	0x01	The transceiver is asked to change its state while receiving.
BUSY_TX	0x02	The transceiver is asked to change its state while transmitting.
FORCE_TRX_OFF	0x03	The transceiver is to be switched off immediately.
IDLE	0x04	The CCA attempt has detected an idle channel.
INVALID_PARAMETER	0x05	A SET/GET request was issued with a parameter in the primitive that is out of the valid range.
RX_ON	0x06	The transceiver is in or is to be configured into the receiver enabled state.
SUCCESS	0x07	A SET/GET, an ED operation, or a transceiver state change was successful.
TRX_OFF	0x08	The transceiver is in or is to be configured into the transceiver disabled state.
TX_ON	0x09	The transceiver is in or is to be configured into the transmitter enabled state.

From IEEE Standard for Radio Interface for White Space Dynamic Spectrum Access Radio Systems Supporting Fixed and Mobile Operation, Dec2015. IEEE.

Parity bits are denoted as p_k, $k = 1...16$. The encoding is performed in a systematic form. Therefore the polynomial

$$a_1 D^{NiCRC+15} + a_2 D^{NiCRC+14} + \ldots + a_{NiCRC} D^{16} + p_1 D^{15} + p_2 D^{14} + \ldots + p_{15} D + p_{16} \quad (19.3)$$

yields a remainder equal to 0 when divided by $g_{CRC16}(D)$.

The bits after CRC attachment are denoted by $b_1, b_1, \ldots, b_{NoCRC}$. The relation between a_i and b_i is defined in Equation (19.4).

$$b_i = a_i, \quad i = 1, \ldots, NiCRC$$
$$b_{NiCRC+k} = p_k, \qquad k = 1, \ldots, 16 \quad (19.4)$$

The bits b_i are transmitted starting from b_1.

It is worth mentioning that $NoCRC = NiCRC + 16$.

19.5.3 Channel Coding (Including Interleaving and Modulation)

19.5.3.1 Scrambling

The data of the input stream shall be randomized in accordance with the configurations depicted in Figure 19.9.

The polynomial for the pseudo random binary sequence (PRBS) generator shall be as defined in Equation (19.5).

$$X^{17} + X^{14} + 1 \quad (19.5)$$

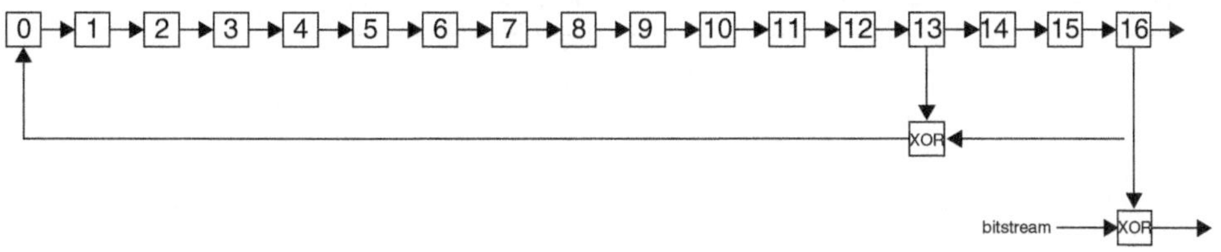

Figure 19.9 Scrambler schematic diagram including initialization vector. From IEEE Standard for Radio Interface for White Space Dynamic Spectrum Access Radio Systems Supporting Fixed and Mobile Operation, Dec 2015. IEEE.

Loading of the sequence "1000000000000000" into the PRBS registers, as indicated in Figure 9, shall be initiated at the start of every CRC coded packet.

19.5.3.2 Convolutional Coding

A convolutional code with constraint length 7 is defined. The input bits of size *NiBCC* are denoted by c_i. First $N_{tb} = 8$ tailbits are added at the end of the input stream. The bit at the input of the convolutional code is donated as d_i. The relation between d_i and c_i is defined in Equation (6).

$$d_i = c_i, \quad i = 1, \ldots, NiBCC$$
$$d_{NiBCC+k} = 0, \qquad k = 1, \ldots, N_{tb} \tag{19.6}$$

The bitstream d_i is encoded by the rate 1/2 convolutional code (generator polynomials are 171 and 133 in octal) with constraint length 7 depicted in Figure 19.10.

Output of the convolutional code shall be done in the order A,B,A,B... The initial value of the shift register of the coder shall be "all 0" when starting to encode the input bits.

19.5.3.3 Puncturing

The system shall allow for a range of punctured convolutional codes, based on a mother convolutional code of rate 1/2. This will allow selection of the most appropriate level of error correction for a given service or data rate. In addition to the mother code of rate 1/2 the system shall allow punctured rates of 2/3, 3/4, and 5/6. The punctured convolutional code specified in Table 19.37 shall be used.

The number *T*

$$T = (NiCRC + 16 + 8)R \tag{19.7}$$

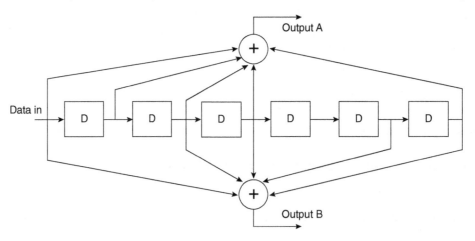

Figure 19.10 Rate 1/2 convolutional code. From IEEE Standard for Radio Interface for White Space Dynamic Spectrum Access Radio Systems Supporting Fixed and Mobile Operation, Dec 2015. IEEE.

Table 19.37 Punctured convolutional code.

Coding rate (R)	Puncturing pattern	Transmitted sequence
1/2	1 1	$A_1 B_1$
2/3	0 1 1 1	$B_1 A_2 B_2$
3/4	1 0 1 1 1 0	$A_1 B_1 B_2 A_3$
5/6	1 0 1 0 1 1 1 0 1 0	$A_1 B_1 B_2 A_3 B_4 A_5$

From IEEE Standard for Radio Interface for White Space Dynamic Spectrum Access Radio Systems Supporting Fixed and Mobile Operation, Dec2015. IEEE.

shall be an integer. For instance, a combination of *NiCRC* = 64 and coding rate $R = 5/6$ is not allowed, while a combination of *NiCRC* = 66 and coding rate $R = 5/6$ is allowed.

19.5.3.4 Bit Interleaving

The interleaving consists of a row column bit interleaving. The input bits of size *NiINT* are denoted as e_i and the output bits are denoted as f_i. In a row-column interleaver bits are written column-wise and read row-wise. The relation between e_i and f_i is defined in Equation (19.8).

$$f_i = e_{1+((i-1)\%m)l+\left\lfloor \frac{i-1}{m} \right\rfloor}, \quad i = 1, \dots, NiINT \tag{19.8}$$

where

l is the row number
m is the column number
%   is modulo function
$\lfloor x \rfloor$ is a function of rounding x to the nearest integer toward minus infinity
m and l are computed using the following procedure:
$t_0 = \lceil \sqrt{NiINT} \rceil$

```
while( 1 )
          if( NiNIT%t₀ == 0 ) then
                    m = t₀ ;
                    l = N/t₀ ;
                    break;
          else
                    t₀ = t₀ - 1 ;
          end if;
end while
```

where

$\lfloor x \rfloor$ is a function of rounding x to the nearest integer toward infinity

19.5.3.5 Bit Padding

The bit padding consists of adding dummy bits to fill a data structure. We denote the size of the bit vector to be mapped into physical channel as *NoPADD*. The size of the bit vector at the input of the padding is denoted as *NiPADD*. *NiPADD* is defined in Equation (19.9).

$$NiPADD = NiINT_1 + NiINT_2 + \dots + NiINT_Q \tag{19.9}$$

where

Q is the number of CRC coded bitstream to map into physical channel

The number of padding bits *NPAD* is derived from *NiPADD* and *NoPADD* using Equation (19.10).

$$NPAD = NoPADD - NiPADD \tag{19.10}$$

The *NPAD* bits are generated using a pseudo random binary sequence. The generator is defined in Equation (19.11).

$$X^{17} + X^{14} + 1 \tag{19.11}$$

Loading sequence "1000000000000000" shall be applied every payload frame, as indicated in Figure 19.9.

The relation between the input bits h_i and the output bits m_i is defined in Equation (19.12).

$$m_i = h_i, \quad i = 1, \dots, NiPADD$$
$$m_{NiPADD+k} = q_k, \quad k = 1, \dots, NPAD \tag{19.12}$$

where

q_k are the bits generated by PRBS.

19.5.3.6 Modulation and Coding Scheme

The binary serial input stream shall be divided into groups of N_{PBS} (1, 2, 4, or 6) bits and converted into complex numbers c_i representing binary phase shift keying (BPSK), quadrature phase shift keying (QPSK), 16-quadrature amplitude modulation (QAM), or 64-QAM constellation complex points. The conversion shall be performed according to Gray-coded constellation mappings, illustrated in Figure 19.11. The bit b0 is the first stream bit received by the QAM modulator.

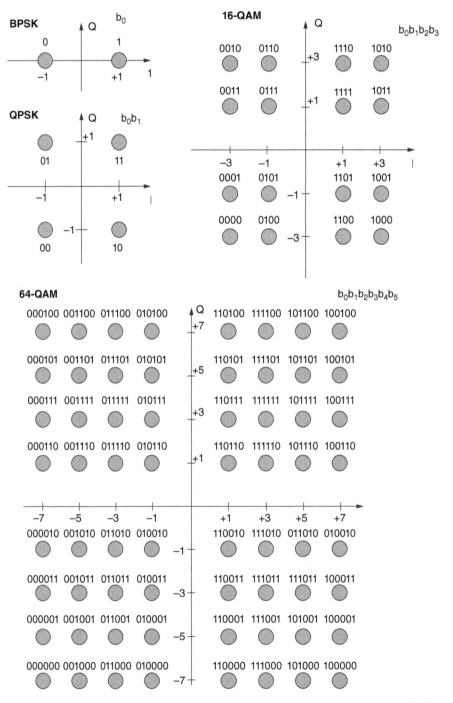

Figure 19.11 BPSK, QPSK, 16-QAM, and 64-QAM constellation bit encoding. From IEEE Standard for Radio Interface for White Space Dynamic Spectrum Access Radio Systems Supporting Fixed and Mobile Operation, Dec 2015. IEEE.

The output values c_i are multiplied by a normalization factor k_{MOD} to generate c_i' as defined in Equation (19.13).

$$c_i' = c_i \cdot k_{MOD} \tag{19.13}$$

The normalization factor k_{MOD}, depends on the base modulation mode, as prescribed in Table 19.38. The purpose of the normalization factor is to achieve the same average power for all constellations.

Various modulation schemes and coding rates are defined. Only a subset of combinations is allowed. Table 19.39 lists the modulation and coding schemes (MCS) defined in this standard.

In case of Beacon, RTS, and CTS frames (see 19.4.3.3, 19.4.3.4, and 19.4.3.5), the MCS 1 (QPSK, coding rate 1/2) should be used. Data and corresponding ACK frame are transmitted using the same MCS (the one specified for data frame, see 19.4.3.4).

19.5.4 Mapping Modulated Symbols to Carriers

19.5.4.1 Payload

In the Case of the Data The stream of complex numbers c_1' is divided into groups of N_C complex numbers, where N_C is the number of non-null modulated carriers, also referred to as active carriers (the range of N_C depends on the mode, as defined in Table 19.41). This shall be denoted by writing the complex number $c_{m,n}'$, which corresponds to index m of the FBMC symbol n , as defined in Equation (19.14).

$$c_i' = c_{m+n \cdot N_C}' = c_{m,n}' \tag{19.14}$$

The block of N_C complex numbers are mapped to N subcarriers according to the carrier mapping function defined by set of parameter duplets that depends on M_C :(Z_l, Δ_l), $l = 0, \ldots, M_C - 1$ as illustrated Figure 19.12. M_C is the number of active subblocks.

Table 19.38 Modulation normalization factor.

Modulation	Normalization factor k_{MOD}
BPSK	1
QPSK	$1/\sqrt{2}$
16-QAM	$1/\sqrt{10}$
64-QAM	$1/\sqrt{42}$

From IEEE Standard for Radio Interface for White Space Dynamic Spectrum Access Radio Systems Supporting Fixed and Mobile Operation, Dec2015. IEEE.

Table 19.39 Modulation and coding scheme index list.

MCS	Modulation	Coding rate
0	BPSK	1/2
1	QPSK	1/2
2	QPSK	3/4
3	16-QAM	1/2
4	16-QAM	3/4
5	64-QAM	2/3
6	64-QAM	3/4
7	64-QAM	5/6

From IEEE Standard for Radio Interface for White Space Dynamic Spectrum Access Radio Systems Supporting Fixed and Mobile Operation, Dec2015. IEEE.

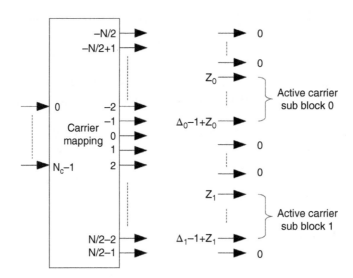

Figure 19.12 Subcarrier mapping. From IEEE Standard for Radio Interface for White Space Dynamic Spectrum Access Radio Systems Supporting Fixed and Mobile Operation, Dec 2015. IEEE.

This mapping is denoted by writing the complex number $d'_{k,n}$, where k is the carrier index of FBMC symbol n, as specified in Equation (19.15).

$$d'_{Z_l+q,n} = c'_{\left(\sum_{u=0}^{l-1} \Delta_u + q\right),n}, \quad q = 0, \ldots, \Delta_l - 1, \quad l = 0, \ldots, M_C - 1, \quad n = 0, \ldots, N_C - 1$$

$$d'_{k,n} = 0, \quad otherwise \tag{19.15}$$

In the Case of the Beacon The stream of complex number c'_1 is divided into groups of Δ_b complex numbers, where Δ_b is the size of the elementary set of active carriers. This shall be denoted by writing the complex number $c'_{m,n}$, which corresponds to index m of the FBMC symbol n, as defined in Equation (19.16).

$$c'_i = c'_{m+n\cdot\Delta_b} = c'_{m,n} \tag{19.16}$$

The block of Δ_b complex numbers is mapped to N_C subcarriers according to the carrier mapping function. This mapping is denoted by writing the complex number $d'_{k,n}$, where k is the carrier index of FBMC symbol n, as defined in Equation (19.17).

$$d'_{Z_l+q,n} = c'_{\left(\left(\sum_{u=0}^{l-1} \Delta_u + q\right)\%\Delta_b\right),n}, \quad q = 0, \ldots, \Delta_l - 1, \quad l = 0, \ldots, M_C, \quad n = 0, \ldots, N_C - 1$$

$$d'_{k,n} = 0, \quad otherwise \tag{19.17}$$

where

% is the modulo operator

In Both Cases Transmission power weights are applied to both data and beacon as defined in Equation (19.18).

$$d_{x,y} = w_x d'_{x,y} \tag{19.18}$$

where

w_x represents power weights that shall be defined in a way to follow the channel mask requirements provided by the ChannelMapMask parameter

Z_l and Δ_l are set to be multiple of 4. According to the chosen mode (see below), constraints on Z_l are defined in Equation (19.19).

$$Z_{l+1} \geq Z_l + \Delta_{l-1}, \quad l = 0, \dots, M_C - 1$$
$$Z_0 \geq Z_{\min}$$
$$Z_{M_C-1} + \Delta_{M_C-1} - 1 \leq Z_{\max}$$
$$N_C = \sum_{l=0}^{M_C-1} \Delta_l \tag{19.19}$$

Range of $Z_{\min}$ and $Z_{\max}$ depends on the mode and is defined in Table 19.41.

A complex-real conversion is performed, where the real and imaginary parts of $d_{k,n}$ are separated to form a real symbol $e_{k,l}$. The complex-real conversion increases the sample rate by a factor of 2 in order to have the complex rate for $d_{k,n}$ and $e_{k,l}$. This corresponds to a parallel to serial conversion of the $d_{k,n}$ with a constant information rate. The selection of the part of the complex symbols that shall be delayed (real part or imaginary part) depends on the parity of k ($k = 0$ is assumed even) as illustrated in Figure 19.13.

The real numbers $e_{k,l}$, are multiplied by an offset QAM sequence to form a new complex symbol $h_{k,l}$, as defined in Equation (19.20).

$$h_{k,l} = (-1)^{k \cdot l}(j)^{k+l} e_{k,l} \tag{19.20}$$

where

j is complex number equal to $\sqrt{-1}$

The complex numbers $h_{k,l}$ are up sampled by a factor $N/2$ and then filtered by the filter defined by $G^k(z)$ and summed to form the baseband output signal r_p as illustrated Figure 19.14.

The filters $G^k(z)$ having $K_{FBMC} \cdot N$ taps are defined in Equation (19.21).

$$G_i^k(z) = P_i W_i^k, \quad i = 0, \dots, K_{FBMC}N - 2, \quad k = -N/2, \dots, N/2 - 1$$
$$G_i^k(z) = 0, \quad i = K_{FBMC}N - 1, \quad k = -N/2, \dots, N/2 - 1 \tag{19.21}$$

where

W_i^k is defined in Equation (19.22)
P_i is defined in Equation (19.23)

$$W_i^k = \exp\left(2\pi j \frac{ki}{N}\right) \tag{19.22}$$

$$P_i = \tilde{P}_0 + 2 \sum_{k=1}^{K_{FBMC}-1} (-1)^k \tilde{P}_k \cos\left(\frac{2\pi k}{K_{FBMC}N}(i+1)\right) \tag{19.23}$$

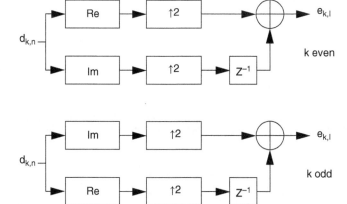

Figure 19.13 FBMC filtering pre-processing. From IEEE Standard for Radio Interface for White Space Dynamic Spectrum Access Radio Systems Supporting Fixed and Mobile Operation, Dec 2015. IEEE.

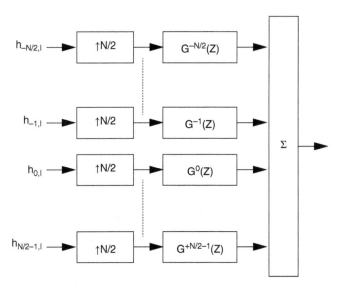

Figure 19.14 FBMC filtering. From IEEE Standard for Radio Interface for White Space Dynamic Spectrum Access Radio Systems Supporting Fixed and Mobile Operation, Dec 2015. IEEE.

The filter coefficient indexation is illustrated Figure 19.15.

The coefficients $\widetilde{P}_k$ are defined Table 19.40. K_{FBMC} may be equal to 3 or 4.

From IEEE Standard for Radio Interface for White Space Dynamic Spectrum Access Radio Systems Supporting Fixed and Mobile Operation, Dec2015. IEEE.

The maximum number of subcarrier N depends on the mode, as prescribed in Table 19.41.

19.5.4.2 Preamble

A preamble is added at the beginning of the payload (see Figure 19.16). M_p FBMC symbols are used for the preamble. M_p is set to 8. It is composed of pilot carriers of type I spaced every 4 active carriers for the whole duration of the preamble and pilots carriers of type II spaced every 16 active carriers and located on symbol 4.

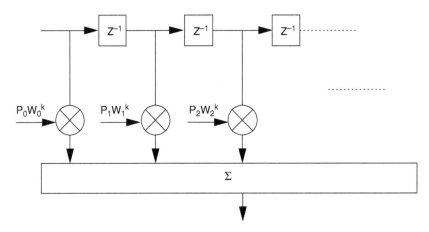

Figure 19.15 Filter coefficient indexation. From IEEE Standard for Radio Interface for White Space Dynamic Spectrum Access Radio Systems Supporting Fixed and Mobile Operation, Dec 2015. IEEE.

Table 19.40 Filter coefficients.

K_{FBMC}	α	$\widetilde{P}_0$	$\widetilde{P}_1$	$\widetilde{P}_2$	$\widetilde{P}_3$
3	0.97195983	1	α	$\sqrt{1-\alpha^2}$	Not used
4	0.97195983	1	α	$1/\sqrt{2}$	$\sqrt{1-\alpha^2}$

Table 19.41 Mode parameters.

Mode	N	Intercarrier spacing	Z_{min}	Z_{max}	Δ_l 2 MHz	Δ_l 8 MHz
Mode 4K	4096	3.75 kHz	−1008	1007	504 (1.86MHz)	2016 (7.56MHz)
Mode 1K	1024	15.00 kHz	−252	251	124 (1.86MHz)	504 (7.56MHz)
Mode 0.5K	512	30.00 kHz	−126	125	64 (1.92MHz)	252 (7.56MHz)
Mode 0.25K	256	60.00 kHz	−62	61	32 (1.92MHz)	124 (7.44MHz)

From IEEE Standard for Radio Interface for White Space Dynamic Spectrum Access Radio Systems Supporting Fixed and Mobile Operation, Dec2015. IEEE.

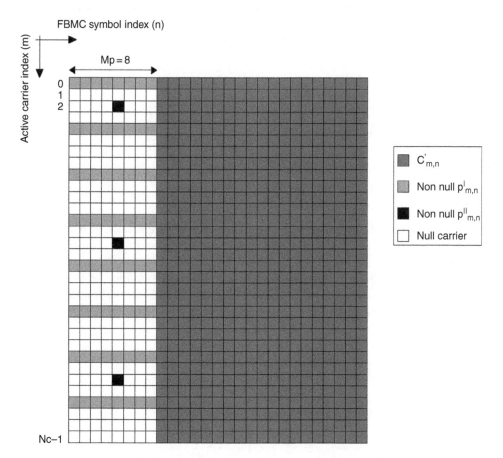

Figure 19.16 Example of FBMC preamble and payload. From IEEE Standard for Radio Interface for White Space Dynamic Spectrum Access Radio Systems Supporting Fixed and Mobile Operation, Dec 2015. IEEE.

This shall be denoted by writing the complex number $p_{m,n}$ derived from $p_{m,n}^I$ (for type I pilots) and $p_{m,n}^{II}$ (for type II pilots), $m = 0, ..., N_C - 1$, $n = 0, ..., 7$ (subcarrier m of FBMC symbol n), as defined in Equation (19.24), Equation (19.25), and Equation (19.26).

$$p_{m,n}^I = (-1)^n \alpha^I (2s_m^I - 1), if \quad m\%4 = 0$$
$$p_{m,n}^I = 0, \quad otherwise \tag{19.24}$$

$$p_{m,n}^{II} = (-1)^n \alpha^{II} (2s_m^{II} - 1), if \quad (m - 2)\%16 = 0 \quad and \quad n = 4$$
$$p_{m,n}^{II} = 0, \quad otherwise \tag{19.25}$$

$$p_{m,n} = p_{m,n}^I + p_{m,n}^{II} \tag{19.26}$$

s_m^I is generated using a pseudo random binary sequence. The generator shall be as defined in Equation (19.27).

$$X^{17} + X^{14} + 1 \tag{19.27}$$

The generator shall use the following initialization sequence "1001100010001010", as indicated in Figure 19.12, initialized for each new FBMC symbol. The pseudo random binary sequence generator is shifted whenever $p_{m.n}^{II} \neq 0$ (i.e., whenever $m\%4 = 0$).

s_m^{II} is generated using a pseudo random binary sequence. The generator shall be as defined in Equation

$$X^{17} + X^{14} + 1 \tag{19.28}$$

The generator shall use the following initialization sequence "0001100010001010", as indicated in Figure 19.12, initialized for each new FBMC symbol. The pseudo random binary sequence generator is shifted whenever $p_{m.n}^{II} \neq 0$ (i.e., whenever $(m-2)\%16 = 0$).

α^I is a normalization factor defined in Equation (19.29).

$$\alpha^I = \sqrt{\frac{N_C}{N_p}} \tag{19.29}$$

where

N_p is the number of non-null pilot of type I
α^{II} is a normalization factor equal to 1
$p_{m,n}$ is then modulated as described in 19.5.4.1.

19.5.5 Transmitter Requirements

Transmitter requirements are characterized by the emission spectrum mask. To enable maximal flexibility, transmissions in non-contiguous portions of the spectrum are considered.

The spectrum mask is constrained by local regulation. In the Ultra High Frequency band in the USA, channels of 6 MHz are considered. Also, the Federal Communication Commission specifies that leakage on adjacent channels should be at least 55 dB under the in-band transmit power. In the example of Figure 19.17, if H_1 is occupied by an incumbent signal, T shall be −55 dBr or less. In Europe, channels are 8 MHz wide.

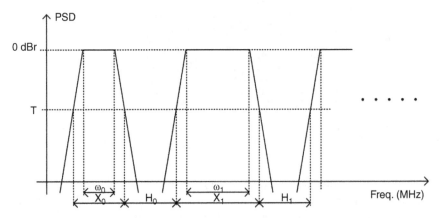

Figure 19.17 Spectrum mask example. From IEEE Standard for Radio Interface for White Space Dynamic Spectrum Access Radio Systems Supporting Fixed and Mobile Operation, Dec 2015. IEEE.

Annex 19A

(informative)

Coexistence Considerations

IEEE 1900.7 devices should sense channels in order to assess the interference levels in the channels that they are potentially going to use. IEEE 1900.7 devices should share the sensed information with the IEEE 1900.7 devices with which they intend to communicate over white space at the other end of the link. A metric for the quality of each channel should be adopted and used as part of the process of deciding which channel to use. One example of such a metric is the allowed power divided by the sensed interference level at the other end of the white space link.

In cases where IEEE 1900.7 devices are aware of sets of contiguous channels that are available and the quality of those channels is approximately equal as determined using the procedures shown in Figure 19A.1, they should choose the channels to use with the objective of maximizing the number of remaining unused contiguous channels.

The choice of metric range as used at point 1 in the above flow chart for which the channel quality is considered as "approximately equal" is not specified in IEEE Std 1900.7 and open to the implementer of the standard.

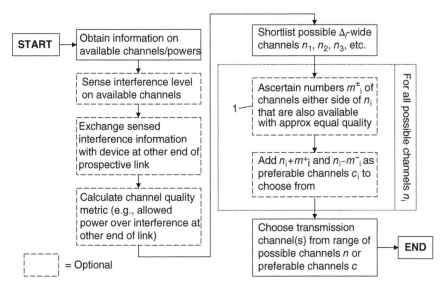

Figure 19A.1 An example of transmission channel selection procedure. From IEEE Standard for Radio Interface for White Space Dynamic Spectrum Access Radio Systems Supporting Fixed and Mobile Operation, Dec 2015. IEEE.

Index

Dynamic Spectrum Access Decisions: Local, Distributed, Centralized, and Hybrid Designs, First Edition. George F. Elmasry.
© 2021 John Wiley & Sons Ltd. Published 2021 by John Wiley & Sons Ltd.
Companion website: www.wiley.com/go/elmasry/dsad